Springer Proceedings in Mathematics & Statistics

Volume 291

Springer Proceedings in Mathematics & Statistics

This book series features volumes composed of selected contributions from workshops and conferences in all areas of current research in mathematics and statistics, including operation research and optimization. In addition to an overall evaluation of the interest, scientific quality, and timeliness of each proposal at the hands of the publisher, individual contributions are all refereed to the high quality standards of leading journals in the field. Thus, this series provides the research community with well-edited, authoritative reports on developments in the most exciting areas of mathematical and statistical research today.

More information about this series at http://www.springer.com/series/10533

Alexey Karapetyants · Vladislav Kravchenko ·
Elijah Liflyand
Editors

Modern Methods in Operator Theory and Harmonic Analysis

OTHA 2018, Rostov-on-Don, Russia, April 22–27

Selected, Revised and Extended Contributions

Springer

Editors
Alexey Karapetyants
Institute of Mathematics, Mechanic and
Computer Sciences and Regional
Mathematical Center
Southern Federal University
Rostov-on-Don, Russia

Vladislav Kravchenko
Department of Mathematics
Cinvestav
Santiago de Querétaro, Querétaro, Mexico

Elijah Liflyand
Department of Mathematics
Bar-Ilan University
Ramat Gan, Israel

ISSN 2194-1009 ISSN 2194-1017 (electronic)
Springer Proceedings in Mathematics & Statistics
ISBN 978-3-030-26747-6 ISBN 978-3-030-26748-3 (eBook)
https://doi.org/10.1007/978-3-030-26748-3

Mathematics Subject Classification (2010): 47Bxx, 46Exx, 35Qxx, 42-xx, 46E15, 46E30, 46E35, 42-00

This Springer imprint is published by the registered company Springer Nature Switzerland AG
The registered company address is: Gewerbestrasse 11, 6330 Cham, Switzerland

Preface

This volume assembles new results concerning the methods, problems, and applications of operator theory and harmonic analysis. Presented by leading researchers, chapters notably touch on recent developments in spectral theory, with applications to PDEs and real-world problems, along with a deep emphasis on the new trends in operator theory and function spaces theory, such as variable exponent spaces and objects of variable nature. The published results will mainly attract researchers who are interested in real and complex variable methods, applications to PDEs, and exploration of new themes and trends in harmonic analysis and operator theory. The relevance of the topics covered in this volume, is stipulated, in particular, by the study of complicated multiparameter objects that require the involvement of operators with variable parameters and functional spaces with fractional and even variable exponents.

This volume consists mainly of the works of speakers at the annual *International Scientific Conference of Modern Methods and Problems of Operator Theory and Harmonic Analysis and Their Applications* (*OTHA-2018*), which took place on April 22–27, 2018, at the Southern Federal University in Rostov-on-Don, Russia. Information on this and preceding conferences is available at the conference website http://otha.sfedu.ru/.

Every year numerous investigators from research and educational institutions worldwide participate in OTHA conferences. Among them are mathematicians from Russia, China, the United States, the United Kingdom, Israel, Finland, Germany, Canada, Italy, Spain, Portugal, Sweden, Mexico, Colombia, Armenia, Azerbaijan, Jordan, Iran, UAE, Belarus, Kazakhstan, Poland, Lithuania, Uzbekistan, and Lithuania. One of the main features of OTHA conferences is the existence of a very special creative atmosphere. This results from both opportunities to form numerous contacts and collaborations and the sharing of inspirational presentations and ideas. The conference in 2018 was supported by the Regional Mathematical Center of the Southern Federal University in collaboration with the Institute of Mathematics, Mechanics and Computer Science of the Southern Federal University. It was also supported by a grant from the Russian Foundation for Basic

Research (https://www.rfbr.ru/) and by the International Society for Analysis and its Applications (ISAAC, isaacmath.org).

We also note that this volume is prepared as part of a special agreement between the ISAAC society and Springer. We thank the ISAAC society for its many years of fruitful collaboration and support, and we thank Springer for providing such a great opportunity. We thank all the authors of this volume for their participation and contributions.

We express our immense regret at the loss of two participants (plenary speakers) who frequently presented at OTHA conferences, our dear friends and great mathematicians, Stasys Rutkauskas (Lithuania) and Garnik Karapetyan (Armenia), both passing in 2018. We have reserved several pages of this volume to honor the memory of these great researchers and friends.

Rostov-on-Don, Russia — Alexey Karapetyants
Santiago de Querétaro, Mexico — Vladislav Kravchenko
Ramat Gan, Israel — Elijah Liflyand

Professor Garnik Karapetyan (1958/02/03–2018/11/29)

With kind permission of Heghine Petrosyan. Copyright remains with the author

Professor Garnik Karapetyan, an outstanding mathematician, a specialist in differential equations and differentiable functions of several variables and other related areas, and a Distinguished Professor of the Russian–Armenian University (RAU), passed away on November 29, 2018.

The diverse scientific activity of Prof. G. Karapetyan meant that he was well-known in the mathematical community. One of the main parts of the theory of differential equations and functional analysis, the general theory of differential equations in the spaces of differentiable functions, is associated with his name. His research played a significant role in the development of a general theory of integral representations and embedding theorems in multianisotropic functional spaces.

Professor Garnik Karapetyan was born on February 3, 1958, in Leninakan (now Gyumri), Armenian SSR. In 1975, he started undergraduate studies at the Faculty of Mechanics and Mathematics of Yerevan State University (YSU) and graduated *cum laude* in 1980. In 1983 he defended his Ph.D. thesis and started work as an assistant at the Department of Numerical Analysis of the Faculty of Informatics and Applied Mathematics at YSU. From 1990 to 1994, he held the position of dean of the

Faculty of Informatics and Applied Mathematics at YSU. He was head of the Mathematics Olympiad Committee of Armenia for 2 years starting in 1997. In 2002, he founded the Department of Mathematics and Mathematical Modeling at RAU. In 2007, he successfully defended his D.Sc. thesis entitled "Properties of the solutions of regular hypoelliptic equations." In 2008, he received the academic rank of professor from the Higher Attestation Commission of the Republic of Armenia and the title of Distinguished Professor at RAU. In 2008, he became a member of the Scientific Council of RAU and the Dissertation Council in Mathematics of the Republic of Armenia. From 2015, he was director of the Applied Mathematics and Informatics program at RAU. He was the first director of the Institute of Mathematics and Informatics at RAU. Professor Garnik Karapetyan was awarded the Medal of Anania Shirakatsi, by decree of the President of the Republic of Armenia.

For many years, he was co-head of the scientific seminar "Functional spaces and their applications in mathematical physics," first at YSU and then RAU. He authored and co-authored 8 textbooks and manuals and more than 60 research articles. His research was mainly devoted to elliptic and hypoelliptic differential equations, the study of the properties of functions in different multianisotropic spaces, and the integral representations and embedding theorems for functions in multianisotropic spaces.

The size limitation applied to this text makes it possible to only briefly describe some of, but not all, the studies he conducted, often with students, which resulted in significant advances and a new deep understanding. His main research activity encompassed:

- The convergence of Galerkin approximations to the solution of the Dirichlet problem for different classes of hypoelliptic equations.
- Liouville and Phragmén–Lindelöf type theorems for general regular equations.
- The behavior of solutions of a certain class of hypoelliptic and μ-elliptic equations in bounded and unbounded domains.
- Estimates for solutions of quasi-elliptic equations in Hölder spaces.
- Schauder-type estimates for solutions of semi-elliptic equations with variable coefficients.
- The study of solutions of hypoelliptic and degenerate equations in a half-space.
- The study of boundary value problems for regular quasilinear equations in unbounded domains.
- Proof of stabilization of solutions for regular and hypoelliptic equations of different classes.
- The study of solutions of nondegenerate equations depending on a parameter.
- The study of eigenvalues of self-adjoint semi-elliptic operators.
- Proof of multiplicative inequalities of Gagliardo–Nirenberg type in multianisotropic Sobolev spaces.
- Approximations of solutions of semi-elliptic equations in a half-space.
- Computation of the index of semi-elliptic operators.
- Proof of the Noetherianness of semi-elliptic and nondegenerate operators.

- Application of the small parameter method for the approximation of eigenvalues of regular hypoelliptic operators.
- Integral representations of functions in different multianisotropic spaces and their embedding theorems.

Professor Garnik Karapetyan was a highly qualified professional in his field and a talented leader. One of his distinguishing qualities was paying particular attention and respect to people and their problems. He was impeccably honest and principled in his work and tactful in relationships with his family members, relatives, friends, colleagues, and students. He was tremendously charismatic; he enjoyed exceptional trust and reverence from students. Not only did he demonstrate genuine interest in everything new in science and life in general, but he also triggered interest in his students and those surrounding him.

Professor G. Karapetyan was not only well-known among scientists and academics; the general public also knew him thanks to the educational program "Open Lesson," broadcast on public TV in Armenia.

What is particularly sad for all his friends and colleagues is that he left this world during his creative ascent. During his last 5–6 years he actively conducted fruitful research on the integral representations of functions in multianisotropic Sobolev spaces and the embedding theorems for such spaces. His most recent work, in the field of isotropic or anisotropic Sobolev spaces, formed part of a body of work that included research by distinguished scientists including S. Sobolev, L. Schwartz, S. Nikolskii, S. Agmon, O. Besov, H. Triebel, E. Gagliardo, P. Lizorkin, I. Slobodetsky, I. Nechas, V. Burenkov, S. Uspenskii, V. Solonnikov, G. Demidenko, A. Kufner, and many others.

Most recently, Prof. G. Karapetyan obtained integral representations for functions in multianisotropic Sobolev spaces generated by regular Newton polyhedra, and, based on these representations, proved embedding theorems for such spaces. Some of the results he obtained are not yet published. His plans for future work included the description of traces of functions in multianisotropic spaces on hypersurfaces of different dimensions and theorems on the continuation of functions outside their domain of definition. We hope that this work will be continued by his disciples.

The family, relatives, friends, colleagues, and students of Garnik Karapetyan, as well as Armenian mathematics, suffered an irretrievable loss due to his untimely death.

Cherished memories of Garnik Karapetyan will forever remain in our hearts.

Adamyan G. V., Aramyan R. H., Avetisyan P. S., Babayan A. H., Baghdasaryan A. G., Berberyan S. L., Darbinyan A. A., Darbinyan A. R., Gevorgyan G. G., Ghazaryan E. M., Ghazaryan H. G., Harutyunyan K. V., Harutyunyan T. N., Margaryan V. N., Petrosyan H. A., Sandoyan E. M., Sargsyan G. Z., Tonoyan R. N., and Tumanyan A. G.

Professor Stasys Rutkauskas (1951/06/06–2018/10/29)

With kind permission of Gintautas Dzemyda. Copyright remains with the author

Professor Stasys Rutkauskas was born on June 6, 1951, in Nečionių in the Širvintai region of Lithuania. He graduated from secondary school in 1968 in Širvintai. In 1973 he finished his studies at the Faculty of Mathematics of Vilnius University (VU) and acquired a diploma as a mathematician. From 1974 to 1977 he was a Ph.D. student at the Institute of Mathematics and Informatics in Vilnius, and in 1978 he defended his Ph.D. theses at Minsk University, receiving a doctoral degree in mathematics. From 1977 to 1980 he worked as a junior research fellow. Starting in 1980 he worked as a senior research fellow at the Institute of Mathematics and Informatics. From 1995 to 2005 he was the Scientific Secretary of the Institute, and in 2005, after defending his habilitation theses and gaining a D.Sc., he became vice-director for research at the Institute of Mathematics and Informatics (nowadays Vilnius University Institute of Data Science and Digital Technologies).

Professor S. Rutkauskas had more than 50 scientific papers published and presented many invited talks at international scientific conferences. The topics of his research included boundary value problems for degenerate elliptic systems and the asymptotic behavior of their solutions near singular points. He actively collaborated with mathematicians from other countries while working on these problems, delivering special lecture courses at foreign universities on multiple occasions.

Professor S. Rutkauskas was a member of the editorial boards of several international mathematical journals including the *Lithuanian Mathematical Journal* and *Mathematical Modelling and Analysis*. He gave lectures to bachelor and master's students in VU, Vilnius Pedagogical University (VPU), and Vilnius Gediminas Technical University (VGTU). His textbook for Ph.D. students entitled *Asymptotic Methods for Ordinary Differential Equations* proved very helpful to young mathematicians. He was the leader of the scientific project entitled "Informatics and Mathematics doctoral studies development (InMaDra)," supported by European structural funds.

He loved his family very much. He had two sons and three granddaughters. He was warm, loved by all, a modest man who left unfinished both works and ideas.

Prof. Dr. Habil. Gintautas Dzemyda
Full member of the Lithuanian Academy of Sciences
Director of the Institute of Mathematics and Informatics
Head of the Systems Analysis Department

Contents

Part I
Function Theory and Approximation Theory

Some General Properties of Operators in Morrey-Type Spaces

Mikhail L. Goldman and Elza Bakhtigareeva

Abstract In the paper we consider general properties of operators acting from rearrangement invariant spaces into generalized Morrey-type spaces. We extract wide class of operators that preserve non-negativity and monotonicity of functions and prove two-sided estimates for their norms. As corollaries we obtain corresponding results for operators of embedding and for Hardy–Littlewood maximal operators. For operators commuting with a shift operator the results are extended to the case of global Morrey-type spaces. As an application of these approaches we establish a criterion of the embedding for a weighted Lorentz space into a Morrey-type space.

Keywords Local and global Morrey-type spaces · Banach function spaces · Decreasing rearrangements · Rearrangement invariant spaces · Lorentz spaces · Estimates of the norms · Criterion of embedding

1 Introduction

In this paper we consider Morrey-type spaces on the basis of Banach function spaces (shortly: BFS) and rearrangement invariant spaces(shortly: RIS). Let us recall the corresponding definitions. We follow here the system of axioms and notation, developed by Bennett and Sharpley [1]. Let $(\mathfrak{R}, \mu)$ be a space with a measure; L_0 be a set of all measurable functions on $(\mathfrak{R}, \mu)$; L_0^+ be a set of all nonnegative measurable functions. Throughout the whole paper, we denote by C a positive constant which is

M. L. Goldman (✉) · E. Bakhtigareeva
Peoples' Friendship University of Russia (RUDN University),
6 Miklukho-Maklaya St, Moscow 117198, Russian Federation
e-mail: seulydia@yandex.ru

E. Bakhtigareeva
e-mail: bakhtigareeva-eg@rudn.ru

M. L. Goldman · E. Bakhtigareeva
Steklov Mathematical Institute of Russian Academy of Sciences,
8 Gubkina St., Moscow 119991, Russian Federation

A. Karapetyants et al. (eds.), *Modern Methods in Operator Theory and Harmonic Analysis*, Springer Proceedings in Mathematics & Statistics 291,
https://doi.org/10.1007/978-3-030-26748-3_1

independent of the main parameters, but it may vary from line to line. The symbol $D \lesssim F$ means that $D \leq CF$.

Definition 1 A mapping $\rho : L_0^+ \to [0, \infty]$ is a Banach function norm, if

$$(A1) \quad \rho(f) = 0 \Leftrightarrow f = 0 \quad \mu\text{-a.e.}; \quad \rho(\alpha f) = \alpha\rho(f), \alpha > 0; \quad \rho(f + g) \leq \rho(f) + \rho(g);$$

$$(A2) \quad 0 \leq g \leq f \quad \mu\text{-a.e.} \Rightarrow \rho(g) \leq \rho(f);$$

$$(A3) \quad 0 \leq f_n \uparrow f \quad \mu\text{-a.e.} \Rightarrow \rho(f_n) \uparrow \rho(f);$$

$$(A4) \quad \mu(\Omega) < \infty \Rightarrow \rho(\chi_\Omega) < \infty;$$

$$(A5) \quad \mu(\Omega) < \infty \Rightarrow \int_\Omega f d\mu < c_\Omega \rho(f).$$

Definition 2 For a Banach function norm ρ the collection $X = X(\rho)$ of all functions f in L_0 for which $\rho(|f|) < \infty$ is called a Banach function space. For each $f \in X$, define

$$||f||_X := \rho(|f|).$$

Remark 1 Note that a Banach function space is complete.

Definition 3 If in a Banach function space X the norm ρ is monotone with respect to decreasing rearrangements:

$$g^* \leq f^* \Rightarrow \rho(g) \leq \rho(f), \tag{1}$$

then X is called a rearrangement invariant space. For a RIS $E = E(R^n) \quad \exists! \quad \tilde{E} = \tilde{E}(R_+)$:

$$||f||_{E(R^n)} = ||f^*||_{\tilde{E}(R_+)} \tag{2}$$

(the Luxemburg representation).

Definition 4 Let $\Omega \subset R^n, \quad \mu(\Omega) = t > 0$. For a RIS $E = E(R^n)$ define a fundamental function:

$$\varphi_E(t) := ||\chi_\Omega||_{E(R^n)} = ||\chi_{[0,t)}||_{\tilde{E}(R_+)}, \quad t \in R_+. \tag{3}$$

Remark 2 Note that $\varphi_E(t) = 0 \Leftrightarrow t = 0, \quad \varphi_E(t) \uparrow, \quad \frac{t}{\varphi_E(t)} \uparrow$, moreover, for associated spaces $E(R^n)$ and $E'(R^n)$ we have

$$\varphi_E(t) \cdot \varphi_{E'}(t) = t, \quad t \in R_+. \tag{4}$$

Remark 3 For $\Re = R^n$ we assume that $\mu = \mu_n$ is a Lebesgue measure in R^n; for $\Re = R_+$ $\mu = \mu_1$ is a Lebesgue measure in R_+.

The outline of this paper is as follows. In Sect. 2 we consider some general properties of operators in local Morrey-type spaces. We prove a nontriviality criterion for the space and obtain order-sharp estimates for a wide class of operators including Hardy–Littlewood maximal operator (Propositions 1–3). Also, we introduce two functions connected with such operators that play an important role in the problem and prove some relations for them (Propositions 4–6). In Sect. 3 for operators commuting with shift operators we prove that their norms coincide into local and global Morrey-type spaces. In Sect. 4 we study the actual problem about relations between weighted Lorentz and Morrey-Lorentz spaces with general weights. These spaces are popular and important in recent theory of function spaces. Section 4 contains the main results about a criterion for embedding of weighted Lorentz spaces into Morrey-type spaces. For all cases of relations between parameters we obtain explicit three-weight order-sharp estimates (in some cases exact values) for the norm of embedding's operator. Finally, in Sect. 5 we prove the results of Sect. 4.

2 Some General Properties of Operators in Local Morrey-Type Spaces

The basic theory of classical Morrey-type spaces and recent development of more general theory about properties of operators in Morrey-type spaces were presented in papers of V. I. Burenkov, A. Gogatishvili, V. S. Guliev, D. Haroske and L. Skrzypczak, P. G. Lemarie-Rieusset, T. Mizuhara, H. Naki, E. D. Nursultanov, R. Mustafaev, Y. Sawano, W. Sickel, T. V. Tararykova, W. Yuan, and D. Yang and others, see, for example, [3, 4, 7–10, 14–17]. We develop here some of these approaches by considering more general basic function spaces and special classes of operators. Concerning our approaches for Hardy-type operators on cones of functions with properties of monotonicity see [12, 13].

Let $E = E(R^n)$ be a rearrangement invariant space (shortly: RIS), $F = F(R_+)$ be a Banach function space (shortly: BFS), $\tilde{E} = \tilde{E}(R_+)$ be a Luxemburg representation of $E(R^n)$ (see (2)). A local Morrey- type space $LM_{EF}(R^n)$ is a set of functions $f \in E^{loc}(R^n)$, such that

$$||f||_{LM_{EF}} \equiv ||f||_{LM_{EF}(R^n)} := || \, ||f||_{E(B_r)} ||_{F(R_+)} < \infty. \tag{5}$$

Here $R_+ = (0, \infty)$, $B_r = B(0, r)$, where

$$B(x, r) = \left\{ y \in R^n : |y - x| < r \right\}; \quad ||f||_{E(B_r)} = ||f \chi_{B_r}||_{E(R^n)}. \tag{6}$$

Further we will use the following short notations for norms in function spaces:

$$E = E(R^n), \tilde{E} = \tilde{E}(R_+), F = F(R_+), \quad \text{so}$$

$$||f||_E := ||f||_{E(R^n)}, ||g||_{\tilde{E}} := ||g||_{\tilde{E}(R_+)}, ||h||_F := ||h||_{F(R_+)}.$$

Proposition 1 *A nontriviality criterion for $LM_{EF}(R^n)$ is the following:*

$$LM_{EF}(R^n) \neq \{0\} \Leftrightarrow \exists r_0 \in R_+ : ||\chi_{(r_0,\infty)}||_F < \infty. \tag{7}$$

Proof Actually, if $f \in E^{loc}(R^n)$, $f \neq 0$, then

$$\exists r_0 \in R_+, c_0 > 0 : ||f||_{E(B_{r_0})} \geq c_0.$$

Then

$$||f||_{E(B_r)} \geq c_0 \chi_{[r_0,\infty)}(r); r \in R_+,$$

so

$$||f||_{LM_{EF}} \geq c_0 ||\chi_{[r_0,\infty)}||_F.$$

In case $||\chi_{[r_0,\infty)}||_F = \infty$ we have $||f||_{LM_{EF}} = \infty$. Otherwise, if $\exists r_0 \in R_+ :$ $||\chi_{[r_0,\infty)}||_F < \infty$, then for $f_0(x) = \chi_{B_{2r_0} \setminus B_{r_0}}$ we have

$$||f_0||_{LM_{EF}} \leq ||\chi_{B_{2r_0}}||_E ||\chi_{[r_0,\infty)}||_F < \infty,$$

so $LM_{EF}(R^n) \neq \{0\}$.

Let Ω_0 be a set of all nonnegative, finite on R_+, decreasing and right continuous functions:

$$\Omega_0 = \{g : R_+ \to [0,\infty); g \downarrow, g(t+0) = g(t), t \in R_+\}. \tag{8}$$

Let Ω be a set of radial functions, belonging to Ω_0 as a function $\rho = |x|$:

$$\Omega = \left\{f(x) = f_0(|x|), x \in R^n; f_0 \in \Omega_0\right\} \subset L_0(R^n). \tag{9}$$

For symmetric rearrangements $f^{\#}$ and decreasing rearrangements f^* we have

$$f \in \Omega \Rightarrow f^{\#}(\rho) = f_0(\rho); f^*(t) = f_0((\frac{t}{v_n})^{\frac{1}{n}}); \quad \rho, t \in R_+. \tag{10}$$

Here v_n is a volume of a unit ball in R^n. Note that for $r \in R_+$ $|B_r| = v_n r^n$. Denote by $\mathcal{A}$ the set of operators $A : \mathcal{D}_{\mathcal{A}} \to L_0(R^n)$, where $\mathcal{D}_{\mathcal{A}} \subset L_0(R^n)$, possessing the following property:

$$A : \mathcal{D}_{\mathcal{A}} \cap \Omega \mapsto L_0(R^n) \cap \Omega. \tag{11}$$

Proposition 2 *Under the notation above for $A \in \mathcal{A}$*

$$||Af||_{LM_{EF}} \leq || \quad ||(Af)^* \chi_{(0,|B_r|)}||_{\tilde{E}}||_F, \tag{12}$$

$\forall f \in \mathcal{D}_{\mathcal{A}}$. *If* $f \in \mathcal{D}_{\mathcal{A}} \cap \Omega$, *then*

$$||Af||_{LM_{EF}} = || \ \ ||(Af)^{*}\chi_{(0,|B_r|)}||_{\tilde{E}}||_{F}. \tag{13}$$

Here for $(Af)^{*}(t)\chi_{(0,|B_r|)}(t)$ *the norm in* $\tilde{E}(R_+)$ *is taken by* t, *and an outer norm in* $F(R_+)$ – *by* r.

Proof 1. Note that

$$(Af\cdot\chi_{B_r})^{*}(t) \le (Af)^{*}(t)\chi_{(0,|B_r|)}(t), t \in R_+. \tag{14}$$

Actually, $|Af\cdot\chi_{B_r}| \le |Af| \Rightarrow (Af\cdot\chi_{B_r})^{*}(t) \le (Af)^{*}(t), t \in R_+$.

Further, $\mu\left\{x \in R^n : \ \ |Af(x)\chi_{B_r}(x)| \ne 0\right\} \le |B_r|$, so $(Af\cdot\chi_{B_r})^{*}(t) = 0$ for $t > |B_r|$. As a result, the estimate (14) holds. It means that for RIS $E(R^n)$ we have

$$||Af||_{E(B_r)} = ||(Af\chi_{B_r})^{*}||_{\tilde{E}} \le ||(Af)^{*}\chi_{(0,|B_r|)}||_{\tilde{E}}.$$

Then, taking into account the monotonicity of the norm in a BFS $F(R_+)$, we get

$$||Af||_{LM_{EF}} = || \ \ ||Af||_{E(B_r)}||_{F} \le || \ \ ||(Af)^{*}\chi_{(0,|B_r|)}||_{\tilde{E}}||_{F} \Rightarrow (12).$$

Next, if $f \in \mathcal{D}_{\mathcal{A}} \cap \Omega$, then $Af \in L_0 \cap \Omega$, so, according to (10), we have

$$(Af)(x) = (Af)_0(\rho), \quad \rho = |x|, (Af)_0 \in \Omega_0 \Rightarrow (Af)^{*}(t) = (Af)_0((\frac{t}{v_n})^{\frac{1}{n}}).$$

Thus,

$$(Af)^{*}(t)\chi_{(0,|B_r|)}(t) = (Af)_0((\frac{t}{v_n})^{\frac{1}{n}})\chi_{(0,|B_r|)}(t). \tag{15}$$

But we also have $Af\chi_{B_r} \in \Omega$, so

$$Af(x)\chi_{B_r}(x) = (Af)_0(\rho)\chi_{(0,r)}(\rho), \quad (Af\chi_{B_r})_0(\rho) = (Af)_0(\rho)\chi_{(0,r)}(\rho), \rho \in R_+.$$

Hence, using the equality (10), we get

$$\begin{aligned}(Af\chi_{B_r})^{*}(t) = (Af\chi_{B_r})_0((\frac{t}{v_n})^{\frac{1}{n}}) = (Af)_0((\frac{t}{v_n})^{\frac{1}{n}})\chi_{(0,r)}((\frac{t}{v_n})^{\frac{1}{n}}) = \\ = (Af)_0((\frac{t}{v_n})^{\frac{1}{n}})\chi_{(0,|B_r|)}(t).\end{aligned} \tag{16}$$

Comparing (15) and (16) shows that for $f \in \mathcal{D}_{\mathcal{A}} \cap \Omega$ we have

$$(Af\chi_{B_r})^{*}(t) = (Af)^{*}(t)\chi_{(0,|B_r|)}(t), \quad t \in R_+. \tag{17}$$

Then, for RIS $E(R^n)$ and $f \in \mathcal{D}_{\mathcal{A}} \cap \Omega$

$$||Af||_{E(B_r)} = ||Af\chi_{B_r}||_E = ||(Af\chi_{B_r})^*)||_{\tilde{E}} = ||(Af)^*\chi_{(0,|B_r|)}||_{\tilde{E}}.$$

From here and from (5) we get (13).

Example 1 The identity operator $I \in \mathcal{A}$; $\mathcal{D}_{\mathcal{I}} = L_0(R^n)$, so that

$$||f||_{LM_{EF}} \leq || \quad ||f^*\chi_{(0,|B_r|)}||_{\tilde{E}}||_F, \quad f \in L_0(R^n); \tag{18}$$

$$||f||_{LM_{EF}} = || \quad ||f^*\chi_{(0,|B_r|)}||_{\tilde{E}}||_F, \quad f \in \Omega. \tag{19}$$

Example 2 The maximal operator $M \in \mathcal{A}$;
$\mathcal{D}_{\mathcal{M}} = \left\{ f \in L_1^{loc}(R^n) : f^{**}(t) < \infty, t \in R_+ \right\} \subset L_0(R^n)$; so that

$$||Mf||_{LM_{EF}} \leq || \quad ||(Mf)^*\chi_{(0,|B_r|)}||_{\tilde{E}}||_F, \quad f \in \mathcal{D}_{\mathcal{M}}; \tag{20}$$

$$||Mf||_{LM_{EF}} = || \quad ||(Mf)^*\chi_{(0,|B_r|)}||_{\tilde{E}}||_F, \quad f \in \Omega \cap \mathcal{D}_{\mathcal{M}}. \tag{21}$$

Here

$$(Mf)(x) = \sup_{r>0} \frac{1}{|B(x,r)|} \int\limits_{B(x,r)} |f(y)|dy, \quad f \in \mathcal{D}_{\mathcal{M}}. \tag{22}$$

The following property of the maximal operator takes place:

$$f \in \mathcal{D}_{\mathcal{M}} \cap \Omega \Rightarrow Mf \in \Omega. \tag{23}$$

Actually, if $f \in \mathcal{D}_{\mathcal{M}}$, then $\int_0^t f^*(\tau)d\tau < \infty, \forall t \in R_+$, so that, taking into account the known estimate (see [1]), we have

$$(Mf)^*(t) \cong f^{**}(t) = t^{-1} \int_0^t f^*(\tau)d\tau < \infty, \forall t \in R_+. \tag{24}$$

So, $|Mf| < \infty$ a.e. Let us show that if $f \in \mathcal{D}_{\mathcal{M}} \cap \Omega$, then $Mf \in \Omega$. Let $x_0, \tilde{x}_0 \in R^n, \quad |x_0| = |\tilde{x}_0|$. For $f \in \mathcal{D}_{\mathcal{M}} \cap \Omega$ we have

$$\int\limits_{B(x_0,r)} |f(y)|dy = \int\limits_{B(\tilde{x}_0,r)} |f(\tilde{y})|d\tilde{y}. \tag{25}$$

Indeed, if $G : R^n \to R^n$ is a rotation, such that $G(B(x_0, r)) = B(\tilde{x}_0, r)$, then $y \in B(x_0, r) \leftrightarrow \tilde{y} = Gy \in B(\tilde{x}_0, r), dy = d\tilde{y}, |y| = |\tilde{y}|; f(y) = f(\tilde{y})$. It leads to (25). So,

$$(Mf)(x_0) = \sup_{r>0} \frac{1}{|B(x_0,r)|} \int\limits_{B(x_0,r)} |f(y)|dy =$$

$$= \sup_{r>0} \frac{1}{|B(\tilde{x}_0, r)|} \int\limits_{B(\tilde{x}_0,r)} |f(\tilde{y})| d\tilde{y} = (Mf)(\tilde{x}_0),$$

i.e. $(Mf)(x)$ is a radial function as well as $f(x)$. Moreover,

$$|\tilde{x}_0| > |x_0| \Rightarrow (Mf)(\tilde{x}_0) \le (Mf)(x_0). \tag{26}$$

So, it leads to (23), i.e. $M \in \mathcal{A}$.

Corollary 1 *For the maximal operator the following estimate takes place*

$$||Mf||_{LM_{EF}} \cong || \;\; ||f^{**}\chi_{(0,|B_r|)}||_{\tilde{E}}||_F, \quad f \in \Omega \cap \mathcal{D}_{\mathcal{M}}. \tag{27}$$

Indeed, (27) follows from (21) and (24).

Now for $c_0 \in [1, \infty)$ introduce a set of operators

$$\mathcal{A}_0(c_0) = \mathcal{A} \cap \left\{A : (Af)^\# \le c_0 A(f^\#)\right\}, \tag{28}$$

i. e.

$$A \in \mathcal{A}_0(c_0) \Leftrightarrow \begin{cases} 1.(A(f^\#))^\# = A(f^\#); \\ 2.(Af)^\# \le c_0 A(f^\#). \end{cases} \tag{29}$$

Proposition 3 *Let $E_i(R^n)$ be a RIS; $i = 1, 2$; $F(R_+)$ be a BFS;*

$$A : E_1(R^n) \to L_0(R^n); \quad A \in \mathcal{A}_0(c_0), \quad c_0 \in [1, \infty). \tag{30}$$

Then, for the norm

$$||A|| = ||A||_{E_1 \to LM_{E_2F}} \tag{31}$$

the following inequality takes place

$$||A||_* \le ||A|| \le c_0 ||A||_*. \tag{32}$$

Here

$$||A||_* = \sup\left\{|| \; ||(Af)^*\chi_{(0,|B_r|)}||_{\tilde{E}_2}||_F : \; f \in E_1 \cap \Omega; ||f||_{E_1} \le 1\right\}. \tag{33}$$

Proof For the norm of the operator $A \in \mathcal{A}$

$$||A|| = \sup\left\{||Af||_{LM_{E_2F}} : \; f \in E_1, ||f||_{E_1} \le 1\right\} \tag{34}$$

we use the estimate (12) and a Luxemburg representation. Then,

$$||A|| \le \sup\left\{|| \; ||(Af)^*\chi_{(0,|B_r|)}||_{\tilde{E}_2}||_F : \; f \in E_1, ||f^*||_{\tilde{E}_1} \le 1\right\}.$$

We also have $A \in \mathcal{A}_0(c_0)$, and, according to (29), for $t \in R_+$

$$(Af)^*(t) = (Af)^\#((\frac{t}{v_n})^{\frac{1}{n}}) \le c_0(A(f^\#))^\#((\frac{t}{v_n})^{\frac{1}{n}}) = c_0(A(f^\#))^*(t). \tag{35}$$

Thus,

$$||A|| \le c_0 \sup\left\{|| \, ||(A(f^\#)^*\chi_{(0,|B_r|)}||_{\tilde{E}_2}||_F : \, f \in E_1, ||f^*||_{\tilde{E}_1} \le 1\right\}.$$

But $f^* = (f^\#)^*$ and

$$f \in E_1, ||f^*||_{\tilde{E}_1} \le 1 \Leftrightarrow h = f^\# \in E_1 \cap \Omega, \; ||h^*||_{\tilde{E}_1} \le 1.$$

Hence,

$$||A|| \le c_0 \sup\left\{|| \, ||(Ah)^*\chi_{(0,|B_r|)}||_{\tilde{E}_2}||_F : \, h \in E_1 \cap \Omega, ||h^*||_{\tilde{E}_1} \le 1\right\}. \tag{36}$$

On the other hand, for $A \in \mathcal{A}$, $h \in E_1(R^n) \cap \Omega$ we have the equality (13) with $f = h$. So, from (34) it follows that

$$\begin{gathered} ||A|| \ge \sup\left\{||Ah||_{LM_{\tilde{E}_2 F}} : \, h \in E_1 \cap \Omega, ||h^*||_{\tilde{E}_1} \le 1\right\} = \\ \sup\left\{|| \quad ||(Ah)^*\chi_{(0,|B_r|)}||_{\tilde{E}_2}||_F : \, h \in E_1 \cap \Omega, ||h^*||_{\tilde{E}_1} \le 1\right\}. \end{gathered} \tag{37}$$

Equations (32), (33) follow from (37) and (36).

Corollary 2 *For the identity operator $A = I$ we have*

$$||I|| = ||I||_* = \sup\left\{|| \, ||g\chi_{(0,|B_r|)}||_{\tilde{E}_2}||_F : g \in \tilde{E}_1 \cap \Omega_0; \, ||g||_{\tilde{E}_1} \le 1\right\}. \tag{38}$$

Proof For $A = I \in \mathcal{A}_0(1)$ the inequality (32) becomes an equality

$$||I|| = ||I||_* = \sup\left\{|| \, ||f^*\chi_{(0,|B_r|)}||_{\tilde{E}_2}||_F : \, f \in E_1 \cap \Omega, ||f^*||_{\tilde{E}_1} \le 1\right\}. \tag{39}$$

But,

$$f \in E_1 \cap \Omega, \quad ||f^*||_{\tilde{E}_1} \le 1 \Leftrightarrow g \equiv f^* \in \tilde{E}_1 \cap \Omega_0, ||g||_{\tilde{E}_1} \le 1.$$

Thus, (39) leads to (38).

Corollary 3 *For the maximal operator $M \in \mathcal{A}_0(c_0)$ we have estimates (32), (33) with $A = M$, $c_0 = c_2/c_1 \in [1, \infty)$, c_1, c_2 are constants in the inequality:*

$$c_1(Mf)^*(t) \le t^{-1}\int_0^t f^*(\tau)d\tau \le c_2(Mf)^*(t), \; t \in R_+ \; (see\ (24)). \tag{40}$$

Moreover,

$$c_1||M||_* \le ||M||_{**} \le c_2||M||_*. \tag{41}$$

Here

$$||M||_{**} = \sup\left\{|| \, ||(t^{-1}\int_0^t g(\tau)d\tau)\chi_{(0,|B_r|)}(t)||_{\tilde{E}_2}||_F : g \in \tilde{E}_1 \cap \Omega_0; \, ||g||_{\tilde{E}_1} \le 1\right\} \tag{42}$$

(the inner norm in $\tilde{E}_2(R_+)$ is taken by t; the outer norm in $F(R_+)$– by r).

Proof $M \in \mathcal{A}$ (see Example 2). We want to get the estimate

$$(Mf)^{\#} \le c_0 M(f^{\#}), \; c_0 = c_2/c_1 \in [1, \infty), \tag{43}$$

using (40). From equalities

$$f^*(\tau) = f^{\#}((\frac{\tau}{v_n})^{\frac{1}{n}}); \quad (Mf)^*(t) = (Mf)^{\#}((\frac{t}{v_n})^{\frac{1}{n}})$$

and the estimate (40), using substitutions $(\frac{t}{v_n})^{\frac{1}{n}} = \rho$; $(\frac{\tau}{v_n})^{\frac{1}{n}} = \xi$ we get

$$c_1(Mf)^{\#}(\rho) \le \frac{n}{\rho^n}\int_0^\rho f^{\#}(\xi)\xi^{n-1}d\xi \le c_2(Mf)^{\#}(\rho), \rho \in R_+. \tag{44}$$

Let us replace f by $f^{\#}$ and take into account that $(f^{\#})^{\#} = f^{\#}$. Then

$$c_1(M(f^{\#}))^{\#}(\rho) \le \frac{n}{\rho^n}\int_0^\rho f^{\#}(\xi)\xi^{n-1}d\xi \le c_2(M(f^{\#}))^{\#}(\rho), \rho \in R_+.$$

But $(M(f^{\#}))^{\#} = M(f^{\#})$, so

$$c_1 M(f^{\#})(\rho) \le \frac{n}{\rho^n}\int_0^\rho f^{\#}(\xi)\xi^{n-1}d\xi \le c_2 M(f^{\#})(\rho), \rho \in R_+. \tag{45}$$

From here and from (44) it follows that

$$(Mf)^{\#}(\rho) \le \frac{n}{c_1\rho^n}\int_0^\rho f^{\#}(\xi)\xi^{n-1}d\xi \le \frac{c_2}{c_1}M(f^{\#})(\rho), \rho \in R_+.$$

We got the estimate (43). As a result,

$$M \in \mathcal{A}_0(c_0) \Rightarrow (32), (33), A = M.$$

Finally, from (40) and (33) with A=M we get (41).

Concerning the boundedness problem of the operator A from a RIS into a Morrey-type space the following functions play an important role:

$$\Psi_A(r) = ||A||_{E_1 \to E_2(B_r)} = \sup\left\{||Af||_{E_2(B_r)} : f \in E_1, ||f||_{E_1} \le 1\right\} \tag{46}$$

$$\Phi_A(r) = ||A||^*_{E_1 \to E_2(B_r)} = \sup\left\{||(Af)^* \chi_{(0,|B_r|)}||_{\tilde{E}_2} : f \in E_1 \cap \Omega, ||f||_{E_1} \le 1\right\}. \tag{47}$$

The following proposition describes the monotonicity properties of Ψ_A, Φ_A.

Proposition 4 *Let E_1, E_2 be RISs. Then for the operator $A : E_1(R^n) \to L_0(R^n)$*

$$0 < \Psi_A, \Phi_A \uparrow \text{ on } R_+; \tag{48}$$

$$\Phi_A(\rho) \le \Phi_A(r) \le \Phi_A(\rho)\left(1 + \frac{\varphi_{E_2}(|B_r|)}{\varphi_{E_2}(|B_\rho|)}\right), \; 0 < \rho < r. \tag{49}$$

Here $\varphi_{E_2}(t)$ is a fundamental function of the RIS $E_2(R^n)$:

$$\varphi_{E_2}(t) = ||\chi_{(0,t)}||_{\tilde{E}_2}, \; t \in R_+.$$

In particular, if $\exists \rho \in R_+ : \Phi_A(\rho) < \infty$, then $\Phi_A(r) < \infty \; \forall r \in R_+$.

Proof The property (48) and the left-side inequality in (49) are obvious. Let us prove the right-side inequality in (49). For $0 < \rho < r$ we have

$$||g\chi_{(0,|B_r|)}||_{\tilde{E}_2} \le ||g\chi_{(0,|B_\rho|)}||_{\tilde{E}_2}\left(1 + \frac{\varphi_{E_2}(|B_r|)}{\varphi_{E_2}(|B_\rho|)}\right), \; g \in \Omega_0. \tag{50}$$

Actually, $g\chi_{(0,|B_r|)} = g\chi_{(0,|B_\rho|)} + g\chi_{[|B_\rho|,|B_r|)}$, so that, according to the Minkowski inequality

$$\begin{aligned} ||g\chi_{(0,|B_r|)}||_{\tilde{E}_2} &\le ||g\chi_{(0,|B_\rho|)}||_{\tilde{E}_2} + ||g\chi_{[|B_\rho|,|B_r|)}||_{\tilde{E}_2} \le \\ &\le ||g\chi_{(0,|B_\rho|)}||_{\tilde{E}_2} + g(|B_\rho|)||\chi_{(0,|B_r|)}||_{\tilde{E}_2}. \end{aligned} \tag{51}$$

But for $g \in \Omega_0$ we have

$$g(|B_\rho|) \le |B_\rho|^{-1}\int_0^{|B_\rho|} g(\tau)d\tau = |B_\rho|^{-1}\int_0^\infty g(\tau)\chi_{(0,|B_\rho|)}\chi_{(0,|B_\rho|)}d\tau.$$

From here, using Hölder's inequality, we get

$$g(|B_\rho|) \le |B_\rho|^{-1}||g\chi_{(0,|B_\rho|)}||_{\tilde{E}_2}||\chi_{(0,|B_\rho|)}||_{\tilde{E}'_2} = |B_\rho|^{-1}||g\chi_{(0,|B_\rho|)}||_{\tilde{E}_2}\varphi_{\tilde{E}'_2}(|B_\rho|).$$

Here $\tilde{E}'_2$ is an associated RIS. For its fundamental function there is a known formula:

$$\varphi_{\tilde{E}_2'}(t) = \frac{t}{\varphi_{E_2}(t)}, \; t \in R_+.$$

Therefore,

$$g(|B_\rho|) \le \frac{||g\chi_{(0,|B_\rho|)}||_{\tilde{E}_2}}{\varphi_{E_2}(|B_\rho|)}.$$

If we substitute this inequality into (51), we get the estimate (50). The estimate (50) with $g = (Af)^* \in \Omega_0 \; \forall f \in E_1(R^n)$ leads to:

$$||(Af)^*\chi_{(0,|B_r|)}||_{\tilde{E}_2} \le ||(Af)^*\chi_{(0,|B_\rho|)}||_{\tilde{E}_2}\left(1 + \frac{\varphi_{E_2}(|B_r|)}{\varphi_{E_2}(|B_\rho|)}\right).$$

From here, proceeding to the supremum by $f \in E_1 \cap \Omega, \; ||f||_{E_1} \le 1$, we get the right-side inequality in (49).

The relation between Ψ_A and Φ_A is the same as the relation between the norms $||A||$ and $||A||_*$.

Proposition 5 *Let $E_1(R^n), \; E_2(R^n)$ be RISs, $A \in \mathcal{A}_0(c_0)$. Then*

$$\Phi_A(r) \le \Psi_A(r) \le c_0\Phi_A(r), \; r \in R_+. \tag{52}$$

Proof For $\Psi_A(r)$ let us use the estimate (14), so that

$$\Psi_A(r) = \sup\left\{||(Af\chi_{B_r})^*||_{\tilde{E}_2} : f \in E_1, \; ||f||_{E_1} \le 1\right\} \le$$

$$\sup\left\{||(Af)^*\chi_{(0,|B_r|)}||_{\tilde{E}_2} : f \in E_1, \; ||f||_{E_1} \le 1\right\} = \Phi_A(r).$$

Next, for $A \in \mathcal{A}_0(c_0)$ the inequality (35) takes place, so that

$$\Psi_A(r) \le c_0 \sup\left\{||(Af^\#)^*\chi_{(0,|B_r|)}||_{\tilde{E}_2} : f \in E_1, \; ||f||_{E_1} \le 1\right\}.$$

But $f \in E_1, \; ||f||_{E_1} \le 1 \Rightarrow h = f^\# \in E_1 \cap \Omega, ||h||_{E_1} \le 1$. Therefore,

$$\Psi_A(r) \le c_0 \sup\left\{||(Ah)^*\chi_{(0,|B_r|)}||_{\tilde{E}_2} : h \in E_1 \cap \Omega, \; ||h||_{E_1} \le 1\right\} = c_0\Phi_A(r).$$

We proved the right-side estimate in (52). Let us prove the left-side estimate. For $A \in \mathcal{A}$ we have the estimate (17). So,

$$\Psi_A(r) \ge \sup\left\{||(Af\chi_{B_r})^*||_{\tilde{E}_2} : f \in E_1 \cap \Omega, \; ||f||_{E_1} \le 1\right\} =$$

$$= \sup\left\{||(Af)^*\chi_{(0,|B_r|)}||_{\tilde{E}_2} : f \in E_1 \cap \Omega, \; ||f||_{E_1} \le 1\right\} = \Phi_A(r).$$

For the maximal operator $A = M$ we have the following result.

Proposition 6 *Let $E_1(R^n)$, $E_2(R^n)$ be RISs. Then for functions Ψ_M and Φ_M ((46) and (47) respectively) the following relations take place*

$$\Psi_M(r) \cong \Phi_M(r) \cong \Phi_M^0(r) = \Phi_M^1(r),\ r \in R_+. \tag{53}$$

Here

$$\Phi_M^0(r) = \sup\left\{||f^{**}\chi_{(0,|B_r|)}||_{\tilde{E}_2} : f \in E_1 \cap \Omega,\ ||f||_{E_1} \le 1\right\};$$

$$\Phi_M^1(r) = \sup\left\{||(t^{-1}\int_0^t g(\tau)d\tau)\chi_{(0,|B_r|)}||_{\tilde{E}_2} : g \in \Omega_0,\ ||g||_{\tilde{E}_1} \le 1\right\}.$$

Proof The first relation in (53) follows from (52) with $A = M$, taking into account that $M \in \mathcal{A}_0(c_0)$ (see Corollary 3). In the second relation we use the definition (47) with $A = M$ and the estimate (24). Finally, we have to prove $\Phi_M^0(r) = \Phi_M^1(r)$. For $f \in E_1$, $||f||_{E_1} \le 1$ we have $g = f^* \in \Omega_0$, $||g||_{\tilde{E}_1} = ||f||_{E_1} \le 1$, $t^{-1}\int_0^t g(\tau)\,d\tau = f^{**}$. So,

$$\Phi_M^0(r) \le \sup\left\{||f^{**}\chi_{(0,|B_r|)}||_{\tilde{E}_2} : f \in E_1,\ ||f||_{E_1} \le 1\right\} \le \Phi_M^1(r).$$

On the other hand, for $\varepsilon \in (0, 1)$ denote by g_ε the function with the following properties:

$$g_\varepsilon \in \Omega_0,\ ||g_\varepsilon||_{\tilde{E}_1} \le 1;\ ||(t^{-1}\int_0^t g_\varepsilon d\tau)\chi_{(0,|B_r|)}||_{\tilde{E}_2} \ge \Phi_M^1(r)(1-\varepsilon).$$

For g_ε $\exists f_\varepsilon \in E_1 \cap \Omega$ such that $f_\varepsilon^* = g_\varepsilon$ and

$$f_\varepsilon^{**}(t) = t^{-1}\int_0^t g_\varepsilon d\tau,\ ||f_\varepsilon||_{E_1} = ||g_\varepsilon||_{\tilde{E}_1} \le 1.$$

Namely, we can take $f_\varepsilon(x) = g_\varepsilon(v_n|x|^n)$. Actually, since $g_\varepsilon \in \Omega_0$, then $f_\varepsilon \in \Omega$, moreover,

$$f_\varepsilon^{\#}(\rho) = g_\varepsilon(v_n\rho^n) \Rightarrow (f_\varepsilon)^*(t) = f_\varepsilon^{\#}((\frac{t}{v_n})^{\frac{1}{n}}) = g_\varepsilon(t),\ \rho, t \in R_+.$$

Thus, $\forall r \in R_+$

$$\Phi_M^0(r) \ge ||f_\varepsilon^{**}\chi_{(0,|B_r|)}||_{\tilde{E}_2} = ||(t^{-1}\int_0^t g_\varepsilon d\tau)\chi_{(0,|B_r|)}||_{\tilde{E}_2} \ge \Phi_M^1(r)(1-\varepsilon).$$

Since $\varepsilon \in (0, 1)$ is arbitrary, it follows from here that $\Phi_M^0(r) \ge \Phi_M^1(r)$. So, we proved that

$$\Phi_M^0(r) = \sup\left\{||f^{**}\chi_{(0,|B_r|)}||_{\tilde{E}_2} : f \in E_1 \cap \Omega,\ ||f||_{E_1} \le 1\right\} = \Phi_M^1(r).$$

The following proposition shows the importance of the functions Ψ_A and Φ_A for assessing the norms $||A|| \equiv ||A||_{E_1 \to LM_{E_2 F}}$ and $||A||_*$ (see (33)).

Proposition 7 *Let $E_1(R^n)$, $E_2(R^n)$ be RISs, $F(R_+)$ be a BFS, A be an operator such that $A : E_1(R^n) \to L_0(R^n)$. Then the following estimates take place:*

$$\sup_{r \in R_+} \left[\Psi_A(r) ||\chi_{[r,\infty)}||_F \right] \le ||A|| \le ||\Psi_A||_F; \tag{54}$$

$$\sup_{r \in R_+} \left[\Phi_A(r) || \frac{\varphi_{E_2}(|B_\rho|)\chi_{(0,r)}(\rho)}{\varphi_{E_2}(|B_\rho|) + \varphi_{E_2}(|B_r|)} + \chi_{[r,\infty)}(\rho) ||_F \right] \le ||A||_* \le ||\Phi_A||_F. \tag{55}$$

Proof 1. The right-side estimate in (54) follows from definitions (31), (5) and (46):

$$||A|| = \sup \left\{ ||\, ||Af||_{E_2(B_r)} ||_F : f \in E_1, ||f||_{E_1} \le 1 \right\} \le ||\Psi_A||_F.$$

Analogously, the right-side estimate in (55) follows from definitions (33) and (47).

2. Let us prove the left-side estimate in (54). For $r \in R_+$ and $\varepsilon \in (0, 1)$ $\exists f_0 \in E_1 : ||f_0||_{E_1} \le 1$ and

$$||Af_0||_{E_2(B_r)} \ge \Psi_A(r)(1 - \varepsilon).$$

Then, $\forall \rho \in [r, \infty)$

$$||Af_0||_{E_2(B_\rho)} \ge (1 - \varepsilon)\Psi_A(r).$$

Thus,

$$||Af_0||_{E_2(B_\rho)} \ge (1 - \varepsilon)\Psi_A(r)\chi_{[r,\infty)}(\rho), \ \rho \in R_+.$$

It follows from here that

$$||Af_0||_{LM_{E_2 F}} \ge (1 - \varepsilon)\Psi_A(r)||\chi_{[r,\infty)}||_F.$$

So, according to the definition (31)

$$||A|| = \sup \left\{ ||Af||_{LM_{E_2 F}} : f \in E_1(R^n) : ||f||_{E_1} \le 1 \right\} \ge ||Af_0||_{LM_{E_2 F}} \ge$$
$$(1 - \varepsilon)\Psi_A(r)||\chi_{[r,\infty)}||_F.$$

The inequality is valid $\forall r \in R_+$, so

$$||A| \ge (1 - \varepsilon) \sup_{r \in R_+} \left[\Psi_A(r) ||\chi_{[r,\infty)}||_F \right].$$

Since $\varepsilon \in (0, 1)$ is arbitrary, the left-side inequality in (54) follows from here.

3. Let us prove the left-side estimate in (55). For $r \in R_+$ and $\varepsilon \in (0, 1)$ $\exists f_1 \in E_1(R^n) : ||f_1||_{E_1} \le 1$ and

$$||(Af_1)^*\chi_{(0,|B_r|)}||_{\tilde{E}_2} \geq (1-\varepsilon)\Phi_A(r). \tag{56}$$

Then, $\forall \rho \in [r, \infty)$

$$||(Af_1)^*\chi_{(0,|B_\rho|)}||_{\tilde{E}_2} \geq (1-\varepsilon)\Phi_A(r). \tag{57}$$

For $\rho \in (0, r)$ we use the estimate (50) with $g = (Af_1)^* \in \Omega_0$. Then

$$||(Af_1)^*\chi_{(0,|B_\rho|)}||_{\tilde{E}_2} \geq \frac{\varphi_{E_2}(|B_\rho|)}{\varphi_{E_2}(|B_\rho|) + \varphi_{E_2}(|B_r|)}||(Af_1)^*\chi_{(0,|B_r|)}||_{\tilde{E}_2}.$$

From here and from (56) it follows that for $\rho \in (0, r)$

$$||(Af_1)^*\chi_{(0,|B_\rho|)}||_{\tilde{E}_2} \geq (1-\varepsilon)\frac{\varphi_{E_2}(|B_\rho|)}{\varphi_{E_2}(|B_\rho|) + \varphi_{E_2}(|B_r|)}\Phi_A(r). \tag{58}$$

$\forall \rho \in R_+$ the estimates (57) and (58) lead to

$$||(Af_1)^*\chi_{(0,|B_\rho|)}||_{\tilde{E}_2} \geq (1-\varepsilon)\Phi_A(r)\left[\frac{\varphi_{E_2}(|B_\rho|)\chi_{(0,r)}(\rho)}{\varphi_{E_2}(|B_\rho|) + \varphi_{E_2}(|B_r|)} + \chi_{[r,\infty)}(\rho)\right].$$

Then, according to (33) (with the norm in $F(R_+)$ taken by ρ)

$$||A||_* \geq ||\ ||(Af_1)^*\chi_{(0,|B_\rho|)}||_{\tilde{E}_2}||_F \geq$$

$$\geq (1-\varepsilon)\Phi_A(r)\left[\frac{\varphi_{E_2}(|B_\rho|)\chi_{(0,r)}(\rho)}{\varphi_{E_2}(|B_\rho|) + \varphi_{E_2}(|B_r|)} + \chi_{[r,\infty)}(\rho)\right].$$

The inequality is valid $\forall r \in R_+$, so

$$||A||_* \geq (1-\varepsilon)\sup_{r\in R_+}\left[\Phi_A(r)||\frac{\varphi_{E_2}(|B_\rho|)\chi_{(0,r)}(\rho)}{\varphi_{E_2}(|B_\rho|) + \varphi_{E_2}(|B_r|)} + \chi_{[r,\infty)}||_F\right].$$

Since $\varepsilon \in (0, 1)$ is arbitrary, the left-side inequality in (55) follows from here.

Corollary 4 *For $\rho \in [r, \infty)$ the following inequality takes place*

$$\frac{1}{2} \leq \frac{\varphi_{E_2}(|B_\rho|)}{\varphi_{E_2}(|B_\rho|) + \varphi_{E_2}(|B_r|)} \leq 1,$$

So, the estimate (55) leads to:

$$\sup_{r\in R_+}\left[\Phi_A(r)||\frac{\varphi_{E_2}(|B_\rho|)}{\varphi_{E_2}(|B_\rho|) + \varphi_{E_2}(|B_r|)}||_F\right] \leq ||A||_* \leq ||\Phi_A||_F, \tag{59}$$

and the left-hand side of (59) has the same order as the right-hand side of (55).

Corollary 5 *The estimate (55) also leads to the following one:*

$$\max\left\{\sup_{r\in R_+}\left[\Phi_A(r)||\frac{\varphi_{E_2}(|B_\rho|)\chi_{(0,r)}(\rho)}{\varphi_{E_2}(|B_\rho|)+\varphi_{E_2}(|B_r|)}||_F\right];\ \sup_{r\in R_+}\Phi_A(r)||\chi_{[r,\infty)}||_F\right\}\le ||A||_*, \tag{60}$$

and

$$||\frac{\varphi_{E_2}(|B_\rho|)\chi_{(0,r)}(\rho)}{\varphi_{E_2}(|B_\rho|)+\varphi_{E_2}(|B_r|)}||_F \cong \frac{1}{\varphi_{E_2}(|B_r|)}||\varphi_{E_2}(|B_\rho|)\chi_{(0,r)}||_F. \tag{61}$$

Remark 4 A comparison of the left-side estimates in (54) and (55) shows that the left-hand side of (55) in some sense majorizes the left-hand side of (54).

Consider the examples, showing that from the estimates (54) and (55) in some cases it is possible to extract the sharp conditions for the boundedness of the operator A.

Example 3 Under the conditions of Proposition 7 let $F(R_+) = L_\infty(w),\ w$ is a positive, measurable function.

$$||g||_{L_\infty(w)} = \operatorname*{ess\,sup}_{\rho\in R_+}\ [w(\rho)|g(\rho)|]. \tag{62}$$

Then,

$$||A|| = ||\Psi_A||_F = \operatorname*{ess\,sup}_{\rho\in R_+}\ [w(\rho)\Psi_A(\rho)] = \sup_{\rho\in R_+}\ [\tilde{w}(\rho)\Psi_A(\rho)]. \tag{63}$$

$$||A||_* = ||\Phi_A||_F = \operatorname*{ess\,sup}_{\rho\in R_+}\ [w(\rho)\Phi_A(\rho)] = \sup_{\rho\in R_+}[\tilde{w}(\rho)\Phi_A(\rho)]. \tag{64}$$

Actually, denote

$$\tilde{w} = \operatorname*{ess\,sup}_{\rho\in[r,\infty)} w(\rho).$$

Since $0 \le \Psi_A \uparrow$, we have:

$$||\Psi_A||_F = \operatorname*{ess\,sup}_{\rho\in R_+}\ [w(\rho)\Psi_A(\rho)] = \operatorname*{ess\,sup}_{r\in R_+}\ [\tilde{w}(r)\Psi_A(r)] \le \sup_{r\in R_+}[\tilde{w}(r)\Psi_A(r)]. \tag{65}$$

Thus, according to the upper bound in (54)

$$||A|| \le ||\Psi_A||_F = \operatorname*{ess\,sup}_{r\in R_+}\ [\tilde{w}(r)\Psi_A(r)] \le \sup_{r\in R_+}[\tilde{w}(r)\Psi_A(r)].$$

On the other hand, the left-side inequality in (54) leads to:

$$||A|| \geq \sup_{r\in R_+} [\Psi_A(r)||\chi_{[r,\infty)}||_F] = \sup_{r\in R_+} \left[\Psi_A(r) \operatorname*{ess\,sup}_{\rho\in[r,\infty)} w(\rho) \right] = \sup_{r\in R_+} [\Psi_A(r)\tilde{w}(r)].$$

From here and from (65) we get (63). Analogously, from (55) we get (64).

Example 4 Under the conditions of Proposition 7 let

$$A : E_1(R^n) \to E_2(R^n) \tag{66}$$

be a bounded operator. Then for its norm as the operator

$$A : E_1(R^n) \to LM_{E_2 F}(R^n)$$

we have

$$||A|| < \infty \Leftrightarrow ||1||_F < \infty. \tag{67}$$

Actually, for the bounded operator in (66)

$$\Psi_A(r) = O(1)(r \to +\infty) \Rightarrow ||\Psi_A||_F \lesssim ||1||_F,$$

so, according to the right-side estimate in (54),

$$||1||_F < \infty \Rightarrow ||A|| < \infty.$$

On the other hand, it follows from the boundedness of the operator (66) that $\Psi_A(r) \cong 1,\ r \in [1,\infty)$, then for $||A|| < \infty$ from the left-side estimate in (54) we have:

$$\sup_{r\in[1,\infty)} \Psi_A(r)||\chi_{[r,\infty)}||_F \cong \sup_{r\in[1,\infty)} ||\chi_{[r,\infty)}||_F = ||\chi_{[1,\infty)}||_F < \infty.$$

But for any BFS $F(R_+)$ the relation $||\chi_{(0,1)}||_F < \infty$ takes place. Thus, $||A|| < \infty \Rightarrow ||1||_F < \infty$.

Remark 5 In particular, for the maximal operator $A = M,\ E_1(R^n) = E_2(R^n) = L_p(R^n),\ 1 < p \leq \infty$ we obtain the equivalence (67) since $M : L_p(R^n) \to L_p(R^n)$, $1 < p \leq \infty$, is bounded.

Remark 6 Left-side inequalities in (54) and (55) show, in particular, that for the finiteness of the norms $||A||$ and $||A||_*$ it is necessary and sufficient that conditions $\Psi_A(r) < \infty$ or $\Phi_A(r) < \infty$ hold respectively $\forall r \in R_+$. It means, in particular, that the condition

$$A(E_1(R^n)) \subset E_2^{loc}(R^n) \tag{68}$$

is necessary for the finiteness of $||A||$ (31). Thus, the most interesting case is the one where the condition (68) holds, but (66) does not.

Proposition 8 *Under the conditions of Proposition 7 let the operator A satisfy the following properties*

1. $A : E_1(R^n) \to E_1(R^n)$ *is bounded, i.e.*

$$||A||_0 := \sup\left\{||Af||_{E_1} : f \in E_1, \ ||f||_{E_1} \leq 1\right\} < \infty; \tag{69}$$

2. *A majorizes the identity operator I, i.e.*

$$\exists d_0 \in R_+ : ||f\chi_{B_r}||_{E_2} \leq d_0||(Af)\chi_{B_r}||_{E_2}, r \in R_+. \tag{70}$$

Then, for the norm (31) the following estimates take place:

$$||A|| \leq ||A||_0||I||, \quad ||I|| \leq d_0||A||. \tag{71}$$

Proof 1. Let $g = Af$, $f \in E_1(R^n)$, $||f||_{E_1} \leq 1$. Then $||g||_{E_1} \leq ||A||_0$, so that (see (31))

$$||A|| = \sup\left\{|| \, ||(Af)\chi_{B_r}||_{E_2}||_F : ||f||_{E_1} \leq 1\right\} \leq$$

$$\leq \sup\left\{|| \, ||g\chi_{B_r}||_{E_2}||_F : ||g||_{E_1} \leq ||A||_0\right\} =$$

$$= ||A||_0 \sup\left\{|| \, ||h\chi_{B_r}||_{E_2}||_F : ||h||_{E_1} \leq 1\right\} = ||A||_0||I||.$$

We took into account here that

$$g \in E_1(R^n), ||g||_{E_1} \leq ||A||_0 \Leftrightarrow h = \frac{g}{||A||_0} \in E_1(R^n), ||h||_{E_1} \leq 1.$$

2. On the other hand,

$$||I|| = \sup\left\{|| \, ||f\chi_{B_r}||_{E_2}||_F : ||f||_{E_1} \leq 1\right\} \leq$$

$$\leq d_0 \sup\left\{|| \, ||(Af)\chi_{B_r}||_{E_2}||_F : ||f||_{E_1} \leq 1\right\} = d_0||A||.$$

Corollary 6 *Under the conditions of Proposition 7 let the operator A be the maximal operator ($A = M$), bounded in $E_1(R^n)$. Then,*

$$||M||_* \cong ||M|| \cong ||I||. \tag{72}$$

Actually, for $f \in E_1(R^n)$ we have $|f(x)| \leq (Mf)(x)$ almost everywhere in R^n, so $(f\chi_{B_r})^* \leq (Mf\chi_{B_r})^*$ and the condition (70) holds with $d_0 = 1$. Thus, according to (71), $||M|| \leq ||M||_0||I||$; $||I|| \leq ||M||$. The first relation in (72) follows from (32).

Example 5 Corollary 6 can be applied in case: $E_1 = L_{p_1}(R^n), \ 1 < p_1 \leq \infty$.

3 Transition to Global Morrey-Type Spaces for Operators, Commuting with a Shift Operator

Let $E = E(R^n)$ be a RIS, $F = F(R_+)$ be a BFS. For $x_0 \in R^n$ define

$$||f||_{LM_{EF}(x_0)} := || \, ||f\chi_{B(x_0,r)}||_E||_F. \tag{73}$$

Note that $||f||_{LM_{EF}} = ||f||_{LM_{EF}(0)}$. The norm in a global Morrey-type space is defined as follows:

$$||f||_{GM_{EF}} := \sup_{x_0 \in R^n} ||f||_{LM_{EF}(x_0)}. \tag{74}$$

Next, let $E_1 = E_1(R^n)$, $E_2 = E_2(R^n)$ be RISs, $F = F(R_+)$ a BFS,

$$A : E_1(R^n) \to E_2^{loc}(R^n), \tag{75}$$

$$|||A||| := ||A||_{E_1 \to GM_{E_2F}} = \sup\left\{||Af||_{GM_{E_2F}} : f \in E_1, ||f||_{E_1} \leq 1\right\}. \tag{76}$$

Let us remind that (see (31))

$$||A|| := ||A||_{E_1 \to LM_{E_2F}} = \sup\left\{||Af||_{LM_{E_2F}} : f \in E_1, ||f||_{E_1} \leq 1\right\}. \tag{77}$$

Obviously, $||A|| \leq |||A|||$. Next, let T_{x_0} be a shift operator by vector $x_0 \in R^n$;

$$(T_{x_0}f)(x) = f(x + x_0), \; x \in R^n.$$

Proposition 9 *If the operator A (75) commutes with the shift operator*

$$AT_{x_0} = T_{x_0}A, \; x_0 \in R^n, \tag{78}$$

then the following equality holds

$$|||A||| = ||A||. \tag{79}$$

Proof 1. Let us show that for $f \in E_1(R^n)$

$$||(Af)\chi_{B(x_0,r)}||_{E_2} = ||(AT_{x_0}f)\chi_{B_r}||_{E_2}. \tag{80}$$

Actually, the equality:

$$\chi_{B(x_0,r)}(x) = \chi_{B_r}(x - x_0), \; x \in R^n$$

means that

$$\chi_{B(x_0,r)} = T_{-x_0}\chi_{B_r} \Leftrightarrow \chi_{B_r} = T_{x_0}\chi_{B(x_0,r)}, \tag{81}$$

thus, taking into account (78),

$$(Af)\chi_{B(x_0,r)} = (T_{-x_0}T_{x_0}Af)(T_{-x_0}\chi_{B_r}) = T_{-x_0}[(T_{x_0}Af)\chi_{B_r}] =$$

$$= T_{-x_0}[(A(T_{x_0}f))\chi_{B_r}].$$

Since the norm in the RIS $E_2 = E_2(R^n)$ is shift invariant, the equality (80) follows from here.

2. It follows from (80) (see also (77)) that

$$||Af||_{LM_{E_2F}(x_0)} = || \quad ||(Af)\chi_{B(x_0,r)}||_{E_2}||_F = || \quad ||(AT_{x_0}f)\chi_{B_r}||_{E_2}||_F =$$

$$||AT_{x_0}f||_{LM_{E_2F}} \leq ||A|| \cdot ||T_{x_0}f||_{E_1} = ||A|| \cdot ||f||_{E_1}.$$

At the last step we took into account that the norm in the RIS $E_1 = E_1(R^n)$ is shift invariant.

It follows from here that

$$||Af||_{GM_{E_2F}} = \sup_{x_0 \in R^n} ||Af||_{LM_{E_2F}(x_0)} \leq ||A|| \cdot ||f||_{E_1}.$$

Thus,

$$|||A||| = \sup\left\{||Af||_{GM_{E_2F}} : f \in E_1(R^n), ||f||_{E_1} \leq 1\right\} \leq ||A||.$$

It leads to the equality (79).

Corollary 7 *The equality (79) holds for the embedding operator $A = I$ and for the maximal operator $A = M$.*

Actually, these operators commute with the shift operator. So, for $f \in E_1(R^n)$ we have

$$(T_{x_0}Mf)(x) = (Mf)(x + x_0) = \sup_{r>0} \frac{1}{|B(x + x_0, r)|} \int_{B(x+x_0,r)} |f(y)|dy =$$

$$= \sup_{r>0} \frac{1}{|B(x,r)|} \int_{B(x,r)} |f(z + x_0)|dz = \sup_{r>0} \frac{1}{|B(x,r)|} \int_{B(x,r)} |T_{x_0}f(z)|dz =$$

$$= M(T_{x_0}f)(x),$$

i.e. $T_{x_0}M = MT_{x_0}$. For $A = I$ the equality (78) is obvious.

Remark 7 For the norm in a global Morrey-type space GM_{EF} (74) the following formula takes place:

$$||f||_{GM_{EF}(R^n)} = \sup_{x_0 \in R^n} ||T_{x_0} f||_{LM_{EF}(R^n)}. \tag{82}$$

Actually, taking into account (81), we have for the right-hand side of (82)

$$\sup_{x_0 \in R^n} ||T_{x_0} f||_{LM_{EF}(R^n)} = \sup_{x_0 \in R^n} || \ ||(T_{x_0} f)\chi_{B_r}||_E||_F =$$

$$= \sup_{x_0 \in R^n} || \ ||(T_{x_0} f) T_{x_0} \chi_{B(x_0,r)}||_E||_F = \sup_{x_0 \in R^n} || \ ||T_{x_0}(f \chi_{B(x_0,r)})||_E||_F =$$

$$= \sup_{x_0 \in R^n} || \ ||f \chi_{B(x_0,r)}||_E||_F.$$

At the last step we took into account that the norm in the RIS $E = E(R^n)$ is shift invariant. Thus, according to (73) and (74), it leads to

$$\sup_{x_0 \in R^n} ||T_{x_0} f||_{LM_{EF}} = \sup_{x_0 \in R^n} ||f||_{LM_{EF}(x_0)} = ||f||_{GM_{EF}}.$$

4 The Criterion of the Embedding of a Weighted Lorentz Space into a Morrey-Type Space. Notation and Main Results

Let $E_1 = E_1(R^n)$, $E_2 = E_2(R^n)$ be RISs, $F = F(R_+)$ be a BFS, $LM_{E_2F}(R^n)$ is a local Morrey-type space (see (5)). Estimates of the norm of the embedding operator

$$I : E_1(R^n) \subset LM_{E_2F}(R^n) \tag{83}$$

are of independent interest. In addition, by Proposition 8, in many cases they allow us to estimate the norms of more general operators. The purpose of this section is to give an order-sharp estimate of the norm

$$||I|| = \sup\left\{||f||_{LM_{E_2F}} : \ f \in E_1, ||f||_{E_1} \leq 1\right\} \tag{84}$$

in case where

$$E_1 = \Lambda_{p_1}(u), \ E_2 = \Lambda_{p_2}(v), \ F = L_\theta(w), 0 < p_1, p_2, \theta \leq \infty. \tag{85}$$

Here u, v, w are weights, i. e. measurable non-negative functions on R_+, and $u > 0$ almost everywhere on R_+. $\Lambda_p(u)$ is a weighted Lorentz space:

$$||f||_{\Lambda_p(u)} := \left(\int_0^\infty f^*(t)^p u(t)dt\right)^{\frac{1}{p}}, \ 0 < p < \infty; \tag{86}$$

$$||f||_{\Lambda_\infty(u)} := \operatorname*{ess\,sup}_{t\in R_+}[f^*(t)u(t)],\ \ p=\infty; \tag{87}$$

$L_\theta(w)$ is a weighted Lebesgue space on R_+ :

$$||f||_{L_\theta(w)} := ||fw||_{L_\theta} = \begin{cases} (\int_0^\infty (|f(r)|w(r))^\theta dr)^{\frac{1}{\theta}},\ 0<\theta<\infty, \\ \operatorname*{ess\,sup}_{r\in R_+}(|f(r)|w(r)),\ \theta=\infty. \end{cases} \tag{88}$$

Introduce the following notation:

$$p = p_1 p_2^{-1},\ q = \theta p_2^{-1}, \text{if}\ \ 0 < p_2 < \infty; \tag{89}$$

$$U(t) = \int_0^t u(\tau)d\tau,\ V(t) = \int_0^t v(\tau)d\tau,\ t\in R_+; \tag{90}$$

$$w_\theta(t) = V(t)^q w((\frac{t}{v_n})^{\frac{1}{n}})^\theta t^{\frac{1}{n}-1},\quad \text{if}\ \ 0<\theta<\infty; \tag{91}$$

$$w_\infty(t) = V(t)^{\frac{1}{p_2}} w((\frac{t}{v_n})^{\frac{1}{n}}),\ \text{if}\,\theta=\infty. \tag{92}$$

$$B_p(t) = \frac{1}{V(t)} \sup_{\tau\in(0,t)} \left[V(\tau)U(\tau)^{-\frac{1}{p}}\right],\ \text{if}\,0<p\le 1; \tag{93}$$

$$B_p(t) = \left(\int_0^\infty \left(\frac{V(\tau)}{V(\tau)+V(t)}\right)^{p'} \frac{u(\tau)d\tau}{U(\tau)^{p'}}\right)^{\frac{1}{p'}},\ \text{if}\,1<p<\infty,\ p'=\frac{p}{p-1}; \tag{94}$$

$$B_\infty(t) = \frac{1}{V(t)}\int_0^t \frac{v(\tau)d\tau}{(\operatorname*{ess\,sup}_{\xi\in(0,\tau]} u(\xi))^{p_2}},\ \text{if}\ p=\infty\ (\text{i. e. } p_1=\infty); \tag{95}$$

$$W_q(t) = \left(\int_0^\infty \left(\frac{V(t)}{V(\tau)+V(t)}\right)^q w_\theta(\tau)d\tau\right)^{\frac{1}{q}},\ \text{if}\ \ 1<q<\infty \tag{96}$$

(i. e. $0<\theta=qp_2<\infty$);

$$W_\infty(t) = \operatorname*{ess\,sup}_{\tau\in R_+}\left(\frac{V(t)w_\infty(\tau)}{V(\tau)+V(t)}\right)\ \text{if}\,q=\infty\ (\text{i. e. }\theta=\infty). \tag{97}$$

First we state the result for the case θ, $p_2<\infty$.

Theorem 1 *Under the notation above let* $0<\theta,\ p_2<\infty$. *Then, for the norm (84) the following estimates hold:*

$$||I||^{p_2} \cong \sup_{t\in R_+}\left[W_q(t)B_p(t)\right], \text{ if } p \le q \text{ (i. e. } p_1 \le \theta); \tag{98}$$

$$||I||^{p_2} \cong \left(\int_0^\infty W_q(t)^{s-q}B_p(t)^s w_\theta(t)dt\right)^{\frac{1}{s}}, \text{ if } p > q \text{ (i. e. } p_1 > \theta). \tag{99}$$

Here

$$s = pq(p-q)^{-1} = p_1\theta p_2^{-1}(p_1-\theta)^{-1};\, s-q = \theta^2 p_2^{-1}(p_1-\theta)^{-1}. \tag{100}$$

Remark 8 In certain cases, the general results (98)–(100) can be simplified.
1. In case $0 < p \le 1;\ p \le q$, (i.e. $0 < p_1 \le p_2;\ p_1 \le \theta$) we have

$$||I||^{p_2} \cong \sup_{t\in R_+}\left[W_q(t)U(t)^{-\frac{1}{p}}\right]. \tag{101}$$

2. In case $p = \infty;\ 0 < q < \infty$, (i.e. $p_1 = \infty;\ 0 < \theta < \infty$) we have

$$||I||^{p_2} \cong \left(\int_0^\infty B_\infty(t)^q w_\theta(t)dt\right)^{\frac{1}{q}}. \tag{102}$$

Actually, in this case $s = q$, so (99) will take the form (102).

Now consider the case $p_2 = \infty$. Denote

$$V_\infty(t) = \operatorname*{ess\,sup}_{\tau\in(0,t]} v(\tau);\quad U_\infty(t) = \operatorname*{ess\,sup}_{\tau\in(0,t]} u(\tau). \tag{103}$$

Suppose that $0 < \theta < \infty$. The case $\theta = \infty$ we will consider later.

Theorem 2 *Under the notation above let $p_2 = \infty$. Then, for the norm (84) the following estimates take place in case $0 < \theta < \infty$.*
1. For $0 < p_1 \le \theta$ we have

$$||I|| \cong \sup_{t\in R_+}\left[V_\infty(t)U(t)^{-\frac{1}{p_1}}||\frac{V_\infty(|B_r|)}{V_\infty(|B_r|)+V_\infty(t)}||_{L_\theta(w)}\right]. \tag{104}$$

2. For $p_1 = \infty$ we have

$$||I|| = ||\operatorname*{ess\,sup}_{\tau\in(0,|B_r|]}\frac{v(\tau)}{U_\infty(\tau)}||_{L_\theta(w)}. \tag{105}$$

In (104) and (105) the norm in $L_\theta(w)$ (see (88)) is taken by $r \in R_+$; recall that $|B_r| = v_n r^n$. The case $\theta < p_1 < \infty$ will be mentioned in Remark 10.

Next, consider the case $\theta = \infty$. By Example 3 with $A = I$, we have for $F(R_+) = L_\infty(w)$

$$||I|| = \sup_{r\in R_+} \left[\tilde{w}(r)\Psi_I(r)\right]; \quad \tilde{w}(r) = \operatorname*{ess\,sup}_{\rho\in[r,\infty)} w(\rho). \tag{106}$$

Here, according to (46), (47), and (52) with $c_0 = 1$,

$$\begin{aligned}\Psi_I(r) &:= ||I||_{E_1\to E_2(B_r)} = \sup\left\{||f||_{E_2(B_r)} : f \in E_1, ||f||_{E_1} \le 1\right\} = \\ &= \Phi_I(r) := \sup\left\{||f^*\chi_{(0,|B_r|)}||_{\tilde{E}_2} : f \in E_1, ||f||_{E_1} \le 1\right\}.\end{aligned} \tag{107}$$

Theorem 3 *Under the notation above let $\theta = \infty$. Then, for the norm (84) the equality (106) holds, where*

$$\Psi_I(r) = \sup_{t\in(0,|B_r|]} \left[V(t)^{\frac{1}{p_2}} U(t)^{-\frac{1}{p_1}}\right], \quad \text{if } 0 < p_1 \le p_2 < \infty; \tag{108}$$

$$\Psi_I(r) = \sup_{t\in(0,|B_r|]} \left[V_\infty(t) U(t)^{-\frac{1}{p_1}}\right], \quad \text{if } 0 < p_1 < p_2 = \infty; \tag{109}$$

$$\Psi_I(r) = \operatorname*{ess\,sup}_{t\in(0,|B_r|]} \left[v(t) U_\infty(t)^{-1}\right], \quad \text{if } p_1 = p_2 = \infty; \tag{110}$$

$$\Psi_I(r) \cong \left(\int_0^{|B_r|} \left[V(t)^{\frac{1}{p_2}} U(t)^{-\frac{1}{p_1}}\right]^\sigma \frac{v(t)dt}{V(t)}\right)^{\frac{1}{\sigma}}, \quad \text{if } 0 < p_2 < p_1 < \infty; \tag{111}$$

$\sigma = \frac{p_1 p_2}{p_1 - p_2}$;

$$\Psi_I(r) = \left(\int_0^{|B_r|} v(t) U_\infty(t)^{-p_2} dt\right)^{\frac{1}{p_2}}, \quad \text{if } 0 < p_2 < p_1 = \infty. \tag{112}$$

5 Proofs of Results Formulated in Sect. 4

The proof of Theorem 1 and Remark 8.

1. According to Corollary 2 for $||I||$ the formula (38) is true, which in case of spaces (85) takes the form:

$$||I|| = \sup\left\{\left(\int_0^\infty \left(\int_0^{|B_r|} g^{p_2} v d\tau\right)^{\frac{\theta}{p_2}} w(r)^\theta dr\right)^{\frac{1}{\theta}} : g \in \Omega_0, \left(\int_0^\infty g^{p_1} u d\tau\right)^{\frac{1}{p_1}} \le 1\right\}, \tag{113}$$

in case $p_1 \in (0, \infty)$; or

$$\|I\| = \sup\left\{\left(\int_0^\infty\left(\int_0^{|B_r|} g^{p_2} v\,d\tau\right)^{\frac{\theta}{p_2}} w(r)^\theta dr\right)^{\frac{1}{\theta}} : g\in\Omega_0,\ \operatorname*{ess\,sup}_{\tau\in R_+}[g(\tau)u(\tau)]\le 1\right\}, \tag{114}$$

in case $p_1 = \infty$.

2. First let $p_1 = \infty$. We have the well-known equality

$$\operatorname*{ess\,sup}_{\tau\in R_+}[g(\tau)u(\tau)] = \operatorname*{ess\,sup}_{\tau\in R_+}[g(\tau)U_\infty(\tau)].$$

Here $U_\infty(\tau) = \operatorname*{ess\,sup}_{\xi\in(0,\tau]} u(\xi)$ is increasing, and $g\in\Omega_0$ is decreasing and right continuous. So, the equality can be continued:

$$\operatorname*{ess\,sup}_{\tau\in R_+}[g(\tau)u(\tau)] = \sup_{\tau\in R_+}[g(\tau)U_\infty(\tau)] \le 1.$$

Then, $g(\tau)\le U_\infty(\tau)^{-1}$, $\tau\in R_+$, $\forall g\in\Omega_0 : \operatorname*{ess\,sup}_{\tau\in R_+}[g(\tau)u(\tau)]\le 1$. Thus,

$$\|I\| \le \left(\int_0^\infty\left(\int_0^{|B_r|} v(\tau)U_\infty(\tau)^{-p_2} d\tau\right)^{\frac{\theta}{p_2}} w(r)^\theta dr\right)^{\frac{1}{\theta}}.$$

On the other hand, let $g_0(\tau) = U_\infty(\tau)^{-1}$. Then $0\le g_0\downarrow$ and g_0 can be considered as a continuous function (the set of points of its discontinuity on any interval is at most countable, so, extending the function at these points by right continuity, we get the equivalent function). Thus, we can suppose $g_0(\tau) = U_\infty(\tau)^{-1}\in\Omega_0$, and

$$\operatorname*{ess\,sup}_{\tau\in R_+}[g_0(\tau)u(\tau)] = \sup_{\tau\in R_+}[g_0(\tau)U_\infty(\tau)] = 1.$$

Therefore,

$$\|I\| \ge \left(\int_0^\infty\left(\int_0^{|B_r|} g_0^{p_2} v\,d\tau\right)^{\frac{\theta}{p_2}} w(r)^\theta dr\right)^{\frac{1}{\theta}} =$$

$$= \left(\int_0^\infty\left(\int_0^{|B_r|} v(\tau)U_\infty(\tau)^{-p_2} d\tau\right)^{\frac{\theta}{p_2}} w(r)^\theta dr\right)^{\frac{1}{\theta}}.$$

Thus, for $p_1 = \infty$ we have

$$||I|| = \left(\int_0^\infty \left(\int_0^{|B_r|} v(\tau) U_\infty(\tau)^{-p_2} d\tau \right)^{\frac{\theta}{p_2}} w(r)^\theta dr \right)^{\frac{1}{\theta}} \cong$$

$$\cong \left(\int_0^\infty \left(\int_0^t v(\tau) U_\infty(\tau)^{-p_2} d\tau \right)^{\frac{\theta}{p_2}} w((\frac{t}{v_n})^{\frac{1}{n}})^\theta t^{\frac{1}{n}-1} dt \right)^{\frac{1}{\theta}}.$$

This result corresponds to (102), (91), (95).

3. Now let $p_1 \in (0, \infty)$. In (113) denote $h = g^{p_2}$; $g \in \Omega_0 \Leftrightarrow h \in \Omega_0$. Then

$$||I||^{p_2} = \sup \left\{ \left(\int_0^\infty \left(\int_0^{|B_r|} hv d\tau \right)^q w(r)^\theta dr \right)^{\frac{1}{q}} : h \in \Omega_0; \left(\int_0^\infty h^p u d\tau \right)^{\frac{1}{p}} \leq 1 \right\}$$

$$= \sup \left\{ \left(\int_0^\infty \left[\frac{1}{V(t)} \int_0^t hv d\tau \right]^q w_\theta(t) dt \right)^{\frac{1}{q}} : h \in \Omega_0; \left(\int_0^\infty h^p u d\tau \right)^{\frac{1}{p}} \leq 1 \right\}$$

$$= ||I||_{\Lambda_p(u) \to \Gamma_{q,v}(w_\theta)}, \tag{115}$$

taking into account notation (89)–(91).

Remark 9 For $p_1 = \infty$ we can make an analogous substitution in (114) and get

$$||I||^{p_2} = \sup \left\{ \left(\int_0^\infty \left[\frac{1}{V(t)} \int_0^t hv d\tau \right]^q w_\theta(t) dt \right)^{\frac{1}{q}} : h \in \Omega_0; \operatorname*{ess\,sup}_{\tau \in R_+} \left[h(\tau) u^{p_2}(\tau) \right] \leq 1 \right\}$$

$$= ||I||_{\Lambda_\infty(u^{p_2}) \to \Gamma_{q,v}(w_\theta)}. \tag{116}$$

Further evaluation is carried out on the basis of equalities (115) for $p_1 < \infty$ and (116) for $p_1 = \infty$. It leads to demanded result (98), (99), which in case $p_1 = \infty$ coincides with the one stated above in step 2.

Thus, let $p_1 < \infty$. For the embedding operator at right-hand side of (115) we can apply results on estimates of Hardy-type operators on the cone Ω_0, obtained in [2, 11, 12]. In a convenient form, the embedding criterion for $\Lambda_p(u) \to \Gamma_{q,v}(w_\theta)$ was obtained in [11]. In the modified form it is given at right-hand sides of (98)–(100). Respectively, for $p_1 = \infty$ we apply the embedding criterion for $\Lambda_\infty(u^{p_2}) \to \Gamma_{q,v}(w_\theta)$ and get (102), which was already justified in step 2.

4. It remains to justify Part 1 of Remark 8. Denote

$$A(t) = W_q(t) V(t)^{-1}; \quad B(t) = V(t) U(t)^{-\frac{1}{p}}, \quad t \in R_+.$$

Then $0 < A(t) \downarrow$; $B(t) \geq 0$ and the known equality takes place:

$$\sup_{t\in R_+} A(t)B(t) = \sup_{t\in R_+} A(t) \sup_{\tau\in(0,t]} B(\tau).$$

It shows that the right-hand side of (101) coincides with the right-hand side of (98).

The proof of Theorem 2.

1. In case $0 < p_1 \leq \theta < \infty$ for

$$||I|| = \sup\left\{\left(\int_0^\infty \left[\operatorname*{ess\,sup}_{\rho\in(0,|B_r|)} (g(\rho)v(\rho))\right]^\theta w(r)^\theta dr\right)^{\frac{1}{\theta}} : g \in \Omega_0; \left(\int_0^\infty g^{p_1} u d\tau\right)^{\frac{1}{p_1}} \leq 1\right\} \tag{117}$$

we can apply the result by V. I. Burenkov and M. L. Goldman (see [5]).

Actually, the operator:

$$T[g](r) = \operatorname*{ess\,sup}_{\rho\in(0,|B_r|)} (g(\rho)v(\rho)), g \in \Omega_0,$$

is l_q-convex for $\forall q \in R_+$, i.e.

$$T[(\sum_k g_k^q)^{\frac{1}{q}}] \leq [\sum_k (Tg_k)^q]^{\frac{1}{q}}.$$

Indeed,

$$T[(\sum_k g_k^q)^{\frac{1}{q}}] = \text{ess sup}(\sum_k (g_k v)^q)^{\frac{1}{q}} = [\text{ess sup} \sum_k (g_k v)^q]^{\frac{1}{q}} \leq$$

$$\leq [\sum_k \text{ess sup}(g_k v)^q]^{\frac{1}{q}} = [\sum_k (\text{ess sup}\, g_k v)^q]^{\frac{1}{q}} = [\sum_k (Tg_k)^q]^{\frac{1}{q}}.$$

In particular, it is l_θ-convex, and $0 < p_1 \leq \theta$. Then, according to Theorem 1 in [5], we get

$$||I|| = \sup_{t\in R_+}\left[||T[\chi_{(0,t)}](r)||_{L_\theta(w)} \left(\int_0^t u d\tau\right)^{-\frac{1}{p_1}}\right] =$$

$$= \sup_{t\in R_+}\left[||\operatorname*{ess\,sup}_{\rho\in(0,|B_r|)} [v(\rho)\chi_{(0,t)}(\rho)]||_{L_\theta(w)} U(t)^{-\frac{1}{p_1}}\right]. \tag{118}$$

But for $|B_r| \geq t$

$$\operatorname*{ess\,sup}_{\rho\in(0,|B_r|)} v(\rho)\chi_{(0,t)}(\rho) = \operatorname*{ess\,sup}_{\rho\in(0,t)} v(\rho) = V_\infty(t),$$

and for $|B_r| < t$

$$\operatorname*{ess\,sup}_{\rho\in(0,|B_r|)} v(\rho)\chi_{(0,t)}(\rho) = \operatorname*{ess\,sup}_{\rho\in(0,|B_r|)} v(\rho) = V_\infty(|B_r|).$$

Thus, for $\forall r \in R_+$

$$\operatorname*{ess\,sup}_{\rho\in(0,|B_r|)} \left(v(\rho)\chi_{(0,t)}(\rho)\right) \cong \frac{V_\infty(|B_r|)V_\infty(t)}{V_\infty(|B_r|) + V_\infty(t)}.$$

Substituting this relation into (118), we obtain the estimate (104).

2. Now let $p_1 = \infty$. Let us reason as in Step 2 of the proof of Theorem 1. For $\forall g \in \Omega_0$, for which $\operatorname*{ess\,sup}_{\rho\in R_+}[g(\rho)u(\rho)] \le 1$ we have $g(\rho) \le U_\infty(\rho)^{-1}$, $\rho \in R_+$, so (see analogue in (117)):

$$||I|| \le \left(\int_0^\infty \left[\operatorname*{ess\,sup}_{\rho\in(0,|B_r|)} (v(\rho)U_\infty(\rho)^{-1})\right]^\theta w(r)^\theta dr\right)^{\frac{1}{\theta}}.$$

On the other hand, $g_0(\rho) = U_\infty(\rho)^{-1} \in \Omega_0$, and

$$\operatorname*{ess\,sup}_{\rho\in R_+}[g_0(\rho)u(\rho)] = \sup_{\rho\in R_+}[g_0(\rho)U_\infty(\rho)] = 1.$$

Thus,

$$||I|| \ge \left(\int_0^\infty \left[\operatorname*{ess\,sup}_{\rho\in(0,|B_r|)} (v(\rho)U_\infty(\rho)^{-1})\right]^\theta w(r)^\theta dr\right)^{\frac{1}{\theta}}.$$

These inequalities lead to (105).

Remark 10 Under the conditions of Theorem 2 for $\theta < p_1 < \infty$ the question of the embedding criterion is open. For its solution it is necessary to obtain a two-sided estimate for (117) under those conditions on the parameters. Since

$$\operatorname*{ess\,sup}_{\rho\in(0,|B_r|)}[g(\rho)v(\rho)] = \operatorname*{ess\,sup}_{\rho\in(0,|B_r|)}[g(\rho)V_\infty(\rho)] = \sup_{\rho\in(0,|B_r|)}[g(\rho)V_\infty(\rho)],$$

we have

$$||I|| = \sup\left\{\left(\int_0^\infty \left[\sup_{\rho\in(0,|B_r|)} (g(\rho)V_\infty(\rho))\right]^\theta w(r)^\theta dr\right)^{\frac{1}{\theta}} : g \in \Omega_0; \left(\int_0^\infty g^{p_1} u d\tau\right)^{\frac{1}{p_1}} \le 1\right\}. \tag{119}$$

Remark 11 For $\theta < p_1 < \infty$ consider a particular case, where the answer about the boundedness of $||I||$ can be obtained immediately. Namely, under those conditions let

$$D := \sup_{\tau \in R_+} [V_\infty(\tau) U(\tau)^{-\frac{1}{p_1}}] < \infty. \tag{120}$$

Then,

$$||I|| < \infty \Leftrightarrow \left(\int_0^\infty w(r)^\theta dr \right)^{\frac{1}{\theta}} < \infty. \tag{121}$$

Actually, (120) is equivalent to the embedding

$$E_1(R^n) := \Lambda_{p_1}(u) \subset \Lambda_\infty(v) =: E_2(R^n),$$

i.e. the condition (120) is equivalent to the boundedness of $I : E_1(R^n) \to E_2(R^n)$ (see (109)). In this case we are under the conditions of Example 4 with $A = I$. Thus, the equivalence (67) takes place. In our case it takes the form (121).

The proof of Theorem 3.

1. The validity of equality (106) was justified in Example 3 (with $A = I$). It is necessary to prove the relations (108)–(112). According to (107) with $E_1(R^n) := \Lambda_{p_1}(u)$, $E_2(R^n) := \Lambda_{p_2}(v)$, we have

$$\Psi_I(r) = \sup \left\{ \left(\int_0^\infty f^*(\rho)^{p_2} v_r(\rho) d\rho \right)^{\frac{1}{p_2}} : f \in E_1, ||f||_{E_1} \le 1 \right\},$$

$$v_r(\rho) = v(\rho) \chi_{(0,|B_r|)}(\rho), \ \rho \in R_+.$$

But $\left\{ f \in E_1, ||f||_{E_1} \le 1 \right\} \Leftrightarrow \left\{ g = f^* \in \Omega_0;\ (\int_0^\infty g^{p_1} u d\rho)^{\frac{1}{p_1}} \le 1 \right\}$ (with corresponding modification for $p_1 = \infty$). So,

$$\Psi_I(r) = \sup \left\{ \left(\int_0^{|B_r|} g^{p_2} v d\rho \right)^{\frac{1}{p_2}} : g \in \Omega_0;\ (\int_0^\infty g^{p_1} u d\rho)^{\frac{1}{p_1}} \le 1 \right\}. \tag{122}$$

The first integral in (122) does not depend on the values of the function $g \in \Omega_0$ in case $\rho > |B_r|$. Therefore, sup$\{...\}$ is achieved on functions $g \in \Omega_0 : g(\rho) = 0, \rho > |B_r|$. Thus,

$$\Psi_I(r) = \sup \left\{ \left(\int_0^{|B_r|} g^{p_2} v d\rho \right)^{\frac{1}{p_2}} : g \in \Omega_0;\ (\int_0^{|B_r|} g^{p_1} u d\rho)^{\frac{1}{p_1}} \le 1 \right\} \tag{123}$$

for $0 < p_1, p_2 \le \infty$ (with corresponding modification in case $p_1 = \infty$ and/or $p_2 = \infty$).

2. In case $0 < p_1 \le p_2 < \infty$ to calculate this quantity, we can apply the result by V. I. Burenkov and M. L. Goldman (see [5]), according to which (for the identity operator $T = I$):

$$\Psi_I(r) = \sup_{t\in(0,|B_r|]} \left[\left(\int_0^t v d\rho \right)^{\frac{1}{p_2}} \left(\int_0^t u d\rho \right)^{-\frac{1}{p_1}} \right].$$

It is the equality (108). In case $0 < p_1 < p_2 = \infty$ (123) takes the form:

$$\Psi_I(r) = \sup \left\{ \operatorname*{ess\,sup}_{\rho\in(0,|B_r|]} (gv) : g \in \Omega_0;\ (\int_0^{|B_r|} g^{p_1} u d\rho)^{\frac{1}{p_1}} \le 1 \right\}.$$

Here, applying the same result of Theorem 2 in [5], we get

$$\Psi_I(r) = \sup_{t\in(0,|B_r|]} \left[\left(\operatorname*{ess\,sup}_{\rho\in(0,t)} v \right) \left(\int_0^t u d\rho \right)^{-\frac{1}{p_1}} \right] = \sup_{t\in(0,|B_r|]} \left[V_\infty(t) U(t)^{-\frac{1}{p_1}} \right].$$

It is the equality (109).

3. In case $p_1 = \infty$ we reason as in Step 2 of the proof of Theorem 1. We have the same estimate:

$$g \in \Omega_0;\ \operatorname*{ess\,sup}_{\rho\in(0,|B_r|]} [g(\rho)u(\rho)] \le 1 \Rightarrow g(\rho) \le U_\infty(\rho)^{-1}, \rho \in (0, |B_r|], \tag{124}$$

so, in (123)

$$\Psi_I(r) \le \left(\int_0^{|B_r|} v(\rho) U_\infty(\rho)^{-p_2} d\rho \right)^{\frac{1}{p_2}}, \quad 0 < p_2 < \infty;$$

$$\Psi_I(r) \le \operatorname*{ess\,sup}_{\rho\in(0,|B_r|]} \left[v(\rho) U_\infty(\rho)^{-1} \right], p_2 = \infty.$$

On the other hand,

$$g_0(\rho) = U_\infty(\rho)^{-1} \in \Omega_0;\ \operatorname*{ess\,sup}_{\rho\in(0,|B_r|]} [g_0(\rho)u(\rho)] = \sup_{\rho\in(0,|B_r|]} [g_0(\rho)U_\infty(\rho)] = 1,$$

thus, according to (123) in case $p_2 < \infty$ we have

$$\Psi_I(r) \ge \left(\int_0^{|B_r|} v(\rho) U_\infty(\rho)^{-p_2} d\rho \right)^{\frac{1}{p_2}};$$

and in case $p_2 = \infty$

$$\Psi_I(r) \geq \operatorname*{ess\,sup}_{\rho\in(0,|B_r|]} \left[v(\rho)U_\infty(\rho)^{-1}\right].$$

It means that we proved (110) and (112).

4. It remains to consider the case $0 < p_2 < p_1 < \infty$. Here, to estimate $\Psi_I(r)$ in (123) we can apply the results by M. L. Goldman (see Theorem 1.1 in [12]).

Denote

$$u_r(\rho) = \begin{cases} u(\rho), & \rho \in (0, |B_r|]; \\ 1, & \rho > |B_r|. \end{cases}$$

Without loss of generality, we can assume

$$0 < U(1) = \int_0^1 u(\rho)d\rho < \infty. \tag{125}$$

If the integral in (125) is diverging, then

$$\left\{g \in \Omega_0; \left(\int_0^\infty g^{p_1} u d\rho\right)^{\frac{1}{p_1}} < \infty\right\} = \{0\},$$

and there is no sense in searching $\Psi_I(r)$ (122). Thus, for $u_r(\rho)$ the nondegeneracy conditions hold:

$$\int_0^1 u_r(\rho)d\rho < \infty, \int_0^\infty u_r(\rho)d\rho = \infty.$$

So, for

$$\Psi_I(r) = \sup\left\{\left(\int_0^\infty g^{p_2} v_r d\rho\right)^{\frac{1}{p_2}} : g \in \Omega_0; \left(\int_0^\infty g^{p_1} u_r d\rho\right)^{\frac{1}{p_1}} \leq 1\right\}$$

we can apply Theorem 1.1 in [12], according to which:

$$\Psi_I(r) \cong \left\{\int_0^\infty \left[\left(\int_0^t v_r d\rho\right)^{\frac{1}{p_2}} \left(\int_0^t u_r d\rho\right)^{-\frac{1}{p_1}}\right]^\sigma \frac{dU_r(t)}{U_r(t)}\right\}^{\frac{1}{\sigma}},$$

i.e.

$$\Psi_I(r) \cong \left\{\int_0^\infty \left[V_r(t)^{\frac{1}{p_2}} U_r(t)^{-\frac{1}{p_1}}\right]^\sigma \frac{dU_r(t)}{U_r(t)}\right\}^{\frac{1}{\sigma}}.$$

Here,

$$V_r(t) = \int_0^t v_r d\rho = \begin{cases} V(t), & t \in (0, |B_r|] \\ V(|B_r|), & t > |B_r|; \end{cases}$$

$$U_r(t) = \int_0^t u_r d\rho = \begin{cases} U(t), & t \in (0, |B_r|] \\ U(|B_r|) + (t - |B_r|), & t > |B_r|. \end{cases}$$

Denote

$$0 < \Psi(t) = U_r(t)^{-\frac{1}{p_1}} \downarrow; 0 < \Phi(t) = V_r(t)^{\frac{1}{p_2}} \uparrow .$$

Then,

$$\Psi_I(r) \cong \left\{ \int_0^\infty [\Phi\Psi]^\sigma \left(-\frac{d\Psi(t)}{\Psi(t)}\right) \right\}^{\frac{1}{\sigma}}, \tag{126}$$

and Φ, Ψ are absolutely continuous on $(0, T]$ for $\forall T \in (0, \infty)$ and

$$\Phi(+0) = 0; \ \Phi(+\infty) = V(|B_r|)^{\frac{1}{p_2}} < \infty; \ \Psi(+\infty) = 0.$$

Therefore, we can apply Theorem 6.1 (see [12]) to (126) (in fact, this is integration by parts with the justification that the substitutions go to zero):

$$\left\{ \int_0^\infty [\Phi\Psi]^\sigma \left(-\frac{d\Psi(t)}{\Psi(t)}\right) \right\}^{\frac{1}{\sigma}} = \left\{ \int_0^\infty [\Phi\Psi]^\sigma \frac{d\Phi(t)}{\Phi(t)} \right\}^{\frac{1}{\sigma}}.$$

Substituting this equality into (126), we get

$$\Psi_I(r) \cong \left\{ \int_0^\infty \left[V_r(t)^{\frac{1}{p_2}} U_r(t)^{-\frac{1}{p_1}} \right]^\sigma \frac{v_r(t)dt}{V_r(t)} \right\}^{\frac{1}{\sigma}}.$$

Taking into account introduced notation, it leads to (111).

Acknowledgements The work for Sects. 2, 3 has been prepared with the support of the "RUDN University Program 5-100" and funded by the Russian Foundation for Basic Research according to the research project 18-51-06005. The work for Sects. 4, 5 has been prepared in Steklov Mathematical Institute of Russian Academy of Sciences and supported by Russian Science Foundation, Project No 19-11-00087.
The authors are sincerely grateful to the reviewers for their careful reading of the paper and valuable advice.

References

1. Bennett, C., Sharpley, R.: Interpolation of Operators in Pure and Applied Mathematics, vol. 129. Academic Press Inc, Boston (1988)
2. Bennett, G., Grosse-Erdmann, K.-G.: Weighted Hardy inequalities for decreasing sequences and functions. Math. Anal. **334**, 489–531 (2006)
3. Burenkov, V.I.: Recent progress in studying the boundedness of classical operators of real analysis in general Morrey-type spaces I. Eurasian Math. J. **3**(3), 11–32 (2012)

4. Burenkov, V.I.: Recent progress in studying the boundedness of classical operators of real analysis in general Morrey-type spaces II. Eurasian Math. J. **4**(1), 21–45 (2013)
5. Burenkov, V.I., Gol'dman, M.L.: Calculation of the norm of a positive operator on the cone of monotone functions. Proc. Steklov Inst. Math. **210**, 47–65 (1995)
6. Burenkov, V.I., Goldman, M.L.: Necessary and sufficient conditions for the boundedness of the maximal operator from Lebesgue spaces to Morrey-type spaces. Math. Inequal. Appl. **17**(2), 401–418 (2014)
7. Burenkov, V.I., Guliev, V.S.: Necessary and sufficient conditions for the boundedness of the Riesz potentials in local Morrey-type spaces. Potential Anal. **30**(3), 211–249 (2009)
8. Burenkov, V.I., Nursultanov, E.D., Chiganbaeva, D.K.: Description of the interpolation spaces for a pair of local Morrey-type spaces and their generalizations. Proc. Steklov Inst. Math. **284**, 97–128 (2014)
9. Burenkov, V.I, Tararykova, T.V.E.: An analog of Young inequality for convolutions of functions for general Morrey-type spaces. Proc. Steklov Inst. Math. **293**, 107–126 (2016)
10. Burenkov, V.I., Gogatishvili, A., Guliev, V.S., Mustafaev, R.: Boundedness of the fractional maximal operator in local Morrey-type spaces. Complex Var. Elliptic Equ. **55**(8–10), 739–758 (2010)
11. Carro, M., Gogatishvili, A., Martin, J., Pick, L.: Inequalities involving two Hardy operators with applications to embeddings of function spaces. J. Oper. Theory **59**(2), 309–332 (2008)
12. Goldman, M.L.: Hardy type inequalities on the cone of quasimonotone functions. Russ. Acad. Sci. Far-East Branch Computer Center, Res. Rep. no 98/31, 1–69 (1998)
13. Goldman, M.L.: On equivalent criteria for the boundedness of Hardy type operators on the cone of decreasing functions. Anal. Math. **37**, 83–102 (2011)
14. Haroske, D.D., Skrzypczak, L.: On Sobolev and Franke-Jawerth embeddings of smoothness Morrey spaces. Rev. Mat. Comput. **27**(2), 541–573 (2014)
15. Kuang J.C.: Generalized Hausdorff operators on weighted Morrey-Herz spaces. Acta Math. Sinica (Chin. Ser.) **55**, 895–902 (2012)
16. Lemarie-Rieusset, P.G.: Multipliers and Morrey spaces. Potential Anal. Springer **38**(3), 741–752 (2013)
17. Yuan, W., Sickel, W., Yang, D.: Morrey and Campanato meet Besov, Lizorkin and Triebel. Lecture Notes in Mathematics, vol. 2005. Springer, Berlin (2010)

Characterization of Parabolic Fractional Maximal Function and Its Commutators in Orlicz Spaces

Vagif S. Guliyev, Ahmet Eroglu and Gulnara A. Abasova

Abstract In this paper, we give a necessary and sufficient condition for the boundedness of the parabolic fractional maximal operator and its commutators in Orlicz spaces.

Keywords Orlicz space · Parabolic fractional maximal function · Commutator · BMO

1 Introduction

The theory of boundedness of classical operators of the real analysis, such as the maximal operator, fractional maximal operator, Riesz potential and the singular integral operators etc, from one Lebesgue space to another one is well studied by now. These results have good applications in the theory of partial differential equations. However, in the theory of partial differential equations, along with Lebesgue spaces, Orlicz spaces also play an important role.

For $x \in \mathbb{R}^n$ and $r > 0$, we denote by $B(x, r)$ the open ball centered at x of radius r. Let $|B(x, r)|$ be the Lebesgue measure of the ball $B(x, r)$.

Let P be a real $n \times n$ matrix, all of whose eigenvalues have positive real part. Let $A_t = t^P \quad (t > 0)$, and set $\gamma = \operatorname{tr} P$. Then, there exists a quasi-distance ρ associated with P such that

V. S. Guliyev (✉)
Institute of Mathematics and Mechanics of NAS of Azerbaijan, Baku, Azerbaijan
e-mail: vagif@guliyev.com

Dumlupinar University, Kutahya 43100, Turkey

A. Eroglu
Omer Halisdemir University, Nigde 51240, Turkey

G. A. Abasova
Azerbaijan State University of Economics, Baku, Azerbaijan

A. Karapetyants et al. (eds.), *Modern Methods in Operator Theory and Harmonic Analysis*, Springer Proceedings in Mathematics & Statistics 291,
https://doi.org/10.1007/978-3-030-26748-3_2

(a) $\rho(A_t x) = t\rho(x)$, $t > 0$, for every $x \in \mathbb{R}^n$;
(b) $\rho(0) = 0$, $\rho(x-y) = \rho(y-x) \geq 0$ and $\rho(x-y) \leq k(\rho(x-z) + \rho(y-z))$;
(c) $dx = \rho^{\gamma-1} d\sigma(w) d\rho$, where $\rho = \rho(x)$, $w = A_{\rho^{-1}} x$
and $d\sigma(w)$ is a measure on the ellipsoid $\{w : \rho(w) = 1\}$.

Then, $\{\mathbb{R}^n, \rho, dx\}$ becomes a space of homogeneous type in the sense of Coifman–Weiss. Thus $\mathbb{R}^n$, endowed with the metric ρ, defines a homogeneous metric space [3, 5]. The balls with respect to ρ, centered at x of radius r, are just the ellipsoids $\mathcal{E}(x,r) = \{y \in \mathbb{R}^n : \rho(x-y) < r\}$ with the Lebesgue measure $|\mathcal{E}(x,r)| = v_\rho r^\gamma$, where v_ρ is the volume of the unit ellipsoid in $\mathbb{R}^n$. If $P = I$, then clearly $\rho(x) = |x|$ and $\mathcal{E}(x,r) = B(x,r)$. Note that in the standard parabolic case $P = (1, \ldots, 1, 2)$ we have

$$\rho(x) = \sqrt{\frac{|x'|^2 + \sqrt{|x'|^4 + x_n^2}}{2}}, \qquad x = (x', x_n).$$

Let $S_\rho = \{w \in \mathbb{R}^n : \rho(w) = 1\}$ be the unit ρ-sphere (ellipsoid) in $\mathbb{R}^n$ $(n \geq 2)$ equipped with the normalized Lebesgue surface measure $d\sigma$. The parabolic fractional maximal function $M_\alpha^P f$, $0 \leq \alpha < \gamma$, of a function $f \in L_1^{\rm loc}(\mathbb{R}^n)$ is given by

$$M_\alpha^P f(x) = \sup_{\mathcal{E} \ni x} |\mathcal{E}|^{-1+\frac{\alpha}{\gamma}} \int_{\mathcal{E}} |f(y)| dy$$

and the parabolic fractional maximal commutator of M_α^P with a locally integrable function b is defined by

$$M_{b,\alpha}^P f(x) = \sup_{\mathcal{E} \ni x} |\mathcal{E}|^{-1+\frac{\alpha}{\gamma}} \int_{\mathcal{E}} |b(x) - b(y)||f(y)| dy,$$

where the supremum is taken over all parabolic balls $\mathcal{E} \subset \mathbb{R}^n$ containing x. If $\alpha = 0$, then $M^P \equiv M_0^P$ is the parabolic maximal operator and $M_b^P \equiv M_{b,0}^P$ is the parabolic maximal commutator operator. If $P = I$, then $M_\alpha \equiv M_\alpha^I$ is the fractional maximal operator and $M \equiv M^I$ is the Hardy–Littlewood maximal operator. It is well known that the parabolic fractional maximal operators play an important role in harmonic analysis (see [6, 19]).

On the other hand, we can define the (nonlinear) commutator of the parabolic fractional maximal operator M_α^P with a locally integrable function b by

$$[b, M_\alpha^P] f(x) f = b(x) M_\alpha^P f(x) - M_\alpha^P (bf)(x).$$

For more details on the operators $M_{b,\alpha}^P$ and $[b, M_\alpha^P]$, where $0 \leq \alpha < \gamma$, we refer to [10] and references therein.

In this work we present the characterization for the boundedness of the parabolic fractional maximal operator M_α^P (Theorem 5) and its commutators $M_{b,\alpha}^P$ (Theorem 4) in Orlicz spaces.

2 Young Functions and Orlicz Spaces

Orlicz space was first introduced by Orlicz in [16, 17] as a generalization of Lebesgue spaces L^p. Since then these spaces have been one of important functional frames in the mathematical analysis, and especially in real and harmonic analysis. Orlicz spaces also appropriately substitute for L^1 space when L^1 space does not work.

First, we recall the definition of Young functions.

Definition 1 A function $\Phi : [0, \infty) \to [0, \infty]$ is called a Young function if Φ is convex, left-continuous, $\lim_{r \to +0} \Phi(r) = \Phi(0) = 0$ and $\lim_{r \to \infty} \Phi(r) = \infty$.

From the convexity and $\Phi(0) = 0$ it follows that any Young function is increasing. If there exists $s \in (0, \infty)$ such that $\Phi(s) = \infty$, then $\Phi(r) = \infty$ for $r \geq s$. The set of Young functions such that

$$0 < \Phi(r) < \infty \quad \text{for} \quad 0 < r < \infty$$

will be denoted by $\mathcal{Y}$. If $\Phi \in \mathcal{Y}$, then Φ is absolutely continuous on every closed interval in $[0, \infty)$ and bijective from $[0, \infty)$ to itself.

For a Young function Φ and $0 \leq s \leq \infty$, let

$$\Phi^{-1}(s) = \inf\{r \geq 0 : \Phi(r) > s\}.$$

If $\Phi \in \mathcal{Y}$, then Φ^{-1} is the usual inverse function of Φ. It is well known that

$$r \leq \Phi^{-1}(r)\widetilde{\Phi}^{-1}(r) \leq 2r \qquad \text{for } r \geq 0, \tag{1}$$

where $\widetilde{\Phi}(r)$ is defined by

$$\widetilde{\Phi}(r) = \begin{cases} \sup\{rs - \Phi(s) : s \in [0, \infty)\} , & r \in [0, \infty) \\ \infty , & r = \infty. \end{cases}$$

A Young function Φ is said to satisfy the Δ_2-condition, denoted also as $\Phi \in \Delta_2$, if

$$\Phi(2r) \leq C\Phi(r), \qquad r > 0$$

for some $C > 1$. If $\Phi \in \Delta_2$, then $\Phi \in \mathcal{Y}$. A Young function Φ is said to satisfy the ∇_2-condition, denoted also by $\Phi \in \nabla_2$, if

$$\Phi(r) \leq \frac{1}{2C}\Phi(Cr), \qquad r \geq 0$$

for some $C > 1$.

Definition 2 (Orlicz Space) For a Young function Φ, the set

$$L^{\Phi}(\mathbb{R}^n) = \left\{ f \in L^1_{\mathrm{loc}}(\mathbb{R}^n) : \int_{\mathbb{R}^n} \Phi(k|f(x)|)dx < \infty \text{ for some } k > 0 \right\}$$

is called the Orlicz space. If $\Phi(r) = r^p$, $1 \le p < \infty$, then $L^{\Phi}(\mathbb{R}^n) = L^p(\mathbb{R}^n)$. If $\Phi(r) = 0$, $(0 \le r \le 1)$ and $\Phi(r) = \infty$, $(r > 1)$, then $L^{\Phi}(\mathbb{R}^n) = L^{\infty}(\mathbb{R}^n)$. The space $L^{\Phi}_{\mathrm{loc}}(\mathbb{R}^n)$ is defined as the set of all functions f such that $f\chi_{\mathcal{E}} \in L^{\Phi}(\mathbb{R}^n)$ for all parabolic balls $\mathcal{E} \subset \mathbb{R}^n$.

The Orlicz space $L^{\Phi}(\mathbb{R}^n)$ is a Banach space with respect to the norm

$$\|f\|_{L^{\Phi}} = \inf\left\{ \lambda > 0 : \int_{\mathbb{R}^n} \Phi\Big(\frac{|f(x)|}{\lambda}\Big)dx \le 1 \right\}.$$

For a measurable set $\Omega \subset \mathbb{R}^n$, a measurable function f and $t > 0$, let $m(\Omega, f, t) = |\{x \in \Omega : |f(x)| > t\}|$. In the case $\Omega = \mathbb{R}^n$, we shortly denote it by $m(f, t)$.

Definition 3 The weak Orlicz space

$$WL^{\Phi}(\mathbb{R}^n) = \{f \in L^1_{\mathrm{loc}}(\mathbb{R}^n) : \|f\|_{WL^{\Phi}} < \infty\}$$

is defined by the norm

$$\|f\|_{WL^{\Phi}} = \inf\left\{ \lambda > 0 \ : \ \sup_{t>0} \Phi(t) m\Big(\frac{f}{\lambda},\ t\Big) \le 1 \right\}.$$

We note that $\|f\|_{WL^{\Phi}} \le \|f\|_{L^{\Phi}}$,

$$\sup_{t>0} \Phi(t)m(\Omega,\ f,\ t) = \sup_{t>0} t\, m(\Omega,\ f,\ \Phi^{-1}(t)) = \sup_{t>0} t\, m(\Omega,\ \Phi(|f|),\ t)$$

and

$$\int_{\Omega} \Phi\Big(\frac{|f(x)|}{\|f\|_{L^{\Phi}(\Omega)}}\Big)dx \le 1, \qquad \sup_{t>0} \Phi(t)m\Big(\Omega,\ \frac{f}{\|f\|_{WL^{\Phi}(\Omega)}},\ t\Big) \le 1, \tag{2}$$

where $\|f\|_{L^{\Phi}(\Omega)} = \|f\chi_{\Omega}\|_{L^{\Phi}}$ and $\|f\|_{WL^{\Phi}(\Omega)} = \|f\chi_{\Omega}\|_{WL^{\Phi}}$.

The following analogue of the Hölder's inequality is well known (see, for example [18]).

Theorem 1 *Let $\Omega \subset \mathbb{R}^n$ be a measurable set and functions f and g measurable on Ω. For a Young function Φ and its complementary function $\widetilde{\Phi}$, the following inequality is valid*

$$\int_{\Omega} |f(x)g(x)|dx \le 2\|f\|_{L^{\Phi}(\Omega)}\|g\|_{L^{\widetilde{\Phi}}(\Omega)}.$$

By elementary calculations we have the following property.

Lemma 1 *Let Φ be a Young function and $\mathcal{E}$ be a parabolic balls in $\mathbb{R}^n$. Then*

$$\|\chi_{\mathcal{E}}\|_{L^\Phi} = \|\chi_{\mathcal{E}}\|_{WL^\Phi} = \frac{1}{\Phi^{-1}\left(|\mathcal{E}|^{-1}\right)}.$$

By Theorem 1, Lemma 1 and (1) we get the following estimate.

Lemma 2 *For a Young function Φ and for the parabolic balls $\mathcal{E} = \mathcal{E}(x, r)$ the following inequality is valid:*

$$\int_{\mathcal{E}} |f(y)|dy \le 2|\mathcal{E}|\Phi^{-1}\left(|\mathcal{E}|^{-1}\right) \|f\|_{L^\Phi(\mathcal{E})}.$$

3 Parabolic Fractional Maximal Function in Orlicz Spaces

In this section, we shall give a necessary and sufficient condition for the boundedness of M_α^P on Orlicz spaces and weak Orlicz spaces. We begin with the boundedness of the parabolic maximal operator on Orlicz spaces.

Theorem 2 ([1]) *Let Φ be a Young function.*

(i) The parabolic maximal operator M^P is bounded from $L^\Phi(\mathbb{R}^n)$ to $WL^\Phi(\mathbb{R}^n)$, and the inequality

$$\|M^P f\|_{WL^\Phi} \le C_0 \|f\|_{L^\Phi} \tag{3}$$

holds with constant C_0 independent of f.

(ii) The parabolic maximal operator M^P is bounded on $L^\Phi(\mathbb{R}^n)$, and the inequality

$$\|M^P f\|_{L^\Phi} \le C_0 \|f\|_{L^\Phi} \tag{4}$$

holds with constant C_0 independent of f if and only if $\Phi \in \nabla_2$.

For proving our main results, we need the following estimate.

Lemma 3 *If $\mathcal{E}_0 := \mathcal{E}(x_0, r_0)$, then for every $x \in \mathcal{E}_0$*

$$M_\alpha^P \chi_{\mathcal{E}_0}(x) \ge c_0 r_0^\alpha,$$

where $c_0 = |\mathcal{E}(0, 1)|$.

Proof Let $x \in \mathcal{E}_0$. Then

$$\begin{aligned} M_\alpha^P \chi_{\mathcal{E}_0}(x) &= \sup_{t>0} |\mathcal{E}(x,t)|^{-1+\frac{\alpha}{\gamma}} |\mathcal{E}(x,t) \cap \mathcal{E}_0| \\ &\ge |\mathcal{E}_0|^{-1+\frac{\alpha}{\gamma}} |\mathcal{E}_0 \cap \mathcal{E}_0| = |\mathcal{E}_0|^{\frac{\alpha}{\gamma}} = c_0 r_0^\alpha. \end{aligned}$$

□

The following result completely characterizes the boundedness of M_α^P on Orlicz spaces.

Theorem 3 *Let $0 < \alpha < \gamma$, Φ, Ψ be Young functions and $\Phi \in \mathcal{Y}$. The condition*

$$r^{-\frac{\alpha}{\gamma}}\Phi^{-1}(r) \le C\Psi^{-1}(r) \tag{5}$$

for all $r > 0$, where $C > 0$ does not depend on r, is necessary and sufficient for the boundedness of M_α^P from $L^\Phi(\mathbb{R}^n)$ to $WL^\Psi(\mathbb{R}^n)$. Moreover, if $\Phi \in \nabla_2$, the condition (5) is necessary and sufficient for the boundedness of M_α^P from $L^\Phi(\mathbb{R}^n)$ to $L^\Psi(\mathbb{R}^n)$.

Proof For arbitrary ball $\mathcal{E} = \mathcal{E}(x, r)$ we represent f as

$$f = f_1 + f_2, \quad f_1(y) = f(y)\chi_{2k\mathcal{E}}(y), \quad f_2(y) = f(y)\chi_{\mathbb{R}^n\setminus\mathcal{E}(x,2kr)}(y), \quad r > 0,$$

and have

$$M_\alpha^P f(x) \le M_\alpha^P f_1(x) + M_\alpha^P f_2(x).$$

Let y be an arbitrary point in $\mathcal{E}$. If $\mathcal{E}(y, t) \cap (\mathbb{R}^n \setminus \mathcal{E}(x, 2kr)) \neq \emptyset$, then $t > r$. Indeed, if $z \in \mathcal{E}(y, t) \cap (\mathbb{R}^n \setminus \mathcal{E}(x, 2kr))$, then

$$t > \rho(y - z) \ge \frac{1}{k}\rho(x - z) - \rho(x - y) > 2r - r = r.$$

On the other hand, $\mathcal{E}(y, t) \cap (\mathbb{R}^n \setminus \mathcal{E}(x, 2kr)) \subset \mathcal{E}(x, 2kt)$. Indeed, if $z \in \mathcal{E}(y, t) \cap (\mathbb{R}^n \setminus \mathcal{E}(x, 2kr))$, then we get $\rho(x - z) \le k\rho(y - z) + k\rho(x - y) < kt + kr < 2kt$.

Hence

$$\begin{aligned}
M_\alpha^P f_2(y) &\le C \sup_{t>0} \frac{1}{|\mathcal{E}(y,t)|^{1-\frac{\alpha}{\gamma}}} \int_{\mathcal{E}(y,t)\cap(\mathbb{R}^n\setminus\mathcal{E}(x,2kr))} |f(z)|dz \\
&\le C \sup_{t>r} \frac{1}{|\mathcal{E}(x,2kt)|^{1-\frac{\alpha}{\gamma}}} \int_{\mathcal{E}(x,2kt)} |f(z)|dz \\
&= C \sup_{t>2kr} \frac{1}{|\mathcal{E}(x,t)|^{1-\frac{\alpha}{\gamma}}} \int_{\mathcal{E}(x,t)} |f(z)|dz \\
&\le C \sup_{r<t<\infty} t^\alpha \Phi^{-1}(|\mathcal{E}(x,t)|^{-1}) \, \|f\|_{L^\Phi(\mathcal{E}(x,t))} \\
&\le C\|f\|_{L^\Phi} \sup_{r<t<\infty} t^\alpha \Phi^{-1}(|\mathcal{E}(x,t)|^{-1}).
\end{aligned}$$

Consequently from Hedberg's trick, see [11], and the last inequality, we have

$$M_\alpha^P f(y) \le Cr^\alpha Mf(y) + \|f\|_{L^\Phi} \sup_{r<t<\infty} t^\alpha \Phi^{-1}(t^{-\gamma}).$$

Thus, by (5) we obtain

$$M_\alpha^P f(x) \le C M^P f(x) \frac{\Psi^{-1}(r^{-\gamma})}{\Phi^{-1}(r^{-\gamma})} + \|f\|_{L^\Phi} \Psi^{-1}(r^{-\gamma}).$$

Choose $r > 0$ so that $\Phi^{-1}(r^{-\gamma}) = \frac{M^P f(x)}{C_0 \|f\|_{L^\Phi}}$. Then

$$\frac{\Psi^{-1}(r^{-\gamma})}{\Phi^{-1}(r^{-\gamma})} = \frac{(\Psi^{-1} \circ \Phi)(\frac{M^P f(x)}{C_0\|f\|_{L^\Phi}})}{\frac{M^P f(x)}{C_0\|f\|_{L^\Phi}}}.$$

Therefore, we get

$$M_\alpha^P f(x) \le C_1 \|f\|_{L^\Phi} (\Psi^{-1} \circ \Phi)\Big(\frac{M^P f(x)}{C_0 \|f\|_{L^\Phi}}\Big).$$

Let C_0 be as in (3). Then by Theorem 2, we have

$$\begin{aligned}
&\sup_{r>0} \Psi(r)\, m\Big(\mathcal{E}, \frac{|M_\alpha^P f(x)|}{C_1\|f\|_{L^\Phi}}, r\Big) = \sup_{r>0} r\, m\Big(\mathcal{E}, \Psi\Big(\frac{|M_\alpha^P f(x)|}{C_1\|f\|_{L^\Phi}}\Big), r\Big) \\
\le\ &\sup_{r>0} r\, m\Big(\mathcal{E}, \Phi\Big(\frac{M^P f(x)}{C_0\|f\|_{L^\Phi}}\Big), r\Big) \le \sup_{r>0} \Phi(r)\, m\Big(\frac{M^P f(x)}{\|M^P f\|_{WL^\Phi}}, r\Big) \le 1,
\end{aligned}$$

i.e.

$$\|M_\alpha^P f\|_{WL^\Psi(\mathcal{E})} \le C \|f\|_{L^\Phi}. \tag{6}$$

By taking the supremum over $\mathcal{E}$ in (6), we get

$$\|M_\alpha^P f\|_{WL^\Psi} \le C \|f\|_{L^\Phi},$$

since the constants in (6) don't depend on x and r.

Let C_0 be as in (4). Since $\Phi \in \nabla_2$, by Theorem 2, we have

$$\int_{\mathcal{E}} \Psi\left(\frac{M_\alpha^P f(x)}{C_1\|f\|_{L^\Phi}}\right) dx \le \int_{\mathcal{E}} \Phi\left(\frac{M^P f(x)}{C_0\|f\|_{L^\Phi}}\right) dx \le \int_{\mathbb{R}^n} \Phi\left(\frac{M^P f(x)}{\|M^P f\|_{L^\Phi}}\right) dx \le 1,$$

i.e.

$$\|M_\alpha^P f\|_{L^\Psi(\mathcal{E})} \le C \|f\|_{L^\Phi}. \tag{7}$$

By taking the supremum over $\mathcal{E}$ in (7), we get

$$\|M_\alpha^P f\|_{L^\Psi} \le C \|f\|_{L^\Phi},$$

since the constants in (7) don't depend on x and r.

We shall now prove the necessity. Let $\mathcal{E}_0 = \mathcal{E}(x_0, r_0)$ and $x \in \mathcal{E}_0$. By Lemma 3, we have $r_0^\alpha \le C M_\alpha^P \chi_{\mathcal{E}_0}(x)$. Therefore, by Lemma 1, we have

$$r_0^\alpha \le C\Psi^{-1}(|\mathcal{E}_0|^{-1})\|M_\alpha^P \chi_{\mathcal{E}_0}\|_{WL^\Psi(\mathcal{E}_0)} \le C\Psi^{-1}(|\mathcal{E}_0|^{-1})\|M_\alpha^P \chi_{\mathcal{E}_0}\|_{WL^\Psi}$$
$$\le C\Psi^{-1}(|\mathcal{E}_0|^{-1})\|\chi_{\mathcal{E}_0}\|_{L^\Phi} \le C\frac{\Psi^{-1}(r_0^{-\gamma})}{\Phi^{-1}(r_0^{-\gamma})}$$

and

$$r_0^\alpha \le C\Psi^{-1}(|\mathcal{E}_0|^{-1})\|M_\alpha^P \chi_{\mathcal{E}_0}\|_{L^\Psi(\mathcal{E}_0)} \le C\Psi^{-1}(|\mathcal{E}_0|^{-1})\|M_\alpha^P \chi_{\mathcal{E}_0}\|_{L^\Psi}$$
$$\le C\Psi^{-1}(|\mathcal{E}_0|^{-1})\|\chi_{\mathcal{E}_0}\|_{L^\Phi} \le C\frac{\Psi^{-1}(r_0^{-\gamma})}{\Phi^{-1}(r_0^{-\gamma})}.$$

Since this is true for every $r_0 > 0$, we are done. □

4 Parabolic Fractional Maximal Commutators in Orlicz Spaces

In this section, as an application of Theorem 3 we consider the boundedness of parabolic fractional maximal commutator operator $M_{b,\alpha}^P$ and commutator of parabolic fractional maximal operator $[b, M_\alpha^P]$ on Orlicz spaces, when b belongs to the parabolic Lipschitz space, by which some new characterizations of the parabolic Lipschitz spaces are given.

Definition 4 Let $0 < \beta < 1$, we say that a function b belongs to the parabolic Lipschitz space $\dot{\Lambda}_{\beta,P}(\mathbb{R}^n)$ if there exists a constant C such that for all $x, y \in \mathbb{R}^n$,

$$|b(x) - b(y)| \le C\rho(x-y)^\beta.$$

The smallest such constant C is called the $\dot{\Lambda}_{\beta,P}(\mathbb{R}^n)$ norm of b and is denoted by $\|b\|_{\dot{\Lambda}_{\beta,P}(\mathbb{R}^n)}$.

To prove the theorems, we need auxiliary results. The first one is the following characterizations of Lipschitz spaces, which is due to DeVore and Sharpley in the isotropic case $P = I$ [4].

Lemma 4 *Let* $0 < \beta < 1$, *we have*

$$\|f\|_{\dot{\Lambda}_{\beta,P}(\mathbb{R}^n)} \approx \sup_{\mathcal{E}} \frac{1}{|\mathcal{E}|^{1+\beta/\gamma}} \int_{\mathcal{E}} |f(x) - f_{\mathcal{E}}|dx,$$

where $f_{\mathcal{E}} = \frac{1}{|\mathcal{E}|}\int_{\mathcal{E}} f(y)dy$.

Lemma 5 *Let* $0 < \beta < 1$, $0 \le \alpha$, $\alpha + \beta < \gamma$ *and* $b \in \dot{\Lambda}_{\beta,P}(\mathbb{R}^n)$, *then the following pointwise estimate holds:*

$$M^P_{b,\alpha} f(x) \leq C\|b\|_{\dot{\Lambda}_{\beta,P}(\mathbb{R}^n)} M^P_{\alpha+\beta} f(x).$$

Proof If $b \in \dot{\Lambda}_{\beta,P}(\mathbb{R}^n)$, then

$$\begin{aligned} M^P_{b,\alpha}(f)(x) &= \sup_{\mathcal{E}\ni x} |\mathcal{E}|^{-1+\frac{\alpha}{\gamma}} \int_{\mathcal{E}} |b(x)-b(y)||f(y)|dy \\ &\leq C\|b\|_{\dot{\Lambda}_{\beta,P}(\mathbb{R}^n)} \sup_{\mathcal{E}\ni x} |\mathcal{E}|^{-1+\frac{\alpha+\beta}{\gamma}} \int_{\mathcal{E}} |f(y)|dy \\ &= C\|b\|_{\dot{\Lambda}_{\beta,P}(\mathbb{R}^n)} M^P_{\alpha+\beta} f(x). \end{aligned}$$

□

Lemma 6 *If $b \in L^1_{\mathrm{loc}}(\mathbb{R}^n)$ and $\mathcal{E}_0 := \mathcal{E}(x_0, r_0)$, then*

$$|\mathcal{E}_0|^{\frac{\alpha}{\gamma}} |b(x) - b_{\mathcal{E}_0}| \leq M^P_{b,\alpha} \chi_{\mathcal{E}_0}(x) \ \text{ for every } \ x \in \mathcal{E}_0.$$

Proof For $x \in \mathcal{E}_0$, we get

$$\begin{aligned} &M^P_{b,\alpha}\, \chi_{\mathcal{E}_0}(x) = \sup_{\mathcal{E}\ni x} |\mathcal{E}|^{-1+\frac{\alpha}{\gamma}} \textstyle\int_{\mathcal{E}} |b(x)-b(y)|\chi_{\mathcal{E}_0}(y)dy \\ &= \ \sup_{\mathcal{E}\ni x} |\mathcal{E}|^{-1+\frac{\alpha}{n}} \textstyle\int_{\mathcal{E}\cap\mathcal{E}_0} |b(x)-b(y)|dy \geq |\mathcal{E}_0|^{-1+\frac{\alpha}{\gamma}} \int_{\mathcal{E}_0} |b(x)-b(y)|dy \\ &\geq \ \Big||\mathcal{E}_0|^{-1+\frac{\alpha}{\gamma}} \textstyle\int_{\mathcal{E}_0} (b(x)-b(y))dy\Big| = |\mathcal{E}_0|^{\frac{\alpha}{\gamma}} |b(x) - b_{\mathcal{E}_0}|. \end{aligned}$$

□

The following theorem is valid.

Theorem 4 *Let $0 < \beta < 1$, $0 \leq \alpha$, $\alpha + \beta < \gamma$, $b \in L^1_{\mathrm{loc}}(\mathbb{R}^n)$, Φ, Ψ be Young functions and $\Phi \in \mathcal{Y}$.*

1. *If $\Phi \in \nabla_2$ and the condition*

$$t^{-\frac{\alpha+\beta}{\gamma}} \Phi^{-1}\big(t\big) \leq C\Psi^{-1}\big(t\big), \tag{8}$$

holds for all $t > 0$, where $C > 0$ does not depend on t, then the condition $b \in \dot{\Lambda}_{\beta,P}(\mathbb{R}^n)$ is sufficient for the boundedness of $M^P_{b,\alpha}$ from $L^\Phi(\mathbb{R}^n)$ to $L^\Psi(\mathbb{R}^n)$.

2. *If the condition*

$$\Psi^{-1}(t) \leq C\Phi^{-1}(t)t^{-\frac{\alpha+\beta}{\gamma}}, \tag{9}$$

holds for all $t > 0$, where $C > 0$ does not depend on t, then the condition $b \in \dot{\Lambda}_{\beta,P}(\mathbb{R}^n)$ is necessary for the boundedness of $M^P_{b,\alpha}$ from $L^\Phi(\mathbb{R}^n)$ to $L^\Psi(\mathbb{R}^n)$.

3. *If $\Phi \in \nabla_2$ and $\Psi^{-1}(t) \approx \Phi^{-1}(t)t^{-\frac{\alpha+\beta}{\gamma}}$, then the condition $b \in \dot{\Lambda}_{\beta,P}(\mathbb{R}^n)$ is necessary and sufficient for the boundedness of $M^P_{b,\alpha}$ from $L^\Phi(\mathbb{R}^n)$ to $L^\Psi(\mathbb{R}^n)$.*

Proof (1) The first statement of the theorem follows from Theorem 3 and Lemma 5.

(2) We shall now prove the second part. Suppose that $M^P_{b,\alpha}$ is bounded from $L^{\Phi}(\mathbb{R}^n)$ to $L^{\Psi}(\mathbb{R}^n)$ and $\Psi^{-1}(t) \le C\Phi^{-1}(t)t^{-(\alpha+\beta)/\gamma}$. Choose any parabolic ball $\mathcal{E} in \mathbb{R}^n$, by Lemmas 1 and 2

$$\begin{aligned}\frac{1}{|\mathcal{E}|^{1+\frac{\beta}{\gamma}}}\int_{\mathcal{E}}|b(y)-b_{\mathcal{E}}|dy &= \frac{1}{|\mathcal{E}|^{1+\frac{\alpha+\beta}{\gamma}}}\int_{\mathcal{E}}\left|\frac{1}{|\mathcal{E}|^{1-\frac{\alpha}{\gamma}}}\int_{\mathcal{E}}(b(y)-b(z))dz\right|dy\\ &\le \frac{1}{|\mathcal{E}|^{1+\frac{\alpha+\beta}{\gamma}}}\int_{\mathcal{E}} M^P_{b,\alpha}\big(\chi_{\mathcal{E}}\big)(y)dy\\ &\le \frac{2\Psi^{-1}(|\mathcal{E}|^{-1})}{|\mathcal{E}|^{\frac{\alpha+\beta}{\gamma}}}\|M^P_{b,\alpha}\big(\chi_{\mathcal{E}}\big)\|_{L^{\Psi}(\mathcal{E})}\\ &\le \frac{C}{|\mathcal{E}|^{\frac{\alpha+\beta}{\gamma}}}\frac{\Psi^{-1}(|\mathcal{E}|^{-1})}{\Phi^{-1}(|\mathcal{E}|^{-1})}\le C.\end{aligned}$$

Thus by Lemma 4 we get $b \in \dot{\Lambda}_{\beta,P}(\mathbb{R}^n)$.

(3) The third statement of the theorem follows from the first and second parts of the theorem. □

If we take $\alpha = 0$ at Theorem 4, we have the following new result.

Corollary 1 *Let* $0 < \beta < 1$, $b \in L^1_{\rm loc}(\mathbb{R}^n)$, Φ, Ψ *be Young functions and* $\Phi \in \mathcal{Y}$.

1. *If* $\Phi \in \nabla_2$ *and the condition* $\Phi^{-1}(t)t^{-\beta/\gamma} \le C\Psi^{-1}(t)$ *holds, then the condition* $b \in \dot{\Lambda}_{\beta,P}(\mathbb{R}^n)$ *is sufficient for the boundedness of* M^P_b *from* $L^{\Phi}(\mathbb{R}^n)$ *to* $L^{\Psi}(\mathbb{R}^n)$.
2. *If* $\Psi^{-1}(t) \le C\Phi^{-1}(t)t^{-\beta/\gamma}$, *then the condition* $b \in \dot{\Lambda}_{\beta,P}(\mathbb{R}^n)$ *is necessary for the boundedness of* M^P_b *from* $L^{\Phi}(\mathbb{R}^n)$ *to* $L^{\Psi}(\mathbb{R}^n)$.
3. *If* $\Phi \in \nabla_2$ *and* $\Psi^{-1}(t) \approx \Phi^{-1}(t)t^{-\beta/\gamma}$, *then the condition* $b \in \dot{\Lambda}_{\beta,P}(\mathbb{R}^n)$ *is necessary and sufficient for the boundedness of* M^P_b *from* $L^{\Phi}(\mathbb{R}^n)$ *to* $L^{\Psi}(\mathbb{R}^n)$.

If we take $\Phi(t) = t^p$ and $\Psi(t) = t^q$ with $1 \le p < \infty$ and $1 \le q < \infty$ at Theorem 4, we have the following result.

Corollary 2 *Let* $0 < \beta < 1$, $0 \le \alpha$, $\alpha + \beta < \gamma$, $b \in L^1_{\rm loc}(\mathbb{R}^n)$, $1 < p < q < \infty$ *and* $\frac{1}{p} - \frac{1}{q} = \frac{\alpha+\beta}{\gamma}$. *Then the condition* $b \in \dot{\Lambda}_{\beta,P}(\mathbb{R}^n)$ *is necessary and sufficient for the boundedness of* $M^P_{b,\alpha}$ *from* $L^p(\mathbb{R}^n)$ *to* $L^q(\mathbb{R}^n)$.

The following theorem is valid.

Theorem 5 *Let* $0 < \beta < 1$, $0 \le \alpha$, $\alpha + \beta < \gamma$, $b \in L^1_{\rm loc}(\mathbb{R}^n)$, Φ, Ψ *be Young functions and* $\Phi \in \mathcal{Y}$.

1. *If condition (8) holds, then the condition* $b \in \dot{\Lambda}_{\beta,P}(\mathbb{R}^n)$ *is sufficient for the boundedness of* $M^P_{b,\alpha}$ *from* $L^{\Phi}(\mathbb{R}^n)$ *to* $WL^{\Psi}(\mathbb{R}^n)$.
2. *If condition (9) holds and* $\frac{t^{1+\varepsilon}}{\Psi(t)}$ *is almost decreasing for some* $\varepsilon > 0$, *then the condition* $b \in \dot{\Lambda}_{\beta,P}(\mathbb{R}^n)$ *is necessary for the boundedness of* $M^P_{b,\alpha}$ *from* $L^{\Phi}(\mathbb{R}^n)$ *to* $WL^{\Psi}(\mathbb{R}^n)$.
3. *If* $\Psi^{-1}(t) \approx \Phi^{-1}(t)t^{-(\alpha+\beta)/\gamma}$ *and* $\frac{t^{1+\varepsilon}}{\Psi(t)}$ *is almost decreasing for some* $\varepsilon > 0$, *then the condition* $b \in \dot{\Lambda}_{\beta,P}(\mathbb{R}^n)$ *is necessary and sufficient for the boundedness of* $M^P_{b,\alpha}$ *from* $L^{\Phi}(\mathbb{R}^n)$ *to* $WL^{\Psi}(\mathbb{R}^n)$.

Proof (1) The first statement of the theorem follows from Theorem 3 and Lemma 5.

(2) For any fixed ball $\mathcal{E}_0$ such that $x \in \mathcal{E}_0$ by Lemma 6 we have

$$|\mathcal{E}_0|^{\alpha/\gamma}|b(x) - b_{\mathcal{E}_0}| \leq M^P_{b,\alpha}\chi_{\mathcal{E}_0}(x).$$

This together with the boundedness of $M^P_{b,\alpha}$ from $L^\Phi(\mathbb{R}^n)$ to $WL^\Psi(\mathbb{R}^n)$ and Lemma 1

$$|\{x \in \mathcal{E}_0 : |\mathcal{E}_0|^{\alpha/n}|b(x) - b_{\mathcal{E}_0}| > \lambda\}| \leq |\{x \in \mathcal{E}_0 : M^P_{b,\mathcal{E}}\chi_{\mathcal{E}_0}(x) > \lambda\}| \\ \leq \frac{1}{\Psi\left(\frac{\lambda}{C\|\chi_{\mathcal{E}_0}\|_{L^\Phi}}\right)} = \frac{1}{\Psi\left(\frac{\lambda\Phi^{-1}(|\mathcal{E}_0|^{-1})}{C}\right)}.$$

Let $t > 0$ be a constant to be determined later, then

$$\begin{aligned}\int_{\mathcal{E}_0}|b(x) - b_{\mathcal{E}_0}|dx =&|\mathcal{E}_0|^{-\alpha/\gamma}\int_0^\infty |\{x \in \mathcal{E}_0 : |b(x) - b_{\mathcal{E}_0}| > |\mathcal{E}_0|^{-\alpha/\gamma}\lambda\}|d\lambda \\ =&|\mathcal{E}_0|^{-\alpha/\gamma}\int_0^t \{x \in \mathcal{E}_0 : |b(x) - b_{\mathcal{E}_0}| > |\mathcal{E}_0|^{-\alpha/\gamma}\lambda\}|d\lambda \\ &+ |\mathcal{E}_0|^{-\alpha/\gamma}\int_t^\infty |\{x \in \mathcal{E}_0 : |b(x) - b_{\mathcal{E}_0}| > |\mathcal{E}_0|^{-\alpha/\gamma}\lambda\}|d\lambda \\ \leq& t|\mathcal{E}_0|^{1-\alpha/\gamma} + |\mathcal{E}_0|^{-\alpha/\gamma}\int_t^\infty \frac{d\lambda}{\Psi\left(\frac{\lambda\Phi^{-1}(|\mathcal{E}_0|^{-1})}{C}\right)} \\ \leq& Ct|\mathcal{E}_0|^{1-\alpha/\gamma} + \frac{|\mathcal{E}_0|^{-\alpha/\gamma}t}{\Psi\left(\frac{t\Phi^{-1}(|\mathcal{E}_0|^{-1})}{C}\right)},\end{aligned}$$

where we use almost decreasingness of $\frac{t^{1+\varepsilon}}{\Psi(t)}$ in the last step.

Set $t = C|\mathcal{E}_0|^{\frac{\alpha+\beta}{\gamma}}$ in the above estimate, we have

$$\int_{\mathcal{E}_0}|b(x) - b_{\mathcal{E}_0}|dx \leq C|\mathcal{E}_0|^{1+\beta/\gamma}.$$

Thus by Lemma 4 we get $b \in \dot{\Lambda}_{\beta,P}(\mathbb{R}^n)$ since $\mathcal{E}_0$ is an arbitrary parabolic ball in $\mathbb{R}^n$.

(3) The third statement of the theorem follows from the first and second parts of the theorem. □

If we take $\alpha = 0$ at Theorem 5, we have the following new result.

Corollary 3 *Let* $0 < \beta < 1$, $b \in L^1_{\mathrm{loc}}(\mathbb{R}^n)$, Φ, Ψ *be Young functions and* $\Phi \in \mathcal{Y}$.

1. *If the condition* $\Phi^{-1}(t)t^{-\beta/\gamma} \leq C\Psi^{-1}(t)$ *holds, then the condition* $b \in \dot{\Lambda}_{\beta,P}(\mathbb{R}^n)$ *is sufficient for the boundedness of* M^P_b *from* $L^\Phi(\mathbb{R}^n)$ *to* $WL^\Psi(\mathbb{R}^n)$.

2. *If* $\Psi^{-1}(t) \le C\Phi^{-1}(t)t^{-\beta/\gamma}$ *and* $\frac{t^{1+\varepsilon}}{\Psi(t)}$ *is almost decreasing for some* $\varepsilon > 0$, *then the condition* $b \in \dot{\Lambda}_{\beta,P}(\mathbb{R}^n)$ *is necessary for the boundedness of* M_b^P *from* $L^\Phi(\mathbb{R}^n)$ *to* $WL^\Psi(\mathbb{R}^n)$.

3. *If* $\Psi^{-1}(t) \approx \Phi^{-1}(t)t^{-\beta/\gamma}$ *and* $\frac{t^{1+\varepsilon}}{\Psi(t)}$ *is almost decreasing for some* $\varepsilon > 0$, *then the condition* $b \in \dot{\Lambda}_{\beta,P}(\mathbb{R}^n)$ *is necessary and sufficient for the boundedness of* M_b^P *from* $L^\Phi(\mathbb{R}^n)$ *to* $WL^\Psi(\mathbb{R}^n)$.

If we take $\Phi(t) = t^p$ and $\Psi(t) = t^q$ with $1 \le p < \imath$ and $1 \le q \le \imath$ at Theorem 5, we have the following result.

Corollary 4 *Let* $0 < \beta < 1$, $0 \le \alpha$, $\alpha + \beta < \gamma$, $b \in L^1_{\text{loc}}(\mathbb{R}^n)$, $1 \le p < q \le \imath$ *and* $\frac{1}{p} - \frac{1}{q} = \frac{\alpha+\beta}{\gamma}$. *Then the condition* $b \in \dot{\Lambda}_{\beta,P}(\mathbb{R}^n)$ *is necessary and sufficient for the boundedness of* $M_{b,\alpha}^P$ *from* $L^p(\mathbb{R}^n)$ *to* $WL^q(\mathbb{R}^n)$.

To state our results, we recall the definition of the maximal operator with respect to a fixed ball. For a fixed ball $\mathcal{E}_0$, the fractional maximal function with respect to $\mathcal{E}_0$ of a function f is given by

$$M_{\alpha,\mathcal{E}_0}(f)(x) = \sup_{\mathcal{E}_0 \supseteq \mathcal{E} \ni x} \frac{1}{|\mathcal{E}_0|^{1-\frac{\alpha}{\gamma}}} \int_{\mathcal{E}} |f(y)|dy, \qquad 0 \le \alpha < \gamma,$$

where the supremum is taken over all the parabolic balls $\mathcal{E}$ with $\mathcal{E} \subseteq \mathcal{E}_0$ and $x \in \mathcal{E}$.

Theorem 6 *Let* $0 < \beta < 1$, $0 \le \alpha$, $\alpha + \beta < \gamma$ *and* b *be a locally integrable non-negative function. Suppose that* Φ, Ψ *be Young functions satisfying* $\Phi \in \mathcal{Y} \cap \nabla_2$ *and* $\Psi^{-1}(t) \approx \Phi^{-1}(t)t^{-\frac{\alpha+\beta}{\gamma}}$. *Then the following statements are equivalent:*

1. $b \in \dot{\Lambda}_{\beta,P}(\mathbb{R}^n)$.
2. $[b, M_\alpha^P]$ *is bounded from* $L^\Phi(\mathbb{R}^n)$ *to* $L^\Psi(\mathbb{R}^n)$.
3. *There exists a constant* $C > 0$ *such that*

$$\sup_{\mathcal{E}} |\mathcal{E}|^{-\beta/\gamma}\Psi^{-1}\big(|\mathcal{E}|^{-1}\big)\|b(\cdot) - |\mathcal{E}|^{-\alpha/\gamma} M_{\alpha,\mathcal{E}}^P(b)(\cdot)\|_{L^\Psi(\mathcal{E})} \le C. \tag{10}$$

Proof (1) $\Rightarrow$ (2): The following estimate was proved in [20]. Let b be any non-negative locally integrable function. Then

$$|[b, M_\alpha^P](f)(x)| \le M_{b,\alpha}^P(f)(x), \qquad x \in \mathbb{R}^n \tag{11}$$

holds for all $f \in L^1_{\text{loc}}(\mathbb{R}^n)$.

It follows from (11) and Theorem 4 that $[b, M_\alpha^P]$ is bounded from $L^\Phi(\mathbb{R}^n)$ to $L^\Psi(\mathbb{R}^n)$ since $b \in \dot{\Lambda}_{\beta,P}(\mathbb{R}^n)$.

(2) $\Rightarrow$ (3): For any fixed ball $\mathcal{E} \subset \mathbb{R}^n$ and all $x \in \mathcal{E}$, we have (see (2.4) in [20]).

$$M_\alpha^P(\chi_{\mathcal{E}})(x) = |\mathcal{E}|^{\alpha/\gamma} \quad \text{and} \quad M_\alpha^P(b\chi_{\mathcal{E}})(x) = M_{\alpha,\mathcal{E}}^P(b)(x).$$

Then,

$$
\begin{aligned}
&|\mathcal{E}|^{-\beta/\gamma}\Psi^{-1}\big(|\mathcal{E}|^{-1}\big)\|b(\cdot)-|\mathcal{E}|^{-\alpha/\gamma}M^P_{\alpha,\mathcal{E}}(b)(\cdot)\|_{L^\Psi(\mathcal{E})}\\
&=|\mathcal{E}|^{-\frac{\alpha+\beta}{\gamma}}\Psi^{-1}\big(|\mathcal{E}|^{-1}\big)\|b(\cdot)M^P_\alpha(\chi_\mathcal{E})(\cdot)-M^P_\alpha(b\chi_\mathcal{E})(\cdot)\|_{L^\Psi(\mathcal{E})}\\
&=|\mathcal{E}|^{-\frac{\alpha+\beta}{\gamma}}\Psi^{-1}\big(|\mathcal{E}|^{-1}\big)\|[b,M^P_\alpha](\chi_\mathcal{E})\|_{L^\Psi(\mathcal{E})}\\
&\le C|\mathcal{E}|^{-\frac{\alpha+\beta}{\gamma}}\Psi^{-1}\big(|\mathcal{E}|^{-1}\big)\|\chi_\mathcal{E}\|_{L^\Phi}\le C
\end{aligned}
$$

which implies (3) since the ball $\mathcal{E}\subset\mathbb{R}^n$ is arbitrary.

(3) $\Rightarrow$ (1): From [20] we have,

$$
\frac{1}{|\mathcal{E}|^{1+\frac{\beta}{\gamma}}}\int_\mathcal{E}|b(x)-b_\mathcal{E}|dx\le\frac{2}{|\mathcal{E}|^{1+\frac{\beta}{\gamma}}}\int_\mathcal{E}|b(x)-|\mathcal{E}|^{-\alpha/\gamma}M^P_{\alpha,\mathcal{E}}(b)(x)|dx.
$$

it follows from Lemma 2 and (10) that

$$
\frac{1}{|\mathcal{E}|^{1+\frac{\beta}{\gamma}}}\int_\mathcal{E}|b(x)-b_\mathcal{E}|dx\le\frac{4}{|\mathcal{E}|^{\frac{\beta}{\gamma}}}\Psi^{-1}\big(|\mathcal{E}|^{-1}\big)\|b(\cdot)-|\mathcal{E}|^{-\alpha/\gamma}M^P_{\alpha,\mathcal{E}}(b)(\cdot)\|_{L^\Psi(\mathcal{E})}\le C.
$$

Thus by Lemma 4 we get $b\in\dot{\Lambda}_{\beta,P}(\mathbb{R}^n)$.

If we take $\alpha=0$ at Theorem 6, we have the following result.

Corollary 5 *Let $0<\beta<1$ and b be a locally integrable non-negative function. Suppose that Φ,Ψ be Young functions, $\Phi\in\mathcal{Y}\cap\nabla_2$ and $\Psi^{-1}(t)\approx\Phi^{-1}(t)t^{-\frac{\beta}{\gamma}}$. Then the following statements are equivalent:*

1. $b\in\dot{\Lambda}_{\beta,P}(\mathbb{R}^n)$.
2. $[b,M^P]$ *is bounded from $L^\Phi(\mathbb{R}^n)$ to $L^\Psi(\mathbb{R}^n)$.*
3. *There exists a constant $C>0$ such that*

$$
\sup_\mathcal{E}|\mathcal{E}|^{-\beta/\gamma}\Psi^{-1}\big(|\mathcal{E}|^{-1}\big)\|b(\cdot)-M^P_\mathcal{E}(b)(\cdot)\|_{L^\Psi(\mathcal{E})}\le C.
$$

If we take $\Phi(t)=t^p$ and $\Psi(t)=t^q$ with $1\le p<\infty$ and $1\le q<\infty$ at Theorem 6, we have the following result.

Corollary 6 *Let $0<\beta<1$, $0\le\alpha$, $\alpha+\beta<\gamma$, $b\in L^1_{\mathrm{loc}}(\mathbb{R}^n)$, b be a locally integrable non-negative function, $1<p<q\le\infty$ and $\frac{1}{p}-\frac{1}{q}=\frac{\alpha+\beta}{\gamma}$. Then the following statements are equivalent:*

1. $b\in\dot{\Lambda}_{\beta,P}(\mathbb{R}^n)$.
2. $[b,M^P_\alpha]$ *is bounded from $L^p(\mathbb{R}^n)$ to $L^q(\mathbb{R}^n)$.*
3. *There exists a constant $C>0$ such that*

$$
\sup_\mathcal{E}\frac{1}{|\mathcal{E}|^{\beta/\gamma}}\left(\frac{1}{|\mathcal{E}|}\int_\mathcal{E}|b(x)-|\mathcal{E}|^{-\alpha/\gamma}M^P_{\alpha,\mathcal{E}}(b)(x)|^q dx\right)^{1/q}\le C.
$$

Remark 1 From the proof of Theorem 6 one can see that the assumption $b \geq 0$ is not used in (2) $\Rightarrow$ (3) and (3) $\Rightarrow$ (1). This means (2) and (3) are sufficient conditions for $b \in \dot{\Lambda}_{\beta,P}(\mathbb{R}^n)$. But we do not know if (2) and (3) are necessary for $b \in \dot{\Lambda}_{\beta,P}(\mathbb{R}^n)$.

Indeed, we have obtained the following result.

Corollary 7 *Let $0 < \beta < 1$, $0 \leq \alpha$, $\alpha + \beta < \gamma$ and b be a locally integrable function. Suppose that Φ, Ψ be Young functions, $\Phi \in \mathcal{Y} \cap \nabla_2$ and $\Psi^{-1}(t) \approx \Phi^{-1}(t)\, t^{-\frac{\alpha+\beta}{\gamma}}$. If one of the following statements is true, then $b \in \dot{\Lambda}_{\beta,P}(\mathbb{R}^n)$:*

1. *$[b, M_\alpha^P]$ is bounded from $L^\Phi(\mathbb{R}^n)$ to $L^\Psi(\mathbb{R}^n)$.*
2. *There exists a constant $C > 0$ such that*

$$\sup_{\mathcal{E}} |\mathcal{E}|^{-\beta/\gamma} \Psi^{-1}\big(|\mathcal{E}|^{-1}\big) \|b(\cdot) - |\mathcal{E}|^{-\alpha/\gamma} M_{\alpha,\mathcal{E}}(b)(\cdot)\|_{L^\Psi(\mathcal{E})} \leq C.$$

Theorem 7 *Let $b \geq 0$ be a locally integrable function, $0 < \beta < 1$, $0 \leq \alpha$, $\alpha + \beta < \gamma$ and $b \in \dot{\Lambda}_{\beta,P}(\mathbb{R}^n)$. Suppose that Φ, Ψ be Young functions, $\Phi \in \mathcal{Y}$ and condition (8) holds. Then $[b, M_\alpha^P]$ is bounded from $L^\Phi(\mathbb{R}^n)$ to $WL^\Psi(\mathbb{R}^n)$.*

Proof Obviously, it follows from (11) and Theorem 5.

If we take $\Phi(t) = t^p$ and $\Psi(t) = t^q$ with $1 \leq p < \infty$ and $1 \leq q < \infty$ at Theorem 7, we have the following result.

Corollary 8 *Let $b \geq 0$ be a locally integrable function, $0 < \beta < 1$, $0 \leq \alpha$, $\alpha + \beta < \gamma$, $b \in \dot{\Lambda}_{\beta,P}(\mathbb{R}^n)$, $1 \leq p < q \leq \infty$ and $\frac{1}{p} - \frac{1}{q} = \frac{\alpha+\beta}{\gamma}$. Then $[b, M_\alpha^P]$ is bounded from $L^p(\mathbb{R}^n)$ to $WL^q(\mathbb{R}^n)$.*

Remark 2 Note that Theorem 6 in the isotropic case $P = I$ were proved in [10].

Acknowledgements The authors thank the referees for careful reading the paper and useful comments. The research of V.S. Guliyev was partially supported by the grant of 1st Azerbaijan-Russia Joint Grant Competition (Grant No. EIF-BGM-4-RFTF-1/2017-21/01/1).

References

1. Abasova, G.A.: Boundedness of the parabolic maximal operator in Orlicz spaces. Trans. Natl. Acad. Sci. Azerb. Ser. Phys.-Tech. Math. Sci. **37**(4), Mathematics 5–11 (2017)
2. Abasova, G.A.: Characterization of parabolic fractional integral and its commutators in Orlicz spaces. Casp. J. Appl. Math. Ecol. Econ. **6**(2), 1–13 (2018)
3. Besov, O.V., Il'in, V.P., Lizorkin, P.I.: The L_p-estimates of a certain class of non-isotropically singular integrals. (Russian) Dokl. Akad. Nauk SSSR **169**, 1250–1253 (1966)
4. De Vore, R.A., Sharpley, R.C.: Maximal functions measuring smoothness. Mem. Am. Math. Soc. **47**(293), viii+115 (1984)
5. Fabes, E.B., Rivère, N.: Singular integrals with mixed homogeneity. Stud. Math. **27**, 19–38 (1966)
6. Folland, G.B., Stein, E.M.: Hardy Spaces on Homogeneous Groups. Mathematical Notes, vol. 28. Princeton University Press, Princeton (1982)

7. Fu, X., Yang, D., Yuan, W.: Boundedness of multilinear commutators of Calderón-Zygmund operators on Orlicz spaces over non-homogeneous spaces. Taiwan. J. Math. **16**, 2203–2238 (2012)
8. Genebashvili, I., Gogatishvili, A., Kokilashvili, V., Krbec, M.: Weight Theory for Integral Transforms on Spaces of Homogeneous Type. Longman, Harlow (1998)
9. Guliyev, V.S., Deringoz, F., Hasanov, S.G.: Riesz potential and its commutators on Orlicz spaces. J. Inequal. Appl. Paper No. 75 (2017), 18 pp
10. Guliyev, V.S., Deringoz, F., Hasanov, S.G.: Fractional maximal function and its commutators on Orlicz spaces. Anal. Math. Phys. **9**(1), 1–16 (2019).
11. Hedberg, L.I.: On certain convolution inequalities. Proc. Am. Math. Soc. **36**, 505–510 (1972)
12. Izuki, M., Sawano, Y.: Characterization of BMO via ball Banach function spaces. Vestn. St.-Petersbg. Univ. Mat. Mekh. Astron. **4**(62)(1), 78–86 (2017)
13. John, F., Nirenberg, L.: On functions of bounded mean oscillation. Commun. Pure Appl. Math. **14**, 415–426 (1961)
14. Long, R., Yang, L.: BMO functions in spaces of homogeneous type. Sci. Sin. Ser. A **27**(7), 695–708 (1984)
15. Nakai, E.: On generalized fractional integrals in the Orlicz spaces on spaces of homogeneous type. Sci. Math. Jpn. **54**, 473–487 (2001)
16. Orlicz, W.: Über eine gewisse Klasse von Räumen. vom Typus B, Bull. Acad. Polon. A 207–220 (1932); reprinted in: Collected Papers, PWN, Warszawa 217–230 (1988)
17. Orlicz, W.: Über Räume (L^M). Bull. Acad. Polon. A 93–107 (1936); reprinted in: Collected Papers, PWN, Warszawa 345–359 (1988)
18. Rao, M.M., Ren, Z.D.: Theory of Orlicz Spaces. M. Dekker Inc., New York (1991)
19. Stein, E.M.: Harmonic Analysis: Real Variable Methods, Orthogonality and Oscillatory Integrals. Princeton University Press, Princeton (1993)
20. Zhang, P., Wu, J.: Commutators of the fractional maximal function on variable exponent Lebesgue spaces. Czechoslov. Math. J. **64**(139)(1), 183–197 (2014)

Finite Trees Inside Thin Subsets of $\mathbb{R}^d$

A. Iosevich and K.Taylor

Abstract Bennett, Iosevich and Taylor proved that compact subsets of $\mathbb{R}^d$, $d \geq 2$, of Hausdorff dimensions greater than $\frac{d+1}{2}$ contain chains of arbitrary length with gaps in a non-trivial interval. In this paper we generalize this result to arbitrary tree configurations.

Keywords Finite point configurations · Falconer problem · Tree graphs · Fourier estimates

1 Introduction

We begin with a seminal result due to Ziegler [10], which generalizes an earlier result due to Furstenberg, Katznelson and Weiss [5]. See also [3].

Theorem 1.1 *Let $E \subset \mathbb{R}^d$, of positive upper Lebesgue density in the sense that*

$$\limsup_{R\to\infty} \frac{\mathcal{L}^d\{E \cap [-R, R]^d\}}{(2R)^d} > 0,$$

where $\mathcal{L}^d$ denotes the d-dimensional Lebesgue measure. Let E_δ denote the δ-neighborhood of E. Let $V = \{\mathbf{0}, v^1, v^2, \ldots, v^{k-1}\} \subset \mathbb{R}^d$, where $k \geq 2$ is a positive integer. Then there exists $l_0 > 0$ such that for any $l > l_0$ and any $\delta > 0$ there exists $\{x^1, \ldots, x^k\} \subset E_\delta$ congruent to $lV = \{\mathbf{0}, lv^1, \ldots, lv^{k-1}\}$.

In particular, this result shows that we can recover every simplex similarity type and sufficiently large scaling inside a subset of $\mathbb{R}^d$ of positive upper Lebesgue density.

A. Iosevich
Department of Mathematics, University of Rochester, Rochester, NY, USA
e-mail: iosevich@math.rochester.edu

K. Taylor (✉)
Department of Mathematics, The Ohio State University, Columbus, OH, USA
e-mail: taylor.2952@osu.edu

A. Karapetyants et al. (eds.), *Modern Methods in Operator Theory and Harmonic Analysis*, Springer Proceedings in Mathematics & Statistics 291,
https://doi.org/10.1007/978-3-030-26748-3_3

It is reasonable to wonder whether the assumptions of Theorem 1.1 can be weakened, but the following result due to Falconer [4] (see also Maga [9]) shows that conclusion may fail even if we replace the upper Lebesgue density condition with the assumption that the set is of dimension d.

Theorem 1.2 ([9]) *For any $d \geq 2$ there exists a full dimensional compact set $A \subset \mathbb{R}^d$ such that A does not contain the vertices of any parallelogram. If $d = 2$, then given any triple of points $x^1, x^2, x^3, x^j \in A$, there exists a full dimensional compact set $A \subset \mathbb{R}^2$ such that A does not contain the vertices of any triangle similar to $\Delta x^1x^2x^3$.*

The general question is to study the distance graph with vertices in a compact set of a given Hausdorff dimension. (For more on graph theory, see [2].) More precisely, let E be a compact subset of $\mathbb{R}^d$, $d \geq 2$, and view its points as vertices of a graph where two vertices x, y are connected by an edge if $|x - y| = t$, with $|\cdot|$ denoting the Euclidean distance and t a positive real number. Denote the resulting graph by $G_t(E)$. Theorem 1.2 says that if $d = 2$ and the Hausdorff dimension of E is equal to 2, then $G_t(E)$ does not in general contain a triangle. The situation changes in higher dimensions, as was demonstrated by the first listed author of this paper and Bochen Liu in [6].

Theorem 1.3 ([6]) *For every $d \geq 4$ there exists $s_0 < d$ such that if the Hausdorff dimension of E is $> s_0$, then E contains vertices of an equilateral triangle.*

Definition 1.4 A *path* in a graph is a finite or infinite sequence of edges that connect a sequence of distinct vertices. A path of length k connects a sequence of $(k + 1)$-vertices, and we refer to this sequence of vertices as a *k-chain*.

Bennett and the two authors of this paper proved in [1] that if the Hausdorff dimension of $E \subset \mathbb{R}^d$, $d \geq 2$, is greater than $\frac{d+1}{2}$, then $G_t(E)$ contains an arbitrarily long path. More generally, they proved the following.

Theorem 1.5 (Theorem 1.7 in [1]) *Suppose that the Hausdorff dimension of a compact set $E \subset \mathbb{R}^d$, $d \geq 2$, is greater than $\frac{d+1}{2}$. Then for any $k \geq 1$, there exists an open interval $\tilde{I}$ such that for any $\{t_i\}_{i=1}^k \subset \tilde{I}$ there exists a non-degenerate k-chain in E with gaps $\{t_i\}_{i=1}^k$.*

One of the key aspects of the proof of Theorem 1.5 is the following estimate.

Theorem 1.6 (Theorem 1.8 in [1]) *Suppose that μ is a compactly supported non-negative Borel measure such that*

$$\mu(B(x, r)) \leq Cr^{s_\mu}, \tag{1.1}$$

where $B(x, r)$ is the ball of radius $r > 0$ centered at $x \in \mathbb{R}^d$, for some $s_\mu \in (\frac{d+1}{2}, d]$. Then for any $t_1, \ldots, t_k > 0$ and $\epsilon > 0$,

$$\mu \times \mu \times \cdots \times \mu\{(x^1, x^2, \ldots, x^{k+1}) : t_i - \epsilon \leq |x^{i+1} - x^i| \leq t_i + \epsilon;\ i = 1, 2, \ldots, k\} \leq C\epsilon^k. \tag{1.2}$$

For the purposes of this paper we are interested in the special case of Theorem 1.5 where all the t_i's are equal. Our goal is to extend Theorem 1.5 to more general configurations.

Definition 1.7 A tree is an (undirected) graph in which any two vertices are connected by exactly one path.

Our main result is the following.

Theorem 1.8 *Let $E \subset \mathbb{R}^d$, compact of Hausdorff dimension greater than $\frac{d+1}{2}$ and let T be a tree on $k+1$ vertices. Then there exists a non-empty interval I such that for all $t \in I$, $G_t(E)$ contains T as a subgraph.*

Remark 1.9 For an analogous result in sets of positive upper Lebesgue density, see a result due to Lyall and Magyar in [8].

2 Proof of Theorem 1.8

The proof of Theorem 1.8 is obtained by streamlining and extending the proof of Theorem 1.5 in a direct and simple way.

Let T be a graph on $k+1$ vertices. Enumerate the vertices of T and let $\mathcal{E}(T)$ denote the set of pairs (i, j), $i < j$, such that the ith vertex is connected with the j'th vertex by an edge. Let μ be a Borel measure supported on E and define

$$\mathcal{T}^{\epsilon}_{T,t}(\mu) = \int \dots \int \left\{ \prod_{(i,j)\in\mathcal{E}(T)} \sigma^{\epsilon}_t(x^i - x^j) \right\} d\mu(x^1)\dots d\mu(x^{k+1}). \tag{2.1}$$

It is not difficult to see that Theorem 1.8 would follow from the following estimates:

$$\mathcal{T}^{\epsilon}_{T,t}(\mu) \le C_k, \tag{2.2}$$

and

$$\liminf_{\epsilon\to 0} \mathcal{T}^{\epsilon}_{T,t}(\mu) \ge c_k > 0, \tag{2.3}$$

where $t \in I$, a non-empty interval, $\epsilon > 0$ is taken sufficiently small, and both the upper and lower bounds of c_k and C_k respectively hold independently of $t \in I$ and $\epsilon < \epsilon_0$. We will prove that these estimates hold when the measure μ is replaced in each variable by the restriction of μ to an appropriate subset of E of positive μ-measure.

In the proof of Theorem 1.5 in [1], the upper bound was established using the observation that if $Tf = \lambda * f$, where λ is a compactly supported measure satisfying $|\widehat{\lambda}(\xi)| \le C|\xi|^{-\alpha}$ for some $\alpha > 0$ and μ is a compactly supported Borel measure satisfying $\mu(B(x,r)) \leq Cr^s$ for some $s > d - \alpha$, then T is a bounded operator from $L^2(\mu)$ to $L^2(\mu)$. The lower bound was established using an inductive procedure generalizing the argument due to the authors of this paper and Mihalis Mourgoglou in [7]. In this paper we streamline the procedure by proving the upper bound and the lower bound at the same time.

The key feature of our argument is the following calculation.

Lemma 2.1 *Set*

$$G = G_{t,\epsilon}(1) = \{x \in E : c < \sigma_t^\epsilon * \mu(x) < 2^{m(1)}\},$$

where $c > 0$ *and* $m(1) \in \mathbb{N}$. *There exists a non-empty open interval* I, *an* $\epsilon_0 > 0$, *and a choice of* c, $m(1)$, *and* $\delta > 0$ *so that*

$$\mu(G_{t,\epsilon}) > \delta > 0$$

whenever $t \in I$ *and* $0 < \epsilon < \epsilon_0$.

To prove this result, let $f_\epsilon^t = \sigma_t^\epsilon * \mu(x)$. It was proved in [1] that there exists a non-empty open interval I and an $\epsilon_0 > 0$ so that simultaneously the $L^1(\mu)$ norm of f_ϵ^t is uniformly bounded below and the $L^2(\mu)$ norm is bounded above for all $t \in I$ and $0 < \epsilon < \epsilon_0$. Denote these uniform lower and upper bounds by $0 < c$ and C respectively. Let ϵ and t be such, and set $f = f_\epsilon^t$.

Set $G = G_{t,\epsilon}(1) = \{x : c < f(x) < 2^{m(1)}\}$, where $m(1) \in \mathbb{N}$ is to be determined. Now,

$$c < \int f d\mu(x) = \int_{\{f \leq c\}} f d\mu(x) + \int_G f d\mu(x) + \sum_{l=m(1)}^{\infty} \int_{\{2^l \leq f \leq 2^{l+1}\}} f d\mu(x). \tag{2.4}$$

It is a straight-forward consequence of Chebyshev's inequality and Cauchy–Schwarz that $\mu(\{2^l \leq f \leq 2^{l+1}\}) < C2^{-2l}$. Plugging this into (2.4) and taking $m(1)$ sufficiently large, it quickly follows that $\mu(G)$ is bounded below away from zero with constants independent of $\epsilon > 0$ and t.

By induction, using an identical argument to the one above, one can find the following nested sequence of sets of positive μ-measure (where the lower bound on the measure is independent of $t \in I$ and ϵ small).

Lemma 2.2 *For* $j \in \mathbb{N}$, *set*

$$G_{t,\epsilon}(j+1) = \{x \in G_{t,\epsilon}(j) : c(j+1) < \sigma_t^\epsilon * \mu_j(x) < 2^{m(j)}\},$$

where $\mu_j(x)$ *denotes restriction of the measure* μ *to the set* $G_{t,\epsilon}(j)$ *and* $c(j+1) > 0$. *Then there exists numbers* $m(j+1) \in \mathbb{N}$, $c(j+1) > 0$, *and* $\delta_{j+1} > 0$, *so that if* $t \in I$ *and* $0 < \epsilon < \epsilon_0$, *then*

$$\mu(G_{t,\epsilon}(j+1)) > \delta_{j+1} > 0.$$

We now demonstrate the pigeon-holing argument that allows us to deduce (2.2) and (2.3) when μ in each variable is appropriately restricted.

Fix $k \in \mathbb{N}$, and let T be a tree on $k+1$ vertices. We say that a vertex is isolated if it is connected to only one other vertex. Let $V(1)$ denote the set of isolated vertices of T, and let $x^1, \cdots, x^{N(1)}$ denote the collection of vertices who are connected to at least on vertex in $V(1)$. Let $k_1, \cdots, k_{N(1)}$ denote the number of isolated vertices connected to $x^1, \cdots, x^{N(1)}$ respectively.

Consider the expression in (2.1). Integrating in each $v^j \in V(1)$, we replace each of the expressions

$$\sigma_t^{\epsilon}(x^i - v^j)d\mu(v^j) \qquad \text{by} \qquad \sigma_t^{\epsilon} * \mu(x^i),$$

whenever x^i is connected to v^j. So, if $v^{j_1}, \cdots v^{j_{k_i}} \in V(1)$ are all connected to x^i, then we get an expression of the form

$$\left(\sigma_t^{\epsilon} * \mu(x^i)\right)^{k_i} \tag{2.5}$$

in the integrand.

The next step is to restrict the vertices $x^1, \cdots, x^{N(1)}$ to the set $G(1)$ as in Lemma 2.1. In this way, for each x^i, the expression in (2.5) is bounded above and below by positive constants independent of $t \in I$ and $0 < \epsilon < \epsilon_0$. Due to the positivity of the integrand, we can now consider the expression in (2.1) with terms of the form in (2.5) removed. Finally, let $T(2)$ denote the tree with all of the vertices in $V(1)$ removed, and let $\mathcal{T}_{T,t}^{\epsilon}(2)$ denote the expression in (2.5) with the above mentioned modifications (so any evidence of the vertices in $V(1)$ has been removed) .

We repeat this process. For $j \in \mathbb{N}$, let $T(j+1)$ denote the tree that is obtained after repeating this process j-times. Let $V(j+1)$ denote the set of isolated vertices of $T(j+1)$, and let $y^1, \cdots, y^{N(j)}$ denote the collection of vertices who are connected to at least on vertex in $V(j+1)$. Let $K_1, \cdots, K_{N(j)}$ denote the number of isolated vertices connected to $y^1, \cdots, y^{N(j)}$ respectively.

Consider the expression in $\mathcal{T}_{T,t}^{\epsilon}(j+1)$. Integrating in each $v^j \in V(j+1)$, we replace each of the expressions

$$\sigma_t^{\epsilon}(y^i - v^j)d\mu_j(v^j) \qquad \text{by} \qquad \sigma_t^{\epsilon} * \mu_j(x^i),$$

whenever y^i is connected to v^j. So, if $v^{j_1}, \cdots v^{j_{K_i}} \in V(j+1)$ are all connected to y^i, then we get an expression of the form

$$\left(\sigma_t^{\epsilon} * \mu_j(x^i)\right)^{k_i} \tag{2.6}$$

in the integrand.

The next step is to restrict the vertices $y^1, \cdots, y^{N(j)}$ to the set $G(j+1)$ as in Lemma 2.2. In this way, for each y^i, the expression in $\mathcal{T}_{T,t}^{\epsilon}(j+1)$ is bounded above and below by positive constants independent of $t \in I$ and $0 < \epsilon < \epsilon_0$. Due to the positivity of the integrand, we can now consider the expression in $\mathcal{T}^{\epsilon}_{T,t}(j+1)$ with terms of the form in (2.6) removed. Finally, let $T(j+2)$ denote the tree with all of the vertices in $V(j+1)$ removed.

This procedure terminates after a finite number of steps, and we are left with an expression of the form

$$\iint \sigma_t^{\epsilon}(z^1 - z^2)d\mu_{J_1}(z^1)d\mu_{J_2}(z^2) = \int \sigma_t^{\epsilon} * \mu_{J_1}(z^2)d\mu_{J_2}(z^2),$$

where we assume with out loss of generality that $J_2 \geq J_1$ (so that $G(J_2) \subset G(J_1)$). If $J_2 > J_1$, then this expression is bounded above and below by positive constants independent of $0 < \epsilon < \epsilon_0$ and independent of $t \in I$. We obtain the same conclusion when $J_2 = J_1$ by simply restricting the variable z^2 to the set $G(J_3)$ defined above in Lemma 2.2.

References

1. Bennett, M., Iosevich, A., Taylor, K.: Finite chains inside thin subsets of $\mathbb{R}^d$. Anal. PDE **9**(3), 597–614 (2016). arXiv:1409.2581.pdf
2. Bollobas, B.: Modern Graph Theory. Springer, New York (1998)
3. Bourgain, J.: A Szemeredi type theorem for sets of positive density. Isr. J. Math. **54**(3), 307–331 (1986)
4. Falconer, K.J.: Some problems in measure combinatorial geometry associated with Paul Erdős. http://www.renyi.hu/conferences/erdos100/slides/falconer.pdf
5. Furstenberg, H., Katznelson, Y., Weiss, B.: Ergodic theory and configurations in sets of positive density. Mathematics of Ramsey Theory. Algorithms and Combinatorics, vol. 5, pp. 184–198. Springer, Berlin (1990)
6. Iosevich, A., Liu, B.: Equilateral triangles in subsets of $\mathbb{R}^d$ of large Hausdorff dimension. Isr. Math. J. (accepted for publication) (2016). arXiv:1603.01907.pdf
7. Iosevich, A., Mourgoglou, M., Taylor, K.: On the Mattila-Sjölin theorem for distance sets. Ann. Acad. Sci. Fenn. Math. **37**(2) (2012)
8. Lyall, N., Magyar, A.: Distance graphs and sets of positive upper density in $\mathbb{R}^d$ (2018). arXiv:1803.08916
9. Maga, P.: Full dimensional sets without given patterns. Real Anal. Exch. **36**, 79–90 (2010)
10. Ziegler, T.: Nilfactors of $\mathbb{R}^d$ actions and configurations in sets of positive upper density in $\mathbb{R}^m$. J. Anal. Math. **99**, 249–266 (2006)

Boundedness of Projection Operator in Generalized Holomorphic and Harmonic Spaces of Functions of Hölder Type

Alexey Karapetyants and Joel E. Restrepo

Abstract We prove boundedness of holomorphic projection in the generalized Hölder type space of holomorphic functions in the unit disc with prescribed modulus of continuity and in the variable exponent generalized Hölder spaces of holomorphic functions in the unit disc. We also prove boundedness of harmonic projection in analogous spaces of harmonic functions.

Keywords Holomorphic and harmonic projections · Hölder spaces · Holomorphic functions · Harmonic functions · Variable exponent spaces · Modulus of continuity

1 Introduction

Let $dA(w)$ stand for the normalized area measure on the unit disc $\mathbb{D}$ in complex plane $\mathbb{C}$ so that the area of $\mathbb{D}$ is 1. The symbol $A^p(\mathbb{D})$ stands for the space of holomorphic functions in $\mathbb{D}$ that belong to $L^p(\mathbb{D}) = L^p(\mathbb{D}; dA(z))$. The holomorphic projection $B_{\mathbb{D}}$, given by the integral representation

$$B_{\mathbb{D}} f(z) = \int_{\mathbb{D}} \frac{f(w)}{(1 - z\overline{w})^2} dA(w), \quad z \in \mathbb{D},$$

is well defined on functions in $L^1(\mathbb{D})$, and is bounded from $L^p(\mathbb{D})$ onto $A^p(\mathbb{D})$ for $1 < p < \infty$. The harmonic projection, $b_{\mathbb{D}}$, from $L^p(\mathbb{D})$ onto the corresponding harmonic subspace, denoted as $h^p(\mathbb{D})$, possesses the integral representation with the

A. Karapetyants
Department of Mathematics and Statistics, SUNY, Albany, NY 12222, USA
e-mail: karapetyants@gmail.com

Department of Mathematics and Regional Mathematical Center, Southern Federal University, Rostov-on-Don 344090, Russia

J. E. Restrepo (✉)
Regional Mathematical Center, Southern Federal University, Rostov-on-Don 344090, Russia
e-mail: cocojoel89@yahoo.es

A. Karapetyants et al. (eds.), *Modern Methods in Operator Theory and Harmonic Analysis*, Springer Proceedings in Mathematics & Statistics 291,
https://doi.org/10.1007/978-3-030-26748-3_4

kernel $2\,\mathrm{Re}\left\{\frac{1}{(1-z\overline{w})^2}\right\} - 1$ valid for $f \in L^1(\mathbb{D})$. For basics on theory of $A^p(\mathbb{D})$ and $h^p(\mathbb{D})$ spaces we refer to [2, 4, 6, 9, 19, 20].

The boundedness results for the above mentioned and for more general operators (weighted projections, fractional operators) in a number of various spaces of holomorphic and harmonic functions are well known. We refer to the above mentioned books, references therein, and also to the following sources [1, 3, 5, 12, 14–16].

The Hölder type spaces attract an attention due to numerous applications. Recently such generalized Hölder type spaces of holomorphic and harmonic functions were studied in [11, 13]. The study of the boundedness of the corresponding holomorphic and harmonic projections is a natural continuation of this work.

In the case of constant λ and functions considered on the ball in $\mathbb{C}^n$ such spaces were studied thoroughly, see [19] (see also [20] for $n = 2$). In the case of a general modulus of continuity such spaces appear in [7, 8, 17].

The paper is organized as follows. In Sect. 2 we give some general definitions and useful facts. In Sect. 3 we collect definitions and some properties of Hölder type spaces of holomorphic functions in the unit disc. Sections 4 and 5 are devoted to boundedness of holomorphic projection in Hölder type spaces of holomorphic and harmonic functions (the definition of harmonic Hölder type spaces is given in Sect. 5).

2 Preliminaries

Let $\mathbb{D}$ stand for the unit disc in complex plane $\mathbb{C}$, and $\mathbb{T}$ be the unit circle. Let $\omega : [0, 2] \to \mathbb{R}_+$ and $\gamma : \mathbb{D} \to (1, \infty)$ measurable functions satisfying the following conditions:
1. $\gamma \in L^\infty(\mathbb{D})$;
2. $\omega \in L^\infty([0, 2])$;
3. ω is almost increasing on $[0, 2]$ and $\frac{\omega(h)}{h}$ is almost decreasing on $[0, 2]$;
4. ω satisfies the following Zygmund type condition:

$$\int_t^2 \frac{\omega(s)}{s^{\gamma(z)}} ds \le k(z) \frac{\omega(t)}{t^{\gamma(z)-1}}, \quad 0 < t < 2, \ z \in \mathbb{D}, \tag{1}$$

where $k(z)$ does not depend on t.

The following two lemmas are intuitive. Their proof can be found in [11].

Lemma 1 *Let ω and γ satisfy the above conditions* **1.-4.** *Then*

$$\int_{\mathbb{T}} \frac{\omega(|\tau - z|)}{|\tau - z|^{\gamma(z)}} |d\tau| \le Ck(z) \frac{\omega(1 - |z|)}{(1 - |z|)^{\gamma(z)-1}}, \quad z \in \mathbb{D},$$

where C does not depend on z, and $k(z)$ is from (1).

Lemma 2 *Let γ be bounded on $\mathbb{D}$ and $\gamma(z) > 1$, for $z \in \mathbb{D}$, then*

$$\int_{\mathbb{T}} \frac{|d\tau|}{|\tau - z|^{\gamma(z)}} \leq \frac{C}{\gamma(z) - 1} \frac{1}{(1 - |z|)^{\gamma(z)-1}},$$

where C does not depend on z.

The function $\omega : [0, 2] \to \mathbb{R}_+$ is called modulus of continuity if

1. ω is continuous in a neighborhood of the origin and $\omega(0) = 0$,
2. ω is almost increasing on $[0, 2]$,
3. $\frac{\omega(h)}{h}$ is almost decreasing on $[0, 2]$.

It is easy to see that under the condition $\frac{\omega(h)}{h}$ is almost decreasing on $[0, 2]$ there follows the semi-additivity property: $\omega(t + s) \leq C(\omega(t) + \omega(s))$, $t, s \in [0, 2]$, and the so-called doubling property: $\omega(2t) \leq C_{(2)}\omega(t)$, $t \in [0, 2]$. Here we assume by definition $\omega(h) = \omega(2)$ for $h > 2$.

We will use the following Zygmund type conditions:

$$\int_0^t \frac{\omega(s)}{s} ds \leq C\omega(t), \quad 0 < t < 2, \tag{2}$$

$$\int_t^2 \frac{\omega(s)}{s^2} ds \leq C\frac{\omega(t)}{t}, \quad 0 < t < 2, \tag{3}$$

where C does not depend on t.

Let $\lambda : \mathbb{D} \to [0, 1]$ be a continuous function. We say that λ satisfies the log-condition (log-Hölder condition) on $\mathbb{D}$ if

$$|\lambda(z) - \lambda(w)| \leq \frac{C}{\ln \frac{1}{|z-w|}}, \quad z, w \in \mathbb{D}, \ |z - w| < \frac{1}{2}, \tag{4}$$

where C does not depend on z, w.

One can prove that the log-condition imposed on the function λ implies that it is bounded and uniformly continuous on $\mathbb{D}$. Hence, it extends to a continuous function on $\overline{\mathbb{D}} := \{z \in \mathbb{D} : |z| \leq 1\}$. We will use the same notation λ for the so extended function. The log-condition also implies the property:

$$C_1 R^{\lambda(z)} \leq R^{\lambda(w)} \leq C_2 R^{\lambda(z)},$$

for all $z, w \in \mathbb{D}$ such that $|z - w| \leq R$, where C_1, C_2 do not depend on z, w. In a particular case, if the line segment between z and w lies in $\mathbb{D}$, then

$$C_1|z - w|^{\lambda(z)} \leq |z - w|^{\lambda(\tau)} \leq C_2|z - w|^{\lambda(z)}, \quad \tau = sw + (1 - s)z, \ 0 \leq s \leq 1,$$

where C_1, C_2 do not depend on z, w and s.

3 Hölder Type Spaces of Holomorphic Functions in the Unit Disc

Here we recall the definitions and some properties of the introduced in [11] spaces and reveal some equivalent description of these spaces which will be used later for the proof of main results on boundedness of the holomorphic projection.

We start with the generalized Hölder spaces with prescribed modulus of continuity. Let $\omega : [0, 2] \to \mathbb{R}_+$ be a modulus of continuity.

By $L^\omega(\mathbb{D})$ denote the space of functions measurable in $\mathbb{D}$ such that

$$|f(z) - f(w)| \leq C\omega(|z - w|), \quad z, w \in \mathbb{D},$$

where C does not depend on z, w. The semi-norm and norm of a function $f \in L^\omega(\mathbb{D})$ are respectively given by

$$\|f\|_{\#, L^\omega(\mathbb{D})} = \sup_{z,w\in\mathbb{D}} \frac{|f(z) - f(w)|}{\omega(|z - w|)}, \quad \|f\|_{L^\omega(\mathbb{D})} = \|f\|_{\#, L^\omega(\mathbb{D})} + \|f\|_{L^\infty(\mathbb{D})}.$$

Since ω is a modulus of continuity, it follows that any $f \in L^\omega(\mathbb{D})$ is continuous in $\overline{\mathbb{D}}$. This implies that $|f(u) - f(v)| \leq C\omega(|u - v|)$, for $u, v \in \mathbb{T}$, where C does not depend on u, v.

The generalized Hölder type space of holomorphic functions in the unit disc with prescribed modulus of continuity, denoted by $A^\omega(\mathbb{D})$, is the space of functions f from $L^\omega(\mathbb{D})$ holomorphic in $\mathbb{D}$, with the notation $\|f\|_{\#, A^\omega(\mathbb{D})} = \|f\|_{\#, L^\omega(\mathbb{D})}$.

As typical example of ω one can take $\omega(h) = h^\lambda \ln^{-\beta} \frac{e}{h}$, where $\lambda \in (0, 1)$, $\beta \in \mathbb{R}$. See [11] for more examples.

Under conditions (2) and (3) the space $A^\omega(\mathbb{D})$ possesses equivalent description in terms of behavior of derivative of a function near the boundary $\mathbb{T}$. Indeed by $B^\omega(\mathbb{D})$ denote the space of functions holomorphic in $\mathbb{D}$ such that

$$|f'(z)| \leq C\frac{\omega(1 - |z|)}{1 - |z|}, \quad z \in \mathbb{D},$$

where C does not depend on z. The semi-norm and norm of a function $f \in B^\omega(\mathbb{D})$ are given by

$$\|f\|_{\#, B^\omega(\mathbb{D})} = \sup_{z\in\mathbb{D}} |f'(z)| \frac{1 - |z|}{\omega(1 - |z|)}, \quad \|f\|_{B^\omega(\mathbb{D})} = \|f\|_{\#, B^\omega(\mathbb{D})} + \|f\|_{L^\infty(\mathbb{D})}.$$

Lemma 3 ([11]) *Let ω satisfy (2) and (3), then the spaces $A^\omega(\mathbb{D})$ and $B^\omega(\mathbb{D})$ coincide up to equivalence of norms.*

By $A_*^\omega(\mathbb{D})$ we denote the space of functions holomorphic in $\mathbb{D}$ such that

$$|f(z) - f(w)| \leq C\omega(|1 - z\overline{w}|), \quad z, w \in \mathbb{D},$$

where C does not depend on z, w. Since ω is a modulus of continuity, it follows that any $f \in A^{\omega}_{*}(\mathbb{D})$ is continuous in $\overline{\mathbb{D}}$. The semi-norm and norm of a function $f \in A^{\omega}_{*}(\mathbb{D})$ are given by

$$\|f\|_{\#,A^{\omega}_{*}(\mathbb{D})} = \sup_{z,w\in\mathbb{D}} \frac{|f(z) - f(w)|}{\omega(|1 - z\overline{w}|)}, \quad \|f\|_{A^{\omega}_{*}(\mathbb{D})} = \|f\|_{\#,A^{\omega}_{*}(\mathbb{D})} + \|f\|_{L^{\infty}(\mathbb{D})}.$$

Theorem 1 *Let ω satisfy (2) and (3). Then the spaces $A^{\omega}(\mathbb{D})$ and $A^{\omega}_{*}(\mathbb{D})$ coincide up to equivalence of norms.*

Proof Let $f \in A^{\omega}(\mathbb{D})$. Since $\left|\frac{z-w}{1-z\overline{w}}\right| \leq 1$ for $z, w \in \mathbb{D}$ and ω is almost increasing we have $\omega(|z - w|) \leq C\omega(|1 - z\overline{w}|)$, where C does not depend on $z, w \in \mathbb{D}$. Therefore $f \in A^{\omega}_{*}(\mathbb{D})$ and $\|f\|_{A^{\omega}_{*}(\mathbb{D})} \leq C\|f\|_{A^{\omega}(\mathbb{D})}$.

Let $f \in A^{\omega}_{*}(\mathbb{D})$. Since f is continuous in $\overline{\mathbb{D}}$ one can see that

$$|f(\tau) - f(\sigma)| \leq C\omega(|1 - \sigma\overline{\tau}|) = C\omega(|\tau - \sigma|),$$

for $\tau, \sigma \in \mathbb{T}$. Hence, by [11, Theorem 3.3] we obtain $\|f\|_{A^{\omega}(\mathbb{D})} \leq C\|f\|_{A^{\omega}_{*}(\mathbb{D})}$.

Now we consider the variable exponent generalized Hölder spaces. Let $\lambda : \mathbb{D} \to [0, 1]$ be a continuous function satisfying the log-condition (4) in $\mathbb{D}$. Note that this condition in application to variable exponent Hölder spaces first appeared in [10] and [18] in real analysis settings.

By $L^{\lambda(\cdot)}(\mathbb{D})$ denote the space of functions f measurable in $\mathbb{D}$ such that

$$|f(z) - f(w)| \leq C|z - w|^{\lambda(z)}, \quad \text{for all} \quad z, w \in \mathbb{D},$$

or what is the same

$$|f(z) - f(w)| \leq C|z - w|^{\lambda(w)}, \quad \text{for all} \quad z, w \in \mathbb{D},$$

where C does not depend on z, w. The semi-norm and norm of a function $f \in L^{\lambda(\cdot)}(\mathbb{D})$ are given by

$$\|f\|_{\#,L^{\lambda(\cdot)}(\mathbb{D})} = \sup_{z,w\in\mathbb{D}} \frac{|f(z) - f(w)|}{|z - w|^{\lambda(z)}}, \quad \|f\|_{L^{\lambda(\cdot)}(\mathbb{D})} = \|f\|_{\#,L^{\lambda(\cdot)}(\mathbb{D})} + \|f\|_{L^{\infty}(\mathbb{D})}.$$

The variable exponent generalized Hölder spaces $A^{\lambda(\cdot)}(\mathbb{D})$ of holomorphic functions in $\mathbb{D}$, denoted as $A^{\lambda(\cdot)}(\mathbb{D})$, is the space of functions f from $L^{\lambda(\cdot)}(\mathbb{D})$ holomorphic in $\mathbb{D}$, with the notation $\|f\|_{\#,A^{\lambda(\cdot)}(\mathbb{D})} = \|f\|_{\#,L^{\lambda(\cdot)}(\mathbb{D})}$.

By $B^{\lambda(\cdot)}(\mathbb{D})$ denote the space of functions f holomorphic in $\mathbb{D}$ such that

$$|f'(z)| \leq C(1 - |z|)^{\lambda(z)-1}, \quad z \in \mathbb{D},$$

where C does not depend on z. The semi-norm and norm of a function $f \in B^{\lambda(\cdot)}(\mathbb{D})$ are given by

$$\|f\|_{\#,B^{\lambda(\cdot)}(\mathbb{D})} = \sup_{z,w\in\mathbb{D}} |f'(z)|(1-|z|)^{1-\lambda(z)}, \quad \|f\|_{B^{\lambda(\cdot)}(\mathbb{D})} = \|f\|_{\#,B^{\lambda(\cdot)}(\mathbb{D})} + \|f\|_{L^\infty(\mathbb{D})}.$$

Lemma 4 ([11]) *Let λ satisfy the log-condition (4) and*

$$0 < \inf_{z\in\mathbb{D}} \lambda(z) \leq \sup_{z\in\mathbb{D}} \lambda(z) < 1, \tag{5}$$

then the spaces $A^{\lambda(\cdot)}(\mathbb{D})$ and $B^{\lambda(\cdot)}(\mathbb{D})$ coincide up to equivalence of norms.

By $A_*^{\lambda(\cdot)}(\mathbb{D})$ we denote the space of functions f holomorphic in $\mathbb{D}$ such that

$$|f(z) - f(w)| \leq C|1 - z\overline{w}|^{\lambda(z)}, \quad z, w \in \mathbb{D},$$

or what is the same

$$|f(z) - f(w)| \leq C|1 - z\overline{w}|^{\lambda(w)}, \quad z, w \in \mathbb{D},$$

where C does not depend on z, w. If $0 < \inf_{z\in\mathbb{D}} \lambda(z)$, then any $f \in A_*^{\omega}(\mathbb{D})$ is continuous in $\overline{\mathbb{D}}$. The semi-norm and norm of a function $f \in A_*^{\lambda(\cdot)}(\mathbb{D})$ are given by

$$\|f\|_{\#,A_*^{\lambda(\cdot)}(\mathbb{D})} = \sup_{z,w\in\mathbb{D}} \frac{|f(z) - f(w)|}{|1 - z\overline{w}|^{\lambda(z)}}, \quad \|f\|_{A_*^{\lambda(\cdot)}(\mathbb{D})} = \|f\|_{\#,A_*^{\lambda(\cdot)}(\mathbb{D})} + \|f\|_{L^\infty(\mathbb{D})}.$$

We conclude this section with the equivalent description of $A^{\lambda(\cdot)}(\mathbb{D})$.

Theorem 2 *Let λ satisfy the log-condition (4) and (5). Then $A^{\lambda(\cdot)}(\mathbb{D})$ and $A_*^{\lambda(\cdot)}(\mathbb{D})$ coincide up to equivalence of norms.*

Proof The proof is similar to the proof of Theorem 1 by using [11, Theorem 4.3].

4 Boundedness of Holomorphic Projection in Generalized Hölder Spaces of Holomorphic Functions in $\mathbb{D}$

Theorem 3 *Let ω satisfy conditions (2) and (3). Then $B_{\mathbb{D}}$ is bounded from $L^\omega(\mathbb{D})$ to $A^\omega(\mathbb{D})$.*

Proof Since $\int_{\mathbb{D}} \frac{\overline{w}}{(1-z\overline{w})^3} dA(w) = 0$, we have

$$(B_{\mathbb{D}} f)'(z) = 2\int_{\mathbb{D}} \frac{\overline{w} f(w)}{(1-z\overline{w})^3} dA(w) = 2\int_{\mathbb{D}} \frac{\overline{w}[f(w) - f(z)]}{(1-z\overline{w})^3} dA(w). \tag{6}$$

Hence, it follows that

$$\begin{aligned}
|(B_{\mathbb{D}}f)'(z)| &\le 2\int_{\mathbb{D}} \frac{|f(w)-f(z)|}{|1-z\overline{w}|^3}dA(w)\\
&\le C\|f\|_{\#,L^{\omega}(\mathbb{D})}\int_{\mathbb{D}} \frac{\omega(|z-w|)}{|1-z\overline{w}|^3}dA(w)\\
&\le C\|f\|_{\#,L^{\omega}(\mathbb{D})}\int_{\mathbb{D}} \frac{\omega(|1-z\overline{w}|)}{|1-z\overline{w}|^3}dA(w)\\
&\le C_1\|f\|_{\#,L^{\omega}(\mathbb{D})}\int_0^1 rdr\int_{\mathbb{T}} \frac{\omega(|1-zr\overline{\sigma}|)}{|1-zr\overline{\sigma}|^3}|d\sigma|\\
&= C_1\|f\|_{\#,L^{\omega}(\mathbb{D})}\int_0^1 rdr\int_{\mathbb{T}} \frac{\omega(|\sigma-zr|)}{|\sigma-zr|^3}|d\sigma|,
\end{aligned}$$

where the constants C and C_1 do not depend on f. In the above estimation we used that $|z-w|<|1-z\overline{w}|$ for all $z, w\in\mathbb{D}$ and the fact that ω is almost increasing on $[0,2]$. Using Lemma 1 with $k(z)$ being a constant from condition (3), we obtain

$$|(B_{\mathbb{D}}f)'(z)| \le C_1\|f\|_{\#,L^{\omega}(\mathbb{D})}\int_0^1 \frac{\omega(1-r|z|)}{(1-r|z|)^2}dr. \tag{7}$$

Fix arbitrary $\delta\in(0,1)$. For $0\le|z|\le\delta$ the integral in the right side of (7) is bounded by a constant depending only on δ and we have

$$|(B_{\mathbb{D}}f)'(z)|\frac{1-|z|}{\omega(1-|z|)} \le C_{\delta}\|f\|_{\#,L^{\omega}(\mathbb{D})},\quad 0\le|z|\le\delta.$$

For $0<\delta\le|z|<1$ changing variable of integration in (7) and using Zygmund condition (3) we obtain

$$\begin{aligned}
|(B_{\mathbb{D}}f)'(z)| &\le \frac{C_1}{|z|}\|f\|_{\#,L^{\omega}(\mathbb{D})}\int_{1-|z|}^1 \frac{\omega(t)}{t^2}dt\\
&\le \frac{C_1}{|z|}\|f\|_{\#,L^{\omega}(\mathbb{D})}\frac{\omega(1-|z|)}{(1-|z|)} \le \frac{C_1}{\delta}\|f\|_{\#,L^{\omega}(\mathbb{D})}\frac{\omega(1-|z|)}{(1-|z|)}.
\end{aligned}$$

The above estimates imply the inequality

$$\|B_{\mathbb{D}}f\|_{\#,B^{\omega}(\mathbb{D})} = \sup_{z\in\mathbb{D}}|(B_{\mathbb{D}}f)'(z)|\frac{1-|z|}{\omega(1-|z|)} \le C\|f\|_{\#,L^{\omega}(\mathbb{D})},$$

where the constant C may be chosen not depending on f. The proof now follows from the equivalence of norms in $A^{\omega}(\mathbb{D})$ and $B^{\omega}(\mathbb{D})$ (see Lemma 3).

Theorem 4 *Let λ satisfy the log-condition (4) and (5). Then $B_{\mathbb{D}}$ is bounded from $L^{\lambda(\cdot)}(\mathbb{D})$ to $A^{\lambda(\cdot)}(\mathbb{D})$.*

Proof Given $f \in B^{\lambda(\cdot)}(\mathbb{D})$ we have the equality (6), and therefore we obtain

$$\begin{aligned}
|(B_{\mathbb{D}} f)'(z)| &\leq C\|f\|_{\#,L^{\lambda(\cdot)}(\mathbb{D})} \int_{\mathbb{D}} \frac{dA(w)}{|1 - z\overline{w}|^{3-\lambda(z)}} \\
&\leq C_1\|f\|_{\#,L^{\lambda(\cdot)}(\mathbb{D})} \int_0^1 r dr \int_{\mathbb{T}} \frac{|d\sigma|}{|1 - zr\overline{\sigma}|^{3-\lambda(z)}} \\
&= C_1\|f\|_{\#,L^{\lambda(\cdot)}(\mathbb{D})} \int_0^1 r dr \int_{\mathbb{T}} \frac{|d\sigma|}{|\sigma - zr|^{3-\lambda(z)}} \\
&\leq \frac{C_2}{2 - \lambda(z)}\|f\|_{\#,L^{\lambda(\cdot)}(\mathbb{D})} \int_0^1 \frac{dr}{(1 - r|z|)^{2-\lambda(z)}},
\end{aligned}$$

where C, C_1, C_2 do not depend on f. Here we used Lemma 2. The rest of the prof is similar to that of Theorem 3 in view of Lemma 4.

5 Generalized Hölder Type Spaces of Harmonic Functions in the Unit Disc and Boundedness of Harmonic Projection

In this section we study the boundedness of the harmonic projection in the framework of the generalized Hölder type spaces of harmonic functions in the unit disc.

The generalized Hölder type space of harmonic functions in the unit disc with prescribed modulus of continuity, denoted by $\mathcal{A}^{\omega}(\mathbb{D})$, is the space of functions f from $L^{\omega}(\mathbb{D})$ harmonic in $\mathbb{D}$, with the notation $\|f\|_{\#,\mathcal{A}^{\omega}(\mathbb{D})} = \|f\|_{\#,L^{\omega}(\mathbb{D})}$.

The variable exponent generalized Hölder spaces $\mathcal{A}^{\lambda(\cdot)}(\mathbb{D})$ of harmonic functions in $\mathbb{D}$ is the space of functions f from $L^{\lambda(\cdot)}(\mathbb{D})$ harmonic in $\mathbb{D}$, with the notation $\|f\|_{\#,\mathcal{A}^{\lambda(\cdot)}(\mathbb{D})} = \|f\|_{\#,L^{\lambda(\cdot)}(\mathbb{D})}$.

Before we formulate and prove theorem for $\mathcal{A}^{\omega}(\mathbb{D})$ and $\mathcal{A}^{\lambda(\cdot)}(\mathbb{D})$, we need to recall the definitions of the spaces $\mathcal{B}^{\omega}(\mathbb{D})$, and $\mathcal{B}^{\lambda(\cdot)}(\mathbb{D})$ (see [13]).

The space $\mathcal{B}^{\omega}(\mathbb{D})$ is defined as to consist of functions harmonic in $\mathbb{D}$ such that

$$|\nabla f(z)| \leq C\frac{\omega(1 - |z|)}{1 - |z|}, \quad z \in \mathbb{D},$$

where C does not depend on z and $\nabla f(z) = \nabla f(x + iy) = \left(\frac{\partial f(z)}{\partial x}, \frac{\partial f(z)}{\partial y}\right)$. The semi-norm and norm of a function $f \in \mathcal{B}^{\omega}(\mathbb{D})$ are given by

$$\|f\|_{\#,\mathcal{B}^{\omega}(\mathbb{D})} = \sup_{z\in\mathbb{D}} |\nabla f(z)|\frac{1 - |z|}{\omega(1 - |z|)}, \quad \|f\|_{\mathcal{B}^{\omega}(\mathbb{D})} = \|f\|_{\#,\mathcal{B}^{\omega}(\mathbb{D})} + \|f\|_{L^{\infty}(\mathbb{D})}.$$

The space $\mathcal{B}^{\lambda(\cdot)}(\mathbb{D})$ is defined in a similar way with the obvious changes: replace $\frac{\omega(1-|z|)}{1-|z|}$ with $(1 - |z|)^{\lambda(z)-1}$ (see [13] for details).

In [13] is was shown that under the conditions (2) and (3) the spaces $\mathcal{A}^{\omega}(\mathbb{D})$ and $\mathcal{B}^{\omega}(\mathbb{D})$ coincide up to norm equivalence; the coincidence of $\mathcal{A}^{\lambda(\cdot)}(\mathbb{D})$ and $\mathcal{B}^{\lambda(\cdot)}(\mathbb{D})$ up to norm equivalence is true provided (5) takes place.

Theorem 5 *The following statements are true.*

1. *Let ω satisfy conditions (2) and (3). Then the operator $b_{\mathbb{D}}$ is bounded from $L^{\omega}(\mathbb{D})$ to $\mathcal{A}^{\omega}(\mathbb{D})$.*
2. *Let λ satisfy the log-condition (4) and (5). Then the operator $b_{\mathbb{D}}$ is bounded from $L^{\lambda(\cdot)}(\mathbb{D})$ to $\mathcal{A}^{\lambda(\cdot)}(\mathbb{D})$.*

Proof The proof follows the lines of the proofs of Theorems 3, 4 with some natural changes. We sketch the proof in the case of the first statement; the second one is similar.

We use the fact under the conditions (2) and (3) the spaces $\mathcal{A}^{\omega}(\mathbb{D})$ and $\mathcal{B}^{\omega}(\mathbb{D})$ coincide up to norm equivalence. Similarly as in Theorem 3 this allows us to obtain the inequality

$$|\nabla(b_{\mathbb{D}} f)(z)| \leq 8 \int_{\mathbb{D}} \frac{|w||f(w) - f(z)|}{|1 - z\overline{w}|^3} dA(w) \leq C \int_{\mathbb{D}} \frac{|f(w) - f(z)|}{|1 - z\overline{w}|^3} dA(w).$$

and the rest of the proof is now straightforward.

Acknowledgements Alexey Karapetyants acknowledges the support of the Fulbright Research Scholarship program and the warm hospitality of the Mathematics Department at the State University of New York at Albany during the time when this research was completed. Alexey Karapetyants is partially supported by the Russian Foundation for Fundamental Research (Grant Number 18-01-00094).

References

1. Aleman, A., Pott, S., Reguera, M.C.: Sarason conjecture on the Bergman space. Int. Math. Res. Not. **2017**(14), 4320–4349. https://doi.org/10.1093/imrn/rnw134
2. Axler, S., Bourdon, P., Ramey, W.: Harmonic Function Theory. Springer, New York (2001)
3. Békollé, D., Bonami, A.: Inégalités á poids pour le noyau de Bergman. C. R. Acad. Sci. Paris Sér. A-B **286**(18), A775–A778 (1978)
4. Djrbashian, A.E., Shamoian, F.A.: Topics in the theory of A^p_α spaces. Teubner-Texte zur Math. **105** (1988). BSB B. G. Teubner Verlagsgesellschaft, Leipzig
5. Dostanić, M.: Boundedness of the Bergman projections on L^p spaces with radial weights. Pub. Inst. Math. **86**, 5–20 (2009)
6. Duren, P., Schuster, A.: Mathematical Surveys and Monographs. Bergman Spaces, vol. 100. American Mathematical Society, Providence (2004)
7. Dyakonov, K.M.: The moduli of holomorphic functions in Lipschitz spaces. Mich. Math. J. **44**(1), 139–147 (1997)
8. Dyakonov, K.M.: Equivalent norms on Lipschitz-type spaces of holomorphic functions. Acta Math. **178**, 143–167 (1997)

9. Hedenmalm, H., Korenblum, B., Zhu, K.: Theory of Bergman Spaces. Springer, New York (2000)
10. Karapetyants, N.K., Ginzburg, A.I.: Fractional integrals and singular integrals in the Hölder classes of variable order. Integral Transform. Spec. Funct. **2**(2), 91–106 (1994)
11. Karapetyants, A., Samko, S.: Generalized Hölder spaces of holomorphic functions in domains in the complex plane. Mediterr. J. Math. **15**, 226 (2018). https://doi.org/10.1007/s00009-018-1272-z
12. Karapetyants, A., Rafeiro, H., Samko, S.: Boundedness of the Bergman projection and some properties of Bergman type spaces. Complex Anal. Oper. Theory (2018). https://doi.org/10.1007/s11785-018-0780-y
13. Karapetyants, A., Restrepo, J.E.: Generalized Hölder type spaces of harmonic functions in the unit ball and half space. Czechoslovak Math. J. (to appear) (2019)
14. Karapetyants, A., Samko, S.: Spaces $BMO_{p(\cdot)}(\mathbb{D})$ of a variable exponent $p(z)$. Georgian Math J. **17**, 529–542 (2010)
15. Karapetyants, A., Samko, S.: Mixed norm spaces of analytic functions as spaces of generalized fractional derivatives of functions in Hardy type spaces. Fract. Calc. Appl. Anal. **5**, 1106–1130 (2017). https://doi.org/10.1515/fca-2017-0059
16. Karapetyants, A., Samko, S.: On boundedness of Bergman projection operators in Banach spaces of holomorphic functions in half plane and harmonic functions in half space. J. Math. Sci. **226**(4), 344–354 (2017). https://doi.org/10.1007/s10958-017-3538-6, ISSN 1072-3374
17. Pavlović, M.: On Dyakonov's paper "Equivalent norms on Lipschitz-type spaces of holomorphic functions". Acta Math. **183**, 141–143 (1999)
18. Ross, B., Samko, S.: Fractional integration operator of variable order in the Hölder spaces $H^{\lambda(x)}$. Int. J. Math. Math. Sci. **18**(4), 777–788 (1995)
19. Zhu, K.: Spaces of Holomorphic Functions in the Unit Ball. Graduate Texts in Mathematics. Springer, Berlin (2004)
20. Zhu, K.: Operator Theory in Function Spaces, Mathematical Surveys and Monographs, vol. 138, 2nd edn. American Mathematical Society, Providence (2007)

Generalized Fourier Series by the Double Trigonometric System

K. S. Kazarian

Abstract Necessary and sufficient conditions are obtained on the function M such that $\{M(x, y)e^{ikx}e^{imy} : (k, m) \in \Omega\}$ is complete and minimal in $L^p(\mathbb{T}^2)$ when $\Omega^c = \{(0, 0)\}$ and $\Omega^c = 0 \times \mathbb{Z}$. If $\Omega^c = 0 \times \mathbb{Z}_0$, $\mathbb{Z}_0 = \mathbb{Z} \setminus \{0\}$ it is proved that the system $\{M(x, y)e^{ikx}e^{imy} : (k, m) \in \Omega\}$ cannot be complete minimal in $L^p(\mathbb{T}^2)$ for any $M \in L^p(\mathbb{T}^2)$. In the case $\Omega^c = \{(0, 0)\}$ necessary and conditions are found in terms of the one dimensional case.

Keywords Generalized Fourier series · Multiplicative completeness · Strong x−singularity

1 Introduction

The present study is motivated by the desire to extend the concept of generalized Fourier series (GFS) for functions of various variables. The concept of GFS can be described as follows. Let (X, Σ, μ) be a measurable space with a positive measure μ, $\mu(X) > 0$ and let $L^2(X, \Sigma, \mu)$ be the space of measurable functions $f : X \to \mathbb{C}$ with the norm $\|f\|_2 = (\int_X |f(t)|^2 dt)^{\frac{1}{2}} < \infty$. For a complete orthonormal system $\Phi = \{\varphi_k\}_{k=1}^{\infty}$

$$\sum_{k=1}^{\infty} a_k(f)\varphi_k = f \quad \text{in} \quad L^2(X, \Sigma, \mu) \tag{1}$$

where for any $k \in \mathbb{N}$

$$a_k(f) = \int_X f(t)\overline{\varphi_k(t)}d\mu. \tag{2}$$

The series (1)–(2) is the Fourier series of the function f with respect to the system Φ. When the system Φ is the trigonometric system it is called the Fourier series of

K. S. Kazarian (✉)
Universidad Autónoma de Madrid, 28049 Madrid, Spain
e-mail: kazaros.kazarian@uam.es

A. Karapetyants et al. (eds.), *Modern Methods in Operator Theory and Harmonic Analysis*, Springer Proceedings in Mathematics & Statistics 291,
https://doi.org/10.1007/978-3-030-26748-3_5

the given function. Representation of a given function by a trigonometric series is a classical topic (see [7, 9, 10] and many others). It is well known that a measurable function can be represented by a $\Phi-$ series where the coefficients are not defined by (2).

It seems something transcendental to find an algorithm that determines the coefficients b_k, $k \in \mathbb{N}$ such that the series $\sum_{k=1}^{\infty} b_k \varphi_k$ represents a given function f when f is not integrable. Such a problem was formulated by Luzin [7]. The following strategy can be an inexhaustible source for the study of the Luzin's problem.

At the first step fix a subset $\mathbb{N}_1 \subset \mathbb{N}$ such that for some $m \in L^2(X, \Sigma, \mu)$ the system

$$\{m\varphi_k\}_{k\in\mathbb{N}\setminus\mathbb{N}_1} \quad \text{is complete in} \quad L^2(X, \Sigma, \mu). \tag{3}$$

Determine those functions m for which $\{m\varphi_k\}_{k\in\mathbb{N}\setminus\mathbb{N}_1}$ is complete and minimal in $L^2(X, \Sigma, \mu)$ if it is possible. Afterwards if we fix any such m then the system $\{m\varphi_k\}_{k\in\mathbb{N}\setminus\mathbb{N}_1}$ will have a unique biorthogonal system $\{\psi_k\}_{k\in\mathbb{N}\setminus\mathbb{N}_1}$ in $L^2(X, \Sigma, \mu)$. When $\{\psi_k\}_{k\in\mathbb{N}\setminus\mathbb{N}_1}$ is total with respect to the space $L^2(X, \Sigma, \mu)$ then for any measurable function g such that $mg \in L^2(X, \Sigma, \mu)$ one can consider the series

$$\sum_{k\in\mathbb{N}\setminus\mathbb{N}_1} b_k(g)\varphi_k \qquad b_k(g) = \int_X g(t)\overline{\psi_k(t)}d\mu, \quad k \in \mathbb{N} \setminus \mathbb{N}_1. \tag{4}$$

The trigonometric system is the best object for testing the described idea because of its importance in various areas of mathematics.

Generalized Fourier series and some applications were studied in [3, 4, 6] when card $\mathbb{N}_1 < \infty$. It is not known if the described strategy is viable for the trigonometric system if card $\mathbb{N}_1 = \infty$ (see [2]). Any essential progress in the problem formulated below will be very helpful to clarify the question.

We denote $\mathbb{T} = \mathbb{R}/2\pi\mathbb{Z}$ and consider the complex form of the trigonometric system $\{e^{ikx} : k \in \mathbb{Z}\}$ defined on the set $\mathbb{T}$, where the set of all integer numbers is denoted by $\mathbb{Z}$. The following theorems were proved in [2].

Let $1 \le n_1 < n_2 < \cdots < n_k < n_{k+1} < \cdots$ and let $\Omega = \{-n_k\}_{k=1}^{\infty} \cup \{n_k\}_{k=1}^{\infty}$ be an infinite set of natural numbers such that $\Omega^c \neq \emptyset$, where

$$\Omega^c = \mathbb{Z} \setminus \Omega = \{-m_k\}_{k=1}^{\infty} \cup \{0\} \cup \{m_k\}_{k=1}^{\infty}.$$

Let $S_\Omega = \overline{\text{span}}_{L^1(\mathbb{T})}\{e^{ikx} : k \in \Omega^c\}$. If $p \ge 1$ then its conjugate number p' is defined by the equation $\frac{1}{p} + \frac{1}{p'} = 1$.

Theorem A. Let $1 \le p < \infty$ and let $M \in L^p(\mathbb{T})$.

Then the system $\{M(x)e^{ikx} : k \in \Omega\}$ is complete in $L^p(\mathbb{T})$ if and only if the following condition holds:

$$\text{If} g[\overline{M}]^{-1} \in L^{p'}(\mathbb{T}) \text{ for some } g \in S_\Omega \text{ then } g(x) = 0 \text{ a.e. on } \mathbb{T} \tag{5}$$

Theorem B. Let $1 \le p < \infty$ and let $M \in L^p(\mathbb{T})$.

The system $\{M(x)e^{ikx} : k \in \Omega\}$ is minimal in $L^p(\mathbb{T})$ if and only if the following condition holds:

$$\text{If for any } k \in \Omega \text{ there exists a function } g_k \in S_\Omega \text{ such that } [e^{ikx} + g_k(x)][\overline{M}]^{-1} \in L^{p'}(\mathbb{T}). \tag{6}$$

The following open problem was formulated in [2].

Problem. Describe pairs (Ω, M) with card $\Omega^c = \infty$ such that conditions (5) and (6) hold simultaneously.

Unfortunately no any subset $\Omega \subset \mathbb{Z}$, card $\Omega^c = \infty$ is known such that the conditions (5) and (6) hold simultaneously. In the present paper it is shown that the similar question for the double trigonometric system has a positive answer. It should be mentioned that for the Haar system the described strategy can be successfully implemented when card $\mathbb{N}_1 = \infty$ (see [3, 5]). First results on multiplicative completion of sets of functions were obtained in [1, 8].

2 Multiplicative Completion of Some Subsystems of the Double Trigonometric System

We will consider the double trigonometric system. The n-multiple case can be studied in a similar way. We suppose that $\Omega \subset \mathbb{Z}^2$ is an infinite set such that Ω^c is not empty, where $\Omega^c = \mathbb{Z}^2 \setminus \Omega$. In this case we modify the definition of the class

$$S_\Omega = \{f \in L^1(\mathbb{T}^2) : \int_{\mathbb{T}^2} f(x, y)e^{-ikx}e^{-imy}dxdy = 0 \quad \forall (k, m) \in \Omega\}.$$

It is clear that S_Ω is a closed subspace of $L^1(\mathbb{T}^2)$.

Theorem 1 *Let $1 \le p < \infty$ and let $M \in L^p(\mathbb{T}^2)$. Then the system*

$$\{M(x, y)e^{ikx}e^{imy} : (k, m) \in \Omega\} \tag{7}$$

is complete in $L^p(\mathbb{T}^2)$ if and only if the following condition holds:

$$\text{If } g[\overline{M}]^{-1} \in L^{p'}(\mathbb{T}^2) \text{ for some } g \in S_\Omega, \text{ then } g(x) = 0 \text{ a.e.} \tag{8}$$

Proof Suppose that (7) is complete in $L^p(\mathbb{T}^2)$ and let $g \in S_\Omega$ be a non trivial function such that $\frac{g}{M} \in L^{p'}(\mathbb{T}^2)$. Then for any $(k, m) \in \Omega$

$$\int_{\mathbb{T}^2} M(x, y)e^{ikx}e^{imy}\overline{\frac{g(x, y)}{M(x, y)}}dxdy = \int_{\mathbb{T}^2} e^{ikx}e^{imy}\overline{g(x, y)}dxdy = 0.$$

Which contradicts the completeness of the system (7). Hence, (8) holds.

Now suppose that (8) holds and for some $\varphi \in L^{p'}(\mathbb{T}^2)$

$$\int_{\mathbb{T}^2} M(x,y)e^{ikx}e^{imy}\overline{\varphi(x,y)}dxdy = 0 \qquad \text{for all} \quad (k,m) \in \Omega.$$

Which yields that $\overline{M}\varphi \in S_\Omega$ and $\varphi(x,y) = 0$ a.e. on $\mathbb{T}^2$.

Theorem 2 *Let $1 \le p < \infty$ and let $M \in L^p(\mathbb{T}^2)$. The system (7) is minimal in $L^p(\mathbb{T}^2)$ if and only if the following condition holds:*

If for any $(k,m) \in \Omega$ there exists a function $g_{k,m} \in S_\Omega$ such that

$$[e^{ikx}e^{imy} + g_{k,m}(x,y)][\overline{M(x,y)}]^{-1} \in L^{p'}(\mathbb{T}^2). \tag{9}$$

Proof Suppose that (7) is minimal in $L^p(\mathbb{T}^2)$. Then there exists a system $\{\varphi_{j,l}\}_{(j,l)\in\Omega} \subset L^{p'}(\mathbb{T}^2)$ such that

$$\int_{\mathbb{T}^2} M(x,y)e^{ikx}e^{imy}\overline{\varphi_{j,l}(x,y)}dxdy = \delta_{kj}\delta_{ml} \qquad \text{for all} \quad (k,j),(m,l) \in \Omega.$$

Hence, for any $(j,l) \in \Omega$ we have that

$$\int_{\mathbb{T}^2} e^{ikx}e^{imy}[M(x,y)\overline{\varphi_{j,l}(x,y)} - \frac{1}{(2\pi)^2}e^{-ijx}e^{-ily}]dxdy = 0 \quad \text{for all}\ (k,m) \in \Omega.$$

Which yields $\overline{M(x,y)}\varphi_{j,l}(x) - \frac{1}{(2\pi)^2}e^{ijx}e^{ily} = g_{j,l}(x,y) \in S_\Omega$. The proof of the necessity is finished.

If (9) holds then it is easy to check that the system $\{\varphi_{k,m}\}_{(k,m)\in\Omega} \subset L^{p'}(\mathbb{T}^2)$, where

$$\varphi_{k,m}(x) = \frac{e^{ikx}e^{imy} + g_{k,m}(x)}{(2\pi)^2\overline{M(x,y)}} \qquad \text{for} \quad (k,m) \in \Omega \tag{10}$$

is biorthogonal to (7).

2.1 The Case $\Omega^c = \{(0,0)\}$

Denote $\mathbb{Z}_0 = \mathbb{Z} \setminus \{0\}$ and $\mathbb{Z}_0^2 = \mathbb{Z}^2 \setminus \{(0,0)\}$.

Theorem 3 *Let $1 \le p < \infty$ and let $M \in L^p(\mathbb{T}^2)$. Then the system*

$$\{M(x,y)e^{ikx}e^{imy} : (k,m) \in \mathbb{Z}_0^2\} \tag{11}$$

is complete and minimal in $L^p(\mathbb{T}^2)$ *if and only if the systems* $\{u(t)e^{int} : n \in \mathbb{Z}_0\}$ *and* $\{v(t)e^{int} : n \in \mathbb{Z}_0\}$ *are complete and minimal in* $L^p(\mathbb{T})$*, where*

$$\frac{1}{u(x)} = \left(\int_{\mathbf{T}} |M(x,y)|^{-p'} dy\right)^{\frac{1}{p'}} \quad \text{and} \quad \frac{1}{v(y)} = \left(\int_{\mathbf{T}} |M(x,y)|^{-p'} dx\right)^{\frac{1}{p'}}. \tag{12}$$

Proof By Theorem 1 it follows that the system (11) is complete in $L^p(\mathbb{T}^2)$ if and only if

$$\int_{\mathbf{T}^2} |M(x,y)|^{-p'} dxdy = +\infty. \tag{13}$$

Hence, by Theorem 2 the system (11) is minimal in $L^p(\mathbb{T}^2)$ if there exist unique numbers $a_{kl} \in \mathbb{C}$, $(k,l) \in \mathbb{Z}_0$ such that

$$\int_{\mathbf{T}^2} |e^{ikx}e^{ily} - a_{kl}|^{p'} |M(x,y)|^{-p'} dxdy < +\infty \quad \text{for any} \quad (k,l) \in \mathbb{Z}_0. \tag{14}$$

We consider (14) respectively for $(k,0)$ and $(0,l)$, where k and l belong to $\mathbb{Z}_0$. By the Fubini-Tonelli theorem it follows that the functions u and v are positive a.e. on $\mathbb{T}$. On the other hand we have that for almost any $x \in \mathbb{T}$

$$2\pi \le \left(\int_{\mathbf{T}} |M(x,y)|^{p} dy\right)^{\frac{1}{p}} \left(\int_{\mathbf{T}} |M(x,y)|^{-p'} dy\right)^{\frac{1}{p'}},$$

which yields

$$\int_{\mathbf{T}} u(x)^p dx \le \frac{1}{(2\pi)^p} \int_{\mathbf{T}^2} |M(x,y)|^p dxdy.$$

Similarly we obtain that $v \in L^p(\mathbb{T})$. Afterwards by (13) and (14) we easily obtain that there exists $x_0 \in \mathbb{T}$ such that

$$\int_{\mathbf{T}} u(x)^{-p'} dx = +\infty \quad \text{and} \quad \int_{\mathbf{T}} |e^{ikx} - e^{ikx_0}|^{p'} u(x)^{-p'} dx < +\infty. \tag{15}$$

By Proposition 3 of [5] it follows that the system $\{u(t)e^{int} : n \in \mathbb{Z}_0\}$ is complete and minimal in $L^p(\mathbb{T})$. Similarly we obtain that $\{v(t)e^{int} : n \in \mathbb{Z}_0\}$ is complete and minimal in $L^p(\mathbb{T})$.

The following theorem gives another characterization.

Theorem 4 *Let* $1 \le p < \infty$ *and let* $M \in L^p(\mathbb{T}^2)$*. Then the system (11) is complete and minimal in* $L^p(\mathbb{T}^2)$ *if and only if holds (13) and*

$$\int_{\mathbf{T}^2} |\sin\frac{x-x_0}{2}|^{p'} |M(x,y)|^{-p'} dxdy < +\infty,$$

$$\int_{\mathbb{T}^2} |\sin \frac{y-y_0}{2}|^{p'} |M(x,y)|^{-p'} dxdy < +\infty.$$

for some $(x_0, y_0) \in \mathbb{T}^2$.

Proof We skip the proof of the necessity because the arguments are similar to those used in the proof of the previous theorem. To finish the proof we have to check the relations (14) for $a_{kl} = e^{ikx_0}e^{ily_0}$. Write

$$\begin{aligned} & \left(\int_{\mathbb{T}^2} |e^{ikx}e^{ily} - e^{ikx_0}e^{ily_0}|^{p'} |M(x,y)|^{-p'} dxdy\right)^{\frac{1}{p'}} \\ \leq \quad & \left(\int_{\mathbb{T}^2} |e^{ily} - e^{ily_0}|^{p'} |M(x,y)|^{-p'} dxdy\right)^{\frac{1}{p'}} \\ + & \left(\int_{\mathbb{T}^2} |e^{ikx} - e^{ikx_0}|^{p'} |M(x,y)|^{-p'} dxdy\right)^{\frac{1}{p'}} < +\infty. \end{aligned}$$

Corollary 1 *Let* $1 \leq p < \infty$ *and let* $M \in L^p(\mathbb{T}^2)$. *Then for any* $(\nu, \mu) \in \mathbb{Z}^2$ *the system*

$$\{M(x,y)e^{ikx}e^{imy} : (k,m) \in \mathbb{Z}^2 \setminus (\nu, \mu)\}$$

is complete and minimal in $L^p(\mathbb{T}^2)$ *if and only if the system*

$$\{M(x,y)e^{ikx}e^{imy} : (k,m) \in \mathbb{Z}_0^2\}$$

is complete and minimal in $L^p(\mathbb{T}^2)$.

The assertion of the corollary is obvious because by multiplying the elements of the system $\{e^{ikx}e^{imy} : (k,m) \in \mathbb{Z}^2 \setminus (\nu, \mu)\}$ by $e^{-i\nu x}e^{-i\mu y}$ we obtain the system $\{e^{ikx}e^{imy} : (k,m) \in \mathbb{Z}_0^2\}$. On the other hand it is easy to observe that in our case the conditions (8), (9) remain true if $M(x,y)$ is multiplied by a function with modulus equal to one almost everywhere.

Example 1 Let $1 \leq p < \infty$, $(x_0, y_0) \in \mathbb{T}^2$ and let $M(x,y) = |\sin \frac{x-x_0}{2}|^\alpha + |\sin \frac{y-y_0}{2}|^\alpha$, where $\frac{1}{p'} \leq \alpha < 1 + \frac{1}{p'}$. Then the system $\{e^{ikx}e^{imy} : (k,m) \in \mathbb{Z}_0^2\}$ is complete and minimal in $L^p(\mathbb{T}^2)$.

2.2 The Case $\Omega^c = 0 \times \mathbb{Z}$

Further in this section it is supposed that $\Omega \subset \mathbb{Z}^2$ is such that $\Omega^c = 0 \times \mathbb{Z}$.

Lemma 1 *Let* $g \in S_\Omega$ *then* $g(x,y) = h(y)$, *where* $h \in L^1(\mathbb{T})$.

Proof Let

$$h(y) = \frac{1}{2\pi} \int_{\mathbb{T}} g(x,y)dx.$$

Then $h \in L^1(\mathbb{T})$ and for any $k \in \mathbb{Z}$

$$c_m(h) = \frac{1}{2\pi}\int_{\mathbb{T}} h(y)e^{-imy}dy = \frac{1}{(2\pi)^2}\int_{\mathbb{T}^2} g(x,y)e^{-imy}dxdy.$$

It is easy to check that for any $(k, m) \in \mathbb{Z}^2$

$$\frac{1}{(2\pi)^2}\int_{\mathbb{T}^2} [g(x,y) - h(y)]e^{-ikx}e^{-imy}dxdy = 0.$$

Definition 1 Let $M \in L^p(\mathbb{T}^2)$ and $1 \le p < \infty$. We say that the function $M(x, y)$ has a strong $x-$singularity of degree p $(1 \le p < \infty)$ if for any measurable set $E \subset \mathbb{T}$, $|E| > 0$

$$M^{-1} \notin L^{p'}(\mathbb{T} \times E), \qquad \frac{1}{p} + \frac{1}{p'} = 1.$$

Proposition 1 *Let $1 \le p < \infty$ and let $M \in L^p(\mathbb{T}^2)$. Then the system*

$$\{M(x,y)e^{ikx}e^{imy} : (k,m) \in \Omega\} \tag{16}$$

is complete in $L^p(\mathbb{T}^2)$ if and only if $M(x, y)$ has a strong $x-$singularity of degree p.

Proof Suppose that the function $M(x, y)$ has a strong $x-$singularity of degree p $(1 \le p < \infty)$. If for some $g \in S_\Omega$ we have that $\frac{g}{M} \in L^{p'}(\mathbb{T}^2)$ then by Lemma 1 it follows that

$$\int_{\mathbb{T}^2} |g(x,y)|^{p'}|M(x,y)|^{-p'}dxdy = \int_{\mathbb{T}} |h(y)|^{p'}\int_{\mathbb{T}} |M(x,y)|^{-p'}dxdy < +\infty. \tag{17}$$

Hence, the set $G = \{y \in \mathbb{T} : |h(y)| > 0\}$ should be of measure zero. Which yields that $g(y) = 0$ a.e. on $\mathbb{T}$ and by Theorem 1 follows that the system (16) is complete in $L^p(\mathbb{T}^2)$.

For the proof of the necessity suppose that the system (16) is complete in $L^p(\mathbb{T}^2)$. Hence, by Theorem 1 we have that for any non trivial $g \in S_\Omega$

$$\int_{\mathbb{T}^2} \frac{|g(y)|^{p'}}{|M(x,y)|^{p'}}dxdy = +\infty.$$

For any measurable set $E \subset \mathbb{T}$, $|E| > 0$ we have that $\chi_E(y) \in S_\Omega$ which yields that $M(x, y)$ has a strong $x-$singularity of degree p.

For our further study we define a class of functions Υ.

Definition 2 We say that $\phi \in \Upsilon$ if $\phi \in L^\infty(\mathbb{T})$ and $\frac{1}{\phi} \in L^\infty(\mathbb{T})$.

Definition 3 We say that a function $M \in L^p(\mathbb{T}^2)$ has an $(x, P)-$singularity of degree p $(1 \le p < \infty)$ if $M^{-1} \notin L^{p'}(\mathbb{T}^2)$ and

$$\int_{\mathbb{T}^2} \frac{|e^{ix} - P(y)|^{p'}}{|M(x,y)|^{p'}} dxdy < +\infty, \tag{18}$$

where $P \in \Upsilon$.

Proposition 2 *Let $1 \leq p < \infty$ and let $M \in L^p(\mathbb{T}^2)$. Then the system (16) is minimal in $L^p(\mathbb{T}^2)$ if and only if one of the following conditions hold:*

$$\int_{\mathbb{T}^2} |M(x,y)|^{-p'} dxdy < +\infty \tag{19}$$

or the function $M(x, y)$ has an $(x, P)-$singularity of degree p.

Proof At first we suppose that (19) holds. Let

$$\psi_{j,l}(x,y) = (2\pi)^{-2} e^{ijx} e^{ily} [\overline{M(x,y)}]^{-1} \quad \text{for} \quad (j,l) \in \Omega.$$

One can easily check that the system $\{\psi_{j,l}(x,y) : (j,l) \in \Omega\} \subset L^{p'}(\mathbb{T}^2)$ is biorthogonal with (16).

Now let us suppose that the function $M(x, y)$ has an $(x, P)-$singularity of degree p. Let

$$\xi_{k,m}(x,y) = (2\pi)^{-2} [e^{ikx} e^{imy} - P^k(y) e^{imy}][\overline{M(x,y)}]^{-1}, \qquad (k,m) \in \Omega. \tag{20}$$

Clearly $\xi_{k,m} \in L^{p'}(\mathbb{T}^2)$ for any $(k,m) \in \Omega$. Moreover, it is easily that the system $\{\xi_{k,m}(x,y) : (k,m) \in \Omega\}$ is biorthogonal with (16).

Suppose that the system (16) is minimal in $L^p(\mathbb{T}^2)$. Then by Theorem 2 we have that the system $\{\varphi_{j,l}\}_{(j,l)\in\Omega}$ biorthogonal with (16) is defined by the Eq. (10) and $g_{k,m} \in S_\Omega$. If $g_{0,1}(x,y) = 0$ a.e. then (19) holds. If $\frac{1}{M} \notin L^{p'}(\mathbb{T}^2)$ then $g_{0,1}$ is a non trivial function and by Lemma 1 it we have that $g_{0,1}(x,y) = h_{0,1}(y)$. Let

$$\begin{aligned} P(y) = -h_{1,0}(y) \ \text{ if } \ \frac{1}{2} < |h_{1,0}(y)| < 2; \\ P(y) = \frac{1}{2} \ \text{ if } \ |h_{1,0}(y)| \leq \frac{1}{2} \ \text{ and} \\ 2 \ \text{ if } \ |h_{1,0}(y)| \geq 2. \end{aligned}$$

Clearly $P \in \Upsilon$ and by the relation

$$\frac{e^{ix} + g_{1,0}(y)}{(2\pi)^2 \overline{M(x,y)}} \in L^{p'}(\mathbb{T}^2).$$

it is easy to check that $M(x,y)$ has an $(x, P)-$singularity of degree p.

Definition 4 We say that $M \in L^p(\mathbb{T}^2)$ has a strong $(x, P)-$singularity of degree $p\,(1 \leq p < \infty)$ if M has a strong $x-$singularity and an $(x, P)-$singularity of degree p for some $P \in \Upsilon$.

Proposition 3 *Let $1 \le p < \infty$ and let $M \in L^p(\mathbb{T}^2)$. Suppose that $\Omega^c = 0 \times \mathbb{Z}$. Then the system (16) is complete and minimal in $L^p(\mathbb{T}^2)$ if and only if the function $M(x, y)$ has a strong $(x, P)-$singularity of degree p with $|P(y)| \equiv 1$ a.e. on $\mathbb{T}$.*

Proof By Propositions 1 and 2 we have to show that if the system (16) is complete and minimal in $L^p(\mathbb{T}^2)$ then the conditions of the proposition hold with $|P(y)| \equiv 1$ a.e. on $\mathbb{T}$. We provide the proof by reduction to absurdity. Suppose that $|P(y)| \neq 1$ if $y \in E, |E| > 0$. Then for some $\delta > 0$ we have that $||P(y)| - 1| > \delta$ if $y \in F \subset E, |F| > 0$. On the other hand we have that (18) holds. Hence,

$$\int_{\mathbb{T}\times F} \frac{1}{|M(x, y)|^{p'}} dxdy < +\infty$$

which contradicts the condition that $M(x, y)$ has a strong $x-$singularity of degree p. The proof of sufficiency is obvious.

Lemma 2 *Let $1 \le p < \infty$ and $\Omega \subset \mathbb{Z}^2$ is such that $\Omega^c = 0 \times \mathbb{Z}$. Suppose that $M \in L^p(\mathbb{T}^2)$ has a strong $(x, P)-$singularity of degree p with $|P(y)| \equiv 1$. Then the system (16) is complete minimal in $L^p(\mathbb{T}^2)$ and its conjugate system $\{\xi_{j,l}(x, y) : (j, l) \in \Omega\}$ is defined by the conditions (20) and for any $(n, m) \in \mathbb{Z}^2$*

$$e^{inx} e^{imy} (e^{ix} - P(y)) = \overline{M(x, y)}\xi_{n+1,m}(x, y) - P(y)\overline{M(x, y)}\xi_{n,m}(x, y).$$

Proof The first part of the lemma follows by Proposition 3 and the proof of Proposition 2.

For any $(n, m) \in \mathbb{Z}^2$ we write

$$\begin{aligned} e^{inx} e^{imy} (e^{ix} - P(y)) \quad & -\overline{M(x, y)}\xi_{n+1,m}(x, y) \\ = -P(y)e^{inx} e^{imy} + P^{n+1}(y)e^{imy} & = -\overline{M(x, y)}P(y)\xi_{n,m}(x, y). \end{aligned}$$

Theorem 5 *Let $1 \le p < \infty$ and $\Omega \subset \mathbb{Z}^2$ is such that $\Omega^c = 0 \times \mathbb{Z}$. Suppose that is such that Then the system (16) is an $M-$basis in $L^p(\mathbb{T}^2)$ if and only if $M \in L^p(\mathbb{T}^2)$ has a strong $(x, P)-$singularity of degree p with $|P(y)| \equiv 1$.*

Proof If the system (16) is an $M-$basis in $L^p(\mathbb{T}^2)$ then by Proposition 3 it follows that M has a strong $(x, P)-$singularity of degree p with $|P(y)| \equiv 1$. On the other hand if the function $M(x, y)$ has a strong $(x, P)-$singularity of degree p with $|P(y)| \equiv 1$ a.e. on $\mathbb{T}$ then by Proposition 3 the system (16) is complete and minimal in $L^p(\mathbb{T}^2)$ and the system $\{\xi_{j,l}(x, y) : (j, l) \in \Omega\}$ conjugate to (16) is defined by the Eq. (20). Let $f \in L^p(\mathbb{T}^2)$ be such that

$$\int_{\mathbb{T}^2} f(x, y)\overline{\xi_{j,l}}(x, y)dxdy = 0, \quad \text{for all} \quad (j, l) \in \Omega.$$

Then by (16) and the Fubini-Tonelli theorem we will have that

$$\int_{\mathbb{T}} e^{-ily} \int_{\mathbb{T}} f(x,y)[e^{-ijx} - \overline{P^j(y)}][\overline{M(x,y)}]^{-1} dxdy = 0, \quad \text{for all} \quad l \in \mathbb{Z}, j \in \mathbb{Z}_0.$$

Which yields $\Phi_j(y) = 0$ a.e. on $\mathbb{T}$ for all $j \in \mathbb{Z}_0$, where

$$\Phi_j(y) = \int_{\mathbb{T}} f(x,y)[e^{-ijx} - \overline{P^j(y)}][\overline{M(x,y)}]^{-1} dx \qquad j \in \mathbb{Z}_0.$$

Let $y_0 \in \mathbb{T}$ be such that the following conditions hold:

$$\int_{\mathbb{T}} |f(x,y_0)|^p dx < +\infty, \ \Phi_j(y_0) = 0 \ \text{for all} \ j \in \mathbb{Z}_0,$$

and

$$\int_{\mathbb{T}} \frac{1}{|M(x,y_0)|^{p'}} dx = +\infty, \quad \int_{\mathbb{T}} \frac{|e^{ix} - P(y_0)|^{p'}}{M(x,y_0)|^{p'}} dx < +\infty.$$

Thus we have that $P(y_0) = e^{ix_0}$ for some $x_0 \in \mathbb{T}$ and

$$\int_{\mathbb{T}} f(x,y_0)[e^{-ijx} - e^{-ijx_0}][\overline{M(x,y_0)}]^{-1} dx = 0 \qquad \text{for all} \quad j \in \mathbb{Z}_0,$$

where $f(\cdot, y_0) \in L^p(\mathbb{T})$. According to the corresponding result in the one dimensional case (see [6]) it follows that $f(x, y_0) = 0$ for almost any $x \in \mathbb{T}$. On the other hand we have that the above conditions are true for almost all $y \in \mathbb{T}$. Which yields that $f = 0$, a.e. on $\mathbb{T}^2$.

Example 2 Let

$$M(x,y) = |\sin \frac{x - x_0}{2}|^{\alpha} \ \text{for} \ (x,y) \in \mathbb{T}^2,$$

where $x_0 \in \mathbb{T}$ and $\frac{1}{p'} \le \alpha < 1 + \frac{1}{p'}$. It is easy to check that m has a strong $(x, P)-$singularity of degree p with $P(y) = e^{ix_0}$ if $y \in \mathbb{T}$. By Theorem 5 it follows that the system (16) is an $M-$basis in in $L^p(\mathbb{T}^2)$ with the conjugate system

$$\xi_{k,m}(x,y) = (2\pi)^{-2}[e^{ikx}e^{imy} - e^{ikx_0}e^{imy}][\overline{M(x,y)}]^{-1}, \qquad (k,m) \in \Omega.$$

2.3 The Case $\Omega^c = 0 \times \mathbb{Z}_0$

In the cases studied above we have that if the system $\{M(x,y)e^{ikx}e^{imy} : (k,m) \in \Omega\}$ is complete and minimal in $L^p(\mathbb{T}^2)$ then it is an $M-$basis in $L^p(\mathbb{T}^2)$. Suppose that $\Omega_0 \subset \mathbb{Z}^2$ is such that $\Omega_0^c = 0 \times \mathbb{Z}_0$. In this section we prove that if the system

$$\{M(x,y)e^{ikx}e^{imy} : (k,m) \in \Omega_0\} \qquad (21)$$

is complete in $L^p(\mathbb{T}^2)$ then it is not minimal.

Theorem 6 *Let $1 \le p < \infty$ and let $M \in L^p(\mathbb{T}^2)$. Suppose that $\Omega_0 \subset \mathbb{Z}^2$ is such that $\Omega_0^c = 0 \times \mathbb{Z}_0$. Then the system (21) is complete in $L^p(\mathbb{T}^2)$ if and only if the function $M(x, y)$ has a strong $x-$singularity of degree p.*

Proof By Proposition 1 it is clear that if the weight function $w(x, y)$ has a strong $x-$singularity of degree p then the system (21) is complete in $L^p(\mathbb{T}^2)$.

For the proof of the necessity suppose that the system (21) is complete in $L^p(\mathbb{T}^2)$. By Theorem 1 we have that for any non trivial $g \in S_{\Omega_0}$

$$\int_{\mathbb{T}^2} \frac{|g(y)|^{p'}}{|M(x, y)|^{p'}} dxdy = +\infty.$$

Let $E \subset \mathbb{T}$, $|E| > 0$ be any measurable set and $E_1 \subset E$ be such that $|E_1| = \frac{1}{2}|E|$. It is easy to observe that we have that $\chi_{E_1}(y) - \chi_{E \setminus E_1}(y) \in S_{\Omega_0}$ which yields that $M(x, y)$ has a strong $x-$singularity of degree p.

Proposition 4 *Let $1 \le p < \infty$ and suppose that $\Omega_0 \subset \mathbb{Z}^2$ is such that $\Omega_0^c = 0 \times \mathbb{Z}_0$. Then for any function $M \in L^p(\mathbb{T}^2)$ the system (7) is not complete minimal in $L^p(\mathbb{T}^2)$.*

Proof Suppose that for a function $M(x, y)$ the system (7) is complete minimal in $L^p(\mathbb{T}^2)$. By Proposition 6 we have that $M(x, y)$ has a strong $x-$singularity of degree p. By Theorem 2 it follows that there exists $g_{0,0} \in S_{\Omega_0}$ such that

$$\frac{1 + g_{0,0}(y)}{M(x, y)} \in L^{p'}(\mathbb{T}^2).$$

Which yields that

$$\int_{\mathbb{T}} |1 + g_{0,0}(y)|^{p'} \int_{\mathbb{T}} \frac{1}{|M(x, y)|^{p'}} dxdy < +\infty.$$

The last condition contradicts the condition that $M(x, y)$ has a strong x-singularity of degree p.

We say that $g \in \Upsilon_0$ if $g \in \Upsilon$ and $\int_{\mathbb{T}} g(t)dt = 0$. By Lemma 1 it easily follows that if $g \in S_{\Omega_0}$ then $g(x, y) = h(y)$, where $h \in L^1(\mathbb{T})$ and $\int_{\mathbb{T}} h(y)dy = 0$.

Proposition 5 *Let $1 \le p < \infty$ and let $M \in L^p(\mathbb{T}^2)$. Then the system (7) is minimal in $L^p(\mathbb{T}^2)$ if and only if or holds the condition (19) or the function $M(x, y)$ has an $(x, P)-$singularity of degree p with $P \in \Upsilon_0$ and for some $Q \in \Upsilon_0$*

$$\int_{\mathbb{T}^2} |1 - Q(y)|^{p'} |M(x, y)|^{-p'} dxdy < \infty. \tag{22}$$

Proof If the condition (19) holds then the proof is similar to the proof of Proposition 2.

Now let us suppose that the function $M(x, y)$ has an $(x, P)-$singularity of degree p with $P \in \Upsilon_0$ and for some $Q \in \Upsilon_0$ holds the condition (22). Clearly $\Upsilon_0 \subset S_{\Omega_0}$. Thus if we put

$$\xi_{0,0}(x, y) = (2\pi)^{-2}[1 - Q(y)e^{imy}][\overline{M(x, y)}]^{-1},$$

and

$$\xi_{k,m}(x, y) = (2\pi)^{-2}[e^{ikx}e^{imy} - P^k(y)e^{imy}][\overline{M(x, y)}]^{-1}, \qquad (k, m) \in \Omega_0 \setminus (0, 0).$$

Clearly $\xi_{k,m} \in L^{p'}(\mathbb{T}^2)$ for any $(k, m) \in \Omega_0$. Moreover, it is easily that the system $\{\xi_{k,m}(x, y) : (k, m) \in \Omega_0\}$ is biorthogonal with (21).

Suppose that the system (21) is minimal in $L^p(\mathbb{T}^2)$. Then by Theorem 2 we have that the system $\{\varphi_{j,l}\}_{(j,l)\in\Omega_0}$ biorthogonal with (21) is defined by the Eq. (10) and $g_{k,m} \in S_{\Omega_0}$. If $g_{0,0}(y) = 0$ a.e. then (19) holds. If $\frac{1}{M} \notin L^{p'}(\mathbb{T}^2)$ then $g_{0,0}$ is a non trivial function and by Lemma 1 it we have that $g_{0,0}(x, y) = h_{0,0}(y)$ and $\int_{\mathbb{T}} h_{0,0}(y)dy = 0$. Let $Q(y) = h_{0,0}(y)$ if $y \in G_0 := \{y \in \mathbb{T} : \frac{1}{2} < |h_{1,0}(y)| < 2\}$ and let $\beta = \int_{G_0} h_{1,0}(y)dy$. If $|G_0| = 2\pi$ then the function Q is defined and $Q \in \Upsilon_0$. If $|G_0| < 2\pi$ then putting $Q(y) = \frac{-\beta}{2\pi-|G_1|}$ if $y \in \mathbb{T} \setminus G_1$. Clearly $Q \in \Upsilon_0$ and the relation (22) holds. In a similar way we define $P \in \Upsilon_0$ so that

$$\frac{e^{ix} + P(y)}{(2\pi)^2\overline{M(x, y)}} \in L^{p'}(\mathbb{T}^2).$$

Thus $M(x, y)$ has an $(x, P)-$singularity of degree p.

References

1. Boas, R.P., Pollard, H.: The multiplicative completion of sets of functions. Bull. Am. Math. Soc. **54**, 518–522 (1948)
2. Kazarian, K.S.: Some open problems related to generalized Fourier series. In: Georgakis, C., Stokolos, A.M., Urbina, W. (eds.) Special Functions, Partial Differential Equations and Harmonic Analysis. Springer Proceedings in Mathematics & Statistics, vol 108, pp. 105–113. Springer, Cham (2014). https://doi.org/10.1007/978-3-319-10545-1_10
3. Kazarian, K.S.: Summability and convergence almost everywhere of generalized Fourier and Fourier-Haar series (in Russian). Izv. Akad. Nauk Arm. SSR, Ser. Mat. **20**(2), 145–162 (1985) (in Russian); English translation in Soviet Jour. Contemp. Math. Anal. **18**, 63–82 (1985)
4. Kazarian, K.S.: Summability of generalized Fourier series in a weighted metric and almost everywhere. Doklady AN USSR **287**(3), 543–546 (1986); English transl. in Soviet Math. Doklady **33**(2), 416–419 (1986)
5. Kazarian, K.S.: On bases and unconditional bases in the spaces $L^p(d\mu)$, $1 \leq p<\infty$. Stud. Math. **71**, 227–249 (1982)

6. Kazarian, K.S.: Summability of generalized Fourier series and Dirichlet's problem in $L^p(d\mu)$ and weighted H^p-spaces ($p>1$). Anal. Math. **13**, 173–197 (1987)
7. Luzin, N.N.: Integral and Trigonometric Series. Gostekhizdat, Moscow (1951). (in Russian)
8. Price, J.J., Zink, R.E.: On sets of functions that can be multiplicatively completed. Ann. Math. **82**(1), 139–145 (1965)
9. Riemann, B.: On the representation of a function by a trigonometric series. In: Riemann Bernhard Collected Papers, Translated from the 1892 German Edition by R. Baker, C. Christensen and H. Orde, pp. 219–256. Kendrick Press, Heber City, UT (2004)
10. Zygmund, A.: Trigonometric Series, vol. 1–2. Cambridge University Press, Cambridge (1959)

Hardy Type Inequalities in the Category of Hausdorff Operators

Elijah Liflyand

Abstract Classical Hardy's inequalities are concerned with the Hardy operator and its adjoint, the Bellman operator. Hausdorff operators in their various forms are natural generalizations of these two operators. In this paper, we try to adjust the scheme used by Bradley for Hardy's inequalities with general weights to the Hausdorff setting. It is not surprising that the obtained necessary conditions differ from the sufficient conditions as well as that both depend not only on weights but also on the kernel that generate the Hausdorff operator. For the Hardy and Bellman operators, the obtained necessary and sufficient conditions coincide and reduce to the classical ones.

Keywords Hardy inequality · Hausdorff operator

1 Introduction

Let $L^p_v := L^p_{v(\cdot)}(Y)$, with

$$\|f\|_{p,v} = \left(\int_Y |f(y)|^p v(y)^p \, dy\right)^{\frac{1}{p}}, \tag{1}$$

be the weighted Lebesgue space, where in our considerations $Y = \mathbb{R}_+ = [0, \infty)$. The weight v (and u) is a non-negative locally integrable function.

Hardy's inequalities are an important part of analysis and frequently used tools. They exist in various forms, there are books on them (see, e.g., [5, 6, 9]). We shall rest on Hardy's inequalities with general weights [1]:

E. Liflyand (✉)
Department of Mathematics, Bar-Ilan University, 52900 Ramat-Gan, Israel
e-mail: liflyand@math.biu.ac.il

S.M. Nikol'skii Institute of Mathematics RUDN University,
6 Miklukho-Maklay St, Moscow 117198, Russia

A. Karapetyants et al. (eds.), *Modern Methods in Operator Theory and Harmonic Analysis*, Springer Proceedings in Mathematics & Statistics 291,
https://doi.org/10.1007/978-3-030-26748-3_6

$$\|P_x f\|_{q,u} \ll \|f\|_{p,v} \tag{2}$$

and

$$\|Q_x f\|_{q,u} \ll \|f\|_{p,v}, \tag{3}$$

with $f \ge 0$, $1 \le p \le q < \infty$, and

$$P_x f = \int_0^x f(y)\,dy, \qquad Q_x f = \int_x^\infty f(y)\,dy$$

being the Hardy and Bellman operators, respectively. Inequality (2) holds if and only if, for each $r > 0$,

$$\left(Q_r u^q\right)^{1/q} \left(P_r v^{-p'}\right)^{1/p'} \ll 1, \tag{4}$$

where here and in similar conditions the constant in $\ll$ on the right does not depend on r. For $p = 1$ and $q < \infty$, see also [9, (5.12)]. Similarly, (3) holds if and only if, for each $r > 0$,

$$\left(P_r u^q\right)^{1/q} \left(Q_r v^{-p'}\right)^{1/p'} \ll 1. \tag{5}$$

We are motivated by the fact that the Hardy and Bellman operators are (simple) partial cases of much more general Hausdorff operators defined, by means of a function φ, in most works as (see, e.g., [7, 8])

$$(\mathcal{H}f)(x) = (\mathcal{H}_\varphi f)(x) = \int_{\mathbb{R}_+} \frac{\varphi(t)}{t} f\left(\frac{x}{t}\right) dt. \tag{6}$$

This is only one of the ways to define Hausdorff operators, for instance, they can be defined on $\mathbb{R}$ or the change of variables in f can be different. This form is convenient because of its symmetry, that is, the right-hand side of (6) is equal to

$$\int_{\mathbb{R}_+} \frac{\varphi\left(\frac{x}{t}\right)}{t} f(t)\,dt. \tag{7}$$

To make it even closer to the Hardy inequalities business, we will deal with the following simplified form of Hausdorff operators

$$(Hf)(x) = (H_\varphi f)(x) = \int_{\mathbb{R}_+} \varphi\left(\frac{x}{t}\right) f(t)\,dt. \tag{8}$$

By this, denoting by χ_A the indicator function of the set A, we have $Hf(x) = P_x f$ provided $\varphi(s) = \chi_{[1,\infty)}(s)$ and $Hf(x) = Q_x f$ provided $\varphi(s) = \chi_{(0,1]}(s)$.

Our goal is to generalize (3) and (2) to a wider family of Hausdorff operators. One cannot say that certain attempts of such type have never been taken. For example, the operator in [10] may be considered as a Hausdorff operator with a compact supported kernel. For the general theory of Hardy's inequalities, besides several books, one of which is already cited [9], the recent survey [2] can be recommended.

Our strategy for extending the part "if" will be repeating calculations in [1] but then stop at a proper moment. In other words, the first part of these calculations is very general while the second one is specific for P_x and Q_x. Definitely, replacing the general H_φ by one of them and continuing the calculations like in [1] will give these two important partial cases.

The necessity part is more specific and assumes additional properties of the Hausdorff operator that automatically hold for the Hardy operator or for the Bellman operator.

The outline of this paper is as follows. In the next section, we prove the claimed results for the operator defined by (8). In Sect. 3, we consider similar problems for its adjoint. This is a natural and necessary step, since the same takes place in the classical situation: the Bellman operator is the adjoint operator of the Hardy operator. In Sect. 4, we reformulate (just be redefining the kernel) the corresponding results for $\mathcal{H}_\varphi$ in (6). Finally, in Sect. 5, we prove the Hardy type inequalities for more general operator. As in the previous section, they are mainly based on the results obtained in Sect. 2. Though the necessary conditions proved to be sharp for the Hardy and Bellman operators, the question of their sharpness in general remains open.

2 Main Results

In fact, "playing" with φ, one can derive a variety of Hardy type sufficient conditions. The question is which one is sharp, that is, becomes necessary for the whole class of functions $f \in L_v^p$. We give one of them, considering also the problem whether the $H_\varphi f$ is well-defined for $f \in L_v^p$.

Theorem 1 *Let* $1 < p \le q < \infty$, $f \in L_v^p$, $\varphi \in L^\infty$, *and* $\varphi, f \ge 0$. *We have*

$$\left(\int_0^\infty u(x)^q \left|\int_0^\infty \varphi(\frac{x}{t}) f(t)\, dt\right|^q dx\right)^{\frac{1}{q}} \ll \left(\int_0^\infty f(t)^p v(t)^p dt\right)^{\frac{1}{p}} \tag{9}$$

provided

$$\begin{aligned} &\int_0^\infty u(x)^q \varphi(\tfrac{x}{t})^{\frac{q}{p}} \left(\int_0^t \varphi(\tfrac{x}{s}) v(s)^{-p'} ds\right)^{\frac{q}{pp'}} \\ \times\quad &\left(\int_0^\infty \varphi(\tfrac{x}{s}) v(s)^{-p'} ds\right)^{\frac{q}{p'p'}} dx \ll 1. \end{aligned} \tag{10}$$

If

$$\varphi(\frac{x}{t}) \int_0^t v(s)^{-p'} ds \ll \int_0^t \varphi(\frac{x}{s}) v(s)^{-p'} ds, \tag{11}$$

then

$$\int_0^\infty u(x)^q \varphi(\frac{x}{t})^{\frac{q}{p}} \left(\int_0^t \varphi(\frac{x}{s})v(s)^{-p'} ds\right)^{\frac{q}{p'}} dx \ll 1 \tag{12}$$

is necessary for (9) to hold.

Proof Let us first check that the Hausdorff operator is well defined. Indeed, by Hölder's inequality we get

$$\int_0^\infty \varphi(\frac{x}{t}) f(t)\,dt \le \left(\int_0^\infty \varphi(\frac{x}{t}) f(t)^p v(t)^p dt\right)^{\frac{1}{p}} \left(\int_0^\infty \varphi(\frac{x}{t}) v(t)^{-p'} dt\right)^{\frac{1}{p'}}.$$

The first integral on the right is finite since $f \in L_v^p$ and $\varphi \in L^\infty$, while the second one is finite since (10) assumes it finiteness almost everywhere.

Let now (10) hold. By Hölder's inequality, we obtain

$$\begin{aligned} &\int_0^\infty u(x)^q \left|\int_0^\infty \varphi(\frac{x}{t}) f(t)\,dt\right|^q dx \\ = &\int_0^\infty u(x)^q \quad \left|\int_0^\infty \varphi(\frac{x}{t})^{\frac{1}{p}} f(t)v(t)h(x,t)\varphi(\frac{x}{t})^{\frac{1}{p'}} v(t)^{-1} h(x,t)^{-1}\,dt\right|^q dx \\ &\le \int_0^\infty u(x)^q \left(\int_0^\infty \varphi(\frac{x}{t}) f(t)^p v(t)^p h(x,t)^p\,dt\right)^{\frac{q}{p}} \\ &\times \left(\int_0^\infty \varphi(\frac{x}{t}) v(t)^{-p'} h(x,t)^{-p'}\,dt\right)^{\frac{q}{p'}} dx, \end{aligned} \tag{13}$$

where

$$h(x,t) = \left(\int_0^t \varphi(\frac{x}{s}) v(s)^{-p'} ds\right)^{\frac{1}{pp'}}.$$

The last integral on the right can be reduced, by substitution $y = h(x,t)^{pp'}$, to

$$\begin{aligned} \int_0^\infty \varphi(\frac{x}{t}) v(t)^{-p'} h(x,t)^{-p'}\,dt &= \int_{h(x,0)^{pp'}}^{h(x,\infty)^{pp'}} z^{-\frac{1}{p}} dz \\ &= p' \left(\int_0^\infty \varphi(\frac{x}{s}) v(s)^{-p'} ds\right)^{\frac{1}{p'}}. \end{aligned}$$

Applying the generalized Minkowski inequality to the right-hand side of (13), which is

$$\int_0^\infty \left(u(x)^p \int_0^\infty \varphi(\frac{x}{t}) f(t)^p v(t)^p h(x,t)^p \, dt \left(\int_0^\infty \varphi(\frac{x}{s}) v(s)^{-p'} ds \right)^{\frac{p}{p'p'}} \right)^{\frac{q}{p}} dx$$

times $(p')^{q/p'}$, we get

$$\left[\int_0^\infty f(t)^p v(t)^p \left(\int_0^\infty u(x)^q \varphi(\frac{x}{t})^{\frac{q}{p}} h(x,t)^q h(x,\infty)^{\frac{qp}{p'}} dx \right)^{\frac{p}{q}} dt \right]^{\frac{q}{p}}$$

times the same constant. By (10), it is dominated by the right-hand side of (9) in the power q, as desired.

Conversely, assume that (9) holds. This means that it is true also in the form

$$\left(\int_0^\infty u(x)^q \left| \int_0^t \varphi(\frac{x}{s}) f(s) \, ds \right|^q dx \right)^{\frac{1}{q}} \ll \left(\int_0^t f(s)^p v(s)^p ds \right)^{\frac{1}{p}}$$

for each t. For $f(s) = v(s)^{-p'}$, we reduce it to

$$\left(\int_0^\infty u(x)^q \varphi(\frac{x}{t})^{\frac{q}{p}} \left| \int_0^t \varphi(\frac{x}{s}) v(s)^{-p'} \, dt \right|^q \left(\varphi(\frac{x}{t}) \int_0^t v(s)^{-p'} ds \right)^{-\frac{q}{p}} dx \right)^{\frac{1}{q}} \ll 1.$$

Applying now (11), we arrive at (12). The proof is complete.

Remark 1 As mentioned, taking $\varphi(s) = \chi_{(1,\infty)}(s)$, we reduce (12) to (4). Condition (11) is satisfied in this case automatically. However, (10) also follows from (4) in this case. This is just the second part of the proof of the sufficiency in [1]. Indeed, for the chosen φ, condition (10) is

$$\int_t^\infty u(x)^q \left(\int_0^t v(s)^{-p'} ds \right)^{\frac{q}{pp'}} \left(\int_0^x v(s)^{-p'} ds \right)^{\frac{q}{p'p'}} dx \ll 1.$$

Applying (4), we estimate the left-hand side by

$$\left(\int_0^t v(s)^{-p'} ds \right)^{\frac{q}{pp'}} \int_t^\infty u(x)^q \left(\int_x^\infty u(s)^q ds \right)^{-\frac{1}{p'}} dx.$$

After integration in x, it becomes

$$\left(\int_0^t v(s)^{-p'} ds \right)^{\frac{q}{pp'}} p \left(\int_t^\infty u(s)^q ds \right)^{-\frac{1}{p}}.$$

Assuming this uniformly bounded gives exactly (4).

Counterparts of the obtained results are related to defining $h(x,t)$ by integrating over (t,∞).

Theorem 2 *Let $1 < p \le q < \infty$, $f \in L^p_v$, $\varphi \in L^\infty$, and $\varphi, f \ge 0$. We have (9) provided*

$$\int_0^\infty u(x)^q \varphi(\frac{x}{t})^{\frac{q}{p}} \left(\int_t^\infty \varphi(\frac{x}{s}) v(s)^{-p'} ds\right)^{\frac{q}{pp'}} \times \left(\int_0^\infty \varphi(\frac{x}{s}) v(s)^{-p'} ds\right)^{\frac{q}{p'p'}} dx \ll 1. \tag{14}$$

If

$$\varphi(\frac{x}{t}) \int_t^\infty v(s)^{-p'} ds \ll \int_t^\infty \varphi(\frac{x}{s}) v(s)^{-p'} ds, \tag{15}$$

then

$$\int_0^\infty u(x)^q \varphi(\frac{x}{t})^{\frac{q}{p}} \left(\int_t^\infty \varphi(\frac{x}{s}) v(s)^{-p'} ds\right)^{\frac{q}{p'}} dx \ll 1 \tag{16}$$

is necessary for (9) to hold.

Remark 2 Similarly to the above, taking $\varphi(s) = \chi_{(0,1)}(s)$, we reduce (16) to (5). Condition (15) is satisfied in this case automatically. However, (14) also follows from (5) in this case.

Remark 3 Condition (10) can be made necessary if we assume

$$\varphi(\tfrac{x}{t}) \int_0^t v(s)^{-p'} ds \ll \left(\int_0^t \varphi(\tfrac{x}{s}) v(s)^{-p'} ds\right)^{p(1-\frac{1}{pp'})} \left(\int_0^\infty \varphi(\tfrac{x}{s}) v(s)^{-p'} ds\right)^{-\frac{p}{p'p'}}$$

instead of (11). However, the class of Hausdorff operators for which this is true does not seem to be rich enough, since it does not include the classical Hardy operator. Similar can be said on (14).

3 Adjoint Operators

Let us figure out which operator is adjoint of (8). We shall formally use the definition of the adjoint T^* of T in the form

$$\int_0^\infty g(x) Tf(x)\, dx = \int_0^\infty f(t) T^* g(t)\, dt.$$

In our notation,

$$\int_0^\infty g(x) H_{\varphi(\cdot)} f(x)\,dx = \int_0^\infty f(t) \int_0^\infty g(x)\varphi\Big(\frac{x}{t}\Big)\,dx\,dt$$
$$= \int_0^\infty f(t)\, H_{\varphi(\frac{1}{\cdot})} g(t)\,dt,$$

and

$$H^* = (H_{\varphi(\cdot)})^* = H_{\varphi(\frac{1}{\cdot})}.$$

Similarly, the adjoint of (6) (or, equivalently, (7)) can be established. It is of the form

$$\mathcal{H}_\varphi^* = \mathcal{H}_\Phi, \qquad \Phi(u) = \frac{1}{u}\varphi\Big(\frac{1}{u}\Big).$$

We can now show that Theorem 2 is just Theorem 1 applied to the adjoint operator, with the same conditions but for the special weights. More precisely, the following holds.

Theorem 3 *Let* $1 < p \le q < \infty$, $f \in L_v^p$, $\varphi \in L^\infty$, *and* $\varphi, f \ge 0$. *Let* $U(t) = u(\frac{1}{t})t^{-\frac{2}{q}}$ *and* $V(t) = v(\frac{1}{t})t^{\frac{2}{p'}}$. *We have*

$$\left(\int_0^\infty u(x)^q \left|\int_0^\infty \varphi(\frac{t}{x}) f(t)\,dt\right|^q dx\right)^{\frac{1}{q}} \ll \left(\int_0^\infty f(t)^p v(t)^p dt\right)^{\frac{1}{p}} \tag{17}$$

provided

$$\int_0^\infty U(x)^q \varphi(\frac{x}{r})^{\frac{q}{p}} \left(\int_0^r \varphi(\frac{x}{s}) V(s)^{-p'} ds\right)^{\frac{q}{pp'}}$$
$$\times \left(\int_0^\infty \varphi(\frac{x}{s}) V(s)^{-p'} ds\right)^{\frac{q}{p'p'}} dx \ll 1. \tag{18}$$

If

$$\varphi(\frac{r}{x}) \int_0^r V(s)^{-p'} ds \ll \int_0^r \varphi(\frac{x}{s}) V(s)^{-p'} ds, \tag{19}$$

then

$$\int_0^\infty U(x)^q \varphi(\frac{x}{r})^{\frac{q}{p}} \left(\int_0^r \varphi(\frac{x}{s}) V(s)^{-p'} ds\right)^{\frac{q}{p'}} dx \ll 1 \tag{20}$$

is necessary for (17) to hold.

Proof We simply apply Theorem 2 to (17). Substitutions $x \to \frac{1}{x}$, $s \to \frac{1}{s}$ and $\frac{1}{t} = r$ along with the use of appropriate notations readily imply Theorem 3.

4 Reformulations for $\mathcal{H}$

The obtained results can easily be reformulated for $\mathcal{H}$. The idea is very simple: Theorems 1 and 2 are applied to the function $\frac{f(t)}{t}$ rather than to $f(t)$, with the weight $tv(t)$ in place of $v(t)$ in order to stay within the condition $f \in L_v^p$. The change of the weight is seen in the conditions.

Theorem 4 *Let $1 < p \le q < \infty$, $f \in L_v^p$, $\varphi \in L^\infty$, and $\varphi, f \ge 0$. We have*

$$\Big(\int_0^\infty u(x)^q \Big|\int_0^\infty \varphi(\frac{x}{t})\frac{1}{t} f(t)\,dt\Big|^q dx\Big)^{\frac{1}{q}} \ll \Big(\int_0^\infty f(t)^p v(t)^p dt\Big)^{\frac{1}{p}} \tag{21}$$

provided

$$\int_0^\infty u(x)^q \varphi(\frac{x}{t})^{\frac{q}{p}} \Big(\int_0^t \varphi(\frac{x}{s})[sv(s)]^{-p'} ds\Big)^{\frac{q}{pp'}} \times \Big(\int_0^\infty \varphi(\frac{x}{s})[sv(s)]^{-p'} ds\Big)^{\frac{q}{p'p'}} dx \ll 1. \tag{22}$$

If

$$\varphi(\frac{x}{t}) \int_0^t [sv(s)]^{-p'} ds \ll \int_0^t \varphi(\frac{x}{s})[sv(s)]^{-p'} ds, \tag{23}$$

then

$$\int_0^\infty u(x)^q \varphi(\frac{x}{t})^{\frac{q}{p}} \Big(\int_0^t \varphi(\frac{x}{s})[sv(s)]^{-p'} ds\Big)^{\frac{q}{p'}} dx \ll 1 \tag{24}$$

is necessary for (21) to hold.

Theorem 5 *Let $1 < p \le q < \infty$, $f \in L_v^p$, $\varphi \in L^\infty$, and $\varphi, f \ge 0$. We have (21) provided*

$$\int_0^\infty u(x)^q \varphi(\frac{x}{t})^{\frac{q}{p}} \left(\int_t^\infty \varphi(\frac{x}{s})[sv(s)]^{-p'} ds\right)^{\frac{q}{pp'}}$$
$$\times \left(\int_0^\infty \varphi(\frac{x}{s})[sv(s)]^{-p'} ds\right)^{\frac{q}{p'p'}} dx \ll 1. \quad (25)$$

If

$$\varphi(\frac{x}{t}) \int_t^\infty [sv(s)]^{-p'} ds \ll \int_t^\infty \varphi(\frac{x}{s})[sv(s)]^{-p'} ds, \quad (26)$$

then

$$\int_0^\infty u(x)^q \varphi(\frac{x}{t})^{\frac{q}{p}} \left(\int_t^\infty \varphi(\frac{x}{s})[sv(s)]^{-p'} ds\right)^{\frac{q}{p'}} dx \ll 1 \quad (27)$$

is necessary for (21) to hold.

5 More General Hausdorff Operators

We consider a somewhat more advanced one-dimensional version of the Hausdorff operator, apparently first introduced in [3] (in a slightly different form and on the whole $\mathbb{R}$; see also [4]):

$$(\mathcal{H}f)(x) = (\mathcal{H}_{\varphi,g}f)(x) = \int_{\mathbb{R}_+} \varphi(t)g(t)f(g(t)x)\,dt\,. \quad (28)$$

If g satisfies certain natural assumptions (those that $\frac{1}{t}$ definitely satisfies), the corresponding results follow from the assertions of the previous section by substitution and change of the averaging function. Indeed, we shall assume that $g : \mathbb{R}_+ \to \mathbb{R}_+$ is monotone decreasing and differentiable except maybe at the point $t = 0$. Substituting $xg(t) = u$ in (28), we arrive at

$$(\mathcal{H}_{\varphi,g}f)(x) = \int_{\mathbb{R}_+} \varphi\Big(g^{-1}(\frac{u}{x})\Big)\frac{u}{x}\frac{1}{xg'(\frac{u}{x})} f(u)\,du$$
$$= \int_{\mathbb{R}_+} \varphi\Big(g^{-1}(\frac{u}{x})\Big)\frac{u^2}{x^2g'(\frac{u}{x})}\frac{f(u)}{u}\,du.$$

Denoting

$$\Phi(t) = \varphi\Big(g^{-1}(\frac{1}{t})\Big)\frac{1}{t^2g'(\frac{1}{t})},$$

we obtain

$$(\mathcal{H}_{\varphi,g}f)(x) = (\mathcal{H}_{\Phi}f)(x),$$

and we are ready to adjust Theorems 4 and 5. The above assumptions on g are assumed without mentioning them in the formulations.

Theorem 6 *Let $1 < p \le q < \infty$, $f \in L^p_v$, $\varphi \in L^\infty$, and $\varphi, f \ge 0$. We have*

$$\left(\int_0^\infty u(x)^q \left|\int_0^\infty \varphi(t)g(t)f(xg(t))\,dt\right|^q dx\right)^{\frac{1}{q}} \ll \left(\int_0^\infty f(t)^p v(t)^p dt\right)^{\frac{1}{p}} \tag{29}$$

provided

$$\int_0^\infty u(x)^q \Phi(\frac{x}{t})^{\frac{q}{p}} \left(\int_0^t \Phi(\frac{x}{s})[sv(s)]^{-p'} ds\right)^{\frac{q}{pp'}} \times \left(\int_0^\infty \Phi(\frac{x}{s})[sv(s)]^{-p'} ds\right)^{\frac{q}{p'p'}} dx \ll 1. \tag{30}$$

If

$$\Phi(\frac{x}{t}) \int_0^t [sv(s)]^{-p'} ds \ll \int_0^t \Phi(\frac{x}{s})[sv(s)]^{-p'} ds, \tag{31}$$

then

$$\int_0^\infty u(x)^q \Phi(\frac{x}{t})^{\frac{q}{p}} \left(\int_0^t \Phi(\frac{x}{s})[sv(s)]^{-p'} ds\right)^{\frac{q}{p'}} dx \ll 1 \tag{32}$$

is necessary for (29) to hold.

Theorem 7 *Let $1 < p \le q < \infty$, $f \in L^p_v$, $\varphi \in L^\infty$, and $\varphi, f \ge 0$. We have (29) provided*

$$\int_0^\infty u(x)^q \Phi(\frac{x}{t})^{\frac{q}{p}} \left(\int_t^\infty \varphi(\frac{x}{s})[sv(s)]^{-p'} ds\right)^{\frac{q}{pp'}} \times \left(\int_0^\infty \Phi(\frac{x}{s})[sv(s)]^{-p'} ds\right)^{\frac{q}{p'p'}} dx \ll 1. \tag{33}$$

If

$$\Phi(\frac{x}{t}) \int_t^\infty [sv(s)]^{-p'} ds \ll \int_t^\infty \Phi(\frac{x}{s})[sv(s)]^{-p'} ds, \tag{34}$$

then

$$\int_0^\infty u(x)^q \Phi(\frac{x}{t})^{\frac{q}{p}} \left(\int_t^\infty \Phi(\frac{x}{s})[sv(s)]^{-p'} ds\right)^{\frac{q}{p'}} dx \ll 1 \tag{35}$$

is necessary for (29) to hold.

Acknowledgements The author thanks M. Goldman, L. Roncal, V. Stepanov and S. Tikhonov for stimulating discussions. The author is also indebted to the anonymous referees for thorough reading of the manuscript and valuable remarks.

References

1. Bradley, J.S.: Hardy inequalities with mixed norms. Canad. Math. Bull. **21**, 405–408 (1978)
2. Gogatishvili, A., Stepanov, V.D.: Reduction theorems for weighted integral inequalities on the cone of monotone functions. Uspekhi Math. Nauk **68**, 3–68 (Russian). English translation Russian Math. Surv. **68**, 597–664 (2013)
3. Kuang, J.C.: Hausdorff, Generalized, operators on weighted Morrey-Herz spaces. Acta Math. Sinica (Chin. Ser.) **55**, 895–902. Chinese; Chinese, English summaries (2012)
4. Kuang, J.C.: Generalized Hausdorff operators on weighted Herz spaces. Math. Vesnik **66**, 19–32 (2014)
5. Kufner, A., Persson, L.-E.: Weighted Inequalities of Hardy Type. World Scientific Publishing Co. Inc, River Edge (2003)
6. Kufner, A., Maligranda, L., Persson, L.-E.: The Hardy Inequality. About Its History and Some Related Results. Vydavatelský Servis, Plzen (2007)
7. Liflyand, E.: Hausdorff operators on Hardy spaces. Eurasian Math. J. **4**(4), 101–141 (2013)
8. Liflyand, E., Móricz, F.: The Hausdorff operator is bounded on the real Hardy space $H^1(\mathbb{R})$. Proc. Am. Math. Soc. **128**, 1391–1396 (2000)
9. Opic, B., Kufner, A.: Hardy-Type Inequalities. Longman (1990)
10. Xiao, J.: L^p and BMO bounds of weighted Hardy-Littlewood averages. J. Math. Anal. Appl. **262**, 660–666 (2001)

Harmonic Analysis and Hypercomplex Function Theory in Co-dimension One

Helmuth R. Malonek, Isabel Cação, M. Irene Falcão and Graça Tomaz

Abstract Fundamentals of a function theory in co-dimension one for Clifford algebra valued functions over $\mathbb{R}^{n+1}$ are considered. Special attention is given to their origins in analytic properties of holomorphic functions of one and, by some duality reasons, also of several complex variables. Due to algebraic peculiarities caused by non-commutativity of the Clifford product, generalized holomorphic functions are characterized by two different but equivalent properties: on one side by local *derivability* (existence of a well defined derivative related to co-dimension one) and on the other side by *differentiability* (existence of a local approximation by linear mappings related to dimension one). As important applications, sequences of harmonic Appell polynomials are considered whose definition and explicit analytic representations rely essentially on both dual approaches.

Keywords Clifford algebras · Hypercomplex differential forms · Hypercomplex derivative · Hypercomplex Appell polynomials

1 Introduction

In 1968, exactly 50 years ago, E. M. Stein and G. Weiss proved in their seminal paper [1], the "*correspondence of irreducible representations of several rotation groups to first order constant coefficient partial differential equations generalizing the Cauchy-Riemann equations.*" They showed how certain properties of complex

H. R. Malonek (✉) · I. Cação
CIDMA and Department of Mathematics, University of Aveiro, 3810-193 Aveiro, Portugal
e-mail: hrmalon@ua.pt

M. I. Falcão
CMAT and Department of Mathematics and Applications, University of Minho,
4710-057 Braga, Portugal

G. Tomaz
CIDMA and Research Unit for Inland Development, Department of Mathematics,
Polytechnic of Guarda, 6300-659 Guarda, Portugal

A. Karapetyants et al. (eds.), *Modern Methods in Operator Theory and Harmonic Analysis*, Springer Proceedings in Mathematics & Statistics 291,
https://doi.org/10.1007/978-3-030-26748-3_7

one-dimensional function theory extend to solutions of those systems, in particular the fact of being harmonic solutions. In their list of systems one can find a generalized Riesz system [2], the Moisil-Theodoresco system [3], spinor systems as n-dimensional generalization of Dirac's equations [4], Hodge–de Rham equations [5] and special cases of them. But the aim of proving that correspondence between representation groups and partial differential equations were merely of qualitative nature and deeply connected with properties of harmonic functions in several real variables. But whereas that paper uses purely real methods, two years later R. Delanghe published *On Regular-analytic Functions with Values in a Clifford Algebra* [6], thereby extending Fueters results [7] about solutions of the generalized Riesz system from the quaternionic to the general Clifford Algebra case.

Nowadays, function theoretic methods over non-commutative algebras are applied in a big variety of mathematical fields, like for example in potential theory, differential geometry, operator theory, BVP of partial differential equations, analytic number theory, discrete and computational mathematics and their corresponding applications in Sciences and Engineering. For a first introduction, particularly dealing with Maxwell's and Schrödinger's equations the reader should consult Kravchenko [8]. An account about the development of this field until 2000 is given by Delanghe in [9].

But it seems to be remarkable that the first attempt for a systematical development of methods based on Clifford algebras started already 40 years before Delanghe's article [6] and was done by the Swiss mathematician Rudolf Fueter (1880–1950). He initiated around 1930 the foundation of a theory of quaternion valued functions of a quaternion variable [10]. Being already a worldwide known number theorist, he was interested in such a tool for the development of new analytical methods in Number Theory (c.f. [11]). But soon he recognized that he had found an new approach to generalize the classical complex function theory in a way different from that of holomorphic functions of several complex variables. It was exactly 70 years ago that in the Fall Semester of 1948/49, Fueter gave one of his last lessons on *Funktionentheory im Hypercomplexen* (Function theory in the hypercomplex) [12]. Since this manuscript of the lessons, still approved by Fueter, were never published as book, it is not well known that it contains as Chapter IV: *Funktionentheorie der Clifford'schen Algebren* on 70 pages (pp. 247–317) a complete description of an approach to function theory in Clifford algebras based on generalized Cauchy integral theorems. As far as we know, only in the book [13] one can find a remark that R. Fueter, who is commonly considered only as one of the founders of Quaternionic Analysis (cf. [14]), made already contributions to what was 30 years later coined *Clifford Analysis* [15, 16].

During 50 years, having been guided by the paper [1], the research on qualitative properties of generalized Cauchy-Riemann or Dirac systems continues to rely heavily on methods of representation theory and its relationship to Harmonic Analysis (see [4, 9, 15, 16]). It legitimates to think about Clifford Analysis as a refinement of Harmonic Analysis. Indeed, in this sense it is a powerful tool for applying those methods to the study of problems in PDE, in particular what concerns their symmetries and the algebraic structures behind (Heisenberg algebras, Lie algebras etc.).

But the aim of this paper is to call attention to a complementary treatment of higher-dimensional generalized Cauchy-Riemann systems, namely to function theoretic methods adapted to the approximation and numerical calculation of its solutions, briefly called *monogenic functions*. The approach we will present is closely connected with the hypercomplex generalization of the Wirtinger calculus (which is the one complex variable case of the Dolbeault calculus in the theory of several complex variables). It is not only the basis for understanding the generalization of the Cauchy approach to holomorphic functions [17] but also clarifies the usefulness of *several hypercomplex variables* and corresponding differential forms [18–20]. The fact that the conjugate hypercomplex Cauchy-Riemann operator acts as derivation operator like in the complex case (cf. [21]) confirmed the observation of S. Semmes (1996), that *Clifford Analysis can be considered as co-dimension one function theory* [22]. Moreover, this non-standard approach opened the way to investigate new analytic, geometric and combinatorial properties of monogenic functions. Even new number theoretic results could be obtained, thereby in some sense coming back to Fueter's motivation for his work [11].

The present paper is a revised and substantially abridged version of [20] including recent results on topics for which the coexistence of both concepts, i.e. hypercomplex derivability as well as hypercomplex differentiability, was essential. After a short introduction to Clifford algebras in Sects. 2 and 3, a calculus of alternative hypercomplex differential forms is considered in Sect. 4. As simple consequences the generalized Stokes' formula as well as a generalized Cauchy integral formula are derived, stressing the relation valid for holomorphic functions in the complex plane, too. At the end of this section the expression of the generalized Cauchy-Riemann operator as areolar derivative in the sense of Pompeiu [23] is given. Naturally, the next Sect. 5 introduces the concept of hypercomplex derivability again based on alternative differential forms. The analogous expression of the hypercomplex derivative as another areolar derivative in the sense of Pompeiu finishes this section. Section 6 refers to two essential theorems proved in [17] resp. [21] which together establish the complete equivalence of the Cauchy and the Riemann approaches for monogenic functions. As direct consequence of differentiability, Sect. 7 introduces the local expansion of those hypercomplex functions in power series of several hypercomplex variables almost analogous to the case of multivariate real series expansion. The last Sect. 8 is dedicated to a first resume of recent results in the theory of generalized sequences of Appell polynomials [24] being a type of application of monogenic functions where all their fundamental properties mentioned before come together.

2 Clifford Algebras

A finite-dimensional algebra with a unit element over the field of real or complex numbers was formerly known as a hypercomplex system. Clifford algebras are examples of them. They possess an isomorphic representation as elements of a $(2^n \times 2^n)$

matrix algebra. The matrix representations of complex numbers and quaternions are well known special cases.

Clifford algebras as associative non-commutative algebras over the field of real numbers can be defined as follows (for more details and other possibilities see, for instance, [20]).

Definition 1 Let $\{e_0, e_1, \ldots, e_n\}$ be a ONB of $\mathbb{R}^{n+1}$ subject to the multiplication rules

$$e_k e_l + e_l e_k = -2\delta_{kl} e_0, \quad (k, l = 1, \ldots, n),$$

where δ_{ij} is the Kronecker symbol. The associative 2^n-dimensional universal Clifford algebra $\mathcal{C}\ell_{0,n}$ over $\mathbb{R}$ with unity is the set of numbers $\alpha \in \mathcal{C}\ell_{0,n}$ written in the form

$$\alpha = \sum_A \alpha_A e_A,$$

with the basis $\{e_A : A \subseteq \{1, \ldots, n\}\}$ formed by

$$e_A = e_{h_1} e_{h_2} \ldots e_{h_r}, 1 \leq h_1 < \ldots < h_r \leq n, \quad e_\emptyset = e_0 = 1,$$

and where the components α_A are real numbers. The conjugate element to α is defined by $\bar{\alpha} = \sum_A \alpha_A \bar{e}_A$, where

$$\bar{e}_A = \bar{e}_{h_r} \bar{e}_{h_{r-1}} \ldots \bar{e}_{h_1}; \bar{e}_k = -e_k \ (k = 1, \ldots, n), \ \bar{e}_0 = e_0 = 1.$$

The norm of $\alpha \in \mathcal{C}\ell_{0,n}$ is defined by $|\alpha| = (\sum_A \alpha\bar{\alpha})^{1/2} = (\sum_A \alpha_A^2)^{1/2}$.

3 Hypercomplex Structures of $\mathbb{R}^{n+1}$

The theory of complex functions of several complex variables uses for the description of the pre-image set the algebra of complex numbers by combining $2n$ real variables $(x_1, \ldots, x_n, y_1, \ldots, y_n) \in \mathbb{R}^{2n}$ to a vector of n complex variables $(z_1, \ldots, z_n) \in \mathbb{C}^n$; $z_k = x_k + iy_k, \quad k = 1, \ldots n$. Then it is possible to realize the inverse variable transformation from the complex to the real variables with the help of the vector of their conjugates $(\bar{z}_1, \ldots, \bar{z}_n)$ in the form

$$x_k = \frac{1}{2}(\bar{z}_k + z_k) \quad \text{and} \quad y_k = \frac{i}{2}(\bar{z}_k - z_k), \quad k = 1, \ldots n.$$

In such a way $\mathbb{R}^{2n}$ is endowed with a complex structure by identifying $\mathbb{R}^{2n}$ with $\mathbb{C}^n$, symbolically $\mathbb{R}^{2n} \cong \mathbb{C}^n$. Like in the ordinary case of two real and one complex variables ($\mathbb{R}^2 \cong \mathbb{C}^1$) this leads to the reduction of the real dimension $2n$ to the half, i.e., to n, compared to $\mathbb{C}^n$. In particular, the local property of complex differentiability of a function $f : \Omega \subset \mathbb{C}^n \rightarrow \mathbb{C}$ implies that f may be considered as a function of $z_k, \ k = 1, \ldots n$ and not of $\bar{z}_k, \ k = 1, \ldots n$. This is due to the fact that it can locally be approximated by a linear mapping (the differential) of the vector $(dz_1, \ldots dz_n)$. One important consequence is the representation of $f : \Omega \subset \mathbb{C}^n \rightarrow \mathbb{C}$ by multiple power series in $(z_1, \ldots, z_n)$ (Weierstrass approach).

If we are looking for a hypercomplex structure of $\mathbb{R}^{n+1}$ in the aforementioned sense, then we can recognize several possibilities for similar procedures.

I. Paravectors. Let the element $x = (x_0, x_1, \ldots, x_n)$ of $\mathbb{R}^{n+1}$ be identified with

$$z = x_0 + x_1 e_1 + \cdots + x_n e_n \in \mathcal{A}_n := \text{span}_{\mathbb{R}}\{1, e_1, \ldots, e_n\}$$

and called *paravector*. The conjugate of z is given by

$$\bar{z} = x_0 - x_1 e_1 - \cdots - x_n e_n$$

and its norm is

$$|z| := \sqrt{z\bar{z}} = (\sum_{i=1}^{n} x_i^2)^{1/2},$$

analogously to the complex case. Therefore each non-zero $z \in \mathcal{A}_n$ is invertible and its inverse is $z^{-1} = \frac{\bar{z}}{|z|^2}$.

The usual approach to hypercomplex function theory considers $\mathcal{C}\ell_{0,n}$-valued functions of the form $f(z) = \sum_A f_A(z) e_A, \ f_A(z) \in \mathbb{R}$, as mappings

$$f : \Omega \subset \mathbb{R}^{n+1} \cong \mathcal{A}_n \longmapsto \mathcal{C}\ell_{0,n}.$$

Of course, the big advantage of this approach is to deal with only one hypercomplex variable z. Compared with the ordinary case of two real and one complex variables ($\mathbb{R}^2 \cong \mathbb{C}^1$) it reflects the reduction of the real dimension $(n + 1)$ *to dimension 1*, i.e., by n, compared to $\mathbb{R}^{n+1}$.

II. Several hypercomplex variables. A second hypercomplex structure of $\mathbb{R}^{n+1}$ different from that given by $\mathcal{A}_n$ consists in the following isomorphism:

$$\mathbb{R}^{n+1} \cong \mathcal{H}^n = \{\mathbf{z} = (z_1, \ldots, z_n) : z_k = x_k - x_0 e_k; x_0, x_k \in \mathbb{R}, \ k = 1, \ldots, n\}.$$

More detailed, this means to take n copies $\mathbb{C}_k$ of $\mathbb{C}$ identifying $i \cong e_k, \ (k = 1, \ldots, n)$; $x_0 \cong \Re z$; $x_k \cong \Im z$; where $z \in \mathbb{C}$, and let $\mathbb{C}_k := -e_k \mathbb{C}$. Then $\mathcal{H}^n$ is the cartesian product $\mathcal{H}^n := \mathbb{C}_1 \times \cdots \times \mathbb{C}_n$ and $\mathcal{C}\ell_{0,n}$-valued functions $f(z) = \sum_A f_A(z) e_A$ are considered as mappings

$$f : \Omega \subset \mathbb{R}^{n+1} \cong \mathcal{H}^n \longmapsto \mathcal{C}\ell_{0,n}.$$

Remark 1 In the following sections we will see that simple algebraic relations allow to consider both approaches as in some sense dual approaches. The qualitatively new point of view is the (not only formal) relationship with $\mathbb{C}^n$. Later we will see that one of the most important facts for the usefulness of hypercomplex function theory is its independence from the parity of the underlying real space and its direct relationship to harmonic function theory.

4 Calculus of Alternative Hypercomplex Differential Forms

In the usual Riemann approach to monogenic functions the hypercomplex form of Stokes' theorem plays a central role. This is due to the fact that for monogenic functions Stokes' theorem can be interpreted as a generalization of Cauchy's integral theorem for holomorphic functions in the plane. We will see that the consideration of $\mathcal{H}^n$ permits a better insight into the nature of the hypercomplex form of Stokes' theorem, too. Indeed, the hypercomplex form of Stokes' theorem can be derived in a very natural way by using $(n+1)$ hypercomplex differential forms defined by:

$$dz_0 = dx_0, \quad dz_k = -e_k dx_0 + dx_k; \quad k = 1, \ldots, n, \tag{1}$$

i.e., in terms of basic real differential 1-forms. In this way we can see that the $\mathcal{H}^n$-approach reflects some formal duality to the case of several complex variables. It is the key to the notion of hypercomplex derivative and also reveals the role of the conjugated Cauchy-Riemann operator as derivation operator. Therefore we define the outer product of the basic differential forms (1) in the following way:

Definition 2 The product of two basic hypercomplex 1-forms is defined by

$$\begin{aligned} dz_k \wedge dz_l &= (-e_k dx_0 + dx_k) \wedge (-e_l dx_0 + dx_l) \\ &= dx_k \wedge dx_l - e_k dx_0 \wedge dx_l - e_l dx_k \wedge dx_0, \\ dz_k \wedge dz_0 &= (-e_k dx_0 + dx_k) \wedge dx_0 = dx_k \wedge dx_0; \quad k, l = 1, \ldots, n. \end{aligned}$$

A basic hypercomplex p-form is the outer product of p different basic hypercomplex 1-forms.

The general form of a basic hypercomplex p-form can easily be written with the help of a multi-index $\nu = (\nu_0, \nu_1, \ldots, \nu_{p-1})$, $0 \le \nu_0 < \nu_1 < \cdots < \nu_{p-1} \le n$ and relies on the fact (property 1 below) that the set of basic hypercomplex 1-forms is an alternative outer algebra (exactly as in the case of real differential forms). Therefore every basic p-form $\omega_p \in \bigwedge_{\mathcal{A}_n}^p$ can be written in a unique way as

$$\omega_p = dz_{\nu_0} \wedge dz_{\nu_1} \wedge \ldots \wedge dz_{\nu_{p-1}} =: dz_\nu.$$

The verification of the following properties is straightforward.

1. The outer product is alternative, i.e.,

$$dz_k \wedge dz_l = -dz_l \wedge dz_k, \quad k \neq l, \quad dz_k \wedge dz_k = 0.$$

2. If the notation $d\hat{x}_m \ (m = 0, \ldots, n)$ means that in the ordered outer product of the 1-forms $dx_k \ (k = 0, \ldots, n)$ the factor dx_m is absent, then the hypercomplex surface element in $\mathbb{R}^{n+1}$ is given by $d\sigma_{(n)} = d\hat{x}_0 - e_1 d\hat{x}_1 + \cdots + (-1)^n e_n d\hat{x}_n$ (cf. [16]). In terms of (1) there holds

$$d\sigma_{(n)} = dz_1 \wedge \cdots \wedge dz_n. \tag{2}$$

3. The volume element dV in $\mathbb{R}^{n+1}$ has in terms of (1) the simple expression

$$dV = dz_0 \wedge d\sigma_{(n)} = (-1)^n d\sigma_{(n)} \wedge dz_0 = dz_0 \wedge dz_1 \wedge \cdots \wedge dz_n. \tag{3}$$

4. With the usual notation $\partial_k = \frac{\partial}{\partial x_k}$, let $df = \partial_0 f dx_0 + \partial_1 f dx_1 \cdots + \partial_n f dx_n$ be the differential of $f \in C^1(\mathbb{R}^{n+1}, \Omega)$. By using (1), the right (resp. left) hypercomplex form of the differential of f has the form:

$$\begin{aligned} df &= (fD)dz_0 + \partial_1 f dz_1 + \cdots + \partial_n f dz_n \ \text{ resp.} \\ df &= dz_0(Df) + dz_1 \partial_1 f + \cdots + dz_n \partial_n f, \end{aligned}$$

where

$$D = \partial_0 + e_1 \partial_1 + \cdots + e_n \partial_n,$$

is the so-called *generalized Cauchy-Riemann operator.*

If $\overline{D} = \partial_0 - e_1 \partial_1 - \cdots - e_n \partial_n$ denotes the conjugate operator to D then $D\overline{D} = \overline{D}D = \Delta$, factorizing the Laplace operator like in the complex case.

5. Consider the hypercomplex $(n-1)$–form

$$d\sigma_{(n-1)} := -e_1 d\hat{x}_{0,1} + e_2 d\hat{x}_{0,2} + \cdots + (-1)^n e_n d\hat{x}_{0,n},$$

where $d\hat{x}_{0,m} \ (m = 1, \ldots, n)$ stands for the ordered outer product of the 1-forms $dx_k \ (k = 0, \ldots, n)$ where the factors dx_0 and dx_m are absent. Then for $n = 1$ the constant 0–form $d\sigma_{(0)} = -e_1 \cong -i$ is obtained, i.e., exactly the factor that is necessary to pass from $z \in \mathbb{C}$ to $z_1 = -iz \in \mathbb{C}_1$. The designation of this $(n-1)$-form by $d\sigma_{(n-1)}$ is based in the following fact. The 1-form of the vector-part of dz, i.e.,

$$\text{Vec } dz = e_1 dx_1 + \cdots + e_n dx_n$$

can be considered as related to a hyperplane $x_0 = c = \text{const}$. Then $d\sigma_{(n-1)}$ is just the corresponding surface-element with respect to the effective variables $(x_1, \ldots, x_n)$.

Now let $\nu = (\nu_0, \nu_1, \ldots, \nu_{p-1}), \ 0 \leq \nu_0 < \nu_1 < \cdots < \nu_{p-1} \leq n$, be a multi-index, and for a given set $f_\nu = f_\nu(z)$ of $\binom{n+1}{p}$ $C\ell_{0,n}$-valued continuously differentiable functions in Ω, consider the forms $\omega_p = \sum_\nu dz_\nu f_\nu(z)$, resp. $\omega_p = \sum_\nu f_\nu(z) dz_\nu$. In the case of a left form we have:

Definition 3 Let $\omega_p = \sum_\nu dz_\nu f_\nu(z)$ be a continuously real differentiable left p–form on Ω. Then its (outer) derivative $d\omega_p$ is defined as the $(p+1)$–form

$$d\omega_p = \sum_\nu (-1)^p dz_\nu \wedge df_\nu(z),$$

where df_ν is the differential of f_ν, i.e., the outer derivative of the 0–form f_ν.

For right p–forms or two-sided p–forms ω_p (i.e., having $C\ell_{0,n}$-valued coefficients on both sides) the definition of $d\omega_p$ is straightforward.

The use of alternative hypercomplex differential forms permits a very lucid deduction of the hypercomplex form of Stokes' integral theorem as well as the generalization of Cauchy's theorem. Therefore let $\omega = f(z)d\sigma_{(n)}$ be a hypercomplex differential n-form built in analogy to the real case as a product of the surface element $d\sigma_{(n)}$ in $\mathbb{R}^{n+1}$, i.e. (2), with a function $f \in C^1(\mathbb{R}^{n+1}, \Omega)$. The general Stokes' theorem in the form

$$\int_{\partial\Omega} \omega = \int_\Omega d\omega$$

applied to a $(n+1)$-dimensional positively oriented domain Ω implies together with the Definition 3 of the outer derivative and with (3) that

$$\int_{\partial\Omega} d\sigma_{(n)} f(z) = \int_\Omega (-1)^n d\sigma_{(n)} \wedge df = \int_\Omega (-1)^n d\sigma_{(n)} \wedge dz_0 Df = \int_\Omega Df dV. \quad (4)$$

The following theorem shows that the generalized Cauchy-Riemann operator D can be characterized as an areolar derivative in the sense of Pompeiu. The concept of *areolar derivative in the sense of Pompeiu* originated from [23] and has been discussed in the hypercomplex context in [13, 20, 25].

Theorem 1 *Let $\mathbf{z}_\star$ be a fixed point in a positively oriented differentiable and contractible domain $\Omega \subset \mathcal{H}^n$ with smooth boundary. Consider a regular sequence of subdomains $\{\Omega_n\}$ which is shrinking to $\mathbf{z}_\star$ if n tends to infinity and whereby $\mathbf{z}_\star$ belongs to all Ω_n. For a real differentiable function f defined in Ω holds*

$$(fD)(\mathbf{z}_\star) = \lim_{n\to\infty} \frac{1}{\text{mes } \Omega_n} \int_{\partial\Omega_n} f(z) dz_1 \wedge \cdots \wedge dz_n$$

i.e. the (right) generalized Cauchy-Riemann operator D is a (right) generalized areolar derivative in the sense of Pompeiu of $f = f(\mathbf{z})$ over Ω at $\mathbf{z}_\star$.

Remember that in our notation, using (3),

$$\text{mes } \Omega_n = \int_{\Omega_n} dV = \int_{\Omega_n} dz_0 \wedge dz_1 \wedge \cdots \wedge dz_n.$$

The proof of this theorem relies on the application of the hypercomplex form of Stokes' theorem and the mean value property (c.f. [20]).

Only one step is left for obtaining a generalized *Cauchy's integral formula*. Therefore we mention

Definition 4 Let $f = f(\mathbf{z})$ be continuously real differentiable in an open set $\Omega \subset \mathbb{R}^{n+1} \cong \mathcal{H}^n \cong \mathcal{A}_n$. Then f is called *left (resp. right) monogenic* in Ω, if and only if $Df = 0$ ($fD = 0$).

The components $f_A(\mathbf{z})$ of a monogenic function $f(\mathbf{z})$ are real harmonic functions not only pluri-harmonic like in the case of several complex variables ("refinement" of harmonic function theory).

The designation of left (resp. right) *monogenic functions* which we are also using has historically been introduced in [16]. For stressing more the complete coincidence with the situation in the complex case, recently in [13] and other papers f is called (Clifford) holomorphic.

Cauchy's integral formula follows immediately from (4): If f is (left) monogenic then $\omega = d\sigma_{(n)}f(\mathbf{z})$ is closed, i.e.

$$\int_{\partial\Omega} d\sigma_{(n)}f(\mathbf{z}) = 0.$$

Now we are able to introduce the already announced notion of hypercomplex derivability and the definition of the hypercomplex derivative (c.f. [21]).

5 Hypercomplex Derivability

The notion of left (L-) or right (R-) hypercomplex derivability relies on the intrinsic relations between two forms of degree n (more concretely on $d\sigma_{(n)}$ and $d(d\sigma_{(n-1)}f)$ or $d(fd\sigma_{(n-1)})$) generalizing an idea of [26].

Definition 5 A function $f : \mathcal{H}^n \mapsto \mathcal{C}\ell_{0,n}$ is L-(R-) derivable at $\mathbf{z} \in \mathcal{H}^n$ if it is real differentiable at $\mathbf{z}$ and there exists $A_{f,L}(\mathbf{z})$ $(A_{f,R}(\mathbf{z})) \in \mathcal{C}\ell_{0,n}$ such that

$$d(d\sigma_{(n-1)}f) = d\sigma_{(n)}A_{f,L}(\mathbf{z}) \ \text{ resp. } \ d(fd\sigma_{(n-1)}) = A_{f,R}(\mathbf{z})d\sigma_{(n)}.$$

$A_{f,L}(\mathbf{z})$ ($A_{f,R}(\mathbf{z})$) are called the left and right derivative of f at $\mathbf{z}$, respectively.

In [21] for the first time could be proven that the conjugate Cauchy-Riemann operator, more concretely $\frac{1}{2}\overline{D}$, acts like a hypercomplex derivation operator as one should expect in analogy to the complex case, since $A_{f,L}(\mathbf{z}) = \frac{1}{2}\overline{D}f$, resp. $A_{f,R}(\mathbf{z}) = \frac{1}{2}f\overline{D}$. Moreover, with this definition one has a generalization of the classical Cauchy approach which relies directly on a limit process like in the classical complex case. In analogy to Theorem 1 one has also its expression in form of an *areolar derivative in the sense of Pompeiu*. In the left (and analogously in the right) case we have

Theorem 2 *Let $\mathcal{S} \subset \Omega$ be an oriented differentiable n-dimensional hypersurface with boundary $\partial\mathcal{S}_m$ and $\mathbf{z}^*$ be a fixed point in $\mathcal{S}$. Consider a sequence of subdomains $\{\mathcal{S}_m\}$ which is shrinking to $\mathbf{z}^*$ if $m \to \infty$. Suppose now that the function f is left monogenic in Ω, i.e. $Df = 0$. Then the left hypercomplex derivative $\frac{1}{2}\overline{D}$ is an areolar derivative in the sense of Pompeiu of the form*

$$\frac{1}{2}\overline{D}f = \lim_{m\to\infty}\left[\int_{\mathcal{S}_m} d\sigma_{(n)}\right]^{-1}\int_{\partial\mathcal{S}_m}(d\sigma_{(n-1)}f).$$

Moreover, following [21], we have

Theorem 3 *Consider $f = f(\mathbf{z})$ with $\mathbf{z} \in \Omega \subset \mathcal{H}^n$. Then f is left (resp. right) monogenic in $\mathbf{z}$ iff f has in $\mathbf{z}$ a uniquely defined hypercomplex areolar derivative $f'_L = \frac{1}{2}\overline{D}f$; ($f'_R = \frac{1}{2}f\overline{D}$) in the sense of Pompeiu.*

For example, for the left case, $\frac{1}{2}\overline{D}f = \frac{1}{2}(\partial_0 f - e_1\partial_1 f - \cdots - e_n\partial_n f) = \partial_0 f$ due to the fact that $Df = \partial_0 f + e_1\partial_1 f + \cdots + e_n\partial_n f = 0$. Obviously, the hypercomplex derivative of a monogenic function f is a differential coefficient between two differential forms (cf. [26]), this time not of degree 1 as usual, but of degree n (which is, of course, the same in the simplest case $n = 1$, i.e., in the complex plane case). Notice also the natural presence of the factor $1/2$ on $\overline{D}$ like in the case of Wirtinger's complex partial derivative $\partial_z = \frac{1}{2}(\partial_x - i\partial_y)$. Several formulas would become more symmetric and compatible with the complex case, if one would use this coefficient also in the hypercomplex case. For simplicity and from now on we use the notation $\partial := \frac{1}{2}\overline{D}$.

6 Hypercomplex Differentiability

The adequate definition of hypercomplex differentiability as generalization of complex differentiability remained an unsolved problem until the end of the 80-ties of the last century (c.f. [27]). Based on the use of the hypercomplex structure expressed by $\mathbb{R}^{n+1} \cong \mathcal{H}^n$ and the fact that differentiability is equivalent with the possibility of local linearization of the increment of the considered function, we have the following

Definition 6 Let f be a continuous mapping from a neighborhood of $\mathbf{z}_* \in \mathcal{H}^n$ into $\mathcal{C}\ell_{0,n}$. Then f is called *left hypercomplex differentiable* (resp. *right hypercomplex differentiable*) at $\mathbf{z}_*$ if there exists a left (resp. right) $\mathcal{C}\ell_{0,n}$-linear mapping ℓ such that

$$\lim_{\Delta\mathbf{z}\to 0}\frac{|f(\mathbf{z}_* + \Delta\mathbf{z}) - f(\mathbf{z}_*) - \ell(\Delta\mathbf{z})|}{\|\Delta\mathbf{z}\|} = 0.$$

We say that a function f is *hypercomplex differentiable* in $\Omega \subset \mathbb{R}^{n+1} \cong \mathcal{H}^n$ if it is hypercomplex differentiable at all points of Ω (c.f. [17]).

Moreover, it has been shown in [17], that the equivalence of the concept of hypercomplex differentiability and monogeneity is guaranteed. Analogously to Theorem 3 the following theorem is valid.

Theorem 4 *Let $f = f(\mathbf{z})$ be continuously real differentiable in an open set $\Omega \subset \mathcal{H}^n$. Then f is hypercomplex L- (R-) differentiable in Ω, if and only if $Df = 0$ ($fD = 0$) in Ω, i.e. iff f is left (resp. right) monogenic.*

Again, but now through the property of hypercomplex differentiability the Cauchy approach to holomorphic functions is generalized. In [28], which discussed the impossibility of adequate generalizations of Cauchy's as well as Weierstrass' approaches (by the concept of convergent power series) to holomorphic functions in the higher-dimensional hypercomplex setting, this approach has been overseen. Both theorems, Theorem 3 together with Theorem 4, show that *hypercomplex derivability*, *hypercomplex differentiability* and *monogeneity* of a hypercomplex function $f : \Omega \subset \mathbb{R}^{n+1} \cong \mathcal{H}^n \longmapsto \mathcal{C}\ell_{0,n}$ are equivalent.

7 Generalized Power Series

The aim of this section is to show that the $\mathcal{H}^n$-approach to monogenic functions leads in a very natural and direct way to power series in several hypercomplex variables (Weierstrass approach). For instance, no auxiliary application of Legendre or Gegenbauer polynomials (cf. [14, 16]) is needed. In general, the non-commutative multiplication in Clifford algebras causes many difficulties in hypercomplex function theory. But the systematical use of n hypercomplex variables allows a deeper structural insight from the beginning and simplifies essentially the calculations, particularly when dealing with monogenic polynomials. For more details see [20]. We begin by illustrating some aforementioned facts in relation to the classical complex case thereby also calling attention to important differences to the hypercomplex case.

1. Consider an open set $\Omega \subset \mathbb{R}^3$ and a continuously differentiable vector field (the velocity field of a flow) $\mathbf{g} = (g_0, g_1, g_2)$ on Ω. Then the flow of a non-compressible fluid without sources nor sinks is the solution of a *Riesz system* (particular case of the Riesz system considered in [1]), namely

$$\begin{cases} \text{div} \;\; \mathbf{g} = 0 \\ \text{curl} \;\; \mathbf{g} = 0 \end{cases}.$$

In the equivalent hypercomplex setting we consider a paravector valued function $f = f_0 + f_1 e_1 + f_2 e_2$ where $(f_0, f_1, f_2) := (g_0, -g_1, -g_2)$. Then the Riesz system can be written with respect to f in the compact form as

$$Df = 0 \quad \text{or} \quad fD = 0$$

and represents obviously a generalized Cauchy-Riemann system in $\mathbb{R}^3$.

2. The hypercomplex variables

$$f_k(z) = z_k := x_k - x_0 e_k = -\frac{1}{2}(ze_k + ze_k), \quad (k = 1, \ldots, n)$$

are left- and right-monogenic functions, also called *totally regular variables* (c.f. [6, 14]).
3. The identity function $f(z) = z \in \mathcal{A}_n$ is *not* monogenic unless $n = 1$ (the classical complex case), since $Df = fD = 1 - n$.
4. Powers of z, i.e. $f(z) = z^n$ and simple products of the totally regular variables like $z_j \cdot z_k, \; j \neq k$, are *not monogenic*. This and the previous property show that it is not possible to deal in the same way as usual with the non-monogenic ordinary power of the underlying variable $z \in \mathcal{A}_n$.
5. Symmetric products of the totally regular variables in the form

$$z_j \times z_k := \frac{1}{2}(z_j z_k + z_k z_j) = x_j x_k - x_0 x_k e_j - x_0 x_j e_k$$

are left- and right-monogenic. More general, it has been proven in [29], that if $\nu = (\nu_1, \ldots, \nu_n)$ is a multi-index, all homogeneous monogenic polynomials of degree $|\nu| = k$ can be obtained as linear combinations (from the left or from the right) of generalized powers given in the form

$$\begin{aligned} \mathbf{z}^{\nu} := z_1^{\nu_1} \times \cdots \times z_n^{\nu_n} &= \underbrace{z_1 \times \cdots \times z_1}_{\nu_1} \times \cdots \times \underbrace{z_n \times \cdots \times z_n}_{\nu_n} \\ &= \frac{1}{k!} \sum_{\pi(i_1, \ldots, i_k)} z_{i_1} \cdots z_{i_k}, \end{aligned} \tag{5}$$

where the sum is taken over *all* permutations of $\{i_1, \ldots, i_k\} \subseteq \{1, \ldots, n\}$ and $z_j = x_j - x_0 e_j, \; j = 1, \ldots, n$.
Moreover, all functions of the form $f(z) = \mathbf{z}^{\nu}$, are left and right monogenic and $Cl_{0,n}$-linear independent. Therefore they can be used as basis for generalized power series. Following [29] it has been shown that the generalized power series of the form

$$P(\mathbf{z}) = \sum_{k=0}^{\infty} \left(\sum_{|\nu|=k} \mathbf{z}^{\nu} c_{\nu} \right), c_{\nu} \in Cl_{0,n}$$

generates in the neighborhood of the origin a monogenic from the left function $f(\mathbf{z})$ and coincides in the interior of its domain of convergence with the Taylor series of $f(\mathbf{z})$, i.e, in a neighborhood of the origin we have

$$f(\mathbf{z}) = \sum_{k=0}^{\infty} \frac{1}{k!} \left(\sum_{|\nu|=k} \mathbf{z}^{\nu} \binom{k}{\nu} \frac{\partial^{|\nu|} f(\mathbf{0})}{\partial \mathbf{x}^{\nu}} \right),$$

where $\mathbf{x} = (x_1, \ldots, x_n)$. Analogously monogenic from the right power series can be considered as we will do in the sequel.

By using classical methods of estimation the following theorem can be proved:

Theorem 5 ([20]) *Let $P(\mathbf{a}, \mathbf{z})$ be a given right power series around $\mathbf{a} \in \mathcal{H}^n$. If $P(\mathbf{a}, \mathbf{z})$ converges in some polycylindric domain of the form*

$$\mathcal{U}(\mathbf{r}, \mathbf{a}) = \{\mathbf{z} \in \mathcal{H}^n : |z_k - a_k| = ((x_0 - \alpha_0)^2 + (x_k - \alpha_k)^2)^{1/2} < r_k, k = 1, \ldots, n\},$$

then $P(\mathbf{a}, \mathbf{z})$ is right hypercomplex differentiable in $\mathcal{U}(\mathbf{r}, \mathbf{a})$ and the partial derivatives with respect to x_k are obtained by formal differentiation as

$$\frac{\partial P(\mathbf{z})}{\partial x_k} = \sum_{\mu} c_\mu \mu_k (\mathbf{z} - \mathbf{a})^{\mu - \tau_k} \tag{6}$$

where τ_k is the multi-index with 1 at place k and zero otherwise.

Formula (6) shows that the derivatives also represent monogenic functions in the same domain of convergence. By induction follows

Theorem 6 *Every R- (L-) generalized power series is infinitely R- (L-) hypercomplex differentiable inside the domain of convergence.*

Calculating the partial derivatives of the generalized powers $(\mathbf{z} - \mathbf{a})^\mu$ in $\mathbf{z} = \mathbf{a}$ leads to

$$\frac{\partial^{|\nu|}}{\partial \mathbf{x}^\nu} (\mathbf{z} - \mathbf{a})^\mu = \begin{cases} \mu!, & \text{if } \nu = \mu \\ 0, & \text{if } \nu \neq \mu \end{cases}$$

and this implies

Theorem 7 *Every convergent R-power series generates in the interior of its domain of convergence a monogenic function $f(\mathbf{z})$ and coincides there with the Taylor series of $f(\mathbf{z})$, i.e., in a neighborhood of $\mathbf{z} = \mathbf{a}$ we have*

$$f(\mathbf{z}) = \sum_{|\mu|=0} \frac{1}{\mu!} \frac{\partial^{|\mu|} f(\mathbf{a})}{\partial \mathbf{x}^\mu} (\mathbf{z} - \mathbf{a})^\mu$$

(analogously for L–series; there the coefficients are on the right side of the powers).

In view of the unique determination of the coefficients of the Taylor series of a monogenic function we can formulate the uniqueness theorem for generalized Taylor series:

Theorem 8 *If the coefficients of two generalized R- (L-) Taylor series coincide in an arbitrarily small neighborhood of the common point of development* **a** *then they coincide identically.*

The uniqueness theorem is the basis for the *Cauchy-Kowalewskaya extension* of a real-analytic $\mathcal{C}\ell_{0,n}$-valued function in $\mathbb{R}^n$:

Theorem 9 *Let* $f(\mathbf{x})$ *be real-analytic in the parallelepiped*

$$\mathcal{V}(\mathbf{r}) = \{\mathbf{x} \in \mathbb{R}^n : |x_k| < r_k, k = 1, \ldots, n\}.$$

Then an analytic continuation of f *to an R- (L-) monogenic function in*

$$\mathcal{U}(\mathbf{r}) = \{\mathbf{z} \in \mathcal{H}^n : |z_k| < r_k, k = 1, \ldots, n\}.$$

is given in a unique way by the function

$$f_R^*(\mathbf{z}) = \sum_{|\mu|=0} \frac{1}{\mu!} \frac{\partial^{|\mu|} f(0)}{\partial \mathbf{x}^\mu} \mathbf{z}^\mu \;\; \textit{resp.} \;\; f_L^*(\mathbf{z}) = \sum_{|\mu|=0} \mathbf{z}^\mu \frac{1}{\mu!} \frac{\partial^{|\mu|} f(0)}{\partial \mathbf{x}^\mu}, \tag{7}$$

and we have

$$f_R^*(\mathbf{z})|_{x_0=0} = f_L^*(\mathbf{z})|_{x_0=0} = f^*(\mathbf{z})|_{\mathbb{R}^n} = f(\mathbf{x}).$$

Proof Inside of $\mathcal{V}(\mathbf{r})$ the function $f(\mathbf{x})$ has the Taylor series representation

$$f(\mathbf{x}) = \sum_{|\mu|=0} \frac{1}{\mu!} \frac{\partial^{|\mu|} f(0)}{\partial \mathbf{x}^\mu} \mathbf{x}^\mu \tag{8}$$

and

$$f_R^*(\mathbf{z})|_{x_0=0} = f_L^*(\mathbf{z})|_{x_0=0} = f^*(\mathbf{z})|_{\mathbb{R}^n} = f(\mathbf{x})$$

becomes obvious. The convergence of (7) in $\mathcal{U}(\mathbf{r})$ is guaranteed by the convergence of (8) in $\mathcal{V}(\mathbf{r})$ and the uniqueness follows from the uniqueness theorem for generalized power series. □

The $f_R^*(\mathbf{z})$ resp. $f_L^*(\mathbf{z})$ are called the right (resp. left) *Cauchy–Kowalewskaya extension* (CK-*extension*) of f. In fact, Theorem 9 together with property 5 mentioned in the begin of this section, immediately indicates the way of passing from a power series in $\mathbf{x}$, i.e. given in $\mathbb{R}^n$, to its monogenic continuation in $\mathcal{H}^n$ as function of $\mathbf{z}$ simply by substituting $\mathbf{x}$ by $\mathbf{z}$ and changing at the same time the ordinary product to the symmetric product (5).

8 An Application: Hypercomplex Appell Polynomials

8.1 Definition and Some Properties

In 1880, Appell [24] considered sequences of polynomials of degree k of one real variable, $(p_k(x))_{k\geq 0}$, satisfying the recurrence relations

$$p_k'(x) = kp_{k-1}(x),\ k = 1, 2, \ldots, \tag{9}$$

where $p_0(x)$ is a non-zero constant.

Nowadays, any such sequence is called an Appell sequence, its members are called Appell polynomials and (9) is usually referred as Appell property. Of course, the basic idea is that the Appell polynomials behave like power-law functions under the differentiation operation. The classical examples of Appell polynomials besides the monomials $(x^k)_{k\geq 0}, x \in \mathbb{R}$ are the Bernoulli, Euler and Hermite polynomials.

The following equivalent conditions characterizing Appell polynomials underpin the different approaches that have been developed to deal with real Appell sequences as well as their generalizations (see e.g. [30–32]).

Theorem 10 *Let $(p_k(x))_{k\geq 0}$ be a sequence of polynomials of one real variable. Then the following conditions are equivalent:*

1. *$(p_k(x))_{k\geq 0}$ is an Appell sequence;*
2. *$(p_k(x))_{k\geq 0}$ satisfies*

$$p_k(x+y) = \sum_{j=0}^{k} \binom{k}{j} x^{k-j} p_j(y); \tag{10}$$

3. *There exists $(a_k)_{k\geq 0}$, with $a_0 \neq 0$, such that*

$$p_k(x) = \sum_{j=0}^{k} \binom{k}{j} a_j x^{k-j}. \tag{11}$$

In hypercomplex analysis context, the hypercomplex derivative of a monogenic function is the key player in the generalization of Appell's concept of power-like polynomial sequences (9) to higher dimensions. This generalization was first introduced in [33] by using several hypercomplex variables (see Theorem 13) and reads as follows.

Definition 7 A sequence of $\mathcal{A}_n$–valued monogenic polynomials $(\mathcal{F}_k(x))_{k\geq 0}$ is called an Appell sequence, if $\mathcal{F}_k(x)$ is of exact degree k and

$$\partial \mathcal{F}_k(x) = k\,\mathcal{F}_{k-1}(x),\quad x \in \mathcal{A}_n,\quad k = 1, 2, \ldots, \tag{12}$$

where $\mathcal{F}_0(x)$ is a non-zero constant.

Based on this concept, it is possible to obtain a binomial-type identity for hypercomplex Appell sequences, which extend the identity (10) to the hypercomplex case [34, Theorem 1].

Theorem 11 *A monogenic polynomial sequence* $(P_k(x))_{k\geq 0}$ *is an Appell sequence if and only if it satisfies*

$$P_k(x) = P_k(x_0 + \underline{x}) = \sum_{j=0}^{k} \binom{k}{j} x_0^{k-j} P_j(\underline{x}).$$

Hypercomplex Appell polynomials have received a lot of attention in the last decade. They have been studied in detail in several papers by different authors and various applications have been considered [33–41]. In what follows we focus on a class of polynomials of the form

$$P_k(x) = \sum_{s=0}^{k} \binom{k}{s} c_s(n) x_0^{k-s} \underline{x}^s, \tag{13}$$

with coefficients $c_s(n) \in \mathbb{R}$ and $c_0(n) \neq 0$, for all n. For such class of polynomials we can derive an hypercomplex analogue of (11) (see [39, Theorem 2] for details). Moreover, we have (cf. [39, 42]):

Theorem 12 *Polynomials* $P_k(x)$ *of the form* (13) *are monogenic if and only if*

$$c_k(n) = \frac{(\frac{1}{2})_{\lfloor \frac{k+1}{2} \rfloor}}{(\frac{n}{2})_{\lfloor \frac{k+1}{2} \rfloor}} c_0(n), \tag{14}$$

where $a_{(r)} = \frac{\Gamma(a+r)}{\Gamma(a)}$, *(r=1,2, ...) denotes the Pochhammer symbol, with* $a_{(0)} = 1$. *In such cases,* $(P_k(x))_{k\geq 0}$ *is an Appell sequence*

Polynomials of the form (13)–(14) with the normalization $c_0(n) = 1$ have been referred as *Standard Appell Polynomials* (SAP) in [43] and denoted by $\mathcal{P}_k^n(x)$. It is clear that this initial value can be changed to any real or hypercomplex constant different from zero. In [44] the concept of Appell sequence was generalized further by considering as first term an arbitrary generalized polynomial constant of a fixed degree.

We stress the fact that $\mathcal{P}_k^1(x) = (x_0 + e_1 x_1)^k$ are the usual powers of the holomorphic variable $z = x_0 + e_1 x_1$ (with the usual identification of e_1 with the complex imaginary unit). Moreover, real values of x lead to $\mathcal{P}_k^n(x_0) = x_0^k$, while for $x = \underline{x}$, we obtain the essential property which characterizes the difference to the complex case $\mathcal{P}_k^n(\underline{x}) = c_k(n)\underline{x}^k$. One straightforward approach to construct SAP is to consider the CK-extension of the powers $c_k(n)\underline{x}^k$, which leads to the following equivalent representation of $\mathcal{P}_k^n(x)$ in terms of generalized powers (5).

Theorem 13 *The SAP can be expressed as*

$$\mathbf{P}_k(\mathbf{z}) = c_k(n) \sum_{|\nu|=k} z_1^{\nu_1} \times \cdots \times z_n^{\nu_n} \binom{k}{\nu} e_1^{\nu_1} \times \cdots \times e_n^{\nu_n}.$$

Since $\mathcal{P}_k^n(1) = \mathbf{P}_k(-e_1, -e_2, \ldots, -e_n) = 1$, Theorems 12–13 allow to derive the following relation involving the generators of the algebra.

$$\sum_{|\nu|=k} (-1)^k \binom{k}{\nu} (e_1^{\nu_1} \times \cdots \times e_n^{\nu_n})^2 = \frac{(\frac{n}{2})_{\lfloor \frac{k+1}{2} \rfloor}}{(\frac{1}{2})_{\lfloor \frac{k+1}{2} \rfloor}}.$$

This identity is one among many others nice relations that have been obtained through the use of the sequences $(c_k(n))_{k\geq 0}$. In fact such sequences, in particular the $n = 2$ case, have important applications in harmonic analysis, theory of stable holomorphic functions and combinatorics. We refer the interested reader to [45, 46] and references therein.

8.2 A Matrix Approach

A matrix representation of real Appell sequences $(p_k(x))_{k\geq 0}$ developed in [32] relies on the matrix H whose entries are given by

$$(H)_{ij} = \begin{cases} i, & i = j+1 \\ 0, & \text{otherwise}, \end{cases} \qquad i, j = 0, 1, \ldots. \tag{15}$$

In this context the role of H is essentially that of a derivation matrix.

For the sake of handle Appell sequences in a closed form, the vector

$$\mathbf{p}(x) = \left[p_0(x)\ p_1(x)\ \cdots\ p_m(x)\right]^T,$$

whose entries are the $m + 1$ first terms of the Appell sequence $(p_k(x))_{k\geq 0}$, is introduced. Also, H is truncated in order to obtain a square matrix of order $m + 1$. In this case, H is a nilpotent matrix of degree $m + 1$.

The differential equation $\mathbf{p}'(x) = H\mathbf{p}(x)$ is the matrix counterpart of (9) for the truncated Appell sequence $(p_k(x))_{0\leq k\leq m}$. Its general solution is

$$\mathbf{p}(x) = e^{Hx}\mathbf{p}(0) \equiv \mathcal{P}(x)\mathbf{p}(0),$$

where $\mathcal{P}(x)$ is the well known generalized Pascal matrix defined by

$$(\mathcal{P}(x))_{ij} = \begin{cases} \binom{i}{j} x^{i-j}, & i \geq j \\ 0, & \text{otherwise}, \end{cases} \qquad i, j = 0, 1, \ldots, m,$$

Note that by using the explicit representation (11), $\mathbf{p}(0) = \left[a_0\ a_1\ \cdots\ a_m\right]^T$, $a_0 \neq 0$.

Aiming a matrix approach of Appell sequences in the hypercomplex context, multivariate homogeneous polynomials of the form (13) are considered. Such polynomials correspond to (11) by replacing of real variable x by x_0, real part of the paravector $x = x_0 + \underline{x}$, and the constant a_j by $a_j\underline{x}^j$.

Denoting by $\mathbf{P}(x)$ and $\mathbf{P}(\underline{x})$ the vectors

$$\mathbf{P}(x) \equiv \mathbf{P}(x_0, \underline{x}) = \left[P_0(x)\ P_1(x)\ \cdots\ P_m(x)\right]^T, \tag{16}$$

and

$$\mathbf{P}(\underline{x}) \equiv \mathbf{P}(0, \underline{x}) = \left[a_0\underline{x}^0\ a_1\underline{x}^1\ \cdots\ a_m\underline{x}^m\right]^T,$$

respectively, a matrix form of the truncated sequence $(P_k(x))_{0\leq k\leq m}$ is

$$\mathbf{P}(x) = e^{Hx_0}\mathbf{P}(\underline{x}). \tag{17}$$

Considering the vector $\xi(\underline{x}) = \left[1\ \underline{x}^1\ \cdots\ \underline{x}^m\right]^T$ and the diagonal matrix

$$\mathcal{D} = \text{diag}[a_0\ a_1\ \cdots a_m], \tag{18}$$

an equivalent representation of (17) is

$$\mathbf{P}(x) = e^{Hx_0}\mathcal{D}\xi(\underline{x}). \tag{19}$$

The relation (see [4])

$$\partial_{\underline{x}}(\underline{x}^k) = \begin{cases} -k\underline{x}^{k-1}, & k \text{ even} \\ -(n+k-1)\underline{x}^{k-1}, & k \text{ odd} \end{cases}$$

allows to achieve in matrix form the action of $\partial_{\underline{x}}$ on the vector $\xi(\underline{x})$. In fact, straightforward calculations lead to $\partial_{\underline{x}}\,\xi(\underline{x}) = \tilde{H}\xi(\underline{x})$, where $\tilde{H}$ is the matrix defined by

$$(\tilde{H})_{ij} = \begin{cases} -(n+i-1), & i = j+1 \text{ and } j \text{ even} \\ -i, & i = j+1 \text{ and } j \text{ odd} \\ 0, & \text{otherwise}, \qquad i, j = 0, 1, \ldots, m \end{cases} \tag{20}$$

(cf. [41], Proposition 3.1).

Theorem 14 *Let $\mathcal{D}$ be the matrix defined in* (18), *with $a_0 \neq 0$, and H, $\tilde{H}$ as defined in* (15), (20), *respectively. If $\mathbf{P}(x)$ is the vector* (16), *then the following conditions are equivalent:*

1. *$\mathbf{P}(x)$ is an Appell vector;*
2. *The matrix $\mathcal{D}$ verifies the relation $H\mathcal{D} + \mathcal{D}\tilde{H} = 0$.*

Proof (1 ⇒ 2) Applying to both sides of (19) the hypercomplex differential operator we obtain

$$\bar{\partial}\mathbf{P}(x) = \frac{1}{2}e^{Hx_0}(H\mathcal{D} + \mathcal{D}\tilde{H})\xi(\underline{x}).$$

The result holds from the monogeneity of the components of $\mathbf{P}(x)$.

(2 ⇒ 1) Under the hypothesis, the monogeneity of the components of $\mathbf{P}(x)$ is obvious. Furthermore, since $H\mathcal{D} = -\mathcal{D}\tilde{H}$,

$$\begin{aligned}\partial\mathbf{P}(x) &= \frac{1}{2}e^{Hx_0}(H\mathcal{D} - \mathcal{D}\tilde{H})\xi(\underline{x})\\ &= He^{Hx_0}\mathcal{D}\xi(\underline{x}) = H\mathbf{P}(x),\end{aligned}$$

which is the corresponding matrix representation of (12) in Definition 7. This completes the proof of the theorem. □

8.3 Orthogonality

Practical problems of Physics and Engineering require to deal with numerical approximations of monogenic functions, like for instance, solutions of complicated differential equations. In this sense, the construction of basis of monogenic functions, in particular, monogenic polynomials, plays a crucial role. As we have seen, there exists equivalence between the existence of a local Taylor series in terms of generalized powers, hypercomplex derivability/differentiability and monogeneity of Clifford-valued functions. The Fourier expansion can be achieved as well if we construct an orthonormal basis. Unfortunately, a basis formed by the generalized powers is not orthogonal in L_2 and the numerical costs for its orthonormalization are enormous for higher degrees. A different approach was followed in [38] in the case of 3 dimensions using the concept of Gelfand-Tsetlin bases. The generalization of this approach to arbitrary dimensions was done in [37] leading to an iterative process where the building blocks are the (in general) non-monogenic $\mathcal{C}\ell_{0,n}$-valued polynomials

$$X^{(k-j)}_{n+1,j}(x) = F^{(k-j)}_{n+1,j}(x) + \frac{j+1}{n+2j}F^{(k-j-1)}_{n+1,j-1}(x)\,\underline{x}, \tag{21}$$

where

$$F^{(k-j)}_{n+1,j}(x) = \frac{(j+1)_{k-j}}{(n-1+2j)_{k-j}}|\underline{x}|^{k-j}C^{\frac{n-1}{2}+j}_{k-j}\left(\frac{x_0}{|x|}\right),$$

with $F^{(-1)}_{n+1,k+1} \equiv 0, x = (x_0, \ldots, x_n) \in \mathbb{R}^{n+1}$, $|\,.\,|$ is the usual Euclidean norm in $\mathbb{R}^{n+1}$, and C^{ν}_m is the Gegenbauer polynomial of degree m and parameter $\nu \neq 0$.

The resulting basis of orthogonal[1] monogenic polynomials is formed by the polynomials

$$f_{k,\mu} = X^{(k-k_n)}_{n+1,k_n} X^{(k_n-k_{n-1})}_{n,k_{n-1}} \cdots X^{(k_3-k_2)}_{3,k_2} \zeta^{k_2},$$

where $\zeta := x_1 - x_2\, e_1 e_2$ and μ is an arbitrary sequence of integers $(k_{n+1}, k_n, \ldots, k_2)$ such that $k = k_{n+1} \geq k_n \geq \ldots \geq k_2 \geq 0$.

Earlier, in the book [4], multiples of the building blocks (21) appeared when the CK-extension was applied to the vector-valued polynomials $\underline{x}^{k-j} P_j(\underline{x})$, where $P_j(\underline{x})$ are arbitrarily fixed monogenic polynomials of degree j $(j = 0, \ldots, k)$. Since the resulting polynomials do not have the Appell property, an appropriate normalization constant was considered by the first three authors in the paper [40]. More explicitly, the CK-extension was applied to the polynomials

$$c_{k,j}(n) \binom{k}{j} \underline{x}^{k-j} P_j(\underline{x}),$$

where

$$c_{k,j}(n) = \begin{cases} \dfrac{(k-j)!!(n+2j-2)!!}{(n+k+j-1)!!}, & \text{if } k, j \text{ have different parities} \\ c_{k-1,j}(n), & \text{if } k, j \text{ have the same parity} \end{cases},$$

for $k \geq 1$, $j = 0, \ldots, k$ and $c_{0,0}(n) = 1$. The result is the sequence of orthogonal monogenic $\mathcal{C}\ell_{0,n}$-valued Appell polynomials $\left(\widetilde{X}^{(k-j)}_{n+1,j}\right)_{k \in \mathbb{N}_0}$, where

$$\widetilde{X}^{(k)}_{n+1,j}(x) := X^{(k-j)}_{n+1,j}(x)\, P_j(\underline{x}), \quad x \in \mathcal{A}_n. \tag{22}$$

The (in general) non-monogenic polynomials $X^{(k-j)}_{n+1,j}$ can be built by a simple shift of the coefficients of the monogenic SAP studied in the Sect. 8.1. In fact, the paper [40] shows that

Theorem 15 *For all $k \in \mathbb{N}_0$ and each fixed j $(j = 0, \ldots, k)$, it holds*

$$X^{(k-j)}_{n+1,j}(x) = \binom{k}{j} \mathcal{P}^{n+2j}_{k-j}(x), \quad x \in \mathcal{A}_n.$$

Notice that for a fixed k and $j = 0$, $X^{(k)}_{n+1,0}$ coincide exactly with the SAP $\mathcal{P}^n_k$, constituting the only monogenic building block in the sequence defined by (22). This fact is reflected in the monogenic sequence $\left(\widetilde{X}^{(k)}_{n+1,j}\right)_{k \in \mathbb{N}_0}$ by the choice $P_0(\underline{x}) \equiv 1$.

As a consequence of Theorem 15, the polynomials $X^{(k-j)}_{n+1,j}$ $(j = 0, \ldots, k)$, explicitly given by (21), admit the simpler representation

[1]The inner product in L_2 is given by $(f, g)_{\mathcal{C}\ell_{0,n}} = \int_{B^{n+1}} \bar{f}\, g\, d\lambda^{n+1}$, where λ^{n+1} is the Lebesgue measure and B^{n+1} is the unit ball in $\mathbb{R}^{n+1}$.

$$X_{n+1,j}^{(k-j)}(x) = \binom{k}{j} \sum_{s=0}^{k-j} \binom{k-j}{s} c_s(n+2j)\, x_0^{k-j-s}\, \underline{x}^s, \quad x \in \mathcal{A}_n,$$

where $c_s(n+2j)$ $(s = 0, \dots k-j, j = 0, \dots, k)$ are given by (14) with $c_0(n) = 1$.

As it is well known from the theory, any sequence of orthogonal polynomials satisfies a three-term recurrence relation and a second order differential equation. The immediate question arises: can we derive similar results for the orthogonal sequence of monogenic polynomials $\left(\widetilde{X}_{n+1,j}^{(k)} : j = 0, \dots, k\right)_{k \in \mathbb{N}_0}$? The affirmative answer was done in the paper [40] leading to the following results.

Theorem 16 *For all $k \in \mathbb{N}_0$ and each fixed j $(j = 0, \dots, k)$, the monogenic polynomials $\widetilde{X}_{n+1,j}^{(k-j)}(x)$, $x \in \mathcal{A}_n$ satisfy the three-term type recurrence*

$$(n+k+1+j)(k+2-j)\widetilde{X}_{n+1,j}^{(k+2)} - \left[(n+2k+2)x_0 + \underline{x}\right](k+2)\widetilde{X}_{n+1,j}^{(k+1)} \\ + (k+2)(k+1)\,|x|^2\,\widetilde{X}_{n+1,j}^{(k)} = 0,$$

$$\widetilde{X}_{n+1,j}^{(j)} = P_j(\underline{x}), \quad \widetilde{X}_{n+1,j}^{(j+1)} = (j+1)(x_0 + \tfrac{1}{n+2j}\underline{x})P_j(\underline{x}).$$

Notice that the appearance of a polynomial of second degree as coefficient of $\widetilde{X}_{n+1,j}^{(k)}$ instead of a constant, like in the classical case, is justified by the fact that we are dealing with homogeneous polynomials.

Theorem 17 *For all $k \in \mathbb{N}_0$ and each fixed j $(j = 0, \dots, k)$, the monogenic polynomials $\widetilde{X}_{n+1,j}^{(k)}(x)$, $x \in \mathcal{A}_n$, satisfy the second order differential equation*

$$|x|^2\partial^2 y(x) - ((n+2k-2)x_0 + \underline{x})\partial y(x) + (n+k+j-1)(k-j)y(x) = 0.$$

Acknowledgements The work of the first, second and fourth authors was supported by Portuguese funds through the CIDMA-Center for Research and Development in Mathematics and Applications, and the Portuguese Foundation for Science and Technology ("FCT-Fundação para a Ciência e Tecnologia"), within project PEst-OE/MAT/UI4106/2013. The work of the second author was supported by Portuguese funds through the CMAT-Centre of Mathematics and FCT within the Project UID/MAT/00013/2013.

References

1. Stein, E.M., Weiss, G.: Generalization of the Cauchy-Riemann equations and representations of the rotation group. Am. J. Math. **90**(1), 163–196 (1968)
2. Brackx, F., Delanghe, R., De Schepper, H.: Hardy spaces of solutions of generalized Riesz and Moisil-Teodorescu systems. Complex Var. Elliptic Equ. **57**(7–8), 771–785 (2012)
3. Moisil, G., Teodorescu, N.: Functions holomorphes dans l'espace. Mathematica Cluj **5**, 142–159 (1931)

4. Delanghe, R., Sommen, F., Souček, V.: Clifford algebra and spinor-valued functions. A function theory for the Dirac operator. Mathematics and its Applications, vol. 53. Kluwer Academic Publishers, Dordrecht (1992)
5. Delanghe, R., Lávička, R., Souček, V.: The Gelfand–Tsetlin bases for Hodge–de Rham systems in Euclidean spaces Math. Methods Appl. Sci. **35**(7), 745–757 (2012)
6. Delanghe, R.: On regular-analytic functions with values in a Clifford Algebra. Math. Ann. **185**, 91–111 (1970)
7. Fueter, R.: Über die analytische Darstellung der regulären Funktionen einer Quaternionenvariablen Comm. Math. Helvetici **8**, 371–378 (1935). https://doi.org/10.1007/BF01199562
8. Kravchenko, V.V.: Applied quaternionic analysis. Research and Experience in Mathematics, vol. 28, Heldermann Verlag, Lemgo (2003)
9. Delanghe, R.: Clifford analysis: history and perspective. Comput. Methods Funct. Theory **1**(1), 107–153 (2001)
10. Fueter, R.: Über Funktionen einer Quaternionenvariablen. Atti Congr. Int. Mat, Bologna (1928)
11. Malonek, H.: Rudolf Fueter and his motivation for hypercomplex function theory. Adv. Appl. Clifford Algebr. **11**(S2), 219–230 (2001)
12. Fueter, R.: Functions of a hyper complex variable. Manuscript of Lecture Notes, Fall Semester 1948/49, written and supplemented by E. Bareiss, ETH Bibliothek Zürich, 318p (1950)
13. Gürlebeck, K., Habetha, K., Sprößig, W.: Holomorphic Functions in the Plane and n-Dimensional Space. Birkhäuser Verlag, Basel (2008)
14. Gürlebeck, K., Sprößig, W.: Quaternionic analysis and elliptic boundary value problems. International Series of Numerical Mathematics, vol. 89, Birkhäuser-Verlag, Basel (1990)
15. Ryan, J.: Clifford analysis with generalize elliptic and quasi-elliptic functions. Appl. Anal. **13**, 151–171 (1982)
16. Brackx, F., Delanghe, R., Sommen, F.: Clifford Analysis. Pitman, Boston, London, Melbourne (1982)
17. Malonek, H.: A new hypercomplex structure of the Euclidean space $\mathbb{R}^{n+1}$ and a concept of hypercomplex differentiability. Complex Variables, Theory Appl. **14**, 25–33 (1990)
18. Malonek, H.: The concept of hypercomplex differentiability and related differential forms. In: Kühnau, R., Tutschke, W. (eds.) Studies in Complex Analysis and its Applications to Partial Differential Equations 1, vol. 256, pp. 193–202. Pitman , Longman (1991)
19. Malonek, H.: Hypercomplex derivability—The characterization of monogenic functions in $\mathbb{R}^{n+1}$ by their derivative. In: Ryan, J. et al. (eds.) Clifford Algebras and their Applications in Mathematical Physics, vol. 2. Progress in Physics, vol. 19, Birkhäuser (2000)
20. Malonek, H.R.: Selected topics in hypercomplex function theory. In: Eriksson, S.-L. (ed.) Clifford Algebras and Potential Theory. Report Series 7, University of Joensuu, pp. 111–150 (2004)
21. Gürlebeck, K., Malonek, H.: A hypercomplex derivative of monogenic functions in $\mathbb{R}^{n+1}$ and its applications. Complex Var. Theory Appl. **39**, 199–228 (1999)
22. Ryan, J.: Clifford algebras in analysis and related topics. Studies in Advanced Mathematics, CRC PRESS (1996)
23. Pompeiu, M.: Sur une classe de fonctions dúne variable complexe. Rend. Circ. Mat. Palermo **33**, 108– 113 (1912), **35**, 277–281 (1913)
24. Appell, P.: Sur une classe de polynômes. Ann. Sci. École Norm. Sup. **9**(2), 119–144 (1880)
25. Mitrea, M., Sabac, F.: Pompeiu's integral representation formula. History and Mathematics. Rev. Roumaine Math. Pures Appl. **43**, 211–226 (1998)
26. Edwards, H.E.: Advanced Calculus: A Differential Forms Approach, 3rd edn. Birkhäuser, Boston, Basel, Berlin (1993)
27. Malonek, H.: On the concept of holomorphy in higher dimensions—Hypercomplex differentiability and series in permutative powers. Habilitation thesis, Halle (1987)
28. Sudbery, A.: Quaternionic analysis. Math. Proc. Camb. Phil. Soc. **85**, 199–225 (1979)
29. Malonek, H.: Power series representation for monogenic functions in $\mathbb{R}^{n+1}$ basead on a permutational product. Complex Var. Theory Appl. **15**, 181–191 (1990)
30. Sheffer, I.M.: Note on Appell polynomials. Bull. Am. Math. Soc. **51**, 739–744 (1945)

31. Lee, D.W.: On multiple Appell polynomials. Proc. Am. Math. Soc. **139**(6), 2133–2141 (2011)
32. Aceto, L., Malonek, H.R., Tomaz, G.: A unified matrix approach to the representation of Appell polynomials. Integral Transf. Spec. Funct. **26**(6), 426–441 (2015)
33. Malonek, H.R., Falcão, M.I.: Special monogenic polynomials—properties and applications. In: Simos, T.S., Psihoyios, G., Tsitouras, C. (eds.) AIP Conference Proceedings, vol. 936, pp. 765–767 (2007)
34. Cação, I., Falcão, M.I., Malonek, H.R.: Laguerre derivative and monogenic Laguerre polynomials: an operational approach. Math. Comput. Model. **53**(5–6), 1084–1094 (2011)
35. Bock, S., Gürlebeck, K.: On a generalized Appell system and monogenic power series. Math. Methods Appl. Sci. **33**(4), 394–411 (2010)
36. Eelbode, D.: Monogenic Appell sets as representations of the Heisenberg algebra. Adv. Appl. Clifford Algebra **22**(4), 1009–1023 (2012)
37. Làvička, R.: Complete orthogonal Appell systems for spherical monogenics. Complex Anal. Oper. Theory **6**, 477–489 (2012)
38. Bock, S., Gürlebeck, K., Làvička, R., Souček, V.: Gelfand-Tstelin bases for spherical monogenics in dimension 3. Rev. Mat. Iberuom. **28**(4), 1165–1192 (2012)
39. Cação, I., Falcão, M.I., Malonek, H.R.: Matrix representations of a basic polynomial sequence in arbitrary dimension. Comput. Methods Funct. Theory **12**(2), 371–391 (2012)
40. Cação, I., Falcão, M.I., Malonek, H.R.: Three term recurrence relations for systems of Clifford algebra-valued polynomial sequences. Adv. Appl. Clifford Algebras **27**, 71–85 (2017)
41. Aceto, L., Malonek, H.R., Tomaz, G.: Matrix approach to hypercomplex Appell polynomials. Appl. Numer. Math. **116**, 2–9 (2017)
42. Cação, I., Falcão, M.I., Malonek, H.R.: Combinatorial identities in the context of hypercomplex function theory. AIP Conference Proceedings 1978, 280004-1-280004-4 (2018). https://doi.org/10.1063/1.5043904
43. Cruz, C., Falcão, M.I., Malonek, H.R.: Monogenic pseudo-complex power functions and their applications. Math. Methods Appl. Sci. **37**, 1723–1735 (2014)
44. Peña Peña, D.: Shifted Appell sequences in Clifford analysis. Results Math. **63**, 1145–1157 (2013)
45. Cação, I., Falcão, M.I., Malonek, H.R.: Hypercomplex polynomials, Vietoris' rational numbers and a related integer numbers sequence. Complex Anal. Oper. Theory **11**, 1059–1076 (2017)
46. Cação, I., Falcão, M.I., Malonek, H.R., Tomaz, G.: Combinatorial identities associated with a multidimensional polynomial sequence. J. Integer Seq. **21**(7), 18.7.4 (2018)

Paraproduct in Besov–Morrey Spaces

Yoshihiro Sawano

Abstract The paraproduct plays a key role in some highly singular partial differential equations. For example, that for Hölder–Zygmund spaces is used to solve stochastic differential equations, as demonstrated by Gubinelli, Imkeller, Perkowski, and Hairer. In this note, which is organized in a self-contained manner, the counterparts for Besov Morrey spaces are obtained.

Keywords Besov–Morrey spaces · Paraproduct

1 Introduction

In this note we investigate the boundedness property of the pointwise multiplier for functions in Hölder–Zygmund spaces and Besov–Morrey spaces, including commutators. These operations are sometimes called bilinear estimates, and use the paraproduct. Bilinear estimates and an analysis of the paraproduct in Besov–Morrey spaces appear in [6, 8] to study the Navier-Stokes and other semi-linear equations. Starting from recent seminal papers [2–4] on the analysis of stochastic differential equations, we investigate these operators from the viewpoint of harmonic analysis.

To describe our first result, we recall some notation. We use the following convention on balls in $\mathbf{R}^n$. The symbol $B(x, r)$ denotes the *ball centered at x of radius r*. Namely, we write

$$B(x, r) \equiv \{y \in \mathbf{R}^n \ : \ |x - y| < r\}$$

when $x \in \mathbf{R}^n$ and $r > 0$. Given a ball B, $c(B)$ denotes its *center* and $r(B)$ represents its *radius*. We write $B(r)$ instead of $B(o, r)$ when $o \equiv (0, 0, \dots, 0)$. Keeping this definition in mind, we define Morrey spaces. Let $1 \leq q \leq p < \infty$. The *Morrey norm* $\| \cdot \|_{\mathcal{M}^p_q}$ is defined by

Y. Sawano (✉)
Department of Mathematical Science, 1-1 Minami-Ohsawa, Hachioji, Tokyo 192-0397, Japan
e-mail: yoshihiro-sawano@celery.ocn.ne.jp

A. Karapetyants et al. (eds.), *Modern Methods in Operator Theory and Harmonic Analysis*, Springer Proceedings in Mathematics & Statistics 291,
https://doi.org/10.1007/978-3-030-26748-3_8

$$\|f\|_{\mathcal{M}_q^p} \equiv \sup_{x\in\mathbf{R}^n, r>0} |B(x,r)|^{\frac{1}{p}-\frac{1}{q}} \|f\|_{L^q(B(x,r))}$$

for a measurable function f. The *Morrey space* $\mathcal{M}_q^p(\mathbf{R}^n)$ is the set of all measurable functions f for which $\|f\|_{\mathcal{M}_q^p}$ is finite. Next, let us recall the definition of Besov–Morrey spaces [6, 12, 16]. Choose $\psi \in C_c^\infty(\mathbf{R}^n)$ so that

$$\chi_{B(\frac{6}{5})} \le \psi \le \chi_{B(\frac{3}{2})}. \tag{1}$$

We write

$$\varphi_0(\xi) = \psi(\xi), \quad \varphi_j(\xi) = \psi(2^{-j}\xi) - \psi(2^{-j+1}\xi), \quad \psi_j(\xi) = \psi(2^{-j}\xi)$$

for $j \in \mathbf{N}$ and $\xi \in \mathbf{R}^n$.

For $f \in L^1(\mathbf{R}^n)$, define the *Fourier transform* and the *inverse Fourier transform* by:

$$\mathcal{F}f(\xi) \equiv (2\pi)^{-\frac{n}{2}} \int_{\mathbf{R}^n} f(x)e^{-ix\cdot\xi}dx, \quad \mathcal{F}^{-1}f(x) \equiv (2\pi)^{-\frac{n}{2}} \int_{\mathbf{R}^n} f(\xi)e^{ix\cdot\xi}d\xi.$$

Here and below we write $\theta(D)f \equiv \mathcal{F}^{-1}[\theta \cdot \mathcal{F}f]$ for $\theta \in \mathcal{S}(\mathbf{R}^n)$ and $f \in \mathcal{S}'(\mathbf{R}^n)$. It is known that $\theta(D)f \in \mathcal{S}'(\mathbf{R}^n) \cap L^1_{\rm loc}(\mathbf{R}^n)$ and it satisfies

$$\theta(D)f(x) = (2\pi)^{-\frac{n}{2}}\langle f, \mathcal{F}^{-1}\theta(x-\cdot)\rangle$$

for all $x \in \mathbf{R}^n$. We define

$$\|f\|_{\mathcal{N}^s_{pqr}} \equiv \left(\sum_{j=0}^{\infty}(2^{js}\|\varphi_j(D)f\|_{\mathcal{M}_q^p})^r\right)^{\frac{1}{r}}$$

for $f \in \mathcal{S}'(\mathbf{R}^n)$.

Let $1 \le q \le p < \infty$, $1 \le r \le \infty$ and $s \in \mathbf{R}$. The space $\mathcal{N}^s_{pqr}(\mathbf{R}^n)$, which we call the *Besov–Morrey space*, is the set of all $f \in \mathcal{S}'(\mathbf{R}^n)$ for which the norm $\|f\|_{\mathcal{N}^s_{pqr}}$ is finite. This notation differs slightly from that used in [6]. We follow the notation in [12, 14, 19]. The parameter s describes the differential property as is indicated by the relations $\mathcal{N}^{s+\varepsilon}_{pqr}(\mathbf{R}^n) \subset \mathcal{N}^s_{pqr}(\mathbf{R}^n)$ and $\partial_j : \mathcal{N}^{s+1}_{pqr}(\mathbf{R}^n) \subset \mathcal{N}^s_{pqr}(\mathbf{R}^n)$ for all $\varepsilon > 0$ and $j = 1, 2, \ldots, n$. It is also clear from the triangle inequality in $\mathcal{M}_q^p(\mathbf{R}^n)$ that $\mathcal{N}^0_{pq1}(\mathbf{R}^n) \subset \mathcal{M}_q^p(\mathbf{R}^n)$. The main results in this note are shown below:

Theorem 1 *Let* $1 \le q_1 \le p_1 < \infty$, $1 \le q_2 \le p_2 < \infty$, $1 \le q \le p < \infty$, $1 \le r \le \infty$, *and* $s > 0$. *Assume that*

$$\frac{1}{p} = \frac{1}{p_1} + \frac{1}{p_2}, \quad \frac{1}{q} = \frac{1}{q_1} + \frac{1}{q_2}. \tag{2}$$

Then for $f \in \mathcal{N}^s_{p_1q_1r}(\mathbf{R}^n)$ *and* $g \in \mathcal{N}^s_{p_2q_2r}(\mathbf{R}^n)$*, the product* $f \cdot g \in \mathcal{N}^s_{pqr}(\mathbf{R}^n)$ *makes sense and satisfies*

$$\|f \cdot g\|_{\mathcal{N}^s_{pqr}} \le C\|f\|_{\mathcal{N}^s_{p_1q_1r}}\|g\|_{\mathcal{N}^s_{p_2q_2r}}.$$

Theorem 1 is an extension of the inequality

$$\|f \cdot g\|_{\mathcal{M}^p_q} \le C\|f\|_{\mathcal{M}^{p_1}_{q_1}}\|g\|_{\mathcal{M}^{p_2}_{q_2}}$$

for $f \in \mathcal{M}^{p_1}_{q_1}(\mathbf{R}^n)$ and $g \in \mathcal{M}^{p_2}_{q_2}(\mathbf{R}^n)$. In [8, Corolary 3.22], Mazzucato investigated the case of $p_0 = p_1 = p$ and $q_0 = q_1 = q$ instead of assuming (2) and the results extend [6]. The proof of Theorem 1 hinges on the paraproduct introduced by Bony [1]. Let $f, g \in \mathcal{S}'(\mathbf{R}^n)$. The (right) paraproduct/low-high decomposition $f \preceq g$ of f and g is defined as

$$f \preceq g = \sum_{j=2}^{\infty} \psi_{j-2}(D)f \cdot \varphi_j(D)g,$$

while the (left) paraproduct/high-low decomposition $f \succeq g$ of f and g is defined as

$$f \succeq g = \sum_{j=2}^{\infty} \varphi_j(D)f \cdot \psi_{j-2}(D)g.$$

It should be noted that Mazzucato investigated these operators in [8, (3.64a)] for Besov–Morrey spaces. Furthermore, the resonant operator/high-high decomposition $f \odot g$ of f and g is defined by

$$f \odot g = \sum_{j=0}^{\infty} \varphi_j(D)f \cdot \varphi_j(D)g + \sum_{j=1}^{\infty} \varphi_{j-1}(D)f \cdot \varphi_j(D)g + \sum_{j=1}^{\infty} \varphi_j(D)f \cdot \varphi_{j-1}(D)g.$$

As a result, we formally obtain the decomposition of the product $f \cdot g$ as

$$f \cdot g = f \preceq g + f \succeq g + f \odot g.$$

This definition goes back to Bony [1]. Mazzucato investigated these operators in [8, (3.64b)] for Besov–Morrey spaces. These three linear operators are the key linear operators used in the proof of Theorem 1.

Another aim of this paper is to extend the results used in [2, 4], which also use these operators, to the Morrey setting.

Theorem 2 *Assume that the parameters* α, β, s *satisfy*

$$0 < \alpha \le 1, \quad s + \beta < 0 < s + \alpha + \beta.$$

Then for $f \in \mathrm{Lip}_\alpha(\mathbf{R}^n)$*,* $g \in \mathcal{C}^\beta(\mathbf{R}^n)$ *and* $h \in \mathcal{N}^s_{pqr}(\mathbf{R}^n)$

$$\|(f \preceq g) \odot h - f(g \odot h)\|_{\mathcal{N}_{pqr}^{s+\alpha+\beta}} \le C\|f\|_{\mathrm{Lip}^\alpha}\|g\|_{\mathcal{C}^\beta}\|h\|_{\mathcal{N}_{pqr}^s}.$$

This result is the counterpart to [2, Lemma 2.4].

Here we briefly recall how Besov–Morrey spaces emerged. See [12, 19] for an exhaustive account. The first paper dates back to 1984. In [9] Netrusov considered Besov–Morrey spaces. Later Kozono and Yamazaki investigated Besov–Morrey spaces and applied them to the Navier–Stokes equations [6]. Mazzucato expanded this application in [8]. Decompositions of Besov–Morrey spaces can be found in [7, 14, 16]. After that, Yang and Yuan defined Besov-type spaces and Triebel–Lizorkin-type spaces in [17, 18]. A close relation between these spaces is pointed out in [15]. Recently, there have been further studies and developments in this area. For example, Haroske and Skrzypczak investigated the embedding relation of Besov–Morrey spaces [5]. One of the important consequences of defining Besov–Morrey spaces is that we have the embedding

$$\mathcal{N}_{pq\infty}^s(\mathbf{R}^n) \hookrightarrow \mathcal{C}^{s-\frac{n}{p}}(\mathbf{R}^n)$$

for $s > \frac{n}{p}$. See [13].

The rest of this paper is organized as follows. Section 2 is devoted to collecting preliminary facts. Sections 3 and 4 prove Theorems 1 and 2.

2 Preliminaries

2.1 Schwartz Distributions and the Fourier Transform

Let us recall the notation of multi-indexes to define the Schwartz space $\mathcal{S}(\mathbf{R}^n)$. By a "multiindex", we mean an element in ${\mathbf{N}_0}^n \equiv \{0, 1, 2, \dots\}^n$. In this paper a tacit understanding is that all functions assume their value in $\mathbf{C}$. For a multi-index $\alpha = (\alpha_1, \alpha_2, \dots, \alpha_n) \in {\mathbf{N}_0}^n$ $x = (x_1, x_2, \dots, x_n) \in \mathbf{R}^n$, we define $x^\alpha \equiv {x_1}^{\alpha_1}{x_2}^{\alpha_2}\cdots{x_n}^{\alpha_n}$. For a multi-index $\beta = (\beta_1, \beta_2, \dots, \beta_n) \in {\mathbf{N}_0}^n$ and $f \in C^\infty(\mathbf{R}^n)$, we set

$$\partial^\beta f \equiv \left(\frac{\partial}{\partial x_1}\right)^{\beta_1}\left(\frac{\partial}{\partial x_2}\right)^{\beta_2}\cdots\left(\frac{\partial}{\partial x_n}\right)^{\beta_n} f.$$

Definition 1 *(Schwartz function space $\mathcal{S}(\mathbf{R}^n)$)* For multi-indexes $\alpha, \beta \in {\mathbf{N}_0}^n$ and a function φ, write $\varphi_{(\alpha,\beta)}(x) \equiv x^\alpha\partial^\beta\varphi(x)$, $x \in \mathbf{R}^n$ temporarily. The *Schwartz function space* $\mathcal{S}(\mathbf{R}^n)$ is the set of functions satisfying

$$\mathcal{S}(\mathbf{R}^n) \equiv \bigcap_{\alpha,\beta\in{\mathbf{N}_0}^n} \left\{\varphi \in C^\infty(\mathbf{R}^n) \,:\, \varphi_{(\alpha,\beta)} \in L^\infty(\mathbf{R}^n)\right\}.$$

The elements in $\mathcal{S}(\mathbf{R}^n)$ are called the *test functions*.

Denote by $\mathcal{S}'(\mathbf{R}^n)$ the set of all continuous linear mappings from $\mathcal{S}(\mathbf{R}^n)$ to $\mathbf{C}$. Denote by $\langle f, \varphi \rangle$ the value of f evaluated at φ. That is, we define $\langle f, \varphi \rangle \equiv f(\varphi)$.

Note that $\mathcal{S}(\mathbf{R}^n)$ is embedded into $L^1(\mathbf{R}^n)$ and that $\mathcal{F}$ mapsto $\mathcal{S}(\mathbf{R}^n)$ isomorphically to itself. By duality, $\mathcal{F}$ mapsto $\mathcal{S}'(\mathbf{R}^n)$ isomorphically to itself.

A function $h \in C^\infty(\mathbf{R}^n)$ is said to *have at most polynomial growth at infinity*, if for all $\alpha \in \mathbf{N}_0{}^n$, there exist $C_\alpha > 0$ and $N_\alpha > 0$ such that:

$$|\partial^\alpha h(x)| \le C_\alpha \langle x \rangle^{N_\alpha}, \quad x \in \mathbf{R}^n. \tag{3}$$

Here we are interested in inclusion:

$$\mathrm{supp}(\mathcal{F}[f \cdot g]) \subset \mathrm{supp}(\mathcal{F}f) + \mathrm{supp}(\mathcal{F}g) \tag{4}$$

for $f, g \in \mathcal{S}'(\mathbf{R}^n)$ with at most polynomial growth at infinity. Usually, we assume that $\mathcal{F}f$ is compactly supported.

Let Ω be a bounded set in $\mathbf{R}^n$. Denote by $\mathcal{S}'_\Omega(\mathbf{R}^n)$ the set of all distributions whose Fourier transform is contained in the closure $\overline{\Omega}$. Define $\mathcal{S}_\Omega(\mathbf{R}^n) \equiv \mathcal{S}'_\Omega(\mathbf{R}^n) \cap \mathcal{S}(\mathbf{R}^n)$.

Lemma 1 *(i) For all $F \in C^\infty_{\mathrm{c}}(\mathbf{R}^n)$, $G \in \mathcal{S}(\mathbf{R}^n)$,*

$$\mathrm{supp}(F * G) \subset \mathrm{supp}(F) + \mathrm{supp}(G). \tag{5}$$

(ii) Let K be a compact set. Then for all $f \in \mathcal{S}_K(\mathbf{R}^n)$, $g \in \mathcal{S}(\mathbf{R}^n)$, (4) holds.

Proof (i) The proof of (5) is standard: Simply write out the convolution $f * g$ completely in terms of the integral to give

$$\begin{aligned}\{x \in \mathbf{R}^n \,:\, F * G(x) \neq 0\} &\subset \{x \in \mathbf{R}^n \,:\, F(x) \neq 0\} + \{x \in \mathbf{R}^n \,:\, G(x) \neq 0\} \\ &\subset \mathrm{supp}(F) + \mathrm{supp}(G).\end{aligned}$$

Since $\mathrm{supp}(F)$ is compact and $\mathrm{supp}(G)$ is closed, $\mathrm{supp}(F) + \mathrm{supp}(G)$ is a closed set. Thus, taking the closure of the above inclusion, we conclude that (5) holds.

(ii) Inclusion (4) is a consequence of $\mathcal{F}[f \cdot g] = (2\pi)^{-\frac{n}{2}} \mathcal{F}f * \mathcal{F}g$ and the fact that $\mathcal{F}$ maps $\mathcal{S}(\mathbf{R}^n)$ isomorphically.

Define the convolution $f * g$ by $f * g(x) \equiv \displaystyle\int_{\mathbf{R}^n} f(x-y)g(y)\mathrm{d}y$ as long as the integral makes sense.

A band-limited distribution is one whose Fourier transform is compactly supported.

Lemma 2 *For all band-limited distributions $f \in \mathcal{S}'(\mathbf{R}^n)$ and all functions $g \in \mathcal{S}(\mathbf{R}^n)$, (4) holds.*

Proof Let $\tau \in C_c^\infty(\mathbf{R}^n)$ be such that $\mathrm{supp}(\tau) \cap (\mathrm{supp}(\mathcal{F}f) + \mathrm{supp}(\mathcal{F}g)) = \emptyset$. We need to show that

$$\langle \mathcal{F}[f \cdot g], \tau \rangle = 0.$$

By the definition of the Fourier transform this is equivalent to showing

$$\langle f \cdot g, \mathcal{F}\tau \rangle = 0.$$

Since $g \in \mathcal{S}(\mathbf{R}^n)$,

$$\langle f \cdot g, \mathcal{F}\tau \rangle = \langle f, g \cdot \mathcal{F}\tau \rangle$$

from the definition of the pointwise multiplication $f \cdot g \in \mathcal{S}'(\mathbf{R}^n)$ for $f \in \mathcal{S}'(\mathbf{R}^n)$ and $g \in \mathcal{S}(\mathbf{R}^n)$. We note that

$$\mathcal{F}^{-1}[g \cdot \mathcal{F}\tau] = (2\pi)^{-\frac{n}{2}} \mathcal{F}^{-1} g * \tau.$$

Thus, by the definition of the Fourier transform $\mathcal{F}$ acting on $\mathcal{S}'(\mathbf{R}^n)$

$$\langle f \cdot g, \mathcal{F}\tau \rangle = (2\pi)^{-\frac{n}{2}} \langle \mathcal{F}f, \mathcal{F}^{-1} g * \tau \rangle.$$

From the definition of the Fourier transform $x \in \mathrm{supp}(\mathcal{F}g)$ if and only if $-x \in \mathrm{supp}(\mathcal{F}^{-1}g)$. Since $\mathrm{supp}(\tau) \cap (\mathrm{supp}(\mathcal{F}f) + \mathrm{supp}(\mathcal{F}g)) = \emptyset$,

$$\mathrm{supp}(\tau * \mathcal{F}^{-1}g) \cap \mathrm{supp}(\mathcal{F}f) \subset (\mathrm{supp}(\tau) + \mathrm{supp}(\mathcal{F}^{-1}g)) \cap \mathrm{supp}(\mathcal{F}f) = \emptyset.$$

thanks to Lemma 1. Thus, $\langle f \cdot g, \mathcal{F}\tau \rangle = 0$ and (4) holds.

Corollary 1 *For all band-limited $f, g \in \mathcal{S}(\mathbf{R}^n)$, (4) holds.*

Proof Let $\tau \in C_c^\infty(\mathbf{R}^n)$ be such that $\mathrm{supp}(\tau) \cap (\mathrm{supp}(\mathcal{F}f) + \mathrm{supp}(\mathcal{F}g)) = \emptyset$. We need to show that $\langle f \cdot g, \mathcal{F}\tau \rangle = 0$. Let $\Phi \in \mathcal{S}(\mathbf{R}^n)$ be such that $\Phi(0) = 1$ and that $\mathrm{supp}(\mathcal{F}\Phi) \subset B(1)$. Then

$$\langle f \cdot g, \mathcal{F}\tau \rangle = \lim_{\varepsilon \downarrow 0} \langle f \cdot g, \Phi(\varepsilon \cdot)^2 \mathcal{F}\tau \rangle,$$

since

$$\lim_{\varepsilon \downarrow 0} \Phi(\varepsilon \cdot)^2 \mathcal{F}\tau = \tau$$

in $\mathcal{S}(\mathbf{R}^n)$. If $\varepsilon > 0$ is chosen so that $\mathrm{supp}(\tau) \cap (\mathrm{supp}(\mathcal{F}f) + \mathrm{supp}(\mathcal{F}g) + \overline{B(2\varepsilon)}) = \emptyset$, then

$$\langle \Phi(\varepsilon \cdot) f \cdot \Phi(\varepsilon \cdot) g, \mathcal{F}\tau \rangle = 0$$

thanks to Lemma 2, since $\Phi(\varepsilon \cdot) f$ is band-limited due to Lemma 1. Thus, $\langle f \cdot g, \mathcal{F}\tau \rangle = 0$.

2.2 Lipschitz Spaces and Hölder–Zygmund Spaces

Let $0 < \alpha \le 1$. Let $\mathrm{Lip}^{\alpha}(\mathbf{R}^n)$ be the set of all bounded continuous functions $f : \mathbf{R}^n \to \mathbf{C}$ for which the quantity $\|f\|_{\mathrm{Lip}^{\alpha}} \equiv \|f\|_{L^\infty} + \sup\limits_{x,y\in\mathbf{R}^n} |x-y|^{-\alpha}|f(x)-f(y)|$ is finite. Let ψ satisfy (1). We write

$$\varphi_0(\xi) \equiv \psi(\xi), \quad \varphi_j(\xi) \equiv \psi(2^{-j}\xi) - \psi(2^{-j+1}\xi), \quad \psi_j(\xi) \equiv \psi(2^{-j}\xi)$$

for $j \in \mathbf{N}$ and $\xi \in \mathbf{R}^n$ as before. The (Besov)–Hölder–Zygmund space $\mathcal{C}^{\beta}(\mathbf{R}^n)$ with $\beta \in \mathbf{R}$. is defined as the set of all $f \in \mathcal{S}'(\mathbf{R}^n)$ for which

$$\|f\|_{\mathcal{C}^\beta} = \sup_{j\in\mathbf{N}_0} 2^{j\beta}\|\varphi_j(D)f\|_{L^\infty}$$

is finite. It is noteworthy that $\mathrm{Lip}^{\alpha}(\mathbf{R}^n)$ and $\mathcal{C}^{\alpha}(\mathbf{R}^n)$ are isomorphic for all $0 < \alpha < 1$ but $\mathrm{Lip}^{1}(\mathbf{R}^n)$ is a proper subset of $\mathcal{C}^{1}(\mathbf{R}^n)$. Although (1) is typically replaced by $\chi_{B(1)} \le \psi \le \chi_{B(2)}$, if a stronger condition (1) is posed on ψ, what we are doing can be quantified. The following is an example of such an attempt. It is used to measure the size of the support of the Fourier transform of the functions.

Example 1 Let $j, k, l \in \mathbf{N}$ satisfy $l \ge 2$.

(i) We note that $\varphi_k \cdot \psi_{l-2} \neq 0$ only if $l \ge k$. In this case,

$$\mathrm{supp}(\mathcal{F}[\varphi_k(D)\psi_{l-2}(D)f \cdot \varphi_l(D)g]) \subset \overline{B\left(\frac{3}{2}\cdot 2^k\right)} + \overline{B\left(\frac{3}{2}\cdot 2^l\right) \setminus B\left(\frac{3}{5}\cdot 2^l\right)}.$$

(ii) Assume $l \ge k+2$. Since

$$\frac{1}{8} < \frac{3}{5} - \frac{3}{8} < \frac{3}{2} + \frac{3}{8} < 2,$$

$$\mathrm{supp}(\mathcal{F}[\varphi_k(D)\psi_{l-2}(D)f \cdot \varphi_l(D)g]) \subset \overline{B\left(2^{l+1}\right) \setminus B\left(\frac{1}{8}\cdot 2^l\right)}.$$

Consequently,

$$\varphi_j(D)[\varphi_k(D)\psi_{l-2}(D)f \cdot \varphi_l(D)g] \neq 0$$

only if $l-3 \le j+1 \le l+1$ or $l-3 \le j-1 \le l+1$, or equivalently $l-4 \le j \le l+2$.

2.3 Estimates in Besov–Morrey Spaces

For the paraproducts, we use the following observation:

Lemma 3 *Let* $1 \le q \le p < \infty$, $1 \le r \le \infty$ *and* $s \in \mathbf{R}$. *Suppose that we have* $\{f_j\}_{j=1}^{\infty} \subset \mathcal{M}_q^p(\mathbf{R}^n)$ *satisfying* $f_0 \in \mathcal{S}'_{B(8)}(\mathbf{R}^n)$, $f_j \in \mathcal{S}'_{B(2^{j+3})\setminus B(2^{j-1})}(\mathbf{R}^n)$, $j \in \mathbf{N}$ *and*

$$\left(\sum_{j=0}^{\infty}(2^{js}\|f_j\|_{\mathcal{M}_q^p})^r\right)^{\frac{1}{r}} < \infty.$$

Then

$$f = \sum_{j=0}^{\infty} f_j \in \mathcal{N}_{pqr}^s(\mathbf{R}^n)$$

with

$$\|f\|_{\mathcal{N}_{pqr}^s} \le C\left(\sum_{j=0}^{\infty}(2^{js}\|f_j\|_{\mathcal{M}_q^p})^r\right)^{\frac{1}{r}}.$$

Let $j \in \mathbf{Z}$ and $\tau \in \mathcal{S}(\mathbf{R}^n)$. Then define $\tau_j \equiv \tau(2^{-j}\cdot)$.

Proof Let $\psi, \varphi_j \in C_c^{\infty}(\mathbf{R}^n)$ be as above for each $j \in \mathbf{N}_0$.
Then

$$\varphi_k(D)f = \sum_{j=\max(0,k-4)}^{k+4} \varphi_k(D)f_j \quad (k \in \mathbf{N}).$$

Thus,

$$\|\varphi_k(D)f\|_{\mathcal{M}_q^p} \le C \sum_{j=\max(0,k-4)}^{k+4} \|f_j\|_{\mathcal{M}_q^p} \quad (k \in \mathbf{N}).$$

Consequently,

$$\begin{aligned}
\|f\|_{\mathcal{N}_{pqr}^s} &= \left(\sum_{k=0}^{\infty}(2^{ks}\|\varphi_k(D)f\|_{\mathcal{M}_q^p})^r\right)^{\frac{1}{r}} \\
&\le C\left(\sum_{k=4}^{\infty}\left(2^{ks}\sum_{j=\max(0,k-4)}^{k+4}\|f_j\|_{\mathcal{M}_q^p}\right)^r\right)^{\frac{1}{r}} \\
&\le C\sum_{j=0}^{8}\|f_j\|_{\mathcal{M}_q^p} + C\left(\sum_{k=4}^{\infty}\left(2^{ks}\sum_{l=-4}^{4}\|f_{k+l}\|_{\mathcal{M}_q^p}\right)^r\right)^{\frac{1}{r}} \\
&\le C\sum_{j=0}^{4}\|f_j\|_{\mathcal{M}_q^p} + C\sum_{l=-4}^{4}\left(\sum_{k=4}^{\infty}\left(2^{ks}\|f_{k+l}\|_{\mathcal{M}_q^p}\right)^r\right)^{\frac{1}{r}}
\end{aligned}$$

$$\le C \left(\sum_{j=0}^{\infty} (2^{js} \| f_j \|_{\mathcal{M}^p_q})^r \right)^{\frac{1}{r}}.$$

3 Paraproduct

3.1 Paraproduct

For the paraproducts, we use the following observation:

Lemma 4 *Let* $1 \le q_1 \le p_1 < \infty$, $1 \le r_1 \le \infty$, $1 \le q_2 \le p_2 < \infty$, $1 \le r_2 \le \infty$, $1 \le q \le p < \infty$, $1 \le r \le \infty$, *and* $s_0, s_1, s \in \mathbf{R}$. *Assume that*

$$\frac{1}{p} = \frac{1}{p_1} + \frac{1}{p_2}, \quad \frac{1}{q} = \frac{1}{q_1} + \frac{1}{q_2}, \quad \frac{1}{r} = \frac{1}{r_1} + \frac{1}{r_2}, \quad s = s_1 + s_2.$$

Suppose that we are given collections $\{f_j\}_{j=1}^{\infty}, \{g_j\}_{j=1}^{\infty} \subset \mathcal{M}^p_q(\mathbf{R}^n)$ *satisfying* $f_j \in \mathcal{S}'_{B(2^{j-1})}(\mathbf{R}^n)$, $g_j \in \mathcal{S}'_{B(2^{j+2}) \setminus B(2^j)}(\mathbf{R}^n)$, $j \in \mathbf{N}$ *and*

$$\left(\sum_{j=1}^{\infty} (2^{js_1} \| f_j \|_{\mathcal{M}^{p_1}_{q_1}})^{r_1} \right)^{\frac{1}{r_1}}, \quad \left(\sum_{j=1}^{\infty} (2^{js_2} \| g_j \|_{\mathcal{M}^{p_2}_{q_2}})^{r_2} \right)^{\frac{1}{r_2}} < \infty.$$

Then

$$\sum_{j=1}^{\infty} f_j \cdot g_j \in \mathcal{N}^s_{pqr}(\mathbf{R}^n)$$

and

$$\left\| \sum_{j=1}^{\infty} f_j \cdot g_j \right\|_{\mathcal{N}^s_{pqr}} \le C \left(\sum_{j=1}^{\infty} (2^{js_1} \| f_j \|_{\mathcal{M}^{p_1}_{q_1}})^{r_1} \right)^{\frac{1}{r_1}} \left(\sum_{j=1}^{\infty} (2^{js_2} \| g_j \|_{\mathcal{M}^{p_2}_{q_2}})^{r_2} \right)^{\frac{1}{r_2}}.$$

Proof Thanks to Corollary 1, we have $\mathrm{supp}(f_j \cdot g_j) \subset \overline{B(2^{j+3}) \setminus B(2^{j-1})}$ for all $j \in \mathbf{N}$. Thus by the equivalent expression (see Lemma 3) and the Hölder inequality, we have

$$\left\| \sum_{j=1}^{\infty} f_j \cdot g_j \right\|_{\mathcal{N}^s_{pqr}} \le C \left(\sum_{j=1}^{\infty} (2^{js} \| f_j \cdot g_j \|_{\mathcal{M}^p_q})^r \right)^{\frac{1}{r}}$$
$$\le C \left(\sum_{j=1}^{\infty} (2^{j(s_1+s_2)} \| f_j \|_{\mathcal{M}^{p_1}_{q_1}} \| g_j \|_{\mathcal{M}^{p_2}_{q_2}})^r \right)^{\frac{1}{r}}$$
$$\le C \left(\sum_{j=1}^{\infty} (2^{js_1} \| f_j \|_{\mathcal{M}^{p_1}_{q_1}})^{r_1} \right)^{\frac{1}{r_1}} \left(\sum_{j=1}^{\infty} (2^{js_2} \| f_j \|_{\mathcal{M}^{p_2}_{q_2}})^{r_2} \right)^{\frac{1}{r_2}}.$$

3.2 Resonant Part

To handle the resonant part, To prove this type of estimates, the atomic decomposition can be used by taking advantage of the assumption $s > 0$ and $p, q, r \ge 1$. Here we estimate the distributions directly. This corresponds to [2, Lemma A3].

Lemma 5 *Let* $1 \le q \le p < \infty$, $1 \le r \le \infty$ *and* $s > 0$. *Suppose that we are given a collection* $\{f_j\}_{j=0}^{\infty} \subset \mathcal{M}^p_q(\mathbf{R}^n)$ *satisfying* $f_j \in \mathcal{S}'_{B(2^{j+2})}(\mathbf{R}^n)$, $j \in \mathbf{N}_0$ *and*

$$\left(\sum_{j=0}^{\infty} (2^{js} \| f_j \|_{\mathcal{M}^p_q})^r \right)^{\frac{1}{r}} < \infty.$$

Then

$$\sum_{j=0}^{\infty} f_j \in \mathcal{N}^s_{pqr}(\mathbf{R}^n).$$

Proof Let $\psi, \varphi_j \in C^\infty_{\rm c}(\mathbf{R}^n)$ be as above for each $j \in \mathbf{N}_0$. We have

$$2^{ks} \sum_{j=0}^{\infty} |\varphi_k(D) f_j| \le \sum_{j=\max(0,k-3)}^{\infty} 2^{(k-j)s} |\varphi_k(D)[2^{js} f_j]|.$$

Consequently, by the translation invariance of $\mathcal{M}^p_q(\mathbf{R}^n)$ and the equality

$$\|\mathcal{F}^{-1}\varphi_k\|_{L^1} = \|\mathcal{F}^{-1}\varphi_1\|_{L^1}$$

for all $k \in \mathbf{N}$

$$\left\| 2^{ks} \sum_{j=0}^{\infty} |\varphi_k(D) f_j| \right\|_{\mathcal{M}^p_q} \le C \sum_{j=\max(0,k-3)}^{\infty} 2^{(k-j)s} (2^{js} \| \varphi_k(D) f_j \|_{\mathcal{M}^p_q})$$
$$\le C \sum_{j=\max(0,k-3)}^{\infty} 2^{(k-j)s} (2^{js} \| f_j \|_{\mathcal{M}^p_q}).$$

Since $s > 0$, by the Hölder inequality

$$\left\| 2^{ks} \sum_{j=0}^{\infty} |\varphi_k(D) f_j| \right\|_{\mathcal{M}^p_q}$$
$$\le C \sum_{j=\max(0,k-3)}^{\infty} 2^{\frac{1}{2}(k-j)s} 2^{\frac{1}{2}(k-j)s} (2^{js} \| f_j \|_{\mathcal{M}^p_q})$$
$$\le C \left(\sum_{j=\max(0,k-3)}^{\infty} 2^{\frac{1}{2}(k-j)sr'} \right)^{\frac{1}{r'}} \left(\sum_{j=\max(0,k-3)}^{\infty} (2^{\frac{1}{2}(k-j)s} (2^{js} \| f_j \|_{\mathcal{M}^p_q}))^r \right)^{\frac{1}{r}}$$
$$\le C \left(\sum_{j=\max(0,k-3)}^{\infty} (2^{\frac{1}{2}(k-j)s} (2^{js} \| f_j \|_{\mathcal{M}^p_q}))^r \right)^{\frac{1}{r}}.$$

Thus, taking the ℓ^r-norm gives

$$\left\| \sum_{j=0}^{\infty} f_j \right\|_{\mathcal{N}^s_{pqr}} \le C \left(\sum_{j=0}^{\infty} (2^{js} \| f_j \|_{\mathcal{M}^p_q})^r \right)^{\frac{1}{r}}.$$

Corollary 2 *Let* $1 \le q_1 \le p_1 < \infty$, $1 \le r_1 \le \infty$, $1 \le q_2 \le p_2 < \infty$, $1 \le r_2 \le \infty$, $1 \le q \le p < \infty$, $1 \le r \le \infty$, *and* $s_0, s_1, s \in \mathbf{R}$. *Assume that*

$$\frac{1}{p} = \frac{1}{p_1} + \frac{1}{p_2}, \quad \frac{1}{q} = \frac{1}{q_1} + \frac{1}{q_2}, \quad \frac{1}{r} = \frac{1}{r_1} + \frac{1}{r_2}, \quad s = s_1 + s_2 > 0.$$

Suppose that we are given collections $\{f_j\}_{j=0}^{\infty}, \{g_j\}_{j=0}^{\infty} \subset \mathcal{M}^p_q(\mathbf{R}^n)$ *satisfying* f_j, $g_j \in \mathcal{S}'_{B(2^{j+1})}(\mathbf{R}^n)$, $j \in \mathbf{N}_0$ *and*

$$\left(\sum_{j=0}^{\infty} (2^{js_1} \| f_j \|_{\mathcal{M}^{p_1}_{q_1}})^{r_1} \right)^{\frac{1}{r_1}}, \quad \left(\sum_{j=0}^{\infty} (2^{js_2} \| g_j \|_{\mathcal{M}^{p_2}_{q_2}})^{r_2} \right)^{\frac{1}{r_2}} < \infty$$

Then

$$\sum_{j=0}^{\infty} f_j \cdot g_j \in \mathcal{N}^s_{pqr}(\mathbf{R}^n).$$

and

$$\left\| \sum_{j=0}^{\infty} f_j \cdot g_j \right\|_{\mathcal{N}^s_{pqr}} \le C \left(\sum_{j=0}^{\infty} (2^{js_1} \| f_j \|_{\mathcal{M}^{p_1}_{q_1}})^{r_1} \right)^{\frac{1}{r_1}} \left(\sum_{j=0}^{\infty} (2^{js_2} \| g_j \|_{\mathcal{M}^{p_2}_{q_2}})^{r_2} \right)^{\frac{1}{r_2}}.$$

Proof In fact, by Corollary 1, we see that $f_j \cdot g_j \in \mathcal{S}'_{B(2^{j+2})}(\mathbf{R}^n)$. Thus, invoking Lemma 5 and using the Hölder inequality twice gives

$$\begin{aligned} \left\| \sum_{j=0}^{\infty} f_j \cdot g_j \right\|_{\mathcal{N}^s_{pqr}} &\le C \left(\sum_{j=0}^{\infty} (2^{js} \| f_j \cdot g_j \|_{\mathcal{M}^p_q})^r \right)^{\frac{1}{r}} \\ &\le C \left(\sum_{j=0}^{\infty} (2^{js_1} \| f_j \|_{\mathcal{M}^{p_1}_{q_1}})^{r_1} \right)^{\frac{1}{r_1}} \left(\sum_{j=0}^{\infty} (2^{js_2} \| g_j \|_{\mathcal{M}^{p_2}_{q_2}})^{r_2} \right)^{\frac{1}{r_2}}. \end{aligned}$$

Remark 1 Under the understanding $\mathcal{M}^\infty_q(\mathbb{R}^n) = L^\infty(\mathbb{R}^n)$ for any $0 < q < \infty$, one may assume $p_1 = \infty$.

3.3 Conclusion of the Proof of Theorem 1

We prove Theorem 1 below. Lemma 4 gives

$$\begin{aligned} \| f \preceq g \|_{\mathcal{N}^s_{pqr}} &= \left\| \sum_{j=2}^{\infty} \psi_{j-2}(D) f \cdot \varphi_j(D) g \right\|_{\mathcal{N}^s_{pqr}} \\ &\le C \sup_{j \in \mathbf{N}_0} \| \psi_j(D) f \|_{\mathcal{M}^{p_1}_{q_1}} \left(\sum_{j=2}^{\infty} (2^{js} \| g_j \|_{\mathcal{M}^{p_2}_{q_2}})^r \right)^{\frac{1}{r}}. \end{aligned}$$

Since $\psi_j(D) f = (2\pi)^{-\frac{n}{2}} \mathcal{F}^{-1}\psi_j * f$ and $\mathcal{F}^{-1}\psi_j = 2^{jn}\mathcal{F}^{-1}\psi(2^j \cdot)$, we have

$$\begin{aligned} \| \psi_j(D) f \|_{\mathcal{M}^{p_1}_{q_1}} &= (2\pi)^{-\frac{n}{2}} \| \mathcal{F}^{-1}\psi_j * f \|_{\mathcal{M}^{p_1}_{q_1}} \\ &\le (2\pi)^{-\frac{n}{2}} \| \mathcal{F}^{-1}\psi_j \|_{L^1} \| f \|_{\mathcal{M}^{p_1}_{q_1}} \\ &= (2\pi)^{-\frac{n}{2}} \| \mathcal{F}^{-1}\psi \|_{L^1} \| f \|_{\mathcal{M}^{p_1}_{q_1}}. \end{aligned}$$

Thus,

$$\|f \preceq g\|_{\mathcal{N}^s_{pqr}} \le C\|f\|_{\mathcal{M}^{p_1}_{q_1}} \left(\sum_{j=2}^{\infty}(2^{js}\|g_j\|_{\mathcal{M}^{p_2}_{q_2}})^r\right)^{\frac{1}{r}}.$$

Recall that $s > 0$. Since $\mathcal{M}^{p_1}_{q_1}(\mathbf{R}^n) \supset \mathcal{N}^s_{p_1q_1r}(\mathbf{R}^n)$, we have

$$\|f \preceq g\|_{\mathcal{N}^s_{pqr}} \le C\|f\|_{\mathcal{N}^s_{p_1q_1r}}\|g\|_{\mathcal{N}^s_{p_2q_2r}}.$$

Likewise,

$$\|f \succeq g\|_{\mathcal{N}^s_{pqr}} \le C\|f\|_{\mathcal{N}^s_{p_1q_1r}}\|g\|_{\mathcal{N}^s_{p_2q_2r}}.$$

Meanwhile, we have

$$\|f \odot g\|_{\mathcal{N}^s_{pqr}} \le \|f \odot g\|_{\mathcal{N}^{2s}_{pqr}} \le C\|f\|_{\mathcal{N}^s_{p_1q_12r}}\|g\|_{\mathcal{N}^s_{p_2q_22r}} \le C\|f\|_{\mathcal{N}^s_{p_1q_1r}}\|g\|_{\mathcal{N}^s_{p_2q_2r}}$$

by Corollary 2. Combining these observations provides the desired result.

4 Commutator Estimate

We recall the following lemma obtained in [2, Lemma 2.2]:

Lemma 6 *Let* $0 < \alpha \le 1$, $j \in \mathbf{N}_0$, *and let* $F \in \mathrm{Lip}^\alpha(\mathbf{R}^n)$, $G \in L^\infty(\mathbf{R}^n)$. *Then*

$$\|\varphi_j(D)[F \cdot G] - F\varphi_j(D)G\|_{L^\infty} \le C2^{-j\alpha}\|F\|_{\mathrm{Lip}^\alpha}\|G\|_{L^\infty}.$$

This is a slight extension of [2, Lemma 2.2] for the case where $\alpha = 1$. For the sake of convenience, we recall the whole proof.

Proof Since $\varphi_j(D)H(x) = (2\pi)^{-\frac{n}{2}}\mathcal{F}^{-1}\varphi_j * H(x)$ for all $H \in \mathcal{S}'(\mathbf{R}^n)$ that grows polynomially at infinity,

$$\begin{aligned}&\varphi_j(D)[F \cdot G](x) - F(x)\varphi_j(D)G(x)\\ &= (2\pi)^{-\frac{n}{2}}\int_{\mathbf{R}^n} 2^{jn}\mathcal{F}^{-1}\varphi(2^j(x-y))(F(y)-F(x))G(y)\mathrm{d}y.\end{aligned}$$

Hence letting

$$C \equiv (2\pi)^{-\frac{n}{2}}\int_{\mathbf{R}^n} |z|^\alpha|\mathcal{F}^{-1}\varphi(z)|\mathrm{d}z$$

gives

$$\|\varphi_j(D)[F \cdot G] - F\varphi_j(D)G\|_{L^\infty} \le C2^{-j\alpha}\|F\|_{\mathrm{Lip}^\alpha}\|G\|_{L^\infty},$$

as required.

Lemma 7 *Let* $0 < \alpha \le 1$, $j \in \mathbf{N}_0$, *and let* $F \in \mathrm{Lip}^\alpha(\mathbf{R}^n)$, $G \in L^\infty(\mathbf{R}^n)$. *Then*

$$\|\varphi_j(D)[F \preceq G] - F\varphi_j(D)G\|_{L^\infty} \le C2^{-j\alpha}\|F\|_{\mathrm{Lip}^\alpha}\|G\|_{L^\infty}.$$

This is a slight extension of [2, Lemma 2.3] to the case where $\alpha = 1$. For the sake of convenience we supply the proof.

Proof Assume $j \gg 1$; otherwise we can mimic the argument below and can readily incorporate the case where j is small. We decompose

$$\begin{aligned}
&\varphi_j(D)[F \preceq G] - F\varphi_j(D)G \\
&= \sum_{k=j-3}^{j+3} \left(\varphi_j(D)[F \preceq \varphi_k(D)G] - F\varphi_j(D)\varphi_k(D)G\right) \\
&= \sum_{k=j-3}^{j+3} \left(\varphi_j(D)[F \cdot \varphi_k(D)G] - F\varphi_j(D)\varphi_k(D)G - \varphi_j(D)[F \succeq \varphi_k(D)G]\right).
\end{aligned}$$

Let k be fixed. Lemma 6 gives

$$\|\varphi_j(D)[F \cdot \varphi_k(D)G] - F\varphi_j(D)\varphi_k(D)G\|_{L^\infty} \le C2^{-j\alpha}\|F\|_{\mathrm{Lip}^\alpha}\|G\|_{L^\infty}.$$

Meanwhile, using

$$\varphi_j(D)[F \succeq \varphi_k(D)G] = \sum_{l=j-5}^{j+5} \varphi_j(D)[\varphi_l(D)F \succeq \varphi_k(D)G]$$

for $k \in [j-3, j+3]$ yields

$$\begin{aligned}
\|\varphi_j(D)[F \succeq \varphi_k(D)G]\|_{L^\infty} &\le \sum_{l=j-5}^{j+5} \|\varphi_j(D)[\varphi_l(D)F \succeq \varphi_k(D)G]\|_{L^\infty} \\
&\le C \sum_{l=j-5}^{j+5} \|\varphi_l(D)F \succeq \varphi_k(D)G\|_{L^\infty} \\
&\le C \sup_{l,l' \in \mathbf{N}_0} \|\varphi_l(D)F \cdot \varphi_{l'}(D)G\|_{L^\infty} \\
&\le C2^{-j\alpha}\|F\|_{\mathrm{Lip}^\alpha}\|G\|_{L^\infty}.
\end{aligned}$$

We prove Theorem 2 to conclude this note.

Proof We decompose

$$(f \preceq g) \odot h - f(g \odot h) = \sum_{j=0}^{\infty}[\varphi_j(D)[f \preceq g] - f \cdot \varphi_j(D)g]\varphi_j(D)h$$
$$+\sum_{j=1}^{\infty}(\varphi_{j-1}(D)[f \preceq g] - f\varphi_{j-1}(D)g)\varphi_j(D)h$$
$$+\sum_{j=1}^{\infty}[\varphi_j(D)[f \preceq g] - f \cdot \varphi_j(D)g]\varphi_{j-1}(D)h.$$

Below we handle the first term only because the other two terms are dealt with in a similar manner. We decompose

$$\sum_{j=0}^{\infty}[\varphi_j(D)[f \preceq g] - f \cdot \varphi_j(D)g]\varphi_j(D)h$$
$$= \sum_{j=0}^{\infty}[\varphi_j(D)[\psi_{j+4}(D)f \preceq g] - \psi_{j+4}(D)f \cdot \varphi_j(D)g]\varphi_j(D)h$$
$$+\sum_{j=0}^{\infty}\sum_{k=j+5}^{\infty} \varphi_j(D)[\varphi_k(D)f \preceq g] \cdot \varphi_j(D)h$$
$$-\sum_{j=0}^{\infty}\sum_{k=j+5}^{\infty} \varphi_k(D)f \cdot \varphi_j(D)g \cdot \varphi_j(D)h.$$

Since

$$\varphi_j(D)[\psi_{j+4}(D)f \preceq g] - \psi_{j+4}(D)f \cdot \varphi_j(D)g|_{L^\infty} = \mathrm{O}(2^{-j(\alpha+\beta)}|f|_{\mathrm{Lip}^\alpha}|g|_{\mathcal{C}^\beta})$$

for all $m = (m_1, m_2, \ldots, m_n) \in \mathbf{N}_0{}^n$, we have

$$\left\|\sum_{j=0}^{\infty}[\varphi_j(D)[\psi_{j+4}(D)f \preceq g] - \psi_{j+4}(D)f \cdot \varphi_j(D)g]\varphi_j(D)h\right\|_{\mathcal{N}_{pqr}^{s+\alpha+\beta}}$$
$$\leq C\|f\|_{\mathrm{Lip}^\alpha}\|g\|_{\mathcal{C}^\beta}\|h\|_{\mathcal{N}_{pqr}^s}.$$

Using Example 1, we estimate the second term as

$$\sum_{j=0}^{\infty}\sum_{k=j+5}^{\infty} \varphi_j(D)[\varphi_k(D)f \preceq g] \cdot \varphi_j(D)h$$

$$
\begin{aligned}
&= \sum_{j=0}^{\infty} \sum_{k=j+5}^{\infty} \varphi_j(D)[\varphi_k(D)\psi_{k-2}(D)f \cdot \varphi_k(D)g] \cdot \varphi_j(D)h \\
&\quad + \sum_{j=0}^{\infty} \sum_{k=j+5}^{\infty} \varphi_j(D)[\varphi_k(D)\psi_{k-1}(D)f \cdot \varphi_{k+1}(D)g] \cdot \varphi_j(D)h.
\end{aligned}
$$

Next, we note that

$$
\begin{aligned}
&\|\varphi_k(D)f \cdot \varphi_j(D)g \cdot \varphi_j(D)h\|_{\mathcal{M}^p_q} \\
&\le C2^{-k\alpha-j(s+\beta)}\|f\|_{\mathrm{Lip}^\alpha}\|g\|_{\mathcal{C}^\beta}\|2^{js}\varphi_j(D)h\|_{\mathcal{M}^p_q}.
\end{aligned}
$$

Adding this estimate over j, k gives

$$
\begin{aligned}
&\left\{\sum_{k=5}^{\infty}\left(2^{k(s+\alpha+\beta)}\left\|\sum_{j=0}^{k-5}\varphi_k(D)f \cdot \varphi_j(D)g \cdot \varphi_j(D)h\right\|_{\mathcal{M}^p_q}\right)^r\right\}^{\frac{1}{r}} \\
&\le C\left\{\sum_{k=5}^{\infty}\left(\sum_{j=0}^{k-5}2^{(k-j)(s+\beta)}\|f\|_{\mathrm{Lip}^\alpha}\|g\|_{\mathcal{C}^\beta}\|2^{js}\varphi_j(D)h\|_{\mathcal{M}^p_q}\right)^r\right\}^{\frac{1}{r}} \\
&= C\left\{\sum_{k=5}^{\infty}\sum_{j=0}^{k-5}\left(2^{\frac{1}{2}(k-j)(s+\beta)}\|f\|_{\mathrm{Lip}^\alpha}\|g\|_{\mathcal{C}^\beta}\|2^{js}\varphi_j(D)h\|_{\mathcal{M}^p_q}\right)^r\right\}^{\frac{1}{r}} \\
&= C\left\{\sum_{j=0}^{\infty}\left(\|f\|_{\mathrm{Lip}^\alpha}\|g\|_{\mathcal{C}^\beta}\|2^{js}\varphi_j(D)h\|_{\mathcal{M}^p_q}\right)^r\right\}^{\frac{1}{r}} \\
&= C\|f\|_{\mathrm{Lip}^\alpha}\|g\|_{\mathcal{C}^\beta}\|h\|_{\mathcal{N}^s_{pqr}}.
\end{aligned}
$$

Acknowledgements The author is thankful to Professor Alexey Karapetyants for his invitation to the conference OTHA 2018. The author is also grateful to Professors Yuzuru Inahama and Masato Hoshino for their encouragement to write this note.

References

1. Bony, J.M.: Calcul symbolique et propagation des singularités pour les équations aux dérivées partielles non linéaires. (French), [Symbolic calculus and propagation of singularities Quantitative analysis in Sobolev imbedding theorems for and applications to spectral theory, nonlinear partial differential equations] Ann. Sci. École Norm. Sup. (4) **14**(2), 209–246 (1981)
2. Gubinelli, M., Imkeller, P., Perkowski, N.: Paracontrolled distributions and singular PDEs. Forum Math. Pi, 3:e6, 75 (2015)
3. Gubinelli, M., Imkeller, P., Perkowski, N.: A fourier approach to pathwise stochastic integration. Electron. J. Probab. **21**, Number 2016, paper no. 2, 37 pp (2016)

4. Hairer, M.: A theory of regularity structures. Invent. Math. **198**(2), 269–504 (2014)
5. Haroske, D.D., Skrzypczak, L.: On Sobolev and Franke-Jawerth embeddings of smoothness Morrey spaces. Rev. Mat. Complut. **27**(2), 541–573 (2014)
6. Kozono, H., Yamazaki, M.: Semilinear heat equations and the Navier-Stokes equation with distributions in new function spaces as initial data. Commun. PDE **19**, 959–1014 (1994)
7. Mazzucato, A.L.: Decomposition of Besov–Morrey spaces. Harmonic analysis at Mount Holyoke, pp. 279–294. South Hadley, MA (2001). Contemp. Math. **320**, Am. Math. Soc. Providence, RI (2003)
8. Mazzucato, A.L.: Besov-Morrey spaces: function space theory and applications to non-linear PDE. Trans. Am. Math. Soc. **355**(4), 1297–1364 (2003)
9. Netrusov, Y.V.: Some imbedding theorems for spaces of Besov–Morrey type. (Russian), Numerical methods and questions in the organization of calculations, 7. Zap. Nauchn. Sem. Leningrad. Otdel. Mat. Inst. Steklov. (LOMI) **139**, 139–147 (1984)
10. Najafov, A.M.: Some properties of functions from the intersection of Besov-Morrey type spaces with dominant mixed derivatives. Proc. A. Razmadze Math. Inst. **139**, 71–82 (2005)
11. Rosenthal, M.: Local means, wavelet bases and wavelet isomorphisms in Besov-Morrey and Triebel-Lizorkin-Morrey spaces. Math. Nachr. **286**(1), 59–87 (2013)
12. Sawano, Y.: Theory of Besov spaces. Development in Mathematics, vol. 56. Springer
13. Sawano, Y., Sugano, S., Tanaka, H.: Identification of the image of Morrey spaces by the fractional integral operators. Proc. A. Razmadze Math. Inst. **149**, 87–93 (2009)
14. Sawano, Y., Tanaka, H.: Decompositions of Besov-Morrey spaces and Triebel-Lizorkin-Morrey spaces. Math. Z. **257**(4), 871–905 (2007)
15. Sawano, Y., Yang, D., Yuan, W.: New applications of Besov-type and Triebel-Lizorkin-type spaces. J. Math. Anal. Appl. **363**, 73–85 (2010)
16. Tang, L., Xu, J.S.: Some properties of Morrey type Besov-Triebel spaces. Math. Nachr. **278**, 904–917 (2005)
17. Yang, D., Yuan, W.: A new class of function spaces connecting Triebel-Lizorkin spaces and Q spaces. J. Funct. Anal. **255**, 2760–2809 (2008)
18. Yang, D., Yuan, W.: New Besov-type spaces and Triebel-Lizorkin-type spaces including Q spaces. Math. Z. **265**, 451–480 (2010)
19. Yuan, W., Sickel, W., Yang, D.: Morrey and Campanato Meet Besov, Lizorkin and Triebel. Lecture Notes in Mathematics, vol. 2005. Springer, Berlin (2010)

Part II
Functional Analysis and Operator Theory

Analogs of the Khintchin—Kolmogorov Inequalities in Discrete Morrey Spaces

Evgenii I. Berezhnoĭ

Abstract We prove certain estimates of the Rademacher sums in the discrete local LM_{l,L^p} and global GM_{l,L^p} Morrey function spaces. Using this estimates, we give necessary and sufficient conditions for the validity of an analogue of the Khinchin-Kolmogorov inequality in the spaces LM_{l,L^p} and GM_{l,L^p}.

Keywords Banach ideal spaces · Rademacher sums · Khintchin—Kolmogorov inequalities · Discrete local and global Morrey spaces · Discrete approximate local and global Morrey space

2010 Mathematics Subject Classification: 46E30 · 46B20 · 46B42

The Rademacher series play a special role in the study of geometric properties of Banach spaces. Investigations of the Rademacher sums in symmetric spaces functions are well presented in the books [1, 8, 11]. On the other hand, the Morrey spaces M_{λ,L^p} [12] and their various generalizations [4–6, 14] have recently come into play in analysis.

The main purpose of this paper is to investigate the behavior of the Rademacher series in the discrete local and global Morrey function spaces, which are not rearrangement invariant.

The first result on bounds for the Rademacher series in Morrey spaces is due to S. Astashkin and L. Maligranda. In [2], they proved an analogue of the Khintchin—Kolmogorov inequality in the classical Morrey space $M_{L^\infty_w,L^p}$. In this paper, we consider discrete local LM_{l,L^p} and global GM_{l,L^p} spaces constructed on the basis of the Lebesgue space L^p while the outer norm is calculated in an arbitrary ideal space of sequences l. We give necessary and sufficient conditions for the validity of an analogue of the Khinchin-Kolmogorov inequality in the spaces LM_{l,L^p} and GM_{l,L^p}. We note that the results thus obtained are also of interest for the classical Morrey spaces.

The reported study was funded by RFBR according to the research project 18-51-06005.

E. I. Berezhnoĭ (✉)
Yaroslavl State University, Yaroslavl, Russia
e-mail: ber@uniyar.ac.ru

A. Karapetyants et al. (eds.), *Modern Methods in Operator Theory and Harmonic Analysis*, Springer Proceedings in Mathematics & Statistics 291,
https://doi.org/10.1007/978-3-030-26748-3_9

1 Introduction and Preliminaries

Let $T = [0, 1]$ be a segment, with 0 and 1 identified. Let μ be the Lebesgue measure on T, let $S(\mu)$ be the space of measurable periodic functions $f : T \to R$, and let $\chi(D)$ stand for the characteristic function of a set $D \subseteq T$. A Banach space X of measurable functions on T is said to be ideal [7, 8] if it follows from the condition $f \in X$, the measurability of g, and the validity of the inequality $|g(t)| \leq |f(t)|$ for almost all $t \in T$ that $g \in X$ and $\|g|X\| \leq \|f|X\|$ (the symbol $\|f|X\|$ denotes the norm of an element f in the space X).

Classical examples of ideal Banach spaces are the spaces of Orlicz, Lorentz, Marcinkiewicz, and symmetric spaces [7, 8, 11].

For every ideal space X, the dual ideal space $X^{'}$ is well defined: it consists of the functionals f, continuous on X and representable in an integral form, with the norm defined by the equation

$$\|f|X^{'}\| = \sup\{\int_T f(t)g(t)dt : \|g|X\| \leq 1\}.$$

Let $v \in S(\mu)$ and $v > 0$ almost everywhere (v is a weight). We denote by the symbol X_v a new ideal space in which the norm is given by the equation $\|f|X_v\| = \|f \cdot v|X\|$. One can readily verify the equation

$$(X_v)^{'} = (X^{'})_{1/v}. \tag{1}$$

When $X = L^p$, our definition of weighted space differs somewhat from the standard one: if $X = L^p$, the weight is usually included in the measure.

Along with function spaces, we need ideal sequence spaces. Let $e^i = \{, ..., 0, 1, 0, ...\}$ (here 1 is the i-th entry) be the standard basis in the space of one- or two-sided sequences. We denote by the symbol l an ideal space of sequences $x = \sum_i x_i e^i$, with the norm $\|x|l\|$.

The definitions of the Orlicz, Lorentz and Marcinkiewicz and other symmetric sequence spaces are similar to the definitions of the corresponding function spaces. All the properties listed above for function spaces, including (1), are valid for sequence spaces. For details concerning the theory of sequence spaces, see [10].

The classical Morrey space M_{λ,L^p}, $(\lambda \in R)$, see [12], consists of the functions $f \in L_1^{loc}(R^n)$ for which the following norm is finite:

$$\|f|M_{\lambda,L^p}\| = \sup_{\tau>0} \tau^{-\lambda}\|f\chi(B(0,\tau))|L^p\|.$$

We note that if $\lambda = 0$, then $M_{\lambda,L^p} = L^p$; if $\lambda = n/p$, then $M_{\lambda,L^p} = L^\infty$. If $\lambda < 0$ or $\lambda > n/p$, then M_{λ,L^p}, $\lambda \in R$ is trivial; more precisely, only the function identically equal to zero belongs to it.

If we now replace the Lebesgue space L^p in the definition of the classical Morrey space by an ideal space X, we obtain the Morrey space $M_{\lambda,X}$, constructed from the ideal space X, with the norm defined by the equation

$$\|f|M_{\lambda,X}\| = \sup_{\tau>0} \tau^{-\lambda} \|f\chi(B(0,\tau))|X\|.$$

The next step in the extension of Morrey spaces is the replacement of the outer sup-norm by the norm in an ideal space L and the replacement of the balls $B(0,\tau)$ by the homothetic sets $U(0,\tau) \subset R^n$. Below, we always assume that $0 \in U(0,1)$, $U(0,1)$ is star-shaped with respect to the point 0, that is, if $t \in U(0,1)$, then $\lambda t \in U(0,1)$ for $\lambda \in (0,1)$, and $\mu(U(0,1)) > 0$.

Thus, the natural definition of local Morrey space constructed from the family $U(0,\tau)$, with continuous parameter τ, has the following form (see [3]).

Definition 1 Given an ideal space X on R^n, an ideal space L on R_+, and a star-shaped set $U(0,1) \subset R^n$, $(\mu(U(0,1)) > 0)$, the local Morrey space $LM_{L,X}$ consists of those $f \in L^{1,loc}(R^n)$ for which the following norm is finite:

$$\|f|LM_{L,X}\| = \| \|f\chi(U(0,\cdot))|X\| |L\|.$$

The global Morrey space $GM_{l,X}$ consists of those $f \in L^{1,loc}(R^n)$ for which the following norm is finite:

$$\|f|GM_{L,X}\| = \sup_h \|f(h+.)|LM_{L,X}\|.$$

The spaces introduced in the definition 1 are called continuous local and global Morrey spaces, respectively. Along with the space $M_{L,X}$, we need the local and global Morrey spaces constructed from a family of sets $U(0,\tau_i)$, with the discrete parameter (see [3]).

We denote by Υ the set of non-negative number sequences $\tau = \{\tau_i\}$ each of which satisfies the conditions $\forall i : \tau_i < \tau_{i+1}$, $\lim_{i\to-\infty} \tau_i = 0$.

For every sequence $\tau = \{\tau_i\}$, we construct a family of sets $U(0,\tau_i)$ and a family of disjoint annuli $R(0,\tau_{i-1},\tau_i) = U(0,\tau_i) \setminus U(0,\tau_{i-1})$.

Definition 2 Given an ideal space X on R^n, a star-shaped set $U(0,1) \subset R^n$ for which $\mu(U(0,1)) > 0$, an ideal space l sequences with the standard basis $\{e^i\}$ and a sequence $\tau \in \Upsilon$, the discrete local Morrey space $LM^{\tau}_{l,X}$ consists of those $f \in L^{1,loc}(R^n)$ for which the following norm is finite:

$$\|f|LM^{\tau}_{l,X}\| = \| \Sigma_i e^i \|f\chi(U(0,\tau_i))|X\| |l\|.$$

The discrete global Morrey space $GM^{\tau}_{l,X}$ consists of those $f \in L^{1,loc}(R^n)$ for which the following norm is finite:

$$\|f|GM^{\tau}_{l,X}\| = \sup_h \|f(h+.)|LM^{\tau}_{l,X}\|.$$

Definition 3 Given an ideal space X on R^n, a star-shaped set $U(0, 1) \subset R^n$ for which $\mu(U(0, 1)) > 0$, an ideal space l of sequences with the standard basis $\{e^i\}$ and a sequence $\tau \in \Upsilon$, by the approximate local Morrey space $\overline{LM^{\tau}_{l,X}}$ we mean the set of functions $f \in L^{1,loc}(R^n)$ for each of which the following norm is finite:

$$\|f|\overline{LM^{\tau}_{l,X}}\| = \|\sum_i e^i \|f\chi(R(0, \tau_{i-1}, \tau_i)|X\|\,|l\|,$$

The approximate global space of Morrey $\overline{GM^{\tau}_{l,X}}$ consist of those $f \in L^{1,loc}(R^n)$ for which the norms is finite:

$$\|f|\overline{GM^{\tau}_{l,X}}\| = \sup_h \|f(h+.)|\overline{LM^{\tau}_{l,X}}\|.$$

From the fact that X and l are ideal spaces, it follows that the spaces $GM^{r}_{l,X}$, $LM^{r}_{l,X}$, $\overline{GM^{r}_{l,X}}$, $\overline{LM^{r}_{l,X}}$ are also ideal and, therefore, Banach spaces.

It follows immediately from the definition that the following embeddings hold:

$$GM^{r}_{l,X} \subseteq LM^{r}_{l,X}, \quad \overline{GM^{r}_{l,X}} \subseteq \overline{LM^{r}_{l,X}};$$

$$\overline{LM^{r}_{l,X}} \subseteq LM^{r}_{l,X}, \quad \overline{GM^{r}_{l,X}} \subseteq GM^{r}_{l,X}, \tag{2}$$

where each constant of the embedding is one.

The following crucial theorem gives natural conditions on the embedding inverse to (2).

Theorem 1 *Let the spaces* $\overline{LM^{r}_{l,X}}$, $LM^{r}_{l,X}$, $\overline{GM^{r}_{l,X}}$ *and* $GM^{r}_{l,X}$ *be constructed from the spaces* X *and* l*, a star-shaped set* $U(0, 1)$ *and a sequence* $\tau \in \Upsilon$*. We introduce an operator* $P : l \to l$ *by the equation*

$$P(\Sigma_i e^i x_i) = \Sigma_i e^i y_i, \quad y_i = \Sigma_{j \le i} x_j.$$

When $\|P|l \to l\| = c_0 < \infty$*, the spaces* $\overline{LM^{r}_{l,X}}$ *and* $LM^{r}_{l,X}$ *consist of the same elements, and the following inequalities hold:*

$$\|f|\overline{LM^{r}_{l,X}}\| \le \|f|LM^{r}_{l,X}\| \le c_0\|f|\overline{LM^{r}_{l,X}}\|. \tag{3}$$

Analogously, if $c_0 < \infty$*, then the spaces* $\overline{GM^{r}_{l,X}}$ *and* $GM^{r}_{l,X}$ *consist of the same elements, and the following inequalities hold:*

$$\|f|\overline{GM^{r}_{l,X}}\| \le \|f|GM^{r}_{l,X}\| \le c_0\|f|\overline{GM^{r}_{l,X}}\|,$$

The proof of the embedding $LM^{r}_{l,X} \subseteq \overline{LM^{r}_{l,X}}$ and the inequality (3) is given in [3]. The counterparts for the global Morrey space are similar.

An important point in Theorem 1 is that the conditions are imposed on the space l and are independent of the sequence $\tau \in \Upsilon$.

If in the Definitions 1–3 one replaces R^n, $L^{1,loc}(R^n)$, and $U(0, \tau_i)$ by Ω, $L^{1,loc}(\Omega)$ and $U(0, \tau_i) \bigcap \Omega$, respectively, then one arrives at the definitions of the Morrey spaces on the set Ω.

Let $r_k(t) = sign\{sin(2^k \pi t)\}, k = 1, 2, ...$ be the Rademacher functions. The definition of the Rademacher functions implies the equalities

$$r_k(t) \equiv r_1(2^{k-1}t); \quad r_k(t) \equiv r_k(t + 2^{-k+1}), \ \ k = 1, 2, ...$$

Consider the set of the Rademacher sums

$$R_m(t) = \Sigma_{k=1}^m a_k r_k(t); \quad R(t) = \sum_{k=1}^{\infty} a_k r_k(t); \quad m \in N, \quad a_k \in R \ \ (\forall k \in N).$$

The behavior of the Rademacher sums in $L^p = L^p[T]$ spaces is well known (cf. [1, 11]). In particular, it follows from the classical Khintchin—Kolmogorov inequality that there exist constants A_p, $B_p > 0$ such that for any sequence $\{a_k\}_{k=1}^{\infty}$ of real numbers and any $m \in N$, we have

$$A_p(\sum_{k=1}^{m} |a_k|^2)^{1/2} \leq \|R_m(.)|L^p(T)\| \leq B_p(\sum_{k=1}^{m} |a_k|^2)^{1/2}, \quad (0 < p < \infty), \tag{4}$$

where the constants do not depend on $m \in N$.

In the case of symmetric spaces X on T, Rodin and Semenov [13] proved that $\|R(.)|X\| \approx (\sum_{k=1}^{\infty} |a_k|^2)^{1/2}$ if and only if $G \subseteq X$, where G is the closure of L^1 in the Orlicz space L^M, generated by the function $M(u) = e^{u^2} - 1$.

The main purpose of this paper is to investigate the behavior of the Rademacher sums in the Morrey function spaces $\overline{LM^r_{l,L^p}}$, LM^r_{l,L^p}, $\overline{GM^r_{l,L^p}}$ and GM^r_{l,L^p}.

In what follows, by c_0, c_1, ... we will denote the constants independent of the essential parameters.

2 Main Results

To obtain lower bounds for the Rademacher series, we will need one embedding theorem for Morrey spaces.

Theorem 2 *Let the Morrey spaces* $\overline{LM^{\tau}_{l,X}}$ *and* $\overline{GM^{\tau}_{l,X}}$ *be constructed by means of the ideal spaces* X, l *and the sequences* $\tau = \{\tau_i\}$. *The continuous embedding*

$$\overline{LM^{\tau}_{l,X}} \subseteq L^1 \tag{5}$$

is valid if and only if the inequality

$$\| \sum_i e^i \| \chi(R(0, \tau_{i-1}, \tau_i) | X' | l' \| = c_1 < \infty \tag{6}$$

holds, and the embedding constant in (5) is not greater than c_1.

If the inequality

$$inf_h \ \| \sum_i e^i \| \chi(R(h, \tau_{i-1}, \tau_i) | X' | l' \| = c_2 < \infty \tag{7}$$

holds, then the continuous embedding

$$\overline{GM^{\tau}_{l,X}} \subseteq L^1 \tag{8}$$

is valid, and the embedding constant in (8) is not greater than c_2.

Using the Theorem 1, we obtain the following

Corollary 1 *Let an operator P be bounded in the space l. The continuous embedding*

$$LM^{\tau}_{l,X} \subseteq L^1 \tag{9}$$

is valid if and only if the inequality (6) holds, and the embedding constant in (9) does not exceed $c_3 = c_1 \|P|l \to l\|$.

Let an operator P be bounded in the space l and let (8) be satisfied. Then we have the continuous embedding

$$GM^{\tau}_{l,X} \subseteq L^1, \tag{10}$$

and the embedding constant in (10) is not greater than $c_4 = c_2 \|P|l \to l\|$.

We now fix the space L^p, $p \in [1, \infty)$, the star-shaped set $U(0, 1) = U(0, \tau_0) = (-0, 5; 0, 5)$ and the sequence $\tau = \{2^{-i}\}_1^{\infty}$ by which we construct the new spaces $LM^{\tau}_{l,X}$, $\overline{LM^{\tau}_{l,X}}$ and $GM^{\tau}_{l,X}$, $\overline{GM^{\tau}_{l,X}}$.

Theorem 3 *For each number* $m \in N$, *the inequalities*

$$\| \sum_{i=1}^{m} e^i 2^{1-(i+2)/p} \max\{|R_{i+2}(-3 \cdot 2^{-(i+2)})|, |R_{i+2}(3 \cdot 2^{-(i+2)})|, 2^{-p} A_p (\sum_{k=i+3}^{m} a_k^2)^{1/2}\} | l \| \leq$$

$$\|R_m(.)|\overline{LM^{\tau}_{l,L^p}}\| \leq$$

$$\| \sum_{i=1}^{m} e^i 2^{-(i+1)/p} \{2B_p (\sum_{j=i+2}^{m} a_j^2)^{1/2} + |R_{i+1}(2^{-(i+1)})| + |R_{i+1}(-2^{-(i+1)})|\} | l \|; \tag{11}$$

$$\sup_h \|\sum_{i=1}^{m} e^i 2^{1-(i+2)/p}\{\max\{|R_{i+2}(-3\cdot 2^{-(i+2)}+h)|, |R_{i+2}(3\cdot 2^{-(i+2)}+h)|, 2^{-p}A_p(\sum_{k=i+3}^{m} a_k^2)^{1/2}\}|l\| \leq$$
$$\|R_m(.)|\overline{GM^{\tau}_{l,L^p}}\| \leq$$
$$\|\sum_{i=1}^{m} e^i 2^{1-(i+1)/p}\{B_p(\sum_{j=i+2}^{m} a_j^2)^{1/2} + \sum_{k=1}^{i+1}|a_k|\}|l\| \tag{12}$$

hold true. Here A_p, B_p are the constants in the Khintchin—Kolmogorov inequality (4).

Applying Theorem 1, we obtain

Corollary 2 *If the operator P is bounded in l, then for each number $m \in N$, the inequalities*

$$\|\sum_{i=1}^{m} e^i 2^{1-(i+2)/p}\max\{|R_{i+2}(-3\cdot 2^{-(i+2)})|, |R_{i+2}(3\cdot 2^{-(i+2)})|, 2^{-p}A_p(\sum_{k=i+3}^{m} a_k^2)^{1/2}\}|l\| \leq$$
$$\|R_m(.)|LM^{\tau}_{l,L^p}\| \leq$$
$$c_0\|\sum_{i=1}^{m} e^i 2^{-(i+1)/p}\{2B_p(\sum_{j=i+2}^{m} a_j^2)^{1/2} + |R_{i+1}(2^{-(i+1)})| + |R_{i+1}(-2^{-(i+1)})|\}|l\|;$$

$$\sup_h \|\sum_{i=1}^{m} e^i 2^{1-(i+2)/p}\{\max\{|R_{i+2}(-3\cdot 2^{-(i+2)}+h)|, |R_{i+2}(3\cdot 2^{-(i+2)}+h)|, 2^{-p}A_p(\sum_{k=i+3}^{m} a_k^2)^{1/2}\}|l\| \leq$$
$$\|R_m(.)|GM^{\tau}_{l,L^p}\| \leq$$
$$c_0\|\sum_{i=1}^{m} e^i 2^{1-(i+1)/p}\{B_p(\sum_{j=i+2}^{m} a_j^2)^{1/2} + \sum_{k=1}^{i+1}|a_k|\}|l\|$$

hold, and the constant c_0 does not exceed $\|P|l \to l\|$.

Theorem 4 *The inequality*

$$\|R(.)|\overline{GM^{\tau}_{l,L^p}}\| \leq c_{p,0}(l)(\sum_{k=1}^{\infty}|a_k|^2)^{1/2}$$

is satisfied for any $\{a_k\} \in l^2$, with a constant $c_{p,0}(l)$, if and only if the inequality

$$\|\sum_{i=1}^{\infty} e^i 2^{-i/p}(\sum_{j=1}^{i} a_j)|l\| \leq c_{p,1}(l)(\sum_{k=1}^{\infty}|a_k|^2)^{1/2} \tag{13}$$

holds, with a constant $c_{p,1}(l)$.

Corollary 3 *If the operator P is bounded in l, then the inequality*

$$\|R(.)|GM^{\tau}_{l,L^p}\| \leq c_{p,2}(l)(\sum_{k=1}^{\infty}|a_k|^2)^{1/2}$$

is satisfied if and only if the inequality (13) is satisfied.

Combining Theorems 2 and 4, we obtain the following result.

Theorem 5 *Let the embedding (8) be valid.*
The inequalities

$$a_p(l)(\sum_{k=1}^{\infty}|a_k|^2)^{1/2} \le \|R(.)|\overline{GM^{T}_{l,L^p}}\| \le b_p(l)(\sum_{k=1}^{\infty}|a_k|^2)^{1/2}$$

are satisfied with constants $a_p(l)$ and $b_p(l)$ if and only if the inequality (13) is satisfied.
If the operator P is bounded in l, then the inequalities

$$a'_p(l)(\sum_{k=1}^{\infty}|a_k|^2)^{1/2} \le \|R(.)|GM^{T}_{l,L^p}\| \le b'_p(l)(\sum_{k=1}^{\infty}|a_k|^2)^{1/2}$$

are satisfied with the constants $a'_p(l)$ and $b'_p(l)$ if and only if the inequality (13) is satisfied.

A result on the Khintchin—Kolmogorov inequalities for the Morrey spaces GM^{T}_{l,L^p} and $\overline{GM^{T}_{l,L^p}}$ in the basic case $l = l^q_\nu$ is in order.
We first note that in this case the inequality (6) is equivalent to the inequality

$$(\sum_{j=1}^{\infty}(2^{jp'}\cdot v_j)^{-q'})^{1/q'} < \infty, \quad 1/p + 1/p' = 1, \quad 1/q + 1/q' = 1, \qquad (14)$$

with the usual modification for $p' = \infty$ or $q' = \infty$.
Further, we give the conditions under which the operator P is bounded in l^q_ν. This problem is well known and the answer can be found, for example, in [9]. It reads as follows.

Theorem 6 *Let $l = l^q_\nu$. The operator P is bounded in the space $l = l^q_\nu$ if and only if the conditions*

$$\sup_{i=-1,-2,\dots} (\sum_{j=-\infty}^{i}(\frac{1}{\nu_j})^{q'})^{1/q'} \cdot (\sum_{j=i}^{0}\nu_j{}^{q})^{1/q} < \infty, \quad \frac{1}{q'} + \frac{1}{q} = 1, \qquad q \in (1,\infty), \ (15)$$

$$\sup_{i=-1,-2,\dots} \frac{1}{\nu_i}\cdot(\sum_{j=i}^{0}\nu_j) < \infty, \quad q = 1; \qquad \sup_{i=-1,-2,\dots} \nu_i(\sum_{j=-\infty}^{i}\frac{1}{\nu_j}) < \infty, \quad q = \infty \ (16)$$

are satisfied.

Finally, we describe the sequence spaces for which the inequality (13) holds. For this, we first reformulate the inequality (13). We define the operator H by the equality

$$H(\sum_{j=1}^{\infty} e^j a_j) = \sum_{i=1}^{\infty} e^i 2^{-i/p} (\sum_{j=1}^{i} a_j). \tag{17}$$

It is not difficult to verify the validity of the following lemma.

Lemma 1 *Inequality (13) is satisfied if and only if the operator H is bounded as an operator taking l^2 into l, and the constant $c_{p,2}(l)$ in (13) coincides with the norm $\|H|l^2 \to l\|$.*

It follows from the Lemma 1 that in the case $l = l^q_\nu$, the problem for the inequality (13) to be valid is again a classical problem and the answer can again be found, for example, in ([9], pp. 57–59).

Theorem 7 *The operator H is bounded as an operator from l^2 to l^q_ν if and only if the relations*

$$\sum_{i=1}^{\infty} \left(i^{r/2} (2^{-i/p} v_i)^q \cdot \left(\sum_{j=i}^{\infty} (2^{-j/p} \nu_j)^q\right)^{r/2}\right) < \infty, \quad \frac{1}{q} - \frac{1}{2} = \frac{1}{r}, \qquad q \in [1, 2); \tag{18}$$

$$\sup_{i \in N} i^{1/2} (\sum_{j=i}^{\infty} 2^{-jq/p} \nu_j^q)^{1/q} < \infty, \qquad q \geq 2; \tag{19}$$

are fulfilled.

Combining now Theorems 2–7, we finally obtain

Theorem 8 *Suppose that the inequality (14) holds true. The inequalities*

$$a_p(l) (\sum_{k=1}^{\infty} |a_k|^2)^{1/2} \leq \|R(.)|\overline{GM^{\tau}_{l^q_\nu, L^p}}\| \leq b_p(l) (\sum_{k=1}^{\infty} |a_k|^2)^{1/2}$$

are satisfied if and only if the inequalities (18) and (19) are satisfied.

Suppose that the inequalities (14), (15), (16) hold. The inequalities

$$a_p^{'}(l) (\sum_{k=1}^{\infty} |a_k|^2)^{1/2} \leq \|R(.)|GM^{\tau}_{l, L^p}\| \leq b_p^{'}(l) (\sum_{k=1}^{\infty} |a_k|^2)^{1/2}$$

are satisfied if and only if the inequalities (18) and (19) are satisfied.

3 Proofs of Theorems 2–4

Proof of Theorem 2.

The equation

$$\int_T |f(t)|dt = \int_T \sum_{i=1}^{\infty} \chi(R(0, r_{i-1}, r_i))|f(t)|\, dt = \sum_{i=1}^{\infty} \int_T \chi(R(0, r_{i-1}, r_i))|f(t)|\, dt \tag{20}$$

and the definition of duality imply the inequalities

$$\int_T |f(t)|\, dt \leq \sum_{i=1}^{\infty} \|\chi(R(0, r_{i-1}, r_i))f(.)|X\| \|\chi(R(0, r_{i-1}, r_i))|X^{'}\| \leq$$

$$\|\sum_{i=1}^{\infty} e^i \|\chi(R(0, r_{i-1}, r_i))f(.)|X\| |l\| \cdot \|\sum_{i=1}^{\infty} e^i \|\chi(R(0, r_{i-1}, r_i))|X^{'}\| |l^{'}\| =$$

$$\|f|\overline{LM^{\tau}_{l,X}}\| \cdot \|\sum_{i=1}^{\infty} e^i \|\chi(R(0, r_{i-1}, r_i))|X^{'}\| |l^{'}\|. \tag{21}$$

Thus the sufficiency of condition (6) is proved.

The necessity of condition (6) also follows from (20)–(21), since all the inequalities are sharp and the norm in L^1 is shift invariant.

The proof of Theorem 2 is complete.

Proof of Theorem 3. The definition implies the equality

$$\|R_m(h+.)(\chi(-2^{-i}, -2^{-(i+1)}) + \chi(2^{-(i+1)}, 2^{-i}))|L^p\| =$$

$$(\int_{-2^{-i}}^{-2^{-(i+1)}} |\sum_{k=1}^{m} a_k r_k(t+h)|^p dt + \int_{2^{-(i+1)}}^{2^{-i}} |\sum_{k=1}^{m} a_k r_k(t+h)|^p dt)^{1/p} = (J_i^- + J_i^+)^{1/p}.$$

Let us prove the right-hand side of the inequality (11). We estimate $(J_i^+)^{1/p}$ from above:

$$(J_i^+)^{1/p} = (\int_{2^{-(i+1)}}^{2^{-i}} |\sum_{k=1}^{m} a_k r_k(t+h)|^p dt)^{1/p} \leq$$

$$(\int_{2^{-(i+1)}}^{2^{-i}} |\sum_{k=1}^{i+1} a_k r_k(t+h)|^p dt)^{1/p} + (\int_{2^{-(i+1)}}^{2^{-i}} |\sum_{k=i+2}^{m} a_k r_k(t+h)|^p dt)^{1/p} = I_{i,1}^+ + I_{i,2}^+.$$

Changing the variable $t = 2^{-(i+1)}s + 2^{-(i+1)}$, we get

$$I^+_{i,2} = (\int_{2^{-(i+1)}}^{2^{-i}} |\sum_{k=i+2}^{m} a_k r_k(t+h)|^p dt)^{1/p} =$$

$$2^{-(i+1)/p}(\int_0^1 |\sum_{j=i+2}^{m} a_j r_j(2^{-(i+1)}s + 2^{-(i+1)} + h)|^p ds)^{1/p} =$$

$$2^{-(i+1)/p}(\int_0^1 |\sum_{j=1}^{m-i-1} a_{j+i+1} r_j(s)|^p ds)^{1/p}.$$

The last equality follows from the periodicity of the function $\sum_{j=1}^{m-i-1} a_{j+i+1} r_j(s)$.

Applying the inequality (4), we obtain the inequalities invariant under the shift h:

$$2^{-(i+1)/p} A_p(\sum_{j=i+2}^{m} a_j^2)^{1/2} \leq I^+_{i,2} \leq 2^{-(i+1)/p} B_p(\sum_{j=i+2}^{m} a_j^2)^{1/2}. \tag{22}$$

Substituting for $t = 2^{-(i+1)}s + 2^{-(i+1)}$, we get, for estimating $I^+_{i,1}$,

$$I^+_{i,1} = (\int_{2^{-(i+1)}}^{2^{-i}} |\sum_{k=1}^{i+1} a_k r_k(t+h)|^p dt)^{1/p} =$$

$$2^{-(i+1)/p}(\int_0^1 |\sum_{j=1}^{i+1} a_j r_j(2^{-(i+1)}s + 2^{-(i+1)} + h)|^p ds)^{1/p}.$$

If the interval $[2^{-(i+1)} + h, 2^{-i} + h)$ is a part of the interval of constancy of the function $r_{i+1}(.)$, then the equality

$$I^+_{i,1} = 2^{-(i+1)/p} |\sum_{j=1}^{i+1} a_j r_j(2^{-(i+1)} + h)| = 2^{-(i+1)/p} |R_{i+1}(2^{-(i+1)} + h)| \tag{23}$$

is fulfilled.

We note immediately that in the case $h = 0$ the function $r_{i+1}(.)$ is constant on the interval $[2^{-(i+1)}, 2^{-(i)})$, and therefore the equality (23) holds. Thus, combining (22) and (23) and letting $h = 0$, we obtain the estimate

$$(J_i^+)^{1/p} \leq 2^{-(i+1)/p}(B_p(\sum_{j=i+2}^{m} a_j^2)^{1/2} + |R_{i+1}(2^{-(i+1)})|). \tag{24}$$

Similar estimates are valid for $(J_i^-)^{1/p}$:

$$(J_i^-)^{1/p} \leq 2^{-(i+1)/p}(B_p(\sum_{j=i+2}^{m} a_j^2)^{1/2} + |R_{i+1}(-2^{-(i+1)})|). \tag{25}$$

The right-hand side of the inequality (11) follows from (24) and (25).

To prove the right-hand side of the inequality (12), it suffices to use (23) and the elementary inequalities

$$I_{i,1}^+ \leq 2^{-(i+1)/p} \sup_{h\in(0,1)} |R_{i+1}(t+h)| = 2^{-(i+1)/p} \sum_{j=1}^{i+1} |a_j|;$$

$$I_{i,1}^- \leq 2^{-(i+1)/p} \sup_{h\in(0,1)} |R_{i+1}(t+h)| = 2^{-(i+1)/p} \sum_{j=1}^{i+1} |a_j|.$$

To estimate $R_m(.)$ from below, we use the elementary inequality

$$\|R_m(h+.)(\chi(-2^{-i}, -2^{-(i+1)}) + \chi(2^{-(i+1)}, 2^{-i}))|L^p\| =$$
$$(\int_{-2^{-i}}^{-2^{-(i+1)}} |\sum_{k=1}^{m} a_k r_k(t+h)|^p dt + \int_{2^{-(i+1)}}^{2^{-i}} |\sum_{k=1}^{m} a_k r_k(t+h)|^p dt)^{1/p} \geq$$
$$\max\{(\int_{-2^{-i}}^{-2^{-(i+1)}} |\sum_{k=1}^{m} a_k r_k(t+h)|^p dt)^{1/p};\quad (\int_{2^{-(i+1)}}^{2^{-i}} |\sum_{k=1}^{m} a_k r_k(t+h)|^p dt)^{1/p}\} =$$
$$\max\{(J_i^-)^{1/p}; (J_i^+)^{1/p}\}.$$

We then estimate each of the integrals $(J_i^-)^{1/p}$, $(J_i^+)^{1/p}$.

We first estimate $(J_i^+)^{1/p}$.

Let us define $j = 3 + \lfloor 2^{(i+2)} h \rfloor$, where $\lfloor t \rfloor$ is the integer part of t. Then the closed interval $I_i(j) = [j2^{-(i+2)}, (j+1)2^{-(i+2)}]$ is a subset of $[2^{-(i+1)} + h, 2^{-i} + h]$. Therefore, from the equality

$$(J_i^+)^{1/p} = (\int_{2^{-(i+1)}+h}^{2^{-i}+h} |\sum_{k=1}^{m} a_k r_k(t)|^p dt)^{1/p},$$

the inequality

$$(J_i^+)^{1/p} \geq (\int_{I_i(j)} |\sum_{k=1}^{m} a_k r_k(t)|^p dt)^{1/p} =$$
$$2^{-(i+2)/p}(\int_0^1 |\sum_{k=1}^{i+2} a_k r_k(j2^{-(i+2)} + 2^{-(i+2)}t) + \sum_{k=i+3}^{m} a_k r_k(j2^{-(i+2)} + 2^{-(i+2)}t)|^p dt)^{1/p}$$

follows. We note that if $t \in [0, 1]$ then the sum

$$d_+(i) = \sum_{k=1}^{i+2} a_k r_k(j2^{-(i+2)} + 2^{-(i+2)}t)$$

is constant. Since $2^{-(i+1)} + 2^{-(i+2)} + h \in [j2^{-(i+2)}, (j+1)2^{-(i+2)})$, the equality

$$d_+(i) = R_{i+2}(3 \cdot 2^{-(i+2)} + h) \tag{26}$$

is satisfied.

Now, using the periodicity of the Rademacher functions and the elementary relations

$$(\int_0^1 |d_+(i) + \sum_{k=i+3}^{m} a_k r_k(i2^{-(i+2)} + s2^{-(i+2)}|^p ds)^{1/p} =$$

$$(\int_0^{1/2} |d_+(i) + \sum_{k=i+3}^{m} a_k r_k(s2^{-(i+2)}|^p ds + \int_0^{1/2} |d_+(i) - \sum_{k=i+3}^{m} a_k r_k(s2^{-(i+2)}|^p ds)^{1/p},$$

$$(\int_T |2f(t)|^p dt)^{1/p} \leq (\int_T |f(t) - g(t)|^p dt)^{1/p} + (\int_T |f(t) + g(t)|^p dt)^{1/p},$$

we obtain the validity of the equalities

$$(J_i^+)^{1/p} \geq$$

$$2^{-(i+2)/p}(\int_0^{1/2} |d_+(i) + \sum_{k=i+3}^{m} a_k r_k(2^{-(i+2)}t)|^p dt + \int_0^{1/2} |d_+(i) - \sum_{k=i+3}^{m} a_k r_k(2^{-(i+2)}t)|^p dt)^{1/p} \geq$$

$$2^{1-(i+2)/p} \max\{|d_+(i)|, (\int_0^{1/2} | \sum_{k=i+3}^{m} a_k r_k(s2^{-(i+2)}|^p ds)^{1/p}\}. \tag{27}$$

Further, we again use the Khintchin—Kolmogorov inequality (4) and obtain the validity of the equalities

$$(\int_0^{1/2} | \sum_{k=i+3}^{m} a_k r_k(s2^{-(i+2)}|^p ds)^{1/p}\} = (\int_0^{1/2} | \sum_{k=i+3}^{m} a_k r_{k-(i+2)}(s)|^p ds)^{1/p}\} =$$

$$(\frac{1}{2}\int_0^1 | \sum_{k=i+3}^{m} a_k r_{k-(i+2)}(s)|^p ds)^{1/p}\} \geq 2^{-p} A_p(\sum_{k=i+3}^{m} a_k^2)^{1/2}. \tag{28}$$

To estimate $(J_i^+)^{1/p}$, we proceed similarly. We get

$$(J_i^+)^{1/p} \geq 2^{1-(i+2)/p} \max\{|R_{i+2}(-3 \cdot 2^{-(i+2)} + h)|, 2^{-p} A_p(\sum_{k=i+3}^{m} a_k^2)^{1/2}\}. \tag{29}$$

The right-hand sides of the inequalities (11) and (12) follow from the relations (26)–(29).

The proof of Theorem 3 is complete.

Proof of Theorem 4.

Let us prove the sufficiency of condition (13).

From the right-hand side of the inequality (12), we derive

$$\|R_m(.)|\overline{GM^{\tau}_{l,L^p}}\| \leq$$
$$\|\sum_{i=1}^{m} e^i 2^{1-(i+1)/p}\{B_p(\sum_{j=i+2}^{m} a_j^2)^{1/2} + \sum_{k=1}^{i+1}|a_k|\}|l\| \leq$$
$$\|\sum_{i=1}^{m} e^i 2^{1-(i+1)/p}\{B_p(\sum_{j=1}^{m} a_j^2)^{1/2} + \sum_{k=1}^{i+1}|a_k|\}|l\| \leq$$
$$B_p(\sum_{j=1}^{m} a_j^2)^{1/2}\|\sum_{i=1}^{m} e^i 2^{1-(i+1)/p}|l\| + \|\sum_{i=1}^{m} e^i 2^{1-(i+1)/p}\sum_{k=1}^{i+1}|a_k||l\| \leq$$
$$B_p(\sum_{j=1}^{\infty} a_j^2)^{1/2}\|\sum_{i=1}^{\infty} e^i 2^{1-(i+1)/p}|l\| + \|\sum_{i=1}^{\infty} e^i 2^{1-(i+1)/p}\sum_{k=1}^{i+1}|a_k||l\|. \tag{30}$$

The last inequality and completeness of the space $\overline{GM^{\tau}_{l,L^p}}$ imply the sufficiency of condition (13).

The necessity of condition (13) can be proved as follows. Let the sequence a_k be positive and decreasing.

If $i \geq 6$, then $|R_{i+2}(3 \cdot 2^{-(i+2)})| \geq \sum_{k=1}^{i} a_k - (a_{i+1} + a_{i+2}) \geq \frac{1}{4}\sum_{k=1}^{i+2} a_k$. Therefore, for $i \geq 6$, we derive from the left-hand side of the inequality (12), with $h = 0$,

$$\|R(.)|\overline{GM^{\tau}_{l,L^p}}\| \geq$$
$$\|\sum_{i=1}^{m} e^i 2^{1-(i+2)/p}|R_{i+2}(3 \cdot 2^{-(i+2)})||l\| \geq \|\sum_{i=6}^{m} e^i 2^{1-(i+2)/p}|\sum_{k=1}^{i+2}|a_k||l\|. \tag{31}$$

The last inequality implies the necessity of condition (13).

The proof of Theorem 4 is complete.

References

1. Astashkin, S.V.: Rademacher system in function spaces. In: Fizmatlit, M. (ed.), p. 584 (2016) (in Russian)
2. Astashkin, S.V., Maligranda, L.: Rademacher function in Morrey spaces. J. Math. Anal. Appl. **444**(2), 1133–1154 (2016)
3. Berezhnoi, E.I.: A discrete version of Morrey spaces. Izvestiya RAN: Ser. Mat. **81**(1), 3–30 (2017). English transl. Izvestiya: Math **81**(1), 1–28 (2017). https://doi.org/10.1070/IM8379
4. Burenkov, V.I.: Recent progress in studying the boundedness of classical operators of real analysis in general Morrey-type spaces. I. Eurasian Math. J. **3**(3), 11–32 (2012)
5. Burenkov, V.I.: Recent progress in studying the boundedness of classical operators of real analysis in general Morrey-type spaces. II. Eurasian Math. J. **4**(1), 21–45 (2013)
6. Burenkov, V.I., Nursultanov, E.D., Chigambayeva, D.K.: Description of the interpolation spaces for a pair of local Morrey-type spaces and thier generalizations. In: Proceedings of the Steklov Institute of Mathematics, vol. 284, pp. 97–128 (2014)
7. Kantorovich, L.V., Akilov, G.P.: Functional Analysis, Nauka, Moscow 1977 (in Russian). English transl. Pergamon Press, New York (1982)
8. Krein, S.G., Petunin, Y.I., Semenov, E.M.: Interpolation of Linear Operators, Nauka, Moscow, (1978) (in Russian); English transl.: American Mathematical Society Providence (1982)
9. Kufner, A., Maligranda, L., Persson, L.-E.: The Hardy Inequality. About its History and Some Related Results, Pilsen (2007)
10. Lindenstrauss, J., Tzafriri, L.: Classical Banach Spaces. Springer, Berlin-New York, I. Sequence Spaces (1977)
11. Lindenstrauss, J., Tzafriri, L.: Classical Banach Spaces. Springer, Berlin-New York, II. Function Spaces (1979)
12. Morrey, C.B.: On the solution of quasi-linear elliptic partial differential equations. Trans. Am. Math. Soc. **43**, 126–166 (1938)
13. Rodin, V.A., Semyonov, E.M.: Rademacher series in symmetric spaces. Anal. Math. **1**, 207–222 (1975)
14. Yuan, W., Sickel, W., Yang, D.: Compact embeddings of radial and subradial subspaces of some Besov-type spaces related to Morrey spaces. J. Approx. Theory **174**, 121–139 (2013)

Mellin Convolution Equations

Roland Duduchava

Abstract In the present paper we collect results on Mellin convolution equations (MCEs), obtained recently. We start with the motivation and in the first section are exposed MCEs which encounter in applications. Further we expose the boundedness results of corresponding operators in the Lebesgue space with weight. Fourier convolution (Wiener-Hopf) equations are defined and their connection to MCEs is described. Solvability and Fredholm properties and the index formulae for MCEs are formulated in terms of the symbol functions assigned to them. Results on the Banach algebra generated by Mellin and Wiener-Hopf operators in the Lebesgue space are exposed: The symbol function is defined and the Fredholm criteria is formulated, the index formula is written. In conclusion we expose relatively new results on Fredholm property and index of MCEs in the Bessel potential spaces. These results are applied to the mixed boundary value problem (Mixed BVP) for the Laplace equation in an angle, which reduces to an equivalent MCE. Results on the Fredholm property, solvability and index of such equations are formulated and, in conclusion, applied to the above formulated mixed BVP for the Laplace equation.

Keywords Mellin convolution · Fourier convolution · Weighted Lebesgue space · Bessel potential space · Banach algebra · Symbol function · Fixed singularity · Fredholm property · Index · Laplace equation · Boundary value problem

1 Mellin Convolution Equations—Are They Useful?

It is well-known that various boundary value problems for PDE in planar domains with angular points on the boundary, e.g. Lamé systems in elasticity (cracks in elastic media, reinforced plates), Maxwell's system and Helmholtz equation in electromagnetic scattering, Cauchy–Riemann systems, Carleman–Vekua systems in generalized analytic function theory etc. can be studied with the help of the Mellin convolution

---

R. Duduchava (✉)
The University of Georgia, 77a, M. Kostava str., 0171 Tbilisi, Georgia
e-mail: roldud@gmail.com

A. Karapetyants et al. (eds.), *Modern Methods in Operator Theory and Harmonic Analysis*, Springer Proceedings in Mathematics & Statistics 291,
https://doi.org/10.1007/978-3-030-26748-3_10

equations of the form

$$\mathbf{A}\varphi(t) := c_0\varphi(t) + \frac{c_1}{\pi i}\int_0^a \frac{\varphi(\tau)\,dt}{\tau - t} + \int_0^a \mathcal{K}\Big(\frac{t}{\tau}\Big)\varphi(\tau)\,\frac{d\tau}{\tau} = f(t), \qquad 0 < t < a \tag{1}$$

for some $0 < a < \infty$ and systems of such equations with the following constraint on the kernel $\mathcal{K}$

$$\int_0^\infty t^{\beta-1}|\mathcal{K}(t)|\,dt < \infty, \quad 0 < \beta < 1 \tag{2}$$

(cf., e.g., [6, 7, 13, 24]). These equation are applied also to uncertainty analysis in hydrosystems engineering to prevent failure of high risk objects such as dams, levees, storm sewers, or pollution control systems (see [28]).

The condition (2) makes the operator $\mathbf{A}$ in (1) a bounded operator in the weighted Lebesgue space $\mathbb{L}_p(I_a, t^\gamma)$,

$$\|\mathbf{A}\varphi|\mathbb{L}_p(I_a, t^\gamma)\| \leqslant C\|\varphi|\mathbb{L}_p(I_a, t^\gamma)\| \qquad \forall\varphi \in \mathbb{L}_p(I_a, t^\gamma),$$

provided $1 \leqslant p \leqslant \infty$, $-1 < \gamma < p - 1$, $\beta := (1+\gamma)/p$ (cf. [6]). Here $I_a := (0, a)$ and the norm $\|\psi|\mathbb{L}_p(\mathbb{R}^+, t^\gamma)\|$ is defined as follows

$$\|\psi|\mathbb{L}_p(I_a, t^\gamma)\| := \left[\int_0^a t^\gamma|\psi(t)|^p dt\right]^{1/p}.$$

In particular, integral equations with fixed singularities in the kernel

$$c_0\varphi(t) + \frac{c_1}{\pi i}\int_0^a \frac{\varphi(\tau)\,dt}{\tau - t} + \sum_{k=0}^n \frac{c_{k+2}t^{k-r_k}}{\pi i}\int_0^a \frac{\tau^{r_k}\varphi(\tau)\,d\tau}{(\tau+t)^{k+1}} = f(t), \quad 0 \leqslant t \leqslant a, \tag{3}$$

where $0 \leqslant r_k \leqslant k$ are of type (1) and encounter rather often in plain theory of elasticity and boundary value problems for elliptic equations in domains with non-smooth (Lipschitz, angular) boundary (see, e.g. [6], for details).

Similar integral equations with fixed singularities in kernels, but on curves, arise in the theory of singular integral equations with the complex conjugation

$$a(t)\varphi(t) + \frac{b(t)}{\pi i}\int_\Gamma \frac{\varphi(\tau)\,dt}{\tau - t} + \frac{e(t)}{\pi i}\overline{\int_\Gamma \frac{\overline{\varphi(\tau)}\,dt}{\tau - t}} = f(t), \quad t \in \Gamma \tag{4}$$

and in more general R-linear equations

$$a(t)\varphi(t)+b(t)\overline{\varphi(t)}+\frac{c(t)}{\pi i}\int\limits_{\Gamma}\frac{\varphi(\tau)\,dt}{\tau-t}+\frac{d(t)}{\pi i}\int\limits_{\Gamma}\frac{\overline{\varphi(\tau)}\,dt}{\tau-t}+$$
$$+\frac{e(t)}{\pi i}\overline{\int\limits_{\Gamma}\frac{\varphi(\tau)\,dt}{\tau-t}}+\frac{g(t)}{\pi i}\overline{\int\limits_{\Gamma}\frac{\overline{\varphi(\tau)}\,dt}{\tau-t}}=f(t),\quad t\in\Gamma,\tag{5}$$

if the contour Γ possesses corner points. Note that a complete theory of such equations is presented in [9, 18]. Equations (4) and (5) are considered in the Lebesgue space $\mathbb{L}_p(\Gamma,\rho)$ with the weight function

$$\rho(t):=\prod_{j=1}^{n}|t-t_j|^{\gamma_j},\qquad 1<p<\infty,\quad -1<\gamma_j<p-1,\quad j=1,\ldots,n$$

and this space is defined as follows: $t_1,\ldots,t_n\in\Gamma$ are the corner points of a piecewise-smooth contour Γ and the weighted Lebesgue space $\mathbb{L}_p(\Gamma,\rho)$ consists of functions $\varphi(t)$ such that the following norm is finite:

$$\|\psi|\mathbb{L}_p(\Gamma,\rho)\|:=\left[\int_{\gamma}\rho(t)|\varphi(t)|^p|dt|\right]^{1/p}.$$

Perhaps the most important application Mellin convolution equations have in boundary value problems for elliptic equations in domains with angular points on the boundary (in Lipschitz domains in general). This topic will be highlighted in Sect. 4.

2 Mellin Convolution Equations: Definition and Fredholm Properties

The integral operator in (1) is a Mellin convolution. More generally, if $g\in\mathbb{L}_\infty(\mathbb{R})$ is an essentially bounded measurable $N\times N$ matrix function, the Mellin convolution operator $\mathfrak{M}_g^0$ is defined by

$$\mathfrak{M}_g^0\varphi(t):=\mathcal{M}_\beta^{-1}g\mathcal{M}_\beta\varphi(t)=\frac{1}{2\pi}\int\limits_{-\infty}^{\infty}g(\xi)\int\limits_{0}^{\infty}\left(\frac{t}{\tau}\right)^{i\xi-\beta}\varphi(\tau)\frac{d\tau}{\tau}\,d\xi,\tag{6}$$

where $\varphi\in\mathbb{S}(\mathbb{R}^+)$ and $\mathbb{S}(\mathbb{R}^+)$ is the Schwartz space of fast decaying smooth functions on $\mathbb{R}^+$, whereas $\mathcal{M}_\beta$ and $\mathcal{M}_\beta^{-1}$ are the Mellin transform and its inverse, defined as follows

$$\mathcal{M}_\beta\psi(\xi) := \int_0^\infty t^{\beta-i\xi}\psi(t)\,\frac{dt}{t}, \quad \xi \in \mathbb{R},$$

$$\mathcal{M}_\beta^{-1}\varphi(t) := \frac{1}{2\pi}\int_{-\infty}^\infty t^{i\xi-\beta}\varphi(\xi)\,d\xi, \quad t \in \mathbb{R}^+.$$

The function $g(\xi)$ is usually referred to as a symbol of the Mellin convolution operator $\mathfrak{M}_g^0$.

If the operator $\mathfrak{M}_g^0$ has a bounded extension to the weighted Lebesgue space

$$\left\|\mathfrak{M}_g^0\varphi \mid \mathbb{L}_p(\mathbb{R}^+, t^\gamma)\right\| \leqslant M\left\|\varphi \mid \mathbb{L}_p(\mathbb{R}^+, t^\gamma)\right\|, \tag{7}$$

where $M > 0$ and the weighted norm in $\mathbb{L}_p(\mathbb{R}^+, t^\gamma)$ is defined by the equality

$$\left\|\varphi \mid \mathbb{L}_p(\mathbb{R}^+, t^\gamma)\right\| := \left[\int_0^\infty t^\gamma|\varphi(t)|^p\,dt\right]^{1/p},$$

then the symbol $g(\xi)$ is called a Mellin $\mathbb{L}_{p,\gamma}$-*multiplier.*

The transformations

$$\mathbf{Z}_\beta : \mathbb{L}_p(\mathbb{R}^+, t^\gamma) \longrightarrow \mathbb{L}_p(\mathbb{R}), \quad \mathbf{Z}_\beta\varphi(\xi) := e^{-\beta t}\varphi(e^{-\xi}), \quad \xi \in \mathbb{R},$$
$$\mathbf{Z}_\beta^{-1} : \mathbb{L}_p(\mathbb{R}) \longrightarrow \mathbb{L}_p(\mathbb{R}^+, t^\gamma), \quad \mathbf{Z}_\beta^{-1}\psi(t) := t^{-\beta}\psi(-\ln t), \quad t \in \mathbb{R}^+,$$

arrange an isometrical isomorphism between the corresponding $\mathbb{L}_p$-spaces. Moreover, the relations

$$\begin{aligned}
&\mathcal{M}_\beta = \mathcal{F}\mathbf{Z}_\beta, \quad \mathcal{M}_\beta^{-1} = \mathbf{Z}_\beta^{-1}\mathcal{F}^{-1},\\
&\mathfrak{M}_g^0 = \mathcal{M}_\beta^{-1}g\mathcal{M}_\beta = \mathbf{Z}_\beta^{-1}\mathcal{F}^{-1}g\mathcal{F}\mathbf{Z}_\beta = \mathbf{Z}_\beta^{-1}\mathbf{W}_g^0\mathbf{Z}_\beta,\\
&-1 < \gamma < p-1, \qquad \beta := \frac{1+\gamma}{p}. \qquad 0 < \beta < 1,
\end{aligned} \tag{8}$$

where $\mathcal{F}$ and $\mathcal{F}^{-1}$ are the Fourier transform and its inverse

$$\mathcal{F}\varphi(\xi) := \int_{-\infty}^\infty e^{i\xi x}\varphi(x)\,dx, \quad \mathcal{F}^{-1}\psi(x) := \frac{1}{2\pi}\int_{-\infty}^\infty e^{-i\xi x}\psi(\xi)\,d\xi, \quad x \in \mathbb{R},$$

show a close connection between a Mellin $\mathfrak{M}_a^0$ and the corresponding Fourier convolution operators

$$W_g^0\varphi := \mathcal{F}^{-1}g\mathcal{F}\varphi, \quad \varphi \in \mathbb{S}(\mathbb{R}),$$

as well as between the corresponding transforms. Here $\mathbb{S}(\mathbb{R})$ denotes the Schwartz space of smooth functions, decaying fast at the infinity.

It is easy to verify that Mellin convolution operators are dilation invariant

$$D_\omega \mathfrak{M}_a^0 = \mathfrak{M}_a^0 D_\omega, \qquad D_\omega\varphi(t) := \varphi(\omega t), \quad \omega > 0,$$

while Fourier convolution operators are translation invariant

$$V_\lambda W_a^0 = W_a^0 V_\lambda, \qquad V_\lambda\varphi(t) := \varphi(t-\lambda), \quad -\infty < \lambda < \infty.$$

Due to the Hörmander's theorem (see [23, Theorem 1, 2]) W_a^0 is then represented as a convolution

$$(W_g^0\varphi)(t) = (k * \varphi)(t), \qquad k \in \mathbb{S}'(\mathbb{R}), \quad \varphi \in \mathbb{S}(\mathbb{R}), \quad t \in \mathbb{R}$$

where $\mathbb{S}'(\mathbb{R})$ denotes the space of slowly growing distributions (the dual space to $\mathbb{S}(\mathbb{R})$) and $(\cdot, \cdot)$ denotes the duality pairing between the dual spaces

$$k * \varphi(t) := (k\,(t-\cdot)\,, \varphi(\cdot)).$$

$k \in \mathbb{S}'(\mathbb{R})$ is a slowly growing distributional kernel and the Fourier transform of k coincides with the symbol $a(\xi) = (\mathcal{F}k)(\xi)$, $\xi \in \mathbb{R}$.

Due to the equalities (8) and Hörmander's theorem we derive that $\mathfrak{M}_g^0$ is then represented as a multiplicative convolution

$$\mathfrak{M}_g^0\psi(t) = r \star \psi(t), \qquad r \in \mathbb{S}'(\mathbb{R}^+), \quad \psi \in \mathbb{S}(\mathbb{R}^+), \quad t \in \mathbb{R}^+,$$

where

$$r \star \psi(t) := (r\left(\frac{t}{\cdot}\right), \psi(\cdot)),$$

with a slowly growing distributional kernel r and the Mellin transform of r coincides with the symbol $g(\xi) = (\mathcal{M}_\beta r)(\xi)$, $\xi \in \mathbb{R}$.

A bounded function $g(\xi)$, $\xi \in \mathbb{R}$ is called a *Fourier* $\mathbb{L}_p$*-multiplier* if the operator $W_g^0 : \mathbb{L}_p(\mathbb{R}) \longrightarrow \mathbb{L}_p(\mathbb{R})$ is bounded. The set of all $\mathbb{L}_p$-multipliers is denoted by $\mathcal{M}_p(\mathbb{R})$.

From (8) we derive easily the following.

Proposition 1 (see [6]) *Let* $1 < p < \infty$. *The class of Mellin* $\mathbb{L}_{p,\gamma}$*-multipliers coincides with the Banach algebra* $\mathcal{M}_p(\mathbb{R})$ *of Fourier* $\mathbb{L}_p$*-multipliers for arbitrary* $-1 < \gamma < p-1$ *and is independent of the parameter* γ.

In other words, a Mellin convolution operator $\mathfrak{M}_a^0$ *in* (6) *is bounded in the weighted Lebesgue space* $\mathbb{L}_p(\mathbb{R}^+, t^\gamma)$ *(cf.* (7)*) if and only if* $a \in \mathcal{M}_p(\mathbb{R})$.

Remark 1 Stechkin theorem on multipliers (see, e.g. [6]) states that the Banach algebra $\mathcal{M}_p(\mathbb{R})$ contains the algebra $\mathbf{V}_1(\mathbb{R})$ of all functions with bounded variation.

Another big class of multipliers is the Wiener algebra $\mathbb{W}(\mathbb{R})$, which consists of Fourier transforms of the absolutely integrable functions extended with constants

$$\mathbb{W}(\mathbb{R}) := \{c + \mathcal{F}\varphi(\xi) \ : \ c = \text{const}, \quad \varphi \in \mathbb{L}_1(\mathbb{R})\}.$$

Note, that the condition (2) on the kernel means that the corresponding symbol $\mathcal{A}(\xi) := \mathcal{M}_\beta \mathcal{K}(\xi)$ of the operator A in (1) belongs to the Wiener algebra $\mathbb{W}(\mathbb{R})$.

Proposition 2 (see [6]) *Let* $1 < p < \infty$, $-1 < \gamma < p - 1$ *and* $a \in \mathcal{M}_p(\mathbb{R})$. *The Mellin convolution operator* $\mathfrak{M}^0_a$ *is Fredholm in the weighted Lebesgue space* $\mathbb{L}_p(\mathbb{R}^+, t^\gamma)$ *if and only if the inverse is a multiplier again* $a^{-1} \in \mathcal{M}_p(\mathbb{R})$.

Moreover, if $\mathfrak{M}^0_a$ *is Fredholm it is invertible and the inverse operator is* $\mathfrak{M}^0_{a^{-1}}$.

Remark 2 Note that the ellipticity condition

$$\inf_{\xi \in \mathbb{R}} |a(\xi)| > 0 \tag{9}$$

for the multiplier $a \in \mathcal{M}_p(\mathbb{R})$ does not guarantee that the inverse is a multiplier again $a^{-1} \in \mathcal{M}_p(\mathbb{R})$, except the case $p = 2$. The class $\mathcal{M}_2(\mathbb{R})$ coincides with the class of all essentially bounded functions $\mathbb{L}_\infty(\mathbb{R})$ and the ellipticity condition (8) for $a \in \mathcal{M}_2(\mathbb{R})$ is equivalent to $a^{-1} \in \mathcal{M}_2(\mathbb{R})$.

The same, equivalence of the invertibility and the ellipticity conditions, hold for the symbols from the subalgebras $\mathcal{M}^0_p(\mathbb{R}) := \bigcap_{p-\varepsilon<r<p+\varepsilon} \mathcal{M}_r(\mathbb{R})$ for some $\varepsilon > 0$.

Let $\ell_{I_a}\varphi$ be the extension by 0 to $\mathbb{R}^+ = (0, \infty)$ of a function $\varphi(t)$ defined initially on the interval $I_a \subset \mathbb{R}^+$ and $r_{I_a}\psi$ be the left inverse to ℓ_{I_a}, which restricts a function $\psi(t)$, defined initially on the semi-axes $\mathbb{R}^+ = (0, \infty)$ to the interval I_a.

The Mellin convolution equation on the interval $I_a = (0, a)$, $a > 0$, is the restriction

$$\mathfrak{M}_g\varphi = r_{I_a}\mathfrak{M}^0_g \ell_{I_a}\varphi, \qquad \varphi \in \mathbb{S}(I_a). \tag{10}$$

The Fredholm theory and the unique solvability of Eq. (1) in the weighted Lebesgue spaces were accomplished in [6]. This investigation was based on the following observation: if $1 < p < \infty$, $-1 < \gamma < p - 1$, $\beta := (1 + \gamma)/p$, the following mutually invertible exponential transformations

$$\begin{aligned} Z_\beta &: \mathbb{L}_p(I_a, t^\gamma) \longrightarrow \mathbb{L}_p(\mathbb{R}^+), \\ Z_\beta\varphi(\xi) &:= a^\beta e^{-\beta\xi}\varphi(ae^{-\xi}), \ \ \xi \in \mathbb{R}^+ := (0, \infty), \\ Z^{-1}_\beta &: \mathbb{L}_p(\mathbb{R}^+) \longrightarrow \mathbb{L}_p(I_a, t^\gamma), \\ Z^{-1}_\beta\psi(t) &:= a^{-\beta}t^{-\beta}\psi(-\ln(a^{-1}t)), \ \ t \in I_a, \end{aligned} \tag{11}$$

transform the Eq. (1), treated in the weighted Lebesgue space $f, \varphi \in \mathbb{L}_p(I_a, t^\gamma)$ into the Fourier convolution equation on the semi-axes (the Wiener-Hopf equation)

$W_{\mathcal{A}_\beta}\psi = g$, $\psi = Z_\beta\varphi$, $g = Z_\beta f \in \mathbb{L}_p(\mathbb{R}^+)$ of the form

$$W^0_{\mathcal{A}_\beta}\psi(x) = c_0\psi(x) + \int_0^\infty \mathcal{K}_1(x-y)\varphi(y)\,dy = g(x), \qquad t \in \mathbb{R}^+,$$

$$\mathcal{K}_1(x) = e^{-\beta x}\Big[\frac{c_1}{1-e^{-x}} + \mathcal{K}(e^{-x})\Big].$$

Note that the symbol of the operator $W_{\mathcal{A}_\beta}$, viz. the Fourier transform of the kernel

$$\mathcal{A}_\beta(\xi) := c_0 + \int_{-\infty}^\infty e^{i\xi x}\mathcal{K}_1(x)\,dx$$

$$:= c_0 - ic_1 \cot\pi(\beta - i\xi) + \int_{-\infty}^\infty e^{(i\xi-\beta)x}\mathcal{K}(e^{-x})\,dx, \quad \xi \in \mathbb{R} \tag{12}$$

is a piecewise continuous function. Let us recall that the theory of Wiener-Hopf convolution equations with discontinuous symbols is well developed, cf. [14–17, 27]. This allows one to investigate various properties of the operators (1), (3). In particular, Fredholm criteria, index formula and conditions of unique solvability of the Eqs. (1) and (3) have been established in [6]. Here is the main result.

Theorem 1 (see Theorem 8.1 in [6]) *Let* $1 < p < \infty$, $-1 < \gamma < p-1$ *and* $g \in \mathcal{M}^0_p(\mathbb{R}) \cap C(\overline{\mathbb{R}})$ *be an* $N \times N$ *matrix* $\mathbb{L}_p$ *multiplier, continuous on* $\mathbb{R}$ *and might have jump point only at the infinity (see e.g., (12)).*

The equation

$$(\mathfrak{M}_g\varphi)(t) = f(t), \qquad f, \varphi \in \mathbb{L}_p(I_a, t^\gamma), \quad t \in I_a \tag{13}$$

is normally solvable (the operator $\mathfrak{M}_g$ *has a closed range in* $\mathbb{L}_p(I_a, t^\gamma)$*) if and only if its symbol*

$$g_p(t,\xi) := \begin{cases} g(t), & for \quad t \in \mathbb{R}, \\ g_+ - ig_- \cot\pi(\beta - i\xi), & for \quad t = \infty, \quad \xi \in \mathbb{R}, \end{cases} \tag{14}$$

$$g_\pm := \frac{g(+\infty) \pm g(-\infty)}{2}$$

is elliptic $\inf_{t,\xi\in\mathbb{R}} |\det g_p(t,\xi)| > 0$.

If $\mathfrak{M}_g$ *is normally solvable, it is Fredholm and the index of the operator is*

$$\operatorname{Ind}\mathfrak{M}_g = -\operatorname{ind}\det g_p = -\frac{1}{2\pi}[\det g_p(t,0)]_{\mathbb{R}} - \frac{1}{2\pi}[\det g_p(\infty,\xi)]_{\mathbb{R}}, \tag{15}$$

where $[h(x)]_{\mathbb{R}}$ *denotes the increment of the function* $h(x)$ *when* x *ranges through* $\mathbb{R}$ *from* $-\infty$ *to* ∞.

Moreover, if the equation is scalar $N = 1$ *the following holds:*

i. *The Eq.* (13) *has a unique solution* $\varphi \in \mathbb{L}_p(I_a, t^\gamma)$ *for arbitrary right hand side* $f \in \mathbb{L}_p(I_a, t^\gamma)$ *if* $\operatorname{ind} g_p = 0$;
ii. *The homogeneous equation* $\mathfrak{M}_g = 0$ *has exactly* $m = -\operatorname{ind}\det g_p$ *linearly independent solutions in* $\mathbb{L}_p(I_a, t^\gamma)$ *and Eq.* (13) *has solutions for arbitrary* $f \in \mathbb{L}_p(I_a, t^\gamma)$ *if* $\operatorname{ind} g_p < 0$;
iii. *The Eq.* (13) *has a unique solution* $\varphi \in \mathbb{L}_p(I_a, t^\gamma)$ *only for those right hand sides* $f \in \mathbb{L}_p(I_a, t^\gamma)$ *which satisfy exactly* $m = -\operatorname{ind} g_p$ *orthogonality conditions with all solutions to the dual homogeneous equation in the dual space, provided* $\operatorname{ind} g_p > 0$.

Concerning the Eqs. (4) and (5): If their coefficients are piecewise-continuous matrix functions, one can construct a function $\mathcal{A}_\beta(t, \xi), t \in \Gamma, \xi \in \mathbb{R}, \boldsymbol{\beta} := (\beta_1, \ldots, \beta_n)$, called the symbol of the equation of the related operator $\mathbf{A}$ responsible for the following properties of the equation:

- The equation is Fredholm in $\mathbb{L}_p(\Gamma, \rho)$ if and only if its symbol is elliptic., i.e. iff $\inf_{(t,\xi)\in\Gamma\times\mathbb{R}} |\det \mathcal{A}_\beta(t, \xi)| > 0$;
- To an elliptic symbol $\mathcal{A}_\beta(t, \xi)$ there corresponds an integer valued index $\operatorname{ind}\det \mathcal{A}_\beta(t, \xi)$, the winding number, and $\operatorname{Ind}\mathbf{A} = -\operatorname{ind}\det \mathcal{A}_\beta(t, \xi)$.

For more detailed survey of the theory and various applications to the problems of elasticity we refer the reader to [6–8, 13–16, 26].

Similar approach to boundary integral equations on curves with corner points based on Mellin transformation has been exploited by M. Costabel and E. Stephan [3, 4]. Integral equations with involutive operators were studied in [25].

3 Banach Algebra of Operators Generated by Mellin and Fourier Convolutions

The proofs of the results on Banach algebra of operators presented here can be found in [19, Section 7].

Let $\dot{\mathbb{R}} := \mathbb{R} \cup \{\infty\}$ denote one point compactification of the real axes $\mathbb{R}$ and $\overline{\mathbb{R}} := \mathbb{R} \cup \{\pm\infty\}$—the two point compactification of $\mathbb{R}$. By $C(\dot{\mathbb{R}})$ (by $C(\overline{\mathbb{R}})$, respectively) we denote the space of continuous functions $g(x)$ on $\mathbb{R}$ which have the equal limits at the infinity $g(-\infty) = g(+\infty)$ (limits at the infinity can be different $g(-\infty) \neq g(+\infty)$. By $PC(\overline{\mathbb{R}})$ is denoted the space of piecewise-continuous functions on $\overline{\mathbb{R}}$, having the limits $a(t \pm 0)$ at all points $t \in \dot{\mathbb{R}}$, including the infinity.

Proposition 3 ([20]) *The Banach algebra, generated by the Cauchy singular integral operator*

$$S_{\mathbb{R}^+}\varphi(x) := \frac{1}{\pi i}\int_0^\infty \frac{\varphi(y)\,dy}{y-x} = W_{-\mathrm{sign}\xi}\varphi(x) \qquad x \in \mathbb{R}, \tag{16}$$

and the identity operator I on the semi-axis $\mathbb{R}^+$, contains Fourier convolution operators with symbols having discontinuity of the jump type only at zero and at the infinity and Mellin convolution operators with continuous symbols on $\dot{\mathbb{R}}$ (including the infinity).

Let us consider the Banach algebra $\mathfrak{A}_p(\mathbb{R}^+)$ generated by Mellin convolution and Fourier convolution operators, i.e. by the operators

$$\mathbf{A} := \sum_{j=1}^{m} \mathfrak{M}^0_{a_j} W_{b_j}, \tag{17}$$

and there compositions, in the Lebesgue space $\mathbb{L}_p(\mathbb{R}^+)$. Here $\mathfrak{M}^0_{a_j}$ are Mellin convolution operators with continuous $N \times N$ matrix symbols $a_j \in C\mathcal{M}_p(\overline{R})$, W_{b_j} are Fourier convolution operators with $N \times N$ matrix symbols $b_j \in C\mathcal{M}_p(\overline{\mathbb{R}} \setminus \{0\}) := C\mathcal{M}_p(\overline{\mathbb{R}}^- \cup \overline{\mathbb{R}}^+)$. The algebra of $N \times N$ matrix $\mathbb{L}_p$-multipliers $C\mathcal{M}_p(\overline{\mathbb{R}} \setminus \{0\})$ consists of those $N \times N$ matrix multipliers $b \in \mathcal{M}_p(\mathbb{R})$ which are continuous on the semi-axis $\mathbb{R}^-$ and $\mathbb{R}^+$ and might have finite jump discontinuities at 0 and at the infinity.

This and more general algebras were studied in [20] and also in earlier works [7, 13, 19, 27].

Note that the algebra $\mathfrak{A}_p(\mathbb{R}^+)$ is actually a subalgebra of the Banach algebra generated by the Cauchy singular integral operator and the identity operator (see Proposition 3). Let $\mathfrak{S}(\mathbb{L}_p(\mathbb{R}^+))$ denote the ideal of all compact operators in $\mathbb{L}_p(\mathbb{R}^+)$. Since the quotient algebra $\mathfrak{F}_p(\mathbb{R}^+)/\mathfrak{S}(\mathbb{L}_p(\mathbb{R}^+))$ is commutative in the scalar case $N = 1$, the following is true.

Lemma 1 (see Corollary 3.9, [19]) *The quotient algebra $\mathfrak{A}_p(\mathbb{R}^+)/\mathfrak{S}(\mathbb{L}_p(\mathbb{R}^+))$ is commutative in the scalar case $N = 1$.*

To describe the symbol of the operator (17), consider the infinite clockwise oriented "rectangle" $\mathfrak{R} := \Gamma_1 \cup \Gamma_2^- \cup \Gamma_2^+ \cup \Gamma_3$, where (cf. Fig. 1)

$$\Gamma_1 := \overline{\mathbb{R}} \times \{+\infty\}, \quad \Gamma_2^\pm := \{\pm\infty\} \times \overline{\mathbb{R}}^+, \quad \Gamma_3 := \overline{\mathbb{R}} \times \{0\}.$$

The symbol $\mathcal{A}_p(\omega)$ of the operator $\mathbf{A}$ in (17) is a function on the set $\mathfrak{R}$, viz.

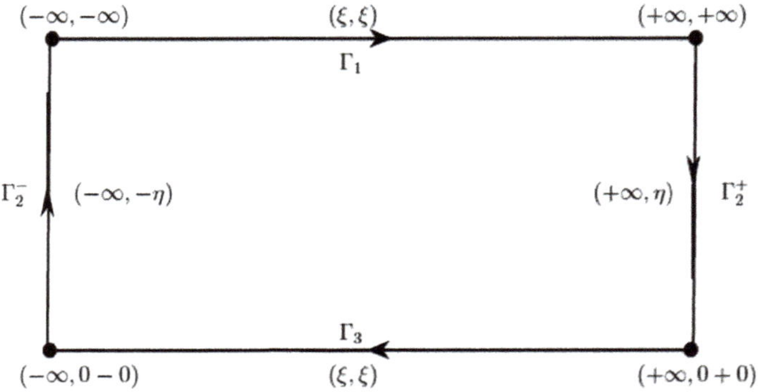

Fig. 1 The domain $\mathfrak{R}$ of definition of the symbol $\mathcal{A}^s_p(\omega)$

$$\mathcal{A}_p(\omega) := \begin{cases} \sum\limits_{j=1}^{m} a_j(\xi)(b_j)_p(\infty,\xi), & \omega = (\xi,\infty) \in \overline{\Gamma_1}, \\ \sum\limits_{j=1}^{m} a_j(+\infty)b_j(-\eta), & \omega = (+\infty,\eta) \in \Gamma_2^+, \\ \sum\limits_{j=1}^{m} a_j(-\infty)b_j(\eta), & \omega = (-\infty,\eta) \in \Gamma_2^-, \\ \sum\limits_{j=1}^{m} a_j(\xi)(b_j)_p(0,\xi), & \omega = (\xi,0) \in \overline{\Gamma_3}. \end{cases} \tag{18}$$

The symbol $\mathcal{A}_p(\omega)$, when $\omega = (\xi,\infty)$ ranges through the infinite interval Γ_1 (see Fig. 1) fills the gap between the following points on the complex plane

$$\infty + i\sum_{j=1}^{m} a_j(\xi)b_j(-\infty) \qquad \text{and} \qquad \infty + i\sum_{j=1}^{m} a_j(\xi)b_j(+\infty),$$

and when $\omega = (\xi,0)$ ranges through the infinite interval Γ_3 (see Fig. 1) it fills the gap between the following points on the complex plane

$$0 + i\sum_{j=1}^{m} a_j(\xi)b_j(0-0) \qquad \text{and} \qquad 0 + i\sum_{j=1}^{m} a_j(\xi)b_j(0+0).$$

The connecting function $g_p(\infty,\xi)$ in (18) for a piecewise continuous function $g \in PC(\overline{\mathbb{R}})$ is defined as follows

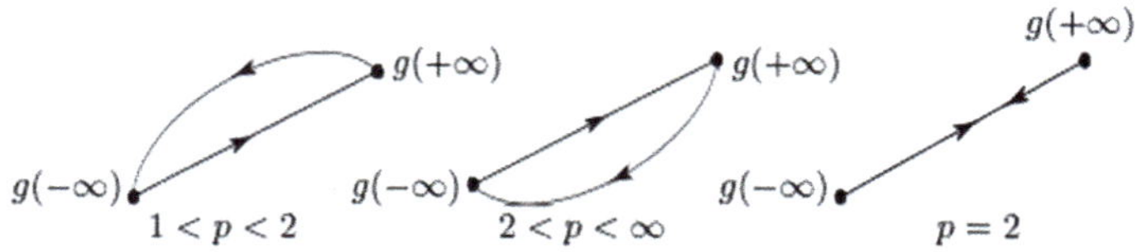

Fig. 2 Arc condition

$$\begin{aligned} g_p(x,\xi) &:= \frac{1}{2}\left[g(x+0)+g(x-0)\right] - \frac{i}{2}\left[g(x+0)-g(x-0)\right]\cot\pi\Big(\frac{1}{p}-i\xi\Big) \\ &= e^{i\pi\frac{g_x^+ + g_x^-}{2}}\,\frac{\cos\pi\Big(\frac{1}{p}+\frac{g_x^+ - g_x^-}{2}-i\xi\Big)}{\sin\pi\Big(\frac{1}{p}-i\xi\Big)}, \qquad \xi\in\mathbb{R}, \qquad (19) \\ g_x^\pm &:= \frac{1}{\pi i}\ln g(x\pm 0), \quad \operatorname{Re} g_x^\pm = \frac{1}{\pi}\arg g(x\pm 0), \quad x\in\dot{\mathbb{R}} := \mathbb{R}\cup\{\infty\} \end{aligned}$$

and fills up the discontinuity gap at ∞ between the following points on the extended complex plane $\infty + ig(-\infty)$ and $\infty + ig(+\infty)$ with an oriented circular arc. From every point of this circular arc the oriented line on the complex plane, connecting the points $\infty + ig(-\infty)$ and $\infty + ig(+\infty)$ is seen under the angle π/p. Moreover, the oriented arc lies above the oriented interval if $1/2 < 1/p < 1$ (i.e., if $1 < p < 2$) and the oriented arc is under the oriented interval if $0 < 1/p < 1/2$ (i.e., if $2 < p < \infty$). For $p = 2$ the oriented arc coincides with the oriented interval (see Fig. 2).

A similar geometric interpretation is valid for the function $g_p(t,\xi)$, which connects the point $t + ig(t-0)$ with $t + ig(t+0)$ for those t for which $g(\tau)$ has a jump discontinuity.

To make the symbol $\mathcal{A}_p(\omega)$ continuous, we endow the rectangle $\mathfrak{R}$ with a special topology: the distance on the curves Γ_1, $\Gamma_2^\pm$, Γ_3 and on $\overline{\mathbb{R}}$ are defined as

$$\rho(x,y) := \left|\arg\frac{x-i}{x+i} - \arg\frac{y-i}{y+i}\right| \quad \text{for arbitrary } x,y\in\overline{\mathbb{R}}.$$

In this topology, the length $|\mathfrak{R}|$ of $\mathfrak{R}$ is 6π, and the symbol $\mathcal{A}_p(\omega)$ is continuous everywhere on $\mathfrak{R}$. The image of the function $\det\mathcal{A}_p(\omega)$, $\omega\in\mathfrak{R}$ ($\det\mathcal{B}_p(\omega)$) is a closed curve on the complex plane. It follows from the continuity of the symbol at the angular points of the rectangle $\mathfrak{R}$ where the one-sided limits coincide. Thus

$$\mathcal{A}_p(\pm\infty,\infty) = \sum_{j=1}^m a_j(\pm\infty)b_j(\mp\infty),$$

$$\mathcal{A}_p(\pm\infty,0) = \sum_{j=1}^m a_j(\pm\infty)b_j(0\mp 0).$$

Hence, if the symbol of the corresponding operator is elliptic, i.e. if

$$\inf_{\omega\in\mathfrak{R}} \big|\det \mathcal{A}_p(\omega)\big| > 0, \tag{20}$$

the increment of the argument $(1/2\pi)\arg \mathcal{A}_p(\omega)$ when ω ranges through $\mathfrak{R}$ in the positive direction; is an integer, is called the winding number or the symbol function and is denoted by $\operatorname{ind}\det \mathcal{A}_p$.

Theorem 2 (see Theorem 3.10, [19]) *Let* $1 < p < \infty$ *and* $\mathbf{A}$ *be defined by* (17). *The operator* $\mathbf{A} : \mathbb{L}_p(\mathbb{R}^+) \longrightarrow \mathbb{L}_p(\mathbb{R}^+)$ *is Fredholm if and only if its symbol* $\mathcal{A}_p(\omega)$ *is elliptic. If* $\mathbf{A}$ *is Fredholm, the index of the operator is*

$$\operatorname{Ind}\mathbf{A} = -\operatorname{ind}\det \mathcal{A}_p. \tag{21}$$

Remark 3 (*see Remark 3.11,* [19]) Let us emphasize that the formula (21) does not contradict the invertibility of "pure Mellin convolution" operators $\mathfrak{M}^0_a : \mathbb{L}_p(\mathbb{R}^+) \longrightarrow \mathbb{L}_p(\mathbb{R}^+)$ with an elliptic matrix symbol $a \in C\mathcal{M}^0_p(\mathbb{R})$, $\inf\limits_{\xi\in\mathbb{R}} |a(\xi)| > 0$, stated in Proposition 2, even if $\operatorname{ind} a \neq 0$.

In fact, computing the symbol of $\mathfrak{M}^0_a$ by formula (18), one obtains

$$(\mathfrak{M}^0_a)_p(\omega) := \begin{cases} a(\xi), & \omega = (\xi, \infty) \in \overline{\Gamma_1}, \\ a(+\infty), & \omega = (+\infty, \eta) \in \Gamma_2^+, \\ a(-\infty), & \omega = (-\infty, \eta) \in \Gamma_2^-, \\ a(\xi), & \omega = (\xi, 0) \in \overline{\Gamma_3}. \end{cases}$$

Noting that on the sets Γ_1 and Γ_3 the variable ω runs in opposite direction, the increment of the argument $[\arg\det(\mathfrak{M}^0_a)_p(\omega)]_{\mathfrak{R}} = 0$ is zero, implying $\operatorname{Ind}\mathfrak{M}^0_a = 0$.

In contrast to the above, a pure Fourier convolution operators $W_b : \mathbb{L}_p(\mathbb{R}^+) \longrightarrow \mathbb{L}_p(\mathbb{R}^+)$ with the elliptic matrix symbol $b \in C\mathcal{M}^0_p(\mathbb{R})$, $\inf\limits_{\xi\in\mathbb{R}} |b_p(\xi, \eta)| > 0$ can have non-zero index: Since

$$b_p(\omega) := \begin{cases} b_p(\infty, \xi), & \omega = (\xi, \infty) \in \overline{\Gamma_1}, \\ b(-\eta), & \omega = (+\infty, \eta) \in \Gamma_2^+, \\ b(\eta), & \omega = (-\infty, \eta) \in \Gamma_2^-, \\ b(0), & \omega = (\xi, 0) \in \overline{\Gamma_3}, \end{cases}$$

one arrives at the well-known formula

$$\operatorname{Ind} W_b = -\operatorname{ind} b_p.$$

Moreover, in the case when the symbol is continuous $b(-\infty) = b(+\infty)$, one has $b_p(\xi, \eta) = b(\xi)$ and if the symbol is elliptic the following formula for the index is valid:

$$\operatorname{Ind} W_b = -\operatorname{ind} b_p = -\operatorname{ind}\det b.$$

If $\mathcal{A}_p(\omega)$ is the symbol of an operator $\mathbf{A}$ of (18), the set $\mathcal{R}(\mathcal{A}_p) := \{\mathcal{A}_p(\omega) \in \mathbb{C} : \omega \in \mathfrak{R}\}$ coincides with the essential spectrum of $\mathbf{A}$. Recall that the essential spectrum $\sigma_{ess}(\mathbf{A})$ of a bounded operator $\mathbf{A}$ is the set of all $\lambda \in \mathbb{C}$ such that the operator $\mathbf{A} - \lambda I$ is not Fredholm in $\mathbb{L}_p(\mathbb{R}^+)$ or, equivalently, the coset $[\mathbf{A} - \lambda I]$ is not invertible in the quotient (Calkin) algebra $\mathfrak{A}_p(\mathbb{R}^+)/\mathfrak{S}(\mathbb{L}_p(\mathbb{R}^+))$. Then, due to Banach theorem, the essential norm $|||\mathbf{A}|||$ of the operator $\mathbf{A}$ can be estimated as follows

$$\sup_{\omega\in\mathfrak{R}} |\mathcal{A}_p(\omega)| \leqslant |||\mathbf{A}||| := \inf_{\mathbf{T}\in\mathfrak{S}(\mathbb{L}_p(\mathbb{R}^+))} \left\|(\mathbf{A}+\mathbf{T}) \mid \mathcal{L}(\mathbb{L}_p(\mathbb{R}^+))\right\|. \tag{22}$$

The inequality (22) enables the continuous extension of the symbol map (18)

$$[\mathbf{A}] \longrightarrow \mathcal{A}_p(\omega), \quad [\mathbf{A}] \in \mathfrak{A}_p(\mathbb{R}^+)/\mathfrak{S}(\mathbb{L}_p(\mathbb{R}^+)) \tag{23}$$

on the whole Banach algebra $\mathfrak{A}_p(\mathbb{R}^+)$. Now, applying Theorem 2 and a standard methods, cf. [20, Theorem 3.2], one can derive the following result.

Corollary 1 (see Corollary 3.12, [19]) *Let $1 < p < \infty$ and $\mathbf{A} \in \mathfrak{A}_p(\mathbb{R}^+)$. The operator $\mathbf{A} : \mathbb{L}_p(\mathbb{R}^+) \longrightarrow \mathbb{L}_p(\mathbb{R}^+)$ is Fredholm if and only if it is symbol $\mathcal{A}_p(\omega)$ is elliptic. If $\mathbf{A}$ is Fredholm, then*

$$\operatorname{Ind}\mathbf{A} = -\operatorname{ind}\mathcal{A}_p.$$

Corollary 2 (see Corollary 3.13, [19]) *The space of maximal ideals of the commutative quotient Banach algebra $\mathfrak{A}_p(\mathbb{R}^+)/\mathfrak{S}(\mathbb{L}_p(\mathbb{R}^+))$ generated by scalar operators in* (17) *(the case $N = 1$), is homomorphic to $\mathfrak{R}$, and the symbol map in* (18), (23) *is a Gelfand homeomorphism of the corresponding Banach algebras.*

Remark 4 (*see Remark 3.14,* [19]) All the above results are valid in a more general setting viz., for the Banach algebra $\mathfrak{PA}^{N\times N}_{p,\alpha}(\mathbb{R}^+)$ generated by the operators

$$\mathbf{A} := \sum_{j=1}^{m} \left[d_j^1 \mathfrak{M}^0_{a_j^1} W_{b_j^1} + d_j^2 \mathfrak{M}^0_{a_j^2} H_{c_j^1} + d_j^3 W^0_{b_j^2} H_{c_j^2} \right] \tag{24}$$

with piecewise-continuous $N \times N$ matrix coefficients $d_j^1, d_j^2, d_j^3 \in PC^{N\times N}(\overline{\mathbb{R}})$ in the weighted Lebesgue space of N-vector-functions $\mathbb{L}_p^N(\mathbb{R}^+, x^\alpha)$.

Symbols of Mellin convolution operators $\mathfrak{M}^0_{a_j^1}$, $\mathfrak{M}^0_{a_j^2}$, Winer–Hopf (Fourier convolution) operators $W_{b_j^1}$, $W_{b_j^2}$ and Hankel operators $H_{c_j^1}$, $H_{c_j^2}$ are $N \times N$ piecewise-continuous matrix $\mathbb{L}_p$-multipliers $a_j^k, b_j^k, c_j^k \in PC^{N\times N}\mathcal{M}_p(\mathbb{R})$.

The spectral set $\Sigma(\mathfrak{PA}^{N\times N}_{p,\alpha}(\mathbb{R}^+))$ of such Banach algebra (viz., the set where the symbols are defined, e.g. $\mathfrak{R}$ for the Banach algebra $\mathfrak{A}_p^{N\times N}(\mathbb{R}^+)$ investigated above) is more sophisticated and described in the papers [16, 17, 20, 27].

Let $\mathfrak{C}\mathfrak{A}_{p,\alpha}(\mathbb{R}^+)\mathfrak{S}(\mathbb{L}_p(\mathbb{R}^+))$ be the sub-algebra of $\mathfrak{P}\mathfrak{A}_{p,\alpha}(\mathbb{R}^+) = \mathfrak{P}\mathfrak{A}^{1\times 1}_{p,\alpha}(\mathbb{R}^+)$ generated by scalar operators (24) with continuous coefficients $c_j, h_j \in C(\overline{\mathbb{R}})$ and scalar piecewise-continuous $\mathbb{L}_p$-multipliers) $a_j, b_j, d_j, g_j \in PC\mathcal{M}_p(\mathbb{R})$. The quotient-algebra $\mathfrak{C}\mathfrak{A}_{p,\alpha}(\mathbb{R}^+)/\mathfrak{S}(\mathbb{L}_p(\mathbb{R}^+))$ with respect to the ideal of all compact operators is a commutative algebra and the spectral set $\Sigma(\mathfrak{P}\mathfrak{A}_{p,\alpha}(\mathbb{R}^+))$ is homeomorphic to the set of maximal ideals.

We drop further details about the Banach algebra $\mathfrak{P}\mathfrak{A}^{N\times N}_{p,\alpha}(\mathbb{R}^+)$ since the above formulated result is sufficient for the purpose of this and subsequent papers dealing with the BVPs in domains with corners at the boundary.

4 Mellin Convolution Equations in the Bessel Potential Spaces

In boundary integral equations for elliptic partial differential equations in the domains with angular points (Lipschitz domains in general) we encounter, after proper localization, Mellin convolution equations as an equivalent reformulation.

For $s \in \mathbb{R}$ and $1 < p < \infty$, the Bessel potential space, known also as a fractional Sobolev space, is the subspace of the Schwartz space $\mathbb{S}'(\mathbb{R})$ of all distributions having the finite norm

$$\|\varphi \mid \mathbb{H}^s_p(\mathbb{R})\| := \left[\int_{-\infty}^{\infty} \left|\mathcal{F}^{-1}\left(1+|\xi|^2\right)^{s/2}(\mathcal{F}\varphi)(t)\right|^p dt\right]^{1/p} < \infty.$$

For an integer parameter $s = m = 1, 2, \dots,$ the space $\mathbb{H}^s_p(\mathbb{R})$ coincides with the usual Sobolev space endowed with an equivalent norm

$$\|\varphi \mid \mathbb{W}^m_p(\mathbb{R})\| := \left[\sum_{k=0}^{m} \int_{-\infty}^{\infty} \left|\frac{d^k\varphi(t)}{dt^k}\right|^p dt\right]^{1/p}.$$

If $s < 0$, one gets the space of distributions. Moreover, $\mathbb{H}^{-s}_{p'}(\mathbb{R})$ is the dual to the space $\mathbb{H}^s_p(\mathbb{R}^+)$, provided $p' := \dfrac{p}{p-1}$, $1 < p < \infty$. Note that $\mathbb{H}^s_2(\mathbb{R})$ is a Hilbert space with the inner product

$$\langle \varphi, \psi\rangle_s = \int_{\mathbb{R}} (\mathcal{F}\varphi)(\xi)\overline{(\mathcal{F}\psi)(\xi)}(1+\xi^2)^s\, d\xi, \ \varphi, \psi \in \mathbb{H}^s(\mathbb{R}).$$

By r_Σ we denote the operator restricting functions or distributions defined on $\mathbb{R}$ to the subset $\Sigma \subset \mathbb{R}$. Thus $\mathbb{H}^s_p(\mathbb{R}^+) = r_+(\mathbb{H}^s_p(\mathbb{R}))$, and the norm in $\mathbb{H}^s_p(\mathbb{R}^+)$ is defined by

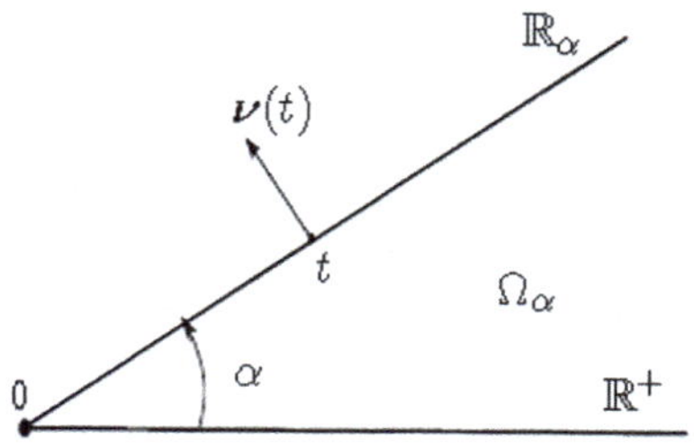

Fig. 3 Model domain Ω_α

$$\left\| f \mid \mathbb{H}^s_p(\mathbb{R}^+) \right\| = \inf_{\ell} \left\| \ell f \mid \mathbb{H}^s_p(\mathbb{R}) \right\|,$$

where ℓf stands for any extension of f to a distribution in $\mathbb{H}^s_p(\mathbb{R})$.

Further, we denote by $\widetilde{\mathbb{H}}^s_p(\mathbb{R}^+)$ the (closed) subspace of $\mathbb{H}^s_p(\mathbb{R})$ which consists of all distributions supported in the closure of $\mathbb{R}^+$.

Note that $\widetilde{\mathbb{H}}^s_p(\mathbb{R}^+)$ is embedded continuously in $\mathbb{H}^s_p(\mathbb{R}^+)$ and if $s \in (1/p - 1, 1/p)$, these two spaces coincide. Moreover, $\mathbb{H}^s_p(\mathbb{R}^+)$ may be viewed as the quotient-space $\mathbb{H}^s_p(\mathbb{R}^+) := \mathbb{H}^s_p(\mathbb{R})/\widetilde{\mathbb{H}}^s_p(\mathbb{R}^-)$, $\mathbb{R}^- := (-\infty, 0)$.

Consider a model domain: Ω_α, which is the angle of magnitude α between the half axes $R_1 := \mathbb{R}^+$ and the beam $R_2 := \mathbb{R}_\alpha$ turned by the angle α from $\mathbb{R}^+$ (see Fig. 3).

In the domain Ω_α we consider the following mixed (Dirichlet-Neumann) BVP:

$$\begin{cases} \Delta u(t) = f(t), & t \in \Omega_\alpha, \\ u^+(s) = g(s), & \text{on } \mathbb{R}^+, \\ (\partial_\nu u)^+(s) = h(s), & \text{on } \mathbb{R}_\alpha, \end{cases} \tag{25}$$

$$f \in \mathbb{L}_p(\Omega_\alpha), \quad g \in \mathbb{H}^{s-1/p}_p(\mathbb{R}^+), \quad h \in \mathbb{H}^{s-1-1/p}_p(\mathbb{R}_\alpha), \quad 1 < p < \infty, \quad s > \frac{1}{p}.$$

By applying Lax-Milgram lemma we easily derive the following.

Theorem 3 (see [21]) *The boundary value problem* (25) *has, at most, one solution in the space* $\mathbb{H}^1(\Omega_{\alpha_j})$ *the classical setting* $p = 2$, $s = 1$.

By applying the potential method to the BVP (25) we derive the following equivalent boundary integral equation, BIE (for details see, e.g., [11]).

$$\mathbf{A}\Phi := \begin{bmatrix} 1 & 0 \\ 0 & 1 \end{bmatrix} \Phi - \frac{1}{2} \begin{bmatrix} 0 & 1 \\ 0 & 1 \end{bmatrix} \mathbf{K}_{e^{i\alpha_j}} \Phi + \frac{1}{2} \begin{bmatrix} 1 & 0 \\ 1 & 0 \end{bmatrix} \mathbf{K}_{e^{-i\alpha_j}} \Phi = F, \tag{26}$$

$$F = \begin{bmatrix} F_+ \\ F_- \end{bmatrix}, \quad \Phi = \begin{bmatrix} \varphi_+ \\ \varphi_- \end{bmatrix}, \quad \varphi_\pm \in \widetilde{\mathbb{H}}^{s-1-1/p}_p(\mathbb{R}^+), \quad F_\pm \in \mathbb{H}^{s-1-1/p}_p(\mathbb{R}^+).$$

where

$$\mathbf{K}^1_{e^{\pm i\alpha}}\psi(t) := \int_0^\infty \frac{\psi(\tau)d\tau}{t - e^{\pm i\alpha}\tau}, \quad -\pi < \alpha < \pi, \tag{27}$$

are Mellin convolution operators with meromorphic kernels (see below).

If we consider the mixed BVP for an elliptic operator on an open hypersurface with Lipschitz (angular) boundary, applying localization we can study solvability of such BVPs in the non-classical setting with the help of local representatives, which are model BVPs in angles Ω_α like the mixed BVP (25) above (see [2] for details). These model BVPs are then reduced to the equivalent BIEs like (26).

As we can see from the example and from the similar examples considered in [11, 12] and also recent publications on nano-photonics [1, 22], while investigating elliptic BVPs in the angular domain we encounter Mellin convolution equations in the Bessel potential spaces. In particular, the operator **A** in (26) should be treated in the following space setting

$$\mathbf{A} \; : \; \widetilde{\mathbb{H}}^\sigma_p(\mathbb{R}^+) \longrightarrow \mathbb{H}^\sigma_p(\mathbb{R}^+) \tag{28}$$

where, in this particular case, $\sigma = s - 1 - 1/p$.

If **A** in (28) is a bounded operator (of order 0), the following diagram is commutative (so called *operator lifting*)

$$\begin{array}{ccc} \widetilde{\mathbb{H}}^\sigma_p(\mathbb{R}^+) & \xrightarrow{\mathbf{A}} & \mathbb{H}^\sigma_p(\mathbb{R}^+) \\ \uparrow \Lambda^{-\sigma}_+ & & \downarrow \Lambda^\sigma_- \\ \mathbb{L}_p(\mathbb{R}^+) & \xrightarrow{\Lambda^\sigma_-\mathbf{A}\Lambda^{-\sigma}_+} & \mathbb{L}_p(\mathbb{R}^+), \end{array} \tag{29}$$

with the following *Bessel potential operators*

$$\begin{array}{l} \Lambda^r_+ = W_{\lambda^r_+} \; : \; \widetilde{\mathbb{H}}^s_p(\mathbb{R}^+) \to \widetilde{\mathbb{H}}^{s-r}_p(\mathbb{R}^+)\,, \\ \Lambda^r_- = W_{\lambda^r_-} \; : \; \mathbb{H}^s_p(\mathbb{R}^+) \to \mathbb{H}^{s-r}_p(\mathbb{R}^+)\,, \\ \lambda^r_\pm(\xi) := (\xi \pm i)^r, \qquad \xi \in \mathbb{R}^+, \end{array} \tag{30}$$

which are Fourier convolution operators (classical ΨDOs).

The diagram (29) provides an *equivalent lifting of the operator* **A** *to the operator* $\Lambda^\sigma_-\mathbf{A}\Lambda^{-\sigma}_+ \; : \; \mathbb{L}_p(\mathbb{R}^+) \longrightarrow \mathbb{L}_p(\mathbb{R}^+)$.

Theorem 4 (see Lemma 5.1 in [6] and [10]) *For arbitrary $r \in \mathbb{R}$ the Bessel potential operators* (30) *arrange isomorphisms of spaces for all $s \in \mathbb{R}$ and $1 < p < \infty$.*

Moreover, if $\mathbf{A} = W_a \; : \; \mathbb{H}^s_p(\mathbb{R}^+) \longrightarrow \mathbb{H}^s_p(\mathbb{R}^+)$ is a bounded Fourier convolution operator, the lifted operator is a Fourier convolution operator again

$$W_{a_s} \; : \; \mathbb{L}_p(\mathbb{R}^+) \longrightarrow \mathbb{L}_p(\mathbb{R}^+), \tag{31}$$

with the symbol

$$a_s(\xi) = \lambda_-^s(\xi)a(\xi)\lambda_+^{-s}(\xi) = \left(\frac{\xi - i}{\xi + i}\right)^s a(\xi). \tag{32}$$

Corollary 3 (see [6]) *The Fourier convolution operators*

$$W_a^0 \,:\, \mathbb{H}_p^s(\mathbb{R}) \longrightarrow \mathbb{H}_p^s(\mathbb{R}) \quad and \quad W_a \,:\, \widetilde{\mathbb{H}}_p^s(\mathbb{R}^+) \longrightarrow \mathbb{H}_p^s(\mathbb{R}^+)$$

are bounded if and only if they are bounded in $\mathbb{L}_p$ spaces, i.e., if a is a Fourier $\mathbb{L}_p$-multiplier $a \in \mathfrak{M}_p(\mathbb{R})$.

We have a completely different reality when we consider Mellin convolution operators in the Bessel potential spaces. First we have the problem with the boundedness. It is easy to write an example of a Mellin operator with the symbol from multiplier class $\mathfrak{M}_p(\mathbb{R})$ which is not bounded in $\mathbb{H}_p^s(\mathbb{R}^+)$ for any $1 < p < \infty$, $s \in \mathbb{R}$, $s \neq 0$. The following boundedness result for the Mellin convolution operator can easily be derived.

Proposition 4 (see [19, Proposition 0.1]) *Let $1 < p < \infty$ and let $m = 1, 2, \ldots$ be an integer. If a function $\mathcal{K}$ satisfies the condition*

$$\int_0^1 t^{\frac{1}{p}-m-1}|\mathcal{K}(t)|\,dt + \int_1^\infty t^{\frac{1}{p}-1}|\mathcal{K}(t)|\,dt < \infty, \tag{33}$$

then the Mellin convolution operator (see (1))

$$\mathbf{A}\varphi(t) = c_0\varphi(t) + \frac{c_1}{\pi i}\int_0^\infty \frac{\varphi(\tau)\,dt}{\tau - t} + \int_0^\infty \mathcal{K}\Big(\frac{t}{\tau}\Big)\varphi(\tau)\,\frac{d\tau}{\tau}, \tag{34}$$

is bounded in the setting of the Bessel potential spaces

$$\mathbf{A} \,:\, \widetilde{\mathbb{H}}_p^s(\mathbb{R}^+) \longrightarrow \mathbb{H}_p^s(\mathbb{R}^+) \tag{35}$$

for any $0 \leqslant s \leqslant m$.

Note that the kernels of operators $\mathbf{K}^1_{e^{\pm i\alpha}}$ in (27) does not satisfy the condition (33) for $s \geqslant 1$. A suitable result was proved in [19] by introducing the class of *admissible kernels*.

Definition 1 (*see Definition 2.1 in* [19]) We call a kernel

$$\mathcal{K}(t) := \sum_{j=0}^{\ell} \frac{d_j}{t - c_j} + \sum_{j=\ell+1}^{\infty} \frac{d_j}{(t - c_j)^{m_j}}, \quad c_j \neq 0, \quad j = 0, 1, \ldots,$$
$$c_0, \ldots, c_\ell \in \mathbb{R}, \quad 0 < \alpha_k := |\arg c_k| \leqslant \pi, \quad k = \ell + 1, \ell + 2, \ldots. \tag{36}$$

admissible iff the points $c_{\ell+1}, c_{\ell+2}, \dots$ do not condense to the positive semi-axes and their real parts are bounded uniformly.

$$0 < \varliminf_{\ell \leqslant j \to \infty} \operatorname{Re} c_j \leqslant \varlimsup_{\ell \leqslant j \to \infty} \operatorname{Re} c_j \leqslant K < \infty.$$

Theorem 5 (see Theorem 2.4 in [19]) *Let $1 < p < \infty$. The Mellin convolution operator $\mathfrak{M}^0_{a_\beta}$ in (34) with an admissible kernel $\mathcal{K}$ (see (36)) is bounded in the Bessel potential space setting (35) for all $s \in \mathbb{R}$ provided the condition*

$$K_\beta := \sum_{j=0}^{\infty} 2^{m_j} |d_j|\, |c_j|^{\beta - m_j} < \infty.$$

holds and $m^0 := \sup_{j=0,1,\dots} m_j < \infty$.

Let us note that the symbol (the Mellin transform of the kernel) of the Mellin convolution operator

$$\mathfrak{M}_{a^m_{c,\beta}} := \frac{1}{\pi} \int_0^\infty \mathbb{K}^m_c\left(\frac{t}{\tau}\right) \varphi(\tau) \frac{d\tau}{\tau} = \frac{1}{\pi} \int_0^\infty \frac{\tau^{m-1} \varphi(\tau)\, d\tau}{(t - c\tau)^m} \tag{37}$$

with the model meromorphic kernel (see (36))

$$\mathbb{K}^m_c(x) := \frac{1}{\pi} \frac{\tau^{m-1}}{(t-c)^m}, \qquad 0 < \arg\, c < 2\pi, \quad m = 1, 2, \dots \tag{38}$$

is calculated in [19, formula (39)]:

$$a^m_{c,\beta}(\xi) = \mathcal{M}_\beta \mathbb{K}^m_c(\xi) = \binom{\beta - i\xi - 1}{m-1} \frac{(-1)^{m-1} e^{-i\pi(\beta - i\xi - m)} c^{\beta - i\xi - m}}{\sin \pi(\beta - i\xi)}. \tag{39}$$

In particular,

$$\begin{aligned}
a^1_{c,\beta}(\xi) &= \frac{e^{-i\pi(\beta - i\xi - 1)} c^{\,\beta - i\xi - 1}}{\sin \pi(\beta - i\xi)}, \quad 0 < \arg c < 2\pi, \\
a^1_{-d,\beta}(\xi) &= \frac{d^{\,\beta - i\xi - 1}}{\sin \pi(\beta - i\xi)}, \quad -\pi < \arg d < \pi, \\
a^1_{-1,\beta}(\xi) &= \frac{1}{\sin \pi(\beta - i\xi)}, \quad \xi \in \mathbb{R}.
\end{aligned} \tag{40}$$

For the Cauchy singular integral operator $S_{\mathbb{R}^+} = iK^1_1 = \mathfrak{s}$ (see (16)) the Mellin symbol is [19, formula (41)]:

$$s_\beta(\xi) = i\mathcal{M}_\beta \mathbb{K}^1_1(t) = \cot \pi(\beta - i\xi) = -i \coth(i\beta - \xi). \tag{41}$$

The following is the key theorem for lifting of Mellin convolution operator in the Bessel potential space to the lebesgue space.

Theorem 6 ([5]) *Let* $-\pi \leqslant \arg c < \pi$, $\arg c \neq 0$, $0 < \arg \gamma < \pi$ *and* $r, s \in \mathbb{R}$, $1 < p < \infty$. *Then the operator* $K^1_c : \widetilde{\mathbb{H}}^s_p(\mathbb{R}^+) \to \mathbb{H}^s_p(\mathbb{R}^+)$ *is lifted equivalently to the operator*

$$\mathbf{A}^{1,s}_c := \mathbf{\Lambda}^s_{-\gamma} \mathbf{K}^1_c \mathbf{\Lambda}^{-s}_\gamma \ : \ \mathbb{L}_p(\mathbb{R}^+) \to \mathbb{L}_p(\mathbb{R}^+) \tag{42}$$

and

$$\mathbf{A}^{1,s}_c = \begin{cases} c^{-s} \mathbf{K}^1_c W_{g^s_{-c\gamma,\gamma}}, & \text{if } \ 0 < \arg(-c\gamma) < \pi, \\ c^{-s} W_{g^s_{-\gamma,-\gamma_0}} \mathbf{K}^1_c W_{g^s_{-c\gamma_0,\gamma}}, & \text{if } \ 0 < \arg(c\gamma) < \pi, \end{cases}$$

$$g^s_{-c\gamma,\gamma}(\xi) := \left(\frac{\xi - c\gamma}{\xi + \gamma}\right)^s, \quad g^s_{-\gamma,-\gamma_0}(\xi) := \left(\frac{\xi - \gamma}{\xi - \gamma_0}\right)^s,$$

for all $\gamma_0 \in \mathbb{C}$ *such that* $0 < \arg \gamma_0 < \pi$ *and* $0 < \arg(-c\,\gamma_0) < \pi$.

Since the lifted operator landed in the Banach algebra generated by Fourier and Mellin convolution operators on $\mathbb{R}^+$, by applying results of [13] we can define the symbol of a Mellin convolution operator with meromorphic symbol

$$\mathfrak{M}^0_a \ : \ \widetilde{\mathbb{H}}^s_p(\mathbb{R}^+) \longrightarrow \mathbb{H}^s_p(\mathbb{R}^+). \tag{43}$$

Consider the infinite "rectangle" $\mathfrak{R} := \Gamma_1 \cup \Gamma^-_2 \cup \Gamma^+_2 \cup \Gamma_3$ (cf. Fig. 1).

Based on Theorems 4, 6 and the description of the symbol of an operator in the Banach algebra generated by Fourier and Mellin convolution operators (see [7, 13]) we can write the symbol of an operator

$$\mathbf{A} := d_0 I + \sum_{j=1}^{n} d_j \mathbf{K}^1_{c_j} \ : \ \widetilde{\mathbb{H}}^s_p(\mathbb{R}^+) \to \mathbb{H}^s_p(\mathbb{R}^+), \tag{44}$$

compiled of the identity operator I and Mellin $\mathbf{K}^1_{c_1}, \ldots, \mathbf{K}^1_{c_n}$ convolution operators.

We assume that $b_1, \ldots, b_n \in C\mathcal{M}_p(\overline{\mathbb{R}} \setminus \{0\})$, $d_0, \ldots, d_n, c_1, \ldots, c_n \in \mathbb{C}$.

The symbol of the operator $\mathbf{A} : \widetilde{\mathbb{H}}^s_p(\mathbb{R}^+) \to \mathbb{H}^s_p(\mathbb{R}^+)$ is defined as follows:

$$\mathcal{A}^s_p(\omega) := d_0 \mathcal{I}^s_p(\omega) + \sum_{j=1}^{n} d_j \mathcal{K}^{1,s}_{c_j,p}(\omega). \tag{45}$$

Here $\mathcal{I}^s_p(\omega)$ and $\mathcal{K}^{1,s}_{c_j,p}(\omega)$ are the symbols of the identity operator I and of the operator $\mathbf{K}^1_{c_j}$, defined by the equalities:

$$\mathcal{I}^s_p(\omega) := \begin{cases} g^s_{-\gamma,\gamma,p}(\infty,\xi), & \omega = (\xi,\xi) \in \overline{\Gamma}_1, \\ \left(\dfrac{\eta-\gamma}{\eta+\gamma}\right)^{\mp s}, & \omega = (\infty,\eta) \in \Gamma_2^{\pm}, \\ e^{\pi s i}, & \omega = (\xi, 0) \in \overline{\Gamma}_3, \end{cases} \tag{46a}$$

$$g^s_{-\gamma,\gamma,p}(\infty,\xi) := \frac{e^{2\pi s i}+1}{2} + \frac{e^{2\pi s i}-1}{2i} \cot \pi\Big(\frac{1}{p} - i\xi\Big)$$
$$= e^{\pi s i} \frac{\sin \pi\Big(\frac{1}{p} + s - i\xi\Big)}{\sin \pi\Big(\frac{1}{p} - i\xi\Big)}, \quad \xi \in \mathbb{R}, \quad \eta \in \mathbb{R}^+, \tag{46b}$$

$$\mathcal{K}^{1,s}_{c,p}(\omega) := \begin{cases} \dfrac{e^{-i\pi\left(\frac{1}{p}-i\xi-1\right)} c^{\frac{1}{p}-i\xi-s-1}}{\sin \pi(\frac{1}{p} - i\xi)}, & \omega = (\xi,\xi) \in \overline{\Gamma}_1, \\ 0, & \omega \in \Gamma_2^{\pm} \cup \Gamma_3, \end{cases} \tag{46c}$$

$$0 < \arg c < 2\pi, \quad -\pi < \arg(c\,\gamma) < 0, \quad 0 < \arg \gamma < \pi, \quad c^s = |c|^s e^{is \arg c}.$$

Theorem 7 (see [5, 19]) *Let $1 < p < \infty$, $s \in \mathbb{R}$. The operator $\mathbf{A}$ in* (44) *is Fredholm if and only if its symbol $\mathcal{A}^s_p(\omega)$ defined in* (45), (46a)–(46c) *is elliptic* $\inf\limits_{\omega \in \mathfrak{R}} |\mathcal{A}^s_p(\omega)| > 0$.

If $\mathbf{A}$ is Fredholm, the index of the operator is

$$\operatorname{Ind} \mathbf{A} = -\operatorname{ind} \det \mathcal{A}^s_p.$$

For the definition of the Sobolev–Slobodeckij (Besov) spaces $\mathbb{W}^s_p(\Omega) = \mathbb{B}^s_{p,p}(\Omega)$, $\widetilde{\mathbb{W}}^s_p(\Omega) = \widetilde{\mathbb{B}}^s_{p,p}(\Omega)$ for an arbitrary domain $\Omega \subset \mathbb{R}^n$, including the half axes $\mathbb{R}^+$, we refer to the monograph [29].

Corollary 4 (see [5, 19]) *Let $1 < p < \infty$, $s \in \mathbb{R}$. If the operator $\mathbf{A} : \widetilde{\mathbb{H}}^s_p(\mathbb{R}^+) \longrightarrow \mathbb{H}^s_p(\mathbb{R}^+)$, defined in* (44)*, is Fredholm (is invertible) for all $a \in (s_0, s_1)$ and $p \in (p_0, p_1)$, where $-\infty < s_0 < s_1 < \infty$, $1 < p_o < p_1 < \infty$, then*

$$\mathbf{A} : \widetilde{\mathbb{W}}^s_p(\mathbb{R}^+) \longrightarrow \mathbb{W}^s_p(\mathbb{R}^+), \quad s \in (s_0, s_1), \quad p \in (p_0, p_1) \tag{47}$$

is Fredholm and has the same index

$$\text{Ind}\mathbf{A} = -\text{ind}\det \mathcal{A}^s_p. \tag{48}$$

(is invertible, respectively) in the Sobolev–Slobodeckij (Besov) spaces $\mathbb{W}^s_p = \mathbb{B}^s_{p,p}$.

Let us come back to BVP (25) and call **regular** those pairs of space parameters $(s, p) \in (1/p, \infty) \times (1, \infty)$ for which the BVP (and the corresponding BIE (26)) are Fredholm.

Theorem 8 (see [5, 19]) *Let* $1 < p < \infty$, $s \in \mathbb{R}$. *The mixed boundary value problem* (25) *and the corresponding boundary integral equation* (25) *are Fredholm if and only if the symbol is elliptic*

$$\frac{e^{4\pi i/p}\sin^2\pi(2/p - i\xi - s) + \cos^2[(\pi-\alpha)(2/p - i\xi - s)]}{\sin^2\pi(1/p - i\xi)} \neq 0 \quad \text{for all} \quad \xi \in \mathbb{R}.$$

If the symbol is elliptic and the BVP (25) *is Fredholm, the subplane* $(1/p, \infty) \times (1, \infty)$ *of pairs* (s, p) *decomposes into an infinite union of non-intersecting connected parts of regular pairs.*

If a regular connected part $\mathcal{R}_0$ *contains the point* $(1, 2)$ *(i.e.* $s = 1$, $p = 2$*) then BVP* (25) *and BIE* (26) *are both uniquely solvable for all pairs* $(s, p) \in \mathcal{R}_0$.

Acknowledgements Support of Shota Rustaveli Georgian National Science Foundation is acknowledged within the grant DI-2016-16.

References

1. Bonnet-Ben Dhia, A.S., Chesnel, L., Claeys, X.: Radiation condition for a non-smooth interface between a dielectric and a metamaterial. Math. Models Methods Appl. Sci. **23**(9), 1629–1662 (2013). http://hal.inria.fr/hal-00651008/
2. Buchukuri, T., Duduchava, R., Kapanadze, D., Tsaava, M.: Localization of a Helmholtz boundary value problem in a domain with piecewise-smooth boundary. Proc. A: Razmadze Math. Inst. **162**, 37–44 (2013)
3. Costabel, M.: Boundary integral operators on curved polygons. Ann. Mat. Pura Appl. (4) **133**, 305–326 (1983)
4. Costabel, M., Stephan, E.: The method of Mellin transformation for boundary integral equations on curves with corners. In: Gerasoulis, A., Vichnevetsky, R. (eds.) Numerical Solutions of Singular Integral Equations, pp. 95–102. IMACS, New Brunswick (1984)
5. Didenko, V.D., Duduchava, R.: Mellin convolution operators in the Bessel potential spaces. J. Anal. Appl. **443**, 707–731 (2016)
6. Duduchava, R.: Integral Equations in Convolution with Discontinuous Presymbols, Singular Integral Equations with Fixed Singularities, and Their Applications to Some Problems of Mechanics. BSB B. G. Teubner Verlagsgesellschaft, Leipzig (1979)
7. Duduchava, R.: On general singular integral operators of the plane theory of elasticity. Rend. Sem. Mat. Univ. Politec. Torino **42**(3), 15–41 (1984)
8. Duduchava, R.: An application of singular integral equations to some problems of elasticity. Integr. Equ. Oper. Theory **5**(4), 475–489 (1982)
9. Duduchava, R., Latsabidze, T., Saginashvili, A.: Singular integral operators with the complex conjugation on curves with cusps. Integr. Equ. Oper. Theory **22**(1), 1–36 (1995)

10. Duduchava, R., Speck, F.O.: Pseudodifferential operators on compact manifolds with Lipschitz boundary. Math. Nachr. **160**, 149–191 (1993)
11. Duduchava, R., Tsaava, M.: Mixed boundary value problems for the Helmholtz equation in a model 2D angular domain. Georgian Math. J. http://arxiv.org/abs/1605.09029
12. Duduchava, R., Tsaava, M.: Mixed boundary value problems for the Laplace-Beltrami equation. Complex Var. Elliptic Equ. **63**(10), 1468–1496 (2018)
13. Duduchava, R.V.: General singular integral equations and fundamental problems of the plane theory of elasticity. Trudy Tbiliss. Mat. Inst. Razmadze Akad. Nauk Gruzin. SSR **82**, 45–89 (1986) (in Russian)
14. Duduchava, R.V.: On Wiener-Hopf integral operators. Math. Nachr. **65**(1), 59–82 (1975)
15. Duduchava, R.V.: Convolution integral operators with discontinuous symbols. Trudy Tbiliss. Mat. Inst. Razmadze **50**, 34–41 (1975) (in Russian)
16. Duduchava, R.V.: Integral operators of convolution type with disconnected coefficients. Math. Nachr. **79**, 75–98 (1977)
17. Duduchava, R.V.: Integral equations of convolution type with discontinuous coefficients. Soobshch. Akad. Nauk Gruzin. SSR **92**(2), 281–284 (1978) (in Russian)
18. Duduchava, R.V., Latsabidze, T.I.: The index of singular integral equations with complex-conjugate functions on piecewise-smooth lines. Trudy Tbiliss. Mat. Inst. Razmadze Akad. Nauk Gruzin. SSR **76**, 40–59 (1985) (in Russian)
19. Duduchava, R.V.: Mellin convolution operators in Bessel potential spaces with admissible meromorphic kernels. Mem. Differ. Equ. Math. Phys. **60**, 135–177 (2013)
20. Duduchava, R.V.: On algebras generated by convolutions and discontinuous functions. Special issue: Wiener–Hopf problems and applications (Oberwolfach, 1986). Integr. Equ. Oper. Theory **10**(4), 505–530 (1987)
21. Duduchava, R.V., Tsaava, M., Tsutsunava, T.: Mixed boundary value problem on hypersurfaces. Int. J. Differ. Equ. **2014**, Article ID 245350, 8 p
22. Gramotnev, D.K., Bozhevolnyi, S.I.: Plasmonics beyond the diffraction limit. Nat. Photon. **4**(2), 83–91 (2010)
23. Hörmander, L.: Linear Partial Differ. Oper. Springer-Verlag, Heidelberg (1963)
24. Kalandya, A.: Mathematical Methods of Two-Dimensional Elasticity. Mir, Moscow (1975)
25. Karapetyants, N., Samko, S.: Equations with Involutive Operators. Birkhauser, Basel, Berlin, Boston (2001), 427 p
26. Schneider, R.: Integral equations with piecewise continuous coefficients in L^p-spaces with weight. J. Integr. Equ. **9**(2), 135–152 (1985)
27. Thelen, G.: Zur Fredholmtheorie singulärer Integrodifferentialoperatoren auf der Halbachse. Dissertation Dr. rer. nat. Darmstadt (1985)
28. Tung, Y.-K., Yen, B.-C.: Hydrosystems Engineering Uncertainty Analysis. McGraw Hill Professional, Technology & Engineering (2005), 285 p
29. Triebel, H.: Interpolation Theory, Function Spaces, Differential Operators. Johann Ambrosius Barth, Heidelberg (1995)

Integral Operators of the $\mathcal{L}$-Convolution Type in the Case of a Reflectionless Potential

Davresh Hasanyan, Armen Kamalyan, Martin Karakhanyan and Ilya M. Spitkovsky

Abstract The notion of the $\mathcal{L}$-convolution operator is introduced by changing the Fourier operator in the definition of the (regular) convolution operator to the operator intertwining the Sturm-Liouville operator $\mathcal{L}$ with the multiplication operator. Along the same lines, the $\mathcal{L}$-Wiener-Hopf operator is introduced. For the latter, the invertibility is studied in the case of a reflectionless potential and piecewise continuous symbols.

Keywords Reflectionless potential · $\mathcal{L}$-Wiener-Hopf operator · $\mathcal{L}$-Symbol

1 Problem Setup

Consider the Sturm-Liouville equation

$$-y''(x) + c(x)y(x) = \lambda^2 y(x), \qquad x \in \mathbb{R}, \tag{1.1}$$

with the real potential c satisfying

D. Hasanyan
Emerson Climate Technologies Inc., Sidney, OH 45365, USA
e-mail: davreshh@yahoo.com

A. Kamalyan (✉) · M. Karakhanyan
Yerevan State University, 375025 Yerevan, Armenia
e-mail: kamalyan_armen@yahoo.com; armen.kamalyan@ysu.am

M. Karakhanyan
e-mail: m_karakhan@yahoo.com

A. Kamalyan
Institute of Mathematics, National Academy of Sciences, 375019 Yerevan, Armenia

I. M. Spitkovsky
New York University Abu Dhabi (NYUAD), Abu Dhabi, United Arab Emirates
e-mail: ims2@nyu.edu; imspitkovsky@gmail.com

A. Karapetyants et al. (eds.), *Modern Methods in Operator Theory and Harmonic Analysis*, Springer Proceedings in Mathematics & Statistics 291,
https://doi.org/10.1007/978-3-030-26748-3_11

$$\int_{-\infty}^{\infty} (1 + |x|)|c(x)|\, dx < \infty, \tag{1.2}$$

and a complex parameter λ (see, e.g., [1–4]).

An important role in the theory of Eq. (1.1) is played by the Jost solutions $e_+(x, \lambda)$ ($x \in \mathbb{R}$, $\operatorname{im} \lambda \geqslant 0$) and $e_-(x, \lambda)$ ($x \in \mathbb{R}$, $\operatorname{im} \lambda \leqslant 0$) characterized by the boundary conditions

$$\lim_{x \to \pm\infty} e^{-i\lambda x} e_\pm(x, \lambda) = 1, \qquad \lim_{x \to \pm\infty} e^{-i\lambda x} e'_\pm(x, \lambda) = i\lambda.$$

For real $\lambda \neq 0$ the pairs $e_+(x, \lambda), e_+(x, -\lambda), e_-(x, \lambda), e_-(x, -\lambda)$ form a fundamental system of solutions to (1.1) [2, 3]. We have therefore

$$e_+(x, \lambda) = b(\lambda) e_-(x, -\lambda) + b_0(\lambda) e_-(x, \lambda).$$

Recall that the potential c satisfying (1.2) is called *reflectionless* [3, 5] if $b = 0$. Such potentials generate families of exact solutions to Korteweg-de Vries equations, the so called N-soliton solutions, see [5] and also [6].

Note that condition $b = 0$ is equivalent to the reflection coefficients $r_\pm(\lambda) = \mp b(\mp\lambda)/b_0(\lambda)$ both being equal to zero. By solving an inverse scattering problem on the line it is possible to obtain the description of all reflectionless potentials satisfying (1.2) (see [1, 3]) parameterized by $\lambda_k = i\mu_k$, $k = 1, \ldots, N$, where μ_k are distinct positive numbers, and $m_k^+ (> 0)$, $k = 1, \ldots, N$.

Namely, the reflectionless potential is defined by the formula

$$c(x) = -2 \frac{d^2}{dx^2} \ln \Delta(x),$$

where

$$\Delta(x) = \det \left[\delta_{ij} + \frac{m_i^+ \exp(-(\mu_i + \mu_j)x)}{\mu_i + \mu_j} \right] \qquad i, j = 1, \ldots, N,$$

and δ_{ij} is the Kronecker symbol.

We suppose in what follows that the potential c is reflectionless. By convention, the zero potential $c = 0$ also is reflectionless.

The operators $\mathcal{L}$ and $\mathcal{L}_0$ defined as $\mathcal{L} = -y'' + cy$ and $\mathcal{L}_0 y = -y''$, respectively, on a dense in $L_2(\mathbb{R})$ Sobolev space $\mathcal{W}_2^2(\mathbb{R})$, are selfadjoint (see [1]). The set $\{\lambda_k, m_k^+, k = 1, \ldots, N\}$ represents the right scattering data. In particular, λ_k^2, $k = 1, \ldots, N$, are the eigenvalues of $\mathcal{L}$.

By $m(a)$ ($a \in L_\infty(\mathbb{R})$) and J we denote the operators acting on $L_p(\mathbb{R})$ ($1 \leqslant p \leqslant \infty$) according to the formulas $m(a)y = ay$, $(Jy)(x) := \hat{y}(x) = y(-x)$. We also let $\mathtt{t}$ stand for the propagation coefficient, in our case defined by

$$\mathtt{t}(\lambda) = \hat{b}_0(\lambda) = b_0^{-1}(\lambda) = \prod_{k=1}^{N} \frac{\lambda + i\mu_k}{\lambda - i\mu_k}\,.$$

Formulas

$$\begin{aligned} u^-(x,\lambda) &= \mathtt{t}(\lambda)e_+(x,\lambda) = e_-(x,\lambda),\\ u^+(x,\lambda) &= \mathtt{t}(\lambda)e_-(x,-\lambda) = e_+(x,-\lambda), \qquad x,\lambda \in \mathbb{R}\,. \end{aligned} \tag{1.3}$$

define a complete set (parameterized by λ) of orthonormal eigenfunctions $u^-(x,\lambda)$, $u^+(x,\lambda)$ for the left and right scattering problem respectively, corresponding to the continuous spectrum of the operator $\mathcal{L}$, in the case of a reflectionless potential.

We define the operators $U_\mp$, $U : L_2(\mathbb{R}) \to L_2(\mathbb{R})$ by the formulas

$$(U_\mp y)(\lambda) = \frac{1}{\sqrt{2\pi}} \int_{-\infty}^{\infty} u^\mp(x,\lambda)\, y(x)\, dx, \qquad \lambda \in \mathbb{R}, \tag{1.4}$$

and

$$U = m(\chi_+)U_- + m(\chi_-)JU_+.$$

Here $\chi_\pm$ is the characteristic function of the half-line $\mathbb{R}_\pm = \{x \in \mathbb{R}\colon\ \pm x \geq 0\}$.

We define also the operators $\ell : L_2^2(\mathbb{R}_+, \frac{1}{2}|x|^{-1/2}) \to L_2(\mathbb{R})$ acting as $(\ell y)(x) = y(x^2)$, and $\pi_\pm : L_p(\mathbb{R}) \to L_p(\mathbb{R}_\pm)$, $\pi_\pm^0 : L_p(\mathbb{R}_\pm) \to L_p(\mathbb{R})$ satisfying
$(\pi_\pm y)(x) = y(x)$, $x \in \mathbb{R}_\pm$,

$$(\pi_+^0 y)(x) = \begin{cases} y(x), & x \in \mathbb{R}_+ \\ 0, & x \in \mathbb{R}_- \end{cases}, \qquad (\pi_-^0 y)(x) = \begin{cases} 0, & x \in \mathbb{R}_+ \\ y(x), & x \in \mathbb{R}_- \end{cases}.$$

According to [1], the operator $T_+ : L_2(\mathbb{R}) \to L_2(\mathbb{R}_+, \frac{1}{2}|x|^{-1/2})$ given by the formula

$$T_+ = \begin{pmatrix} \ell^{-1}\pi_+ U_+ \\ \ell^{-1}\pi_+ U_- \end{pmatrix}$$

is a partial isometry:

$$T_+^* T_+ = I - P, \qquad T_+ T_+^* = I. \tag{1.5}$$

Here P is the orthogonal projection of $L_2(\mathbb{R})$ onto the invariant subspace H_d of the operator $\mathcal{L}$ spanned by its eigenvectors ($\dim H_d = N$).

Observing that

$$U = \pi_+^0 \pi_+ U_- + J\pi_+^0 \pi_+ U_+ = (J\pi_+^0 \ell, \pi_+^0 \ell)T_+$$

and taking into consideration obvious equalities $J^* = J$, $\pi_\pm^* = \pi_\pm^0$, $\ell^* = \ell^{-1}$ holding on $L_2(\mathbb{R})$ along with (1.5), we conclude that

$$U^*U = I - P, \qquad UU^* = I. \tag{1.6}$$

Along with T_+, let us define the operator

$$T_0 = \begin{pmatrix} \ell^{-1}\pi_+ F^{-1} \\ \ell^{-1}\pi_+ F \end{pmatrix} : L_2(\mathbb{R}) \to L_2^2\Big(\mathbb{R}_+, \frac{1}{2}|x|^{-1/2}\Big),$$

where F is the Fourier transform:

$$(Fy)(\lambda) = \frac{1}{\sqrt{2\pi}} \int_{-\infty}^{\infty} e^{i\lambda x} y(x)\, dx, \qquad y \in L_2(\mathbb{R}).$$

It was shown in [1] that the equality

$$\mathcal{L} T_+^* T_0 = T_+^* T_0 \mathcal{L}_0 \,. \tag{1.7}$$

holds on $\mathcal{W}_2^2(\mathbb{R})$.

Since $T_+^* = U_+^* \pi_+^0 \ell$, $U_-^* \pi_+^0 \ell$, we have $T_+^0 T_0 = U^* F$ and so (1.6) and (1.7) imply

$$U \mathcal{L} U^* = m(\lambda^2)$$

on the set of functions $y \in L_2(\mathbb{R})$ satisfying $\int_{-\infty}^{\infty} |y(x)|^2 (1 + |x|^2)^2 dx < \infty$.

The restriction of the operator U onto the subspace $H_c = L_2(\mathbb{R}) \ominus H_d$ acts unitarily from H_c to $L_2(\mathbb{R})$. In order to construct a unitary operator on $L_2(\mathbb{R})$ intertwining $\mathcal{L}$ with the multiplication operator, we proceed as follows.

Consider the space $L_2(\Delta, \delta)$, where $\Delta = \{1, \ldots, N\}$, and δ is the Dirac measure on Δ. Let $\varphi_1, \ldots, \varphi_N$ be an orthonormal basis of H_d. Define a unitary operator $\tilde{U} : H_d \to L_2(\Delta, \delta)$ by the rule $\tilde{U}\varphi_k = \xi_k$, where $\xi_k(j) = \delta_{kj}$. Then the operator $\hat{U} : H_c \oplus H_d \to L_2(\mathbb{R}) \oplus L_2(\Delta, \delta)$ with the block matrix representation

$$\hat{U} = \begin{pmatrix} U|_{H_c} & 0 \\ 0 & \tilde{U} \end{pmatrix}$$

also is unitary.

It is easy to see that the equality $\tilde{U} \mathcal{L} \tilde{U}^* \xi = \lambda^2 \xi$ holds for any $\xi \in L_2(\Delta, \delta)$. In other words, $\hat{U} \mathcal{L} \hat{U}^* = m(\lambda^2)$ on a dense submanifold of $L_2(\mathbb{R})$.

Note that the multiplication operator $\tilde{m}(d)$ on a finite dimensional space $L_2(\Delta, \delta)$ is completely characterized by the vector $d = (d_1, \ldots, d_N)^T \in \mathbb{C}^N$, since it is acting as $\hat{m}(d)\xi = (d_1\xi_1, \ldots, d_N\xi_N)^T$.

In the particular case $c = 0$ the operator $\hat{U}$ is nothing but the Fourier transform.

By analogy, the $\mathcal{L}$-convolution operator with the $\mathcal{L}$-symbol (a, d) on the space $L_2(\mathbb{R})$, where $a \in L_\infty(\mathbb{R})$, $d = (d_1, \ldots, d_N)^T \in \mathbb{C}^N \cong L_\infty(\Delta, \delta)$, is defined as follows:

$$W^0_{\mathcal{L}}(a,d) = \hat{U}^{+} \begin{pmatrix} m(a) & 0 \\ 0 & \hat{m}(d) \end{pmatrix} \hat{U}.$$

Equivalently,

$$W^0_{\mathcal{L}}(a,d)y = U^* m(a) U y + \sum_{i=1}^{N} d_i \varphi_i(x) \int_{-\infty}^{\infty} \varphi_i(t)\, y(t)\, dt, \quad y \in L_2(\mathbb{R}). \tag{1.8}$$

Denote by $\mathcal{M}_{p,\mathcal{L}}$, $1 \leqslant p < \infty$, the set of all functions $a \in L_\infty(\mathbb{R})$ such that the operator $W^0_{\mathcal{L}}(a,d)$ is bounded in L_p-norm on $L_p(\mathbb{R}) \cap L_2(\mathbb{R})$. For such a, the extension of $W^0_{\mathcal{L}}(a,d)$ by continuity to the whole $L_p(\mathbb{R})$ will be denoted by the same symbol $W^0_{\mathcal{L}}(a,d)$, and called the $\mathcal{L}$-convolution operator with the $\mathcal{L}$-symbol (a,d).

Respectively, $W_{\mathcal{L}}(a,d) := \pi_+ W^0_{\mathcal{L}}(a,d)\pi^0_+$ will be called the $\mathcal{L}$-Wiener-Hopf operator with the $\mathcal{L}$-symbol (a,d).

In the case $\mathcal{L} = \mathcal{L}_0$ in place of $\mathcal{M}_{p,\mathcal{L}_0}$, $W^0_{\mathcal{L}_0}(a,d)$, $W_{\mathcal{L}_0}(a,d)$ we will write simply $\mathcal{M}_p$, $W^0(a)$, and $W(a)$.

This paper is devoted to the invertibility problem for operators of the form $W_{\mathcal{L}}(a,d)$. The case $N=1, d=0$ was treated earlier in [7]. On the other hand, in [8] the $\mathcal{L}$-convolution operator was defined as the integral operator with the kernel $K(x,t)$ satisfying

$$-\frac{\partial^2}{\partial x^2}K + c(x)K = -\frac{\partial^2}{\partial t^2}K + c(t)K$$

along with some additional conditions. Such an operator indeed admits representation (1.8), though under severe restrictions on the $\mathcal{L}$-symbol.

2 Properties of the Operators $U_\pm$, U

Jost solutions of (1.1) admit the representations [1–3]

$$e_\pm(x,\lambda) = (I + \mathcal{K}_\pm)(e^{i\lambda x}), \tag{2.1}$$

where the transformation operators

$$((I+\mathcal{K}_+)y)(x) = y(x) + \int_x^{\infty} K_+(x,\tau)y(\tau)\, d\tau,$$

$$((I+\mathcal{K}_-)y)(x) = y(x) + \int_{-\infty}^{x} K_-(x,\tau)y(\tau)\, d\tau$$

are bounded on $L_p(\gamma, \infty)$ and $L_p(-\infty, \gamma)$, respectively, for all $1 \leqslant p \leqslant \infty$ and $\gamma \in \mathbb{R}$.

The kernels $K_+(x, \tau)$ $(-\infty < x \leqslant \tau < \infty)$ and $K_-(x, \tau)$ $(-\infty < \tau \leqslant x < \infty)$ satisfy the Gelfand-Levitan-Marchenko equations [1–3]

$$\sum_{k=1}^{N}(m_k^+)^2 e^{-\mu_k x} + K_+(x,\tau) + \int_x^\infty K_+(x,s)\sum_{k=1}^{N}(m_k^+)^2 e^{-\mu_k(\tau+s)}ds = 0, \quad \tau > x,$$

$$\sum_{k=1}^{N}(m_k^-)^2 e^{\mu_k x} + K_-(x,\tau) + \int_{-\infty}^x K_-(x,s)\sum_{k=1}^{N}(m_k^-)^2 e^{\mu_k(\tau+s)}ds = 0, \quad \tau < x.$$

Here the positive numbers m_k^-, $k = 1, \dots, N$, along with λ_k, $k = 1, \dots, n$, comprise the left scattering data, and are related to m_k^+ by the formulas [2]

$$\frac{1}{(m_k^-)^2} = (m_k^+)^2 b_0'(i\mu_k), \qquad k = 1, \dots, N.$$

It is easy to see that the kernels $K_\pm$ are defined by the formulas

$$K_\pm(x,\tau) = -\sum_{k=1}^{N}\varphi_k^\pm(x)\cdot\psi_k^\pm(\tau), \tag{2.2}$$

where $\psi_k^\pm(\tau) = m_k^\pm e^{\mp\mu_k\tau}$, $k = 1, \dots, N$, and $\varphi_k^\pm$ are determined by the system of linear equations

$$\varphi_k^\pm(x) + \sum_{s=1}^{N}\frac{\psi_k^\pm(x)\psi_s^\pm(x)}{\mu_k+\mu_s}\varphi_s^\pm(x) = \psi_k^\pm(x), \quad k = 1, \dots, N. \tag{2.3}$$

We define the operators $I - \Gamma_\pm : L_p(\mathbb{R}_\pm) \to L_p(\mathbb{R}_\pm)$, $I - \Gamma : L_p(\mathbb{R}) \to L_p(\mathbb{R})$, $1 \leqslant p \leqslant \infty$, by the formulas

$$((I-\Gamma_+)y)(x) = y(x) - \sum_{k=1}^{N}\psi_k^+(x)\int_0^x \varphi_k^+(t)\,y(t)\,dt, \quad x \in \mathbb{R}_+$$

$$((I-\Gamma_-)y)(x) = y(x) - \sum_{k=1}^{N}\psi_k^-(x)\int_x^0 \varphi_k^-(t)\,y(t)\,dt, \quad x \in \mathbb{R}_-$$

$$I - \Gamma = \pi_+^0(I-\Gamma_+)\pi_+ + \pi_-^0(I-\Gamma_-)\pi_-\,.$$

Also, denote by S the singular integral operator:

$$(Sy)(x) = \frac{1}{\pi i} \int_{-\infty}^{\infty} \frac{y(\tau)}{\tau - x} \, d\tau, \qquad x \in \mathbb{R},$$

with the Cauchy principal value integral. This operator is bounded on $L_p(\mathbb{R})$, $1 < p < \infty$ (see e.g. [9]). Set $P_\pm = \frac{1}{2}(I + S)$.

Lemma 2.1 *The operators $U_\pm$, U admit the following representations:*

$$U_- = (m(\mathtt{t})P_+ + P_-)F(I - \Gamma), \tag{2.4}$$

$$U_+ = J(P_+ + m(\hat{\mathtt{t}})P_-)F(I - \Gamma), \tag{2.5}$$

$$U = (m(\mathtt{t}\chi_+ + \chi_-)P_+ + m(\chi_+ + \hat{\mathtt{t}}\chi_-)P_-)F(I - \Gamma). \tag{2.6}$$

Proof Using (1.3) and (2.1) we obtain:

$$\begin{aligned}
(U_- y)(\lambda) &= \int_0^\infty u^-(x, \lambda)\, y(x)\, dx + \int_{-\infty}^0 u^-(x, \lambda)\, y(x)\, dx \\
&= \mathtt{t}(\lambda) \int_0^\infty \left\{ e^{i\lambda x} + \int_x^\infty K_+(x, \tau) e^{i\lambda\tau} d\tau \right\} y(x)\, dx \\
&+ \int_{-\infty}^0 \left\{ e^{i\lambda x} + \int_{-\infty}^x K_-(x, \tau) e^{i\lambda\tau} d\tau \right\} y(x)\, dx \\
&= \mathtt{t}(\lambda) \int_0^\infty e^{i\lambda\tau} \left\{ y(\tau) + \int_0^\tau K_+(x, \tau) y(x)\, dx \right\} d\tau \\
&+ \int_{-\infty}^0 e^{i\lambda\tau} \left\{ y(\tau) + \int_\tau^0 K_-(x, \tau) y(x)\, dx \right\} d\tau.
\end{aligned}$$

From the definition of $I - \Gamma_\pm$, $I - \Gamma$ and the equalities $Fm(\chi_\pm) = P_\pm F$ (see [10]) it further follows:

$$\begin{aligned}
U_- = m(\mathtt{t}) F m(\chi_+) \pi_+^0 (I - \Gamma_+) \pi_+ \\
+ F m(\chi_-) \pi_-^0 (I - \Gamma_-) \pi_- = (m(\mathtt{t}) P_+ + P_-) F (I - \Gamma).
\end{aligned}$$

Similarly,

$$(U_+ y)(\lambda) = \int_0^\infty u^+(x, \lambda)\, y(x)\, dx + \int_{-\infty}^0 u^+(x, \lambda)\, y(x)\, dx$$

$$= \int_0^\infty \left\{ e^{-i\lambda x} + \int_x^\infty K_+(x,\tau) e^{-i\lambda\tau} d\tau \right\} y(x)\, dx$$

$$+ \mathtt{t}(\lambda) \int_{-\infty}^0 \left\{ e^{-i\lambda x} + \int_{-\infty}^x K_-(x,\tau) e^{-i\lambda\tau} d\tau \right\} y(x)\, dx$$

$$= \int_0^\infty e^{-i\lambda\tau} \left\{ y(\tau) + \int_0^\tau K_+(x,\tau) y(x)\, dx \right\} d\tau$$

$$+ \mathtt{t}(\lambda) \int_{-\infty}^0 e^{-i\lambda\tau} \left\{ y(\tau) + \int_\tau^0 K_-(x,\tau) y(x)\, dx \right\} d\tau,$$

i.e. $U_+ = F^{-1} m(\chi_+) \pi_+^0 (I - \Gamma_+) \pi_+$
$+ m(\mathtt{t}) F^{-1} m(\chi_-) \pi_-^0 (I - \Gamma_-) \pi_- = J(P_+ + m(\hat{\mathtt{t}}) P_-) F(I - \Gamma).$

Equality (2.6) now follows from (2.4), (2.5), and (1.4). □

Corollary 2.1 *For the operators $U_\pm$, U, the following relations hold:*

$$U = m(\chi_+ + \hat{\mathtt{t}}\chi_-) U_- = J m(\chi_+ + \hat{\mathtt{t}}\chi_-) U_+,$$
$$U_- = m(\mathtt{t}) J U_+ .$$

3 The $\mathcal{L}$-Convolution Operator

From (2.1) it follows that

$$e_\pm(x,\lambda) = e^{i\lambda x} \left(1 - \sum_{s=1}^N \frac{1}{\mu_s \mp i\lambda} \varphi_s^\pm(x) \psi_s^\pm(x) \right). \tag{3.1}$$

In particular, using (2.2):

$$e_\pm(x, \pm i\mu_k) = \frac{1}{m_k^\pm} \varphi_k^\pm(x).$$

Since $\{\lambda_k, m_k^+; k = 1, \ldots, N\}$ and $\{\lambda_k, m_k^-; k = 1, \ldots, N\}$ are the right and left scattering data, respectively, the latter equalities imply that φ_k^+ and φ_k^-, $k = 1, \ldots, N$ are (also the right and left) sets of the normalized eigenfunctions corresponding to the discrete spectrum of the operator $\mathcal{L}$ (see [2]). From (1.3) and (3.1) we obtain:

$$u^-(x,\lambda) = \mathrm{t}(\lambda)e^{i\lambda x}\left(1 - \sum_{s=1}^{N}\frac{1}{\mu_s - i\lambda}\varphi^+(x)\psi^+(x)\right)$$
$$= e^{i\lambda x}\left(1 - \sum_{s=1}^{N}\frac{1}{\mu_s + i\lambda}\varphi^-(x)\psi^-(x)\right), \quad (3.2)$$

$$u^+(x,\lambda) = t(\lambda)e^{-i\lambda x}\left(1 - \sum_{k=1}^{N}\frac{1}{\mu_k - i\lambda}\varphi_k^-(x)\psi_k^-(x)\right)$$
$$= e^{-i\lambda x}\left(1 - \sum_{k=1}^{N}\frac{1}{\mu_k + i\lambda}\varphi_k^-(x)\psi_k^-(x)\right).$$

Define the operators $\overset{\circ}{V}{}_k^{\pm}, \overset{\circ}{V}_k : L_p(\mathbb{R}) \to L_p(\mathbb{R}), 1 \leqslant p \leqslant \infty, k = 1, \ldots, N$, according to the formulas

$$(\overset{\circ}{V}{}_k^+ y)(x) = (\psi_k^+)^{-1}(x)\int_x^{\infty}\varphi_k^+(\tau)\,y(\tau)\,d\tau,$$

$$(\overset{\circ}{V}{}_k^- y)(x) = \psi_k^+(x)\int_{-\infty}^{x}\varphi_k^+(\tau)\,y(\tau)\,d\tau,$$

$$(\overset{\circ}{V}_k y)(x) = \varphi_k^+(x)\int_{-\infty}^{\infty}\varphi_k^+(\tau)\,y(\tau)\,d\tau.$$

The following theorem holds.

Theorem 3.1 *Let $a \in \mathcal{M}_p$, $1 \leqslant p < \infty$. Then $a \in \mathcal{M}_{p,\mathcal{L}}$, and the $\mathcal{L}$-convolution operator $W_{\mathcal{L}}(a,d)$ admits the representation*

$$W_{\mathcal{L}}^0(a,d) = W^0(a) - \sum_{k=1}^{N} m(\varphi_k^+\psi_k^+)W^0(a)\overset{\circ}{V}{}_k^+$$
$$- \sum_{k=1}^{N} m(\varphi_k^+(\psi_k^+)^{-1})W^0(a)\overset{\circ}{V}{}_k^- + \sum_{k=1}^{N} d_k\overset{\circ}{V}_k. \quad (3.3)$$

Proof Due to Corollary 2.1 for $a \in L_\infty(\mathbb{R})$ we have $U^*m(a)U = U_-^*m(a)U_-$. From here and equality (3.2) for $y \in L_p(\mathbb{R}) \cap L_2(\mathbb{R})$ it follows:

$$U^*m(a)Uy = W^0(a)y - \sum_{j=1}^{N} m(\psi_j^+\varphi_j^+)F^{-1}m(af_j^+)Fy$$

$$
\begin{aligned}
&- \sum_{k=1}^{N} F^{-1} m(af_k^-) F m(\psi_k^+ \varphi_k^+) y \\
&+ \sum_{j=1}^{N} \sum_{k=1}^{N} m(\psi_j^+ \varphi_j^+) F^{-1} m(af_j^+ f_k^-) F m(\psi_k^+ \varphi_k^+) y, \qquad (3.4)
\end{aligned}
$$

where

$$
f_k^{\pm}(\lambda) = \frac{1}{\mu_k \pm \lambda}, \qquad \lambda \in \mathbb{R},\ k = 1, \ldots, N.
$$

Observing that

$$
f_j^+(\lambda) f_k^-(\lambda) = \frac{1}{\mu_j + \mu_k} (f_j^+(\lambda) + f_k^-(\lambda)),
$$

the latter summand can be evaluated at $x \in \mathbb{R}$ as

$$
\begin{aligned}
&\frac{1}{2\pi} \sum_{j=1}^{N} \sum_{k=1}^{N} \psi_j^+(x) \varphi_j^+(x) \int_{-\infty}^{\infty} e^{-i\lambda x} \frac{a(\lambda) f_j^+(\lambda)}{\mu_j + \mu_k} \int_{-\infty}^{\infty} e^{i\lambda t} \psi_k^+(t) \varphi_k^+(t) y(t)\, dt\, d\lambda \\
&+ \frac{1}{2\pi} \sum_{j=1}^{N} \sum_{k=1}^{N} \psi_j^+(x) \varphi_j^+(x) \int_{-\infty}^{\infty} e^{-i\lambda x} \frac{a(\lambda) f_j^-(\lambda)}{\mu_j + \mu_k} \int_{-\infty}^{\infty} e^{i\lambda t} \psi_k^+(t) \varphi_k^+(t)\, dt\, d\lambda \\
&= \frac{1}{2\pi} \sum_{j=1}^{N} \varphi_j^+(x) \cdot \\
&\cdot \int_{-\infty}^{\infty} e^{-i\lambda x} a(\lambda) f_j^+(\lambda) \int_{-\infty}^{\infty} e^{i\lambda t} e^{\mu_j (t-x)} \left(\sum_{k=1}^{N} \frac{\psi_j^+(t) \psi_k^+(t) \varphi_k^+(t)}{\mu_j + \mu_k} \right) y(t)\, dt\, d\lambda \\
&+ \frac{1}{2\pi i} \sum_{k=1}^{N} \left(\sum_{j=1}^{N} \frac{\psi_j^+(x) \psi_k^+(x) \varphi_k^+(x)}{\mu_k + \mu_j} \right) \cdot \\
&\cdot \int_{-\infty}^{\infty} e^{-i\lambda x} a(\lambda) f_k^-(\lambda) \int_{-\infty}^{\infty} e^{i\lambda t} \varphi_k(t) e^{\mu_k (x-t)} y(t)\, dt\, d\lambda.
\end{aligned}
$$

From here, using (2.3) we obtain that the last summand (3.4) can be rewritten as

$$
\begin{aligned}
&\sum_{j=1}^{N} m(\varphi_j^+ \psi_j^+) F^{-1} m(af_j^+) F y - \sum_{j=1}^{N} m(\varphi_j^+ \psi_j^+) F^{-1} m(af_j^+) F m(\varphi_j^+ (\psi_j^+)^{-1}) y \\
&+ \sum_{k=1}^{N} F^{-1} m(af_k^-) F m(\varphi_k^+ \psi_k^+) y - \sum_{k=1}^{N} m(\varphi_k^+ (\psi_k^+)^{-1}) F^{-1} m(af_k^-) F m(\varphi_k^+ \psi_k^+) y.
\end{aligned}
$$

Substituting this into (3.4) and using the obvious equalities $W^0(af_k^{\pm}) = W^0(a)$ $W(f_k^{\pm})$:

$$U^*m(a)Uy = W^0(a)y - \sum_{k=1}^{N} m(\varphi_k^+\psi_k^+)\, W^0(a)\, W^0(f_k^+)\, m(\varphi_k^+(\psi_k^+)^{-1})y$$
$$- \sum_{k=1}^{N} m(\varphi_k^+(\psi_k^+)^{-1})\, W^0(a)\, W^0(f_k^-)\, m(\varphi_k^+\psi_k^+)\, y.$$

In its turn, from

$$f_k^+(\lambda) = \int_{-\infty}^{0} e^{\mu_k\tau} e^{i\lambda\tau}\, d\tau, \quad f_k^-(\lambda) = \int_{-\infty}^{0} e^{-\mu_k\tau} e^{i\lambda\tau}\, d\tau, \qquad k = 1, \ldots, N,$$

it is easy to see that

$$W^0(f_k^+)\, m(\varphi_k^+(\psi_k^+)^{-1})y = \overset{\circ}{V}{}_k^+ y, \;\; W^0(f_k^-)\, m(\varphi_k^+\psi_k^+)y = \overset{\circ}{V}{}_k^- y, \quad k = 1, \ldots, N.$$

Let now $a \in \mathcal{M}_p$. Using the latter equalities along with (1.8) we obtain (3.3). In particular, $a \in \mathcal{M}_{p,\mathcal{L}}$. □

Repeating the reasoning from the proof, it is easy to see that (3.4) implies

$$(U^*Uy)(x) = (U_-^*U_-y)(x) = y(x) - \sum_{k=1}^{N} \varphi_k^+(x) \int_{-\infty}^{\infty} \varphi_k^+(\tau)\, y(\tau)\, d\tau.$$

for all $y \in L_2(\mathbb{R})$. This is essentially the first equality in (1.6).

Various additional constraints on the $\mathcal{L}$-symbol yield more specific versions of formula (3.3).

Example 1 Let a belong to the Wiener algebra $\mathtt{W}(\mathbb{R})$, i.e. $a = \alpha + Fk$ where $\alpha \in \mathbb{C}$ and $k \in L_1(\mathbb{R})$. Let also $d = 0$. It is well known that $\mathtt{W}(\mathbb{R}) \subset \mathcal{M}_p$ for all $p \in [1; \infty)$, see e.g. [10]. Formula (3.3) then takes the form

$$W_{\mathcal{L}}^0(a, 0) = \alpha y(x) + W^0(Fk)$$
$$- \sum_{k=1}^{N} \varphi_k^+(x) \int_{-\infty}^{\infty} k(\tau) e^{-\mu_k\tau} \int_{x-\tau}^{\infty} \varphi_k^+(s)\, y(s)\, ds\, d\tau$$
$$- \sum_{k=1}^{N} \varphi_k^+(x) \int_{-\infty}^{\infty} k(\tau) e^{\mu_k\tau} \int_{-\infty}^{x-\tau} \varphi_k^+(s)\, y(s)\, ds\, d\tau\,.$$

Changing the order of integration, we obtain

$$(W^0_{\mathcal{L}}(a,0)y)(x) = \alpha y(x) + \int\limits_{-\infty}^{\infty} k(x-\tau)\,y(\tau)\,d\tau$$

$$-\sum_{k=1}^{N} \varphi_k^+(x) \int\limits_{-\infty}^{\infty} \left\{ \int\limits_{-\infty}^{\infty} k(\tau) e^{\mu_k \tau \,\mathrm{sgn}(x-\tau-s)} d\tau \right\} \varphi_k^+(s)\,y(s)\,ds, \ x \in \mathbb{R}. \tag{3.5}$$

Example 2 Let $a(x) = e^{i\nu x}$, $\nu \in \mathbb{R}$, and $d = 0$. Then $a \in M_p(\mathbb{R})$ for all $p \in [1, \infty)$. Taking into consideration that $(W^0(a)y)(x) = y(x-\nu)$, $x \in \mathbb{R}$, it is not hard to see that formula (3.3) in this case takes the form

$$(W^0(a,0)y)(x) = y(x-\nu) - \sum_{k=1}^{N} e^{\mu_k \nu} \varphi_k^+(x) \int\limits_{x-\nu}^{\infty} \varphi_k^+(\tau)\,y(\tau)\,d\tau$$

$$-\sum_{k=1}^{N} e^{-\mu_k \nu} \varphi_k^+(x) \int\limits_{-\infty}^{x-\nu} \varphi_k^+(\tau)\,y(\tau)\,d\tau, \quad x \in \mathbb{R}. \tag{3.6}$$

Example 3 Let $a(x) = \mathrm{sgn}\,x$. Without dwelling on the justification, we just point out that for $y \in L_p(\mathbb{R})$, $1 < p < \infty$:

$$W^0_{\mathcal{L}}(\mathrm{sgn}, 0) = -(Sy)(x) + \sum_{s=1}^{N} \varphi_s(x) \int\limits_{-\infty}^{\infty} (Ei(\mu_s(\tau - x)) - Ei(\mu_s(x-\tau)))\varphi_s(\tau)\,d\tau,$$

where $Ei(x) = v.p. \int\limits_{-\infty}^{x} \frac{e^t}{t}\,dt$ is the integral exponential function.

4 The $\mathcal{L}$-Wiener-Hopf Operator

We now turn to operators $W_{\mathcal{L}}(a, d)$.

In the case $a = \alpha + Fk \in \mathrm{W}(\mathbb{R})$ according to (3.5) they take the form

$$(W_{\mathcal{L}}(a,d)y)(x) = \alpha y(x) + \int\limits_{0}^{\infty} k(x-\tau)\,y(\tau)\,d\tau$$

$$-\sum_{k=1}^{N} \varphi_k^+(x) \int\limits_{0}^{\infty} \left\{ \int\limits_{-\infty}^{\infty} k(\tau) e^{\mu_k \tau \,\mathrm{sgn}(x-\tau-s)} d\tau \right\} \varphi_k^+(s)\,y(s)\,ds$$

$$+\sum_{k=1}^{N} d_k \varphi_k^+(x) \int\limits_{0}^{\infty} \varphi_k^+(\tau)\,y(\tau)\,d\tau, \qquad x \in \mathbb{R}_+. \tag{4.1}$$

Similarly, (3.6) for $a(x) = e^{-i\nu x}$, $\nu > 0$ implies

$$(W(a,d)y)(x) = y(x+\nu) - \sum_{k=1}^{N} e^{-\mu_k \nu} \varphi_k^+(x) \int_{x+\nu}^{\infty} \varphi_k^+(\tau)\, y(\tau)\, d\tau$$

$$- \sum_{k=1}^{N} e^{\mu_k \nu} \varphi_k^+(x) \int_0^{x+\nu} \varphi_k^+(\tau)\, y(\tau)\, d\tau + \sum_{k=1}^{N} d_k \varphi_k^+(x) \int_0^{\infty} \varphi_k^+(\tau)\, y(\tau)\, d\tau, \ x \in \mathbb{R}_+ . \tag{4.2}$$

For $a(x) = e^{i\nu x}$, $\nu > 0$, the operator $W(a,d)$ acts as

$$(W(a,d)y)(x) = y(x-\nu) - \sum_{k=1}^{N} e^{\mu_k \nu} \varphi_k^+(x) \int_{x-\nu}^{\infty} \varphi_k^+(\tau)\, y(\tau)\, d\tau$$

$$- \sum_{k=1}^{N} e^{-\mu_k \nu} \varphi_k^+(x) \int_0^{x-\nu} \varphi_k^+(\tau)\, y(\tau)\, d\tau + \sum_{k=1}^{N} d_k \varphi_k^+(x) \int_0^{\infty} \varphi_k^+(\tau)\, y(\tau)\, d\tau. \tag{4.3}$$

if $x > \nu$, and as

$$(W(a,d)y)(x) = \sum_{k=1}^{N} (d_k - e^{\mu_k \nu}) \varphi_k^+(x) \int_0^{\infty} \varphi_k^+(\tau)\, y(\tau)\, d\tau \tag{4.3'}$$

if $0 \leqslant x \leqslant \nu$.

Theorem 4.1 *Let $a \in \mathcal{M}_p$, $1 \leqslant p < \infty$. Then on $L_p(\mathbb{R}_+)$ we have*

$$W_{\mathcal{L}}(a, 0) = (I + \mathcal{K})\, W(a)\, (I - \Gamma_+). \tag{4.4}$$

Proof Since $P_\pm^* = P_\pm$ (see [9]), and $\bar{\mathtt{t}} = \hat{\mathtt{t}}$, $|\mathtt{t}| = 1$, for $y \in L_p(\mathbb{R})$ we have

$$\begin{aligned} &(m(\mathtt{t})P_+ + P_-)^* m(a)\, (m(\mathtt{t})P_+ + P_-)y \\ &= (P_+ m(a) P_+ + P_+ m(\hat{\mathtt{t}}a) P_- + P_- m(\mathtt{t}a) P_+ + P_- m(a) P_-)y. \end{aligned}$$

In its turn, invoking $P_\pm F = F m(\chi_\pm)$, for $y \in L_p(\mathbb{R}) \cap L_2(\mathbb{R})$:

$$\begin{aligned} &F^{-1}(m(\mathtt{t})P_+ + P_-)^* m(a)\, (m(\mathtt{t})P_+ + P_-)Fy \\ &= (m(\chi_+), m(\chi_-)) \begin{pmatrix} W^0(a) & W^0(\hat{\mathtt{t}}a) \\ W^0(\mathtt{t}a) & W^0(a) \end{pmatrix} \begin{pmatrix} m(\chi_+) \\ m(\chi_-) \end{pmatrix} y . \end{aligned}$$

Further, due to (1.8), (2.4) and the equalities $U^* m(a) U = U_-^* m(a) U$, $I + \mathcal{K} = (I - \Gamma)^* = \pi_+^0 (I + \mathcal{K}_+) \pi_+ + \pi_-^0 (I + \mathcal{K}_-) \pi_-$ valid on $L_p(\mathbb{R}) \cap L_2(\mathbb{R})$:

$$W^0_{\mathcal{L}}(a,0) = (I+\mathcal{K})(m(\chi_+), m(\chi_-)) \begin{pmatrix} W^0(a) & W^0(\hat{\mathtt{t}}a) \\ W^0(\mathtt{t}a) & W^0(a) \end{pmatrix} (I-\Gamma). \tag{4.5}$$

Since $\mathtt{t}a$, $\hat{\mathtt{t}}a$ lie in $\mathcal{M}_p$ along with a, formula (4.5) extends to the whole $L_p(\mathbb{R})$ by continuity. Taking into consideration the equalities $(I-\Gamma)\pi^0_+ = \pi^0_+(I-\Gamma_+)$, $\pi_+(I+\mathcal{K}) = (I+\mathcal{K}_+)\pi_+$, $\pi_+ m(\chi_+) = \pi_+$, $\pi_+ m(\chi_-) = 0$, $m(\chi_+)\pi^0_+ = \pi^0_+$, $m(\chi_+)\pi^0_- = 0$, we conclude that (4.5) implies (4.4). □

The operators $I+\mathcal{K}_+, I-\Gamma_+ : L_p(\mathbb{R}) \to L_p(\mathbb{R})$ are invertible, since $\mathcal{K}_+$ and Γ_+ are Volterra operators. The explicit formulas for the respective inverses are delivered by the next lemma.

Lemma 4.1 *The operators $(I+\mathcal{K}_+)^{-1}$, $(I-\Gamma_+)^{-1}$ are acting on $L_p(\mathbb{R})$, $1 \leqslant p < \infty$, according to the formulas*

$$((I+\mathcal{K}_+)^{-1}y)(x) = y(x) + \sum_{k=1}^{N} \psi_k^+(x) \int_x^\infty \varphi_k^+(\tau) y(\tau)\, d\tau, \tag{4.6}$$

$$((I-\Gamma_+)^{-1}y)(x) = y(x) + \sum_{k=1}^{N} \varphi_k^+(x) \int_0^x \psi_k^+(\tau) y(\tau)\, d\tau. \tag{4.7}$$

Proof We start with formula (4.7). Consider the system

$$\frac{dz}{dx} = \varphi^+(x)\,(\psi^+(x))^T z(x) \tag{4.8}$$

of ordinary differential equations.

It is well known that any solution z of (4.8) has a finite limit at infinity $\lim_{x\to+\infty} z(x) = z_0$, and the respective Cauchy problem $z(\infty) = z_0$ ($z_0 \in \mathbb{C}^N$) has a unique solution (see [11, §§ 16, 22]).

Let Φ be the fundamental matrix of the system (4.8). Consider the operator $I + L_+ : L_p(\mathbb{R}_+) \to L_p(\mathbb{R}_+)$, $1 \leqslant p < \infty$, defined by the formula

$$((I+L_+)y)(x) = y(x) + (\psi^+(x))^T \Phi(x) \int_0^x \Phi^{-1}(\tau)\, \varphi^+(\tau)\, y(\tau)\, d\tau,$$

where $\varphi^+ = (\varphi_1^+, \dots, \varphi_N^+)^T$, $\psi^+ = (\psi_1^+, \dots, \psi_N^+)^T$.

From the continuity of the matrix function Φ on $[0, \infty)$ and its behavior at infinity it follows that the operator $I + L_+$ is bounded. Based on the results of [12] it is natural to expect that $I + L_+$ is the inverse of the integral operator $I - \Gamma_+$ with a semi-separable kernel. Formally speaking, however, we cannot refer to [12] directly since the domain of the operator under consideration is different from the one treated there. We resort to a direct computation instead.

Let $y \in L_p(\mathbb{R}_+)$, $1 \leqslant p < \infty$. From the equality

$$\varphi^+(\tau)\,(\psi^+)^T(\tau)\,\Phi(\tau) = \frac{d\Phi}{d\tau}(\tau), \qquad \tau \in \mathbb{R}_+,$$

changing the order of integration we obtain

$$(\psi^+)^T(x)\int_0^x \varphi^+(\tau)(\psi^+)^T(\tau)\,\Phi(\tau)\int_0^\tau \Phi^{-1}(s)\varphi^+(s)\,y(s)\,ds\,d\tau$$

$$= (\psi^+)^T(x)\int_0^x\int_\tau^x \frac{d\Phi}{ds}(s)\,ds\,\Phi^{-1}(\tau)\,\varphi^+(\tau)\,y(\tau)\,d\tau$$

$$= (\psi^+)^T(x)\int_0^x [\Phi(x) - \Phi(\tau)]\,\Phi^{-1}(\tau)\,\varphi^+(\tau)\,y(\tau)\,d\tau.$$

It is easy to see from here that $(I - \Gamma_+)(I + L_+)y = y$.

Similarly, from

$$\varphi^+(\tau)(\psi^+)^T(\tau) = \frac{d\Phi}{d\tau}(\tau)\,\Phi^{-1}(\tau),\ -\Phi^{-1}(\tau)\frac{d\Phi}{d\tau}(\tau)\,\Phi^{-1}(\tau) = \frac{d\Phi^{-1}}{d\tau}(\tau),$$

also changing the order of integration:

$$(\psi^+)^T(x)\Phi(x)\int_0^x \Phi^{-1}(\tau)\,\varphi^+(\tau)\,(\psi^+)^T(\tau)\int_0^\tau \varphi^+(s)\,y(s)\,ds$$

$$= (\psi^+)^T(x)\Phi(x)\int_0^x \Phi^{-1}(\tau)\,\frac{d\Phi}{d\tau}(\tau)\,\Phi^{-1}(\tau)\int_0^\tau \varphi^+(s)\,y(s)\,ds$$

$$= (\psi^+)^T(x)\Phi(x)\int_0^x [\Phi^{-1}(s) - \Phi^{-1}(x)]\,\varphi^+(s)\,y(s)\,ds.$$

The latter equality implies that $(I + L_+)(I - \Gamma_+)y = y$, and so the operator $I + L_+$ is indeed the inverse of $(I - \Gamma_+)$.

Let us introduce an $N \times N$ matrix function Ψ the (k, j)-entry of which is given by

$$\delta_{kj} + \frac{m_k^+ m_j^+}{\mu_k + \mu_j} - \int_0^x \psi_k^+(\tau)\,\psi_j^+(\tau)\,d\tau, \qquad k, j = 1, \ldots, N. \tag{4.9}$$

From (2.3) it follows that

$$\Psi\varphi^+ = \psi^+ , \tag{4.10}$$

while (4.9) implies $\lim_{x\to\infty} \Psi(x) = E_N$, the identity $N \times N$ matrix. Moreover, from (4.9) and (4.10):

$$(\Psi^{-1})' = -\Psi^{-1}\Psi'\Psi^{-1} = \Psi^{-1}\psi^+(\psi^+)^T\Psi^{-1} = \varphi^+(\psi^+)^T\Psi^{-1}.$$

So, Ψ^{-1} is the fundamental matrix of the system (4.8). From (4.10) and the equality $(\Psi^{-1})^T = \Psi^{-1}$ it follows that $(\varphi^+)^T = (\psi^+)^T(\Psi^{-1})^T = (\psi^+)^T\Psi^{-1}$. Letting $\Phi = \Psi^{-1}$ and plugging the latter equality and (4.10) into the definition of $I + L_+$ we see that (4.7) holds.

In the case $1 < p < \infty$ equality (4.6) follows from the observation that the operator $I + \mathcal{K}_+$ acting on $L_p(\mathbb{R}_+)$ coincides with $(I - \Gamma_+)^*$, where the operator $I - \Gamma_+$ is acting on $L_q(\mathbb{R}_+)$, $q = p/(p-1)$.

If $p = 1$, then on a dense in $L_1(\mathbb{R}_+)$ subset we have

$$(I + \mathcal{K}_+)^{-1}(I + \mathcal{K}_+)y = y, \qquad (I + \mathcal{K}_+)(I + \mathcal{K}_+)^{-1}y = y,$$

where $(I + \mathcal{K}_+)^{-1}$ is defined by (4.6). From the existence of a bounded inverse of the operator $I + \mathcal{K}_+$ it follows that these equalities hold for all $y \in L_1(\mathbb{R}_+)$. □

By the same token one can show that the operators $(I - \mathcal{K}_-)^{-1}$ and $(I - \Gamma_-)^{-1}$ act on $L_p(\mathbb{R}_-)$, $1 \leqslant p < \infty$, according to the formulas

$$((I + \mathcal{K}_-)^{-1}y)(x) = y(x) + \sum_{k=1}^{N} \psi_k^-(x) \int_{-\infty}^{x} \varphi_k^-(\tau)y(\tau)\,d\tau,$$

$$((I - \Gamma_-)^{-1}y)(x) = y(x) + \sum_{k=1}^{N} \varphi_k^-(x) \int_{x}^{0} \psi_k^-(\tau)y(\tau)\,d\tau.$$

It is known (see e.g. [10]) that any piecewise constant function with finitely many discontinuities on $\mathbb{R}$ lies in $\mathcal{M}_p$, $1 < p < \infty$. The closure of the algebra of all such functions in $\mathcal{M}_p$ is denoted PC_p. Recall that $PC_p \subset PC_2 = PC$ (also see [10]), where PC stands for the set of the functions a having one-sided limits $a(x \pm 0)$ at all $x \in \mathbb{R}$ and also at infinity: $a(\infty \pm 0) = \lim_{x\to\mp\infty} a(x)$. In its turn, $C_p \subset PC_p$, where C_p is the closure of the Wiener algebra $\mathbb{W}(\mathbb{R})$ in $\mathcal{M}_p$. Also, PC_p contains all the functions of bounded variation (see [10]).

For $a \in PC_p$ let us introduce, as in [10], the function $a_p : \dot{\mathbb{R}} \times \bar{\mathbb{R}} \to \mathbb{C}$ (here $\dot{\mathbb{R}} = \mathbb{R} \cup \{\infty\}$ and $\bar{\mathbb{R}} = \mathbb{R} \cup \{\pm\infty\}$ are the one- and two-point compactifications of $\mathbb{R}$, respectively) by the following rule:

$$a_p(x,\xi) = \frac{1}{2}(a(x-0)+a(x+0)) + \frac{1}{2}(a(x-0)-a(x+0)) \operatorname{cth}\left(\frac{i}{p}+\xi\right).$$

The function a has at most countably many discontinuities $x_k (\in \mathbb{R})$ at which $a(x_k - 0) \neq a(x_k + 0)$, and the construction of $a_p(x,\xi)$ can be interpreted as follows (again, see [10]): $a(x_k - 0)$ and $a(x_k + 0)$ are being connected by a circular arc from the points of which the line segment with the endpoints $a(x_k - 0)$, $a(x_k + 0)$ is seen at the angle $2\pi/\max\{p, q\}$ and located to the right/left of this segment if $p > 2$ (respectively, $p < 2$), and merging with it if $p = 2$.

For $a_p(x,\xi) \neq 0$ the index $\operatorname{ind} a_p$ of a_p is understood as the normalized increment of the argument $(2\pi)^{-1} \arg a_p(x,\xi)$ while x traverses $\dot{\mathbb{R}}$ and, at the discontinuities of a in $\dot{\mathbb{R}}$, the parameter ξ traverses $\bar{\mathbb{R}}$ from $-\infty$ to ∞.

Given an integer $\varkappa$, introduce the function $r_\varkappa$ as

$$r_\varkappa(\lambda) = \left(\frac{\lambda - i}{\lambda + i}\right)^{\varkappa}.$$

By definition, a generalized p-factorization $(1 < p < \infty)$ of a function $a \in L_\infty(\mathbb{R})$ is its representation in the form

$$a(x) = a_-(x) r_\varkappa(\lambda) a_+(\lambda),$$

where

$$(\lambda - i)^{-2/p} a_- \in L_p^-, \quad (\lambda - i)^{-2/q} a_-^{-1} \in L_q^-, \quad q = \frac{p}{p-1},$$
$$(\lambda + i)^{-2/q} a_+ \in L_q^+, \quad (\lambda + i)^{-2/p} a_+^{-1} \in L_p^+,$$

and the operator $m(\tilde{a}_+^{-1}) S m(\tilde{a}_-^{-1})$ is bounded on $L_p(\mathbb{R})$. Here $L_p^{\pm} = P_{\pm}(L_p(\mathbb{R}))$, and $\tilde{a}_+(\lambda) = (\lambda + i)^{\frac{2-p}{p}} a_+(\lambda)$, $\tilde{a}_-(\lambda) = (\lambda - i)^{\frac{p-2}{p}} a_-(\lambda)$.

The generalized p-factorization, when exists, is defined uniquely up to constant multiples in $a_\pm$, and $\varkappa$ is called the p-index of a.

Condition

$$\inf |a_p(\lambda,\xi)| > 0, \qquad \lambda \in \dot{\mathbb{R}}, \quad \xi \in \bar{\mathbb{R}}$$

is necessary and sufficient for the generalized p-factorization to exist. Note that a continuous function $a \in C(\dot{\mathbb{R}})$ admits a generalized p-factorization for any (equivalently, all) $p \in (1,\infty)$ if and only if it is invertible. If this is the case, then $\varkappa = \operatorname{ind} a$ and $a_\pm$ can be chosen as $a_\pm = \exp P_\pm b$, where $b = \ln(a r_{-\varkappa})$. In particular, for $a \in \mathtt{W}(\mathbb{R})$ we have $a_+^{\pm 1}, a_-^{\pm 1} \in \mathtt{W}(\mathbb{R})$ (see [13]). Finally, see [14, part 4] for the description of $a_\pm$ in the case when a is not (necessarily) continuous but $\varkappa$ is known a priori. We also note that $C_p \subset C(\dot{\mathbb{R}})$.

From Theorem 4.1 and the results of [10] it follows

Theorem 4.2 *Let $a \in PC_p$, $1 < p < \infty$. Then the operator $W_{\mathcal{L}}(a, 0)$ is normally solvable if and only if*

$$\inf |a_q(\lambda, \xi)| > 0, \qquad \lambda \in \dot{\mathbb{R}}, \quad \xi \in \bar{\mathbb{R}}, \tag{4.11}$$

where $q = \frac{p}{p-1}$. If this condition holds, then $W_{\mathcal{L}}(a, 0)$ is two-side, left, or right invertible if $\varkappa = \operatorname{ind} a_q$ is respectively zero, non-negative, or non-positive. Moreover, $W_{\mathcal{L}}(a, 0)$ is Fredholm, and $\operatorname{Ind} W(a, 0) = -\operatorname{ind} a_q$. Condition (4.11) also implies that the function a admits a generalized q-factorization $a = a_- r_\varkappa a_+$, and for $\varkappa \geqslant 0$ a left inverse of $W_{\mathcal{L}}(a, 0)$ can (on a dense subset of $L_p(\mathbb{R}_+)$) be written as

$$W_{\mathcal{L}}^{(-1)}(a, 0) = (I - \Gamma_+)^{-1} W(a_+^{-1})\, W(a_-^{-1})\, W(r_{-\varkappa})\, (I - \mathcal{K}_+)^{-1}.$$

Similarly, if $\varkappa \leqslant 0$, then a right inverse of $W_{\mathcal{L}}(a, 0)$ on a dense subset of $L_p(\mathbb{R}_+)$ can be written as

$$W_{\mathcal{L}}^{(-1)}(a, 0) = (I - \Gamma_+)^{-1} W(r_{-\varkappa})\, W(a_+^{-1})\, W(a_-^{-1})\, (I + \mathcal{K}_+)^{-1}.$$

For $a \in C_p$, both formulas hold on the whole $L_p(\mathbb{R}_+)$.

If $\varkappa < 0$, then

$$\ker W_{\mathcal{L}}(a, 0) = \operatorname{span}\{(I - \Gamma_+)^{-1} \pi_+ F^{-1} m(a_+^{-1}) g_k,\ k = 1, \ldots, -\varkappa\},$$

where $g_k(\lambda) = (1 - i\lambda)^{-k}$, $\lambda \in \mathbb{R}$.

If $\varkappa > 0$, then the equation $W_{\mathcal{L}}(a, 0)y = f$ has a solution in $L_p(\mathbb{R})$ if and only if

$$\int_0^\infty ((I + \mathcal{K}_+)^{-1} f)(t)\, \overline{h_k(t)}\, dt = 0, \qquad k = 1, \ldots, \varkappa,$$

where $h_k = F^{-1}(\bar{a}_-^{-1}\, g_{-k})$.

In order to tackle operators $W_{\mathcal{L}}(a, d)$, the following lemma will be useful.

Lemma 4.2 *Let X, Y be linear spaces, $B : X \to Y$, $B_0 : X \to Y$ be linear mappings, and let $B^{(-1)} : Y \to X$ be a* generalized inverse *of B, i.e. $BB^{(-1)}B = B$, $B^{(-1)}BB^{(-1)} = B^{(-1)}$. Define $\varepsilon_0 \colon \operatorname{im} B_0 \to Y$, $\varepsilon_1 \colon \ker B^{(-1)} \to Y$, $\varepsilon_2 : \ker B \to X$, $\sigma_0 \colon X \to \operatorname{im} B_0$, $\sigma_1 \colon Y \to \ker B^{(-1)}$, $\sigma_2 : X \to \ker B$ as follows: $\varepsilon_i y = y$ $(i = 0, 1, 2)$, $\sigma_0 x = B_0 x$, $\sigma_1 x = x - BB^{(-1)}x$, $\sigma_2 x = x - B^{(-1)}Bx$. Then the block-matrix mappings*

$$\Xi_1 = \left(\begin{array}{c|cc} B + B_0 & -\varepsilon_0 & \varepsilon_1 \\ \hline -\sigma_0 & I & 0 \\ \sigma_2 & 0 & 0 \end{array}\right) : \begin{array}{c} X \\ \oplus \\ \operatorname{im} B_0 \\ \oplus \\ \ker B^{(-1)} \end{array} \to \begin{array}{c} Y \\ \oplus \\ \operatorname{im} B_0 \\ \oplus \\ \ker B \end{array}$$

and

$$\Xi_2 = \left(\begin{array}{c|cc} B^{(-1)} & B^{(-1)}\varepsilon_0 & \varepsilon_2 \\ \hline -\sigma_0 B^{(-1)} & I + \sigma_0 B^{(-1)}\varepsilon_0 & \sigma_0\varepsilon_2 \\ \sigma_1 & \sigma_1\varepsilon_0 & 0 \end{array}\right) : \begin{array}{c} Y \\ \oplus \\ \operatorname{im} B_0 \\ \oplus \\ \ker B \end{array} \to \begin{array}{c} X \\ \oplus \\ \operatorname{im} B_0 \\ \oplus \\ \ker B^{(-1)} \end{array}$$

satisfy $\Xi_1^{-1} = \Xi_2$.

The proof is by direct computation. Similar results were proved earlier in [15], see also [16].

In the terminology of [12], relation $\Xi_1^{-1} = \Xi_2$ represents the coupling of the operators $B + B_0$ and

$$\mathcal{B} = \begin{pmatrix} I + \sigma_0 B^{(-1)}\varepsilon_0 & \sigma_0\varepsilon_2 \\ \sigma_1\varepsilon_0 & 0 \end{pmatrix} : \begin{array}{c} \operatorname{im} B_0 \\ \oplus \\ \ker B \end{array} \to \begin{array}{c} \operatorname{im} B_0 \\ \oplus \\ \ker B^{(-1)}. \end{array}$$

From the coupling properties it follows that a generalized inverse of $B + B_0$ can be computed by the formula

$$(B + B_0)^{(-1)} = B^{(-1)} - (B^{(-1)}\varepsilon_0 \ \varepsilon_2)\, \mathcal{B}^{(-1)} \begin{pmatrix} \sigma_0 B^{(-1)} \\ \sigma_1 \end{pmatrix}.$$

Moreover,

$$\dim \ker(B + B_0) = \dim \ker \mathcal{B}, \quad \ker(B + B_0) = (B^{(-1)}\varepsilon_0, \varepsilon_2)(\ker \mathcal{B}).$$

Since $W_{\mathcal{L}}(a, d)$ and $W_{\mathcal{L}}(a, 0)$ differ by a finite dimensional operator

$$V_d = \sum_{k=1}^{N} d_k \pi_+ \overset{\circ}{V}_k \pi_+^0,$$

they are (or are not) normally solvable or Fredholm only simultaneously.

Now define the operators $\mathcal{B}_0(a, d)\colon \operatorname{im} V_d \to \operatorname{im} V_d$, $\mathcal{B}_-(a, d)\colon \operatorname{im} V_d \oplus \ker W_{\mathcal{L}}(a, 0) \to \operatorname{im} V_d$, $\mathcal{B}_+(a, d)\colon \operatorname{im} V_d \to \operatorname{im} V_d \oplus \ker W_{\mathcal{L}}^{(-1)}(a, 0)$ by the formulas

$$\mathcal{B}_0(a, d)y = y + V_d W_{\mathcal{L}}^{(-1)}(a, 0)y\,,$$

$$\mathcal{B}_-(a, d)\begin{pmatrix} y \\ z \end{pmatrix} = (\mathcal{B}_0(a, d), V_d)\begin{pmatrix} y \\ z \end{pmatrix} = \mathcal{B}_0(a, d)y + V_d z\,,$$

$$\mathcal{B}_+(a, d)y = \begin{pmatrix} \mathcal{B}_0(a, d)y \\ y - W_{\mathcal{L}}(a, 0)W_{\mathcal{L}}^{(-1)}(a, 0)y \end{pmatrix}.$$

According to [10], if $\varkappa > 0$ for $y \in L_p(\mathbb{R}_+)$ we have

$$y(x) - (W(r_{\varkappa})\,W(r_{-\varkappa})y)(x) = \sum_{k=1}^{\varkappa} \ell_k(x) \int_0^{\infty} y(\tau)\,\ell_k(\tau)\,d\tau\,,$$

where $\ell_k(t) = \sqrt{2}\,\Lambda_{k-1}(2t)e^{-t}$, $k = 1, \ldots, \varkappa$, and Λ_{k-1} are Laguerre polynomials. So:

$$y(x) - (W_{\mathcal{L}}(a,0)\,W_{\mathcal{L}}^{(-1)}(a,0)y)(x) = \\ = \sum_{k=1}^{\varkappa}((I+\mathcal{K}_+)\,W(a_-)\ell_k)(x) \int_0^{\infty} \ell_k(\tau)\,W(a_-^{(-1)})(I+\mathcal{K}_+)^{-1}y(\tau)\,d\tau.$$

Setting $X = Y = L_p(\mathbb{R}_+)$, $B = W_{\mathcal{L}}(a,0)$, $B_0 = V_d$, we see that under condition (4.11) the Fredholm operator $W_{\mathcal{L}}(a,d)$ is coupled with one of the finite dimensional operators $\mathcal{B}_0(a,d)$, $\mathcal{B}_-(a,d)$, or $\mathcal{B}_+(a,d)$, in accordance with $\varkappa$ being equal zero, negative, or positive.

Similarly, the operator $W_{\mathcal{L}}^* = W_{\mathcal{L}}(\bar{a},\bar{d})$ is coupled with the finite dimensional operator $\mathcal{B}_0(\bar{a},\bar{d})$, $\mathcal{B}_+(\bar{a},\bar{d})$, or $\mathcal{B}_-(\bar{a},\bar{d})$, depending on whether $\varkappa = 0$, $\varkappa < 0$, or $\varkappa > 0$.

Let $d_{i_1}, \ldots, d_{i_n}$ be all the non-zero elements among $d_1, \ldots, d_N$. Then of course $\operatorname{im} V_d = \operatorname{span}\{\varphi_{i_1}^+, \ldots, \varphi_{i_n}^+\}$ and $\dim \operatorname{Im} V_d = n$.

We now introduce the matrices $\mathcal{A}_0(a,d)$, $\mathcal{A}_-(a,d)$, $\mathcal{A}_+(a,d)$ as follows:

$$\mathcal{A}_0(a,d) = \left(\delta_{k,s} + \sum_{s=1}^{n} d_{i_k} \int_0^{\infty} \varphi_{i_k}^+(\tau)(W_{\mathcal{L}}^{(-1)}(a,0)\,\varphi_{i_s}^+)(\tau)\,d\tau\right)_{k,s=1,\ldots,n},$$

$$\mathcal{A}_-(a,d) = \left(d_{i_k} \int_0^{\infty} \varphi_{i_k}^+(\tau)((I-\Gamma_+)^{-1}\pi_+F^{-1}m(a_+^{-1})\,g_s)(\tau)\,d\tau\right)_{\substack{k=1,\ldots,n\\ s=1,\ldots,-\varkappa}},$$

$$\mathcal{A}_+(a,d) = \left(\int_0^{\infty} \ell_k(\tau)(W(a_-^{-1})\,(I+\mathcal{K}_+)^{-1}\varphi_{i_s})(\tau)\,d\tau\right)_{\substack{k=1,\ldots,\varkappa\\ s=1,\ldots,n}}.$$

Note that $\mathcal{A}_0$ is a square $n \times n$ matrix, while the $n \times (-\varkappa)$ matrix $\mathcal{A}_-$ is defined only if $\varkappa < 0$ and the $\varkappa \times n$ matrix $\mathcal{A}_+$ exists if $\varkappa > 0$.

It is easy to see that $\sum\limits_{s=1}^{n} \alpha_s\varphi_{i_s}^+ \in \ker \mathcal{B}_0(a,d)$ if and only if $(\alpha_1, \ldots, \alpha_n)^T \in \ker \mathcal{A}_0(a,d)$. Similarly, a vector $\left(\sum\limits_{s=0}^{n} \alpha_s\varphi_{i_s}^+, \sum\limits_{k=1}^{-\varkappa} \beta_k(I-\Gamma_+)^{-1}\pi_+F^{-1}m(a_+^{-1})\,g_s\right)^T$ lies in $\ker \mathcal{B}_-(a,d)$ if and only if $(\alpha_1, \ldots, \alpha_n, \beta_1, \ldots, \beta_{-\varkappa})^T \in \ker[\mathcal{A}_0(a,d), \mathcal{A}_-(a,d)]$, and a function $\sum\limits_{s=1}^{n} \alpha_s\varphi_{i_s}^+$ lies in $\ker \mathcal{B}_+(a,d)$ if and only if

$$(\alpha_1, \dots, \alpha_n)^T \in \ker \begin{pmatrix} \mathcal{A}_0(a,d) \\ \mathcal{A}_+(a,d) \end{pmatrix}.$$

Putting the pieces together, we arrive at

Theorem 4.3 *Let $a \in PC_p$ for some $1 < p < \infty$. Then the operator $W_{\mathcal{L}}(a,d)$ is normally solvable in $L_p(\mathbb{R})$ if and only if* (4.11) *holds. Under this condition $W_{\mathcal{L}}(a,d)$ is in fact Fredholm, with the index equal*

$$\operatorname{Ind} W_{\mathcal{L}}(a,d) = -\varkappa = -\operatorname{ind} a_q\,, \qquad q = \frac{p}{p-1}\,.$$

For $\varkappa = 0$, a generalized inverse of $W_{\mathcal{L}}(a,d)$ is given by

$$W_{\mathcal{L}}^{(-1)}(a,d) = W_{\mathcal{L}}^{(-1)}(a,0) - W_{\mathcal{L}}^{(-1)}(a,0)\mathcal{B}_0^{(-1)}(a,d)W_{\mathcal{L}}^{(-1)}(a,0).$$

Moreover,

$$\dim\ker W_{\mathcal{L}}(a,d) = n - \operatorname{rank}\mathcal{A}_0(a,d),$$

and $\ker W_{\mathcal{L}}(a,d)$ *coincides with*

$$\Big\{W_{\mathcal{L}}^{(-1)}(a,0)\Big(\sum_{k=1}^{n}\alpha_k\varphi_{i_k}^+\Big) \mid (\alpha_1,\dots,\alpha_n)^T \in \ker\mathcal{A}_0(a,d)\Big\}.$$

Equation $W_{\mathcal{L}}(a,d) = f$ has a solution for $f \in L_p(\mathbb{R}_+)$ if and only if

$$\int_0^\infty \overline{f(\tau)}\, W_{\mathcal{L}}^{(-1)}(\bar a,0)\Big(\sum_{k=1}^{n}\alpha_k\varphi_{i_k}^+\Big)(\tau)\,d\tau = 0$$

for all $(\alpha_1,\dots,\alpha_n)^T \in \ker\mathcal{A}_0(\bar a,\bar d)$.
If $\varkappa < 0$, then

$$W_{\mathcal{L}}^{(-1)}(a,d) = W_{\mathcal{L}}^{(-1)}(a,0) - (W_{\mathcal{L}}^{(-1)}(a,0);\, I)\mathcal{B}_-^{(-1)}V_d\,W_{\mathcal{L}}^{(-1)}(a,0),$$
$$\dim\ker W_{\mathcal{L}}(a,d) = n - \varkappa - \operatorname{rank}(\mathcal{A}_0(a,d), \mathcal{A}_-(a,d)),$$
$$\ker W_{\mathcal{L}}(0,d) = \Big\{W_{\mathcal{L}}^{(-1)}(a,0)\Big(\sum_{k=1}^{n}\alpha_k\varphi_{i_k}^+\Big) + (I - \Gamma_+)^{-1}\pi_+F^{-1}m(a_+^{-1})\Big(\sum_{k=1}^{-\varkappa}\beta_k g_k\Big) \mid (\alpha_1,\dots,\alpha_n,\beta_1,\dots,\beta_{-\varkappa})^T \in \ker\{\mathcal{A}_0(a,d), \mathcal{A}_-(a,d)\}\Big\}.$$

Also, equation $W_{\mathcal{L}}(a,d) = f$ with the right hand side $f \in L_p(\mathbb{R}_+)$ has a solution if and only if

$$\int_0^\infty \overline{f(\tau)}\, W_{\mathcal{L}}^{(-1)}(\bar a,0)\Big(\sum_{k=1}^{n}\alpha_k\varphi_{i_k}^+\Big)(\tau)\,d\tau = 0$$

for all $(\alpha_1, \dots, \alpha_n)^T \in \ker \begin{pmatrix} \mathcal{A}_0(\bar{a}, \bar{d}) \\ \mathcal{A}_-(\bar{a}, \bar{d}) \end{pmatrix}$.

Finally, for $\varkappa > 0$*:*

$$W_{\mathcal{L}}^{(-1)}(a, d) = W_{\mathcal{L}}^{(-1)}(a, 0) - W_{\mathcal{L}}^{(-1)}(a, 0)\mathcal{B}_+^{(-1)}(a, d) \begin{pmatrix} V_d W_{\mathcal{L}}^{(-1)}(a, 0) \\ I - W_{\mathcal{L}}(a, 0) W_{\mathcal{L}}^{(-1)}(a, 0) \end{pmatrix},$$

$$\dim \ker W_{\mathcal{L}}(a, d) = n - \operatorname{rank} \begin{pmatrix} \mathcal{A}_0(a, d) \\ \mathcal{A}_+(a, d) \end{pmatrix},$$

$$\ker W_{\mathcal{L}}(a, d) = \left\{ W_{\mathcal{L}}^{(-1)}\Big(\sum_{k=1}^{n} \alpha_k \varphi_{i_k}^+ \Big) \,|\, (\alpha_1, \dots, \alpha_n)^T \in \ker \begin{pmatrix} \mathcal{A}_0(a, d) \\ \mathcal{A}_+(a, d) \end{pmatrix} \right\}.$$

Equation $W_{\mathcal{L}}(a, d) = f$ *with the right hand side* $f \in L_p(\mathbb{R}_+)$ *has a solution if and only if*

$$\int_0^\infty \overline{f(\tau)}\, W_{\mathcal{L}}^{(-1)}(\bar{a}, 0) \Big(\sum_{k=1}^{n} \alpha_k \varphi_{i_k}^+ + \sum_{k=1}^{\varkappa} \beta_k (I - \Gamma_+)^{-1} \pi_+ F^{-1} m(\bar{a}_-^{-1})\, g_k \Big)(\tau)\, d\tau = 0$$

for all $(\alpha_1, \dots, \alpha_n, \beta_1, \dots, \beta_\varkappa)^T \in \ker(\mathcal{A}_0(\bar{a}, \bar{d}), \mathcal{A}_-(\bar{a}, \bar{d})$.

As a particular case, the following statement holds.

Theorem 4.4 *Under conditions* (4.11)*, the operator* $W_{\mathcal{L}}(a, d)$ *is invertible if and only if*

$$\varkappa = 0 \text{ and } \operatorname{rank} \mathcal{A}_0(a, d) = n\,,$$

left (but not right) invertible if and only if

$$\varkappa < 0 \text{ and } \operatorname{rank}\, (\mathcal{A}_0(a, d), \mathcal{A}_-(a, d)) = n\,,$$

and right (but not left) invertible if and only if

$$\varkappa > 0 \text{ and } \operatorname{rank} \begin{pmatrix} \mathcal{A}_0(a, d) \\ \mathcal{A}_+(a, d) \end{pmatrix} = n\,.$$

Note that $\mathcal{M}_1 = \mathtt{W}(\mathbb{R})$. So, based on classical results of [13] under the condition $a \in \mathtt{W}(\mathbb{R})$ Theorems 4.2–4.4 can be extended to the case $p = 1$.

Acknowledgements This work was supported by the Science Committee, Ministry of Education and Science of Armenia within the framework of the joint research given to YSU-SFU-16/1, and by the Faculty Research funding from Division of Science and Mathematics, New York University Abu Dhabi.

References

1. Faddeev, L.D.: The inverse problem in the quantum theory of scattering. II. Itogi Nauki i Tekhniki. Ser. Sovrem. Probl. Mat., 3. VINITI, Moscow, 93–180 (1974) (in Russian) [English translation: J. Soviet Math. 5:3, 334–396 (1976)]
2. Marchenko, V.A.: Sturm-Liouville Operators and Their Applications. Naukova Dumka, Kiev (1977) (in Russian) [English translation: Revised edition: AMS Chelsea Publishing, Providence, RI (2011)]
3. Yurko, V.A.: Introduction to the Inverse Spectral Problems Theory. Fizmatlit, Moscow (2007) (in Russian)
4. Dodd, R.K., Eilbeck, J.C., Gibbon, J.D., Morris, H.C.: Solitons and Nonlinear Wave Equations. Academic Press Inc., London (1984)
5. Zakharov, V.E., Manakov, S.V., Novikov, S.P., Pitaevskii, L.P.: Theory of Solitons. The Inverse Scattering Method. Nauka, Moscow (1980) (in Russian) [English translation: Consultants Bureau, New York (1984)]
6. Calogero, F., Degasperis, A.: Spectral Transform and Solitons: Tools to Solve and Investigate Nonlinear Evolution Equations. North-Holland Publishing, Germany (1982)
7. Kamalyan, A.G., Karakhanyan, I.M., Oganesyan, A.O.: On a class of $\mathcal{L}$-Wiener-Hopf operations. Izv. NAN Armenii, Matem. **53:3**, 21–27 (2018) (in Russian) [English translation: J. Contemp. Math. Anal. 53:3, 121–125 (2018)]
8. Kamalyan, A.G., Khachatryan, I.G., Nersesyan, A.B.: Solvability of integral equations with $\mathcal{L}$-convolution type operators. Izv. NAN Armenii, Matem. **29:6**, 31–81 (1994) (in Russian) [English translation: J. Contemp. Math. Anal. 29:6, 22–66 (1994)]
9. Gohberg, I., Krupnik, N.: One-Dimensional Linear Singular Integral Equations. I. Operator theory: Advances and Applications, vol. 53. Birkhäuser Verlag, Basel (1992)
10. Duduchava, R.V.: Convolution Integral Equations with Discontinuous Presymbols, Singular Integral Equations with Fixed Singularities, and Their Applications to Some Problems in Mechanics. Teubner Texts on Mathematics, Leipzig (1979)
11. Naimark, M.A.: Linear Differential Operators. Frederil Ungar Publishing Co., New York (1967)
12. Bart, H., Gohberg, I., Kaashoek, M.A.: The coupling method for solving integral equations. Operator Theory: Advances and Applications, vol. 12. Birkhäuser Verlag, Basel (1984)
13. Krein, M.G.: Integral equations on the half-line with a kernel depending on the difference of the arguments. Uspehi Mat. Nauk **13:5**(83), 3–120 (1958) (in Russian) [English translation: Amer. Math. Soc. Transl. 22(2), 163–288 (1962)]
14. Litvinchuk, G.S., Spitkovskii, I.M.: Factorization of Measurable Matrix Functions. Operator Theory: Advances and Applications, vol. 25. Birkhäuser Verlag, Basel (1987)
15. Kamalian, A.G., Oganian, V.A.: Constructive method of factorization for a certain class of matrix functions. Izv. NAN Armenii **28:3**, 46–62 (1993) [English translation: J. Contemp. Math. Anal. 18:3, 40–54 (1993)]
16. Lebre, V.G.: Factorization in the Wiener algebra of a class of 2×2 matrix functions. IEOT **12**, 408–423 (1989)

Spectral Theory for Nonlinear Operators: Quadratic Case

Yakov Krasnov

Abstract In many cases, given a non-linear map, linearized systems near its fixed points do qualitatively capture its topological and algebraic properties. This suggests to extend the linear spectral theory to non-linear operators by considering spectra of linearizations in small neighborhoods of the fixed points. In the present paper, we develop this approach for quadratic maps. Several standard concepts such as asymptotic laws for splitting/gluing zeros of polynomial maps) are considered from new (and, possibly, unexpected) angles.

Keywords Nonlinear spectral theory · Cumulative spectrum · Bilinear operator

1 Introduction

1.1 Statement of the Problem

Many attempts have been made to define a concept of spectrum for nonlinear operators. Clearly, a reasonable spectral theory of nonlinear operators should: (a) inherit as many properties of the linear case as possible, and (b) be compatible with the standard spectral theory in the case of linear operators. However, as it was indicated in many sources (see, for example, [2, 13, 14]) a reasonable concept of "nonlinear spectrum" may be different from the classical eigenvalues, which is an important part of the spectrum in the linear case.

Hereafter, we will restrict ourselves to the finite-dimensional case. As a matter of fact, the Jacobian matrix of a linear operator is independent of the choice of a point meaning that the concept of spectrum in this case is *global*. In the non-linear case, spectral properties of linearizations at different points may be quite different. This way one arrives at the following question: given a nonlinear map, what is a set of points (both well-defined and important from the application viewpoint)

Y. Krasnov (✉)
Bar-Ilan University, Ramat Gan, Israel
e-mail: krasnov@math.biu.ac.il

A. Karapetyants et al. (eds.), *Modern Methods in Operator Theory and Harmonic Analysis*, Springer Proceedings in Mathematics & Statistics 291,
https://doi.org/10.1007/978-3-030-26748-3_12

at which spectral properties of the linearizations essentially determine the global behavior of the considered map? Many important results and examples suggest to consider the set of fixed points of the considered map. In what follows, the union of spectra of linearizations of a non-linear operator F at its fixed points will be called *cumulative spectrum* (denoted $\sigma(F)$). Studying cumulative spectrum of nonlinear operators seems to be a problem of formidable complexity and great importance.

1.2 Setting, Goals and Tools

In what follows, all the vector spaces are considered over a field $\mathbb{F}$ which is assumed to be a subfield of $\mathbb{C}$ - complex numbers. Given a vector space V and a map $P : V \to V$, denote by $Null_{\mathbb{C}}(P)$ the set of all *complex* solutions to the equation $P(x) = 0$.

We mainly restrict ourselves to the so-called quadratic homogeneous maps. More precisely, given a finite-dimensional vector space V over a filed $\mathbb{F}$, a polynomial map $Q : V \to V$ is called *homogeneous quadratic* if $Q(\alpha x) = \alpha^2 Q(x)$ for all $x \in V$ and all $\alpha \in \mathbb{F}$.

Essentially, coordinate functions of Q are quadratic forms in n variables. Using the standard polarization formula

$$\Psi_Q(x, y) := \frac{1}{2}\Big(Q(x + y) - Q(x) - Q(y)\Big), \quad x, y \in V, \tag{1}$$

one defines the bi-linear operator as multiplication $\Psi_Q : V \times V \to V$ such that $Q(x) = x * x = x^2$ for all $x \in V$. Thus, any quadratic homogeneous map is induced by a "diagonal" restriction of some multiplication in a commutative (in general, non-associative) algebra (denoted A_Q). In what follows, Q always stands for a homogeneous quadratic map. Some results will be stated for quadratic (in general, non-homogeneous) maps as well as for arbitrary polynomial maps.

Similarly to the linear operator case, a vector $x \neq 0$ may be considered as an eigenvector of Q if there exists an eigenvalue $\lambda \in \mathbb{F}$ such that $Qx = \lambda x$. However, only the following two cases are substantial:

(i) $\lambda = 1$ (in which case every eigenvector x is a fixed point of Q, i.e. $Qx = x$), or

(ii) $\lambda = 0$ (in which case every eigenvector $x \neq 0$ is a zero of Q, i.e. $Qx = 0$). Denote the set of non-zero fixed points of Q and zeros of Q respectively by

$$Fix_V(Q) := \{x \neq 0 \in V : \ Q(x) = x\}, \quad Null_V(Q) := \{x \neq 0 \in V : \ Q(x) = 0\},$$

and put

$$Fix_V^0(Q) := Fix_V(Q) \cup \{0\}.$$

Following the standard notations, given a differentiable map $P : V \to V$, where $\mathbb{F} = \mathbb{R}$ or $\mathbb{C}$, denote by $J_P(x)$ the Jacobian matrix of P at $x \in V$. Due to the Euler formula,

$$J_Q(x)x = 2Q(x) \quad \text{for any homogeneous quadratic map } Q. \tag{2}$$

Regarding $\frac{1}{2}J_Q(x)$ as an endomorphism in the vector space V, we define the corresponding characteristic polynomial by

$$p_{Q(x)}(t) := \det\left(\frac{1}{2}J_Q(x) - tI\right), \quad t \in \mathbb{F}. \tag{3}$$

Given $x \in V$, denote by $\sigma(x)$ the set of all roots (possibly, complex) of the characteristic equation $p_{Q(x)}(t) = 0$ at x (counting multiplicity). In algebra, $\sigma(x)$ is known under the name *Peirce spectrum* of Q at x. Thus, any $t \in \sigma(x)$ is actually an eigenvalue of $\frac{1}{2}J_Q(x)$. Moreover, if $a \in Fix_V(Q)$, then $t = 1$ is an obvious eigenvalue of $\frac{1}{2}J_Q(a)$ (equivalently, $1 \in \sigma(a)$ for $a \in Fix_V(Q)$).

Note that, by the well-known Bezóut theorem (see, for example, [9, 18]), for any quadratic map $Q : V \to V$ admitting finitely many fixed points, one has

$$\text{card}(Fix_V(Q)) \le 2^{\dim V} - 1 \tag{4}$$

We are now in a position to formulate the two-fold *goal* of this paper:

(i) to further develop the geometrical approach to generic quadratic maps initiated in [19];
(ii) to develop a spectral theory for quadratic maps and to study a parallelism between the theory developed and Peirce decomposition widely used in algebra (see [19]).

Our approach rests on the following two ingredients:

(a) Euler-Jacobi formula (giving an algebraic relation between the spectrum at (simple) fixed points of a polynomial map and the corresponding residues);
(b) Resolution of singularities (decomposing a compound singularity into the simpler ones with further application of asymptotic methods (see [7])).

Observe that for a generic quadratic map Q (i.e., $1/2$ is not in the cumulative spectrum of Q), the Euler-Jacobi formula yields several restrictions both on the spectrum and configuration of the fixed points set (expressed in the form of the so-called syzygies, cf. [8, 15, 19]). In particular, an information on the spectrum of sufficiently many fixed points completely determines the rest part of the spectrum as well as prescribes the configurations of fixed points.

1.3 Overview and Related Topics

After the Introduction, the paper is organized as follows. In Sect. 2, we discuss several properties of the cumulative spectrum of generic quadratic homogeneous maps.

Section 3 is devoted to general facts related to a desingularization of a multiple root of a polynomial map (for a quadratic map, the existence of a multiple root means that the map is not generic). In Sect. 4, a desingularization of a double root is discussed in detail (see also [7]). In Sect. 5, we briefly discuss a desingularization of a triple root and non-isolated fixed points.

Many results presented in the paper may be extended to polynomial ordinary differential equations (cf. [12, 13]). Several results which are related to the resolution of general singularities have been discussed in [7] in the context relevant to the so-called *good deformations* of singularities (cf. [18]).

2 Generic Quadratic Homogeneous Maps

2.1 Main Syzygy

We start with the following

Definition 2.1 (*cf.* [19]) Let V be a vector space over a field $\mathbb{F}$. A quadratic homogeneous map $Q : V \to V$ is called *generic* if

$$\det(J_Q(a) - \mathrm{I}) \neq 0 \quad \forall a \in Fix_V(Q). \tag{5}$$

As it was observed in [19], if Q is generic, then $1/2 \notin \sigma(Q)$. Also (see [19]), if $1/2 \notin \sigma(A)$ and the commutative algebra A_Q (cf. (1)) is free from 2-nilpotents, then Q is generic. It turns out that if Q is generic, then elements of $\sigma(Q)$ satisfy a syzygy determined by the following version of the celebrated Euler-Jacobi formula (see [4, p. 106] (see also Theorem 4.3 in [8])).

Theorem 2.2 (Euler-Jacobi Formula) *Let $P = (P_1, \ldots, P_n) : \mathbb{C}^n \to \mathbb{C}^n$ be a polynomial map whose components P_i are polynomials of degree m_i and let $\widetilde{P}$ be the polynomial map, whose components are the highest homogeneous terms of P. Denote by $Null_{\mathbb{C}}(P)$ the set of all complex roots of P and suppose that all $M = m_1 * m_2 * \ldots * m_n$ roots $a_i \in Null_{\mathbb{C}}(P)$ are simple and, in addition, $Null_{\mathbb{C}}(\widetilde{P}) = \{0\}$. Then, for any polynomial $h : \mathbb{C}^n \to \mathbb{C}$ of degree less than the degree of the Jacobian determinant of P, i.e.* $\deg h < m_1 + m_2 + \ldots + m_n - n$, *one has*

$$\sum_{a \in Null(P)} \frac{h(a)}{\det[J_P(a)]} = 0 \tag{6}$$

where $J_P(\cdot)$ denotes the Jacobian matrix of P.

In order to show that all the hypotheses in Theorem 2.2 must be necessarily fulfilled, consider several examples.

Example 2.3 Define $P : \mathbb{R}^2 \to \mathbb{R}^2$ by $P(x_1, x_2) := \{2x_1^2 - x_1x_2 - x_1, x_1^2 - x_2\}$. There are exactly two isolated zeros of P, namely: $\{0, 0\}$ and $\{1, 1\}$. But, the highest homogeneous part $\widetilde{P}$ has a nontrivial kernel, therefore, Theorem 2.2 is not applicable for P.

Example 2.4 Consider a polynomial map $P : \mathbb{R}^2 \to \mathbb{R}^2$ given by $P(x_1, x_2) := \{x_1^2 - x_2^2, 2x_1x_2\}$. In this case, $\widetilde{P}$ has trivial kernel. But, the only zero of P is not simple. Theorem 2.2 is not applicable for this map as well.

To apply Theorem 2.2 to studying the cumulative spectrum of a quadratic homogeneous map $Q : V \to V$, put $P(x) := Q(x) - x$ and take $a \in Fix_V(Q)$. Then (see (2) and (3)),

$$\det[J_P(a)] = 2^n p_{Q(a)}(1/2). \tag{7}$$

Combining Theorem 2.2 with (7) yields the following main result on $\sigma(Q)$ obtained in [19].

Theorem 2.5 *Let $Q : \mathbb{C}^n \to \mathbb{C}^n$ be a generic homogeneous quadratic map and let $p_{Q(a)}(t)$ be a characteristic polynomial of Q at $a \in \mathbb{C}^n$. Then,*

$$\sum_{a \in Fix_V^0(Q)} \frac{p_{Q(a)}(t)}{p_{Q(a)}(1/2)} \equiv 2^n, \quad \forall t \in \mathbb{C}. \tag{8}$$

In particular,

$$\sum_{a \in Fix_V^0(Q)} \frac{p_{Q(a)}^{(k)}(t)}{p_{Q(a)}(1/2)} = 0, \quad k = 1, 2, \ldots, n \tag{9}$$

where $p^{(k)} = \frac{d^k p(t)}{dt^k}$ denotes the k-th derivative of $p(t)$.

For several corollaries of Theorem 2.5 specifying formula (9) for selected classes of quadratic homogeneous maps, we refer to [19].

2.2 Thin Spectrum

A generic homogeneous quadratic map $Q : V \to V$ may a priori have

$$(2^{\dim V} - 1)(\dim V - 1) \tag{10}$$

distinct eigenvalues different from 0 and 1, in which case $\sigma(Q)$ will be called *full*. As a matter of fact, many important classes of generic homogeneous quadratic maps share the following property: their cumulative spectrum consists only of a few distinct prescribed eigenvalues (in particular, the cardinality of the spectrum is less than the maximal one provided by (10)). Following [19], such a spectrum will be called *thin*. The following two simple examples illustrate this concept.

Example 2.6 Define $Q : \mathbb{C}^2 \to \mathbb{C}^2$ by $Q(x_1, x_2) = \{x_1^2 - x_2^2, 2x_1x_2\}$. By direct computation, $Fix_V^0(Q)$ consists of four points: $c_0 = (0, 0)$, $c_1 = (1, 0)$, $c_{2,3} = (1/2, \pm i/2)$ with $\sigma(c_0) = 0$, $\sigma(c_1) = \{1\}$, $\sigma(c_2) = \sigma(c_3) = \{1, 0\}$. Therefore, $\sigma(Q) = \{1, 0\}$, i.e., $\sigma(Q)$ is thin.

Example 2.7 Define $Q : \mathbb{C}^2 \to \mathbb{C}^2$ by $Q(x_1, x_2) = \{x_1^2 + 4x_1x_2, 16x_1x_2 + x_2^2\}$. By direct computation, $Fix_V^0(Q)$ consists of four points: $c_0 = (0, 0)$, $c_1 = (1, 0)$, $c_2 = (0, 1)$, $c_3 = (\frac{1}{21}, \frac{5}{21})$ with $\sigma(c_0) = 0$, $\sigma(c_1) = \{1, 8\}$, $\sigma(c_2) = \{1, 2\}$, and $\sigma(c_3) = \{1, 1/7\}$. Therefore, (in contrast to Example 2.6), in this case, $\sigma(Q)$ is full.

A class of (generic) quadratic homogeneous maps having a thin spectrum is the one satisfying the following property: at any non-zero fixed point, the spectrum is the same. In this case, the map will be called *isospectral*. Below we present a simple necessary condition for a map to be isospectral.

Theorem 2.8 *If a quadratic homogeneous map $Q : V \to V$ is generic and isospectral, then its characteristic polynomial is of the form*

$$p_{Q(x)}(t) = (-1)^n t^n + \det\left(\frac{1}{2} J_Q(x)\right) \quad \forall x \in V. \tag{11}$$

Proof Take the characteristic polynomial (3):

$$p_{Q(x)}(t) = \alpha_0 t^n + \alpha_{n-1}(x)t^{n-1} + \ldots + \alpha_1(x)t + \alpha_n(x), \tag{12}$$

where

$$\alpha_0 = (-1)^n, \quad \alpha_n(x) = \det\left(\frac{1}{2} J_Q(x)\right), \tag{13}$$

and each coefficient $\alpha_k(x)$, $k = 1, \ldots, n-1$, is a homogeneous map of degree $(n - k)$.

Let us show first that for any $a \in Fix_V(Q)$, all the coefficients $\alpha_k(x)$ with $k = 1, \ldots, n-1$, vanish at $x = a$. In fact, the isospectrality implies that $p_{Q(a)}(t) \equiv p(t)$ is independent of the choice of a fixed point a. Substituting $p_{Q(a)}(t) = p(t)$ into (8) yields:

$$\sum_{a \in Fix_V^0(Q)} \frac{p_{Q(a)}(t)}{p_{Q(a)}(1/2)} = \sum_{a \in Fix_V(Q)} \frac{p_{Q(a)}(t)}{p_{Q(a)}(1/2)} + 2^n t^n = (2^n - 1)\frac{p(t)}{p(1/2)} + 2^n t^n$$

From this, we obtain

$$(2^n - 1)\frac{p(t)}{p(1/2)} + 2^n t^n = 2^n,$$

and therefore,

$$p(t) = p(1/2)\frac{2^n}{2^n - 1}(1 - t^n). \tag{14}$$

Combining (14) with (12)–(13) implies

$$p_{Q(a)} \equiv p(t) = (-1)^n(t^n - 1)$$

for all $a \in Fix_V(Q)$. Moreover, using the $(n-k)$-homogeneity of $\alpha_k(x)$, one can show that $p_Q(x) = (-1)^n(t^n - C(x))$ with the scalar function $C(x)$ such that $C(a) = 1$ at any $a \in Fix_V(Q)$. □

The following example illustrates Theorem 2.8.

Example 2.9 Take $Q : \mathbb{C}^2 \to \mathbb{C}^2$ defined by $Q(x_1, x_2) = \{x_1^2 - 2x_1x_2, x_2^2 - 2x_1x_2\}$. By direct computation, (11) is satisfied. On the other hand, $Fix_V^0(Q)$ consists of four points: $c_0 = (0, 0)$, $c_1 = (1, 0)$, $c_2 = (0, 1)$, and $c_3 = (-1, -1)$ with $\sigma(c_0) = 0$ and $\sigma(c_1) = \sigma(c_2) = \sigma(c_3) = \{1, -1\}$. The characteristic polynomial of $Q(x)$ coincide with (11), namely $p_{Q(x)}(t) = t^2 - x_1^2 - x_2^2 + 3x_1x_2$.

In the above examples, we considered quadratic homogeneous maps $Q : V \to V$ with thin spectrum assuming that $\mathrm{card}(Fix_V(Q)) = 2^{\dim V} - 1$. If the above equality is not satisfied, then one can also talk about the maps with a thin spectrum.

Example 2.10 The map $Q : \mathbb{C}^2 \to \mathbb{C}^2$ given by $Q(x_1, x_2) = \{x_1^2 - x_2^2, x_1x_2\}$ has only one non-zero fixed point, namely $c_1 = (1, 0)$. Also, $\sigma(c_1) = \{1, 1/2\}$. Since $1/2 \in \sigma(Q)$ (equivalently, condition (5) is not satisfied), Q is not generic and one cannot use syzygy (8) in this case. In what follows, we will consider several non-generic settings to which formula (8) can be extended.

3 Non-generic Maps: Desingularization

3.1 Multiple Roots

In what follows, if the opposite is not stated explicitly, all the maps (not necessarily quadratic) considered in this section are supposed to be *real polynomial with (possibly complex and non-simple) isolated fixed points.*

Recall that given a polynomial map $P : \mathbb{C}^n \to \mathbb{C}^n$, its root a is called *multiple* if the Jacobian determinant of P at a vanishes. A simple criterium for a fixed point a of a homogeneous quadratic map to be multiple was established in [19].

Theorem 3.1 *Let $Q : \mathbb{C}^n \to \mathbb{C}^n$ be a quadratic homogeneous map, $a \in Fix_V^0(Q)$ and $P(x) := Q(x) - x$. Then, a is a multiple or continuous root of P if and only if $p_{Q(a)}(1/2) = 0$.*

Below we present a criterium for a root of a polynomial map (in general, not necessarily homogeneous quadratic) to be multiple. Observe that in the scalar case, multiple roots appeal to divisibility of polynomials. Our criterium can be traced back to the following concept [16] known as the Gleason B property: a domain $D \subset \mathbb{C}^n$ is said to have the Gleason B property at a point $z = a \in D$ if for every $f \in B(D)$ such that $f(a) = 0$ there exist functions $f_1, f_2, \ldots, f_n \in B(D)$ such that $f(z) = \sum_{j=1}^{n}(z_j - a_j) f_j(z)$.

Lemma 3.2 (Factorization Lemma (see [7])) *Let $P = (p_1, \ldots, p_n) : \mathbb{C}^n \to \mathbb{C}^n$ be a polynomial map with $P(a) = 0$. Then, a is an isolated multiple root of P if and only if there exist coordinates $z = (z_1, \ldots z_n)$ in $\mathbb{C}^n$ and natural $m \geq 2$ such that*

$$p_k(z) = (z_1 - a_1)^m q_{k1}(z_1) + \sum_{i=2}^{n}(z_i - a_i) q_{ki}(z) \quad (k = 1, \ldots, n), \tag{15}$$

where $R(z) = \{q_{ki}(z)\}_{k,i=1}^{n}$ is a polynomial matrix such that the column vector $q_1(z_1) := \{q_{11}(z_1), \ldots, q_{n1}(z_1)\}^t$ depends only on z_1 and is not equal to zero identically. Representation (15) *is unique up to the order of the Jordan blocks of linearization of P.*

3.2 *Desingularization Through Deformation*

In general, it is not an easy task to study a type of a multiple root of a polynomial map. The commonly used method is known as a *resolution of singularity* or *desingularization* (see [3–6, 12, 18]). In what follows, given a polynomial map $P : \mathbb{C}^n \to \mathbb{C}^n$ with an isolated root $a \in V$, by a *desingularization* we mean a small one-parameter deformation of P near a converting P to the map having (in a small neighborhood of a) roots which are simpler than a (in particular, these roots have multiplicities strictly less than the multiplicity of a; we refer to [4, 18] for the definition of multiplicity).

The Factorization Lemma combined with the scalar interpolation polynomial suggests the following formula for a deformation of P near a (cf. [7]):

$$P_\varepsilon(z) = \prod_{j=0}^{m-1}(z_1 - a_1 - \varepsilon j) q_1(z_1) + \sum_{i=2}^{n}(z_i - a_i) q_i(z), \tag{16}$$

where ε is a *sufficiently small* real number. Clearly,

$$P_\varepsilon(a) = P_\varepsilon(a + \varepsilon e_1) = \ldots = P_\varepsilon(a + \varepsilon(m-1)e_1) = 0, \tag{17}$$

where $e_1 = (1, 0, \ldots 0)$. Moreover,

$$P_\varepsilon(z) = P(z) - \varepsilon \frac{m(m-1)}{2}(z_1 - a_1)^{m-1} q_1(z_1) + o(\varepsilon). \tag{18}$$

Remark 3.3 In general, formula (16) does not decompose a into simple roots. However, (16) decreases the multiplicity of a at least by one. Therefore, a step-by-step application of this deformation can be used in order to decompose a multiple root into the simpler ones. Also, if a is a *double* root, then formula (16) does decompose a into simple roots.

Example 3.4 Suppose $a = 0$ is a multiple zero of a quadratic map $P_2 : \mathbb{C}^n \to \mathbb{C}^n$ (in general, *non-homogeneous*). Then, the application of the Factorization Lemma yields:

$$P_2(z) = z_1^2 V_1 + \sum_{i=2}^{n} z_i(V_i + L_i z), \tag{19}$$

where V_i is a vector in $\mathbb{C}^n$ and $L_i : \mathbb{C}^n \to \mathbb{C}^n$ is a linear operator. Furthermore, by splitting $z_1^2 \to z_1(z_1 + \varepsilon)$ [see formula (16)], one obtains the following partial desingularization:

$$P_{2,\varepsilon}(z) = L_0 z + z_1^2 V_1 + \sum_{i=2}^{n} z_i L_i z \tag{20}$$

(here L_0 is an $n \times n$ matrix composed of the following columns $L_0 = [\varepsilon V_1, V_2, \ldots, V_n]$.)

Remark 3.5 Using (partial) desingularizations, one can associate to a multiple singularity a geometric graph reflecting its complexity (see [7] for details). The idea behind the concept of geometric graph can be traced back to the pioneering work of V. Arnold [6] (see also, [4, 5]), where a deep connection between the hierarchy of singularities of *gradient* maps based on their possible decomposition on the one hand, and a decomposition of Dynkin diagrams on the other hand, was established.

3.3 *A*-Equivalence and Topological Equivalence

In general, given a polynomial map P with an isolated multiple root a, constructing the geometric graph associated with a is a problem of formidable complexity. This problem can be simplified by considering several equivalences preserving the graph. In this subsection, we focus on two of them: A-equivalence and topological equivalence.

Definition 3.6 Let $P, Z : \mathbb{C}^n \to \mathbb{C}^n$ be two polynomial maps. Suppose there exists a polynomial matrix $R(z) = \{q_{ki}(z)\}_{k,i=1}^n$ such that

$$P(z) = R(z)Z(z) \quad \text{and} \quad \det R(a) \neq 0. \tag{21}$$

Then, P and Z are said to be A-equivalent near a.

Remark 3.7 (i) If P and Z are A-equivalent, then $\det(R(z)) \neq 0$ for all z close to a and $R^{-1}(z)$ is a C^{∞}-map from a small disc centered at a to $GL(n, \mathbb{C})$. Therefore, Definition 3.6 is compatible with [4].

(ii) If P and Z are A-equivalent near a, then a desingularization of a for P and Z can be achieved by the same formulas (in particular (cf. [4]), the multiplicities of a for P and Z coincide). Clearly, A-equivalence preserves the index up to a sign. Also, we refer to [18] for the so-called transformation law linking residues of A-equivalent maps.

Example 3.8 Keeping in mind Remark 3.7(ii), consider the map (15) more intently. Assume that $\det(R(a) \neq 0$. Then, P is A-equivalent to the map

$$Z_m := ((z_1 - a_1)^m, z_2 - a_2, ..., z_n - a_n) \tag{22}$$

and m coincides with the multiplicity of a for P. Moreover, formula (16) defines a desingularization of a for P and $P_\varepsilon(z) = R(z)Z_{m,\varepsilon}$, where

$$Z_{m,\varepsilon} = \Big(\prod_{j=0}^{m-1}(z_1 - a_1 - \varepsilon j), z_2 - a_2, ..., z_n - a_n\Big)$$

is a desingularization of a for Z_m. More generally, take a polynomial map

$$Z_m = ((z_1 - a_1)^{m_1}, (z_2 - a_2)^{m_2}, \ldots, (z_n - a_n)^{m_n})^t, \quad m = (m_1, \ldots, m_n) \tag{23}$$

(formula (23) considered in $\mathbb{C}^n$ defines the so-called Pham map (see [4]). Assume that P and Z_m are A-equivalent near a. Then, the desingularization of a for both P and Z_m can be achieved with the same zeroes filling out the 0-skeleton of the corresponding rectangular parallelepiped (cf. (17)), given by:

$$(a_1 + k_1\varepsilon, a_2 + k_2\varepsilon, \ldots, a_n + k_n\varepsilon)^t, \quad k_i \in \{0, ..., m_i - 1\}. \tag{24}$$

The A-equivalence relation allows one to reduce the algebraic complexity of a considered map while the topological equivalence we are going to define allows to perform very rough change of variables near a singularity a.

Definition 3.9 Two polynomial maps $P, S : \mathbb{C}^n \to \mathbb{C}^n$ with $P(a) = S(a) = 0$ are called *topologically equivalent near* a, if there exist neighborhoods $U \ni a$ and $V \ni 0$ and homeomorphisms $i : U \to U$, $j : V \to V$ such that $i(a) = a$, $j(0) = 0$ and $P \circ i \equiv j \circ S$ on U (see [4]).

Clearly, if P and S are topologically equivalent near a, then a desingularization of a for P and S can be achieved by the same formulas. In the sequel, we will systematically use the following obvious statement.

Proposition 3.10 *Let $P, S : \mathbb{C}^n \to \mathbb{C}^n$ be polynomial maps with $P(a) = S(a) = 0$ and assume that P is obtained from S (near a) by a composition of a topological* *equivalence and A-equivalence. Then, the geometric graphs associated to a for P and S are isomorphic.*

The A-equivalence and topological equivalence come together in the case of the so-called singularities of type A_m (cf. [4, 6]).

Definition 3.11 Let $P : \mathbb{C}^n \to \mathbb{C}^n$ be a polynomial maps with $P(a) = 0$. We say that a is a singularity of type A_m if $J_P(a)$ contains exactly one Jordan block corresponding to zero eigenvalue.

The following statement is well-known in two-dimension (see [12]).

Proposition 3.12 *Let $P : \mathbb{C}^2 \to \mathbb{C}^2$ be a polynomial map with $P(a) = 0$ and let a be a singularity of type A_m. Then, P can be represented as a composition of a topological equivalence, A-equivalence and the map*

$$Z_m = ((z_1 - a_1)^m, z_2 - a_2, ..., z_n - a_n).$$

4 Desingularization of Double Points

In this section, following [7], we present several results related to the desingularization of a double point. An example of a quadratic homogeneous map supporting these results is also considered.

4.1 *Basic Formulas*

Proposition 4.1 (see [1, 7]) *Under the notations and assumptions of Lemma 3.2, suppose, in addition, that a is a double root of P. Take the coordinates provided by Lemma 3.2 (in particular, $m = 2$ in (15)). Let P_ε be a desingularization provided by* (16). *Then:*

(i) for $k = 0, 1$, one has:

$$\det(J_{P_\varepsilon}(a + k\varepsilon e_1)) = (-1)^{1-k}\varepsilon \det(R(a + \varepsilon k e_1)); \tag{25}$$

(ii)

$$\frac{\partial}{\partial x_1} \det(J_P(x))\Big|_{x=a} = 2 \det(R(a)). \tag{26}$$

4.2 *Patterns*

Under the notations and assumptions of Proposition 4.1, denote by ξ the unit eigenvector corresponding to zero eigenvalue of the linearization of P at a and let

$\mathcal{P} : [0, \varepsilon_o] \times \mathbb{C}^n \to \mathbb{C}^n$ be a *good deformation* of P near a (cf. [7, 18]). Assume that $\mathcal{P}(\varepsilon, a) = 0$ for all ε. In this case, Lemma 3.2 and Proposition 4.1 allow us to describe several patterns that $\mathcal{P}$ must satisfy:

- *decomposition of double root:* $\frac{\partial}{\partial \xi} \det(J_{\mathcal{P}}(z))\Big|_{z=a} \neq 0$, moreover, $\mathcal{P}$ decomposes the root a into two simple roots only along ξ;
- *interaction of invariants of linearization at the regular roots:*
  - determinants at the simple roots have asymptotically the same magnitude but are opposite in signs (in particular, $\operatorname{ind}(a, P) = 0$);
  - traces at the simple roots are asymptotically equal to the trace of the linearization of P at a;
- *coalescing of simple singularities*: explicit asymptotic formulas for the invariants of linearizations at the simple roots allow one to effectively compute invariants of the original double root as well as to determine its topological type, for example,
  - one has
  $$\operatorname{res}_P(r, a) = \frac{\partial}{\partial \xi}\left(\frac{r(z)}{\det(R(z))}\right)\Big|_{z=a}, \tag{27}$$
  where $r : \mathbb{C}^n \to \mathbb{C}$ is an arbitrary polynomial (see [7]).
  - for $n = 2$, if a is semi-hyperbolic for P, then a is a saddle-node, while if a is nilpotent for P, then a is a cusp (cf. [12], Theorems 2.19 and 3.5).

4.3 Generalized Euler-Jacobi Formula

Let $P : \mathbb{C}^n \to \mathbb{C}^n$ be a polynomial map. Denote by A_s (resp. A_d) the set of simple (resp. double) roots of P. The statement following below (it is an immediate consequence of the Global Residue Theorem (see [18]), Euler-Jacobi formula for simple singularities (see [18]), Bezout Theorem (see, for example, [9]) and computational formula (27)) is a generalization of the Euler-Jacobi formula to polynomial maps with simple and double singularities.

Theorem 4.2 (see [7]) *Let $P = (P_1, ..., P_n) : \mathbb{C}^n \to \mathbb{C}^n$ be a polynomial map ($\deg(P_i) = d_i$, $i = 1, ..., n$) with $Null_{\mathbb{C}}(P) = A_s \cup A_d$. Assume that*

$$\operatorname{card}(Null_{\mathbb{C}}(P)) = \prod_{i=1}^{n} d_i$$

(each root counted according to its multiplicity). Denote by $\widetilde{P}$ the polynomial map, whose components are the highest homogeneous terms of the components of P and assume that $Null_{\mathbb{C}}(\widetilde{P}) = \{0\}$. Finally, let $r : \mathbb{C}^n \to \mathbb{C}$ be a polynomial with $\deg(r) \leq \sum_{m=1}^{n} d_m - (n+1)$. *Then,*

$$\sum_{a_j \in A_s} \frac{r(a_j)}{\det(J_p(a_j))} + \sum_{a_j \in A_d} \frac{\partial}{\partial \xi_{a_j}} \left(\frac{r(z)}{\det(R(z))} \right) \Bigg|_{z=a_j} = 0, \tag{28}$$

where $z = (z_1, ..., z_n)$, $\frac{\partial}{\partial \xi_{a_j}}(\cdot)$ *stands for the directional derivative along the unit eigenvector of* $J_P(a_j)$ *corresponding to the zero eigenvalue, and* $R(z)$ *is a polynomial matrix given by* (15) *as "factorization" of* P *near a double point* a_j.

Remark 4.3 (i) We refer to [15], where a variant of Euler-Jacobi formula including simple and double roots was presented for $n = 2$.

(ii) One can easily extend (28) to the case when P has only simple roots and roots of type A_m.

4.4 Example

Consider the quadratic map $Q : \mathbb{R}^2 \to \mathbb{R}^2$ given by $Q(x) = (x_1x_2 + ax_1^2, bx_1^2 + x_2^2)$, where $a, b \neq 0$. By inspection, $Fix^0_{\mathbb{R}^2}(Q)$ consists of:

$$c_0 = (0, 0),\ c_{1,2} = (0, 1),\ c_3 = \left(\frac{a}{a^2 + b}, \frac{b}{a^2 + b} \right),$$

and

$$\sigma(c_0) = \{0\},\ \sigma(c_{1,2}) = \{1, 1/2\},\ \sigma(c_3) = \left\{1, \frac{b}{2(a^2 + b)}\right\} \tag{29}$$

It follows immediately from (29) that $c_{1,2}$ is a double point for $Q - I$. In order to study $Q - I$ near the double fixed point $x = c_{1,2} = (0, 1)$, rewrite $Q - I$ in the following form:

$$Q(x) - x = \begin{bmatrix} x_1x_2 + ax_1^2 - x_1 \\ x_2^2 + bx_1^2 - x_2 \end{bmatrix} = x_1^2 \begin{bmatrix} a \\ b \end{bmatrix} + (x_2 - 1) \begin{bmatrix} x_1 \\ x_2 \end{bmatrix} = R(x)X_2(x), \tag{30}$$

where

$$R(x) = \begin{bmatrix} a & x_1 \\ b & x_2 \end{bmatrix} \quad \text{and} \quad X_2 = \begin{bmatrix} (x_1 - 0)^2 \\ x_2 - 1 \end{bmatrix} \tag{31}$$

Clearly, $\det(R(c_{1,2})) = a \neq 0$, so that the desingularization of $c_{1,2}$ can be carried out along the x_1-axis. In addition, $\frac{\partial}{\partial x_1} R(x) \equiv -b \neq 0$. Hence, using (28), one can calculate the residue at the double fixed point $c_{1,2}$ as follows:

$$\operatorname{res}(r, Q, c_{1,2}) = \frac{\partial}{\partial x_1} \left(\frac{r(x)}{\frac{\partial}{\partial x_1} R(x)} \right) \Bigg|_{x=c_{1,2}} = -\frac{\partial}{\partial x_1} \left(\frac{r(x)}{b} \right) \Bigg|_{x=c_{1,2}} \tag{32}$$

Remark 4.4 Observe also that if $b = 0$, then $c_{1,2} = c_3$ so that c_3 is a triple fixed point.

5 Concluding Remarks: Triple and Non-isolated Fixed Points

We complete the paper with several remarks related to a desingularization of triple and non-isolated fixed points.

5.1 Triple Points: Two-Step Algorithm

Below we present an algorithm allowing (at least, in principal) to completely desingularize triple roots (i.e. of multiplicity three). Let $P : \mathbb{C}^n \to \mathbb{C}^n$ be a polynomial map and let $a \in \mathbb{C}^n$ be a *triple* root of P. In contrast to the case of double root, the conclusion of Proposition 3.12 is not true anymore in this case. To be more specific, formula (15) with $m = 2$ still takes place, while there is the following alternative: either (i) P is A-equivalent to X_3 near a, in which case a complete desingularization of a (i.e., splitting into three simple roots) can be achieved using formula (16), or (ii) $\det(Q(a)) = 0$, and then necessarily a is a *double* root for P_ε while $a_\varepsilon := a + \varepsilon e_1$ is a simple root (in particular, $\mathrm{ind}(a, P) = \mathrm{ind}(a_\varepsilon, P_\varepsilon) = \pm 1$). Clearly, in the case (ii), there are no more roots for P_ε near a. Hence, applying a formula similar to (16), to P_ε near a, one can completely desingularize a in the case (ii).

Before giving an illustration of the above 2-step algorithm, let us present a simple analytic criterium for a root to be triple.

Proposition 5.1 *Let $P : \mathbb{C}^n \to \mathbb{C}^n$ be a polynomial map with $P(a) = 0$. Then, a is a triple root of P if and only if the following conditions are satisfied:*

(a) $\det(J_P(a)) = 0$ *and the eigenspace of $J_P(a)$ corresponding to zero eigenvalue is one-dimensional;*
(b) there exist coordinates $(x_1, ..., x_n)$ in which representation (15) *takes place with $m = 2$ (here x_1 corresponds to zero eigenspace);*
(c) $\frac{\partial}{\partial x_1} \det(J_P(x))\Big|_{x=a} = 0$;
(d) there exists a direction y_1 such that $\frac{\partial^2}{\partial x_1 \partial y_1} \det(J_P(x))\Big|_{x=a} \neq 0$.

Example 5.2 Consider the map

$$P(x, y) = \begin{bmatrix} y - x^2 + y^2 \\ xy \end{bmatrix} = x^2 \begin{bmatrix} -1 \\ 0 \end{bmatrix} + y \begin{bmatrix} 1 + y \\ x \end{bmatrix}. \tag{33}$$

By inspection, the origin is a triple root for P (cf. Proposition 5.1; here the eigenspace of $J_P(0,0)$ corresponding to zero eigenvalue coincides with the space span(e_1), $e_1 = \begin{bmatrix}1\\0\end{bmatrix}$. Formula (16) takes (33) to

$$P_\varepsilon(x, y) = P(x, y) - \varepsilon(x - 0)\begin{bmatrix}-1\\0\end{bmatrix} = \begin{bmatrix}y - x^2 + \varepsilon x + y^2\\xy\end{bmatrix}. \tag{34}$$

By inspection,

$$J_{P_\varepsilon}(0,0) = \begin{bmatrix}\varepsilon & 1\\0 & 0\end{bmatrix} \quad\text{and}\quad J_{P_\varepsilon}(\varepsilon, 0) = \begin{bmatrix}-\varepsilon & 1\\0 & \varepsilon\end{bmatrix}, \tag{35}$$

meaning that $(\varepsilon, 0)$ is a regular root while the origin is a double root for P_ε. Let

$$M := \begin{bmatrix}-1 & 1\\\varepsilon & 0\end{bmatrix} \tag{36}$$

be the matrix of eigenvectors of $J_{P_\varepsilon}(0,0)$ corresponding to the eigenvalues 0 and ε, respectively. Combining (36) with Lemma 3.2 suggests new coordinates (u, v) given by $x = v - u$; $y = \varepsilon u$. This change of coordinates takes P_ε to $\widetilde{P}_\varepsilon$:

$$\widetilde{P}_\varepsilon(u, v) = M^{-1}P_\varepsilon M\begin{bmatrix}u\\v\end{bmatrix} = u^2\begin{bmatrix}-1\\\varepsilon^2 - 2\end{bmatrix} + v\begin{bmatrix}u\\\varepsilon + 3u - v\end{bmatrix}. \tag{37}$$

Applying an analog of formula (16) one obtains:

$$P_{\varepsilon\delta}(u, v) = \widetilde{P}_\varepsilon(u, v) - \delta(u - 0)\begin{bmatrix}-1\\\varepsilon^2 - 2\end{bmatrix} = \begin{bmatrix}uv - u^2 + \delta u\\\delta(2 - \varepsilon^2)u + \varepsilon v - 2u^2 + 3uv - v^2 + \varepsilon^2 u^2\end{bmatrix}. \tag{38}$$

It is easy to see that (38) is a good perturbation of $\widetilde{P}_\varepsilon$ near the origin (provided $\delta << \varepsilon$). Indeed, by direct computation, $P_{\varepsilon\delta}$ admits the following four roots:

$$\begin{bmatrix}0\\0\end{bmatrix}, \begin{bmatrix}0\\\varepsilon\end{bmatrix}, \begin{bmatrix}\delta\\0\end{bmatrix}, \begin{bmatrix}-(\varepsilon + \delta)/\varepsilon^2\\-(\varepsilon + \delta + \delta\varepsilon^2)/\varepsilon^2\end{bmatrix}. \tag{39}$$

The determinants of $J_{f_{\varepsilon\delta}}$ at these points are

$$\varepsilon\delta, \quad -\varepsilon^2 - \varepsilon\delta, \quad -\varepsilon\delta - \delta^2 - \delta^2\varepsilon^2, \quad -(\varepsilon^2 + 2\varepsilon\delta + \varepsilon^3\delta + \delta^2 + \delta^2\varepsilon^2)/\varepsilon^2, \tag{40}$$

respectively. Since $\varepsilon, \delta > 0$, it follows that these determinants are different from zero, meaning that all the roots are regular.

5.2 Spectrum at Non-isolated Fixed Points

In contrast to the linear operators, quadratic maps may have the continuous set of fixed points. Consider a map $Q : \mathbb{R}^3 \to \mathbb{R}^3$ given by

$$Q(x) = \begin{bmatrix} x_1^2 + \frac{1}{2}(x_1x_2 - x_2x_3 + x_3x_1) \\ x_2^2 + \frac{1}{2}(x_1x_2 + x_2x_3 - x_3x_1) \\ x_3^2 - \frac{1}{2}(x_1x_2 - x_2x_3 - x_3x_1) \end{bmatrix}. \tag{41}$$

Then, a simple analysis reveals that there exist exactly three sets of fixed points:

$$c_0 = (0, 0, 0), \quad c_1 = \frac{2}{3}\Big(1, 1, 1\Big),$$

and a continuous set of fixed points belonging to the circle given by

$$c = \{x_1, x_2, x_3 : \left(x_1 - \frac{1}{3}\right)^2 + \left(x_2 - \frac{1}{3}\right)^2 + \left(x_3 - \frac{1}{3}\right)^2 = \frac{2}{3}, \quad x_1 + x_2 + x_3 = 1\}. \tag{42}$$

Also,

$$\sigma(c_0) = \{0, 0, 0\}, \qquad \sigma(c_1) = \{1, 1, 1\}.$$

In addition, for any element e of c, one has

$$\sigma(e) = \{0, 1, 1/2\}. \tag{43}$$

The presence of $1/2$ in the spectrum of c is not by chance. It turns out that

Proposition 5.3 (see [19]) *Let $Q : \mathbb{R}^n \to \mathbb{R}^n$ be a homogeneous quadratic map admitting a continuous set $c(t)$ of fixed points. Then, $1/2 \in \sigma(c(t))$ for all t.*

One can show that there exists a desingularization of (43) converting $Fix_{\mathbb{C}}(Q)$ to six isolated fixed points $c_2, c_3, c_4, c_5, c_6, c_7$ belonging to a small neighborhood of the circle, while $c_0 = 0$ and c_1 lie outside it (see the Fig. 1).

Let Q_α be a small deformation of Q near $\alpha = 1/2$:

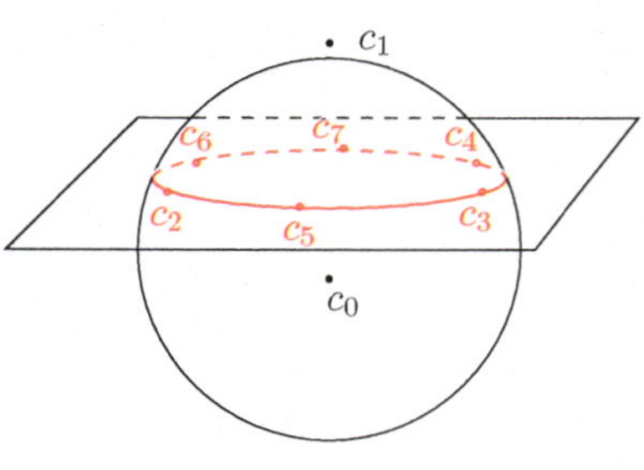

Fig. 1 Desingularization of non isolated fixed points layout along circle

$$Q_\alpha(x) = \begin{bmatrix} x_1^2 + \alpha(x_1x_2 - x_2x_3 + x_3x_1) \\ x_2^2 + \alpha(x_1x_2 + x_2x_3 - x_3x_1) \\ x_3^2 - \alpha(x_1x_2 - x_2x_3 - x_3x_1) \end{bmatrix}. \tag{44}$$

It is easy to see that the deformed map Q_a has seven (isolated) fixed points $c_0, \ldots, c_7$, of which $c_2, \ldots c_7$ belong to a small neighborhood of the circle (42) and are given by:

$$c_2 = \{1/(\alpha+1), 1/(\alpha+1), -\alpha/(\alpha+1)\}, \quad c_1 = \{1/(\alpha+1), 1/(\alpha+1), 1/(\alpha+1)\},$$
$$c_3 = \{0, 0, 1\}, \quad c_4 = \{-\alpha/(\alpha+1), 1/(\alpha+1), 1/(\alpha+1)\}, \quad c_5 = \{1, 0, 0\},$$
$$c_6 = \{1/(\alpha+1), -\alpha/(\alpha+1), 1/(\alpha+1)\}, \quad c_7 = \{0, 1, 0\}$$

Acknowledgements The author wishes to cordially thank the referee for various suggestions and comments that enable to make many important corrections and to significantly improve the presentation of the paper.

References

1. Aizenberg, L.A., Yuzhakov, A.P.: Integral Representations and Residues in Multidimensional Complex Analysis. Translations of Mathematical Monographs, 58th edn. American Mathematical Society, Providence R.I. (1983)
2. Appell, J., Dörfner, M.: Some spectral theory for nonlinear operators. Nonlinear Anal. **28**, 1955–1976 (1997)
3. Arnold, V., Goryunov, V., Lyashko, O., Vasiliev, V.: Singularity Theory II: Classification and Applications. Encyclopedia of Mathematical Sciences, vol. 39. Springer, Berlin (1992). (Russian edition 1988)
4. Arnold, V.I., Varchenko, A.N., Gusein-Zade, S.M.: Singularities of differentiable maps, vol. I. Birkhäuser, Boston (1985)
5. Arnold, V.I.: Singularities, bifurcations, and catastrophes. *Usp. Fiz. Nauk***141**, 569–590
6. Arnold, V.: Normal forms for functions near degenerate critical points, the Weyl groups of A_k, D_k, E_k and Lagrangian singularities. Funktsional'nyi Analiz i Ego Prilozheniya **6**, 3–25 (1972)
7. Balanov, Z., Krasnov, Y.: On good deformations of A_m -singularities. Discrete Cont. Dyn. Syst. Ser. S **12**(7) (2019)
8. Balanov, Z., Kononovich, A., Krasnov, Y.: Projective dynamics of homogeneous systems: local invariants, syzygies and the global residue theorem. Proc. Edinb. Math. Soc. **II**(55), 577–589 (2012)
9. Bezout, E.: General Theory of Algebraic Equations. Princeton University Press, Princeton (2006)
10. Coppel, W.A.: A survey of quadratic systems. J. Differ. Equ. **2**, 293–304 (1966)
11. Dumortier, F.: Singularities of vector fields on the plane. J. Differ. Equ. **23**, 53–106 (1977)
12. Dumortier, F., Llibre, J., Artés, J.C.: Qualitative Theory of Planar Differential Systems. Springer, Berlin (2006)
13. Edmunds, D.E., Evans, W.D.: Spectral theory and differential operators. Oxford University Press, New York (1987)

14. Furi, M., Martelli, M., Vignoli, A.: Contributions to the spectral theory for nonlinear operators in Banach spaces. Ann. Mat. Pura. Appl. **118**, 229–294 (1978)
15. Gasull, A., Torregrosa, J.: Euler-Jacobi formula for double points and applications to quadratic and cubic systems. Bull. Belg. Math. Soc. Simon Stevin **6**, 337–346 (1999)
16. Gleason, A.M.: Finitely generated ideals in Banach algebras. J. Math. Mech. **13**, 125–132 (1964)
17. Griffiths, P.: Variations on a theorem of Abel. Inven. Math. **35**, 321–390 (1976)
18. Griffiths, P., Harris, J.: Principles of Algebraic Geometry. Wiley, New York (1978)
19. Krasnov, Y., Tkachev, V.G.: Idempotent geometry in generic algebras. Adv. Appl. Clifford Algebr. **28**(5), 84–98 (2018)

Factorization of Order Bounded Disjointness Preserving Multilinear Operators

Anatoly G. Kusraev and Zalina A. Kusraeva

Abstract Given a finite collection of disjointness preserving linear operators with values in an f-algebra, the mapping defined as their pointwise product is a disjointness preserving multilinear operator. The central result asserts that an arbitrary order bounded disjointness preserving multilinear operator from the Cartesian product of vector lattices into an arbitrary vector lattice has a similar structure. Using this fact, we establish a multilinear version of the Hart theorem on the associated lattice homomorphism, and also give some consequences on the structure of disjointness preserving homogeneous polynomials.

Keywords Vector lattice · f-algebra · Disjointness preserving operator · Multilinear operator · Homogeneous polynomial · Factorization

1 Introduction

The study of disjointness preserving linear operators is an interesting area of the general theory of regular operators in vector lattices. A systematic study of this class of operators begins with the seminal work of Abramovich et al. [4]. There is an extensive literature devoted to various aspects of disjointness preserving operators. We indicate here only several monographs: Abramovich et al. [2], Abramovich and Kitover [3], Aliprantis and Burkinshow [5], Kusraev and Kutateladze [26], Meyer-

A. G. Kusraev · Z. A. Kusraeva (✉)
Southern Mathematical Institute of Vladikavkaz Scientific Center of Russian Academy of Sciences, 362027 Vladikavkaz, Russia
e-mail: zali13@mail.ru

A. G. Kusraev
e-mail: kusraev@smath.ru

Z. A. Kusraeva
Regional Mathematical Center of Southern Federal University, 344006 Rostov-on-Don, Russia

A. Karapetyants et al. (eds.), *Modern Methods in Operator Theory and Harmonic Analysis*, Springer Proceedings in Mathematics & Statistics 291,
https://doi.org/10.1007/978-3-030-26748-3_13

Nieberg [36]; Ph.D. Thesis [38]; and surveys: Boulabiar [9], Boulabiar et al. [12], Gutman [21], Huijsmans [23].

Disjointness preserving multilinear operators have been studied a little. Bilinear disjointness preserving operators were apparently first considered by Kusraev and Tabuev [27], see also the survey Bu et al. [15]. The structural properties of bilinear lattice homomorphisms were studied in Boulabier and Toumi [14] and then the main results of this work were extended to lattice multimorphisms by Boulabiar [7].

The main purpose of this work is to establish some results on the factorization of order bounded disjointness preserving multilinear operators in vector lattices and to give some new application to disjointness preserving multilinear operators and homogeneous polynomials.

In Sect. 2 we briefly sketch the needed information concerning order bounded disjointness preserving operators including multilinear versions of Meyer's theorem, Fremlin's fundamental construction of the tensor product of vector lattices with the observation that linearization of regular multilinear operators via Fremlin tensor product preserves disjointness, and test for verifying that a multlinear operator is order bounded and disjointness preserving.

An example of a disjointness preserving multilinear operator can be constructed as follows. Given a finite collection of disjointness preserving linear operators $T_1, \dots, T_n$ defined on respective vector lattices $E_1, \dots, E_n$, but taking values from the same f-algebra $(A, \odot)$ and $a \in A$, one can construct a disjointness preserving multilinear operator T from $E_1 \times \cdots \times E_n$ to A by putting

$$T(x_1, \dots, x_n) = a \odot T(x_1) \odot \cdots \odot T(x_n) \ \ (x_1 \in E_1, \dots, x_n \in E_n).$$

The central result of Sect. 4 asserts that each order bounded disjointness preserving multilinear operator from the Cartesian product of vector lattices into an arbitrary vector lattice has a similar structure. Moreover, as shown in Sect. 3, if T is a lattice multimorphism then one can choose $a \in A$ to be the unit element. A special case of symmetric multilinear operators is considered in Sect. 5. In Sect. 6 we deduce from this fact multilinear and polynomial versions of Hart's theorem on the associated lattice multimorphism and the associated polymorphism.

Throughout the sequel $\mathbb{B}(E)$ and $\mathbb{P}(E)$ stand respectively for the complete Boolean algebras of all bands and all band projections in a vector lattice E. A *partition of unity* in a Boolean algebra $\mathbb{B}$ is a family $(b_\xi)_{\xi\in\Xi} \subset \mathbb{B}$ such that $\bigvee_{\xi\in\Xi} b_\xi = \mathbb{1}$ and $b_\xi \wedge b_\eta = 0$ whenever $\xi \neq \eta$.

We use the standard notation and terminology of Aliprantis and Burkinshaw [5] for the theory of vector lattices. In the present paper we assume that all vector spaces are defined over the field of reals and all vector lattices are Archimedean. We let := denote the assignment by definition, while $\mathbb{N}$ and $\mathbb{R}$ symbolize the naturals and the reals.

2 Preliminaries

In this section, we briefly sketch the needed information concerning multilinear operators in vector lattices. In the sequel we fix a natural $n \in \mathbb{N}$ and vector lattices $E_1, \dots, E_n$ and F.

Say that a n-linear operator $T : E_1 \times \cdots \times E_n \to F$ is *positive* and write $T \geq 0$ if $T(x_1, \dots, x_n) \geq 0$ for all $0 \leq x_1 \in E_1, \dots, 0 \leq x_n \in E_n$; *regular* if T can be written as a difference of two positive n-linear operators; *order bounded* if $T(A)$ is order bounded in F for each order bounded set A in $E_1 \times \cdots \times E_n$; *lattice multimorphism* or *lattice n-morphism* if the linear operator $x \mapsto T(x_1, \dots, x_{j-1}, x, x_{j+1}, \dots, x_n)$ is a lattice homomorphism for any choice of $1 \leq j \leq n$ and $0 \leq x_k \in E_k$, $j \neq k \leq n$. Here and in what follows we agree that x_{-1} and x_{n+1} are omitted.

It can be easily seen that T is a multimorphism if and only if $|T(x_1, \dots, x_n)| = T(|x_1|, \dots, |x_n|)$ for all $x_1 \in E_1, \dots, x_n \in E_n$, while T is positive if and only if $|T(x_1, \dots, x_n)| \leq T(|x_1|, \dots, |x_n|)$ for all $x_1 \in E_1, \dots, x_n \in E_n$.

A remarkable tool for studying positive and regular multilinear operators is the *Fremlin tensor product* of vector lattices. In the next two results we collect the most important properties of the Fremlin tensor product, following Fremlin [19] and Schep [39].

Theorem 1 *There exists a unique up to isomorphism vector lattice $E_1 \bar{\otimes} \cdots \bar{\otimes} E_n$ and a lattice n-morphism $\otimes : E_1 \times \cdots \times E_n \to E_1 \bar{\otimes} \cdots \bar{\otimes} E_n$ such that:*

(1) *Whenever F is a vector lattice and $T : E_1 \times \cdots \times E_n \to F$ is a lattice n-morphism, there is a unique lattice homomorphism $T^{\otimes} : E_1 \bar{\otimes} \cdots \bar{\otimes} E_n \to F$ with $T^{\otimes} \circ \otimes = T$.*

(2) *$\otimes$ induces an embedding of the algebraic tensor product $E_1 \otimes \cdots \otimes E_n$ into $E_1 \bar{\otimes} \cdots \bar{\otimes} E_n$.*

(3) *$E_1 \otimes \cdots \otimes E_n$ is dense in $E_1 \bar{\otimes} \cdots \bar{\otimes} E_n$ in the sense that for every $v \in E_1 \bar{\otimes} \cdots \bar{\otimes} E_n$ there exist $0 \leq x_k \in E_k$ $(k = 1, \dots, n)$ such that for every $\varepsilon > 0$ there is an element $u \in E_1 \otimes \cdots \otimes E_n$ with $|v - u| \leq \varepsilon x_1 \otimes \cdots \otimes x_n$.*

(4) *If $0 < v \in E_1 \bar{\otimes} \cdots \bar{\otimes} E_n$, then here exist $0 \leq x_1 \in E_1, \dots, 0 \leq x_n \in E_n$ such that $0 < x_1 \otimes \cdots \otimes x_n \leq v$.*

(5) *If $v \in E_1 \bar{\otimes} \cdots \bar{\otimes} E_n$, then there exist $x_k \in E_k$ $(k = 1, \dots, n)$ such that $|v| \leq x_1 \otimes \cdots \otimes x_n$.*

The universal property of the Fremlin tensor product, stated in Theorem 1(1) can be extended to a wider class of operators, depending on the qualification of the lattice F. Denote by $L^r(E_1, \dots, E_n; F)$ the vector space of all regular n-linear operators from $E_1 \times \cdots \times E_n$ to F ordered by the cone of positive operators, that is $T_1 \geq T_2$ means that $T_1 - T_2$ is positive.

Theorem 2 *If F is a uniformly complete vector lattice, then there is an order preserving one-to-one correspondence between regular n-linear operators $T : E_1 \times \cdots \times E_n \to F$ and regular linear operators $T^{\otimes} : E_1 \bar{\otimes} \cdots \bar{\otimes} E_n \to F$ such that $T = T^{\otimes} \otimes$. If F is Dedekind complete then $L^r(E_1, \dots, E_n; F)$ is a Dedekind* *complete vector lattice and the mapping $T \mapsto T^{\otimes}$ is a a lattice isomorphism of $L^r(E_1, \dots, E_n; F)$ onto $L^r(E_1 \bar{\otimes} \cdots \bar{\otimes} E_n, F)$.*

A n-linear operator $T : E_1 \times \cdots \times E_n \to F$ is said to be *disjointness preserving* if the linear operator $x \mapsto T(x_1, \dots, x_{j-1}, x, x_{j+1}, \dots, x_n)$ is disjointness preserving for any choice of $1 \leq j \leq n$ and $x_k \in E_k$, $j \neq k \leq n$. It can be easily seen that T is disjointness preserving if and only if $|T(x_1, \dots, x_n)| = |T(|x_1|, \dots, |x_n|)|$ for all $x_k \in E_k$ $(k = 1, \dots, n)$.

The classical result by Meyer [37] dealing with the existence of positive part, negative part, and the modulus of order bounded disjointness preserving linear operators has the following multilinear version, which was observed in Kusraev and Tabuev [27, Theorem 3.4].

Theorem 3 *Let $E_1, \dots, E_n$ and F be vector lattices and T be an order bounded disjointness preserving n-linear operator from $E_1 \times \cdots \times E_n$ to F. Then the following assertions hold:*

(1) *There exist lattice n-morphisms T^+ and T^- from $E_1 \times \cdots \times E_n$ to F such that $T = T^+ - T^-$ and for all $0 \leq x_1 \in E_1, \dots, 0 \leq x_n \in E_n$ we have*

$$T^+(x_1, \dots, x_n) = (T(x_1, \dots, x_n))^+,$$

$$T^-(x_1, \dots, x_n) = (T(x_1, \dots, x_n))^-.$$

(2) *T is regular and* $\mathrm{Im}(T^+) \perp \mathrm{Im}(T^-)$.

(3) *The modulus $|T|$ exists, $|T|$ is a lattice n-morphism, $|T| = T^+ + T^-$ and*

$$|T(x_1, \dots, x_n)| = |T(|x_1|, \dots, |x_n|)| = ||T|(x_1, \dots, x_n)| = |T|(|x_1|, \dots, |x_n|) \quad (1)$$

for all $x_1 \in E_1, \dots, x_n \in E_n$

Combining Theorem 3 with Theorem 1(1) yields that factorization via Fremlin tensor product takes disjointness preserving n-linear operators into disjointness preserving linear operators. This fact was observed for bilinear operators (that is, $n = 2$) in Kusraev and Tabuev [27, Proposition 3.6].

Corollary 1 *There is a one-to-one correspondence between order bounded disjointness preserving n-linear operator $T : E_1 \times \cdots \times E_n \to F$ and regular disjointness preserving linear operator $T^{\otimes} : E_1 \bar{\otimes} \cdots \bar{\otimes} E_n \to F$ such that $T = T^{\otimes} \circ \otimes$. Moreover, $T^+ = (T^{\otimes})^+ \circ \otimes$, $T^- = (T^{\otimes})^- \circ \otimes$, and $|T| = |T^{\otimes}| \circ \otimes$.*

Proof Take an order bounded disjointness preserving n-linear operator T from $E_1 \times \cdots \times E_n$ to F. By Theorem 1(1) there exist lattice n-morphisms S_1 and S_2 from $E_1 \bar{\otimes} \cdots \bar{\otimes} E_n$ to F such that $T^+ = S_1 \otimes$ and $T^- = S_2 \otimes$. Observe that $\mathrm{Im}(S_1|_E) \perp \mathrm{Im}(S_2|_E)$ with $E := E_1 \otimes \cdots \otimes E_n$, since $\mathrm{Im}(T^+) \perp \mathrm{Im}(T^-)$ by Theorem 3(1). It follows from Theorem 1(3) that $\mathrm{Im}(S_1) \perp \mathrm{Im}(S_2)$ and hence the regular operator

$S = S_1 - S_2$ is disjointness preserving. It remains to observe that $T = S\otimes$ and $|T| = (S_1 + S_2)\otimes = |S|\otimes$ and take into account that S_1 and S_2 are unique by Theorem 1(1).

The next result is a convenient test for verifying that a multlinear operator preserves disjointness and is order bounded. For linear operators between Banach lattices the equivalence (1) $\Longleftrightarrow$ (2) was proved by Arendt [6, Theorem 2.4]; it was extended to linear operators between arbitrary vector lattices in Abramovich, Arenson, and Kitover [2, Theorem 3.3] and Huijsmans and de Pagter [25, Proposition 1.2].

Theorem 4 *For a n-linear operator T from $E_1 \times \cdots \times E_n$ to F the following assertions are equivalent:*

(1) *T is order bounded and disjointness preserving.*

(2) *If $|y_j| \leq |x_j|$ in E_j for all $j = 1, \dots, n$, then $|T(y_1, \dots, y_n)| \leq |T(x_1, \dots, x_n)|$ in F.*

Proof The implication (1) $\Longrightarrow$ (2) is immediate from Theorem 3(3). It is also evident that (2) implies the order boundedness of T. Putting $y_j := |x_j|$ in (2) we get that $|T(|x_1|, \dots, |x_n|)| \leq |T(x_1, \dots, x_n)| \leq |T(|x_1|, \dots, |x_n|)|$ for all $x_j \in E_j$ $(j = 1, \dots, n)$. It follows that T is disjointness preserving and hence (2) $\Longrightarrow$ (1).

McPollin and Wickstead [35] characterized the order boundedness of disjointness preserving linear operators in terms of uniformly convergent sequences: *A disjointness preserving linear operator $S : E \to F$ between vector lattices is order bounded* (regular) *if and only if $\bigwedge_{k=1}^{\infty} |S(u_k)| = 0$ for every uniformly convergent to zero sequence (u_k) in E.* An elementary proof as well as the history of this result can be found in Abramovich and Kitover [3]. Next we state a multilinear version of McPollin–Wickstead theorem.

Theorem 5 *A disjointness preserving multilinear operator T from $E_1 \times \cdots \times E_n$ to F is regular if and only if, given $1 \leq j \leq n$ and $x_l \in E_l$, $j \neq l \leq n$, whenever a sequence $(x_{j,k})_{k\in\mathbb{N}}$ is (decreasing and) uniformly convergent to zero in E_j then*

$$\bigwedge_{k=1}^{\infty} \big|T(x_1, \dots, x_{j-1}, x_{j,k}, x_{j+1}, \dots, x_n)\big| = 0.$$

Proof Define a disjointness preserving linear operator S_j from E_j to F by putting $S_j(x) := T(x_1, \dots, x_{j-1}, x, x_{j+1}, \dots, x_n)$ for all $x \in E_j$. By hypothesis, for every decreasing sequence (u_k) uniformly convergent to zero in E_j we have $\bigwedge_{k=1}^{\infty} |S_j(u_k)| = 0$. According to McPollin–Wickstead theorem mentioned above, S_j is regular. It follows from Theorem 4 that $|S_j(u)| \leq |S_j(v)|$ whenever $|u| \leq |v|$ in E_j. Now, assuming $|y_1| \leq |x_1|, \dots, |y_n| \leq |x_n|$ and applying the latter repeatedly we get

$$\begin{aligned} |T(y_1, \dots, y_n)| &\leq |T(x_1, y_2, \dots, y_n)| \leq \\ \leq |T(x_1, x_2, y_3, \dots, y_n)| &\leq \cdots \leq |T(x_1, \dots, x_n)|. \end{aligned}$$

3 Factorization of Lattice Multimorphisms

In this section we show that a lattice n-morphism decomposes into a pointwise f-algebra product of n (linear) lattice homomorphism. In other words, the lattice multimorphism is in a sense a distorted f-algebra multiplication. Recall that any vector lattice F has a *universal completion* F^u, so that F can be identified with an order dense vector sublattice of F^u. Let ι_u be the inclusion operator of F into its universal completion F^u. Moreover, for any weak order unit $\mathbb{1} \in F^u$ there exists a unique multiplication $\odot : (x, y) \mapsto x \odot y$ in F^u such that $(F^u, \odot)$ is an f-algebra and $\mathbb{1}$ is a ring unit, see [5]. Given $a \in F^u$, the multiplication operator $\pi_a : x \mapsto a \odot x$ is an orthomorphism and the mapping $a \mapsto \pi_a$ is an f-algebra isomorphism from F^u onto $\mathrm{Orth}(F^u)$. Thus, we will identify the orthomorphism π_a with the multiplier a and $\mathrm{Orth}(F^u)$ with F^u.

Lemma 1 *For an upward directed set $\mathcal{S}$ of lattice homomorphisms from E to F there exists a lattice homomorphism $T : E \to F^u$ such that $\iota_u \circ \mathcal{S} \subset T^{\perp\perp}$.*

Proof There is no loss of generality in assuming $F = \left(\bigcup_{S \in \mathcal{S}} S(E)\right)^{\perp\perp}$. First we observe that two members S_1 and S_2 of $\mathcal{S}$ are disjoint if and only if their ranges $S_1(E)$ and $S_2(E)$ are disjoint. Indeed, because $\mathcal{S}$ is upward directed, S_1 and S_2 have a common majorant $S \in \mathcal{S}$ and by Kutateladze theorem [5, Theorem 2.50] $S_1 = \pi_1 S$ and $S_2 = \pi_2 S$ for some positive orthomorphisms $\pi_1, \pi_2 \in \mathrm{Orth}(F^u)$. Now, if $S_1 \perp S_2$, then $|\pi_1 - \pi_2|S = |S_1 - S_2| = |S_1 + S_2| = |\pi_1 + \pi_2|S$ so that positive orthomorphisms $|\pi_1 - \pi_2|$ and $|\pi_1 + \pi_2|$ coincide on $S(E)$ and hence on $S(E)^{\perp\perp}$ (see [5, Theorem 2.48]). It follows that $\pi_1 \perp \pi_2$ or, equivalently, $\pi_1(F^u) \perp \pi_2(F^u)$, as an orthomorphism in F^u is a multiplication operator, see [5, Theorem 2.62], and we get $S_1(E) \perp S_2(E)$. The converse is trivial.

By our hypothesis $F = \bigvee_{S \in \mathcal{S}} S(E)^{\perp\perp}$ and hence there exists a partition of unity $(B_\alpha)_{\alpha \in \mathrm{A}}$ in $\mathbb{B}(F)$ with the property that for every $\alpha \in \mathrm{A}$ there exists a lattice homomorphism in $\mathcal{S}$, say $S_\alpha \in \mathcal{S}$, such that $B_\alpha \subset S_\alpha(E)^{\perp\perp}$. Let π_α stand for the band projection in F^u onto the band $B_\alpha^{\perp\perp}$, where disjoint complements are taken in F^u. Now, define an operator $T : E \to F^u$ by putting

$$Tx = o\text{-}\sum_{\alpha \in \mathrm{A}} \pi_\alpha S_\alpha(x) \quad (x \in E).$$

The definition is sound, since $(\pi_\alpha)_{\alpha \in \mathrm{A}}$ is a partition of unity in $\mathbb{P}(F^u)$ and the sum of a family of lattice homomorphism with pairwise disjoint range spaces is again a lattice homomorphism. Assume that $\iota_u \circ S \notin T^{\perp\perp}$ for some $S \in \mathcal{S}$. Then there is a lattice homomorphism $S_0 : E \to F$ with $0 < S_0 \leq S$ and $\iota_u \circ S_0 \in T^\perp$. It follows that $\pi_\alpha(\iota_u \circ S_0) \perp \pi_\alpha S_\alpha$ and hence $\pi_\alpha(\iota_u \circ S_0)(E) \perp B_\alpha$ or, equivalently, $\pi_\alpha(\iota_u \circ S_0)(E) = \{0\}$ for all $\alpha \in \mathrm{A}$. Taking into account that $(\pi_\alpha)_{\alpha \in \mathrm{A}}$ is a partition of unity in $\mathbb{P}(F)$ we get $S_0 = 0$, a contradiction.

Lemma 2 *Let E and F be vector lattices with F Dedekind complete, let $T : E \to F$ be a lattice homomorphism, and $F = T(E)^{\perp\perp}$. Then for any $S \in T^{\perp\perp}$ there exists a* *unique* $\rho_S \in \mathrm{Orth}(F)$ *such that* $S = \rho_S \circ T$. *Moreover,* $S \mapsto \rho_S$ *is a lattice isomorphism from* $T^{\perp\perp}$ *onto* $\{\rho \in \mathrm{Orth}(F^u) : \rho(\mathrm{Im}(T)) \subset F\}$.

Proof If $|S| \le cT$ for some $c \in \mathbb{R}_+$ then the claim follows from Kutateladze theorem [5, Theorem 2.50]. For an arbitrary $S \in T^{\perp\perp}$ there exist a partition of unity (π_k) in $\mathbb{P}(F)$ and a sequence of reals (c_k) such that $\pi_k|S| \le c_k T$ for all $k \in \mathbb{N}$. If $\rho_k \in \mathrm{Orth}(F^u)$ satisfy $\pi_k S = \rho_k T$, then $\rho_S := o\text{-}\sum_{k\in\mathbb{N}} \pi_k \rho_k$ is the desired orthomorphism. It remains to observe that if $\rho_1 \circ T = \rho_2 \circ T$ then the orthomorphisms ρ_1 and ρ_2 agree on the set $T(E)$ they also agree on the band $T(E)^{\perp\perp} = F$, see [5, Theorem 2.48].

Lemma 3 *If a mapping* T *from* $(E_1 \times \ldots \times E_n)_+$ *to* F_+ *is additive and positively homogeneous in each variable, then* T *extends uniquely to a positive n-linear operator* $\widetilde{T}$ *from* $E_1 \times \ldots \times E_n$ *to* F. *The unique extension is given by*

$$\widetilde{T}(x_1, \ldots, x_n) := \sum\nolimits_{\nu\in\mathbb{S}} (-1)^{|\nu|} T(x_{1,k_1}, \ldots, x_{n,k_n}),$$

where $x_{j,0} = x_j^+$, $x_{j,1} = x_j^-$, $\mathbb{S} := \{\nu = (k_1, \ldots, k_n) : k_j = 0, 1\}$, *and* $|\nu| := k_1 + \cdots + k_n$. *Moreover, if*

$$T(x_1, \ldots, x_{k-1}, x, x_{k+1}, \ldots, x_n) \wedge T(x_1, \ldots, x_{k-1}, x, x_{k+1}, \ldots, x_n) = 0$$

for all $0 \le x, y \in E_k$ *with* $x \wedge y = 0$ *and all* $0 \le x_j \in E_j$ *with* $k \ne j \le n$, *then* $\widetilde{T}$ *is a lattice multimorphism.*

Proof See, van Gaans [20, Lemma 2.2] for bilinear operators and Loane [34, Theorem 23] for the general case.

Lemma 4 *Let* $E_1, \ldots, E_n$ *and* F *be vector lattices. For any lattice n-morphism* T *from* $E_1 \times \cdots \times E_n$ *to* F *there exist a lattice* $(n-1)$*-morphism* $T_0 : E_1 \times \cdots \times E_{n-1} \to F^u$ *and a lattice homomorphisms* $T_n : E_n \to F^u$ *such that*

$$T(x_1, \ldots, x_n) = T_0(x_1, \ldots, x_{n-1}) \odot T_n(x_n) \quad (x_1 \in E_1, \ldots, x_n \in E_n).$$

Proof Consider the set $\mathcal{S} \subset L^r(E_n, F)$ consisting of all lattice homomorphisms $T(x_1, \ldots, x_{n-1}, \cdot)$ with $0 \le x_j \in E_j$ $(j = 1, \ldots, n-1)$. Clearly, $\mathcal{S}$ is upward directed, since $T(x_1, \ldots, x_{n-1}, \cdot)$ and $T(y_1, \ldots, y_{n-1}, \cdot)$ are majorized by $T(x_1 \vee y_1, \ldots, x_{n-1} \vee y_{n-1}, \cdot)$. By Lemma 1, there exists a lattice homomorphism $T_n : E_n \to F^u$ such that $\iota_u \circ \mathcal{S} \subset T_n^{\perp\perp}$. We can assume without loss of generality that $F = T_n(E_n)^{\perp\perp}$. By Lemma 2, there is a a lattice isomorphism $S \mapsto \rho_S$ from $T_n^{\perp\perp}$ onto F^u. Denote $G := E_1 \times \cdots \times E_{n-1}$ and define $T_0 : G \to F^u$ as follows: given $(x_1, \ldots, x_{n-1}) \in G_+$, put $T_0(x_1, \ldots, x_{n-1}) := \rho_S$, where $S := T(x_1, \ldots, x_{n-1}, \cdot)$. Then

$$T(x_1, \ldots, x_{n-1}, x) = T_0(x_1, \ldots, x_{n-1}) \odot T_n(x), \tag{2}$$

for all $(x_1, \ldots, x_{n-1}) \in G_+$ and $x \in E_n$. It follows that for fixed $0 \le x_1, y_1 \in E_1$, $0 \le u := (x_2, \ldots, x_{n-1}) \in E_2 \times \cdots \times E_{n-1}$, and arbitrary $x \in E_n$ we have

$$T_0(x_1 + y_1, u) \odot T_n(x) = [T_0(x_1, u) + T_0(y_1, u)] \odot T_n(x),$$
$$T_0(x_1 \wedge y_1, u) \odot T_n(x) = [T_0(x_1, u) \wedge T_0(y_1, u)] \odot T_n(x).$$

Taking into account the equation $F = T_n(E_n)^{\perp\perp}$, we conclude that T_0 is additive and preserves infima in the first variable. The same reasoning applies to any other variable. By Lemma 3, T_0 extends to a lattice $(n-1)$-morphism (also denoted by T_0) on the whole lattice G. Clearly, this extension T_0 satisfies (2) and the proof is complete.

Now we are able to state and prove the main result of the section.

Theorem 6 *Let $E_1, \dots, E_n$ and F be vector lattices and let T be a lattice n-morphism from $E_1 \times \dots \times E_n$ to F. Then there exist n lattice homomorphisms $T_k : E_k \to F^u$ $(k := 1, \dots, n)$ such that*

$$T(x_1, \dots, x_n) = T_1(x_1) \odot \dots \odot T_n(x_n) \quad (x_1 \in E_1, \dots, x_n \in E_n). \tag{3}$$

This representation is unique in the sense that if (3) holds with T_k replaced by some lattice homomorphism $S_k : E_k \to F^u$ for all $k := 1, \dots, n$, and π is the band projection onto $\mathrm{Im}(T)^{\perp\perp}$, then there exist $\tau_1, \dots, \tau_n \in \mathrm{Orth}(F^u)$ with

$$\tau_1 \odot \dots \odot \tau_n = \pi\mathbb{1}, \quad T_k = \tau_k \odot S_k \quad (k = 1, \dots, n). \tag{4}$$

Proof The proof is by induction on $n \in \mathbb{N}$. When $n = 1$ the result is obvious, so let us suppose that the claim is true for $n-1$. According to Lemma 4 there exist a lattice $(n-1)$-morphism T_0 from $E_1 \times \dots \times E_{n-1}$ to F^u and a lattice homomorphisms $T_n : E_n \to F^u$ such that

$$T(x_1, \dots, x_n) = T_0(x_1, \dots, x_{n-1}) \odot T_n(x_n) \quad (x_1 \in E_1, \dots, x_n \in E_n). \tag{5}$$

By the induction hypothesis there exist lattice homomorphisms $T_j : E_j \to F^u$ such that $T_0(x_1, \dots, x_{n-1}) = T_1(x_1) \odot \dots \odot T_{n-1}(x_{n-1})$ for all $x_1 \in E_1, \dots, x_{n-1} \in E_{n-1}$. Putting together the latter and (5) yields the desired representation (3).

Assume now that for some finite collection of lattice homomorphisms $S_k : E_k \to F^u$ $(k := 1, \dots, n)$ the representation (3) holds with T_k replaced by S_k. Then $T_1(x_1) \odot b(g) = S_1(x_1) \odot a(g)$ for all $x_1 \in E_1$ and $g \in G := E_2 \times \dots \times E_n$, where

$$g = (x_2, \dots, x_n), \quad a(g) = S(x_2) \odot \dots \odot S(x_n), \quad b(g) = T(x_2) \odot \dots \odot T(x_n).$$

Now, if $\pi := [a(g)] \circ [b(g)] \neq 0$ then $(\pi b(g))^{-1}$ exists in πF^u and hence $\pi T_1(x_1) = (\pi b(g))^{-1} \odot a(g) \odot S_1(x_1)$ for all $x_1 \in E_1$. Since

$$\bigvee_{g \in G} [a(g)] = \bigvee_{g \in G} [b(g)] = (\mathrm{Im}\, T)^{\perp\perp},$$

we can find a partition of unity (π_ξ) in $\mathbb{P}(F)$ such that $\pi_\xi T_1(x_1) = (\pi_\xi b(g_\xi))^{-1} \odot a(g_\xi) \odot S_1(x_1)$. Putting $\tau_1 := o\text{-}\sum_\xi (\pi_\xi b(g_\xi))^{-1} \odot a(g_\xi)$ yields $T_1 = \tau_1 \odot S_1$. The same argument shows the existence of $\tau_k \in \mathrm{Orth}(F^u)$ with $T_k = \tau_k \odot S_k$ $(1 < k \le n)$. If $\tau := \tau_1 \odot \cdots \odot \tau_n$, then $\tau \odot T = T$ and so τ coincides with the identity operator on $\mathrm{Im}(T)$. It follows that $\tau = \pi\mathbb{1}$.

Corollary 2 *For an arbitrary lattice n-morphism T from $E_1 \times \cdots \times E_n$ to F there exists a unital f-algebra $(A, \odot)$ containing* $\mathrm{Im}(T)$ *as an order dense subset such that the representation* (3) *holds with* $\mathrm{Im}(T_k) \subset A$ *for all $k = 1, \ldots, n$ and the smallest unital f-subalgebra of A containing* $\mathrm{Im}(T_k)$ *is A itself. Moreover such f-algebra $(A, \odot)$ is unique up to lattice isomorphism.*

Proof Apply Theorem 6 with $F := \mathrm{Im}(T^\otimes)$, fix an f-algebra multiplication $\odot$ in F^u, and define A as the smallest f-subalgebra of F^u containing $\bigcup_{k=1}^n \mathrm{Im}(T_k)$. Then A is a desired f-algebra, since $\mathrm{Im}(T)$ as an order dense subset of F by Theorem 1(4). If $(A_1, \circledcirc)$ is another unital f-algebra with the same properties then A_1 is an order dense f-subalgebra of A_1^u [5, Theorem 2.64]. Both vector lattices A^u and A_1^u are universal completion of F^u and hence they are lattice isomorphic. It follows that A and A_1 are lattice isomorphic.

Corollary 3 *Let F and G be vector lattices with fixed f-algebra multiplications $\odot$ and $\circledcirc$ in F^u and G^u, respectively. Then for any vector lattices $E_1, \ldots, E_n$ and lattice homomorphisms $S : F \to G$ and $T_k : E_k \to F^u$ $(k = 1, \ldots, n)$ with* $\mathrm{Im}(T_1) \odot \cdots \odot \mathrm{Im}(T_n) \subset F$ *there exist lattice homomorphisms $S_k : E_k \to G^u$ $(k = 1, \ldots, n)$ such that*

$$S(T_1(x_1) \odot \cdots \odot T_n(x_n)) = S_1(x_1) \circledcirc \cdots \circledcirc S_n(x_n) \quad (x_1 \in E_1, \ldots, x_n \in E_n).$$

Proof Apply Theorem 6 to n-morphism T from $E_1 \times \cdots \times E_n$ to G defined as $T(x_1, \ldots, x_n) = S(T_1(x_1) \odot \cdots \odot T_n(x_n))$ for all $x_1 \in E_1, \ldots, x_n \in E_n$.

The bilinear version of the representation (3) was proved by Kusraev and Tabuev in [28, Theorem 3.2] using Boolean valued analysis approach. This means that, first, the statement of the theorem is established for the scalar case ($F = \mathbb{R}$), and then the general result is derived by interpreting the scalar result in an appropriate Boolean valued model (see also Kusraev and Kutateladze [26, Theorem 3.12.A.3]). If $E_1, \ldots, E_n$ are f-algebras with unit elements, F is an algebra, and a lattice n-morphism T is multiplicative, then the representation (3) is trivial, see Boulabiar [7, Proposition 1.5].

4 Factorization of Order Bounded Disjointness Preserving Operators

In this section we show that Theorem 6 can be extended to order bounded disjoint preserving multilinear operators. To do this we need a multilinear version of polar decomposition theorem for linear operators by Boulabiar and Buskes [11,

Theorem 7]). But first we give the following general result extending another result by Boulabiar and Buskes [11, Theorem 6]). Denote by $\langle\!\langle T\rangle\!\rangle$ the set of all $S \in L^r(E_1, \dots, E_n; F)$ with $S^{\otimes}(u) \in T^{\otimes}(u)^{\perp\perp}$ for all $u \in E_1 \bar{\otimes} \cdots \bar{\otimes} E_n$.

Theorem 7 *Let $E_1, \dots, E_n$ and F be vector lattices with F uniformly complete, T an order bounded disjointness preserving n-linear operator from $E_1 \times \cdots \times E_n$ to F, and the order ideal in F generated by* $\operatorname{Im}(T)$ *is F itself. Then for every $S \in \langle\!\langle T\rangle\!\rangle$ there exists a unique $U := U_S \in$* $\operatorname{Orth}(F)$ *with $S = U|T|$. Moreover, the mapping $S \mapsto U_S$ is a lattice isomorphism of $\langle\!\langle T\rangle\!\rangle$ onto* $\operatorname{Orth}(F)$.

Proof The regular linear operator $T^{\otimes}$ from $E_1 \bar{\otimes} \cdots \bar{\otimes} E_n$ to F in Theorem 2 is disjointness preserving, since $|T|$ exists and is a lattice n-morphism by Theorem 3, $|T^{\otimes}| = |T|^{\otimes}$ by Theorem 2, and $|T|^{\otimes}$ is a lattice homomorphism by Theorem 1(1). Note that the order dense ideal in F generated by $\operatorname{Im}(T^{\otimes})$ is again F by Theorem 1(5). The polar decomposition theorem for order bounded disjointness preserving linear operators due to Boulabiar and Buskes [11, Theorem 6] tells us that for every $\bar{S} \in L^r(E_1 \bar{\otimes} \cdots \bar{\otimes} E_n, F)$ with $\bar{S}(u) \in \{T^{\otimes}(u)\}^{\perp\perp}$ for all $u \in E_1 \bar{\otimes} \cdots \bar{\otimes} E_n$ there exists a unique $U \in \operatorname{Orth}(F)$ for which $\bar{S} = U|T|^{\otimes}$. By Theorem 2 $\bar{S} = S^{\otimes}$ for some $S \in L^r(E_1, \dots, E_n; F)$, so that $S^{\otimes} = U|T|^{\otimes}$. Applying again Theorem 2 we get $S = S^{\otimes}\otimes = U|T|^{\otimes}\otimes = U|T|$. The mapping $S \mapsto U_S$ is obviously a linear bijection; moreover, it is modulus preserving as $|S| = |U_S \circ |T|\,| = |U_S||T|$.

Combining Theorems 6 and 7 yields the following.

Corollary 4 *Let $E_1, \dots, E_n$ and F be vector lattices with F uniformly complete, T an order bounded disjointness preserving n-linear operator from $E_1 \times \cdots \times E_s$ to F, and G the order ideal in F generated by* $\operatorname{Im}(T)$. *Then there exist n lattice homomorphisms $T_k : E_k \to F^u$ $(k := 1, \dots, n)$ such that every $S \in \langle\!\langle T\rangle\!\rangle$ admits the representation*

$$S(x_1, \dots, x_n) = U(T_1(x_1) \odot \cdots \odot T_n(x_n)) \quad (x_1 \in E_1, \dots, x_n \in E_n) \qquad (6)$$

with a unique $U \in$ $\operatorname{Orth}(G)$. *The mapping $S \mapsto U_S$ is a lattice isomorphism of $\langle\!\langle T\rangle\!\rangle$ onto* $\operatorname{Orth}(G)$.

Applying Corollary 4 to T itself yields the following *polar decomposition theorem* (cf. Boulabiar and Buskes [11, Theorem 7]).

Corollary 5 *Let $E_1, \dots, E_n$ and F be vector lattices with F uniformly complete, T an order bounded disjointness preserving n-linear operator from $E_1 \times \cdots \times E_n$ to F, and G the order ideal in F generated by* $\operatorname{Im}(T)$. *Then the following hold:*

(1) *There exists a unique bijective $U \in$* $\operatorname{Orth}(G)$ *for which $|U| = I_G$, $T = U|T|$, and $|T| = UT$.*

(2) *There exist an invertible $U \in$* $\operatorname{Orth}(G)$, *and n lattice homomorphisms $T_k : E_k \to F^u$ $(k := 1, \dots, n)$ such that $|U| = I_G$ and*

$$T(x_1, \dots, x_n) = U(T_1(x_1) \odot \ldots \odot T_n(x_n)) \textit{ for all } x_1 \in E_1, \dots, x_n \in E_n.$$

As an immediate application of Corollary 4, we obtain the following generalization of Kutateladze's theorem (see [5, Theorem 2.50]). Denote by $I(M)$ the order ideal generated by a set M.

Corollary 6 *Suppose F is uniformly complete, $T : E_1 \times \cdots \times E_n \to F$ is a lattice n-morphism, and S is a n-linear operator from $E_1 \times \cdots \times E_n$ to F such that $0 \le S \le T$. Then there exists a unique $w \in \mathrm{Orth}(I(\mathrm{Im}(T))$ for which $S = wT$.*

Proof The claim is easily verified by means of Fremlin's tensor product enabling one to transfer results from linear operators to multilinear operators. As for linear operators, we use the following version of Kutateladze theorem due to Boulabiar and Buskes [11, Corollary 3]: *Given two linear operators U and V from a vector lattice E to a uniformly complete vector lattice F such that $0 \le U \le V$ and V is a lattice homomorphism, there exists a unique orthomorphism $w \in \mathrm{Orth}(I(T(E)))$ for which $0 \le w \le I$ and $U = wV$.*

The next two results deal with the factorization of disjointness preserving multilinear operators between f-algebras with factor operators being algebra homomorphisms and are generalizations of the corresponding linear results by Abid et al. [1, Theorem 4.2 and Corollary 4.3].

Consider some algebras $A_1, \ldots, A_n, A$. A n-linear operator T from $A_1 \times \cdots \times A_n$ to A is *an algebra n-morphism* if $T(x_1y_1, \ldots, x_ny_n) = T(x_1, \ldots, x_n)T(y_1, \ldots, y_n)$ for all $x_j, y_j \in A_j$ $(j = 1, \ldots, n)$. In case $n = 1$ we will speak of *an algebra homomorphism*.

Say that an element w of an f-algebra B is *von Neumann regular* whenever the equality $w^2v = w$ holds for some $v \in B$, see Boulabiar and Hager [13]. An f-algebra B is said to be *von Neumann regular* if all elements in B are von Neumann regular, see [32].

Theorem 8 *Let $A_1, \ldots, A_n$ be f-algebras with unit elements $e_1, \ldots, e_n$, respectively, $(B, \odot)$ be a semiprime f-algebra, and $T : A_1 \times \cdots \times A_n \to B$ be an order bounded n-linear operator with $T(e_1, \ldots, e_n)$ von Neumann regular. Then the following conditions are equivalent:*

(1) *T is disjointness preserving.*

(2) *There is an algebra n-morphism $S : A_1 \times \cdots \times A_n \to B$ such that*

$$T = T(e_1, \ldots, e_n)S.$$

(3) *There exist algebra homomorphisms $S_k : A_k \to B$ $(k = 1, \ldots, n)$ with*

$$T(x_1, \ldots, x_n) = T(e_1, \ldots, e_n) \odot S_1(x_1) \odot \cdots \odot S_n(x_n)$$
$$(x_1 \in A_1, \ldots, x_n \in A_n).$$

Proof The implication (3) $\Longrightarrow$ (2) is trivial, while (2) $\Longrightarrow$ (1) is immediate from the fact that every algebra n-morphism defined on a Cartesian product of semiprime f-algebras $A_1, \ldots, A_n$ is disjointness preserving, since $x \perp y$ is equivalent to $xy = 0$

for all $x, y \in A_k$ and $k \leq n$. To verify the implication (1) $\Longrightarrow$ (3), we first use the same trick as in the proof of Theorem 4.2 in [1], and then apply Theorem 6.

By hypothesis we can pick $v \in B$ such that $T(e_1, \dots, e_n)^2 v = T(e_1, \dots, e_n)$. Define a n-linear operator S from $A := A_1 \times \cdots \times A_n$ to B by $S = vT$, where the juxtaposition vT stands for the operator $u \mapsto v \odot Tu$ $(u \in A)$. Clearly, S is disjointness preserving. Moreover, denoting $\mathbf{e} := (e_1, \dots, e_n)$ we have

$$(S\mathbf{e})^2 = (vT\mathbf{e})^2 = v(T\mathbf{e})^2 v = vT(\mathbf{e}) = S\mathbf{e}.$$

In particular, $S\mathbf{e}$ is positive and, using Meyer's Theorem 3, we conclude that S^- is a lattice n-morphism and $0 = (S\mathbf{e})^- = S^-\mathbf{e}$. From this it is easily seen that $S^-(x_1 \wedge ke_1, \dots, x_n \wedge ke_n) \leq k^n S^-\mathbf{e} = 0$ for all positive $(x_1, \dots, x_n) \in A$, and $k \in \mathbb{N}$. By de Pagter theorem (see [5, Theorem 2.57]) the sequence $(x_j \wedge ke_j)_{k\in\mathbb{N}}$ converges uniformly to x_j for all $j \leq n$ and hence $S^-(x_1, \dots, x_n) = 0$. Consequently, S is a lattice n-morphism. The same kind of reasoning shows that $\mathrm{Im}(S) \subset pB$ with the idempotent $p := S\mathbf{e}$. Observe also that pB itself is an f-algebra with the unit element p, see [1, Lemma 2.1]. Thus, B can be considered as an f-subalgebra of the unital f-algebra $(B^u, \odot)$ with p being a component of the ring unit in B^u, see [5, Theorem 2.58].

Now, applying Theorem 6, we can choose n lattice homomorphisms $T_1, \dots, T_n$ from A_k to B^u such that

$$S(x_1, \dots, x_n) = T_1(x_1) \odot \cdots \odot T_n(x_n) \quad (x_1 \in A_1, \dots, x_n \in A_n). \tag{7}$$

The factors T_k in this representation need to be slightly corrected to get the desired algebra homomorphisms. To this end, note that $p = \bigodot_{j=1}^n T_j(e_j)$ and put

$$p_k := \bigodot_{j \neq k} T_j(e_j), \quad S_k = p_k T_k \quad (k = 1, \dots, n). \tag{8}$$

and take into account that $p = p^{n-1} = p_1 \odot \cdots \odot p_n$. Then according to formula (7) we have $S_k(x_k) = S(e_1, \dots, e_{k-1}, x_k, e_{k+1}, \dots, e_n) \in B$. Observe also that $S_k(e_k)$ is an idempotent, since $S_k(e_k) = S\mathbf{e} = p$. Consequently, $S_k : A_k \to B$ is multiplicative by the result of Huijsmans and de Pagter [24, Theorem 5.4], stating that a positive operator from A_k to B is an algebra homomorphism if and only if it is a lattice homomorphism sending the unit element to an idempotent. Moreover, two easy relations

$$pS(x_1, \dots, x_n) = p(p_1 T_1(x_1) \odot \cdots \odot p_n T_n(x_n)) = p(S_1(x_1) \odot \cdots \odot S_n(x_n)),$$
$$|S(x_1, \dots, x_n) - S_1(x_1) \odot \cdots \odot S_n(x_n)| \leq 2p \quad (|x_k| \leq e_k,\ k = 1, \dots, n)$$

imply that $S(x_1, \dots, x_n) = S_1(x_1) \odot \cdots \odot S_n(x_n)$ whenever $|x_k| \leq e_k$ for all $k = 1, \dots, n$. However, the latter holds for all $x_k \in A_k$, since the order ideal in A_k generated by e_k is uniformly dense in A_k and a positive multilinear operator is uniformly continuous. The same reasoning shows that an order bounded disjointness preserving n-linear operator from A to B vanishing at $\mathbf{e} \in A$ is the zero operator, cf. [1, Lemma 2.2]. Thus, it remains to prove that $T - T\mathbf{e}S$ is disjointness preserving and $(T - T\mathbf{e}S)\mathbf{e} = 0$. The latter is obvious:

$$(T - T\mathbf{e}S)\mathbf{e} = T\mathbf{e} - T\mathbf{e}S\mathbf{e} = T\mathbf{e} - T\mathbf{e}vT\mathbf{e} = T\mathbf{e} - (T\mathbf{e})^2 v = T\mathbf{e} - T\mathbf{e} = 0.$$

To ensure that $T - T\mathbf{e}S$ is disjointness preserving, take $\mathbf{x} = (x_1, \dots, x_n)$ and $\mathbf{y} = (y_1, \dots, y_n)$ in A with $x_j \perp y_j$ for some $1 \le j \le n$ and note that $T\mathbf{x} \perp T\mathbf{y}$ or equivalently $T\mathbf{x} \odot T\mathbf{y} = 0$ as T is disjointness preserving. Then $(T - T\mathbf{e}S)\mathbf{x}$ and $(T - T\mathbf{e}S)\mathbf{y}$ are also disjoint, since

$$\begin{aligned}(T - T\mathbf{e}S)\mathbf{x} \odot (T - T\mathbf{e}S)\mathbf{y} =& (T - uT\mathbf{e}T)\mathbf{x} \odot (T - uT\mathbf{e}T)\mathbf{y} \\ =& (T\mathbf{x} - uT\mathbf{e}T\mathbf{x}) \odot (T\mathbf{y} - uT\mathbf{e}T\mathbf{y}) \\ =& T\mathbf{x} \odot T\mathbf{y} - 2uT\mathbf{e}T\mathbf{x} \odot T\mathbf{y} + (uTe)^2 T\mathbf{x} \odot T\mathbf{y} \\ =& 0,\end{aligned}$$

and this completes the proof.

Given an Archimedean semiprime f-algebra B, denote by $Q(B)$ the *complete* (or *maximal*) *ring of quotients* of B (for the definition and properties of $Q(B)$ see Lambek [32]). Then $Q(B)$ is also an Archimedean semiprime f-algebra, while B is an f-subalgebra of $Q(B)$ (see, Wickstead [42]). The linear version $n = 1$ of the following result is due to Boulabiar [8, Theorem 3.4].

Corollary 7 *An order bounded n-linear operator T from $A_1 \times \cdots \times A_n$ to B is disjointness preserving if and only if there exists n algebra homomorphisms $S_k : A_k \to Q(B)$ $(k = 1, \dots, n)$ such that $T = T(e_1, \dots, e_n)(S_1 \odot \cdots \odot S_n)$.*

Proof The complete ring of quotients $Q(B)$ always is von Neumann regular (see Lambek [32, Sect. 2.4, Proposition 1]).

5 Factorization of Symmetric Lattice Multimorphisms

Next we consider the special case of symmetric multimorphisms and for this purpose we will use the concept of the power of a vector lattice introduced in Boulabier and Buskes [10, Definition 3.1] (for the case $n = 2$ see Buskes and van Rooij [18]).

Let $2 \le n \in \mathbb{N}$ and E be a vector lattice. A pair $(E^{n\circledcirc}, \circledcirc_n)$ is called a *n-power* of E if the following conditions are fulfilled:

(1) $E^{n\circledcirc}$ is a vector lattice and $\circledcirc_n : E^n \to E^{n\circledcirc}$ is a symmetric lattice n-morphism (called the *canonical n-morphism* of E);

(2) for every vector lattice F and every symmetric lattice n-morphism Φ from E^n to F there is a unique lattice homomorphism $S : E^{n\circledcirc} \to F$ such that $\Phi = S \circ \circledcirc_n$.

We put also $E^{1\circledcirc} = E$ and $\circledcirc_1 = I_E$ for convenience.

The following result was obtained by Boulabier and Buskes [10, Theorem 3.2].

Theorem 9 *Let $n \in \mathbb{N}$ and E be a vector lattice. Then E has a unique (up to lattice isomorphism) n-power $(E^{n\circledcirc}, \circledcirc_n)$.*

Proof The idea of the proof is as follows. Let $\bar{\bigotimes}_n E$ be the n-fold Fremlin tensor product of E and I the uniformly closed order ideal in $\bar{\bigotimes}_n E$ generated by the set of all elements $x_1 \otimes \cdots \otimes x_n$ with $x_1, \dots, x_n \in E_+$ and $x_k \wedge x_l = 0$ for some $1 \leq k, l \leq 1$. Then the quotient vector lattice $E^{n\circledcirc} := \bar{\bigotimes}_n E/I$ with the quotient mapping $\circledcirc_n : \bar{\bigotimes}_n E/I \to E^{n\circledcirc}$ is a n-power of E.

The fixed f-algebra multiplication in F^u we denote by $\odot$ and instead of $\circledcirc_n$ we write $\circledcirc$.

Theorem 10 *If $T : E^n \to F$ is a symmetric lattice n-morphism, then there exists a lattice homomorphism $S : E \to F^u$ such that*

$$T(x_1, \dots, x_n) = S(x_1) \odot \cdots \odot S(x_n) \quad (x_1, \dots, x_n \in E).$$

Proof Arguing by induction, note that there is nothing to prove if $n = 1$ and assume that the required representation is true for $n-1$, $n \geq 2$. According to Lemma 4 we have $T(x_1, \dots, x_n) = T_0(x_1, \dots, x_{n-1}) \odot T_n(x_n)$ for all $x_1 \in E_1, \dots, x_n \in E_n$. Clearly, T_0 is a symmetric lattice $(n-1)$-morphism from E^{n-1} to F^u and by induction hypothesis we arrive at the representation $T(x_1, \dots, x_n) = S(x_1) \odot \cdots \odot S(x_{n-1}) \odot T_n(x_n)$. Taking into account the symmetry of T we conclude that $S(x)^{n-1} \odot T_n(y) = S(x)^{n-2} \odot S(y) \odot T_n(x)$ for all $x, y \in E$ (with the convention that $S(x)^0$ is the unit element of F^u). We can assume without loss of generality that $S(E)^{\perp\perp} = F^u$. Therefore, there exists a partition of unity $(\pi_x)_{x \in E_+}$ in $\mathbb{P}(F^u)$ such that $\pi_x \leq [S(x)]$ for all $x \in E_+$, where $[S(x)]$ denotes the band projection in F^u onto the band $S(x)^{\perp\perp}$. Define $\sigma, \tau \in F^u$ by

$$\sigma := \sum_{x \in E_+} \pi_x S(x)^{n-1}, \quad \tau := \sum_{x \in E_+} \pi_x S(x)^{n-2} T_n(x).$$

and observe that σ is a weak order unit and so invertible. Putting $\rho := \sigma^{-1} \odot \tau$ we get $T_n = \rho S$ and hence $T(x_1, \dots, x_n) = \rho \odot S(x_1) \odot \cdots \odot S(x_n)$. Replacing S by $\rho^{1/n} S$ in the last equation we arrive at the required representation.

Theorem 11 *Let E and F be vector lattices. For any symmetric lattice n-morphism $T : E^n \to F$ there exist a vector lattice G and a lattice homomorphism $S : E \to G$ such that $G^{n\circledcirc}$ is a sublattice of F and the representation holds*

$$T(x_1, \dots, x_n) = S(x_1) \circledcirc \cdots \circledcirc S(x_n) \quad (x_1, \dots, x_n \in E). \tag{9}$$

Moreover, for every symmetric order bounded disjointness preserving n-linear operator $S : E^n \to F$ the modulus $T := |S|$ is representable as in (9) and if, in addition, F is uniformly complete, then there exists an invertible $U \in \mathrm{Orth}(I(G^{n\circledcirc}))$ such that

$$S(x_1, \dots, x_n) = U(S(x_1) \circledcirc \cdots \circledcirc S(x_n)) \quad (x_1, \dots, x_n \in E). \tag{10}$$

Proof Let $S : E \to F^u$ be a lattice homomorphism in Theorem 10 and denote $G := S(E)$. Then G is a vector sublattice of $(F^u, \odot)$, while the mapping Φ from G^n to F defined as $\Phi : (u_1, \dots, u_n) \mapsto u_1 \odot \cdots \odot u_n$ is a symmetric lattice n-morphism. In view of Theorem 9 there exists a unique lattice homomorphism $h : G^{n\circledcirc} \to F$ such that $\Phi = h \circ \circledcirc_n$. If $h(u) = 0$ and $0 < |u| \in G^{n\circledcirc}$, then there is $v \in G_+$ with $0 < v \circledcirc \cdots \circledcirc v \le u$ (see [16, Theorem 2.1 (4)]). It follows that $v \odot \cdots \odot v = h(v \circledcirc \cdots \circledcirc v) \le h(|u|) = |h(u)| = 0$ and hence $v = 0$, a contradiction. Thus, h is injective and, since the power of a vector lattice is uniquely defined, we can identify $G^{n\circledcirc}$ with a vector sublattice of F. If F is uniformly complete then the same line of reasoning together with Corollary 4 establishes the second part of the desired result.

Consider a uniformly complete vector lattice and a positive real number p. Using the homogeneous functional calculus, we can introduce new vector operations on E by putting $x \oplus y = (x^p + y^p)^{1/p}$ and $\lambda \circledast x = \lambda^{1/p} x$, where $x, y \in E$ and $\lambda \in \mathbb{R}$. Recall that, given $\lambda, p \in \mathbb{R}$ with $p > 0$, by λ^p we mean $|\lambda|^p \mathrm{sgn}(\lambda)$. Accordingly, $x \oplus y := \phi_p(x, y)$, where $\phi_p : \mathbb{R}^2 \to \mathbb{R}$ is defined as $\phi_p(r, t) = |u|^p \mathrm{sgn}(u)$ with $u := |r|^p \mathrm{sgn}(r) + |t|^p \mathrm{sgn}(t)$ $(r, t \in \mathbb{R})$, see [33]. Endowed with these new operations, the original order and lattice structures, E becomes a vector lattice.

This new vector lattice is called the *p-concavification* of E and is denoted by $E_{(p)}$. Denote by ι_p the identity mapping of $(E, \le)$ considered as an operator from E onto $E_{(p)}$. Clearly, ι_p is order isomorphism of E onto $E_{(p)}$, since the vector lattices E and $E_{(p)}$ have the same underlying ordered set $(E, \le)$. If $n \in \mathbb{N}$ then $E_{(n)} = E^{n\circledcirc}$ and $\iota_n(x) = x^{n\circledcirc}$ for all $x \in E_+$, see Boulabier and Buskes [10].

Lemma 5 *If F is a uniformly complete vector lattice then $F = (F_{(1/n)})^{n\circledcirc}$, where the identity is meant in the sense of lattice isomorphism.*

Proof Applying the Proposition 4.8(ii) in Szulga [40] to F, we obtain $(F_{(1/n)})^{n\circledcirc} = (F_{(1/n)})_{(n)} = F$.

Below $\circledcirc$ denotes the canonical n-morphism both of E and $F_{(1/n)}$ and $\mathrm{Hom}(E, F)$ stands for the set of all lattice homomorphisms from E to F.

Corollary 8 *If F is uniformly complete then there is a one-to-one correspondence between $\mathrm{Hom}(E_{(n)}, F)$ and $\mathrm{Hom}(E, F_{(1/n)})$. Moreover, the lattice homomorphisms $R \in \mathrm{Hom}(E_{(n)}, F)$ and $S \in \mathrm{Hom}(E, F_{(1/n)})$ correspond to each other if and only if*

$$R(x_1 \circledcirc \cdots \circledcirc x_n) = S(x_1) \circledcirc \cdots \circledcirc S(x_n) \quad (x_1, \dots, x_n \in E). \tag{11}$$

Proof Given $S \in \mathrm{Hom}(E, F_{(1/n)})$, define a symmetric lattice n-morphism from E^n to $(F_{(1/n)})_{(n)}$ by $T : (x_1, \dots, x_n) \mapsto S(x_1) \circledcirc \cdots \circledcirc S(x_n)$. By Theorem 9 there is a unique lattice homomorphism $R \in \mathrm{Hom}\left(E_{(n)}, (F_{(1/n)})_{(n)}\right)$ such that $T(x_1, \dots, x_n) = R(x_1 \circledcirc \cdots \circledcirc x_n)$. By Lemma 5, the vector lattice $(F_{(1/n)})_{(n)}$ is identified with F, so that $R \in \mathrm{Hom}(E_{(n)}, F)$. It remains to ensure that the mapping $h : S \mapsto R$ is bijective.

Take an arbitrary lattice homomorphism $R \in \mathrm{Hom}(E_{(n)}, F)$ and define a symmetric lattice n-morphism T from E^n to F as $T : (x_1, \dots, x_n) \mapsto R(x_1 \circledcirc \cdots \circledcirc x_n)$. By Theorem 11, there are a vector G and a lattice homomorphism $S : E \to G$ such that $G^{n\circledcirc}$ is a sublattice of F and $T(x_1, \dots, x_n) = S(x_1) \circledcirc \cdots \circledcirc S(x_n)$ for all $x_k \in E_k$ $(k = 1, \dots, n)$. Again, by Lemma 5, G is identified with a sublattice of $F_{(1/n)}$ and S can be considered as an operator from E to $F_{(1/n)}$. Thus, h is surjective. If $S_1, S_2 \in \mathrm{Hom}(E, F_{(1/n)})$ and $h(S_1) = h(S_2)$, then $S_1(x)^{n\circledcirc} = S_2(x)^{n\circledcirc}$ for all $x \in E$. For a positive $x \in E$ we have $\iota_n(S_1(x)) = S_1(x)^{n\circledcirc} = \iota_n(S_2(x)) = S_2(x)^{n\circledcirc}$ and hence $S_1(x) = S_2(x)$ as ι_n is a bijection. It follows that $S_1 = S_2$ and this completes the proof.

In conclusion of the section we indicate a corollary on the structure of homogeneous disjointness preserving polynomials introduced in Kusraeva [31], see also [29, 30]. Recall that a mapping $P : E \to F$ is called a *homogeneous polynomial of degree* n (or n-*homogeneous polynomial*) if there exists a n-linear operator $T : E^n \to F$ such that $P = T \circ \Delta_n$, where $\Delta_n : E \ni x \mapsto (x, \dots, x) \in E^n$ is the *diagonal mapping*.

There exists a unique symmetric n-linear operator φ with $P = \varphi \circ \Delta_n$ which is denoted by $\check{P}$, so that $P(x) = \check{P}(x, \dots, x)$ for all $x \in E$. Clearly, P is order bounded if and only if so is $\check{P}$.

Say that a n-homogeneous polynomial P is *disjointness preserving* if the associated n-linear operator $\check{P}$ is disjointness preserving; a lattice n-*polymorphism* (or simply lattice *polymorphism*) if $\check{P}$ is a lattice multimorphism; *order bounded*, if P sends order bounded sets in E into order bounded sets in F.

Corollary 9 *Let E and F be vector lattices and $P : E \to F$ a n-homogeneous polynomial. Then there exist a vector lattice G and a lattice homomorphism $S : E \to G$ such that $G^{s\circledcirc}$ is a vector sublattice of F and $Px = (Sx)^{s\circledcirc}$ for all $x \in E$. If, in addition, F is uniformly complete, then for every order bounded disjointness preserving n-homogeneous polynomial $Q : E \to F$ there exists an invertible $U \in \mathrm{Orth}(I(G^{s\circledcirc}))$ such that $Q(x) = U((Sx)^{s\circledcirc})$ for all $x \in E$.*

6 The Associate f-Algebra Multimorphism

Now we consider a multilinear version of Hart's result according to which an f-algebra homomorphism can naturally be associated with each disjointness preserving linear operator, see Hart [22, Theorem 2.1]. In the special case of Banach lattices such an associated lattice homomorphism was previously considered by Wicksted in [41, Theorem 3.1]. We start with the following two lemmas.

Lemma 6 *Let D be a linear subspace of a vector lattice G, and let F be the vector sublattice generated by D. Suppose $\tau : D \to D$ is a linear mapping such that $x^+ \perp y$ $(x \in D,\ y \in G)$ implies $(\tau x)^+ \perp y$. Then τ has a unique extension to a positive orthomorphism $\hat{\tau} : F \to F$.*

Proof See, Hart [22, Lemma 1.5].

Recall that a semiprime f-algebra A can naturally be embedded (by f-algebra isomorphism $a \mapsto \pi_a$ $(a \in A)$ with $\pi_a \in \mathrm{Orth}(A)$ defined as $\pi_a : x \mapsto a \odot x$) into the unital f-algebra $\mathrm{Orth}(A)$ of all orthomorphisms of A as an f-subalgebra and a ring ideal. Moreover, if A has a multiplicative unit then A and $\mathrm{Orth}(A)$ are f-isomorphic, see Aliprantis and Burkishaw [5, Theorem 2.62]. This identification is taken into consideration in the next lemma and the proof of Theorem 12 below.

Lemma 7 *If F is a vector sublattice of a universally complete f-algebra $(A, \odot)$ and $A = F^{\perp\perp}$ then*

$$\mathrm{Orth}(F) = \{a \in A : \ a \odot F \subset F\}.$$

Proof Denote by $I(F)$ the order ideal in A generated by F. Then F is a majorizing sublattice of a uniformly complete vector lattice $I(F)$. By Theorem 4.1 in Buskes and van Rooij [17] every orthomorphism $\pi \in \mathrm{Orth}(F)$ admits a unique extension to $I(F)$. It follows that $\mathrm{Orth}(F) \subset \mathrm{Orth}\ (I(F))$ and the claim follows from Aliprantis and Burkinshaw [5, Theorem 2.63], since $I(F)^u = A$.

Theorem 12 *Let $E_1, \dots, E_n$ and F be vector lattices, T an order bounded disjointness preserving n-linear operator from $E_1 \times \dots \times E_n$ to F, and F_0 a vector sublattice of F generated by* $\mathrm{Im}(T)$. *Then there exists a unique lattice n-morphism $\widetilde{T}$ from* $\mathrm{Orth}(E_1) \times \dots \times \mathrm{Orth}(E_n)$ *to* $\mathrm{Orth}(F_0)$ *such that*

$$\begin{gathered} T(\pi_1 x_1, \dots, \pi_n x_n) = \widetilde{T}(\pi_1, \dots, \pi_n)(T(x_1 \dots, x_n)), \\ \widetilde{T}(\pi_1 \rho_1, \dots, \pi_n \rho_n) = \widetilde{T}(\pi_1, \dots, \pi_n)\widetilde{T}(\rho_1, \dots, \rho_n) \\ (x_j \in E_j;\ \pi_j, \rho_j \in \mathrm{Orth}(E_j);\ j = 1, \dots, n). \end{gathered}$$

Moreover, the restriction of $\widetilde{T}(\pi_1, \dots, \pi_n)$ to the vector sublattice generated by $\mathrm{Im}(|T|)$ *coincides with $|\widetilde{T}|(\pi_1, \dots, \pi_n)$ for all $\pi_j \in$* $\mathrm{Orth}(E_j)$ $(j = 1, \dots, n)$.

Proof Denote $A := F^u$ and let G and $\odot$ stand for the ideal in A generated by F and a fixed f-algebra multiplication in A. Then G is uniformly complete and by Corollary 5(2) there exist n lattice homomorphisms $T_k : E_k \to A$ $(k := 1, \dots, n)$ and an invertible $U \in \mathrm{Orth}(G)$, and such that $|U| = I_G$ and the representation holds:

$$T(x_1, \dots, x_n) = U(T_1(x_1) \odot \dots \odot T_n(x_n)) \quad (x_1 \in E_1, \dots, x_n \in E_n).$$

Clearly, U may be identified with the multiplicative operator $g \mapsto u \odot g$ $(g \in G)$ for some $u \in A$. Denote $F_j := T_j(E_j)$ and note that F_j is a sublattice of A and F_0 coincides with the sublattice in A generated by $\mathrm{Im}(T) = U(F_1 \odot \dots \odot F_n)$. By Hart's theorem [22, Theorem 2.1] there exists a unique lattice homomorphism $\widetilde{T}_j$ from $\mathrm{Orth}(E_j)$ to $\mathrm{Orth}(F_j)$ such that $\widetilde{T}_j(\pi)T_j(x) = T_j(\pi x)$ for all $x \in E_j$ and $\pi \in \mathrm{Orth}(E_j)$. Lemma 7 enables us to identify the orthomorphism $\widetilde{T}_j(\pi)$ with the multiplication operator $x \mapsto a_j \odot y \in A$ $(y \in F_j)$ for some $a_j \in A$. Denote $a := a_1 \odot \dots \odot a_n \in A$. Let D be a vector subspace of A generated by $F_1 \odot \dots \odot F_n$.

Now, for the collection $0 \le \pi_1 \in \mathrm{Orth}(E_1), \dots, 0 \le \pi_n \in \mathrm{Orth}(E_n)$ define the mapping $\tau : D \to A$ by putting $\tau : u \mapsto a \odot u$ $(u \in D)$. Then τ is linear and, moreover, $\tau(D) \subset D$, since for $y_{j,1}, \dots, y_{j,m} \in F_j$ $(j = 1, \dots, n)$ we have

$$v = \sum_{k=1}^{m} y_{1,k} \odot \dots \odot y_{n,k} \implies \tau(v) := \sum_{k=1}^{m} (a_1 \odot y_{1,k}) \odot \dots \odot (a_n \odot y_{n,k}) \in D.$$

Evidently, $v^+ \perp y$ implies $(\tau(v))^+ \perp y$ for all $v \in D$ and $y \in A$. Applying Lemma 6 with $G := A^u$ yields a positive orthomorphism $\hat{\tau} \in \mathrm{Orth}(F_0)$; moreover $\hat{\tau}(y) = a \odot y$ for all $y \in F_0$, that is, $a \odot F_0 \subset F_0$. Define the mapping $\widetilde{T}$ from $\mathrm{Orth}(E_1) \times \dots \times \mathrm{Orth}(E_n)$ to $\mathrm{Orth}(F_0)$ by putting $\widetilde{T}(\pi_1, \dots, \pi_n) := \hat{\tau}$ for positive π_j and extend $\widetilde{T}$ to arbitrary π_j by differences. Then we have by definition

$$\begin{gathered} \widetilde{T}(\pi_1, \dots, \pi_n) = \widetilde{T}_1(\pi_1) \odot \dots \odot \widetilde{T}_n(\pi_n) \\ (\pi_1 \in \mathrm{Orth}(E_1), \dots, \pi_n \in \mathrm{Orth}(E_n)) \end{gathered} \tag{12}$$

and hence $\widetilde{T}$ is lattice n-morphism. Moreover, for all $\pi_1 \in \mathrm{Orth}(E_1), \dots, \pi_n \in \mathrm{Orth}(E_n)$ and $x_1 \in E_1, \dots, x_n \in E_n$ we deduce

$$\begin{aligned} T(\pi_1 x_1, \dots, \pi_n x_n) =& U\big(T_1(\pi_1 x_1) \odot \dots \odot T_n(\pi_n x_n)\big) \\ =& U\big(\widetilde{T}_1(\pi_1) T_1(x_1) \odot \dots \odot \widetilde{T}_n(\pi_n) T_n(x_n)\big) \\ =& u \odot (a_1 \odot T_1(x_1) \odot \dots \odot a_n \odot T_n(x_n)) \\ =& a \odot (u \odot T_1(x_1) \odot \dots \odot T_n(x_n)) \\ =& \widetilde{T}(\pi_1, \dots, \pi_n)(T(x_n \dots, x_n)). \end{aligned}$$

The fact that $\widetilde{T}$ is an algebra n-morphism can be easily seen from (12) taking into account that $\widetilde{T}_k$ are algebra homomorphisms: for arbitrary $\pi_j, \rho_j \in \mathrm{Orth}(E_j)$ $(j = 1, \dots, n)$ we deduce

$$\begin{aligned} \widetilde{T}(\pi_1 \rho_1, \dots, \pi_n \rho_n) &= \widetilde{T}_1(\pi_1 \rho_1) \odot \dots \odot \widetilde{T}_n(\pi_n \rho_n) \\ &= \widetilde{T}_1(\pi_1) \widetilde{T}_1(\rho_1) \odot \dots \odot \widetilde{T}_n(\pi_n) \widetilde{T}_n(\rho_n) \\ &= \big(\widetilde{T}_1(\pi_1) \odot \dots \odot \widetilde{T}_n(\pi_n)\big)\big(\widetilde{T}_1(\rho_1) \odot \dots \odot \widetilde{T}(\rho_n)\big) \\ &= \widetilde{T}(\pi_1, \dots, \pi_n) \widetilde{T}_n(\rho_1, \dots, \rho_n). \end{aligned}$$

To prove the last assertion we use again Hart's theorem [22, Theorem 2.1] saying that the restriction of $T_j(\pi_j)$ to $|T_j|(E_j)$ coincides with $|\widetilde{T_j}|(\pi_j)$ for all $\pi_j \in \mathrm{Orth}(E_j)$. Note that F_0 is wider then the vector sublattice $F_{00} \subset F$ generated by $\mathrm{Im}(|T|)$. Since $\widetilde{T}(\pi_1, \dots, \pi_n) \in \mathrm{Orth}(F_0)$ and $|\widetilde{T}|(\pi_1, \dots, \pi_n) \in \mathrm{Orth}(F_{00})$, it suffices that these orthmorphisms coincide on $\mathrm{Im}(|T|)$. Thus, it remains to observe that

$$\begin{aligned}\widetilde{T}(\pi_1, \ldots, \pi_n)(|T|(x_n \ldots, x_n)) &= |\widetilde{T}(\pi_1, \ldots, \pi_n)(T(x_n \ldots, x_n))| \\ &= |T(\pi_1 x_1, \ldots, \pi_n x_n)| = |T|(\pi_1 x_1, \ldots, \pi_n x_n)\end{aligned}$$

for all $0 \leq x_1 \in E_1, \ldots, 0 \leq x_n \in E_n$ and $0 \leq \pi_1 \in \mathrm{Orth}(E_1), \ldots, 0 \leq \pi_n \in \mathrm{Orth}$ (E_n) and the proof is complete.

Theorem 12 leads to one more version of Hart's theorem for homogeneous polynomials.

Corollary 10 *Let E and F be vector lattices, $P : E \to F$ a lattice polymorphism, and F_0 a vector sublattice of F generated by $P(E)$. Then there exists a unique lattice polymorphism $\widetilde{P}$ from* $\mathrm{Orth}(E)$ *to* $\mathrm{Orth}(F_0)$ *such that*

$$\begin{aligned} P(\pi x) &= \widetilde{P}(\pi)(Px), \\ \widetilde{P}(\pi\rho) &= \widetilde{P}(\pi)\widetilde{P}(\rho), \\ (x \in E;\ \pi,&\ \rho \in \mathrm{Orth}(E)). \end{aligned}$$

Proof Apply Theorem 12 to order bounded disjointness preserving n-linear operator $T := \check{P}$ and define $\widetilde{P}$ by $\widetilde{P}(\pi) := \widetilde{T}(\pi, \ldots, \pi)$ $(\pi \in \mathrm{Orth}(E))$.

Acknowledgements The second author was supported by Russian Foundation for Basic Research (project No 18-31-00205).

References

1. Abid, R., Ben Amor, M.A., Boulabiar, K.: Composition operators on f-algebras. Positivity **21**, 521–537 (2017)
2. Abramovich, Y.A., Arenson, E.L., Kitover, A.K.: Banach $C(K)$-modules and operators preserving disjointness. Pitman Research Notes in Mathematics Series, vol. 277. Longman Scientific & Technical, Harlow (1992)
3. Abramovich, Y.A., Kitover, A.K.: Inverses of Disjointness Preserving Operators. Mem. Amer. Math. Soc. **143**(679). Providence, R.I. (2000)
4. Abramovich, YuA, Veksler, A.I., Koldunov, A.V.: Operators preserving disjointness. Dokl. Akad. Nauk SSSR. **248**, 1033–1036 (1979)
5. Aliprantis, C.D., Burkinshaw, O.: Positive Operators. Springer, Netherlands (2006)
6. Arendt, W.: Spectral theory of Lamperti operators. Indiana Univ. Marh. J. **32**, 199–215 (1983)
7. Boulabiar, K.: Some aspects of Riesz multimorphisms. Indag. Mathem., N.S. 13(4), 419-432 (2002)
8. Boulabiar, K.: Order bounded separating linear maps on Φ-algebras. Houst. J. Math. **30**, 1143–1155 (2004)
9. Boulabiar, K.: Recent trends on order bounded disjointness preserving operators. Ir. Math. Soc. Bulletin. **62**, 43–69 (2008)
10. Boulabiar, K., Buskes, G.: Vector lattice powers: f-algebras and functional calculus. Comm. Algebra. **34**(4), 1435–1442 (2006)
11. Boulabiar, K., Buskes, G.: Polar decomposition of order bounded disjointness preserving operators. Proc. Amer. Math. Soc. **132**(3), 799–806 (2003)
12. Boulabiar, K., Buskes, G., Triki, A.: Results in f-algebras. In: Boulabiar, K., Buskes, G., Triki, A. (eds.) Positivity, pp. 73–96. Birkhäuser, Basel (2007)

13. Boulabiar, K., Hager, A.W.: ℓ-group homomorphisms between reduced archimedean f-rings. Algebr. Univ. **62**, 329–337 (2009)
14. Boulabiar, K., Toumi, M.A.: Lattice bimorphisms on f-algebras. Algebr. Univ. **48**, 103–116 (2002)
15. Bu, Q., Buskes, G., Kusraev, A.G.: Bilinear maps on product of vector lattices. In: Boulabiar, K., Buskes, G., Triki, A. (eds.) Positivity, pp. 97–126. Birkhäuser, Basel (2007)
16. Buskes, G., Kusraev, A.G.: Extension and representation of orthoregular maps. Vladikavkaz Math. J. **9**(1), 16–29 (2007)
17. Buskes, G., van Rooij, A.: Small vector lattices. Math. Proc. Cambr. Philos. Soc. **105**, 523–536 (1989)
18. Buskes, G., van Rooij, A.: Squares of Riesz spaces. Rocky Mt. J. Math. **31**(1), 45–56 (2001)
19. Fremlin, D.H.: Tensor products of Archimedean vector lattices. Amer. J. Math. **94**, 778–798 (1972)
20. van Gaans, O.: The Riesz part of a positive bilinear form. Circumspice. Nijmegen: Katholieke Universiteit Nijmegen, 19–30 (2001)
21. Gutman, A.E.: Disjointness preserving operators. In: Kutateladze, S.S. (ed.) Vector Lattices and Integral Operators, pp. 361–454. Kluwer Academic Publishers, Dordrecht (1996)
22. Hart, D.R.: Some properties of disjointness preserving operators. Indag. Math. **47**, 183–197 (1985)
23. Huijsmans, C.B.: Disjointness preserving operators on Banach lattices. Oper. Theory Adv. Appl. **75**, 173–189 (1999)
24. Huijsmans, C.B., de Pagter, B.: Subalgebras and Riesz subspaces of an f-algebra. Proc. London, Math. Soc. **48**, 161–174 (1984)
25. Huijsmans, C.B., de Pager, B.: Invertible disjointness preserving operators. Proc. Edinb. Math. Soc. **37**, 125–132 (1993)
26. Kusraev, A.G., Kutateladze, S.S.: Boolean Valued Analysis: Selected Topics. Vladikavkaz Scientific Center Press, Vladikavkaz (2014)
27. Kusraev, A.G., Tabuev, S.N.: On disjointness preserving bilinear operators. Vladikavkaz Math. J. **6**(1), 58–70 (2004)
28. Kusraev, A.G., Tabuev, S.N.: On multiplicative representation of disjointness preserving bilinear operators. Sib. Math. J. **49**(2), 357–366 (2008)
29. Kusraeva, Z.A.: Representation of orthogonally additive polynomials. Sib. Math. J. **52**(2), 248–255 (2011)
30. Kusraeva, Z.A.: Orthogonally additive polynomials on vector lattices. Math. Notes **91**(5), 704–710 (2012)
31. Kusraeva, Z.A.: Characterization and multiplicative representation of homogeneous disjointness preserving polynomials. Vladikavkaz Math. J. **18**(1), 51–62 (2016)
32. Lambek, J.: Lectures on Rings and Modules. Blaisdell Publishing Company, Waltham (1966)
33. Lindenstrauss, J., Tzafriri, L.: Classical Banach Spaces. Springer, Berlin (1979). Functional analysis
34. Loane, J.: Polynomials on Riesz spaces. Thesis, Department of Mathematics National University of Ireland, Galway (2007)
35. McPolin, P.T.N., Wickstead, A.W.: The order boundedness of band preserving operators on uniformly complete vector lattices. Math. Proc. Cambridge Philos. Soc. **97**(3), 481–487 (1985)
36. Meyer-Nieberg, P.: Banach Lattices. Springer, Berlin (1991)
37. Meyer, M.: Le stabilisateur d'un espace vectoriel réticulé. C. R. Acad. Sci. Paris. **283**, 249–250 (1976)
38. de Pagter, B.: f-Algebras and orthomorphisms. Ph.D. Thesis, Leiden University (1981)
39. Schep, A.R.: Factorization of positive multilinear maps. Ill. J. Math. **28**(4), 579–591 (1984)
40. Szulga, J.: (p, r)-convex functions on vector lattices. Proc. Edinburg Math. Soc. **37**(2), 207–226 (1994)
41. Wickstead, A.W.: The ideal center of a Banach lattice. Proc. Royal Irish Acad. **76**(4), 15–23 (1976)
42. Wickstead, A.W.: The injective hull of an Archimedean f-algebra. Compos. Math. **62**, 329–342 (1987)

Robbins–Monro Conditions for Persistent Exploration Learning Strategies

Dmitry B. Rokhlin

Abstract We formulate simple assumptions, implying the Robbins–Monro conditions for the Q-learning algorithm with the local learning rate, depending on the number of visits of a particular state (local clock) and the number of iteration (global clock). It is assumed that the Markov decision process is communicating and the learning policy ensures the persistent exploration. The restrictions are imposed on the functional dependence of the learning rate on the local and global clocks. The result partially confirms the conjecture of Bradkte (1994).

Keywords Robbins–Monro conditions · Reinforcement learning · Learning rate · Learning strategy · Persistent exploration · Communicating Markov decision processes

1 Problem Description

Reinforcement learning is aimed at the solution of the Markov decision problems without the exact knowledge of an underlying model. In this paper we address only the case of finite state-action Markov decision processes (MDP). Moreover, for concreteness we discuss only the discounted optimality criterion and the Q-learning algorithm. However, this is not essential since we consider only the Robbins–Monro conditions for the learning rates, and not the convergence of the algorithms. So, the result is applicable to other reinforcement learning algorithms, based on asynchronous stochastic approximation.

The Q-learning can be regarded as an asynchronous version of the classical value iteration algorithm for the Q-function. Recall that a Q-function $Q(x, a)$ is the optimal gain for fixed initial state x and initial action a. The Q-learning algorithm updates

---

D. B. Rokhlin (✉)
Institute of Mathematics, Mechanics and Computer Sciences, Southern Federal University, Mil'chakova str., 8a, 344090 Rostov-on-Don, Russia
e-mail: rokhlin@math.rsu.ru

A. Karapetyants et al. (eds.), *Modern Methods in Operator Theory and Harmonic Analysis*, Springer Proceedings in Mathematics & Statistics 291,
https://doi.org/10.1007/978-3-030-26748-3_14

the current approximation Q_t to Q along a trajectory (x_t, a_t) of states x_t and actions a_t, generated by a selected learning (or exploration) strategy.

A learning strategy is a sequence of probability distributions $\pi_t(a)$ on the action set A (we assume that A does not depend on x). As e.g. in [12], we distinguish between persistent exploration and decaying exploration learning strategies. Persistent exploration (in contrast to the decaying one) means the existence of a uniform lower bound of the form $\pi_t(a) \geq c > 0$.

Besides the learning strategy, a particular instance of the Q-learning algorithm is determined by a learning rate $\gamma_t(x, a)$ which controls the influence of the new information on the update rule. Usually the learning rate is of the form

$$\gamma_t(x, a) = \alpha_t I_{\{x_t = x, a_t = a\}}. \tag{1}$$

The sequence (α_t) will be also called a learning rate. The standard results assert the pointwise convergence $Q_t \to Q$ with probability 1 under the Robbins–Monro conditions (see Theorem 1):

$$\sum_{t=0}^{\infty} \gamma_t = \infty, \quad \sum_{t=0}^{\infty} \gamma_t^2 < \infty. \tag{2}$$

Clearly, (2) implies that each state-action pair (x, a) is visited infinitely often. Assuming that this property is satisfied, it is easy to construct a sequence (α_t) depending on a "local clock" and verifying (2). By a local clock we mean the number of visits of a particular point (x, a) by the sequence (x_t, a_t). Indeed, consider a function $\varphi : \mathbb{Z}_+ \mapsto (0, \infty)$ satisfying the Robbins–Monro conditions, that is,

$$\sum_{t=0}^{\infty} \frac{1}{\varphi(t)} = \infty, \quad \sum_{t=0}^{\infty} \frac{1}{\varphi^2(t)} < \infty.$$

Put $\alpha_t = 1/\varphi(n_t(x, a))$, where

$$n_t(x, a) = \sum_{k=0}^{t} I_{\{x_k = x, a_k = a\}} \tag{3}$$

is the number of visits of (x, a) by the sequence $(x_k, a_k)_{k=0}^t$, and denote by $t_j(x, a)$ the time of j-th visit, $j \geq 1$. Then $n_{t_j}(x, a) = j$ and

$$\sum_{t=0}^{\infty} \gamma_t(x, a) = \sum_{j=1}^{\infty} \alpha_{t_j(x,a)} = \sum_{j=1}^{\infty} \frac{1}{\varphi(n_{t_j}(x, a))} = \sum_{j=1}^{\infty} \frac{1}{\varphi(j)} = \infty.$$

Similarly,

$$\sum_{t=0}^{\infty} \gamma_t^2(x,a) = \sum_{j=1}^{\infty} \frac{1}{\varphi^2(j)} < \infty.$$

If the learning rate α_t explicitly depends on the "global clock", that is, the iteration number t, then the situation becomes more difficult. Let α_t be a deterministic sequence. In his Ph.D. thesis Bradtke ([3], see also [4]) in somewhat different situation, involving function approximation, conjectured that if (α_t) satisfies the Robbins–Monro conditions:

$$\sum_{t=0}^{\infty} \alpha_t = \infty, \quad \sum_{t=0}^{\infty} \alpha_t^2 < \infty,$$

then the same is true for γ_t. In [13] it was mentioned that this conjecture is true if the inter-arrival times $t_{j+1} - t_j$ have a common upper bound or, more specifically, are eventually exponentially distributed with common parameters. However, these conditions are difficult to verify and they depend on the learning strategy.

In this note we show that the Bradtke conjecture holds true for persistent exploration learning strategies. This assertion follows from the main result: Theorem 2.

2 Markov Decision Processes and Q-Learning

Let X and A be finite state and action spaces. Consider the canonical space $\Omega = (X \times A)^\infty$ with the σ-algebra $\mathscr{F}$ generated by projections

$$(x_0, a_0, x_1, a_1, \dots) \mapsto (x_t, a_t).$$

Denote by $\mathscr{F}_t = \sigma(x_0, a_0, \dots, s_t, a_t)$ the natural filtration of the coordinate process. The probabilistic structure of the process (x_t, a_t) is determined by a fixed transition kernel $q(y|x,a)$:

$$\sum_{y \in X} q(y|x,a) = 1, \qquad q(y|x,a) \geq 0$$

and a control (or learning) strategy, which is a sequence $\pi = (\pi_t)$ of probability distributions on the action set A. These objects uniquely determine a unique probability measure $\mathsf{P}_{z,\pi}$ on Ω such that

$$\mathsf{P}_{z,\pi}(x_{t+1} = y|\mathscr{F}_t, a_t) = q(y|x_t, a_t), \quad \mathsf{P}_{z,\pi}(a_t = a|\mathscr{F}_{t-1}, x_t) = \pi_t(a),$$
$$\mathsf{P}_{z,\pi}(x_0 = z) = 1$$

(see, e.g., [7]). Note, that $\pi_t(a)$ is $\sigma(\mathscr{F}_{t-1}, x_t)$-measurable.

Given a reward function $r(x, a, y)$ and a discounting factor $\beta \in [0, 1)$, the total discounted gain is defined by the value function

$$V(z) = \sup_{\pi} \mathsf{E}_{z,\pi} \sum_{t=0}^{\infty} \beta^t r(x_t, a_t, x_{t+1}),$$

where $\mathsf{E}_{z,\pi}$ is the expectation with respect to $\mathsf{P}_{z,\pi}$. As is well known, this function is a unique solution of the Bellman (or dynamic programming) equation:

$$V(x) = \max_{a \in A} \sum_{y \in X} q(y|x, a)(r(x, a, y) + \beta V(y)).$$

The Q-function is the total discounted gain for fixed initial state and initial action:

$$Q(x, a) = \sum_{y \in X} q(y|x, a)(r(x, a, y) + \beta V(y)).$$

This function is a unique solution of the equation

$$Q(x, a) = \sum_{y \in X} q(y|x, a)(r(x, a, y) + \beta \max_{a \in A} Q(y, a)).$$

The Q-learning algorithm proposed in [15] recursively defines the sequence Q_t:

$$\begin{aligned} Q_{t+1}(x, a) = {} & (1 - \alpha_t I_{\{x_t = x, a_t = a\}}) Q_t(x, a) \\ & + \alpha_t I_{\{x_t = x, a_t = a\}} (r(x_t, a_t, x_{t+1}) + \beta \max_{a' \in A} Q_t(x_{t+1}, a')) \end{aligned}$$

for a strictly positive $\mathscr{F}_t$-measurable random variables α_t and an arbitrary initial guess $Q_0(x, a)$.

Let us recall a basic result on the convergence of Q_t to Q with probability 1: see [8, 14].

Theorem 1 *Assume that the learning rate α_t satisfies the Robbins–Monro conditions*

$$\sum_{t=0}^{\infty} \alpha_t I_{\{x_t = x, a_t = a\}} = \infty, \quad \sum_{t=0}^{\infty} \alpha_t^2 I_{\{x_t = x, a_t = a\}} < \infty \quad \mathsf{P}_{z,\pi}\text{-a.s.} \tag{4}$$

for all $(x, a) \in X \times A$. Then

$$\lim_{t \to \infty} Q_t(x, a) = Q(x, a) \quad \mathsf{P}_{z,\pi}\text{-a.s.}$$

In this paper we study only conditions (4) and not the proof of Theorem 1. Under the assumption that each pair (x, a) is visited infinitely often, one simple construction of the learning rate α_t, depending on the local clock (3) and satisfying (4), was given is Sect. 1. In the sequel we solely consider another version of a local clock, defined as the number of visits of a particular state x:

$$N_t(x) = \sum_{k=0}^{t} I_{\{x_k=x\}}. \tag{5}$$

Assume that all states are visited infinitely often $\mathsf{P}_{z,\pi}$-a.s., the learning strategy satisfies the lower bound $\pi_t(a) \geq c(N_t) > 0$, the learning rate is of the form $\alpha_t = 1/\varphi(N_t)$ and

$$\sum_{t=1}^{\infty} \frac{c(t)}{\varphi(t)} = \infty, \qquad \sum_{t=1}^{\infty} \frac{1}{\varphi^2(t)} < \infty, \tag{6}$$

then the Robbins–Monro conditions (4) are satisfied.

Indeed, by the conditional Borel–Cantelli lemma [10] (Chap. 1, Theorem 21), the first condition (4) is implied by (6), since

$$\sum_{t=0}^{\infty} \mathsf{E}_{z,\pi}(\alpha_t I_{\{x_t=x,a_t=a\}}|\mathscr{F}_{t-1}, x_t) = \sum_{t=0}^{\infty} \frac{1}{\varphi(N_t)} I_{\{x_t=x\}} \mathsf{E}_{z,\pi}(I_{\{a_t=a\}}|\mathscr{F}_{t-1}, x_t)$$
$$= \sum_{t=0}^{\infty} \frac{1}{\varphi(N_t)} I_{\{x_t=x\}} \pi_t(a) \geq \sum_{t=0}^{\infty} \frac{1}{\varphi(N_t)} I_{\{x_t=x\}} c(N_t) = \sum_{j=1}^{\infty} \frac{c(j)}{\varphi(j)} = \infty$$

$\mathsf{P}_{z,\pi}$-a.s. For the second condition (4) the argumentation is even easier:

$$\sum_{t=0}^{\infty} \alpha_t^2 I_{\{x_t=x,a_t=a\}} \leq \sum_{t=0}^{\infty} \frac{1}{\varphi^2(N_t)} I_{\{x_t=x\}} = \sum_{j=1}^{\infty} \frac{1}{\varphi^2(j)} < \infty \quad \mathsf{P}_{z,\pi}\text{-a.s.}$$

Note, that the decaying exploration is allowed, but the learning strategy should ensure infinitely many visits of every state and the lower bounds $c(t)$ should be consistent with learning rate: see the first condition (6).

In the next section we allow an explicit dependence of α_t on the global clock t, but consider only persistent exploration learning strategies. Two main examples of persistent exploration learning strategies are

- the Boltzmann exploration:

$$\pi_t(a) = \frac{\exp(Q_t(x_t, a)/\tau)}{\sum_{a'} \exp(Q_t(x_t, a')/\tau)}, \quad \tau > 0.$$

  The required inequality $\pi_t(a) \geq c > 0$ follows from the boundedness of the sequence (Q_t), which is proved in [6].
- ε-greedy exploration which takes a "greedy" action $a_t \in \arg\max Q_t(x_t, a_t)$ with probability $1 - \varepsilon$ and a random action with probability ε.

3 Robbins–Monro Conditions for Persistent Exploration Learning Strategies

A distribution $a \mapsto g(a|x)$ on A, defined for all $x \in X$, is called a stationary randomized strategy. If $g(b(x)|x) = 1$ for some function $b : X \mapsto A$, then the strategy is called deterministic. Such strategy can be identified with the function b. A stationary randomized strategy is called completely mixed if $g(a|x) > 0$ for all $x \in X$, $a \in A$. Any stationary randomized strategy g induces a Markov chain with the transition matrix

$$P(g)(x, y) = \sum_{a \in A} q(y|x, a)g(a|x).$$

An MDP is called communicating (see [1, 5, 9]), if for any $x, y \in X$ there exists a stationary deterministic strategy g such that y is accessible from x in the Markov chain $P(g)$. In other words, there exists $n \in \mathbb{N}$, depending on x, y, such that $P^n(g)(x, y) > 0$. We will use the fact that an MDP is communicating if and only $P(g)$ is irreducible for every completely mixed stationary randomized strategy: see [5, Theorem 2.1].

Define the completely mixed strategy $\overline{g}(a|x) = 1/|A|$, where $|A|$ is the cardinality of A. Let us recall (see [2, Lemma 7.3(i)]) that a Markov chain $P(\overline{g})$ is irreducible if and only if there exist $n \in \mathbb{N}$ such that all entries in the matrix $\sum_{j=1}^{n} P^j(\overline{g})$ are strictly positive. Let $\delta > 0$ be the minimal element of this matrix. Then

$$\sum_{j=1}^{n} P^j(\overline{g})(x, y) \geq \delta. \tag{7}$$

Lemma 1 *Assume that an MDP is communicating and the learning strategy* π *ensures the persistent exploration:* $\pi_t(a) \geq c > 0$. *Then for any function* $f : X \mapsto [0, \infty)$ *we have*

$$\sum_{j=1}^{n} \mathsf{E}_{z,\pi}(f(x_{t+j+1})|\mathscr{F}_t) \geq c^n |A|^n \delta \sum_{y \in X} f(y), \tag{8}$$

where the constants n, δ *satisfy (7).*

Proof Put

$$P^n(g)f(x) = \sum_{y \in X} P^n(g)(x, y)f(y), \quad n \geq 1.$$

Let $k \geq 2$, $f \geq 0$. Then

$$\begin{aligned}
\mathsf{E}_{z,\pi}(f(x_{t+k})|\mathscr{F}_t) &= \sum_x f(x)\mathsf{P}_{z,\pi}(x_{t+k} = x|\mathscr{F}_t)\\
&= \sum_x f(x)\mathsf{E}_{z,\pi}(\mathsf{P}_{z,\pi}(x_{t+k} = x|\mathscr{F}_{t+k-1})|\mathscr{F}_t)\\
&= \sum_x f(x)\mathsf{E}_{z,\pi}(q(x|x_{t+k-1}, a_{t+k-1})|\mathscr{F}_t)\\
&= \sum_x f(x)\mathsf{E}_{z,\pi}(\mathsf{E}_{z,\pi}(q(x|x_{t+k-1}, a_{t+k-1})|\mathscr{F}_{t+k-2}, x_{t+k-1})|\mathscr{F}_t)\\
&= \sum_x f(x)\mathsf{E}_{z,\pi}\left(\sum_a q(x|x_{t+k-1}, a)\pi_{t+k-1}(a)|\mathscr{F}_t\right)\\
&\geq c\sum_x f(x)\mathsf{E}_{z,\pi}\left(\sum_a q(x|x_{t+k-1}, a)|\mathscr{F}_t\right)\\
&= c|A|\sum_x f(x)\mathsf{E}_{z,\pi}\left(P(\overline{g})(x_{t+k-1}, x)|\mathscr{F}_t\right)\\
&= c|A|\mathsf{E}_{z,\pi}\left(P(\overline{g})f(x_{t+k-1})|\mathscr{F}_t\right) \geq c^{k-1}|A|^{k-1}\mathsf{E}_{z,\pi}\left(P^{k-1}(\overline{g})f(x_{t+1})|\mathscr{F}_t\right).
\end{aligned}$$

It follows that

$$\begin{aligned}
&\sum_{j=1}^n \mathsf{E}_{z,\pi}(f(x_{t+j+1})|\mathscr{F}_t) \geq \sum_{j=1}^n c^j|A|^j\mathsf{E}_{z,\pi}\left(P^j(\overline{g})f(x_{t+1})|\mathscr{F}_t\right)\\
&= \sum_{j=1}^n c^j|A|^j\sum_z P^j(\overline{g})f(z)q(z|x_t, a_t) \geq c^n|A|^n\sum_z\sum_{j=1}^n P^j(\overline{g})f(z)q(z|x_t, a_t)\\
&\geq c^n|A|^n\min_z\sum_{j=1}^n P^j(\overline{g})f(z) = c^n|A|^n\min_z\sum_y\sum_{j=1}^n P^j(\overline{g})(z, y)f(y)\\
&\geq c^n|A|^n\delta\sum_y f(y),
\end{aligned}$$

where we used the fact that $c \leq 1/|A|$. □

Under the assumptions of Lemma 1 every state $x \in X$ is visited infinitely often. It is even possible to give a lower bound for the growth rate of the local clock N_t. Namely, we claim that

$$\liminf_{t\to\infty}\frac{N_t(x)}{t} \geq \frac{c^n|A|^n\delta}{n} \qquad \mathsf{P}_{z,\pi}\text{-a.s.} \tag{9}$$

To prove (9) let us represent N_{kn+1}, $k \geq 1$ in the form

$$N_{kn+1}(x) = I_{\{x_0=x\}} + I_{\{x_1=x\}} + \sum_{j=1}^{k} \xi_j, \quad \xi_j = \sum_{l=(j-1)n+2}^{jn+1} I_{\{x_l=x\}}.$$

Furthermore, consider the Doob decomposition

$$\sum_{j=1}^{k} \xi_j = A_k + M_k, \quad k \geq 1$$

with respect to the filtration $\overline{\mathscr{F}}_k = \mathscr{F}_{kn}$, $k \geq 0$. Here (A_k) is a predictable process (compensator):

$$A_k = \sum_{j=1}^{k} \mathsf{E}_{z,\pi}(\xi_j | \overline{\mathscr{F}}_{j-1})$$

and (M_k) is a martingale. By Lemma 1 we have

$$\begin{aligned}\mathsf{E}_{z,\pi}(\xi_j | \overline{\mathscr{F}}_{j-1}) &= \sum_{l=(j-1)n+2}^{jn+1} \mathsf{E}_{z,\pi}\left(I_{\{x_l=x\}} | \mathscr{F}_{(j-1)n}\right) \\ &= \sum_{r=1}^{n} \mathsf{E}_{z,\pi}\left(I_{\{x_{(j-1)n+r+1}=x\}} | \mathscr{F}_{(j-1)n}\right) \geq c^n |A|^n \delta.\end{aligned}$$

It follows that $A_k \geq c^n |A|^n \delta k$. Furthermore,

$$\frac{M_k}{k} \to 0, \quad k \to \infty \qquad \mathsf{P}_{z,\pi}\text{-a.s.}$$

by the law of large numbers for martingales: [11, Chap. 7, Sect. 3, Corollary 2]. Thus,

$$\liminf_{k\to\infty} \frac{N_{kn+1}(x)}{k} \geq c^n |A|^n \delta \qquad \mathsf{P}_{z,\pi}\text{-a.s.} \tag{10}$$

For any $t \in \mathbb{N}$ there exists a unique $k \in \mathbb{N}$ such that $t \in [kn, (k+1)n)$. So, the inequality (9) easily follows from (10):

$$\begin{aligned}\liminf_{t\to\infty} \frac{N_t(x)}{t} &\geq \liminf_{k\to\infty} \frac{N_{kn}(x)}{(k+1)n} = \liminf_{k\to\infty} \frac{N_{(k+2)n}(x)}{(k+3)n} \\ &\geq \liminf_{k\to\infty} \frac{N_{(k+1)n+1}(x)}{k(1+3/k)n} \geq \frac{c^n |A|^n \delta}{n} \qquad \mathsf{P}_{z,\pi}\text{-a.s.}\end{aligned}$$

In Theorem 2, which is the main result of this note, the learning rate will be determined by a function $\varphi : \mathbb{Z}_+ \times \mathbb{Z}_+ \mapsto (0, \infty)$. Assume that

(i) the functions $t \mapsto \varphi(t, j)$, $j \mapsto \varphi(t, j)$ are non-decreasing;

(ii) the function φ satisfies the Robbins–Monro conditions on the diagonal:

$$\sum_{t=0}^{\infty} \frac{1}{\varphi(t,t)} = \infty, \qquad \sum_{t=0}^{\infty} \frac{1}{\varphi^2(t,t)} < \infty. \tag{11}$$

Theorem 2 *Assume that the MDP is communicating and φ satisfies conditions (i), (ii) above. Then the learning rate $\alpha_t = 1/\varphi(t, N_t)$ satisfies the Robbins–Monro conditions (4) for a persistent exploration learning strategy: $\pi_t(a) \geq c > 0$.*

Proof (a) Let us check the first property (4). We will use the notation (1). By the conditional Borel–Cantelli lemma the series

$$\gamma_0 + \gamma_1 + \sum_{j=1}^{k} \zeta_j, \qquad \zeta_j = \sum_{l=(j-1)n+2}^{jn+1} \gamma_l$$

diverges $\mathsf{P}_{z,\pi}$-a.s. if and only if

$$\sum_{j=1}^{\infty} \mathsf{E}_{z,\pi}(\zeta_j | \overline{\mathscr{F}}_{j-1}) = \infty \qquad \mathsf{P}_{z,\pi}\text{-a.s.}, \tag{12}$$

where $\overline{\mathscr{F}}_j = \mathscr{F}_{jn}$. Using the monotonicity properties of φ and the inequality (8), we get

$$\begin{aligned}
\mathsf{E}_{z,\pi}(\zeta_j | \overline{\mathscr{F}}_{j-1}) &= \sum_{l=(j-1)n+2}^{jn+1} \mathsf{E}_{z,\pi}\left(\gamma_l | \mathscr{F}_{(j-1)n}\right) \\
&= \sum_{l=(j-1)n+2}^{jn+1} \mathsf{E}_{z,\pi}\left(\frac{1}{\varphi(l, N_l)} I_{\{x_l = x\}} \mathsf{E}_{z,\pi}(I_{\{a_l = a\}} | \mathscr{F}_{l-1}, x_l) | \mathscr{F}_{(j-1)n}\right) \\
&\geq c \sum_{l=(j-1)n+2}^{jn+1} \mathsf{E}_{z,\pi}\left(\frac{1}{\varphi(l,l)} I_{\{x_l = x\}} | \mathscr{F}_{(j-1)n}\right) \\
&\geq \frac{c}{\varphi(jn+1, jn+1)} \sum_{l=(j-1)n+2}^{jn+1} \mathsf{E}_{z,\pi}\left(I_{\{x_l = x\}} | \mathscr{F}_{(j-1)n}\right) \\
&\geq \frac{c}{\varphi(jn+1, jn+1)} \sum_{r=1}^{n} \mathsf{E}_{z,\pi}\left(I_{\{x_{(j-1)n+r+1} = x\}} | \mathscr{F}_{(j-1)n}\right) \\
&\geq \frac{c^{n+1} |A|^n \delta}{\varphi(jn+1, jn+1)}.
\end{aligned}$$

So, to prove (12), and hence the first relation (4), it is enough to show that

$$\sum_{j=1}^{\infty} \frac{1}{\varphi(jn+1, jn+1)} = \infty.$$

But this is clear, since $\varphi(jn+1, jn+1) \leq \varphi(jn+k, jn+k)$, $k = 1, \dots, n$ and

$$\infty = \sum_{t=1}^{\infty} \frac{1}{\varphi(t,t)} \leq \sum_{j=0}^{\infty} \frac{n}{\varphi(jn+1, jn+1)}.$$

(b) Denote by $\tau_j(x)$ the time of j-th visit, $j \geq 1$ of the point x by the sequence (x_t). Then $N_{\tau_j}(x) = j$ and

$$\sum_{t=0}^{\infty} \frac{1}{\varphi^2(t, N_t)} I_{\{x_t = x, a_t = a\}} \leq \sum_{t=0}^{\infty} \frac{1}{\varphi^2(t, N_t)} I_{\{x_t = x\}}$$
$$= \sum_{j=1}^{\infty} \frac{1}{\varphi^2(\tau_j(x), j)} \leq \sum_{j=1}^{\infty} \frac{1}{\varphi^2(j, j)}$$

since $\tau_j(x) \geq j$ and the function $t \mapsto \varphi(t, j)$ is non-decreasing. Thus, the second condition (4) is implied by the second condition (11). □

For instance, the learning rates

$$\varphi(t, N_t) = \frac{a_1}{(b_1+t)^\alpha} \frac{a_2}{(b_2+N_t)^\beta}, \quad \alpha + \beta \in (1/2, 1], \quad a_i, b_i, \alpha, \beta > 0,$$

$$\varphi(t, N_t) = \frac{a_1}{(b_1+\ln t)^\alpha} \frac{a_2}{(b_2+N_t)^\beta}, \quad \alpha \in (1/2, 1], \quad \beta \in [1/2, 1], \quad a_i, b_i > 0$$

satisfy the conditions of Theorem 2.

For the learning rate depending only on the global clock:

$$\gamma_t = \frac{1}{\varphi(t)} I_{\{x_t = x, a_t = a\}}$$

Theorem 2 partially confirms the mentioned conjecture of Bradtke:

$$\sum_{t=1}^{\infty} \frac{1}{\varphi(t)} = \infty \implies \sum_{t=1}^{\infty} \frac{1}{\varphi(t)} I_{\{x_t = x, a_t = a\}} = \infty$$

for finite state-action communicating MDP, persistent exploration learning strategies and non-decreasing functions φ.

It would be interesting to investigate the case of decaying exploration learning strategies. It is clear that the Robbins–Monro conditions (4) can be ensured only by joint conditions on the learning rate and the randomized learning strategy (π_t). A simple illustration was given by (6).

Acknowledgements The research is supported by the Russian Science Foundation, project 17-19-01038.

References

1. Bather, J.: Optimal decision procedures for finite Markov chains. Part II: communicating systems. Adv. Appl. Probab. **5**(3), 521–540 (1973). https://doi.org/10.2307/1425832
2. Behrends, E.: Introduction to Markov Chains, with Special Emphasis on Rapid Mixing. Vieweg & Sohn, Braunschweig, Wiesbaden, Germany (2000). https://doi.org/10.1007/978-3-322-90157-6
3. Bradtke, S.J.: Incremental dynamic programming for on-line adaptive optimal control. Ph.D. Thesis, University of Massachusetts, Amherst, MA, USA (1994)
4. Bradtke, S.J., Barto, A.G.: Linear least-squares algorithms for temporal difference learning. Mach. Learn. **22**, 33–57 (1996). https://doi.org/10.1023/A:1018056104778
5. Filar, J.A., Schultz, T.A.: Communicating MDPs: equivalence and LP properties. Oper. Res. Lett. **7**(6), 303–307 (1988). https://doi.org/10.1016/0167-6377(88)90062-4
6. Gosavi, A.: Boundedness of iterates in Q-learning. Syst. Control Lett. **55**(4), 347–349 (2006). https://doi.org/10.1016/j.sysconle.2005.08.011
7. Hernández-Lerma, O., Lasserre, J.B.: Discrete-Time Markov Control Processes: Basic Optimality Criteria. Springer, Berlin (1996). https://doi.org/10.1007/978-1-4612-0729-0
8. Jaakkola, T., Jordan, M.I., Singh, S.: On the convergence of stochastic iterative dynamic programming algorithms. Neural Comput. **6**(6), 1185–1201 (1994). https://doi.org/10.1162/neco.1994.6.6.1185
9. Kallenberg L.C.M.: Classification problems in MDPs. In: Hou Z., Filar J.A., Chen A. (eds.) Markov Processes and Controlled Markov Chains. Springer, Boston (2002). https://doi.org/10.1007/978-1-4613-0265-0_9
10. Meyer, P.-A.: Martingales and Stochastic Integrals I. Lecture Notes in Math, vol. 284. Springer, Berlin (1972)
11. Shiryaev, A.N.: Probability. Springer, New York (1996). https://doi.org/10.1007/978-1-4757-2539-1
12. Singh, S., Jaakkola, T., Littman, M.L., Szepesvári, Cs.: Convergence results for single-step on-policy reinforcement-learning algorithms. Mach. Learn. **38**(3), 287–308 (2000). https://doi.org/10.1023/A:1007678930559
13. Szepesvári, Cs., Littman, M.L.: Generalized Markov decision processes: dynamic-programming and reinforcement-learning algorithms. Technical Report CS-96-11, Brown University, Providence, RI
14. Tsitsiklis, J.N.: Asynchronous stochastic approximation and Q-learning. Mach. Learn. **16**(3), 185–202 (1994). https://doi.org/10.1023/A:1022689125041
15. Watkins, C.J.C.H.: Learning from delayed rewards. Ph.D. Thesis, King's College, Cambridge, UK (1989)

On Widths of Invariant Sets

Ekaterina Shulman

Abstract Many problems of approximation theory and operator theory can be reduced to the computation or estimation of n-widths. By definition, the n-width $w_n(A)$ of a subset A in a Banach space X is the minimal distance of A from n-dimensional subspaces. If in X an isometric representation $g \mapsto T_g$ of a group G acts and A is invariant under operators T_g, then it is reasonable to estimate the minimal distances from A to *invariant* subspaces; in this case the *invariant width* $w_n^G(A)$ is defined as the minimal distance to invariant subspaces of dimension *at most* n. This approach, initiated in the author's work [5], allowed to study stability of Levi-Civita functional equations on amenable groups in classes of bounded functions. To catch more general classes of "approximate solutions" one has to consider representations on linear G-spaces that contain invariant subspaces on which the representation acts isometrically—this is the subject of the present work. Some partial results are obtained for general (not necessarily amenable) groups if X is reflexive; for Hilbert spaces the result can be formulated in the form $w_n^G(A) \leq 7\sqrt{n+1}\; w_n(A)$. New applications to the stability of functional equations are given.

Keywords n-widths · Linear G-spaces · Levi-Civita functional equations on groups · Stability of functional equations

1 Introduction

We consider some geometrical problems in linear G-spaces that are concerned with approximation of invariant subsets by finite-dimensional invariant subspaces. To formulate them, recall that the n-width $w_n(A)$ of a bounded subset A in a Banach space X is defined by the formula

E. Shulman (✉)
University of Silesia, 14 Bankowa Str., Katowice 40-007, Poland
e-mail: shulmanka@gmail.com

Vologda State University, 6 S. Orlova Str., Vologda 160035, Russia

A. Karapetyants et al. (eds.), *Modern Methods in Operator Theory and Harmonic Analysis*, Springer Proceedings in Mathematics & Statistics 291,
https://doi.org/10.1007/978-3-030-26748-3_15

$$w_n(A) = \inf\{dist(A, L) : L \in \mathcal{L}_n(X)\},$$

where $\mathcal{L}_n(X)$ is the set of all n-dimensional subspaces in X, and

$$dist(A, L) = \sup_{x \in A} \inf_{y \in L} \|x - y\|.$$

If $T : g \mapsto T_g$ is a representation of a group G on X leaving A invariant, then a covariant analogue of n-widths is defined [5] by

$$w_n^G(A) = \inf\{dist(A, L) : L \in \mathcal{L}_n^G(X)\},$$

where $\mathcal{L}_n^G(X)$ is the set of all *invariant* subspaces of dimension at most n in X.

Recall that a representation T of a topological group G on X is *dual* if $X = Y^*$, for some Banach space Y, all operators T_g are weak*-continuous and the map $g \mapsto T_g$ is continuous in the point-weak* topology of $B(X)$.

It turns out that for amenable groups and dual representations, invariant n-widths can be estimated via n-widths. We will formulate this result from [5] in a modified form; its validity easily follows from the proofs of [5, Theorems 1, 2].

Theorem 1 ([5]) *If a group G is amenable, then there is a function $n \mapsto \beta_G(n)$ such that, for any isometric dual representation of G on a Banach space X and any bounded G-invariant subset $A \subset X$,*

$$w_n^G(A) \leq \beta_G(n) w_n(A), \qquad \textit{for all} \quad n \in \mathbb{N}. \tag{1}$$

Moreover, if X is a Hilbert space then for any (not necessarily amenable) group G and all isometric (not necessarily weakly continuous) representations of G,

$$w_n^G(A) \leq (n+1)^{1/2} w_n(A), \qquad \textit{for all} \quad n \in \mathbb{N}. \tag{2}$$

It should be mentioned that Theorem 1 immediately implies that if an isometric representation has an invariant set whose n-widths decrease sufficiently quickly then the representation has invariant finite-dimensional subspaces of arbitrarily large dimension. We give the precise formulation only for Hilbert spaces.

Corollary 1 *If a unitary representation T of a group has an invariant set A with $w_n(A) = o(n^{-1/2})$, then for each $N \in \mathbb{N}$ there is a finite-dimensional T-invariant subspace L with* $\dim L > N$.

The interest to covariant widths in [5] arose from the study of stability for Levi-Civita functional equations [4, 6, 7] on groups:

$$f(gh) = \sum_{i=1}^{n} a_i(g) b_i(h), \qquad \text{for all} \quad g, h \in G; \tag{3}$$

the geometric results discussed above has appeared sufficiently convenient in consideration of bounded "approximate solutions" of these equations. To remove the restriction of boundedness one has to extend the setting by considering a more general class of representations than the isometric ones. This approach was also used in [5] but the obtained results were qualitative rather than quantitative. More precisely, it was proved that if a measurable function f on an amenable group G satisfies the condition

$$\sup_{g,h} |f(gh) - \sum_{i=1}^{n} a_i(g)b_i(h)| < \infty, \tag{4}$$

with some functions $\{a_i\}_{i=1}^n$ and $\{b_i\}_{i=1}^n$, then f is a sum of a solution of (3) and a bounded function. Here we wish to get a quantitative result. Moreover, we avoid the condition of amenability, as far as possible, obtaining on this way some geometric information for general groups.

2 Results

Let $T : g \mapsto T_g$ be a representation of a group G on a linear space X, and let an invariant subspace Y of X be a Banach space with respect to a norm $\|\cdot\|_Y$. We assume that the restriction of T to Y is isometric: $\|T_g|_Y\| = 1$, for all $g \in G$; for brevity we denote $T_g|_Y$ by $\pi(g)$.

In what follows we fix a finite-dimensional subspace L of X and a vector $\xi \in X$ such that its orbit $O(\xi) = \{T_g\xi : g \in G\}$ is contained in $L + Y$. For each $g \in G$, we choose a decomposition

$$T_g\xi = l(g) + r(g) \tag{5}$$

with $l(g) \in L$, $r(g) \in Y$, and we are interested in the case that the Y-components $r(g)$ are "small" in some sense; in other words, $O(\xi)$ is "close" to L. We will assume that L is linearly generated by $\{l(g) : g \in G\}$, otherwise L can be changed by $span\{l(g) : g \in G\}$. This assumption implies easily that the subspace $L + Y$ is invariant for all T_g.

In formulations below we preserve the introduced notations.

Proposition 1 *Suppose that* $\sup_{g\in G} \|r(g)\| < \infty$ *in* (5). *If L does not contain non-zero G-invariant subspaces then there is a norm on $Y + L$ coinciding with $\|\cdot\|_Y$ on Y and such that the restriction of the representation T to $Y + L$ is bounded in this norm.*

It will be convenient to extend $\|\cdot\|_Y$ to a quasinorm $\|\cdot\|'$ on X by setting $\|x\|' = \infty$ if $x \notin Y$. For arbitrary $x \in X$ and $Q \subset X$, we set

$$d(x, Q) = \inf\{\|x - z\|' : z \in Q\}.$$

Let us call a vector $x \in X$ *finite of order at most n*, if it is contained in an invariant subspace of dimension at most n.

Theorem 2 *If the space Y is reflexive,* $\dim L = n$ *and the vector ξ satisfies the condition*

$$d(T_g\xi, L) \leq \delta, \text{ for all } g \in G, \tag{6}$$

then $\xi = v_0 + p_0$, *where v_0 is finite of order at most n,* $p_0 \in Y$ *and*

$$w_n(O(p_0)) \leq 7\delta.$$

This result reduces the study of vectors with "almost finite-dimensional orbits" to the same problem for isometric representations: if an orbit of a vector in X is close (with respect to the quasinorm $\|\cdot\|'$) to an n-dimensional subspace then it can be decomposed into the sum of a finite vector (of order at most n) and a vector in Y whose orbit is close to an n-dimensional subspace in the norm $\|\cdot\|_Y$.

Corollary 2 *If, in Theorem 2, the space* $(Y, \|\cdot\|_Y)$ *is Hilbert then*

$$w_n^G(O(p_0)) \leq 7\sqrt{n+1}\,\delta.$$

Thus $\xi = \eta_0 + y_0$, *where η_0 is a finite vector of order at most* $2n$, $y_0 \in Y$ *and* $\|y_0\| \leq 7\sqrt{n+1}\,\delta$.

Corollary 3 *If, in assumptions of Theorem 2, G is amenable and the restriction π of the representation T to Y is weakly continuous, then $\xi = \eta + y$, where η is a finite vector of order at most* $2n$, $y \in Y$ *and* $\|y\| \leq 7\beta_G(n)\delta$.

Now we indicate some applications to functional equations.

Let G be a locally compact group. Complex-valued functions on G satisfying the Levi-Civita equation (3) can be described (see [5]) as matrix elements of finite-dimensional representations of G:

$$f(g) = \langle \pi(g)x, y\rangle,$$

for some $x \in X_\pi$, $y \in X_\pi^*$. The minimal possible dimension of X_π coincides with the minimal n in (3) and is called the *order* of the matrix element f. We study functions that satisfy (3) "approximately". This can be formalized by the condition (4), or one can consider a right-invariant Banach space $(Y, \|\cdot\|_Y)$ of functions on G and impose the condition

$$f(gh) - \sum_{i=1}^{n} a_i(g)b_i(h) := \phi_h(g) \in Y \qquad \text{and} \qquad \sup_{h\in G} \|\phi_h\|_Y \leq \infty, \tag{7}$$

or a more detailed one

$$f(gh) - \sum_{i=1}^{n} a_i(g)b_i(h) := \phi_h(g) \in Y \qquad \text{and} \qquad \|\phi_h\|_Y \leq \delta, \quad \text{for each } h \in G. \tag{8}$$

Clearly one can regard this setting as a special case of the geometric situation considered above if for T we take the representation of G on a right-invariant space X of functions on G by right shifts: $T_h f(g) = f(gh)$. The role of Y is played by a T-invariant subspace of X, supplied with a norm $\|\cdot\|_Y$ which is invariant under right shifts: $\|T_h f\|_Y = \|f\|_Y$, for all $f \in Y$ and $h \in G$. So the T-orbit $O(f)$ of an arbitrary function $f \in X$ consists of all right shifts of f.

Corollary 4 *If an unbounded function f on G satisfies* (4) *then the linear span of functions $a_1, \ldots, a_n$ contains a non-zero matrix element of G.*

It should be noted that even for amenable groups, Corollary 4 cannot be strengthened to the statement that f is a sum of a bounded function and a matrix element *in the linear span of functions* $a_1, \ldots, a_n$. Indeed let for example $G = \mathbb{R}$ and $f(t) = t + \arctan t$. Then $|f(t+s) - f(t) - f(s)| < 2$, $a_1(t) = f(t)$, $a_2(t) = 1$ and the only finite vectors in *span*$(\{a_1, a_2\})$ are constants.

The following easy consequence of Corollary 4 is known (at least in the case that $a(g) = b(g) = f(g)$) as the *superstability* of the multiplicative Cauchy functional equation. It was established in [1, 2, 8].

Corollary 5 *If a function f on G is unbounded and*

$$\sup_{g,h \in G} |f(gh) - a(h)b(g)| < \infty,$$

for some functions $a(g)$ and $b(g)$ on G, then $f(g) = \lambda\chi(g)$ where $\chi(gh) = \chi(g)\chi(h)$, $\lambda \in \mathbb{C}$.

Corollary 6 *If a function f satisfies* (8) *with $Y = L_2(G)$, then $f(g) = v(g) + y(g)$ where $v(g)$ is a matrix element of order $\leq 2n$, and $y(g) \in L_2(G)$ with $\|y\|_2 \leq 7\sqrt{n+1}\,\delta$.*

Corollary 7 *If a function f satisfies* (7) *with $Y = L_p(G)$, $1 < p < \infty$, then $f(g) = v(g) + y(g)$ where $v(g)$ is a matrix element of order $\leq 2n$, and $y(g) \in L_p(G)$ with $w_n(O(y)) \leq 7\delta$.*

Clearly instead of L_p there can be chosen any reflexive shift-invariant Banach space of functions.

Corollary 8 *If G is amenable, Y is reflexive, the regular representation is weakly continuous on Y and f satisfies* (7), *then $f = v + y$ where v is a matrix element of order $\leq 2n$, and $y \in Y$ with $\|y\|_Y \leq 7\beta(_G n)\delta$.*

3 Proofs

Proof of Proposition 1
By our assumptions, there is $D > 0$ with

$$T_g\xi - l(g) \in Y \text{ and } \|T_g\xi - l(g)\|_Y = \|r(g)\|_Y \leq D,$$

where $l(g) \in L$. Therefore $T_h l(g) - T_{hg}\xi \in Y$, $T_h l(g) - l(hg) \in Y$; setting $l(G) = \{l(g) : g \in G\}$ we have that $T_h l(G) \subset l(G) + Y$. So the linear span of the set $l(G) + Y$ is invariant for T and we may assume that it coincides with X (otherwise one can change X by this subspace). Moreover we may assume that L is the linear span of $l(G)$, so one can choose $g_1, \ldots, g_n \in G$ such that the vectors $l(g_1), \ldots, l(g_n)$ form a basis of L.

Let us denote $L \cap Y$ by S, and let us choose a complement Y_0 of S in Y. Let us denote by P_0 and $I - P_0$ the projections onto Y_0 and L respectively corresponding to the direct sum $X = L + Y_0$. The restriction $P_0|_Y$ of P_0 to Y is bounded in the norm $\| \cdot \|_Y$.

For any $\eta \in L$, we write

$$T_g\eta = P_0 T_g\eta + (I - P_0)T_g\eta. \tag{9}$$

Let us show that the first summand in (9) is $\| \cdot \|_Y$-bounded on G. For an arbitrary $\eta = \sum_{i=1}^n \lambda_i l(g_i) \in L$, we have

$$\begin{aligned}
&\|P_0 T_g\eta\|_Y = \|\sum_{i=1}^n \lambda_i P_0 T_g l(g_i)\|_Y = \|\sum_{i=1}^n \lambda_i P_0 T_g (T_{g_i}\xi - r(g_i))\|_Y = \\
&\|\sum_{i=1}^n \lambda_i (P_0 T_{gg_i}\xi - P_0\pi(g)r(g_i))\|_Y = \|\sum_{i=1}^n \lambda_i P_0 (l(gg_i) + r(gg_i) - \pi(g)r(g_i))\|_Y = \\
&\|\sum_{i=1}^n \lambda_i P_0 (r(gg_i) - \pi(g)r(g_i))\|_Y \leq 2D \sum_{i=1}^n |\lambda_i| \|P_0|_Y\|.
\end{aligned} \tag{10}$$

This allows us to introduce a seminorm ρ in L as follows

$$\rho(\eta) = \sup\{\|P_0 T_g\eta\|_Y : g \in G\}. \tag{11}$$

Let $N(\rho)$ be the null-space of ρ:

$$N(\rho) = \{\eta \in L : \rho(\eta) = 0\} = \{\eta \in L : P_0 T_g\eta = 0 \, \forall g \in G\} = \{\eta \in L : T_g\eta \in L \, \forall g \in G\} \tag{12}$$

It follows easily that the subspace $N(\rho)$ of L is invariant for T. By the hypothesis we have that $N(\rho) = 0$, hence ρ is a norm on L.

Now we will show that the L-component of the orbit $T_G\eta$ (the second summand in (9)) is bounded with respect to this norm. Indeed,

$$\sup_{g\in G} \rho((I-P_0)T_g\eta) = \sup_{g\in G}\sup_{h\in G} \|P_0T_h(I-P_0)T_g\eta\|_Y \le$$
$$\sup_{g,h\in G} \|P_0T_{hg}\eta\|_Y + \sup_{g,h\in G} \|P_0T_hP_0T_g\eta\|_Y \le \tag{13}$$
$$\rho(\eta) + \sup_{g\in G} \|P_0\pi(h)\|_Y \sup_{g\in G} \|P_0T_g\eta\|_Y \le \rho(\eta)(1+\|P_0|_Y\|).$$

This implies boundedness of T in the norm $\|\cdot\|$ defined by

$$\|\eta + y\| := \|y\|_Y + \rho(\eta), \text{ for } y \in Y_0, \eta \in L.$$

Since Y_0 has finite codimension in Y, this norm is equivalent to $\|\cdot\|_Y$ on Y and we can change it by an equivalent one by choosing a complement W of $L\cap Y$ in L and setting

$$\|w + y\| = \rho(w) + \|y\|_Y,$$

for $w \in W$, $y \in Y$. ■

Proof of Theorem 2

As above we can assume that $X = L + Y$ and L is the linear span of $l(G)$.

Step 1. Suppose that L does not have non-zero T-invariant subspaces. Then, by Proposition 1, the representation T is bounded with respect to some norm $\|\cdot\|$ on X that coincides with $\|\cdot\|_Y$ on Y. Sometimes we will write $\|y\|_Y$ instead of $\|y\|$, to underline that $y \in Y$.

Let as above $S = L\cap Y$ and denote $\dim S$ by m. Choosing a complement W of S in L we write

$$l(g) = w(g) + s(g), \text{ for each } g \in G. \tag{14}$$

We set $\xi_0 = l(e)$, so $\xi - \xi_0 \in Y$ and $\|\xi - \xi_0\| \le \delta$. It follows that

$$\|T_g\xi_0 - l(g)\|_Y = \|\pi(g)(\xi_0 - \xi) + T_g\xi - l(g)\| \le 2\delta \tag{15}$$

whence

$$d(T_g\xi_0, L) \le 2\delta, \text{ for all } g \in G. \tag{16}$$

Since $\xi_0 \in L$ we write $\xi_0 = w_0 + s_0$, where $w_0 = w(e) \in W$, $s_0 = s(e) \in S$.

Since $X = W + Y$ and Y is invariant, operators T_g can be written in the matrix form

$$T_g = \begin{pmatrix} \lambda(g) & 0 \\ p(g) & \pi(g) \end{pmatrix},$$

where λ is a representation of G on W and $p : g \mapsto p(g) \in B(W, Y)$ is a (λ, π)-cocycle:

$$p(gh) = p(g)\lambda(h) + \pi(g)p(h).$$

Since T is bounded the maps λ and p are bounded on G.

For each $g \in G$, the map $F_g : B(W, Y) \to B(W, Y)$ defined by the formula

$$F_g(K) = \pi(g)K\lambda(g)^{-1} + p(g)\lambda(g)^{-1}$$

is affine, and it is easy to check that $F_{gh}(K) = F_g(F_h(K))$, for all $g, h \in G$. It follows that the set

$$\mathcal{E} = \{p(g)\lambda(g)^{-1} : g \in G\} = \{F_g(0) : g \in G\}$$

is invariant for all maps F_g. Clearly $\mathcal{E}$ is a bounded subset of $B(W, Y)$; since $B(W, Y)$ is reflexive, the closed convex span $\mathcal{F} = \overline{conv(\mathcal{E})}$ of $\mathcal{E}$ is a weakly compact subset of $B(W, Y)$. It follows easily from the boundedness of representations π and λ that the action $g \mapsto F_g$ of G on $\mathcal{F}$ is uniformly continuous. Applying the Ryll-Nardzevski theorem (see [3]) we get that there is an operator $K_0 \in \mathcal{F}$ with $F_g(K_0) = K_0$, for all $g \in G$. This means that

$$p(g) = K_0\lambda(g) - \pi(g)K_0, \text{ for all } g \in G. \tag{17}$$

Since $L = span(l(G))$, the linear span of the λ-orbit of w_0 is dense in W.

Now let us consider the set

$$\mathcal{M} = \overline{conv(\{\pi(g)^{-1}K_0\lambda(g) : g \in G\})}.$$

Clearly, $\mathcal{M}$ is a weakly compact convex subset in $B(W, Y)$, invariant under the action $(g, K) \mapsto \pi(g)^{-1}K\lambda(g)$. Arguing as above we obtain a fixed point of this action, that is an operator $K_1 \in \mathcal{M}$ satisfying the condition

$$K_1\lambda(g) = \pi(g)K_1.$$

By (14) and (15),

$$\|T_g\xi_0 - w(g) - s(g)\|_Y \leq 2\delta, \tag{18}$$

hence the matrix representation of $T(g)$ implies

$$w(g) = \lambda(g)w_0$$

and

$$\|p(g)w_0 + \pi(g)s_0 - s(g)\|_Y \leq 2\delta.$$

Taking in account (17) we obtain

$$\|K_0\lambda(g)w_0 - \pi(g)K_0w_0 + \pi(g)s_0 - s(g)\|_Y \leq 2\delta.$$

Hence

$$\|\pi(g)^{-1}K_0\lambda(g)w_0 - K_0w_0 + s_0 - \pi(g)^{-1}s(g)\| \leq 2\delta. \tag{19}$$

Let us denote by $\mathcal{B}$ the ball $\{y \in Y : \|y - K_0 w_0 + s_0\| \leq 2\delta\}$, then

$$\pi(g)^{-1} K_0 \lambda(g) w_0 - \pi(g)^{-1} s(g) \in \mathcal{B}, \text{ for every } g \in G. \tag{20}$$

Since $K_1 \in \mathcal{M}$, it is a norm-limit of convex combinations of operators $\pi(g)^{-1} K_0 \lambda(g)$, so, for every $k \in \mathbb{N}$, there is $R_k \in conv(\{\pi(g)^{-1} K_0 \lambda(g) : g \in G\})$ with $\|K_1 - R_k\| < 1/k$. By definition

$$R_k = \sum_{i=1}^{n_k} \alpha_i^{(k)} \pi(h_i^{(k)})^{-1} K_0 \lambda(h_i^{(k)}),$$

for some $h_i^{(k)} \in G$ and $\alpha_i^{(k)} > 0$ with $\sum_{i=1}^{n_k} \alpha_i^{(k)} = 1$. Setting $z_k = \sum_{i=1}^{n_k} \alpha_i^{(k)} \pi(h_i^{(k)})^{-1} s(h_i^{(k)})$, we get from (20) that $R_k w_0 - z_k \in \mathcal{B}$. Coming to a subsequence we may assume that $z_k \to s_1$ in the weak topology, where $s_1 \in Y$. Since $R_k \to K_1$ when $k \to \infty$, we obtain that $K_1 w_0 - s_1 \in \mathcal{B}$. Thus

$$\|K_1 w_0 - K_0 w_0 + s_0 - s_1\| \leq 2\delta, \tag{21}$$

We claim that the subspace

$$V = \{w + (K_0 - K_1)w : w \in W\}$$

is invariant for T_g.

Indeed let $w = \lambda(h) w_0$, then

$$T_g(w + (K_0 - K_1)w) = T_g(\lambda(h) w_0 + (K_0 - K_1)\lambda(h) w_0)$$

$$= \lambda(g)\lambda(h) w_0 + (K_0 \lambda(g) - \pi(g) K_0)\lambda(h) w_0 + \pi(g)(K_0 - K_1)\lambda(h) w_0$$

$$= \lambda(gh) w_0 + (K_0 - K_1)\lambda(gh) w_0 \in V$$

Since $\{\lambda(h) w_0 : h \in G\}$ generates W, the claim is proved.

Set now

$$\eta = w_0 + (K_0 - K_1) w_0 + s_1.$$

Then

$$\|\xi_0 - \eta\|_Y = \|w_0 + s_0 - (w_0 + (K_0 - K_1) w_0 + s_1)\| = \|K_1 w_0 - K_0 w_0 + s_0 - s_1\| < 2\delta. \tag{22}$$

It follows that $\|T_g \xi_0 - T_g \eta\|_Y \leq 2\delta$, for all g.

Clearly

$$T_g \eta = v(g) + \pi(g) s_1, \tag{23}$$

where $v(g) = T_g(w_0 + (K_0 - K_1) w_0) \in V$.

Furthermore

$$T_g\eta = T_g\xi_0 + T_g(\eta - \xi_0) = l(g) + (T_g\xi_0 - l(g)) + \pi(g)(\eta - \xi_0) = l(g) + z(g),$$

therefore (23) implies

$$\pi(g)s_1 = l(g) - v(g) + z(g). \tag{24}$$

where $\|z(g)\|_Y \leq 4\delta$ by (15) and (22).

Let $M = (L + V) \cap Y$ then $\dim(M) \leq \dim L$. Indeed, setting $Z = (K_0 - K_1)W$ we have $V \subset W + Z$ whence

$$(L + V) \cap Y \subseteq (L + W + Z) \cap Y = (L + Z) \cap Y = L \cap Y + Z = S + Z$$

and

$$\dim M \leq \dim S + \dim Z \leq \dim S + \dim W = \dim L.$$

By (24), $l(g) - v(g) \in Y$, so $l(g) - v(g) \in M$.

It follows now from (24), that the π-orbit of s_1 is 4δ-close to the subspace M.

It remains to set $v_0 = v(e) = w_0 + (K_0 - K_1)w_0$, $p_0 = \xi - v_0$.

Indeed, $v_0 \in V$, the subspace V is invariant and $\dim V = \dim W = n - m$, so v_0 is a finite vector of order at most $n - m$. Furthermore, $p_0 = \xi - \xi_0 + \xi_0 - \eta + \eta - v_0 = (\xi - \xi_0) + (\xi_0 - \eta) + s_1$, so $\|p_0 - s_1\|_Y \leq 3\delta$ whence

$$d(\pi(g)p_0, M) \leq \|\pi(g)(p_0 - s_1)\| + d(\pi(g)s_1, M) \leq 3\delta + 4\delta = 7\delta,$$

for all $g \in G$.

Step 2 Now we consider the general case, removing the assumption that L does not contain non-zero invariant subspaces.

So let $0 \neq L_0$ be a maximal invariant subspace of L, and let $q : X \to X/L_0$ be the standard epimorphism.

The space $q(Y)$ is isomorphic to the space $Y/(Y \cap L_0)$; supplied with the quotient norm of the Banach space it is reflexive. The quotient representation $\widetilde{T} : G \to \mathcal{L}(X/L_0)$, defined by the formula $\widetilde{T}_g(\widetilde{x}) = T_g x + L_0$, preserves $q(Y)$, and the restriction $\widetilde{\pi}$ of $\widetilde{T}$ to $q(Y)$ is isometric. It is easy to check that $d_{q(Y)}(\widetilde{T}_g q(\xi), q(L)) < \delta$ for all $g \in G$. Since $\widetilde{T}$ does not have non-trivial invariant subspaces in $q(L)$, we may apply the result of Step 1 and conclude that there are a $\widetilde{T}$-invariant subspace $\widetilde{V}$ with $\dim \widetilde{V} \leq \dim q(L)$ and a subspace $\widetilde{M} \subset q(Y)$ with $\dim \widetilde{M} \leq \dim q(L)$, such that

$$q(\xi) = \widetilde{v_0} + \widetilde{p_0}$$

for some $\widetilde{v_0} \in \widetilde{V}$ and $\widetilde{p_0} \in q(Y)$ satisfying condition

$$d_{q(Y)}(\widetilde{\pi}(g)\widetilde{p_0}, \widetilde{M}) \leq 7\delta, \text{ for all } g \in G. \tag{25}$$

Now we take $p_0 \in q^{-1}(\widetilde{p})$ and set $v_0 = \xi - p_0$, then $v_0 \in q^{-1}(\widetilde{v_0})$. Setting $V = q^{-1}(\widetilde{V})$ and $M = q^{-1}(\widetilde{M})$ we have that

$$\dim(V) \le \dim(\widetilde{V}) + \dim(L_0) \le \dim(q(L)) + \dim(L_0) = \dim(L) = n$$

and similarly $\dim(M) \le n$.

Furthermore, it follows from (25) that, for each $g \in G$,

$$d_Y(\pi(g)p_0, M) = \inf_{a\in\widetilde{M}} \inf_{s\in a} \|\pi(g)p_0 - s\| = \inf_{a\in\widetilde{M}} \|\widetilde{\pi}(g)\widetilde{p_0} - a\|_{q(Y)} = d_{q(Y)}(\widetilde{\pi}(g)\widetilde{p_0}, \widetilde{M}) \le 7\delta. \blacksquare$$

Proof of Corollary 2. Applying Theorem 2 of [5] to the set $O(p_0)$ where p_0 is the vector constructed in Theorem 2, we obtain the inequality

$$w_n^G(O(p_o)) \le 7\sqrt{n+1}\,\delta.$$

Hence there exists an invariant subspace E of Y of dimension at most n with $d(p_0, E) \le 7\sqrt{n+1}\,\delta$. That means that there is an element $z_0 \in E$ with $\|p_0 - z_0\| \le 7\sqrt{n+1}\,\delta$. Now we set $y_0 = p_0 - z_0$, $\eta_0 = v_0 + z_0$. Since $\eta_0 \in V + E$ we conclude that η_0 is a finite vector of order at most $n + n = 2n$, and $\xi = \eta_0 + y_0$ is a required decomposition. ■

For the convenience of readers we include a brief proof of the cited result from [5], simultaneously correcting some misprints there.

Theorem 3 ([5, Theorem 2]) *If π is a unitary representation of a group G on a Hilbert space H then*

$$w_n^G(A) \le \sqrt{n+1}\, w_n(A),$$

for every G-invariant subset $A \subset H$.

Proof We fix $A \subset H$ and $\delta > w_n(A)$. Choosing $L \in \mathcal{L}_n(H)$ with $dist(A, L) < \delta$, we denote by P the orthogonal projection on L, and let Ω be the closed convex span of the orbit $\{\pi(g)^* P\pi(g) : g \in G\}$ of P. Since $rank(P) < \infty$, it is not difficult to see that Ω is a weakly compact convex subset of $B(H)$, invariant under all maps $F \mapsto \pi(g)^* F\pi(g)$; by the Ryll-Nardzevsky Theorem, these maps have a common fixed point $M_0 \in \Omega$.

The operator $K_0 = 1 - M_0$ belongs to the orbit of operator $1 - P$, and $\pi(g)K_0 = K_0\pi(g)$. It follows that all spectral subspaces of K_0 are invariant under all $\pi(g)$. Let E be the spectral subspace of K_0 corresponding to the interval $[0, \frac{1}{n+1})$. We will show that $\dim E \le n$ and $dist(A, E) \le (n+1)^{1/2}\delta$.

If $\dim E > n$ let us choose an $(n+1)$-dimensional subspace E_1 of E and denote by Q the projection onto E_1. Clearly $trace(K_0 Q) < (n+1)\frac{1}{n+1} = 1$. On the other hand, for each $g \in G$, the operator $M(g) = \pi(g)^*(1 - P)\pi(g)$ is a projection onto a subspace of codimension n, and therefore $(M(g)H) \cap E_1 \ne 0$ whence $trace(M(g)Q) \ge 1$. Since K_0 belongs to the closed convex span of operators $M(g)$, we conclude that $trK_0 Q \ge 1$. The contradiction shows that $\dim E \le n$.

In order to estimate $dist(A, E)$ note that, for each $\xi \in A$,

$$(M(g)\xi, \xi) = \|(1 - P)\pi(g)\xi\|^2 = (dist(\pi(g)\xi, L))^2 \leq \delta^2,$$

so $(K_0\xi, \xi) \leq \delta^2$. It follows easily from the definition of E that $K_0 \geq \frac{1}{n+1}(1 - R)$, where R is the projection onto E. Therefore

$$(dist(\xi, E))^2 = \|(1 - R)\xi\|^2 \leq (n + 1)(K_0\xi, \xi) \leq (n + 1)\delta^2,$$

whence $dist(A, E) \leq (n + 1)^{1/2}\delta$.

It follows that $w_n^G(A) \leq (n + 1)^{1/2}\delta$. Since δ can be chosen arbitrarily close to $w_n(A)$ we obtain the needed inequality. ■

Proof of Corollary 3. Applying Theorem 1 to the set $O(p_0)$, we deduce from the inequality $w_n(O(p_o)) \leq 7\delta$, established in Theorem 2, the inequality $w_n^G(O(p_o)) \leq 7\beta_G(n)\delta$. It remains to repeat the arguments from the proof of Corollary 2. ■

Let us come to the results on functional equations. When the conditions (7) or (4) hold we will always denote by L the linear span of the family $\{a_i : 1 \leq i \leq n\}$. It is clear that $O(f) \subset L + Y$, so we may restrict our attention to $L + Y$ (or assume that $X = L + Y$).

Proof of Corollary 4. We denote by X the space of all functions on G, by Y the space of all bounded functions supplied with the sup-norm. If $f \in X$ is unbounded and satisfies (4) then at least one of the functions b_i is unbounded. Since (4) can be written in the form $\|T_hf - \sum_{i+1}^n b_i(h)a_i\|_Y \leq C$, for some $C > 0$, it follows that $O(f)$ cannot be bounded in a norm on $Y + L$ coinciding with the sup-norm on Y. It remains to apply Proposition 1. ■

Proof of Corollary 5. In our assumptions the space L is one-dimensional: $L = \mathbb{C}a$. So a non-zero invariant subspace $V \subset L$, which exists by Corollary 4, must coincide with L. This means that $a(g)$ is proportional to a multiplicative function on G. ■

Corollaries 6–8 are direct consequences of Corollary 2, Theorem 2 and Corollary 3, respectively.

References

1. Baker, J.: The stability of the cosine equation. Proc. Am. Math. Soc. **80**(3), 411–416 (1980)
2. Baker, J., Lawrence, J., Zorzito, F.: The stability of the equation $f(x + y) = f(x)f(y)$. Proc. Am. Math. Soc. **74**(2), 242–246 (1972)
3. Greenleaf, F.: Invariant Means on Topological Groups and Their Applications. Van Nostrand, New York, Toronto, London (1969)
4. Levi-Cività, T.: Sulle funzioni che ammettono una formula d'addizione del tipo $f(x + y) = \sum_{i=1}^n X_i(x)Y_i(y)$. Atti. Accad. Naz. Lincei Rend. **5**(22), 181–183 (1913)
5. Shulman, E.: Group representations and stability of functional equations. J. Lond. Math. Soc. **54**, 111–120 (1996)
6. Stäkel, P.: Sulla equazione funzionale $f(x + y) = \sum_{i=1}^n X_i(x)Y_i(y)$. Atti. Accad. Naz. Lincei Rend. **5**(22), 392–393 (1913)

7. Stéphanos, C.: Sur une catégorie déquations fonctionnelles. Rend. Circ. Math. Palermo **18**, 360–362 (1904)
8. Szekelyhidi, L.: On a theorem of Baker, Lawrence and Zorzito. Proc. Am. Math. Soc. **84**(1), 95–96 (1982)

The Distance Function and Boundedness of Diameters of the Nearest Elements

Igor' G. Tsar'kov

Abstract The paper is concerned with approximative properties of sets versus the rate of change of the distance function. We solve a number of problems posed recently by W. B. Moor on sets with bounded diameters of sets of nearest points.

Keywords Best approximation · Distance function · Almost convexity · Diameter of the set of nearest points

Given an arbitrary nonempty subset M of a normed linear space X, we let $\varrho(y, M)$ $(y \in X, M \subset X)$ denote the distance to the set M, i.e., $\varrho(y, M) := \inf_{z \in M} \|z - y\|$. By $P_M x$ we denote the set of all nearest points from M to $x \in X$; i.e., $P_M x := \{y \in M \mid \|y - x\| = \varrho(x, M)\}$. The mapping P_M is called the metric projection onto the set M. Next, $B(x, r) = \{y \in X \mid \|y - x\| \leqslant r\}$ and $S(x, r) = \{y \in X \mid \|y - x\| = r\}$ denote, respectively, the ball and the sphere with centre at x and radius $r \geqslant 0$. For brevity, we write B and S instead of $B(0, 1)$ and $S(0, 1)$, respectively. By S^* we denote the dual unit sphere. Given an arbitrary set $M \subset X$, we denote by $\operatorname{conv} M$ and $\overline{\operatorname{conv}}\, M$, the convex hull and closed convex hull of M, respectively.

For brevity, we write $r(x) = \varrho(x, M)$, where $x \in X$. By estimating the maximum rate of change $\varlimsup_{\Delta x \to 0} \frac{r(x+\Delta x) - r(x)}{\|\Delta x\|}$ of the distance function $r(x)$, we will determine approximative and geometric properties of an approximating set. In particular, in a Hilbert space we shall obtain a sharper estimate (than that given by Moor in [1]) for the almost convexity of the approximating set for which $\operatorname{diam} P_M x \leqslant \varepsilon$ for all $x \in X$. We also give estimates for the almost convexity in a similar problem in uniformly smooth and uniformly convex spaces.

To estimate from below the maximum rate of change of the distance function, we shall use a continuous function $f : [a, +\infty) \to [0, 1]$ which is positive on $(a, +\infty)$, where $a \geqslant 0$. We shall assume that the estimate

I. G. Tsar'kov (✉)
Faculty of Mechanics and Mathematics, Moscow State University, Moscow, Russia
e-mail: tsar@mech.math.msu.su

A. Karapetyants et al. (eds.), *Modern Methods in Operator Theory and Harmonic Analysis*, Springer Proceedings in Mathematics & Statistics 291,
https://doi.org/10.1007/978-3-030-26748-3_16

$$\varlimsup_{\Delta x \to 0} \frac{r(x+\Delta x)-r(x)}{\|\Delta x\|} \geqslant f(r(x))$$

holds for all $x \in X$ for which $r(x) > a$.

In the case $f \equiv 1$ and $a = 0$, this condition means the so-called δ-solarity, which in complete spaces implies the almost convexity of approximating sets. A more detailed account and overview of these subjects and related properties can be found in the surveys [2–4].

Let $x \in X$ such that $r(x) = a_0 > a$. Then, for any number $R > a_0$, there exists $\delta_0 > 0$ such that $\inf_{[a_0,R]} f > 12\mu$, where $\mu = \omega(f, \delta_0)$. Given $x \in X$ such that $r(x) > a_0$, let $\delta(x) \in$ be the supremum of $\delta \in (0, \delta_0)$ for which there exists a vector Δx: $\|\Delta x\| = \delta$ satisfying the inequality

$$r(x+\Delta x) - r(x) \geqslant f((r(x)) - \mu)\|\Delta x\|.$$

We have $|r(x+\Delta x) - r(x'+\Delta x)|, |r(x) - r(x')| \leqslant \|x - x'\|$, and hence if $\|x - x'\| \leqslant \mu\|\Delta x\|/4$, then $r(x'+\Delta x) - r(x') \geqslant (f(r(x')) - 3\mu)\|\Delta x\|$ for all x' such that $\|x - x'\| \leqslant \mu\|\Delta x\|/4$.

Theorem 1 *Let X be a Banach space and let $M \subset X$ be such that the inequality*

$$\varlimsup_{\Delta x \to 0} \frac{r(x+\Delta x)-r(x)}{\|\Delta x\|} \geqslant f(r(x))$$

holds for all $x \in X$ for which $r(x) := \varrho(x, M) > a$.

Then, for any $R > r(x)$ and any $x \in X$ such that $r(x) \geqslant a$,

$$\inf\big\{\|x_R - x\| \mid r(x_R) = R\big\} \leqslant \int_{r(x)}^{R} \frac{dr}{f(r)}.$$

Proof Let $\Delta > 0$ be an arbitrary small number. We argue by induction. Given an arbitrary point $x \in X$ such that $r(x) \geqslant a_0 > a$, we put $x_0 = x$.

1°. We find a vector Δx_1 such that $\|\Delta x_1\| = \min\{\Delta, \delta(x_0)/2\}$ and $r(x_0 + \Delta x_1) - r(x_0) \geqslant (f(r(x_0)) - 3\mu)\|\Delta x_1\| \geqslant \mu\|\Delta x_1\|$.

2°. Assuming that the point $x_n = x_{n-1} + \Delta x_n$ is constructed, we find a vector Δx_{n+1} such that $\|\Delta x_{n+1}\| = \min\{\Delta, \delta(x_n)/2\}$ and $r(x_n + \Delta x_{n+1}) - r(x_n) \geqslant (f(r(x_n)) - 3\mu)\|\Delta x_{n+1}\| \geqslant \mu\|\Delta x_{n+1}\|$.

3°. We claim that there exists a number $N \in \mathbb{N}$ such that $r(x_N) \geqslant R - \Delta$. Indeed, if $r(x_n) \leqslant R - \Delta \leqslant R$ for all $n \in \mathbb{N}$, then $\sum_{n=1}^{\infty} \|\Delta x_n\| \leqslant \frac{1}{\mu} \sum_{n=1}^{\infty} (r(x_n) - r(x_{n-1})) \leqslant \frac{R}{\mu}$. Hence, the sequence $\{x_n\}$ converges to some point y and $r(y) \leqslant R - \Delta$. Besides, there exists a vector Δx such that $\|\Delta x\| = \min\{\Delta, \delta(y)/2\}$ and $r(y + \Delta x) - r(y) \geqslant (f(r(y)) - \mu)\|\Delta x\|$. Therefore, the inequality

$$r(x' + \Delta x) - r(x') \geqslant (f(r(x')) - 3\mu)\|\Delta x\|$$

is satisfied in some neighbourhood $O_\eta(y)$. Next, we have $x_n \in O_\eta(y)$ for all n exceeding some $n_0 \in \mathbb{N}$, and so $\lim\limits_{n\to\infty} \|\Delta x_n\| \geqslant \|\Delta x\|$ by the construction of the corresponding sequence, which is impossible. Hence there exists a number $N \in \mathbb{N}$ such that $r(x_N) \geqslant R - \Delta$. It can be assumed that $r(x_N) \in [R - \Delta, R]$.

4°. Thus, we have

$$\frac{\Delta r_i}{f(r_i) - 3\mu} \geqslant \|\Delta x_i\|,$$

where $\Delta r_i = r(x_i) - r(x_{i-1})$, $i = \overline{1, N}$. It follows that

$$\sum_{i=1}^{N} \frac{\Delta r_i}{f(r_i) - 3\mu} \geqslant \sum_{i=1}^{N} \|\Delta x_i\| \geqslant \|x_N - x_0\| = \|x_N - x\|.$$

From any $\varepsilon > 0$ one can find a sufficiently small $\Delta > 0$ such that

$$\int_{r(x)}^{R} \frac{dr}{f(r) - 3\mu} + \varepsilon \geqslant \inf\{\|x_R - x\| \mid r(x_R) = R\}.$$

Since μ can be made arbitrarily small by reducing δ_0, it follows from the arbitrariness of ε that

$$\inf\{\|x_R - x\| \mid r(x_R) = R\} \leqslant \int_{r(x)}^{R} \frac{dr}{f(r)}.$$

Letting a_0 tend to a, we prove the required estimate also for the case $r(x) \geqslant a$, proving the theorem. □

Remark 1 The case $f > 0$ on $(a, +\infty)$ can be extended to the case $f \geqslant 0$ on $[a, +\infty)$. In this case, if the integral $\int_{r(x)}^{R} \frac{dr}{f(r)}$ diverges, we assume that its value is $+\infty$.

Our next aim is to study the conditions which imply the inequality

$$\inf\{\|x_R - x\| \mid r(x_R) = R\} \leqslant \int_{r(x)}^{R} \frac{dr}{f(r)}.$$

Given an arbitrary set $A \subset S^*$ and a vector $\ell \in X$ such that $\|\ell\| = 1$, $x^*(\ell) > 0$ $(x^* \in A)$ we define

$$\alpha(A) = \alpha(A, \ell) := \inf\{x^*(\ell) \mid x^* \in A\}.$$

We let (US) denote the class of all uniformly smooth Banach spaces—these are the spaces in which the modulus of smoothness

$$\Omega(t) := \sup\Big\{\frac{1}{2}\Big(\|x + h\| + \|x - h\|\Big) - 1 \;\Big|\; \|h\| \leqslant t, x \in S\Big\}$$

behaves as $o(t)$ as $t \to 0$. Next, for an arbitrary set $T \subset S(x, r)$ $(r > 0)$, consider the set

$$\mathbb{A}(T) := \{x^* \in S^* \mid x^*(t - x) = \|t - x\|, \ \ t \in T\}.$$

Definition 1 Let M be a nonempty subset of a normed linear space $(X, \|\cdot\|)$. A point $x \in X$ is called a *point of approximative compactness* (for the set M) if, for any sequence $\{y_n\} \subset M$ such that $\|x - y_n\| \to \varrho(x, M)$ $(n \to \infty)$, there exists a subsequence $\{y_{n_k}\}$ converging to some point $y_0 \in M$. By $\mathrm{AC}(M)$ we denote the set of all points of approximative compactness for M. A set M is called *approximatively compact* if $\mathrm{AC}(M) = X$ (see, for example, [2, Sect. 2]).

Theorem 2 *Let* $X \in (\mathrm{US})$, $M \subset X$, $x \in \mathrm{AC}(M) \setminus M$, *and let* $\ell \in X$, $\|\ell\| = 1$. *Then*

$$\overline{\lim_{\Delta x \to 0}} \frac{r(x + \Delta x) - r(x)}{\|\Delta x\|} \geqslant \alpha\big(\mathbb{A}(P_M x)\big).$$

Proof We assume without loss of generality that $x = 0$ and $r(x) = \varrho(x, M) = 1$. Consider an arbitrary point $y \in P_M(0)$ and a functional $y^* \in S^*$ such that $y^*(y) = 1$. Let $\Gamma = \{z \in X \mid y^*(z) = 1\}$ be the tangent hyperplane to the sphere S at the point y. We have $\varrho(-u\ell, \Gamma) - \varrho(0, \Gamma) = uy^*(\ell)$, and besides, $z_u := y(1 + u) - u\ell \in P_\Gamma(-u\ell)$. Since $X \in (\mathrm{US})$, there exists a function $\varphi(\tau) = o(\tau)$ $(\tau \to 0)$ such that any point a lying between the sphere S and the hyperplane Γ satisfies the inequality $\varrho(a, \Gamma) = \varphi(\|a - y\|) = o(\|a - y\|)$.

For an arbitrary $\delta \in (0, y^*(\ell))$ and $x^* \in S^*$ such that $\|x^* - y^*\| \leqslant \delta$, consider the hyperplane $\Gamma_{x^*} = \{z \in X \mid x^*(z) = x^*(y)\}$. We have

$$\begin{aligned}\varrho(-u\ell, \Gamma_{x^*}) - \varrho(0, \Gamma_{x^*}) &= |x^*(y) + x^*(u\ell)| - |x^*(y)| = x^*(y) + x^*(u\ell) - x^*(y) \\ &= ux^*(\ell) \geqslant uy^*(\ell) - u\|x^* - y^*\| = u(y^*(\ell) - \delta).\end{aligned}$$

Next, there exists a number $\varepsilon > 0$ (independent of $y \in P_M 0$) such that the set

$$C_\varepsilon(y) := \{z \in X \mid \|z\| \geqslant 1, \|z - y\| \leqslant \varepsilon\}$$

lies in the set $\{z \in X \mid \exists x^* \in S^* : x^*(z) \geqslant x^*(y), \|x^* - y^*\| \leqslant \delta\}$. Hence $\varrho(-u\ell, C_\varepsilon(y)) - \varrho(0, C_\varepsilon(y)) \geqslant u(y^*(\ell) - \delta)$. We have $x \in \mathrm{AC}(M)$, and so there exists a number $u_0 > 0$ such that $P_M(-u\ell) \subset O_\varepsilon(P_M(0))$ for all $u \in [0, u_0]$. As a result, $\varrho(-u\ell, M) - \varrho(0, M) = \varrho(-u\ell, M) - 1 \geqslant u(\alpha(\mathbb{A}(P_M 0)) - \delta)$ for all $u \in [0, u_0]$; i.e.,

$$\frac{\varrho(-u\ell, M) - \varrho(0, M)}{u} \geqslant \alpha(\mathbb{A}(P_M 0)) - \delta.$$

The required fact follows by making $\delta \to 0$. Theorem 2 is proved. □

Corollary 1 *Let H be a Hilbert space, $M \subset H$, $x \in \mathrm{AC}(M) \setminus M$. Assume that, for some $\ell \in S$, the set $P_M x$ lies in the half-space $\{y \in H \mid (\ell, y - x) \geqslant \alpha\varrho(x, M)\}$. Then*

$$\overline{\lim_{\Delta x \to 0}} \frac{r(x+\Delta x)-r(x)}{\|\Delta x\|} \geqslant \alpha.$$

Proof Given any point $t \in P_M x$ and a vector $e_t := \frac{t-x}{\|t-x\|}$ we have $(e_t, \ell) = \frac{(\ell, t-x)}{\|t-x\|} \geqslant \frac{\alpha\rho(x,M)}{\|t-x\|} = \alpha$. Hence $\alpha(\mathbb{A}(P_M(x))) \geqslant \alpha$. The required fact now follows from Theorem 2. □

In a Banach space X, the quantity

$$J(X) = \sup\left\{\frac{R(M,X)}{\operatorname{diam} M} \;\middle|\; M : 0 < \operatorname{diam} M < \infty\right\},$$

where $R(M,X) := \inf\{R > 0 \mid M \subset B(z,R)\}$ is the Chebyshev radius, is called the *Jung constant* of the space X (see, for example, [6, Sect. I.7]). It is well-known that $J_m = J(H) = \sqrt{\frac{m}{2(m+1)}}$ for a Hilbert space H of finite dimension m. In an infinite-dimensional Hilbert space H one has $J_\infty = J(H) = \sqrt{1/2}$.

Corollary 2 *Let H be a Hilbert space of dimension $n \in \overline{\mathbb{N}}$, $n \geqslant 2$, $M \subset H$, $x \in \mathrm{AC}(M) \setminus M$, $r = \varrho(x, M) > \varepsilon J_{n-1} > 0$, $\operatorname{diam} P_M(x) = \varepsilon$. Then*

$$\overline{\lim_{\Delta x \to 0}} \frac{r(x+\Delta x)-r(x)}{\|\Delta x\|} \geqslant \frac{\sqrt{r^2 - \varepsilon^2 J_{n-1}^2}}{r},$$

where J_{n-1} is the Jung constant of the Hilbert space of dimension $n-1 \in \overline{\mathbb{N}}$.

Proof Let $N = \overline{\operatorname{conv}}\, P_M x$. Then $\operatorname{diam} N \leqslant \varepsilon$. We assume without loss of generality that $\operatorname{diam} N = \varepsilon$. Let y_0 be a unique nearest point for x in N. There exists $x^* \in S$ for which $(x^*, y) \geqslant (x^*, z)$ for all $y \in N$ and $z \in B(x, \varrho(x, N))$, and hence, $(x^*, y - x) \geqslant \varrho(x, N)$. By Carathéodory's theorem, there exists an inclusion-minimal set of points $\{x_i\}_{i=1}^{m+1} \subset P_M x \cap \Gamma$ $(m \leqslant n-1)$ such that $y_0 \in K := \overline{\operatorname{conv}}\, \{x_i\}_{i=1}^{m+1}$, where $\Gamma = \{y \in H \mid (x^*, y) = \varrho(x, N)\}$. By Garkavi's theorem (see [5]), y_0 is the Chebyshev centre of the set K. Hence, since $\|x_k - x_l\| = \varepsilon$ $(k \neq l)$, we see that the radius of the ball $B(x, r) \cap \Gamma$ is equal to $\varepsilon J_m \leqslant \varepsilon J_{n-1}$. By the previous assertion, $\|x - y_0\| = \sqrt{r^2 - \varepsilon^2 J_m^2}$. Hence

$$\overline{\lim_{\Delta x \to 0}} \frac{r(x+\Delta x)-r(x)}{\|\Delta x\|} \geqslant \frac{\|x - y_0\|}{r} \geqslant \frac{\sqrt{r^2 - \varepsilon^2 J_{n-1}^2}}{r},$$

the result required. □

Theorem 3 *Let H be a Hilbert space, $\dim H = n \in \overline{\mathbb{N}}$, $n \geqslant 2$, $M \subset H$ be an approximatively compact set such that $\operatorname{diam} P_M x \leqslant \varepsilon$ for all $x \in H$ for which $\varrho(x, M) \geqslant a > \varepsilon J_{n-1}$. Then $N = \overline{\operatorname{conv}}\, M \subset \overline{O_a(M)}$.*

Proof Indeed, consider an arbitrary point $x \in H$ such that $r(x) := \varrho(x, M) \geqslant a$. By Theorem 1 and Corollary 2, for all $R > a$ and $\delta > 0$ there exists a point $x_R \in H$ such that $\varrho(x_R, M) = R$ and

$$\|x - x_R\| \leqslant \int_{r(x)}^{R} \frac{r\,dr}{\sqrt{r^2 - \varepsilon^2(J_{n-1} + \delta)^2}} = \sqrt{r^2 - \varepsilon^2(J_{n-1} + \delta)^2}\,\Big|_{r(x)}^{R}$$
$$= \sqrt{R^2 - \varepsilon^2(J_{n-1} + \delta)^2} - \sqrt{r^2(x) - \varepsilon^2(J_{n-1} + \delta)^2}.$$

This inequality means that the ball $B(x_R, R)$ contains the ball $B(x, r_\delta)$, where $r_\delta = \sqrt{r^2(x) - \varepsilon^2(J_{n-1} + \delta)^2}$. We claim that the set N has no common points with the interior of the ball $B(x, r_\delta)$. Assume on the contrary that there exist points $\{a_i\}_{i=1}^N \subset M$ and numbers $\{\lambda_i\}_{i=1}^N \subset [0, 1]$, $\sum_{i=1}^N \lambda_i = 1$, such that the point $z = \sum_{i=1}^N \lambda_i a_i$ lies in the ball $B(x, r_\delta - \varepsilon')$ for some $\varepsilon' \in (0, r_\delta)$. Let $d := \max_{i=\overline{1,N}} \|a_i - z\| + r_\delta$ and let $\Gamma_0 = \{v \in H \mid z_0^*(v) = z_0^*(z_0)\}$ ($z_0^* \in S^*$) be the tangent hyperplane to the ball $B(x_R, R)$ at some point $z_0 \in B(x, R - \|x - x_R\|) \cap S(x_R, R)$ (here $R - \|x - x_R\| \geqslant r_\delta$). For sufficiently large R, the tangent plane Γ_0 to the ball $B(x_R, R)$ can be approximated in a d-neighbourhood of the tangent plane by the boundary of the ball $B(x_R, R)$ with accuracy not exceeding $\varepsilon'/2$. Hence, for the half-space $P = \{v \in H \mid z_0^*(v) \leqslant z_0^*(z_0) - \varepsilon'/2\}$, which contains the ball $B(x, r_\delta - \varepsilon'/2)$, the interior of P has no common points with the set $M \cap O_d(z_0)$, and therefore, it does not meet the set $\operatorname{conv}(M \cap O_d(z_0))$. But this contradicts the condition $z \in B(x, r_\delta - \varepsilon')$. Thus, we have proved that N is disjoint from the interior of the ball $B(x, r_\delta)$. Since $\delta > 0$ is arbitrary, the set N has no common points with the interior of the ball $B(x, r_0)$, where $r_0 = \sqrt{r^2(x) - \varepsilon^2 J_{n-1}^2}$. By Hahn–Banach's theorem, there exists a functional $x^* \in S^*$ separating the set N and the ball $B(x, r_0)$ and such that $x^*(v - x) \geqslant r_0$ for all $v \in N$. The half-space $P_x = \{v \in H \mid x^*(v) \geqslant r_0 + x^*(x)\}$ separates from the ball $B(x, r(x))$ its segment $Q = \{v \in B(x, r(x)) \mid x^*(v) \geqslant r_0 + x^*(x)\}$. The base D of this segment is the section of the ball $B(x, r(x))$ by the hyperplane lying at the distance r_0 from the point x, i.e., D is a hyperball of radius εJ_{n-1}.

Note that $\varrho(v, M) \leqslant a$ for all $v \in N$, for otherwise, arguing for v as for x, we would obtain that the point v can be strictly separated by a hyperplane from N. Thus, $N \subset \overline{O_a(M)}$, proving Theorem 3. □

The following result, which refines Theorem 3 of [1], follows directly from Theorem 3.

Corollary 3 *Let H be a Hilbert space,* $\dim H = n \in \overline{\mathbb{N}}$, $n \geqslant 2$, *$M \subset H$ be an approximatively compact set, and let* $\operatorname{diam} P_M(x) \leqslant \varepsilon$ *for all x for which $\varrho(x, M) > \varepsilon J_{n-1}$. Then $N = \overline{\operatorname{conv}}\, M \subset \overline{O_{\varepsilon J_{n-1}}(M)}$.*

Note that the estimate of the almost convexity (the radius of the neighbourhood of M) is sharp on the class of all approximatively compact sets $M \subset H$ for which $\operatorname{diam} P_M(x) \leqslant \varepsilon$ for all $x \in H$.

Corollary 4 *Let H be a Hilbert space,* $\dim H = n \in \overline{\mathbb{N}}$, $n \geqslant 2$, *$M \subset H$ be an approximatively compact set such that* $\operatorname{diam} P_M(x) \leqslant \varepsilon$ *for all x such that $\varrho(x, M) \geqslant a \geqslant \varepsilon J_{n-1}$. Then the boundary $N = \overline{\operatorname{conv} M}$ is contained in $\overline{O_{\varepsilon J_{n-1}}(M)}$.*

Proof Given any point $y_0 \in \partial N$ and an arbitrary $\delta > 0$, there exists a point $x_0 \in \partial N$, $\|y_0 - x_0\| < \delta$, which is a nearest point to some point $x \in H$ such that $1 \leqslant \varrho(x, N) \leqslant r(x) := \varrho(x, M)$. Let $r_0 = \sqrt{r^2(x) - \varepsilon^2 J_{n-1}^2}$. As was shown in the proof of Theorem 3, the interior of the ball $B(x, r_0)$ has no common points with the set N. Hence $\varrho(x, N) = \|x_0 - x\| \geqslant r_0$. Let $x^* \in S^*$ be a functional separating the ball $B(x, \|x_0 - x\|)$ and the set N (we assume that the half-space $P_x = \{v \in H \mid x^*(v) \geqslant \|x_0 - x\| + x^*(x)\}$ contains the set N). This half-space separates from the ball $B(x, r(x))$ its segment $Q = B(x, r(x)) \cap P_x$. The base D of this segment is a hyperball of radius $\leqslant \varepsilon J_{n-1}$ with centre x_0. The distance from the point x_0 to any point of the segment Q is at most εJ_{n-1}. Since $P_M x \subset Q$, we have $\varrho(x_0, M) = \varrho(x_0, P_M x) \leqslant \varepsilon J_{n-1}$. Hence $\varrho(y_0, M) \leqslant \varrho(x_0, M) + \delta \leqslant \varepsilon J_{n-1} + \delta$. Moreover, since δ is arbitrary, we have $\varrho(y_0, M) \leqslant \varepsilon J_{n-1}$, proving the corollary. □

Lemma 1 *Let* $X \in (\mathrm{US})$, $M \subset X$, $x \in \mathrm{AC}(M) \setminus M$, $\operatorname{diam} P_M x \leqslant \varepsilon \leqslant r(x) = \varrho(x, M)$. *Then*

$$\varlimsup_{\Delta x \to 0} \frac{r(x + \Delta x) - r(x)}{\|\Delta x\|} \geqslant 1 - 2\Omega\Big(\frac{\varepsilon}{r(x)}\Big).$$

Proof We assume without loss of generality that $r(x) = 1$. Let $y, y_0 \in P_M x$, $\ell_0 = y_0 - x$, $\ell = y - x$ and let $y^*, y_0^* \in S^*$ be such that $y^*(\ell) = 1$, $y_0^*(\ell_0) = 1$. Then $y^*(\ell_0) = y^*(\ell) + y^*(\ell_0 - \ell) \geqslant 1 - 2\Omega(\varepsilon)$. It follows that

$$\alpha\big(\mathbb{A}(P_M x)\big) \geqslant 1 - 2\Omega\Big(\frac{\varepsilon}{r(x)}\Big).$$

Now the conclusion of the lemma follows from Theorem 2. □

Corollary 5 *Let $X \in (\mathrm{US})$, $M \subset X$ be an approximatively compact set such that* $\operatorname{diam} P_M(x) \leqslant \varepsilon$ *for all $x \in X$. Let $\tau_0 \in (0, 1/2)$ be such that $\Omega(\tau_0) < 1/4$ and that the integral $K(\tau) := \int_0^\tau \frac{\Omega(t)}{t^2} dt$ is convergent. Then for all $x \in X \setminus M$ such that $r(x) = \varrho(x, M) \geqslant r_0 > \varepsilon/\tau_0$ and an arbitrary $R > r(x)$, there exists a point $x_R \in X \setminus M$ such that $\varrho(x_R, M) = R$, $\|x_R - x\| < R - r(x) + 4\varepsilon K(\varepsilon/r_0)$.*

Proof By Theorem 1

$$\|x - x_R\| \leqslant \int_{r(x)}^R \frac{dr}{1 - 2\Omega\big(\frac{\varepsilon}{r}\big)} = \int_{r(x)}^R \left(1 + \frac{2\Omega\big(\frac{\varepsilon}{r}\big)}{1 - 2\Omega\big(\frac{\varepsilon}{r}\big)}\right) dr$$

$$= R - r(x) + \int_{r(x)}^R \frac{2\Omega\big(\frac{\varepsilon}{r}\big)\, dr}{1 - 2\Omega\big(\frac{\varepsilon}{r}\big)}.$$

Hence

$$\left|\int_{r(x)}^{R} \frac{2\Omega\left(\frac{\varepsilon}{r}\right) dr}{1 - 2\Omega\left(\frac{\varepsilon}{r}\right)}\right| \leqslant \int_{r(x)}^{+\infty} 4\Omega\left(\frac{\varepsilon}{r}\right) dr = \left|t = \frac{\varepsilon}{r}\right|$$

$$= 4\varepsilon \int_{0}^{\frac{\varepsilon}{r(x)}} \frac{\Omega(t)}{t^2} dt = 4\varepsilon K\left(\frac{\varepsilon}{r(x)}\right) \leqslant 4\varepsilon K(\varepsilon/r_0),$$

proving the corollary. □

Remark 2 Corollary 5 means that the points $x_0 := x + r(x)\frac{x - x_R}{\|x - x_R\|}$ and $z_0 := x_R + R\frac{x - x_R}{\|x - x_R\|}$, which lie on the spheres $S(x, r(x))$ and $S(x_R, R)$, respectively, lie at the distance $\Delta := \|x - x_R\| + r(x) - R < 4\varepsilon K\left(\frac{\varepsilon}{r_0}\right)$ from each other.

Since $r(x) - 4\varepsilon K\left(\frac{\varepsilon}{r_0}\right) < R - \|x - x_R\|$, we have $B(x_R, R) \supset B\big(x, r(x) - 4\varepsilon K\left(\frac{\varepsilon}{r_0}\right)\big)$. Moreover, $r(x) \geqslant r_0 > \frac{\varepsilon}{\tau_0} > 2\varepsilon$, and so $4\varepsilon K\left(\frac{\varepsilon}{r_0}\right) < \varepsilon$ and $r(x) - K\left(\frac{\varepsilon}{r_0}\right) > \varepsilon$.

Lemma 2 *The set $N = \overline{\mathrm{conv}}\, M$ is disjoint from the ball $B(x, \varepsilon/4) \subset B\big(x, r(x) - 4\varepsilon K\left(\frac{\varepsilon}{r_0}\right)\big)$.*

Proof Assume on the contrary that there exist points $\{a_i\}_{i=1}^N \subset M$ and numbers $\{\lambda_i\}_{i=1}^N \subset [0, 1]$, $\sum\limits_{i=1}^{N} \lambda_i = 1$, such that the point $z = \sum\limits_{i=1}^{N} \lambda_i a_i$ lies in the ball $B(x, \varepsilon/4)$. Let $d := \max\limits_{i=\overline{1,N}} \|a_i - z\| + r(x)$ and let $\Gamma_0 = \{v \in X \mid z_0^*(v) = z_0^*(z_0)\}$ $(z_0^* \in S^*)$ be the tangent hyperplane to the ball $B(x_R, R)$ at some point $z_0 \in B(x, r(x)) \cap S(x_R, R)$. There exists a sufficiently large $R_0 > 0$ such that, for all $R \geqslant R_0$, the tangent plane Γ_0 in the d-neighbourhood of the point z_0 is approximated by the boundary of the ball $B(x_R, R)$ with the accuracy at most $\varepsilon/2$. Then, for the half-space $P = \{v \in X \mid z_0^*(v) \leqslant z_0^*(z_0) - \varepsilon/2\}$, which contains the ball $B(x, \varepsilon/2)$, the interior of P is disjoint from the set $M \cap O_d(z_0)$, and hence, it has no common points with the set $\mathrm{conv}(M \cap O_d(z_0))$. However, the last fact contradicts the condition $z \in B(x, \varepsilon/4)$. Hence N does not meet $B(x, \varepsilon/4)$, the result required. □

From Lemma 2 it follows that $\varrho(x, N) \leqslant r_0$ for all $x \in N = \overline{\mathrm{conv}}\,(M)$, for otherwise N would be disjoint from ball $B(x, \varepsilon/4)$, which is impossible.

Corollary 6 *Let $X \in$ (US). Then $N = \overline{\mathrm{conv}}\,(M) \subset \overline{O_{r_0}(M)}$. In particular, this inclusion holds for $r_0 = \frac{2\varepsilon}{\tau_0}$.*

Definition 2 Given a Banach space X, we denote by $\omega(t) = \omega_X(t)$ the modulus of convexity of X, which is defined as

$$\omega(t) = \inf\left\{1 - \left\|\frac{x + y}{2}\right\| \mid x, y \in S,\ \ \|x - y\| \leqslant t\right\}.$$

By (UR) we denote the class of all uniformly convex spaces X, i.e., the spaces X for which $\omega_X(t) > 0$ for all $t \in (0, 2]$.

Let $\Gamma = \{z \in X \mid x^*(z-x) = \|z_0 - x\|\}$ $(x^* \in S^*)$ be the tangent hyperplane to the ball $B(x_R, R)$ at a point z_0. If $R \geqslant R_0$, where R_0 is a sufficiently large number, then the half-space

$$\Pi_\varepsilon = \left\{z \in X \;\middle|\; x^*(z-x) \geqslant x^*(z_0 - x) - \left(4\varepsilon K\left(\frac{\varepsilon}{r_0}\right) - \Delta\right)\right\}$$

contains the set $B(x, r(x)) \setminus B(x_R, R)$. Hence

$$\operatorname{diam}\big(B(x, r(x)) \setminus B(x_R, R)\big) \leqslant r(x)\omega^{-1}\left(\frac{4\varepsilon K(\frac{\varepsilon}{r_0})}{r(x)}\right),$$

where $\omega^{-1}(t)$ is the inverse function of the modulus of convexity ω_X.

Remark 3 In the case $X = L_p := L_p(T, \Sigma, \mu)$ $(1 < p < \infty)$, there exist $k_p, K_p > 0$ such that $\Omega(t) \leqslant K_p t^\alpha$ and $\omega^{-1}(t) \leqslant k_p t^\beta$, where $\alpha = \min\{p, 2\}$, $\beta = \min\{\frac{1}{p}, \frac{1}{2}\}$. Hence $K(\tau) \leqslant K_p \int_0^\tau t^{\alpha-2} dt = K_p \frac{\tau^{\alpha-1}}{\alpha-1}$, and for $r(x) \geqslant r_0$, we have

$$\begin{aligned}
&r(x)\omega^{-1}\left(4\frac{\varepsilon}{r(x)} K\left(\frac{\varepsilon}{r_0}\right)\right) \leqslant r(x)\omega^{-1}\left(4\frac{\varepsilon}{r(x)}\frac{K_p}{\alpha-1}\left(\frac{\varepsilon}{r_0}\right)^{\alpha-1}\right)\\
=&r(x)\omega^{-1}\left(\frac{4K_p}{r(x)r_0^{\alpha-1}(\alpha-1)}\varepsilon^\alpha\right) \leqslant r(x)k_p\left(\frac{4K_p}{r(x)r_0^{\alpha-1}(\alpha-1)}\right)^\beta \varepsilon^{\alpha\beta}\\
=&A_p r(x)^{1-\beta} r_0^{\beta-\alpha\beta}\varepsilon^{\alpha\beta} = A_p r(x)^{1-\beta} r_0^{\beta-\gamma}\varepsilon^\gamma,
\end{aligned}$$

where $A_p = k_p\left(\frac{4K_p}{(\alpha-1)}\right)^\beta$, $\gamma = \alpha\beta = \min\{\frac{2}{p}, \frac{p}{2}\}$. Hence, for $r(x) = r_0$ and $R \geqslant R_0$,

$$\operatorname{diam}\big(B(x, r(x)) \setminus B(x_R, R)\big) \leqslant A_p r_0^{1-\gamma}\varepsilon^\gamma.$$

The proof of the following result is similar to that of Corollary 4.

Theorem 4 *Let $X \in$ (US) $\cap$ (UR), $M \subset X$ be an approximatively compact set. Assume that* diam $P_M x \leqslant \varepsilon$ *for all $x \in X$ for which $r(x) \geqslant r_0 > \frac{\varepsilon}{\tau_0}$. Then the boundary of the set $N = \overline{\operatorname{conv}}\, M$ is contained in $\overline{O_d(M)}$, where $d = \min\left\{r_0\omega^{-1}\left(4\frac{\varepsilon}{r_0}K\left(\frac{\varepsilon}{r_0}\right)\right), r_0\right\}$.*

In particular, for $X = L_p$ $(1 < p < \infty)$, the set ∂N lies in $\overline{O_d(M)}$, where $\gamma = \min\{\frac{2}{p}, \frac{p}{2}\}$, $d = \min\left\{r_0, A_p r_0^{1-\gamma}\varepsilon^\gamma\right\}$.

Acknowledgements This research was carried out with the financial support of the Russian Foundation for Basic Research (grant nos. 16-01-00295-a and 19-01-00332-a).

References

1. Moors W.B.: Nearly Chebyshev sets are almost convex. Set-Valued Var. Anal., 1–10 (2017). https://doi.org/10.1007/s1128-017-0445-4.
2. Alimov, A.R., Tsar'kov, I.G.: Connectedness and solarity in problems of best and near-best approximation. Russ. Math. Surv. **71**(1), 1–77 (2016). https://doi.org/10.1070/RM9698
3. Balaganskii, V.S., Vlasov, L.P.: The problem of convexity of Chebyshev sets. Russ. Math. Surv. **51**(6), 1127–1190 (1996). https://doi.org/10.1070/RM1996v051n06ABEH003002
4. Alimov, A.R., Tsar'kov, I.G.: Connectedness and other geometric properties of suns and Chebyshev sets. J. Math. Sci. (New York) **217**(6), 683–730 (2016). https://doi.org/10.1007/s10958-016-3000-1
5. Garkavi A.L.: On the Chebyshev center and convex hull of a set. Usp. Matem. Nauk **19**(6), 139–145 (1964). http://mi.mathnet.ru/umn6276
6. Leichtweiß, K.: Konvexe Mengen. VEB, Berlin (1980). (in German)

Part III
Differential Equations and Mathematical Physics

The Influence of Oscillations on Energy Estimates for Damped Wave Models with Time-dependent Propagation Speed and Dissipation

Halit Sevki Aslan and Michael Reissig

Abstract The aim of this paper is to derive higher order energy estimates for solutions to the Cauchy problem for damped wave models with time-dependent propagation speed and dissipation. The model of interest is

$$\begin{cases} u_{tt} - \lambda^2(t)\omega^2(t)\Delta u + \rho(t)\omega(t)u_t = 0, & (t,x) \in [0,\infty) \times \mathbb{R}^n, \\ u(0,x) = u_0(x), \quad u_t(0,x) = u_1(x), & x \in \mathbb{R}^n, \end{cases}$$

The coefficients $\lambda = \lambda(t)$ and $\rho = \rho(t)$ are shape functions and $\omega = \omega(t)$ is a bounded oscillating function. If $\omega(t) \equiv 1$ and $\rho(t)u_t$ is an effective dissipation term, then $L^2 - L^2$ energy estimates are proved in Bui and Reissig (Fourier analysis, trends in mathematics. Birkhäuser, Basel, [2]). In contrast, the main goal of the present paper is to generalize the previous results to coefficients including an oscillating function in the time-dependent coefficients. We will explain how the interplay between the shape functions and oscillating behavior of the coefficient $\omega = \omega(t)$ will influence energy estimates.

Keywords Damped wave model · Effective dissipation · Very fast oscillations · Stabilization condition · WKB analysis · Energy estimate

1 Introduction

Let us consider the following Cauchy problem for the damped wave equation with time-dependent propagation speed and dissipation:

H. S. Aslan · M. Reissig (✉)
Faculty for Mathematics and Computer Science, Technical University Bergakademie Freiberg, Prüferstr 9, 09596 Freiberg, Germany
e-mail: reissig@math.tu-freiberg.de

H. S. Aslan
e-mail: halitsevkiaslan@gmail.com

A. Karapetyants et al. (eds.), *Modern Methods in Operator Theory and Harmonic Analysis*, Springer Proceedings in Mathematics & Statistics 291,
https://doi.org/10.1007/978-3-030-26748-3_17

$$\begin{cases} u_{tt} - a^2(t)\Delta u + b(t)u_t = 0, & (t,x) \in [0,\infty) \times \mathbb{R}^n, \\ u(0,x) = u_0(x), \quad u_t(0,x) = u_1(x), & x \in \mathbb{R}^n. \end{cases} \tag{1}$$

The total energy of the solution to (1) is defined by

$$\mathbb{E}(u,t) = \frac{1}{2}\int_{\mathbb{R}^n} \big(a^2(t)|\nabla u(t,x)|^2 + |u_t(t,x)|^2\big)dx.$$

The properties of the time-dependent propagation speed $a = a(t)$ and the coefficient $b = b(t)$ in the damping term have different effects on the behavior of $\mathbb{E}(u,t)$ as $t \to \infty$.

First we consider the properties of the energy in the case $b(t) \equiv 0$ in (1). If $0 < a_0 \le a(t) \le a_1$, in general one cannot expect the so-called *generalized energy conservation* property (GEC), that is, the both sided estimate

$$C_0\mathbb{E}(u,0) \le \mathbb{E}(u,t) \le C_1\mathbb{E}(u,0)$$

holds for the wave energy with positive constants C_0 and C_1. The main reason is, that an oscillating behavior of $a = a(t)$ may have a very deteriorating influence on the energy behavior of the solutions to (1) (see [5, 13]). However, if we assume for the oscillating behavior of $a = a(t)$ the assumptions

$$|a^{(k)}(t)| \le C_k(1+t)^{-k} \text{ for } k = 0, 1, 2,$$

then (GEC) holds (see [12]). In this case only very slow oscillations are allowed. In general, very fast oscillating coefficients one might expect to destroy the estimates, which are valid for very slow oscillating coefficients. However, in [8] even though the oscillations are very fast, the author has proved (GEC) to (1) under the following additional assumptions to the coefficient:

$$|a^{(k)}(t)| \le C_k(1+t)^{-kr} \text{ for } k = 0, 1, \cdots, M, \tag{2}$$

and

$$\int_0^t |a(\tau) - a_\infty| d\tau \le C(1+t)^q, \tag{3}$$

for some real a_∞ and $q \in (0,1)$ with $r > q + \dfrac{1-q}{M}$. Here (3) is called the *stabilization condition* (see [8]) and by this condition one can get some benefit of higher order regularity of $a = a(t)$. Moreover, in the case $b(t) \equiv 0$ in the paper [10] the authors studied the Cauchy problem (1) after introducing $a(t) = \lambda(t)\omega(t)$ with the monotonously increasing shape function $\lambda = \lambda(t)$ and the (bounded) oscillating function $\omega = \omega(t)$. By using the C^M property of $\lambda = \lambda(t)$ and $\omega = \omega(t)$ and the idea of stabilization condition they proved a two sided estimate

$$C_0 \leq \frac{1}{\lambda(t)} \mathbb{E}_\lambda(u,t) \leq C_1,$$

where the non-negative constants C_0 and C_1 depend on the data and

$$\mathbb{E}_\lambda(u,t) = \frac{1}{2} \int_{\mathbb{R}^n} \big(\lambda^2(t)|\nabla u(t,x)|^2 + |u_t(t,x)|^2\big)dx.$$

Now we discuss the Cauchy problem (1) in the case of constant speed of propagation $a(t) \equiv 1$. Assuming a suitable control of the oscillations in $b = b(t)$, the following classification of damping terms $b(t)u_t$ is proposed in [15, 16]: *non-effective dissipation, effective dissipation, scattering producing* and *over-damping producing*.

Finally, let us consider the Cauchy problem (1) with time-dependent propagation speed and dissipation. In [2] the authors studied the Cauchy problem (1) assuming a suitable control of the oscillations of $a = a(t)$ and $b = b(t)$. They proposed a classification of the damping term $b(t)u_t$ in terms of an increasing speed of propagation $a = a(t)$, given in [15, 16]. For the effective dissipation case, the authors proved the energy of the solution to (1) satisfies the following $L^2 - L^2$ estimate for all $t > 0$:

$$\big\|u_t(t,\cdot), a(t)\nabla u(t,\cdot)\big\|_{L^2} \lesssim a(t)\Big(1 + \int_0^t \frac{a^2(\tau)}{b(\tau)}d\tau\Big)^{-\frac{1}{2}} \big(\|u_0\|_{H^1} + \|u_1\|_{L^2}\big).$$

The main goal of this paper is to derive higher order energy estimates for solutions of the Cauchy problem (1) with time-dependent speed of propagation and time-depending damping term both having a time-dependent oscillation term. For this reason, we assume that the coefficients $a = a(t)$ and $b = b(t)$ in (1) can be represented by the following products:

$$a(t) = \lambda(t)\omega(t) \quad \text{and} \quad b(t) = \rho(t)\omega(t).$$

Here the time-dependent functions $\lambda = \lambda(t)$, $\rho = \rho(t)$ and $\omega = \omega(t)$ are smooth and strictly positive functions. Thus, in this paper we devote ourselves to the following Cauchy problem:

$$\begin{cases} u_{tt} - \lambda^2(t)\omega^2(t)\Delta u + \rho(t)\omega(t)u_t = 0, & (t,x) \in [0,\infty) \times \mathbb{R}^n, \\ u(0,x) = u_0(x), \quad u_t(0,x) = u_1(x), & x \in \mathbb{R}^n, \end{cases} \tag{4}$$

where λ is a monotonously increasing nontrivial shape function in the propagation speed, ρ is a nontrivial shape function in the damping term and ω is a bounded oscillating function in both propagation speed and damping term. Throughout this paper, we restrict ourselves to "effective-like" (due to oscillating term $\omega = \omega(t)$) damping $\rho(t)\omega(t)u_t$ in the sense of [2, 16]. Here effective means that the solution behaves like that of a corresponding heat equation.

Remark 1 We assume that the damped wave equation (4) has the same oscillating function $\omega = \omega(t)$ in both speed of the propagation and damping term. Only for this model we are able to develop a suitable approach basing on tools from WKB-analysis. In particular, the division of the extended phase space into regions and zones (see Sect. 3) with reasonable properties of the separating lines depends heavily on the coincidence of the oscillating part in the propagation speed and the dissipation term.

Then, the main results of this paper are the following estimates for Sobolev solutions to the Cauchy problem (4).

Theorem 1 (Main Theorem) *We assume that $\lambda = \lambda(t)$, $\rho = \rho(t)$ and $\omega = \omega(t)$ satisfy the conditions (A1) to (A5) and (B1) to (B6) (see below). Then, the Sobolev solutions to the Cauchy problem* (4) *satisfy the following estimates with $m \in [1, 2)$ and $\sigma \geq 0$:*

$$\|u(t,\cdot)\|_{\dot{H}^\sigma} \lesssim \big(1 + B_\lambda(0,t)\big)^{-\frac{\sigma}{2}-\frac{n}{2}(\frac{1}{m}-\frac{1}{2})}\big(\|u_0\|_{L^m\cap H^\sigma} + \|u_1\|_{L^m\cap H^{[\sigma-1]_+}}\big),$$
$$\|u_t(t,\cdot)\|_{\dot{H}^\sigma} \lesssim \lambda(t)\big(1 + B_\lambda(0,t)\big)^{-\frac{\sigma}{2}-\frac{n}{2}(\frac{1}{m}-\frac{1}{2})-\frac{1}{2}}\big(\|u_0\|_{L^m\cap H^{\sigma+1}} + \|u_1\|_{L^m\cap H^\sigma}\big),$$

where $B_\lambda(0,t) := \int_0^t \frac{\lambda^2(\tau)}{\rho(\tau)}d\tau$.

Remark 2 The estimates for $\|u(t,\cdot)\|_{\dot{H}^\sigma}$ from Theorem 1 coincide with the ones for solutions to the Cauchy problem (4) with $\omega \equiv 1$. For this reason different influences of the conditions for ω on the estimates for solutions and derivatives (bad influence of the oscillating behavior (A2) and improving influence of the stabilization condition (A3)) are in equilibrium. Some differences in the decay rate appear in the estimates for $\|u_t(t,\cdot)\|_{\dot{H}^\sigma}$. Here we feel some bad influence of ω.

The content of the paper is organized as follows:

- In Sect. 2 we will introduce our assumptions for the coefficients $\lambda = \lambda(t)$, $\rho = \rho(t)$ and $\omega = \omega(t)$.
- In Sect. 3 we apply (hyperbolic and elliptic) WKB-analysis to get representations of solutions. For this reason we divide the extended phase space into some zones and transform the second order equation to a system of first order for suitable micro-energies. This gives us precise information on the structure of fundamental solutions in different parts of the extended phase space.
- Sect. 4 is devoted to glue the estimates for the fundamental solutions from different parts of the extended phase space.
- In Sect. 5 we derive some estimates for large and small frequencies, which prove the main theorem of the Cauchy problem (4).
- In Sect. 6 we explain our assumptions and the main results for some examples of admissible coefficients.
- Some concluding remarks and open problems in Sect. 7 complete the paper.

Notations In this paper we use $f \lesssim g$ for positive functions f and g if there exists a constant $C > 0$ such that the estimate $f \leq Cg$ is valid. Moreover, $f \approx g$ denotes if f

and g satisfy $f \lesssim g$ and $g \lesssim f$. $f = o\big(g\big)$ denotes that $\limsup_{x\to\infty} \big|\frac{f(x)}{g(x)}\big| = 0$. $(|\cdot|)$ denotes for a matrix the matrix of the absolute values of its entries. We introduce $[x]_+ := \max\{x, 0\}$. Finally, by H^σ and $\dot{H}^\sigma$, we denote Bessel and Riesz potential spaces based on L^2, respectively.

2 Assumptions

We assume the following conditions for $\lambda = \lambda(t)$ and $\omega = \omega(t)$ belonging to $C^M(\mathbb{R}_+)$ with $M \geq 2$ (here we follow some ideas of [10]):

(A1) $\lambda(t) > 0$ and $\lambda'(t) > 0$ for all times $t > 0$ and the derivatives of λ satisfy the conditions

$$\lambda_0 \frac{\lambda(t)}{\Lambda(t)} \leq \frac{\lambda'(t)}{\lambda(t)} \leq \lambda_1 \frac{\lambda(t)}{\Lambda(t)}, \quad |d_t^k \lambda(t)| \leq \lambda_k \lambda(t) \Big(\frac{\lambda(t)}{\Lambda(t)}\Big)^k, \quad k = 1, 2, \cdots, M,$$

where λ_0 and all λ_k are positive constants and $\Lambda(t) = 1 + \int_0^t \lambda(\tau)d\tau$ is a primitive of $\lambda(t)$;

(A2) $0 < c_0 \leq \omega(t) \leq c_1$ and the derivatives of ω satisfy the conditions

$$|d_t^k \omega(t)| \leq \omega_k \Xi^{-k}(t), \quad k = 1, 2, \cdots, M,$$

where all ω_k are positive constants and $\Xi = \Xi(t)$ is a positive, monotonous and continuous function satisfying the compatibility condition

$$C_1 \Theta(t) \leq \lambda(t)\Xi(t) \leq C_2 \Lambda(t),$$

where $\Theta = \Theta(t)$ is a strictly increasing continuous function with $\Theta(0) = 1$, $\Theta(t) < \Lambda(t)$ for $t > 0$ and $\Theta(t) = o\big(\Lambda(t)\big)$;

(A3) $\omega = \omega(t)$ is λ-stabilizing towards 1, that is,

$$\int_0^t \lambda(\tau)|\omega(\tau) - 1|d\tau \leq C_3 \Theta(t);$$

(A4) for $M \geq 2$ the following estimate holds:

$$\int_t^\infty \lambda^{-M}(\tau)\Xi^{-M-1}(\tau)d\tau \leq C_4 \Theta^{-M}(t);$$

(A5) the function $F = F(\Lambda(t))$ is defined by

$$\Xi(t) = \frac{F(\Lambda(t))}{\lambda(t)\sqrt{F'(\Lambda(t))}}.$$

This implies that

$$\frac{1}{F\big(\Lambda(t)\big)} = \int_t^\infty \lambda^{-1}(\tau)\Xi^{-2}(\tau)d\tau,$$

where $F(\Lambda(t)) \to \infty$ as $t \to \infty$. Here we suppose that the right-hand side exists for all $t \geq 0$.

Cauchy problems with the increasing speed of propagation have been considered in [14]. Following their approach the assumption (A1) is standard. The assumption (A3), which is introduced as a *stabilization condition*, allows us to control a certain amount of very fast oscillations. Especially, due to the deteriorating influence of oscillations on solutions by the aid of this stabilization condition and higher order regularity of the time-dependent coefficients may be compensated "bad behavior" of the very fast oscillations. For this reason, this stabilization condition describes an error made from the oscillating behavior of the coefficient $\omega = \omega(t)$.

Remark 3 If we consider very slow oscillations (according to the definitions in [13]), that is, $\Theta(t) \equiv \Lambda(t)$ and $F(\Lambda(t)) \equiv \Lambda(t)$ the assumptions (A3) to (A5) trivially hold and by these choices the stabilization condition disappears. Hence, the stabilization condition (A3) has a meaning only in the case $M \geq 2$.

Now motivated by the considerations from [1, 2], in order to study the interaction between the shape functions $\lambda = \lambda(t)$, $\rho = \rho(t)$ and the oscillating function $\omega = \omega(t)$ we assume the following conditions:

(B1) $\rho(t) > 0, \quad \rho(t) = \mu(t)\dfrac{\lambda(t)}{\Lambda(t)}$;

(B2) $\big|d_t^k \mu(t)\big| \leq \mu_k \mu(t)\Big(\dfrac{\lambda(t)}{\Lambda(t)}\Big)^k$ for $k = 1, 2, \cdots, M$, where all μ_k are positive constants;

(B3) $\dfrac{\mu(t)}{\Lambda(t)}$ is monotonic and $\mu(t) \to \infty$ for $t \to \infty$;

(B4) $\dfrac{\lambda^2(t)}{\rho(t)} = \dfrac{\lambda(t)\Lambda(t)}{\mu(t)} \notin L^1(\mathbb{R}_+)$;

(B5) $\big|\big(\rho(t)\omega(t)\big)'\big| = o\big(\,(\rho(t)\omega(t))^2\,\big)$ as $t \to \infty$, this implies that $\mu(t)\dfrac{\Theta(t)}{\Lambda(t)} \to \infty$ as $t \to \infty$;

(B6) $\displaystyle\int_0^t \frac{\lambda^2(\tau)}{\rho(\tau)}d\tau \leq C_5 F^2(\Lambda(t))$, where C_5 is a positive constant.

The assumptions (B3) and (B4) describe the effective damping case related to a given increasing propagation speed. In particular, (B4) excludes the over-damping case (see [2]). The assumption (B5) allows us to control a certain amount of very fast oscillations in the damping term $\rho(t)\omega(t)u_t$.

3 Representation of Solutions

We perform the partial Fourier transformation $\hat{u}(t,\xi) = F_{x\to\xi}(u)(t,\xi)$ to (4) with respect to spatial variables. Then, we have

$$\begin{cases} \hat{u}_{tt} + \lambda^2(t)\omega^2(t)|\xi|^2\hat{u} + \rho(t)\omega(t)\hat{u}_t = 0, & (t,\xi) \in [0,\infty)\times\mathbb{R}^n, \\ \hat{u}(0,\xi) = \hat{u}_0(\xi), \quad \hat{u}_t(0,\xi) = \hat{u}_1(\xi), & \xi \in \mathbb{R}^n. \end{cases} \tag{5}$$

Applying the transformation

$$\hat{u}(t,\xi) = \exp\Big(-\frac{1}{2}\int_0^t \rho(\tau)\omega(\tau)d\tau\Big)v(t,\xi),$$

transfers the Cauchy problem (5) into

$$\begin{cases} v_{tt} + m(t,\xi)v = 0, & (t,\xi) \in [0,\infty)\times\mathbb{R}^n, \\ v(0,\xi) = v_0(\xi), \quad v_t(0,\xi) = v_1(\xi), & \xi \in \mathbb{R}^n, \end{cases} \tag{6}$$

where

$$v_0(\xi) = \hat{u}_0(\xi) \quad \text{and} \quad v_1(\xi) = \frac{\rho(0)\omega(0)}{2}\hat{u}_0(\xi) + \hat{u}_1(\xi).$$

The coefficient $m = m(t,\xi)$ of the mass term is defined by

$$m(t,\xi) := \lambda^2(t)\omega^2(t)|\xi|^2 - \frac{1}{4}\big(\rho(t)\omega(t)\big)^2 - \frac{1}{2}\big(\rho(t)\omega(t)\big)'. \tag{7}$$

Due to the assumptions (B2), (B3) and (B5) we see that $\big(\rho(t)\omega(t)\big)'$ is a negligible term in (7), that is, it holds $\big|\big(\rho(t)\omega(t)\big)'\big| = o\big(\,(\rho(t)\omega(t))^2\,\big)$ as $t\to\infty$. Hence, we can introduce a separating curve Γ as follows:

$$\Gamma := \Big\{(t,\xi)\in\mathbb{R}_+\times\mathbb{R}^n : |\xi| = \frac{1}{2}\frac{\mu(t)}{\Lambda(t)}\Big\}.$$

This curve divides the extended phase space into two regions, the hyperbolic region Π_{hyp} and the elliptic region Π_{ell}, as follows:

$$\Pi_{hyp} = \Big\{(t,\xi) : |\xi| > \frac{1}{2}\frac{\mu(t)}{\Lambda(t)}\Big\} \quad \text{and} \quad \Pi_{ell} = \Big\{(t,\xi) : |\xi| < \frac{1}{2}\frac{\mu(t)}{\Lambda(t)}\Big\}.$$

Let us define the auxiliary weight function

$$\langle\xi\rangle_{\lambda(t),\omega(t)} := \sqrt{\Big|\lambda^2(t)\omega^2(t)|\xi|^2 - \frac{\rho^2(t)\omega^2(t)}{4}\Big|}.$$

Division of the extended phase space. We divide both regions of the extended phase space into zones in order to organize the necessary steps of WKB-analysis. This procedure was developed in [14, 17]. Here we follow some ideas of [2]. However, we shall restrict the considerations to a smaller hyperbolic zone and elliptic zone in the extended phase space to cope with stronger oscillations in $\omega = \omega(t)$ in our approach. The zones are defined as follows:

- hyperbolic zone:

$$Z_{hyp}(N) = \Big\{(t,\xi) : \langle\xi\rangle_{\lambda(t),\omega(t)} \geq N\frac{\rho(t)\omega(t)}{2}\Big\} \cap \Pi_{hyp},$$

  and $\Theta(t)|\xi| \geq N$;
- oscillation subzone:

$$Z_{osc}(N,\varepsilon) = \Big\{(t,\xi) : \varepsilon\frac{\rho(t)\omega(t)}{2} \leq \langle\xi\rangle_{\lambda(t),\omega(t)} \leq N\frac{\rho(t)\omega(t)}{2}\Big\} \cap \Pi_{hyp},$$

  $\Theta(t)|\xi| \leq N$ and $\Lambda(t)|\xi| \geq N$;
- reduced zone:

$$Z_{red}(\varepsilon) = \Big\{(t,\xi) : \langle\xi\rangle_{\lambda(t),\omega(t)} \leq \varepsilon\frac{\rho(t)\omega(t)}{2}\Big\};$$

- elliptic zone:

$$Z_{ell}(d_0,\varepsilon) = \Big\{(t,\xi) : |\xi| \geq \frac{d_0}{F\big(\Lambda(t)\big)}\Big\} \cap \Big\{\langle\xi\rangle_{\lambda(t),\omega(t)} \geq \varepsilon\frac{\rho(t)\omega(t)}{2}\Big\} \cap \Pi_{ell};$$

- dissipative zone:

$$Z_{diss}(d_0) = \Big\{(t,\xi) : |\xi| \leq \frac{d_0}{F\big(\Lambda(t)\big)}\Big\} \cap \Pi_{ell}.$$

Here in general, N is a large positive constant and ε is a small positive constant. Both will be chosen later (Fig. 1).

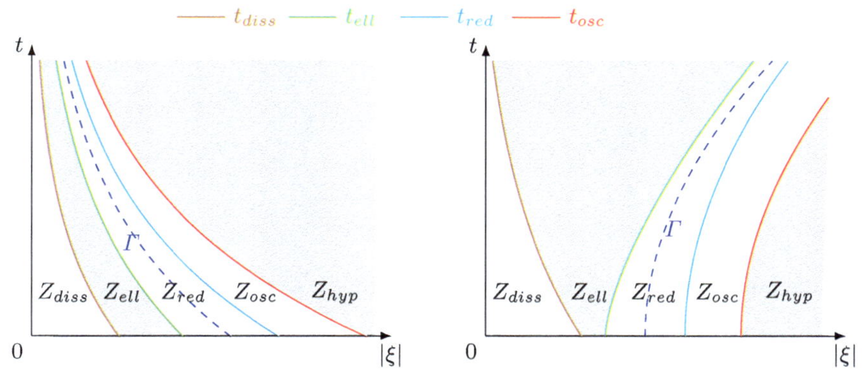

(a) The case that $\mu(t)/\Lambda(t)$ is decreasing (b) The case that $\mu(t)/\Lambda(t)$ is increasing

Fig. 1 Division of the extended phase space into zones

Let us introduce the separating lines between the dissipative zone and the elliptic zone by $t_{diss}(|\xi|) =: t_{diss}$, between the elliptic zone and the reduced zone by $t_{ell}(|\xi|) =: t_{ell}$, between the reduced zone and oscillation subzone by $t_{red}(|\xi|) =: t_{red}$, between the oscillation subzone and the hyperbolic zone by $t_{osc}(|\xi|) =: t_{osc}$. By the definitions of the zones we can see that these separating lines really exist and can be described by functions due to the monotonicity of the functions $\frac{\mu(t)}{\Lambda(t)}$ and $F(\Lambda(t))$.

We define

$$h_1(t,\xi) := \chi\big(|\xi|F(\Lambda(t))\big)\frac{\lambda(t)}{F(\Lambda(t))} + \Big(1-\chi\big(|\xi|F(\Lambda(t))\big)\Big)\lambda(t)|\xi| \tag{8}$$

and

$$h_2(t,\xi) := \chi\left(\frac{\langle\xi\rangle_{\lambda(t),\omega(t)}}{\varepsilon\frac{\rho(t)\omega(t)}{2}}\right)\varepsilon\frac{\rho(t)\omega(t)}{2} + \left(1-\chi\left(\frac{\langle\xi\rangle_{\lambda(t),\omega(t)}}{\varepsilon\frac{\rho(t)\omega(t)}{2}}\right)\right)\langle\xi\rangle_{\lambda(t),\omega(t)}, \tag{9}$$

where $\chi \in C^\infty[0,\infty)$ such that $\chi(t) = 1$ for $0 \le t \le \frac{1}{2}$ and $\chi(t) = 0$ for $t \ge 1$. Introducing the micro-energy $U(t,\xi) := \big(h_1(t,\xi)\hat{u}(t,\xi), D_t\hat{u}(t,\xi)\big)^T$ we obtain from (5) the system of first order

$$D_tU(t,\xi) = \underbrace{\begin{pmatrix} \frac{D_th_1(t,\xi)}{h_1(t,\xi)} & h_1(t,\xi) \\ \frac{\lambda^2(t)\omega^2(t)|\xi|^2}{h_1(t,\xi)} & i\rho(t)\omega(t) \end{pmatrix}}_{A(t,\xi)} U(t,\xi), \tag{10}$$

with the initial condition $U(0,\xi) = \big(h_1(0,\xi)\hat{u}(0,\xi), D_t\hat{u}(0,\xi)\big)^T$.

On the other hand, we define the micro-energy $V(t,\xi) := \big(h_2(t,\xi)v(t,\xi), D_t v(t,\xi)\big)^T$. Then, by (6) we obtain that $V = V(t,\xi)$ satisfies the following system of first order:

$$D_t V(t,\xi) = \underbrace{\begin{pmatrix} \frac{D_t h_2(t,\xi)}{h_2(t,\xi)} & h_2(t,\xi) \\ \frac{m(t,\xi)}{h_2(t,\xi)} & 0 \end{pmatrix}}_{A_V(t,\xi)} V(t,\xi), \tag{11}$$

with the initial condition $V(0,\xi) = \big(h_2(0,\xi)v(0,\xi), D_t v(0,\xi)\big)^T$.

Definition 1 For any $t \geq s \geq 0$, we denote by $E = E(t,s,\xi)$ and $E_V = E_V(t,s,\xi)$ the fundamental solutions of (10) and (11), respectively, that is, the matrix-valued functions solving the Cauchy problems

$$D_t E(t,s,\xi) = A(t,\xi)E(t,s,\xi), \quad E(s,s,\xi) = I,$$

and

$$D_t E_V(t,s,\xi) = A_V(t,\xi)E_V(t,s,\xi), \quad E_V(s,s,\xi) = I.$$

Hence, it is easy to prove that $U(t,\xi) = E(t,0,\xi)\big(h_1(0,\xi)\hat{u}(0,\xi), D_t\hat{u}(0,\xi)\big)^T$ and $V(t,\xi) = E_V(t,0,\xi)\big(h_2(0,\xi)v(0,\xi), D_t v(0,\xi)\big)^T$.

Remark 4 By the previous considerations, after obtaining estimates for $E_V = E_V(t,s,\xi)$ it is sufficient to apply the backward transformation to the original Cauchy problem. That is, we transform back $E_V = E_V(t,s,\xi)$ to estimate the fundamental solution $E = E(t,s,\xi)$ which is related to a system of first order for the micro-energy $\big(\lambda(t)|\xi|\hat{u}, D_t\hat{u}\big)^T$, which gives the representation

$$E(t,s,\xi) = T(t,\xi)E_V(t,s,\xi)T^{-1}(s,\xi), \tag{12}$$

where the matrix $T(t,\xi)$ is defined in the following way:

$$\begin{pmatrix} \lambda(t)|\xi|\hat{u} \\ D_t\hat{u} \end{pmatrix} = \underbrace{\begin{pmatrix} \frac{\lambda(t)|\xi|}{\delta(t)h_2(t,\xi)} & 0 \\ i\frac{\rho(t)\omega(t)}{2\delta(t)h_2(t,\xi)} & \frac{1}{\delta(t)} \end{pmatrix}}_{T(t,\xi)} \begin{pmatrix} h_2(t,\xi)v \\ D_t v \end{pmatrix}$$

with the inverse matrix

$$T^{-1}(t,\xi) = \begin{pmatrix} \frac{\delta(t)h_2(t,\xi)}{\lambda(t)|\xi|} & 0 \\ -i\frac{\rho(t)\omega(t)\delta(t)}{2\lambda(t)|\xi|} & \delta(t) \end{pmatrix},$$

where the auxiliary function

$$\delta(t) := \exp\left(\frac{1}{2}\int_0^t \rho(\tau)\omega(\tau)d\tau\right)$$

is related to the transformed damping term $\rho(t)\omega(t)\hat{u}_t$.

3.1 Considerations in the Hyperbolic Zone

In the hyperbolic zone after M steps of diagonalization procedure we can guarantee that the remainder part is uniformly integrable over the hyperbolic zone. Here we follow some ideas of [8, 10, 17]. First of all let us introduce the following family of symbol classes in the hyperbolic zone.

Definition 2 A function $f = f(t,\xi)$ belongs to the hyperbolic symbol class $S_N^l\{m_1, m_2\}$ of limited smoothness if the estimates

$$\left|D_t^k f(t,\xi)\right| \leq C_k \langle\xi\rangle_{\lambda(t),\omega(t)}^{m_1} \Xi(t)^{-m_2-k}$$

are valid for all $(t,\xi) \in Z_{hyp}(N)$ and all $k = 0, 1, \cdots, l$ with $l \leq M$. Here M is the order of the regularity of the time-dependent coefficients as well as the number of steps of the diagonalization procedure.

We note that in $Z_{hyp}(N)$ the auxiliary symbol $\langle\xi\rangle_{\lambda(t),\omega(t)}$ can be estimated by

$$\langle\xi\rangle_{\lambda(t),\omega(t)} \approx \lambda(t)|\xi|. \tag{13}$$

From the definition of the symbol classes we may conclude the following rules.

Proposition 1 *The following statements are true:*

1. $S_N^l\{m_1, m_2\}$ *is a vector space for all non-negative integers* l;
2. $S_N^l\{m_1, m_2\} \cdot S_N^{l'}\{m_1', m_2'\} \hookrightarrow S_N^{\tilde{l}}\{m_1 + m_1', m_2 + m_2'\}$ *for all non-negative integers* l *and* l' *with* $\tilde{l} = \min\{l, l'\}$;
3. $D_t^k S_N^l\{m_1, m_2\} \hookrightarrow S_N^{l-k}\{m_1, m_2 + k\}$ *for all non-negative integers* l *with* $k \leq l$;
4. $S_N^0\{-M, M+1\} \hookrightarrow L_\xi^\infty L_t^1\big(Z_{hyp}(N)\big)$ *with* M *from the assumption (A4).*

Proof We only verify the fourth property. Indeed, if $f(t,\xi) \in S_N^0\{-M, M+1\}$, then we have

$$\begin{aligned}\int_{t_{osc}}^{\infty} |f(\tau,\xi)| d\tau &\lesssim \int_{t_{osc}}^{\infty} \langle\xi\rangle_{\lambda(\tau),\omega(\tau)}^{-M} \Xi^{-M-1}(\tau) d\tau \\ &\lesssim \int_{t_{osc}}^{\infty} |\xi|^{-M} \lambda^{-M}(\tau) \Xi^{-M-1}(\tau) d\tau \\ &\lesssim |\xi|^{-M} \Theta^{-M}(t_{osc}) \leq \frac{1}{N^M} < \infty,\end{aligned}$$

where we used (13), the assumption (A4) and the definition of $Z_{hyp}(N)$. □

Proposition 2 *Assume the conditions (A1), (A2) and (B1), (B2). Then, the following inequalities hold:*

1. $\left|D_t^k \langle\xi\rangle_{\lambda(t),\omega(t)}\right| \lesssim \langle\xi\rangle_{\lambda(t),\omega(t)} \Xi^{-k}(t)$ *for all* $k = 0, 1, \cdots, l$ *with* $l \leq M$;
2. $\left|D_t^k\big(\rho(t)\omega(t)\big)\right| \lesssim \langle\xi\rangle_{\lambda(t),\omega(t)} \Xi^{-k}(t)$ *for all* $k = 0, 1, \cdots, l$ *with* $l \leq M$.

In the hyperbolic zone we have $h_2(t,\xi) = \langle\xi\rangle_{\lambda(t),\omega(t)}$. So, we introduce the micro-energy $V = \big(\langle\xi\rangle_{\lambda(t),\omega(t)} v, D_t v\big)^T$. Then, it holds

$$D_t V = \begin{pmatrix} 0 & \langle\xi\rangle_{\lambda(t),\omega(t)} \\ \langle\xi\rangle_{\lambda(t),\omega(t)} & 0 \end{pmatrix} V + \begin{pmatrix} \frac{D_t \langle\xi\rangle_{\lambda(t),\omega(t)}}{\langle\xi\rangle_{\lambda(t),\omega(t)}} & 0 \\ -\frac{(\rho(t)\omega(t))'}{2\langle\xi\rangle_{\lambda(t),\omega(t)}} & 0 \end{pmatrix} V.$$

Let us carry out the first step of the diagonalization procedure. The eigenvalues of the first matrix are $\pm\langle\xi\rangle_{\lambda(t),\omega(t)}$. Thus, the matrix of eigenvectors P and its inverse P^{-1} are

$$P = \begin{pmatrix} 1 & -1 \\ 1 & 1 \end{pmatrix}, \quad P^{-1} = \frac{1}{2}\begin{pmatrix} 1 & 1 \\ -1 & 1 \end{pmatrix}.$$

Defining $V^{(0)} := P^{-1} V$, we get the transformed system

$$D_t V^{(0)} = \big(\mathcal{D}_0(t,\xi) + \mathcal{R}_0(t,\xi)\big) V^{(0)},$$

where

$$\mathcal{D}_0(t,\xi) = \begin{pmatrix} \langle\xi\rangle_{\lambda(t),\omega(t)} + \frac{D_t \langle\xi\rangle_{\lambda(t),\omega(t)}}{2\langle\xi\rangle_{\lambda(t),\omega(t)}} - \frac{(\rho(t)\omega(t))'}{4\langle\xi\rangle_{\lambda(t),\omega(t)}} & 0 \\ 0 & -\langle\xi\rangle_{\lambda(t),\omega(t)} + \frac{D_t \langle\xi\rangle_{\lambda(t),\omega(t)}}{2\langle\xi\rangle_{\lambda(t),\omega(t)}} + \frac{(\rho(t)\omega(t))'}{4\langle\xi\rangle_{\lambda(t),\omega(t)}} \end{pmatrix}$$

and

$$\mathcal{R}_0(t,\xi) = \begin{pmatrix} 0 & -\frac{D_t \langle\xi\rangle_{\lambda(t),\omega(t)}}{2\langle\xi\rangle_{\lambda(t),\omega(t)}} + \frac{(\rho(t)\omega(t))'}{4\langle\xi\rangle_{\lambda(t),\omega(t)}} \\ -\frac{D_t \langle\xi\rangle_{\lambda(t),\omega(t)}}{2\langle\xi\rangle_{\lambda(t),\omega(t)}} - \frac{(\rho(t)\omega(t))'}{4\langle\xi\rangle_{\lambda(t),\omega(t)}} & 0 \end{pmatrix}.$$

Note that $\mathcal{R}_0(t,\xi) \in S_N^{M-1}\{0,1\}$. Now we want to carry out further steps of the diagonalization procedure. The goal is to transform the previous system such that the new matrix has diagonal structure and the new remainder belongs to a "better" hyperbolic symbol class.

The diagonalization procedure for the following lemma is essentially based on the approach used in [13, 17] for wave equations with variable speed of propagation.

Lemma 1 *There exists a zone constant $N > 0$ such that for any $k = 0, 1, \cdots, M$ we can find matrices with the following properties:*

- *the matrices $N_k(t,\xi) \in S_N^{M-k}\{0,0\}$ are invertible and $N_k^{-1}(t,\xi) \in S_N^{M-k}\{0,0\}$;*
- *the matrices $\mathcal{D}_k(t,\xi) \in S_N^{M-k}\{1,0\}$ are diagonal and*

$$\mathcal{D}_k(t,\xi) = \operatorname{diag}\left(\tau_k^+(t,\xi), \tau_k^-(t,\xi)\right)$$

  with $|\tau_k^+(t,\xi) - \tau_k^-(t,\xi)| \geq C_k \langle\xi\rangle_{\lambda(t),\omega(t)}$;
- *the matrices $\mathcal{R}_k(t,\xi) \in S_N^{M-k}\{-k, k+1\}$ are antidiagonal;*

all these matrices are defined in $Z_{hyp}(N)$ such that the operator identity

$$\big(D_t - \mathcal{D}_k(t,\xi) - \mathcal{R}_k(t,\xi)\big)N_k(t,\xi) = N_k(t,\xi)\big(D_t - \mathcal{D}_{k+1}(t,\xi) - \mathcal{R}_{k+1}(t,\xi)\big)$$

is valid.

Finally, we obtain for $k = M$ the remainder $\mathcal{R}_M = \mathcal{R}_M(t,\xi) \in S_N^0\{-M, M+1\}$ which is uniformly integrable over the hyperbolic zone by Proposition 1.

To complete the derivation of our representation we need more information on the diagonal matrices $\mathcal{D}_k = \mathcal{D}_k(t,\xi)$. An improvement of the diagonalization procedure was developed in [8]. The author performed more diagonalization steps in order to use structural properties of the coefficient matrices by assuming higher regularity of the entries of the matrix. Following [8, 9] we arrive at the following lemma.

Lemma 2 *The difference of the diagonal entries of $\mathcal{D}_k(t,\xi)$ is real for all $k = 0, 1, \cdots, M-1$.*

Now we want to construct the fundamental solution $E_{hyp}^V = E_{hyp}^V(t,s,\xi)$ with $0 \leq s \leq t$, for the operator

$$D_t - \mathcal{D}_0(t,\xi) - \mathcal{R}_0(t,\xi).$$

For this reason after M steps of diagonalization it is sufficient to construct the fundamental solution satisfying the system

$$D_t E_M(t,s,\xi) = \big(\mathcal{D}_M(t,\xi) + \mathcal{R}_M(t,\xi)\big)E_M(t,s,\xi), \quad E_M(s,s,\xi) = I.$$

At first we solve the diagonal system

$$D_t \mathcal{E}_M(t,s,\xi) = \mathcal{D}_M(t,s,\xi)\mathcal{E}_M(t,s,\xi), \quad \mathcal{E}_M(s,s,\xi) = I, \quad 0 \leq s \leq t.$$

Its fundamental solution is given by

$$\mathcal{E}_M = \mathcal{E}_M(t, s, \xi) = \exp\Big(i \int_s^t \mathcal{D}_M(\theta, \xi) d\theta\Big) = \operatorname{diag}\Big(e^{i \int_s^t \tau_M^+(\theta,\xi)d\theta}, e^{i \int_s^t \tau_M^-(\theta,\xi)d\theta}\Big).$$

We make the *ansatz*

$$E_M(t, s, \xi) = \mathcal{E}_M(t, s, \xi)\mathcal{Q}_M(t, s, \xi)$$

with a uniformly bounded and invertible matrix $\mathcal{Q}_M = \mathcal{Q}_M(t, s, \xi)$. It follows that the matrix $\mathcal{Q}_M$ satisfies the Cauchy problem

$$D_t \mathcal{Q}_M(t, s, \xi) = R_M(t, s, \xi)\mathcal{Q}_M(t, s, \xi), \quad \mathcal{Q}_M(s, s, \xi) = I$$

with the coefficient matrix

$$R_M(t, s, \xi) = \mathcal{E}_M(s, t, \xi)\mathcal{R}_M(t, \xi)\mathcal{E}_M(t, s, \xi).$$

Taking account of $\mathcal{R}_M(t, \xi) \in S_N^0\{-M, M + 1\}$ we obtain

$$\big|R_M(t, s, \xi)\big| = \big|\mathcal{R}_M(t, \xi)\big| \lesssim \langle\xi\rangle_{\lambda(t),\omega(t)}^{-M} \Xi^{-M-1}(t) \lesssim |\xi|^{-M}\lambda^{-M}(t)\Xi^{-M-1}(t).$$

The solution $\mathcal{Q}_M = \mathcal{Q}_M(t, s, \xi)$ can be represented as Peano-Baker series

$$\begin{aligned}&\mathcal{Q}_M(t, s, \xi)\\ &= I + \sum_{k=1}^{\infty} i^k \int_s^t R_M(t_1, s, \xi) \int_s^{t_1} R_M(t_2, s, \xi) \cdots \int_s^{t_{k-1}} R_M(t_k, s, \xi) dt_k \cdots dt_1.\end{aligned}$$

Then, we obtain the following statement.

Lemma 3 *The fundamental solution $E_{hyp}^V = E_{hyp}^V(t, s, \xi)$ is representable in the following form:*

$$E_{hyp}^V(t, s, \xi) = P\Big(\prod_{k=0}^{M-1} N_k(t, \xi)\Big)\mathcal{E}_M(t, s, \xi)\mathcal{Q}_M(t, s, \xi)\Big(\prod_{k=0}^{M-1} N_k^{-1}(s, \xi)\Big)P^{-1}$$

for all $(t, \xi), (s, \xi) \in Z_{hyp}(N)$, where

- *the matrices $N_k = N_k(t, \xi)$ and $N_k^{-1} = N_k^{-1}(t, \xi)$ are uniformly bounded and invertible;*
- *the matrices $\mathcal{Q}_M = \mathcal{Q}_M(t, s, \xi)$ and $\mathcal{Q}_M^{-1} = \mathcal{Q}_M^{-1}(t, s, \xi)$ are uniformly bounded and invertible.*

Finally, an estimate for the fundamental solution $E_{hyp}^V = E_{hyp}^V(t, s, \xi)$ is given by the following statement.

Lemma 4 *Assume the conditions (A1) to (A4) and (B1) to (B3). Then, the fundamental solution* $E^V_{hyp} = E^V_{hyp}(t,s,\xi)$ *satisfies the estimate*

$$\big(|E^V_{hyp}(t,s,\xi)|\big) \lesssim \frac{\sqrt{\lambda(t)}}{\sqrt{\lambda(s)}} \begin{pmatrix} 1 & 1 \\ 1 & 1 \end{pmatrix}$$

uniformly for all $(s,\xi), (t,\xi) \in Z_{hyp}(N)$.

After constructing the fundamental solution $E^V_{hyp} = E^V_{hyp}(t,s,\xi)$ we use the backward "dissipative" transformation to the Fourier transformed original Cauchy problem (5). Thus, we get the following estimate for the fundamental solution $E_{hyp} = E_{hyp}(t,s,\xi)$ in the hyperbolic zone.

Corollary 1 *The fundamental solution* $E_{hyp} = E_{hyp}(t,s,\xi)$ *satisfies the estimate*

$$\big(|E_{hyp}(t,s,\xi)|\big) \lesssim \frac{\sqrt{\lambda(t)}}{\sqrt{\lambda(s)}} \exp\Big(-\frac{1}{2}\int_s^t \rho(\theta)\omega(\theta)d\theta\Big) \begin{pmatrix} 1 & 1 \\ 1 & 1 \end{pmatrix}$$

for all $(s,\xi), (t,\xi) \in Z_{hyp}(N)$.

3.2 Considerations in the Oscillation Subzone

We have already chosen the hyperbolic zone as large as possible to cope with the stronger oscillating behavior of $\omega = \omega(t)$. For this reason we have a zone between the reduced zone $Z_{red}(\varepsilon)$ and the hyperbolic zone $Z_{hyp}(N)$, the so-called *oscillation subzone* $Z_{osc}(N,\varepsilon)$. The basic approach in this zone bases on [10]. Essentially, in $Z_{osc}(N,\varepsilon)$ we relate the fundamental solution $E_{osc} = E_{osc}(t,s,\xi)$ to the fundamental solution $E_\lambda = E_\lambda(t,s,\xi)$ to the corresponding model with $\omega(t) \equiv 1$ and effective dissipation case. Note that $Z_{osc}(N,\varepsilon) \subset Z^\lambda_{hyp}(N)$ such that we can use the known estimates for $E_\lambda = E_\lambda(t,s,\xi)$ from [2] (see Lemma 3.5 of [2]). This estimate reads as follows:

$$\big(|E_\lambda(t,s,\xi)|\big) \lesssim \frac{\sqrt{\lambda(t)}}{\sqrt{\lambda(s)}} \exp\Big(-\frac{1}{2}\int_s^t \rho(\theta)d\theta\Big) \begin{pmatrix} 1 & 1 \\ 1 & 1 \end{pmatrix}.$$

Let us introduce the micro-energy $U(t,\xi) = \big(\lambda(t)|\xi|\hat{u}, D_t\hat{u}\big)^T$. Then, we obtain from (5) the system of first order

$$D_tU = \underbrace{\begin{pmatrix} \frac{D_t\lambda(t)}{\lambda(t)} & \lambda(t)|\xi| \\ \lambda(t)\omega^2(t)|\xi| & i\rho(t)\omega(t) \end{pmatrix}}_{A(t,\xi)} U. \tag{14}$$

Our aim is to construct the corresponding fundamental solution, that is, the matrix-valued solution of the system

$$D_t E_{osc}(t,s,\xi) = A(t,\xi)E_{osc}(t,s,\xi), \quad E_{osc}(s,s,\xi) = I, \ 0 \le s \le t.$$

If we set formally $\omega(t) \equiv 1$ and define the micro-energy to the corresponding model by $U_\lambda(t,\xi) = \big(\lambda(t)|\xi|\hat{u}, D_t\hat{u}\big)^T$, then it satisfies the system of first order

$$D_t U_\lambda = \underbrace{\begin{pmatrix} \frac{D_t\lambda(t)}{\lambda(t)} & \lambda(t)|\xi| \\ \lambda(t)|\xi| & i\rho(t) \end{pmatrix}}_{A_\lambda(t,\xi)} U_\lambda.$$

We denote the corresponding fundamental solution as $E_\lambda = E_\lambda(t,s,\xi)$, i.e., the solution to

$$D_t E_\lambda(t,s,\xi) = A_\lambda(t,\xi)E_\lambda(t,s,\xi), \quad E_\lambda(s,s,\xi) = I, \ 0 \le s \le t.$$

Hence, in $Z_{osc}(N,\varepsilon)$ we relate $E_{osc}(t,s,\xi)$ to $E_\lambda(t,s,\xi)$ and use the *stabilization condition* (A3).

Corollary 2 *Assume the conditions (A1) to (A3) and (B1). Then, the fundamental solution* $E_{osc} = E_{osc}(t,s,\xi)$ *satisfies the estimate*

$$\big(|E_{osc}(t,s,\xi)|\big) \lesssim \frac{\sqrt{\lambda(t)}}{\sqrt{\lambda(s)}} \exp\Big(-\frac{1}{2}\int_s^t \rho(\theta)d\theta\Big)$$

uniformly for $(s,\xi), (t,\xi) \in Z_{osc}(N,\varepsilon),\ 0 \le s \le t.$

3.3 Considerations in the Elliptic Zone

In a similar manner as in the hyperbolic zone we will try to cope with the stronger oscillating behavior of $\omega = \omega(t)$ by shrinking the elliptic zone. For this reason, we shall enlarge the dissipative zone in the extended phase space, since we do not propose any oscillation subzone between these two zones.

In the elliptic zone we can follow the standard diagonalization procedure. That is, contrary to the hyperbolic zone $Z_{hyp}(N)$, we will perform two diagonalization steps to derive an estimate for the fundamental solution. The considerations are based on the papers [2, 16].

Now let us introduce the following family of symbol classes in $Z_{ell}(d_0,\varepsilon)$.

Definition 3 A function $f = f(t,\xi)$ belongs to the elliptic symbol class $S^l\{m_1,m_2\}$ of limited smoothness if the derivatives of f satisfy the estimates

$$\left|D_t^k f(t,\xi)\right| \le C_k \langle\xi\rangle_{\lambda(t),\omega(t)}^{m_1} \Xi(t)^{-m_2-k}$$

for all $(t,\xi) \in Z_{ell}(d_0,\varepsilon)$ and all $k \le l$ with $l \in \mathbb{N}_0$.

Note that the auxiliary symbol $\langle\xi\rangle_{\lambda(t),\omega(t)}$ can be estimated in $Z_{ell}(d_0,\varepsilon)$ by

$$\langle\xi\rangle_{\lambda(t),\omega(t)} \approx \frac{\rho(t)}{2} = \mu(t)\frac{\lambda(t)}{2\Lambda(t)}. \tag{15}$$

Some useful properties of the symbolic calculus are collected in the following proposition.

Proposition 3 *The following statements are true:*

1. *$S^l\{m_1,m_2\}$ is a vector space for all non-negative integers l;*
2. *$S^l\{m_1,m_2\} \cdot S_N^{l'}\{m_1',m_2'\} \hookrightarrow S^{\tilde{l}}\{m_1+m_1', m_2+m_2'\}$ for all non-negative integers l and l' with $\tilde{l} = \min\{l,l'\}$;*
3. *$D_t^k S^l\{m_1,m_2\} \hookrightarrow S^{l-k}\{m_1, m_2+k\}$ for all non-negative integers l with $k \le l$;*
4. *$S^{l-2}\{-1,2\} \hookrightarrow L_\xi^\infty L_t^1\big(Z_{ell}(d_0,\varepsilon)\big)$ for $l \ge 2$.*

Proof We only verify the integrability statement. Indeed, if $f = f(t,\xi) \in S^{l-2}\{-1,2\}$, then it holds

$$\begin{aligned}\int_{t_{diss}}^{t_{ell}} |f(\tau,\xi)|d\tau &\lesssim \int_{t_{diss}}^{t_{ell}} \frac{\Xi^{-2}(\tau)}{\langle\xi\rangle_{\lambda(\tau),\omega(\tau)}}d\tau \lesssim \int_{t_{diss}}^{t_{ell}} \frac{\Lambda(\tau)}{\mu(\tau)\lambda(\tau)}\frac{\lambda^2(\tau)F'(\Lambda(\tau))}{F^2(\Lambda(\tau))}d\tau \\ &\lesssim \frac{1}{|\xi|}\int_{t_{diss}}^{t_{ell}} \frac{\lambda(\tau)F'(\Lambda(\tau))}{F^2(\Lambda(\tau))}d\tau = -\frac{1}{|\xi|}\frac{1}{F\big(\Lambda(\tau)\big)}\Big|_{t_{diss}}^{t_{ell}} \lesssim \frac{1}{|\xi|F(\Lambda(t_{diss}))} \lesssim 1,\end{aligned}$$

where we used (15), the definition of the elliptic region, $\Xi(t) = \frac{F(\Lambda(t))}{\lambda(t)\sqrt{F'(\Lambda(t))}}$ and $|\xi|F\big(\Lambda(t_{diss})\big) = d_0$, respectively. □

In the elliptic zone we introduce the micro-energy $V = \big(\langle\xi\rangle_{\lambda(t),\omega(t)}v, D_t v\big)^T$ for all $t \ge s \ge 0$ and $(t,\xi), (s,\xi) \in Z_{ell}(d_0,\varepsilon)$. Then, the corresponding first order system to the Cauchy problem (6), with respect to the micro-energy V, is stated as

$$D_t V = \begin{pmatrix} 0 & \langle\xi\rangle_{\lambda(t),\omega(t)} \\ -\langle\xi\rangle_{\lambda(t),\omega(t)} & 0 \end{pmatrix} V + \begin{pmatrix} \frac{D_t\langle\xi\rangle_{\lambda(t),\omega(t)}}{\langle\xi\rangle_{\lambda(t),\omega(t)}} & 0 \\ -\frac{(\rho(t)\omega(t))'}{2\langle\xi\rangle_{\lambda(t),\omega(t)}} & 0 \end{pmatrix} V.$$

Performing the diagonalization procedure we get after the second step of the diagonalization that the entries of the remainder matrix are uniformly integrable over the elliptic zone.

Step 1. We denote by P the matrix consisting of eigenvectors of the first matrix and its inverse P^{-1}. So, we have

$$P = \begin{pmatrix} i & -i \\ 1 & 1 \end{pmatrix}, \quad P^{-1} = \frac{1}{2}\begin{pmatrix} -i & 1 \\ i & 1 \end{pmatrix}.$$

Then, defining $V^{(0)} := \tilde{M}^{-1}V$ we get the system

$$D_t V^{(0)} = \big(\mathcal{D}(t,\xi) + \mathcal{R}(t,\xi)\big)V^{(0)},$$

where

$$\mathcal{D}(t,\xi) = \begin{pmatrix} -i\langle\xi\rangle_{\lambda(t),\omega(t)} & 0 \\ 0 & i\langle\xi\rangle_{\lambda(t),\omega(t)} \end{pmatrix},$$

and

$$\mathcal{R}(t,\xi) = \frac{1}{2}\begin{pmatrix} \frac{D_t\langle\xi\rangle_{\lambda(t),\omega(t)}}{2\langle\xi\rangle_{\lambda(t),\omega(t)}} - i\frac{(\rho(t)\omega(t))'}{4\langle\xi\rangle_{\lambda(t),\omega(t)}} & -\frac{D_t\langle\xi\rangle_{\lambda(t),\omega(t)}}{2\langle\xi\rangle_{\lambda(t),\omega(t)}} + i\frac{(\rho(t)\omega(t))'}{4\langle\xi\rangle_{\lambda(t),\omega(t)}} \\ -\frac{D_t\langle\xi\rangle_{\lambda(t),\omega(t)}}{2\langle\xi\rangle_{\lambda(t),\omega(t)}} - i\frac{(\rho(t)\omega(t))'}{4\langle\xi\rangle_{\lambda(t),\omega(t)}} & \frac{D_t\langle\xi\rangle_{\lambda(t),\omega(t)}}{2\langle\xi\rangle_{\lambda(t),\omega(t)}} + i\frac{(\rho(t)\omega(t))'}{4\langle\xi\rangle_{\lambda(t),\omega(t)}} \end{pmatrix}.$$

Then, we obtain

$$\mathcal{D}(t,\xi) \in S^M\{1,0\}, \quad \mathcal{R}(t,\xi) \in S^{M-1}\{0,1\}.$$

Step 2. Let us introduce $F_0(t,\xi) = \operatorname{diag}\mathcal{R}(t,\xi)$. Now we carry out the next step(s) of the diagonalization. The difference of the diagonal entries of the matrix $\mathcal{D}(t,\xi) + F_0(t,\xi)$ is

$$i\alpha(t,\xi) = 2\langle\xi\rangle_{\lambda(t),\omega(t)} + \frac{\big(\rho(t)\omega(t)\big)'}{2\langle\xi\rangle_{\lambda(t),\omega(t)}} \approx 2\langle\xi\rangle_{\lambda(t),\omega(t)} + \frac{o\big(\rho^2(t)\omega^2(t)\big)}{2\langle\xi\rangle_{\lambda(t),\omega(t)}} \approx \langle\xi\rangle_{\lambda(t),\omega(t)}$$

for $t \geq t_0$ with a sufficiently large $t_0 = t_0(\varepsilon)$ by using $|\big(\rho(t)\omega(t)\big)'| = o\big(\rho^2(t)\omega^2(t)\big)$. Now we can follow the usual diagonalization procedure. Therefore, we choose a matrix $N^{(1)} = N^{(1)}(t,\xi)$. Let

$$N^{(1)}(t,\xi) = \begin{pmatrix} 0 & -\frac{\mathcal{R}_{12}}{\alpha} \\ \frac{\mathcal{R}_{21}}{\alpha} & 0 \end{pmatrix} \approx \begin{pmatrix} 0 & i\frac{D_t\langle\xi\rangle_{\lambda(t),\omega(t)}}{4\langle\xi\rangle^2_{\lambda(t),\omega(t)}} - \frac{(\rho(t)\omega(t))'}{8\langle\xi\rangle^2_{\lambda(t),\omega(t)}} \\ i\frac{D_t\langle\xi\rangle_{\lambda(t),\omega(t)}}{4\langle\xi\rangle^2_{\lambda(t),\omega(t)}} + \frac{(\rho(t)\omega(t))'}{8\langle\xi\rangle^2_{\lambda(t),\omega(t)}} & 0 \end{pmatrix}.$$

Taking into consideration the rules of the symbolic calculus we have

$$N^{(1)}(t,\xi) \in S^{M-1}\{-1,1\} \text{ and } N_1(t,\xi) = I + N^{(1)}(t,\xi) \in S^{M-1}\{0,0\}.$$

For a sufficiently large time $t \geq t_0$ the matrix $N_1 = N_1(t,\xi)$ is invertible with uniformly bounded inverse $N_1^{-1} = N_1^{-1}(t,\xi)$. Indeed, in the elliptic zone it holds

$$\big|N_1(t,\xi) - I\big| \leq \frac{\Xi^{-1}(t)}{\langle\xi\rangle_{\lambda(t),\omega(t)}} \lesssim \frac{\Lambda(t)}{\mu(t)\Theta(t)} \to 0 \text{ for } t \to \infty$$

due to the assumption (B5). Let

$$B^{(1)}(t,\xi) = D_t N^{(1)}(t,\xi) - \big(\mathcal{R}(t,\xi) - F_0(t,\xi)\big)N^{(1)}(t,\xi) \in S^{M-2}\{-1,2\},$$

$$\mathcal{R}_1(t,\xi) = -N_1^{-1}(t,\xi)B^{(1)}(t,\xi) \in S^{M-2}\{-1,2\}.$$

Then, we have the following operator identity:

$$\big(D_t - \mathcal{D}(t,\xi) - \mathcal{R}(t,\xi)\big)N_1(t,\xi) = N_1(t,\xi)\big(D_t - \mathcal{D}(t,\xi) - F_0(t,\xi) - \mathcal{R}_1(t,\xi)\big).$$

Hence, the previous steps of the diagonalization procedure give us the following lemma.

Lemma 5 *Assume that $\lambda = \lambda(t)$, $\omega = \omega(t)$ satisfy the conditions (A1), (A2) and (A5) and $\rho = \rho(t)$ satisfies the conditions (B1), (B2) and (B5). Then, there exists a sufficiently large t_0 such that in $Z_{ell}(d_0,\varepsilon)$ the following statements hold:*

- $N_1 \in S^{M-1}\{0,0\}$, *invertible for* $(t,\xi) \in Z_{ell}(d_0,\varepsilon)$ *with* $N_1^{-1} \in S^{M-1}\{0,0\}$;
- $F_0 = \operatorname{diag}\Big(\frac{D_t\langle\xi\rangle_{\lambda(t),\omega(t)}}{2\langle\xi\rangle_{\lambda(t),\omega(t)}} - \frac{i(\rho(t)\omega(t))'}{4\langle\xi\rangle_{\lambda(t),\omega(t)}}, \frac{D_t\langle\xi\rangle_{\lambda(t),\omega(t)}}{2\langle\xi\rangle_{\lambda(t),\omega(t)}} + \frac{i(\rho(t)\omega(t))'}{4\langle\xi\rangle_{\lambda(t),\omega(t)}}\Big) \in S^{M-1}\{0,1\}$;
- $\mathcal{R}_1 \in S^{M-2}\{-1,2\}$ *with* $M \geq 2$.

Moreover, the operator identity

$$\big(D_t - \mathcal{D}(t,\xi) - \mathcal{R}(t,\xi)\big)N_1(t,\xi) = N_1(t,\xi)\big(D_t - \mathcal{D}(t,\xi) - F_0(t,\xi) - \mathcal{R}_1(t,\xi)\big)$$

holds for all $(t,\xi) \in Z_{ell}(d_0,\varepsilon)$.

Step 3. Construction of the fundamental solution. In order to solve the transformed system and construct its fundamental solution we can not follow the considerations from the theory of the hyperbolic zone, since the main diagonal entries are purely imaginary.

Lemma 6 *Assume the conditions (A1), (A2), (A5) and (B1), (B2), (B5). Then, the fundamental solution $E_{ell}^V = E_{ell}^V(t,s,\xi)$ to the transformed operator*

$$D_t - \mathcal{D}(t,\xi) - F_0(t,\xi) - \mathcal{R}_1(t,\xi)$$

can be estimated by

$$\big(|E_{ell}^V(t,s,\xi)|\big) \lesssim \frac{\langle\xi\rangle_{\lambda(t),\omega(t)}}{\langle\xi\rangle_{\lambda(s),\omega(s)}} \exp\Big(\int_s^t \langle\xi\rangle_{\lambda(\tau),\omega(\tau)}d\tau\Big)\begin{pmatrix} 1 & 1 \\ 1 & 1 \end{pmatrix},$$

with $(t,\xi), (s,\xi) \in Z_{ell}(d_0,\varepsilon) \cap \{t \geq t_0(\varepsilon)\}$, $0 \leq s \leq t$.

In order to prove Lemma 6 one can proceed in the same manner as in the papers [2, 16].

Step 4. Transforming back to the original Cauchy problem. Now we want to obtain an estimate for the energy of the solution to our original Cauchy problem. For this reason we need to transform back to get an estimate of the fundamental solution $E_{ell} = E_{ell}(t, s, \xi)$ which is related to a system of first order for the micro-energy $\big(\lambda(t)|\xi|\hat{u}, D_t\hat{u}\big)^T$.

Lemma 7 *Under the assumptions (B1) to (B3) the following holds:*

1. *In the elliptic zone it holds* $\langle\xi\rangle_{\lambda(t),\omega(t)} - \dfrac{\rho(t)\omega(t)}{2} \le -\dfrac{\lambda^2(t)\omega(t)|\xi|^2}{\rho(t)}$,
2. $\dfrac{\delta(s)}{\delta(t)} \exp\Big(\displaystyle\int_s^t \langle\xi\rangle_{\lambda(\tau),\omega(\tau)} d\tau\Big) \le \exp\Big(-|\xi|^2 \displaystyle\int_s^t \frac{\lambda^2(\tau)\omega(\tau)}{\rho(\tau)} d\tau\Big)$,

where $\delta = \delta(t) = \exp\big(\frac{1}{2}\int_0^t \rho(\tau)\omega(\tau)d\tau\big)$.

Proof By using the property

$$\sqrt{x+y} \le \sqrt{x} + \frac{y}{2\sqrt{x}}$$

for any $x \ge 0$ and $y \ge -x$, the first statement is equivalent to the following inequality:

$$\sqrt{\frac{\rho^2(t)\omega^2(t)}{4} - \lambda^2(t)\omega^2(t)|\xi|^2} - \frac{\rho(t)\omega(t)}{2} \le -\frac{\lambda^2(t)\omega(t)|\xi|^2}{\rho(t)}.$$

After integration the second statement follows directly from the first one together with the definition of $\delta = \delta(t)$. □

From Lemma 6 we get for $(t, \xi), (s, \xi) \in Z_{ell}(d_0, \varepsilon)$ the estimate

$$\big(|E_{ell}^V(t, s, \xi)|\big) \lesssim \frac{\rho(t)\omega(t)}{\rho(s)\omega(s)} \exp\Big(\int_s^t \langle\xi\rangle_{\lambda(\tau),\omega(\tau)} d\tau\Big) \begin{pmatrix} 1 & 1 \\ 1 & 1 \end{pmatrix}.$$

This yields in combination with (12) the estimate

$$\begin{aligned} &\big(|E_{ell}(t, s, \xi)|\big) \\ &\lesssim \begin{pmatrix} \lambda(t)|\xi| & 0 \\ \rho(t) & \rho(t) \end{pmatrix} \exp\Big(\int_s^t \Big(\langle\xi\rangle_{\lambda(\tau),\omega(\tau)} - \frac{\rho(\tau)\omega(\tau)}{2}\Big) d\tau\Big) \begin{pmatrix} 1 & 1 \\ 1 & 1 \end{pmatrix} \begin{pmatrix} \frac{1}{\lambda(s)|\xi|} & 0 \\ \frac{1}{\lambda(s)|\xi|} & \frac{1}{\rho(s)} \end{pmatrix} \\ &\lesssim \exp\Big(-|\xi|^2 \int_s^t \frac{\lambda^2(\tau)\omega(\tau)}{\rho(\tau)} d\tau\Big) \begin{pmatrix} \frac{\lambda(t)}{\lambda(s)} & \frac{\lambda(t)|\xi|}{\rho(s)} \\ \frac{\rho(t)}{\lambda(s)|\xi|} & \frac{\rho(t)}{\rho(s)} \end{pmatrix}. \end{aligned} \tag{16}$$

Remark 5 Taking into account the estimates (16), we see that the estimates for the first row are reasonable. However, the estimates for the second row seem to be not reasonable. Because, the estimate for $|E_{ell}^{(22)}|$ is only reasonable for decreasing coefficients $\rho = \rho(t)$ and the estimate for $|E_{ell}^{(21)}|$ is not optimal since the upper bound for $|E_{ell}^{(21)}|$ is not bounded in the elliptic zone. This "contradicts somehow" the damping effect in our model. For this reason we derive a refined estimate which we present in the next step.

Step 5. A refined estimate for the fundamental solution in the elliptic zone.

Corollary 3 *The fundamental solution $E_{ell} = E_{ell}(t, s, \xi)$ satisfies the following estimate:*

$$\big(|E_{ell}(t,s,\xi)|\big) \lesssim \exp\Big(-|\xi|^2\int_s^t \frac{\lambda^2(\tau)\omega(\tau)}{\rho(\tau)}d\tau\Big)\begin{pmatrix} \frac{\lambda(t)}{\lambda(s)} & \frac{\lambda(t)|\xi|}{\rho(s)} \\ \frac{\lambda^2(t)|\xi|}{\lambda(s)\rho(t)} & \frac{\lambda^2(t)|\xi|^2}{\rho(s)\rho(t)} \end{pmatrix} + \frac{\delta^2(s)}{\delta^2(t)}\begin{pmatrix} 0 & 0 \\ 0 & 1 \end{pmatrix}$$

for all $t \geq s$ and $(t,\xi), (s,\xi) \in Z_{ell}(d_0, \varepsilon)$.

Proof Let us assume that $\Phi_k = \Phi_k(t,s,\xi)$, $k = 1, 2$, are solutions to the equation

$$\Phi_{tt} + \lambda^2(t)\omega^2(t)|\xi|^2\Phi + \rho(t)\omega(t)\Phi_t = 0$$

with initial values $\Phi_k(s,s,\xi) = \delta_{1k}$ and $\partial_t\Phi_k(s,s,\xi) = \delta_{2k}$. Then, we have

$$\begin{pmatrix} \lambda(t)|\xi|v(t,\xi) \\ D_t v(t,\xi) \end{pmatrix} = \begin{pmatrix} \frac{\lambda(t)}{\lambda(s)}\Phi_1(t,s,\xi) & i\lambda(t)|\xi|\Phi_2(t,s,\xi) \\ \frac{D_t\Phi_1(t,s,\xi)}{\lambda(s)|\xi|} & iD_t\Phi_2(t,s,\xi) \end{pmatrix}\begin{pmatrix} \lambda(s)|\xi|v(s,\xi) \\ D_t v(s,\xi) \end{pmatrix}.$$

Our basic idea is to relate the entries of the above given estimates to the multipliers $\Phi_k = \Phi_k(t,s,\xi)$ and use Duhamel's formula to improve the estimates for the second row using estimates from the first one (see [16]). □

Remark 6 We are able to derive a refined estimate for the fundamental solution, because in the proof of Corollary 3 we use only estimates for $E_{ell}^{(11)}$ and $E_{ell}^{(12)}$ and both estimates seem to be optimal with our analytical tools.

Remark 7 If we choose a fixed s, then the second summand in Corollary 3 is dominated by the first one. Indeed, if we set $s = t_{ell}$, then by using $\lambda(t_{ell})|\xi| \approx \rho(t_{ell})$ we get the following estimate:

$$\big(|E_{ell}(t,s,\xi)|\big) \lesssim \exp\Big(-|\xi|^2\int_{t_{ell}}^t \frac{\lambda^2(\tau)\omega(\tau)}{\rho(\tau)}d\tau\Big)\begin{pmatrix} \frac{\lambda(t)}{\lambda(t_{ell})} & \frac{\lambda(t)}{\lambda(t_{ell})} \\ \frac{\lambda^2(t)|\xi|}{\lambda(t_{ell})\rho(t)} & \frac{\lambda^2(t)|\xi|}{\lambda(t_{ell})\rho(t)} \end{pmatrix}.$$

3.4 Considerations in the Dissipative Zone

In the dissipative zone we define the micro-energy $U = U(t, \xi)$ by

$$U = \big(\gamma(t)\hat{u}, D_t\hat{u}\big)^T, \quad \gamma(t) := \frac{\lambda(t)}{F(\Lambda(t))}.$$

This seems to be reasonable because we will later need to estimate $\lambda(t)|\xi|\hat{u}$ and it holds $\lambda(t)|\xi| \lesssim \dfrac{\lambda(t)}{F(\Lambda(t))}$ due to the definition of the dissipative zone. Then, the Fourier transformed Cauchy problem (5) leads to the system of first order

$$D_t U = \underbrace{\begin{pmatrix} \frac{D_t\gamma(t)}{\gamma(t)} & \gamma(t) \\ \frac{\lambda^2(t)\omega^2(t)|\xi|^2}{\gamma(t)} & i\rho(t)\omega(t) \end{pmatrix}}_{A(t,\xi)} U. \tag{17}$$

We are interested in the fundamental solution

$$E_{diss} = E_{diss}(t, s, \xi) = \begin{pmatrix} E_{diss}^{(11)} & E_{diss}^{(12)} \\ E_{diss}^{(21)} & E_{diss}^{(22)} \end{pmatrix}$$

to the system (17), that is, the solution of

$$D_t E_{diss}(t, s, \xi) = A(t, \xi) E_{diss}(t, s, \xi), \quad E_{diss}(s, s, \xi) = I,$$

for all $0 \leq s \leq t$ and $(t, \xi), (s, \xi) \in Z_{diss}(d_0)$. Thus, the solution $U = U(t, \xi)$ is represented as

$$U(t, \xi) = E_{diss}(t, s, \xi) U(s, \xi).$$

We will use the auxiliary function

$$\delta(t) = \exp\Big(\frac{1}{2}\int_0^t \rho(\tau)\omega(\tau)d\tau\Big)$$

which is related to the entry $i\rho(t)\omega(t)$ of the coefficient matrix.

The entries $E_{diss}^{(kl)}(t, s, \xi)$, $k, l = 1, 2$, of the fundamental solution $E_{diss}(t, s, \xi)$ satisfy the following system of Volterra integral equations for $k = 1, 2$:

$$D_t E_{diss}^{(1l)}(t, s, \xi) = \frac{D_t\gamma(t)}{\gamma(t)} E_{diss}^{(1l)}(t, s, \xi) + \gamma(t) E_{diss}^{(2l)}(t, s, \xi),$$

$$D_t E_{diss}^{(2l)}(t, s, \xi) = \frac{\lambda^2(t)\omega^2(t)|\xi|^2}{\gamma(t)} E_{diss}^{(1l)}(t, s, \xi) + i\rho(t)\omega(t) E_{diss}^{(2l)}(t, s, \xi)$$

together with their initial conditions

$$\begin{pmatrix} E^{(11)}_{diss}(s,s,\xi) & E^{(12)}_{diss}(s,s,\xi) \\ E^{(21)}_{diss}(s,s,\xi) & E^{(22)}_{diss}(s,s,\xi) \end{pmatrix} = \begin{pmatrix} 1 & 0 \\ 0 & 1 \end{pmatrix}.$$

Then, by direct calculations we get

$$\begin{aligned}
E^{(11)}_{diss}(t,s,\xi) &= \frac{\gamma(t)}{\gamma(s)} + i\gamma(t)\int_s^t E^{(21)}_{diss}(\tau,s,\xi)d\tau,\\
E^{(21)}_{diss}(t,s,\xi) &= \frac{i|\xi|^2}{\delta^2(t)}\int_s^t \frac{\lambda^2(\tau)\omega^2(\tau)}{\gamma(\tau)}\delta^2(\tau)E^{(11)}_{diss}(\tau,s,\xi)d\tau,\\
E^{(12)}_{diss}(t,s,\xi) &= i\gamma(t)\int_s^t E^{(22)}_{diss}(\tau,s,\xi)d\tau,\\
E^{(22)}_{diss}(t,s,\xi) &= \frac{\delta^2(s)}{\delta^2(t)} + \frac{i|\xi|^2}{\delta^2(t)}\int_s^t \frac{\lambda^2(\tau)\omega^2(\tau)}{\gamma(\tau)}\delta^2(\tau)E^{(12)}_{diss}(\tau,s,\xi)d\tau.
\end{aligned}$$

The next lemma is important for deriving suitable estimates for the entries $E^{(kl)}_{diss}(t,s,\xi)$, $k,l=1,2$.

Lemma 8 *The assumption (B3) implies* $\frac{\lambda(t)}{\delta^2(t)} \in L^1(\mathbb{R}_+)$ *with*

$$\int_t^\infty \frac{\lambda(\tau)}{\delta^2(\tau)}d\tau \lesssim \frac{\Lambda(t)}{\delta^2(t)}.$$

Moreover, $\frac{\Lambda(t)}{\delta^2(t)}$ *is monotonously decreasing for large* t.

Proof From $\mu(t)\to\infty$ as $t\to\infty$ it follows $\mu(t)\geq\frac{1+\varepsilon}{\omega(t)}$. Then, we may conclude

$$\delta^2(t) = \exp\Big(\int_0^t \rho(\tau)\omega(\tau)d\tau\Big) \gtrsim \exp\Big((1+\varepsilon)\int_0^t \frac{\lambda(\tau)}{\Lambda(\tau)}d\tau\Big) = \Lambda^{1+\varepsilon}(t),$$

which implies the integrability of $\frac{\lambda(t)}{\delta^2(t)}$. Furthermore, for large t we have

$$\begin{aligned}
\frac{1}{C_\varepsilon}\int_t^\infty \frac{\lambda(\tau)}{\delta^2(\tau)}d\tau &\leq \int_t^\infty \frac{\varepsilon}{\omega(\tau)}\frac{\lambda(\tau)}{\delta^2(\tau)}d\tau \leq \int_t^\infty \Big(\mu(\tau)-\frac{1}{\omega(\tau)}\Big)\frac{\lambda(\tau)}{\delta^2(\tau)}d\tau\\
&= \int_t^\infty \frac{\mu(\tau)\lambda(\tau)\omega(\tau)-\lambda(\tau)}{\omega(\tau)\delta^2(\tau)}d\tau \lesssim \int_t^\infty \frac{\mu(\tau)\lambda(\tau)\omega(\tau)-\lambda(\tau)}{\delta^2(\tau)}d\tau = \frac{\Lambda(t)}{\delta^2(t)}.
\end{aligned}$$

Moreover, we have

$$\frac{d}{dt}\frac{\Lambda(t)}{\delta^2(t)} = \frac{\lambda(t)-\rho(t)\omega(t)\Lambda(t)}{\delta^2(t)} = \frac{\lambda(t)\big(1-\mu(t)\omega(t)\big)}{\delta^2(t)}$$

and $\mu(t) \geq \frac{1+\varepsilon}{\omega(t)}$ for large t, which implies that $\dfrac{\Lambda(t)}{\delta^2(t)}$ is decreasing for large t. This completes the proof. □

In order to estimate the modulus $\big|E_{diss}^{(kl)}(t,s,\xi)\big|$, $k,l=1,2$, of the entries $E_{diss}^{(kl)}(t,s,\xi)$ we will use the assumption (B6).

Corollary 4 *Assume the conditions (A1) for $\lambda(t)$, (A2) for $\omega(t)$, (B3) and (B6) for $\rho(t)$. Then, we have the following estimate in the dissipative zone:*

$$\big(|E_{diss}(t,s,\xi)|\big) \lesssim \frac{\lambda(t)}{F(\Lambda(t))}\begin{pmatrix} \frac{F(\Lambda(s))}{\lambda(s)} & \frac{\Lambda(s)}{\lambda(s)} \\ \frac{F(\Lambda(s))}{\lambda(s)} & \frac{\Lambda(s)}{\lambda(s)} \end{pmatrix},$$

with $(s,\xi),(t,\xi) \in Z_{diss}(d_0)$ and $0 \leq s \leq t$.

Proof First let us consider the first column. Plugging the representation for $E_{diss}^{(21)}(t,s,\xi)$ into the integral equation for $E_{diss}^{(11)}(t,s,\xi)$ gives

$$\frac{\gamma(s)}{\gamma(t)}E_{diss}^{(11)}(t,s,\xi) = 1 - |\xi|^2\int_s^t\int_s^\tau \lambda^2(\theta)\omega^2(\theta)\frac{\delta^2(\theta)}{\delta^2(\tau)}\frac{\gamma(s)}{\gamma(\theta)}E_{diss}^{(11)}(\theta,s,\xi)d\theta d\tau.$$

By setting $y(t,s,\xi) := \dfrac{\gamma(s)}{\gamma(t)}E_{diss}^{(11)}(t,s,\xi)$ and applying partial integration we obtain

$$\begin{aligned} y(t,s,\xi) &= 1 + |\xi|^2\int_s^t \lambda^2(\theta)\omega^2(\theta)\Big(\frac{1}{\rho(\tau)\omega(\tau)}\frac{\delta^2(\theta)}{\delta^2(\tau)}\Big|_\theta^t \\ &\qquad + \int_\theta^t \underbrace{\frac{\big(\rho(\tau)\omega(\tau)\big)'}{\big(\rho(\tau)\omega(\tau)\big)^2}}_{=o(1)\text{ for }\tau\to\infty}\frac{\delta^2(\theta)}{\delta^2(\tau)}d\tau\Big)y(\theta,s,\xi)d\theta \\ &\approx 1 + |\xi|^2\int_s^t \lambda^2(\theta)\omega^2(\theta)\Big(\frac{1}{\rho(t)\omega(t)}\frac{\delta^2(\theta)}{\delta^2(t)} - \frac{1}{\rho(\theta)\omega(\theta)}\Big)y(\theta,s,\xi)d\theta. \end{aligned}$$

Then, we have

$$|y(t,s,\xi)| \lesssim 1 + |\xi|^2\int_s^t \Big(\frac{\lambda^2(\theta)\omega^2(\theta)}{\rho(t)\omega(t)}\frac{\delta^2(\theta)}{\delta^2(t)} + \frac{\lambda^2(\theta)\omega(\theta)}{\rho(\theta)}\Big)|y(\theta,s,\xi)|d\theta.$$

Applying Gronwall's inequality and partial integration, respectively, we get

$$\begin{aligned} |y(t,s,\xi)| &\lesssim \exp\Big(\frac{\lambda^2(t)|\xi|^2}{\rho(t)}\frac{1}{\delta^2(t)}\int_s^t \delta^2(\theta)d\theta + |\xi|^2\int_s^t\frac{\lambda^2(\theta)}{\rho(\theta)}d\theta\Big) \\ &= \exp\Big(\frac{\lambda^2(t)|\xi|^2}{\rho(t)}\frac{1}{\delta^2(t)}\Big(\frac{\delta^2(\theta)}{\rho(\theta)\omega(\theta)}\Big|_s^t \end{aligned}$$

$$+\int_s^t \frac{\big(\rho(\theta)\omega(\theta)\big)'}{\big(\rho(\theta)\omega(\theta)\big)^2}\delta^2(\theta)d\theta\Big) + |\xi|^2 \int_s^t \frac{\lambda^2(\theta)}{\rho(\theta)}d\theta\Big)$$

$$\lesssim \exp\Big(C\frac{\lambda^2(t)|\xi|^2}{\rho^2(t)} + C\frac{d_0^2}{F^2\big(\Lambda(t)\big)}\int_s^t \frac{\lambda^2(\theta)}{\rho(\theta)}d\theta\Big) \lesssim 1.$$

Here we have used $\frac{(\rho(\theta)\omega(\theta))'}{(\rho(\theta)\omega(\theta))^2} = o\big(1\big)$ from the assumption (B5), $\lambda(t)|\xi| \lesssim \rho(t)$ from the definition of the elliptic region, $|\xi| \leq \frac{d_0}{F(\Lambda(t))}$ from the definition of the dissipative zone and the assumption (B6), respectively. Hence, we may conclude

$$\big|E_{diss}^{(11)}(t,s,\xi)\big| \lesssim \frac{\gamma(t)}{\gamma(s)} = \frac{\lambda(t)}{F(\Lambda(t))}\frac{F(\Lambda(s))}{\lambda(s)}.$$

Now we consider $E_{diss}^{(21)}(t,s,\xi)$. By using the estimate for $\big|E_{diss}^{(11)}(t,s,\xi)\big|$ we obtain

$$\big|E_{diss}^{(21)}(t,s,\xi)\big| \lesssim |\xi|^2 \int_s^t \frac{\lambda^2(\tau)}{\gamma(\tau)}\frac{\delta^2(\tau)}{\delta^2(t)}\big|E_{diss}^{(11)}(\tau,s,\xi)\big|d\tau \lesssim \frac{1}{\gamma(s)}\frac{\lambda^2(t)|\xi|^2}{\delta^2(t)}\int_s^t \delta^2(\tau)d\tau$$

$$= \frac{1}{\gamma(s)}\frac{\lambda^2(t)|\xi|^2}{\delta^2(t)}\Big(\frac{1}{\rho(\tau)\omega(\tau)}\delta^2(\tau)\Big|_s^t + \int_s^t \frac{\big(\rho(\tau)\omega(\tau)\big)'}{\big(\rho(\tau)\omega(\tau)\big)^2}\delta^2(\tau)d\tau\Big)$$

$$\lesssim \frac{1}{\gamma(s)}\frac{\lambda^2(t)|\xi|^2}{\rho(t)\omega(t)} \lesssim \frac{1}{\gamma(s)}\lambda(t)|\xi| \lesssim \frac{F\big(\Lambda(s)\big)}{\lambda(s)}\frac{\lambda(t)}{F\big(\Lambda(t)\big)}.$$

Here we used $\frac{(\rho(\theta)\omega(\theta))'}{(\rho(\theta)\omega(\theta))^2} = o\big(1\big)$ from the assumption (B5), $\lambda(t)|\xi| \lesssim \rho(t)$ from the definition of the elliptic region and $|\xi| \leq \frac{d_0}{F(\Lambda(t))}$ from the definition of the dissipative zone, respectively.

Next we consider the entries of the second column. Plugging the representation for $E_{diss}^{(22)}(t,s,\xi)$ into the integral equation for $E_{diss}^{(12)}(t,s,\xi)$ gives

$$\begin{aligned}&E_{diss}^{(12)}(t,s,\xi)\\ &= i\gamma(t)\delta^2(s)\int_s^t \frac{d\tau}{\delta^2(\tau)} - |\xi|^2\gamma(t)\int_s^t\int_s^\tau \lambda^2(\theta)\omega^2(\theta)\frac{\delta^2(\theta)}{\delta^2(\tau)}\frac{1}{\gamma(\theta)}E_{diss}^{(12)}(\theta,s,\xi)d\theta d\tau.\end{aligned}$$

By setting $y(t,s,\xi) := \dfrac{1}{\gamma(t)}E_{diss}^{(12)}(t,s,\xi)$ and proceeding in the same manner as with $E_{diss}^{(11)}(t,s,\xi)$, after integration by parts we obtain

$$\begin{aligned}|y(t,s,\xi)| &\lesssim \delta^2(s)\int_s^t \frac{\lambda(\tau)}{\delta^2(\tau)}\frac{1}{\lambda(\tau)}d\tau\\ &\quad + |\xi|^2\int_s^t \Big(\frac{\lambda^2(\theta)\omega^2(\theta)}{\rho(t)\omega(t)}\frac{\delta^2(\theta)}{\delta^2(t)} + \frac{\lambda^2(\theta)\omega^2(\theta)}{\rho(\theta)\omega(\theta)}\Big)|y(\theta,s,\xi)|d\theta.\end{aligned}$$

Now we can use again Gronwall's inequality, since the first integral can be estimated by $\frac{\Lambda(s)}{\delta^2(s)}\frac{1}{\lambda(s)}$ due to Lemma 8. Then, we get

$$|y(t,s,\xi)| \lesssim \frac{\Lambda(s)}{\lambda(s)}.$$

Thus, we obtain

$$\big|E_{diss}^{(12)}(t,s,\xi)\big| \lesssim \gamma(t)\frac{\Lambda(s)}{\lambda(s)} = \frac{\lambda(t)}{F\big(\Lambda(t)\big)}\frac{\Lambda(s)}{\lambda(s)}.$$

Finally, let us consider $E_{diss}^{(22)}(t,s,\xi)$ by using the estimate for $\big|E_{diss}^{(12)}(t,s,\xi)\big|$. In the same way as in the estimate for $\big|E_{diss}^{(21)}(t,s,\xi)\big|$ we obtain

$$\begin{aligned}\big|E_{diss}^{(22)}(t,s,\xi)\big| &\lesssim \frac{\delta^2(s)}{\delta^2(t)} + |\xi|^2\int_s^t \lambda^2(\tau)\frac{\delta^2(\tau)}{\delta^2(t)}\frac{1}{\gamma(\tau)}\big|E_{diss}^{(12)}(\tau,s,\xi)\big|d\tau\\ &\lesssim \frac{\delta^2(s)}{\delta^2(t)} + \frac{\lambda^2(t)|\xi|^2}{\delta^2(t)}\frac{\Lambda(s)}{\lambda(s)}\int_s^t \delta^2(\tau)d\tau \lesssim \frac{\delta^2(s)}{\delta^2(t)} + \frac{\Lambda(s)}{\lambda(s)}\lambda(t)|\xi|.\end{aligned}$$

We rewrite the last inequality as

$$\frac{\Lambda(t)}{\lambda(t)}\big|E_{diss}^{(22)}(t,s,\xi)\big| \lesssim \frac{\Lambda(s)}{\lambda(s)} + \frac{\Lambda(s)}{\lambda(s)}\Lambda(t)|\xi| \lesssim \frac{\Lambda(s)}{\lambda(s)} + \frac{\Lambda(s)}{\lambda(s)}\frac{\Lambda(t)}{F\big(\Lambda(t)\big)},$$

where we used that $\frac{\Lambda(t)}{\delta^2(t)}$ is decreasing for large t due to Lemma 8. Thus, we get

$$\big|E_{diss}^{(22)}(t,s,\xi)\big| \lesssim \frac{\Lambda(s)}{\lambda(s)}\frac{\lambda(t)}{\Lambda(t)} + \frac{\Lambda(s)}{\lambda(s)}\frac{\Lambda(t)}{F\big(\Lambda(t)\big)}\frac{\lambda(t)}{\Lambda(t)} \lesssim \frac{\lambda(t)}{F\big(\Lambda(t)\big)}\frac{\Lambda(s)}{\lambda(s)},$$

where we used $\Lambda(t) \geq F\big(\Lambda(t)\big)$. This completes the proof. □

3.5 Considerations in the Reduced Zone

In the reduced zone we introduce the micro-energy $V = V(t,\xi)$ by

$$V = \Big(\varepsilon\frac{\rho(t)\omega(t)}{2}v, D_t v\Big)^T.$$

Then, by (6) the function V satisfies the following system:

$$D_t V = \underbrace{\begin{pmatrix} \frac{D_t(\rho(t)\omega(t))}{\rho(t)\omega(t)} & \varepsilon\frac{\rho(t)\omega(t)}{2} \\ \frac{\lambda^2(t)\omega^2(t)|\xi|^2 - \frac{1}{4}(\rho(t)\omega(t))^2 - \frac{1}{2}(\rho(t)\omega(t))'}{\varepsilon\frac{\rho(t)\omega(t)}{2}} & 0 \end{pmatrix}}_{A_V(t,\xi)} V. \tag{18}$$

We want to estimate the fundamental solution $E_{red}^V = E_{red}^V(t,s,\xi)$ to (18), that is, the solution to

$$D_t E_{red}^V(t,s,\xi) = A_V(t,\xi) E_{red}^V(t,s,\xi), \quad E_{red}^V(s,s,\xi) = I.$$

Due to $\big|\big(\rho(t)\omega(t)\big)'\big| = o\big((\rho(t)\omega(t))^2\big)$ for sufficiently large $t \geq t_0$, it holds

$$\frac{\big|D_t\big(\rho(t)\omega(t)\big)\big|}{\rho(t)\omega(t)} \lesssim \varepsilon\rho(t)\omega(t).$$

Moreover, we have the estimate

$$\langle\xi\rangle_{\lambda(t),\omega(t)} \lesssim \varepsilon\frac{\rho(t)\omega(t)}{2}.$$

Hence, we obtain the following estimate:

$$\frac{\big|\lambda^2(t)\omega^2(t)|\xi|^2 - \frac{1}{4}(\rho(t)\omega(t))^2 - \frac{1}{2}(\rho(t)\omega(t))'\big|}{\varepsilon\frac{\rho(t)\omega(t)}{2}} \lesssim \varepsilon\rho(t)\omega(t),$$

where we used $\big|\big(\rho(t)\omega(t)\big)'\big| = o\big(\,(\rho(t)\omega(t))^2\,\big)$. Finally, the norm of the coefficient matrix of (18) can be estimated by $\varepsilon\rho(t)\omega(t)$ for sufficiently large t.

Remark 8 From the backward transformation we may conclude that the fundamental solution $E_{red} = E_{red}(t,s,\xi)$ can be estimated as follows:

$$\big(|E_{red}(t,s,\xi)|\big) \lesssim \exp\Big(-\frac{1}{2}\int_s^t \rho(\tau)\omega(\tau)d\tau\Big)\big(|E_{red}^V(t,s,\xi)|\big).$$

Corollary 5 *Under the assumptions (B1), (B2) and (B5) the fundamental solution $E_{red} = E_{red}(t,s,\xi)$ satisfies the following estimate in the reduced zone:*

$$\big(|E_{red}(t,s,\xi)|\big) \lesssim \exp\Big(\Big(\varepsilon - \frac{1}{2}\Big)\int_s^t \rho(\tau)\omega(\tau)d\tau\Big)\begin{pmatrix} 1 & 1 \\ 1 & 1 \end{pmatrix}$$

for $t \geq s \geq t_0$ with a sufficiently large $t_0 = t_0(\varepsilon)$ and $(t,\xi), (s,\xi) \in Z_{red}(\varepsilon)$.

4 Gluing Procedure

In the previous sections we have derived estimates for fundamental fundamental solutions in different zones. Now we have to glue these estimates from Corollaries 1, 2, 3, 4 and 5.

Taking into account of the estimates in $Z_{hyp}(N)$, $Z_{osc}(N,\varepsilon)$ and $Z_{red}(\varepsilon)$ we can uniformly estimate $\big(|E_{hyp}(t,s,\xi)|\big)$ and $\big(|E_{osc}(t,s,\xi)|\big)$ by the upper bound from the estimate for $\big(|E_{red}(t,s,\xi)|\big)$. Therefore, we can glue $Z_{red}(\varepsilon)$ to the hyperbolic region and we define new regions by

$$\Pi_{hyp}(N,\varepsilon)=Z_{red}(\varepsilon)\cup Z_{osc}(N,\varepsilon)\cup Z_{hyp}(N),$$
$$\Pi_{ell}(d_0,\varepsilon)=Z_{ell}(d_0,\varepsilon)\cup Z_{diss}(d_0).$$

We denote by $t_{diss}(|\xi|)=:t_{diss}$ the separating line between $Z_{ell}(d_0,\varepsilon)$ and $Z_{diss}(d_0)$ and by $t(|\xi|)=:t_{|\xi|}$ the separating curve between $\Pi_{ell}(d_0,\varepsilon)$ and $\Pi_{hyp}(N,\varepsilon)$. Due to the definitions of the zones these separating lines really exist and can be described by functions due to the definitions of the zones and the monotonicity of these functions. Indeed, denoting $\eta(t):=\frac{\mu(t)}{2\Lambda(t)}$, this curve is given by

$$\eta^2(t_{|\xi|})-|\xi|^2=\varepsilon^2\eta^2(t_{|\xi|}),\quad \text{i.e.,}\quad t_{|\xi|}=\eta^{-1}\Big(\frac{|\xi|}{\sqrt{1-\varepsilon^2}}\Big).$$

Definition 4 We denote by $B_\lambda=B_\lambda(s,t)$, $0\le s\le t$, the primitive of $\dfrac{\lambda^2(t)}{\rho(t)}$ which vanishes at $t=s$. So, it is defined by

$$B_\lambda(s,t):=\int_s^t\frac{\lambda^2(\tau)}{\rho(\tau)}d\tau=B_\lambda(0,t)-B_\lambda(0,s).$$

In order to obtain energy estimates, first we establish some auxiliary estimates.

Lemma 9 *Under the assumptions (A1), (B1), (B2) and (B6) the following estimates hold:*

1. *Supposing* $|\xi|F(\Lambda(t_{diss}))=d_0$ *it holds*

$$\exp\Big(-C|\xi|^2\int_0^{t_{diss}}\frac{\lambda^2(\tau)}{\rho(\tau)}d\tau\Big)\approx 1.$$

2. *Supposing* $|\xi|=\eta(t_{|\xi|})\sqrt{1-\varepsilon^2}$ *it holds*

$$\big|d_{|\xi|}t_{|\xi|}\big|\gtrsim\frac{\mu(t_{|\xi|})}{|\xi|\rho(t_{|\xi|})}.$$

Case 1: The function $\eta = \eta(t)$ is Monotonously Decreasing

Now we distinguish between two cases related to the setting of the zones in the extended phase space.

Large Frequencies

In this case the large frequencies are in $Z_{hyp}(N)$ only. Then, it follows

$$\big(|E(t,0,\xi)|\big) \lesssim \Big(\frac{1}{\delta(t)}\Big)^{1-2\varepsilon} \begin{pmatrix} 1 & 1 \\ 1 & 1 \end{pmatrix},$$

where $\delta(t) = \exp\big(\frac{1}{2}\int_0^t \rho(\tau)\omega(\tau)d\tau\big)$.

Small Frequencies

In this case, if $\eta = \eta(t)$ is decreasing, in general all zones appear for small frequencies (see Fig. 1, Case a). Then, we have the following three cases:

Case 1.1: $t \le t_{diss}$

In this case (t,ξ) belongs to $Z_{diss}(d_0)$. Then, we have the following estimate from Corollary 4:

$$\big(|E(t,0,\xi)|\big) \lesssim \frac{\lambda(t)}{F\big(\Lambda(t)\big)} \begin{pmatrix} 1 & 1 \\ 1 & 1 \end{pmatrix}.$$

Case 1.2: $t_{diss} \le t \le t_{|\xi|}$

Now we will glue the estimates in $Z_{ell}(d_0,\varepsilon)$ from Corollary 3 with those in $Z_{diss}(d_0)$ from Corollary 4.

Lemma 10 *The following estimates hold for all $t \in [t_{diss}, t_{|\xi|}]$ and $|\xi| \ge \frac{d_0}{F(\Lambda(t))}$:*

$$\big(|E(t,0,\xi)|\big) \lesssim \exp\big(-C|\xi|^2 B_\lambda(0,t)\big) \begin{pmatrix} \lambda(t)|\xi| & \lambda(t)|\xi| \\ \frac{\lambda^2(t)|\xi|^2}{\rho(t)} & \frac{\lambda^2(t)|\xi|^2}{\rho(t)} \end{pmatrix}.$$

Proof The fundamental solution $E = E(t,0,\xi)$ can be represented as

$$E(t,0,\xi) = E_{ell}(t,t_{diss},\xi)E_{diss}(t_{diss},0,\xi).$$

Then, we have

$$\begin{aligned} \big(|E(t,0,\xi)|\big) &\lesssim \big(|E_{ell}(t,t_{diss},\xi)|\big)\big(|E_{diss}(t_{diss},0,\xi)|\big) \\ &\lesssim \exp\big(-C|\xi|^2 B_\lambda(t_{diss},t)\big) \begin{pmatrix} \frac{\lambda(t)}{\lambda(t_{diss})} & \frac{\lambda(t)|\xi|}{\rho(t_{diss})} \\ \frac{\lambda^2(t)|\xi|}{\lambda(t_{diss})\rho(t)} & \frac{\lambda^2(t)|\xi|^2}{\rho(t_{diss})\rho(t)} \end{pmatrix} \frac{\lambda(t_{diss})}{F\big(\Lambda(t_{diss})\big)} \begin{pmatrix} 1 & 1 \\ 1 & 1 \end{pmatrix} \\ &\lesssim \exp\big(-C|\xi|^2 B_\lambda(0,t)\big) \begin{pmatrix} \lambda(t)|\xi| & \lambda(t)|\xi| \\ \frac{\lambda^2(t)|\xi|^2}{\rho(t)} & \frac{\lambda^2(t)|\xi|^2}{\rho(t)} \end{pmatrix}, \end{aligned}$$

where we used $\lambda(t_{diss})|\xi| \lesssim \rho(t_{diss})$ and $|\xi|F\big(\Lambda(t_{diss})\big) = d_0$. Due to the first statement of Lemma 9 we can extend $B_\lambda(t_{diss}, t)$ to $B_\lambda(0, t)$. This completes the proof.

□

Case 1.3: $t \geq t_{|\xi|}$

To derive the corresponding estimates for $t \in [t_{|\xi|}, \infty)$ we shall estimate the term

$$S(t, |\xi|) := \exp\Big(- C|\xi|^2 \int_0^{t_{|\xi|}} \frac{\lambda^2(\tau)\omega(\tau)}{\rho(\tau)} d\tau\Big) \exp\Big(- \frac{1}{2} \int_{t_{|\xi|}}^t \rho(\tau)\omega(\tau) d\tau\Big).$$

This term explains the competition of influences from different zones. We use the decreasing behavior of the function $S = S(t, |\xi|)$ in $|\xi|$. So, the function $S = S(t, |\xi|)$ takes its maximum for $|\tilde{\xi}|$ satisfying $t = t_{|\tilde{\xi}|}$, that is, the second integral vanishes in $S(t, |\xi|)$ (see [2]).

Lemma 11 *For any $t \geq t_{|\xi|}$ and for a sufficiently small positive constant C the function $S = S(t, |\xi|)$ satisfies the following estimate:*

$$S(t, |\xi|) \leq \max_{\xi \in \mathbb{R}^n} \Big\{ \exp\Big(- C|\xi|^2 \int_0^t \frac{\lambda^2(\tau)\omega(\tau)}{\rho(\tau)} d\tau\Big)\Big\}.$$

Now we will glue the estimates from $\Pi_{hyp}(N, \varepsilon)$ and $Z_{ell}(d_0, \varepsilon)$ for $|\xi| \geq \frac{d_0}{F(\Lambda(t))}$.

Lemma 12 *The following estimates hold for all $|\xi| \geq \frac{d_0}{F(\Lambda(t))}$ and $t \in [t_{|\xi|}, \infty)$:*

$$\big(|E(t, 0, \xi)|\big) \lesssim \exp\big(- C'|\xi|^2 B_\lambda(0, t)\big) \begin{pmatrix} \lambda(t) & \lambda(t)|\xi| \\ \lambda(t) & \lambda(t)|\xi| \end{pmatrix}.$$

Proof Using the representation of the fundamental solution $E(t, 0, \xi)$ as

$$E(t, 0, \xi) = E_{hyp}(t, t_{osc}, \xi) E_{osc}(t_{osc}, t_{red}, \xi) E_{red}(t_{red}, t_{ell}, \xi) E_{ell}(t_{ell}, 0, \xi)$$

we arrive at the estimate

$$\begin{aligned} \big(|E(t, 0, \xi)|\big) &\lesssim \big(|E_{red}(t, t_{|\xi|}, \xi)|\big)\big(|E_{ell}(t_{|\xi|}, 0, \xi)|\big) \\ &\lesssim \Big(\frac{\delta(t_{|\xi|})}{\delta(t)}\Big)^{C_2} \begin{pmatrix} 1 & 1 \\ 1 & 1 \end{pmatrix} \exp\big(- C_1|\xi|^2 B_\lambda(0, t_{|\xi|})\big) \begin{pmatrix} \frac{\lambda(t_{|\xi|})}{\lambda(0)} & \frac{\lambda(t_{|\xi|})|\xi|}{\rho(0)} \\ \frac{\lambda^2(t_{|\xi|})|\xi|}{\lambda(0)\rho(t_{|\xi|})} & \frac{\lambda^2(t_{|\xi|})|\xi|^2}{\rho(0)\rho(t_{|\xi|})} \end{pmatrix} \\ &\lesssim \exp\big(- C'|\xi|^2 B_\lambda(0, t)\big) \begin{pmatrix} \lambda(t) & \lambda(t)|\xi| \\ \lambda(t) & \lambda(t)|\xi| \end{pmatrix}, \end{aligned}$$

where we used $\rho(t_{|\xi|}) \approx \lambda(t_{|\xi|})|\xi|$. Defining $C' := \min\{C_1, C_2\}$ we used Lemma 11 with

$$\exp\big(- C_1|\xi|^2 B_\lambda(0, t_{|\xi|})\big)\Big(\frac{\delta(t_{|\xi|})}{\delta(t)}\Big)^{C_2} \leq \exp\big(- C'|\xi|^2 B_\lambda(0, t)\big).$$

This completes the proof. □

Finally, for small frequencies it remains to glue the estimates in $\Pi_{hyp}(N,\varepsilon)$ with those in $Z_{ell}(d_0,\varepsilon)$ and in $Z_{diss}(d_0)$. We remark that this case comes into play only if $\eta=\eta(t)$ is decreasing. We have already obtained in Lemma 10 the desired estimate after gluing the estimates in $Z_{ell}(d_0,\varepsilon)$ with those in $Z_{diss}(d_0)$. Denoting the glued propagator by $\mathcal{E}=\mathcal{E}(t,0,\xi)$ we will only glue the estimates in $Z_{red}(\varepsilon)$ with the estimate from Lemma 10.

Lemma 13 *The following estimates hold for all $t\in[t_{|\xi|},\infty)$:*

$$\big(|E(t,0,\xi)|\big)\lesssim\exp\big(-C'|\xi|^2B_\lambda(0,t)\big)\lambda(t)|\xi|\begin{pmatrix}1&1\\1&1\end{pmatrix}.$$

Proof We have the following estimate in $\Pi_{hyp}(N,\varepsilon)$:

$$\big(|E_{red}(t,s,\xi)|\big)\lesssim\Big(\frac{\delta(s)}{\delta(t)}\Big)^{1-2\varepsilon}\begin{pmatrix}1&1\\1&1\end{pmatrix}.$$

Then, by using the estimate from Lemma 10 we get

$$\begin{aligned}\big(|E(t,0,\xi)|\big)&\lesssim\big(|E_{red}(t,t_{|\xi|},\xi)|\big)\big(|\mathcal{E}(t_{|\xi|},0,\xi)|\big)\\&\lesssim\Big(\frac{\delta(t_{|\xi|})}{\delta(t)}\Big)^{C_2}\begin{pmatrix}1&1\\1&1\end{pmatrix}\exp\big(-C_1|\xi|^2B_\lambda(0,t_{|\xi|})\big)\begin{pmatrix}\lambda(t_{|\xi|})|\xi| & \lambda(t_{|\xi|})|\xi|\\ \frac{\lambda^2(t_{|\xi|})|\xi|^2}{\rho(t_{|\xi|})} & \frac{\lambda^2(t_{|\xi|})|\xi|^2}{\rho(t_{|\xi|})}\end{pmatrix}\\&\lesssim\exp\big(-C'|\xi|^2B_\lambda(0,t)\big)\lambda(t)|\xi|\begin{pmatrix}1&1\\1&1\end{pmatrix}\end{aligned}$$

for all $t\geq t_{|\xi|}$. Here we used $\rho(t_{|\xi|})\approx\lambda(t_{|\xi|})|\xi|$. Defining $C':=\min\{C_1,C_2\}$ by Lemma 11 we also used

$$\exp\big(-C_1|\xi|^2B_\lambda(0,t_{|\xi|})\big)\Big(\frac{\delta(t_{|\xi|})}{\delta(t)}\Big)^{C_2}\leq\exp\big(-C'|\xi|^2B_\lambda(0,t)\big).$$

This completes the proof. □

Case 2: The Function $\eta=\eta(t)$ is Monotonously Increasing

In this case the elliptic zone lies on the top of the hyperbolic zone. Then, we have two different parts of the phase space to glue (see Fig. 1, Case b).

Small Frequencies

Small frequencies lie completely inside $\Pi_{ell}(d_0,\varepsilon)$. For this reason we can use the estimates that we obtained in *Case 1.1* and *Case 1.2*.

Large Frequencies

Case 2.1: $t\leq t_{|\xi|}$

In this case (t,ξ) belongs to $\Pi_{hyp}(N,\varepsilon)$. Then, we have

$$\big(|E(t,0,\xi)|\big) \lesssim \Big(\frac{1}{\delta(t)}\Big)^{1-2\varepsilon}\begin{pmatrix}1 & 1\\ 1 & 1\end{pmatrix}.$$

Case 2.2: $t \geq t_{|\xi|}$

First let us consider the case $(t,\xi) \in Z_{ell}(d_0,\varepsilon)$. Then, from Corollary 3 we get

$$\big(|E(t,0,\xi)|\big) \lesssim \exp\big(-C'|\xi|^2 B_\lambda(0,t)\big)\begin{pmatrix}\frac{\lambda(t)}{\lambda(0)} & \frac{\lambda(t)|\xi|}{\rho(0)}\\ \frac{\lambda^2(t)|\xi|}{\lambda(0)\rho(t)} & \frac{\lambda^2(t)|\xi|^2}{\rho(0)\rho(t)}\end{pmatrix}.$$

Now we have to glue the estimates in $Z_{ell}(d_0,\varepsilon)$ with those in $\Pi_{hyp}(N,\varepsilon)$.

Lemma 14 *The following estimates hold for all* $t \in [t_{|\xi|},\infty)$*:*

$$\big(|E(t,0,\xi)|\big) \lesssim \exp\big(-C'|\xi|^2 B_\lambda(0,t)\big)\frac{\lambda(t)}{\lambda(t_{|\xi|})}\begin{pmatrix}1 & 1\\ 1 & 1\end{pmatrix}.$$

Proof Taking account of the representation

$$E(t,0,\xi) = E_{ell}(t,t_{ell},\xi)E_{red}(t_{ell},t_{red},\xi)E_{osc}(t_{red},t_{osc},\xi)E_{hyp}(t_{osc},0,\xi),$$

we have

$$\begin{aligned}\big(|E(t,0,\xi)|\big) &\lesssim \big(|E_{ell}(t,t_{|\xi|},\xi)|\big)\big(|E_{red}(t_{|\xi|},0,\xi)|\big)\\ &\lesssim \exp\big(-C_1|\xi|^2 B_\lambda(t_{|\xi|},t)\big)\begin{pmatrix}\frac{\lambda(t)}{\lambda(t_{|\xi|})} & \frac{\lambda(t)|\xi|}{\rho(t_{|\xi|})}\\ \frac{\lambda^2(t)|\xi|}{\lambda(t_{|\xi|})\rho(t)} & \frac{\lambda^2(t)|\xi|^2}{\rho(t_{|\xi|})\rho(t)}\end{pmatrix}\Big(\frac{1}{\delta(t_{|\xi|})}\Big)^{C_2}\begin{pmatrix}1 & 1\\ 1 & 1\end{pmatrix}\\ &\lesssim \exp\big(-C'|\xi|^2 B_\lambda(0,t)\big)\begin{pmatrix}\frac{\lambda(t)}{\lambda(t_{|\xi|})} & \frac{\lambda(t)}{\lambda(t_{|\xi|})}\\ \frac{\lambda(t)}{\lambda(t_{|\xi|})} & \frac{\lambda(t)}{\lambda(t_{|\xi|})}\end{pmatrix},\end{aligned}$$

where we used $\rho(t_{|\xi|}) \approx \lambda(t_{|\xi|})|\xi|$. Defining $C' := \min\{C_1, C_2\}$ and Lemma 13 we also used

$$\exp\big(-C_1|\xi|^2 B_\lambda(t_{|\xi|},t)\big)\Big(\frac{1}{\delta(t_{|\xi|})}\Big)^{C_2} \leq \exp\big(-C'|\xi|^2 B_\lambda(0,t)\big).$$

This completes the proof. □

4.1 Preliminaries

Now we introduce $\hat{K}_0 = \hat{K}_0(t,0,\xi)$ as the solution of the Cauchy problem (5) with the initial conditions $\hat{u}_0(\xi) = 1$ and $\hat{u}_1(\xi) = 0$. Then, we have the following identity for $k = 1, 2$:

$$\begin{pmatrix} \frac{\lambda(t)|\xi|}{h_k(t,\xi)} & 0 \\ 0 & 1 \end{pmatrix} E_k(t,0,\xi) \begin{pmatrix} h_k(0,\xi) \\ 0 \end{pmatrix} = \begin{pmatrix} \frac{\lambda(t)|\xi|}{h_k(t,\xi)} & 0 \\ 0 & 1 \end{pmatrix} \begin{pmatrix} h_k(t,\xi)\hat{K}_0(t,0,\xi) \\ D_t\hat{K}_0(t,0,\xi) \end{pmatrix}$$
$$= \begin{pmatrix} \lambda(t)|\xi|\hat{K}_0(t,0,\xi) \\ D_t\hat{K}_0(t,0,\xi) \end{pmatrix}.$$

Moreover, it holds

$$\begin{pmatrix} \frac{\lambda(t)|\xi|}{h_k(t,\xi)} & 0 \\ 0 & 1 \end{pmatrix} E_k(t,0,\xi) \begin{pmatrix} h_k(0,\xi) \\ 0 \end{pmatrix} = \begin{pmatrix} \frac{\lambda(t)|\xi|}{h_k(t,\xi)} & 0 \\ 0 & 1 \end{pmatrix} \begin{pmatrix} h_k(0,\xi)E_k^{(11)}(t,0,\xi) \\ h_k(0,\xi)E_k^{(21)}(t,0,\xi) \end{pmatrix}$$
$$= \begin{pmatrix} \frac{h_k(0,\xi)}{h_k(t,\xi)}\lambda(t)|\xi|E_k^{(11)}(t,0,\xi) \\ h_k(0,\xi)E_k^{(21)}(t,0,\xi) \end{pmatrix},$$

where $h_1 = h_1(t,\xi)$ and $h_2 = h_2(t,\xi)$ are defined in (8) and (9), respectively. Moreover, $E_1(t,0,\xi) := E(t,0,\xi)$ and $E_2(t,0,\xi) := E_V(t,0,\xi)$ are defined in Definition 1. The above relations allows us to transfer properties of $E_k = E_k(t,0,\xi)$ to $\hat{K}_0 = \hat{K}_0(t,0,\xi)$ for $k = 1, 2$. Thus, we obtain

$$\hat{K}_0(t,0,\xi) = \frac{h_k(0,\xi)}{h_k(t,\xi)} E_k^{(11)}(t,0,\xi),$$
$$D_t\hat{K}_0(t,0,\xi) = h_k(0,\xi)E_k^{(21)}(t,0,\xi).$$

In the same way, we consider $\hat{K}_1 = \hat{K}_1(t,0,\xi)$ as the solution of the Cauchy problem (5) with initial conditions $\hat{u}_0(\xi) = 0$ and $\hat{u}_1(\xi) = 1$. Then, we get

$$\hat{K}_1(t,0,\xi) = \frac{1}{h_k(t,\xi)} E_k^{(12)}(t,0,\xi),$$
$$D_t\hat{K}_1(t,0,\xi) = E_k^{(22)}(t,0,\xi).$$

4.2 Final Estimates

Let us define for any $t \geq 0$ the function

$$\Omega(0,t) := \max\{\eta(0), \eta(t)\}\sqrt{1-\varepsilon^2}.$$

Remark 9 We distinguish between small and large frequencies. Small frequencies satisfy the condition $|\xi| \leq \Omega(0,t)$, while, large frequencies satisfy the condition $|\xi| \geq \Omega(0,t)$.

Summarizing, we arrive at the following estimates for $|\partial_t^l \hat{K}_j(t,0,\xi)|$ with $j, l = 0, 1$.

Corollary 6 *If* $|\xi| \geq \Omega(0,t)$*, then we have the following estimates for* $j, l = 0, 1$*:*

$$\big|\partial_t^l \hat{K}_j(t,0,\xi)\big| \lesssim \lambda(t)^{l-1}|\xi|^{l-j}\Big(\frac{1}{\delta(t)}\Big)^{1-2\varepsilon}. \tag{19}$$

If $\dfrac{d_0}{F(\Lambda(t))} \leq |\xi| \leq \Omega(0,t)$*, then we have the following estimates* $j, l = 0, 1$*:*

$$\big|\partial_t^l \hat{K}_j(t,0,\xi)\big| \lesssim \lambda^l(t)|\xi|^l \exp\big(-C|\xi|^2 B_\lambda(0,t)\big). \tag{20}$$

If $|\xi| \leq \dfrac{d_0}{F(\Lambda(t))}$*, then we have the following estimates* $j, l = 0, 1$*:*

$$\big|\partial_t^l \hat{K}_j(t,0,\xi)\big| \lesssim \frac{\lambda^l(t)}{F^l(\Lambda(t))}. \tag{21}$$

5 Energy Estimates for Solutions to Damped Wave Models with Additional Regularity of the Data

The solution $u = u(t,x)$ to the Cauchy problem

$$\begin{cases} u_{tt} - \lambda^2(t)\omega^2(t)\Delta u + \rho(t)\omega(t)u_t = 0, & (t,x) \in [0,\infty) \times \mathbb{R}^n, \\ u(0,x) = u_0(x), \quad u_t(0,x) = u_1(x), & x \in \mathbb{R}^n, \end{cases} \tag{22}$$

can be represented as

$$u(t,x) = K_0(t,0,x) *_{(x)} u_0(x) + K_1(t,0,x) *_{(x)} u_1(x).$$

Thus, we may conclude the following estimate for the solution $u = u(t,x)$:

$$\|u(t,\cdot)\|_{L^2} = \|\hat{u}(t,\cdot)\|_{L^2} \leq \big\|\hat{K}_0(t,0,\xi)\hat{u}_0(\xi)\big\|_{L^2} + \big\|\hat{K}_1(t,0,\xi)\hat{u}_1(\xi)\big\|_{L^2}.$$

In order to estimate the L^2 norm of $\partial_t^l \partial_x^\sigma K_j(t,0,x) *_{(x)} u_j(x)$ for any $\sigma \geq 0$ and $j, l = 0, 1$, we can follow the techniques used in [6, 11]. Then, we have the following statements for large and small frequencies.

Lemma 15 *The following estimates hold for large frequencies* $|\xi| \geq \Omega(0,t)$*:*

$$\big\||\xi|^\sigma \partial_t^l \hat{K}_j(t,0,\cdot)\hat{u}_j\big\|_{L^2_{\{|\xi| \geq \Omega(0,t)\}}} \lesssim \lambda^l(t)\Big(\frac{1}{\delta(t)}\Big)^{1-2\varepsilon} \|u_j\|_{H^{|\sigma|+l-j}} \tag{23}$$

with $\sigma + l \geq j$ *and for any* $\sigma \geq 0$ *and* $j, l = 0, 1$*. Moreover, if* $\sigma = l = 0$ *and* $j = 1$ *we distinguish between the following two cases:*

1. *If $\eta = \eta(t)$ is increasing, then we have the estimate*

$$\big\| \hat{K}_1(t,0,\cdot)\hat{u}_1 \big\|_{L^2_{\{|\xi| \geq \Omega(0,t)\}}} \lesssim \frac{1}{\eta(t)}\Big(\frac{1}{\delta(t)}\Big)^{1-2\varepsilon} \|u_1\|_{L^2}. \tag{24}$$

2. *If $\eta = \eta(t)$ is decreasing we have the estimate*

$$\big\| \hat{K}_1(t,0,\cdot)\hat{u}_1 \big\|_{L^2_{\{|\xi| \geq \Omega(0,t)\}}} \lesssim \Big(\frac{1}{\delta(t)}\Big)^{1-2\varepsilon} \|u_1\|_{L^2}. \tag{25}$$

Proof We have the following estimate for $\sigma + l \geq j$:

$$\begin{aligned}&\big\| |\xi|^\sigma \partial_t^l \hat{K}_j(t,0,\cdot)\hat{u}_j \big\|_{L^2_{\{|\xi| \geq \Omega(0,t)\}}} \\ &\quad\leq \big\| |\xi|^{j-l} \partial_t^l \hat{K}_j(t,0,\xi) \big\|_{L^\infty_{\{|\xi| \geq \Omega(0,t)\}}} \big\| |\xi|^{\sigma+l-j} \hat{u}_j \big\|_{L^2_{\{|\xi| \geq \Omega(0,t)\}}}.\end{aligned}$$

The second term on the right-hand side can be estimated by $\|u_j\|_{H^{\sigma+l-j}}$. Now let us consider the L^∞ norm of $\partial_t^l K_j(t,0,\xi)$. Indeed, by using the estimate (19) we get

$$|\xi|^{j-l} \big|\partial_t^l \hat{K}_j(t,0,\xi)\big| \lesssim \lambda(t)^{l-1}\Big(\frac{1}{\delta(t)}\Big)^{1-2\varepsilon} \lesssim \lambda^l(t)\Big(\frac{1}{\delta(t)}\Big)^{1-2\varepsilon}.$$

Let $\sigma = l = 0$ and $j = 1$. Then, we arrive at the estimate

$$\big\| \hat{K}_1(t,0,\cdot)\hat{u}_1 \big\|_{L^2_{\{|\xi| \geq \Omega(0,t)\}}} \lesssim \frac{1}{\Omega(0,t)}\Big(\frac{1}{\delta(t)}\Big)^{1-2\varepsilon} \|u_1\|_{L^2}.$$

This completes the proof. □

Lemma 16 *The following estimates hold for small frequencies $\frac{d_0}{F(\Lambda(t))} \leq |\xi| \leq \Omega(0,t)$:*

$$\begin{aligned}&\big\| |\xi|^\sigma \partial_t^l \hat{K}_j(t,0,\cdot)\hat{u}_j \big\|_{L^2_{\{\frac{d_0}{F(\Lambda(t))} \leq |\xi| \leq \Omega(0,t)\}}} \\ &\qquad\lesssim \lambda^l(t)\big(B_\lambda(0,t)\big)^{-\frac{l}{2}}\big(B_\lambda(0,t)\big)^{-\frac{\sigma}{2}-\frac{n}{2}(\frac{1}{m}-\frac{1}{2})}\|u_j\|_{L^m}\end{aligned} \tag{26}$$

for any $\sigma \geq 0$ and $j, l = 0, 1$, where $m \in [1,2)$.

Lemma 17 *The following estimates hold for small frequencies $|\xi| \leq \frac{d_0}{F(\Lambda(t))}$:*

$$\begin{aligned}&\big\| |\xi|^\sigma \partial_t^l \hat{K}_j(t,0,\cdot)\hat{u}_j \big\|_{L^2_{\{|\xi| \leq \frac{d_0}{F(\Lambda(t))}\}}} \\ &\qquad\lesssim \lambda^l(t)\big(F^2(\Lambda(t))\big)^{-\frac{l}{2}}\big(F^2(\Lambda(t))\big)^{-\frac{\sigma}{2}-\frac{n}{2}(\frac{1}{m}-\frac{1}{2})}\|u_j\|_{L^m}\end{aligned} \tag{27}$$

for any $\sigma \geq 0$ *and* $j, l = 0, 1$, *where* $m \in [1, 2)$.

Proof The proofs of Lemmas 16 and 17 use Hölder's inequality and Hausdorff-Young inequality. □

Proof of Theorem 1. Due to condition (B6), from (26) and (27) it follows

$$\big(F^2(\Lambda(t))\big)^{-\frac{\sigma}{2}-\frac{n}{2}(\frac{1}{m}-\frac{1}{2})-\frac{l}{2}} \lesssim \big(B_\lambda(0,t)\big)^{-\frac{\sigma}{2}-\frac{n}{2}(\frac{1}{m}-\frac{1}{2})-\frac{l}{2}}$$

for $l = 0, 1$. For this reason the right-hand sides in the estimates of Lemmas 15 and 17 decay faster than that one in the estimates of Lemma 16. On the other hand, the regularity of the data is coming from the large frequencies from the estimates (23) to (25). In this way we have proved the desired statements. □

Remark 10 We note that in our estimates we replace $B_\lambda(0, t)$ by $1 + B_\lambda(0, t)$. This can be done modulo a compact set in the extended phase space. Such a compact set will never influence the desired estimates.

6 Some Examples

Example 1 (*Polynomial case*) Let $\lambda(t) = (\alpha + 1)(1 + t)^\alpha$, $\alpha > 0$. So, we have

$$\Lambda(t) = (1 + t)^{\alpha+1} \text{ and } \Theta(t) = (1 + t)^{\gamma+1}, \quad -1 < \gamma < \alpha.$$

Moreover, we choose

$$\rho(t) = \frac{(\alpha + 1)^2}{2\alpha - \beta + 1}(1 + t)^\beta, \quad \alpha - \gamma - 1 < \beta < 2\alpha + 1.$$

If we take $\Xi(t) = (1 + t)^\kappa$ from the condition (A2), then the condition (A4) holds with

$$1 > \kappa \geq \kappa_M = 1 - \alpha + \gamma + \frac{\alpha - \gamma}{M + 1},$$

where M denotes the order of regularity for the coefficients. On the other hand, by the condition (A5) we obtain

$$F(\Lambda(t)) \approx (1 + t)^{\alpha+2\kappa-1}, \quad \alpha + 2\kappa - 1 > 0.$$

Finally, from the condition (B6) we have the estimate

$$(1 + t)^{2\alpha-\beta+1} \leq (1 + t)^{2\alpha+4\kappa-2}, \quad \kappa \geq \frac{3 - \beta}{4}.$$

Summarizing, the above relations yield

$$\kappa \geq \frac{3-\beta}{4} > \frac{1-\alpha}{2} > 1-\alpha+\gamma+\frac{\alpha-\gamma}{M+1}.$$

Therefore, we obtain that the range for admissible κ is given by $\kappa \geq \frac{3-\beta}{4}$. Hence, the hypotheses of Theorem 1 are satisfied.

Let us introduce $C_\sigma = \frac{\sigma}{2} + \frac{n}{2}\big(\frac{1}{m} - \frac{1}{2}\big)$. Then, we have the following estimates:

$$\|u(t,\cdot)\|_{\dot{H}^\sigma} \lesssim (1+t)^{-(2\alpha-\beta+1)C_\sigma}\big(\|u_0\|_{L^m\cap H^\sigma} + \|u_1\|_{L^m\cap H^{[\sigma-1]_+}}\big),$$
$$\|u_t(t,\cdot)\|_{\dot{H}^\sigma} \lesssim (1+t)^{-(2\alpha-\beta+1)C_\sigma+\frac{\beta-1}{2}}\big(\|u_0\|_{L^m\cap H^{\sigma+1}} + \|u_1\|_{L^m\cap H^\sigma}\big).$$

Example 2 (*Exponential case*) Let us choose $\lambda(t) = e^t$. So, we have

$$\Lambda(t) = e^t \ \text{ and } \ \Theta(t) = e^{rt}, \ \ 0 < r < 1.$$

Moreover, we choose

$$\rho(t) = \frac{1}{2-q}e^{qt}, \ \ 1-r < q < 2.$$

Now if we take $\Xi(t) = e^{\kappa t}$ from the condition (A2), then the condition (A4) is satisfied with

$$0 > \kappa \geq \kappa_M = r-1+\frac{1-r}{M+1}.$$

On the other hand, by the condition (A5) we get

$$F(\Lambda(t)) \approx e^{(2\kappa+1)t}, \ \ \kappa \geq -\frac{q}{4}.$$

Finally, by the condition (B6) we have the estimate

$$e^{(2-q)t} \leq e^{(2+4\kappa)t}, \ \ \kappa \geq -\frac{q}{4}.$$

The above relations yield

$$\kappa \geq -\frac{q}{4} > -\frac{1}{2} > r-1+\frac{1-r}{M+1}.$$

Therefore, we obtain that the range for admissible κ is given by $\kappa \geq -\frac{q}{4}$. Hence, the hypotheses of Theorem 1 are satisfied.

Introducing $C_\sigma = \frac{\sigma}{2} + \frac{n}{2}\big(\frac{1}{m} - \frac{1}{2}\big)$, we have the following estimates:

$$\|u(t,\cdot)\|_{\dot{H}^\sigma} \lesssim e^{-(2-q)C_\sigma t}\big(\|u_0\|_{L^m\cap H^\sigma} + \|u_1\|_{L^m\cap H^{[\sigma-1]_+}}\big),$$
$$\|u_t(t,\cdot)\|_{\dot{H}^\sigma} \lesssim e^{-(2-q)C_\sigma t} e^{\frac{q}{2}t}\big(\|u_0\|_{L^m\cap H^{\sigma+1}} + \|u_1\|_{L^m\cap H^\sigma}\big).$$

Example 3 (*Super-exponential case*) Let us choose $\lambda(t) = e^t e^{e^t}$. So, we have

$$\Lambda(t) = e^{e^t} \quad \text{and } \Theta(t) = e^{re^t}, \;\; 0 < r < 1.$$

Moreover, we choose

$$\rho(t) = \frac{1}{2-q} e^t e^{qe^t}, \;\; 1 - r < q < 2.$$

Finally, we take $\Xi(t) = e^{-t} e^{\kappa e^t}$ from condition (A2). Then, the condition (A4) holds with

$$0 > \kappa \geq \kappa_M = r - 1 + \frac{1-r}{M+1}.$$

On the other hand, by condition (A5) we have

$$F(\Lambda(t)) \approx e^{(1+2\kappa)e^t}, \;\; 1 + 2\kappa > 0.$$

From condition (B6) we have the estimate

$$e^{(2-q)e^t} \leq e^{(2+4\kappa)e^t}, \;\; \kappa \geq -\frac{q}{4}.$$

For this reason we obtain that the range of admissible κ is given by $\kappa \geq -\frac{q}{4}$. Summarizing, the above relations yield

$$\kappa \geq -\frac{q}{4} > -\frac{1}{2} > r - 1 + \frac{1-r}{M+1}.$$

Hence, the hypotheses of Theorem 1 are satisfied.

Introducing $C_\sigma = \frac{\sigma}{2} + \frac{n}{2}\big(\frac{1}{m} - \frac{1}{2}\big)$, we get the following estimates:

$$\|u(t,\cdot)\|_{\dot{H}^\sigma} \lesssim e^{-(2-q)C_\sigma e^t}\big(\|u_0\|_{L^m\cap H^\sigma} + \|u_1\|_{L^m\cap H^{[\sigma-1]_+}}\big),$$
$$\|u_t(t,\cdot)\|_{\dot{H}^\sigma} \lesssim e^t e^{-(2-q)C_\sigma e^t} e^{\frac{q}{2}e^t}\big(\|u_0\|_{L^m\cap H^{\sigma+1}} + \|u_1\|_{L^m\cap H^\sigma}\big).$$

Construction of an admissible oscillating function. Now we present an admissible non-trivial oscillating function $\omega = \omega(t)$ in (22) satisfying the hypotheses of Theorem 1. The construction of the function $\omega = \omega(t)$ was proposed in [10].

In order to construct a non-trivial function $\omega = \omega(t)$ satisfying (A2), (A3) and (B5) let us choose the positive sequences $\{t_j\}_j$, $\{\delta_j\}_j$ and $\{\eta_j\}_j$ as follows:

$$t_j \to \infty, \quad \delta_j \le \Delta t_j := t_{j+1} - t_j \quad \text{and} \quad \eta_j \le 1, \tag{28}$$

and a function $\psi \in C_0^M(\mathbb{R})$ with

$$\operatorname{supp} \psi \subseteq [0,1], \quad -1 < \psi(t) < 1 \quad \text{and} \quad \int_0^1 |\psi(t)| dt = \frac{1}{2}.$$

Then, we define

$$\omega(t) = 1 + \sum_{j=1}^{\infty} \eta_j \psi\Big(\frac{t - t_j}{\delta_j}\Big).$$

The last sum is convergent, since by (28) for each t at most one term is present. Furthermore, if $c_0 = \min \psi(t)$ and $c_1 = \max \psi(t)$, then we get the bounds

$$0 < 1 + c_0 \le \omega(t) \le 1 + c_1.$$

For (A2) we can take

$$C_1^{-1} \Xi(t_j) \le \delta_j \le C_1 \Xi(t_j) \tag{29}$$

and the sequence $\{t_j\}_j$ satisfying

$$C_2^{-1} \lambda(t_{k+1}) \le \lambda(t_k) \le C_2 \lambda(t_{k+1}) \ \text{ and } \ C_3^{-1} \Lambda(t_{k+1}) \le \Lambda(t_k) \le C_3 \Lambda(t_{k+1}) \tag{30}$$

with positive constants C_j, $j = 1, 2, 3$, which are independent of k. Indeed, by (28) and the definition of $\omega = \omega(t)$ we have

$$\omega(t) = 1 + \eta_k \psi\Big(\frac{t - t_k}{\delta_k}\Big) \quad \text{for all} \quad t \in [t_k, t_{k+1}].$$

Taking account of (29) and (30) it follows

$$|d_t^i \omega(t)| \le C_i \frac{\eta_k}{\delta_k^i} \le \tilde{C}_i \Xi^{-i}(t) \quad \text{for all } t \in [t_k, t_{k+1}], \quad i = 1, \cdots, M.$$

Using the definition of $\omega = \omega(t)$ we may conclude for all $t \in [t_k, t_{k+1}]$ the estimate

$$\int_0^t \lambda(s)|\omega(s) - 1| ds = \sum_{j=1}^{k} \eta_j \int_{t_j}^{t_{j+1}} \lambda(s) \Big|\psi\Big(\frac{s - t_j}{\delta_j}\Big)\Big| ds \le \frac{1}{2} \sum_{j=1}^{k} \eta_j \delta_j \lambda(t_{j+1}).$$

This implies that the stabilization condition (A3) is ensured if we assume $\eta_j \delta_j$ are small enough. Indeed,

$$\Theta(t_{k+1}) \approx \sum_{j=1}^{k} \eta_j \delta_j \lambda(t_{j+1}) = o\Big(\sum_{j=1}^{k} \lambda(t_j)\Delta(t_j)\Big).$$

Example 4 (*Polynomial case*) We consider $\lambda(t) = (\alpha + 1)(1 + t)^{\alpha}, \alpha > 0$. To define one admissible function $\omega = \omega(t)$ let us choose the parameters α, γ and κ from Example 1 and positive sequences $\{t_j\}_j$, $\{\delta_j\}_j$ and $\{\eta_j\}_j$ as follows:

$$t_j = 2^j, \quad \delta_j = 2^{\kappa j} \leq \Delta t_j = t_{j+1} - t_j = 2^j \quad \text{and} \quad \eta_j = 2^{j(\gamma-\alpha-\kappa)}.$$

From Example 1 we have

$$1 > \kappa \geq \frac{3-\beta}{4} > \kappa_M = 1 - \alpha + \gamma + \frac{\alpha - \gamma}{M+1}.$$

Therefore, we get $\gamma - \alpha - \kappa < 0$ and this implies that we have $0 < \eta_j \leq 1$. Moreover, the stabilization condition (A3) holds, since $\eta_j \delta_j = 2^{j(\gamma-\alpha)} < 1$, where $\gamma - \alpha < 0$.

Example 5 (*Exponential case*) We consider $\lambda(t) = e^t$. Let us choose the parameters r and κ from Example 2 and the positive sequences $\{t_j\}_j$, $\{\delta_j\}_j$ and $\{\eta_j\}_j$ by

$$t_j = j, \quad \delta_j = e^{\kappa j} \leq \Delta t_j = t_{j+1} - t_j \quad \text{and} \quad \eta_j = e^{j(r-\kappa-1)}.$$

From Example 2 we have

$$0 > \kappa \geq -\frac{q}{4} > \kappa_M = r - 1 + \frac{1-r}{M+1}.$$

Therefore, we have $\kappa < 0$ and $r - \kappa - 1 < 0$. These imply that we have $\delta_j < 1$ and $0 < \eta_j \leq 1$, respectively. Moreover, the stabilization condition (A3) is satisfied, because $\eta_j \delta_j = e^{j(r-1)} < 1$, where $r - 1 < 0$.

Example 6 (*Super-exponential case*) We consider $\lambda(t) = e^t e^{e^t}$. Let us choose the parameters r and κ from Example 3 and positive sequences $\{t_j\}_j$, $\{\delta_j\}_j$ and $\{\eta_j\}_j$ as follows:

$$t_j = e^j, \quad \delta_j = e^{-j} e^{\kappa e^j} \leq \Delta t_j = t_{j+1} - t_j \quad \text{and} \quad \eta_j = e^{(r-\kappa-1)e^j}.$$

From Example 3 we have

$$0 > \kappa \geq -\frac{q}{4} > \kappa_M = r - 1 + \frac{1-r}{M+1}.$$

These imply that we have $\delta_j < 1$ and $0 < \eta_j \leq 1$, respectively. Moreover, the stabilization condition (A3) holds, since $\eta_j \delta_j = e^{j(r-1)} < 1$.

7 Concluding Remarks and Open Problems

Remark 11 The recent papers [3, 4] are devoted to studying the following Cauchy problem for semi-linear wave models with effective dissipation:

$$\begin{cases} u_{tt} - a^2(t)\Delta u + b(t)u_t = |u|^p, & (t,x) \in [0,\infty) \times \mathbb{R}^n, \\ u(0,x) = u_0(x), \quad u_t(0,x) = u_1(x), & x \in \mathbb{R}^n. \end{cases} \tag{31}$$

Here the considerations are divided into two cases depending on the behavior of the propagation speed: *the super-exponential case* and *the sub-exponential case*, respectively, for the global (in time) existence of small data solutions to (31).

Remark 12 An interesting application of the results of this paper is to study the following Cauchy problem for semi-linear damped wave models with time-dependent speed of propagation and "effective-like" damping term $\rho(t)\omega(t)u_t$ in combination with very-fast oscillations and stabilization condition:

$$\begin{cases} u_{tt} - \lambda^2(t)\omega^2(t)\Delta u + \rho(t)\omega(t)u_t = |u|^p, & (t,x) \in [0,\infty) \times \mathbb{R}^n, \\ u(0,x) = u_0(x), \quad u_t(0,x) = u_1(x), & x \in \mathbb{R}^n. \end{cases}$$

This will be discussed in a forthcoming paper to understand the influence of the oscillations on the global (in time) existence of small data solutions. The key tools are the Gagliardo-Nirenberg inequalities, the fractional chain rule, the fractional Leibniz rule and the fractional powers rules which have been extensively discussed in Harmonic Analysis (cf. the book [7]).

References

1. Bui, T.B.N.: Semi-linear waves with time-dependent speed and dissipation, 154 pp. Ph.D. thesis, TU Bergakademie Freiberg (2013)
2. Bui, T.B.N., Reissig, M.: The interplay between time-dependent speed of propagation and dissipation in wave models. In: Fourier Analysis, Trends in Mathematics, pp. 9–45, Birkhäuser, Basel (2014)
3. Bui, T.B.N., Reissig, M.: Global existence of small data solutions for wave models with super-exponential propagation speed. Nonlinear Anal. **121**, 82–100 (2015). https://doi.org/10.1016/j.na.2014.10.035
4. Bui, T.B.N., Reissig, M.: Global existence of small data solutions for wave models with sub-exponential propagation speed. Nonlinear Anal. **129**, 173–188 (2015). https://doi.org/10.1016/j.na.2015.09.003
5. Colombini, F.: Energy estimates at infinity for hyperbolic equations with oscillating coefficients. J. Diff. Equ. **231**, 598–610 (2006). https://doi.org/10.1016/j.jde.2006.05.014
6. D'Abbicco, M., Lucente, S., Reissig, M.: Semi-linear wave equations with effective damping. Chin. Ann. Math. **34B**(3), 345–380 (2013). https://doi.org/10.1007/s11401-013-0773-0
7. Ebert M.R., Reissig, M.: Methods for Partial Differential Equations. Birkhäuser, Basel, Germany, (2018). https://doi.org/10.1007/978-3-319-66456-9

8. Hirosawa, F.: On the asymptotic behavior of the energy for the wave equation with time depending coefficients. Math. Ann. **339**, 819–838 (2007). https://doi.org/10.1007/s00208-007-0132-0
9. Hirosawa, F., Wirth, J.: C^m-theory of damped wave equations with stabilisation. J. Math. Anal. Appl. **343**, 1022–1035 (2008). https://doi.org/10.1016/j.jmaa.2008.02.024
10. Hirosawa, F., Wirth, J.: Generalised energy conservation law for wave equations with variable propagation speed. J. Math. Anal. Appl. **358**, 56–74 (2009). https://doi.org/10.1016/j.jmaa.2009.04.048
11. Matsumura, A.: On the asymptotic behavior of solutions of semi-linear wave equations. Publ. Res. Inst. Math. Sci. **12**(1), 169–189 (1976)
12. Reissig, M., Smith, J.: $L^p - L^q$ estimate for wave equation with bounded time-dependent coefficient. Hokkaido Math. J. **34**(3), 541–586 (2005)
13. Reissig, M., Yagdjian, K.: About the influence of oscillations on Strichartz-type decay estimates. Rem. Sem. Mat. Torino **58**(3), 375–388 (2000)
14. Reissig, M., Yagdjian, K.: $L_p - L_q$ decay estimates for the solutions of strictly hyperbolic equations of second order with increasing in time coefficients. Math. Nachr. **214**, 71–104 (2000)
15. Wirth, J.: Wave equations with time-dependent dissipation I. Non-Effective dissipation. J. Diff. Equ. **222**, 487–514 (2006). https://doi.org/10.1016/j.jde.2005.07.019
16. Wirth, J.: Wave equations with time-dependent dissipation II. Effective dissipation. J. Diff. Equ. **232**, 74–103 (2007). https://doi.org/10.1016/j.jde.2006.06.004
17. Yagdjian, K.: The Cauchy Problem for Hyperbolic Operators: Multiple Characteristics, Micro-local Approach. Mathematical Topics. Akademie, Berlin (1997)

On a Dirichlet Problem for One Improperly Elliptic Equation

Armenak H. Babayan and Seyran H. Abelyan

Abstract The Dirichlet problem for sixth order improperly elliptic equation is considered. The functional class of boundary functions, where this problem is normally solvable is determined. If the roots of the characteristic equation satisfy some conditions, the number of linearly independent solutions of homogeneous problem and the number of linearly independent solvability conditions of in-homogeneous problem are obtained. Solutions of homogeneous problem and solvability conditions of in-homogeneous problem are obtained in explicit form.

Keywords Dirichlet problem · Correct boundary value problem · Defect numbers · Improperly elliptic equation

1 Introduction. Formulation of the Results

Let $D = \{z : |z| < 1\}$ be the unit disk of the complex plane and $\Gamma = \partial D$ its boundary. We consider sixth order differential equation

$$\sum_{k=0}^{6} G_k \frac{\partial^6 U}{\partial x^k \partial y^{6-k}}(x, y) = 0, \ (x, y) \in D. \tag{1}$$

Here G_k are such complex numbers ($G_0 \neq 0$) that the characteristic equation

$$\sum_{k=0}^{6} G_k \lambda^{6-k} = 0 \tag{2}$$

A. H. Babayan (✉) · S. H. Abelyan
National Polytechnic University of Armenia, Teryan 105, Yerevan 0009, Armenia
e-mail: barmenak@gmail.com

S. H. Abelyan
e-mail: seyran.abelyan@gmail.com

A. Karapetyants et al. (eds.), *Modern Methods in Operator Theory and Harmonic Analysis*, Springer Proceedings in Mathematics & Statistics 291,
https://doi.org/10.1007/978-3-030-26748-3_18

has no real roots. In the paper we suppose that all roots have positive imaginary part, that is the Eq. (1) is improperly elliptic. The paper is mainly devoted to the case when i—the imaginary unit, is the only multiple root of the Eq. (2), but other cases are also briefly considered. The solution of the Eq. (1) from the class $C^6(D) \cap C^{(2,\alpha)}(\overline{D})$ on the boundary Γ has to satisfy Dirichlet boundary conditions

$$\left.\frac{\partial^j U}{\partial r^j}\right|_\Gamma = f_j(\theta), \quad e^{i\theta} \in \Gamma, \quad j = 0, 1, 2. \tag{3}$$

Here $\frac{\partial}{\partial r}$ is a differentiation by the radius r (here and then $z = re^{i\theta}$) and $f_j \in C^{(2-j,\alpha)}(\Gamma)$ are the given functions. As usual $C^{(m,\alpha)}(D)$ is a class of the functions, Hölder continuous in the domain D with all derivatives up to the order m ($C^{(0,\alpha)}(D) \equiv C^{(\alpha)}(D)$). The special functional class of the boundary functions will be defined later.

1.1 Historical Remarks

It is well known (see, for example [1]) that the Dirichlet problem for properly elliptic equation (1) (when the numbers of the roots of the Eq. (2) with positive and negative imaginary parts are equal) with real coefficients is Fredholmian and the defect numbers of the problem (1), (3) are equal. Following [2], we denote defect numbers as the number of linearly independent solutions of homogeneous problem (1), (3) (when $f_j \equiv 0$) and the number of linearly independent conditions provided existence of the solution of in-homogeneous problem. For a long time it was considered that the Dirichlet problem for the Eq. (1) with complex coefficients has the same property. But in 1948 A. V. Bicadze showed that it is not the case [3]. He proved that the homogeneous Dirichlet problem for bi-analytic equation in the unit disk has infinite number of linearly independent solutions and infinite number of linearly independent conditions to the boundary function are necessary for the solvability of the corresponding in-homogeneous problem. After that in works of A. V. Bicadze and his disciples (see, for example [4]) it was shown that for improperly elliptic equation all classical boundary value problems (Dirichlet, Neumann, Robin, etc.) are not correct. We want to mention papers [5, 6], where the difference between strongly connected and weakly connected (terminology of A. V. Bicadze) second order systems of partial differential equations for the correct posing boundary conditions was shown. In works [7, 8] using the generalization of analytic functions—notion of J-analytic function, general boundary value problem for arbitrary elliptic system was investigated. The conditions for correctness of the boundary value problem and the index formula for domains with smooth and piece-wise smooth boundary were obtained. From the functional analysis point of view the boundary conditions, provided the correctness of the corresponding boundary value problem, were obtained in [9, 10]. These conditions (Shapiro–Lopatinskii condition [10]) are necessary for the correct posing of the boundary value problem. The question of the efficient determination of the defect numbers is essentially more difficult than the determination of the index of boundary value problem. In the unit disk for properly elliptic equations the formulas for the defect numbers of the Dirichlet problem of the higher order equation of the form (1) were found in the works [11–13]. In the [11] the problem (1), (3) was investigated also in arbitrary multiply connected domain (it was reduced to the second order Fredholm equation). It was supposed in these papers, that the characteristic equation has only simple roots. In [14] this problem was solved in general case, for arbitrary multiplicity of the roots of characteristic equation.

In the case of properly elliptic equation (1) the Dirichlet problem is Fredholmian. In [15] it was shown, that for strongly connected elliptic systems the Dirichlet conditions do not satisfy the Shapiro–Lopatinskii condition. Thus, the Dirichlet problem for improperly elliptic equation is a non correct boundary value problem. Therefore, investigation of such problem is connected with difficulties. For example, in [16] it was shown, that for the solvability of the boundary value problem for polyanalitic equation a continuum number linearly independent conditions is necessary. In [17] (see [18] also) it was shown, that for the correct formulation of the Dirichlet problem for second order improperly elliptic equation in the unit disk it is necessary to modify space of boundary functions. It was proved in this paper, that if the boundary function has analytic continuation inside the disk and homogeneous problem has only zero solution, then the corresponding in-homogeneous problem is uniquely solvable. In the same paper it was proved, that homogeneous Dirichlet problem has only zero solution, or, for some values of the roots of characteristic equation, has countable number of linearly independent solutions. These solutions are determined in explicit form. The Dirichlet problem for the fourth order improperly elliptic equation was investigated in [19–21].

Unique solvability of the Dirichlet problem was considered not only for elliptic equation (1). In the papers [22–24] the conditions of nontrivial solvability of the homogeneous Dirichlet problem for the second order hyperbolic equation in the different domains were found. Non-trivial solvability of the homogeneous Dirichlet problem for even order differential equations (may be non-elliptic) was considered in [13]. In the recent paper [25] the Cauchy problem for hyperbolic system in the bounded domain was considered. In these cases it is important to take the geometry of the domain under consideration. More detailed historical survey of the problem may be found in [26]. In this paper we will consider the Dirichlet problem for the sixth order improperly elliptic equation.

1.2 Formulation of the Results

For the exact formulation of the obtained results, using operators of complex differentiation $\frac{\partial}{\partial \bar{z}} = \frac{1}{2}\left(\frac{\partial}{\partial x} + i\frac{\partial}{\partial y}\right)$ and $\frac{\partial}{\partial z} = \frac{1}{2}\left(\frac{\partial}{\partial x} - i\frac{\partial}{\partial y}\right)$, let's represent the Eq. (1) in the form:

$$\prod_{p=1}^{6}\left(\frac{\partial}{\partial \bar{z}} - \mu_p \frac{\partial}{\partial z}\right) U = 0. \tag{4}$$

Here $\mu_p = \frac{i-\lambda_p}{i+\lambda_p}$ are complex numbers that may coincide. From the conditions to the roots, we have $|\mu_p| < 1$. We will use both the boundary conditions (3) and the following equivalent boundary conditions

$$\left.\frac{\partial^2 U}{\partial \bar{z}^{2-k} \partial z^k}\right|_{\Gamma} = F_k(\theta),\ z = e^{i\theta} \in \Gamma,\ k = 0, 1, 2; \tag{5}$$

$$u(1, 0) = f_0(1),\ u_r(1, 0) = f_1(1),\ u_\theta(1, 0) = f_0'(1).$$

Here $F_k \in C^{(\alpha)}(\Gamma)$ are uniquely determined by the boundary functions f_k:

$$F_0(\theta) = 0.25 e^{2i\theta}(f_2 + 2i f_1' - f_0'' - f_1 - 2i f_0'),\ F_1(\theta) = 0.25(f_2 + f_1 + f_0''),$$

$$F_2(\theta) = 0.25 e^{-2i\theta}(f_2 - 2i f_1' - f_0'' - f_1 + 2i f_0'), \tag{6}$$

Now, let's introduce the functional class, necessary for further considerations:

Definition 1 Let $\rho \in (0, 1)$ be the given number. We denote $A^{(m,\alpha)}(\rho)$ the class of functions analytic in the annulus $\rho < |z| < 1$ and Hölder continuous up to the boundary with mth order derivatives.

Using these notions we formulate the obtained results.

Theorem 1 *We consider the Dirichlet boundary problem (1), (3) in the case, when all roots of the characteristic Eq. (2) are equal to imaginary unit i. In this case the general solution of the homogeneous problem may be represented in the form*

$$U_0(x, y) = (1 - z\bar{z})^3 v_3(x, y), \tag{7}$$

where the function v_3 is an arbitrary three-analytic function (that is $v_{3\bar{z}\bar{z}\bar{z}} = 0$). Inhomogeneous problem has a solution if and only if the following conditions to the boundary functions f_j $(j = 0, 1, 2)$ hold:

$$\int_{-\pi}^{\pi} f_j(\theta) e^{-li\theta} d\theta = 0, \quad l = -6, -7, \ldots, \quad j = 0, 1, 2,$$

$$20 \int_{-\pi}^{\pi} f_0(\theta) e^{5i\theta} d\theta = 4 \int_{-\pi}^{\pi} f_1(\theta) e^{5i\theta} d\theta = \int_{-\pi}^{\pi} f_2(\theta) e^{5i\theta} d\theta,$$

$$5 \int_{-\pi}^{\pi} f_1(\theta) e^{4i\theta} d\theta - 8 \int_{-\pi}^{\pi} f_0(\theta) e^{4i\theta} d\theta = \int_{-\pi}^{\pi} f_2(\theta) e^{4i\theta} d\theta. \tag{8}$$

Theorem 2 *If the five roots of the characteristic Eq. (2) are equal to imaginary unit* i *and* $\lambda_6 \neq i$ *and* $\Im\lambda_6 > 0$*, then the general solution of the homogeneous problem (1), (5) may be represented in the form*

$$U_0(x, y) = (1 - z\bar{z})^3 v_2(x, y), \tag{9}$$

where the function v_2 *is an arbitrary bi-analytic function (that is* $v_{2\bar{z}\bar{z}} = 0$*). In-homogeneous Dirichlet problem (1), (5) has a solution if and only if the boundary functions* F_j $(j = 0, 1, 2)$ *belong to the class* $A^{(2,\alpha)}(|\mu_6|)$ *(*μ_6 *is defined in (4)) and equality:*

$$\mu_6^2 G_2^-(\theta) \equiv \mu_6 G_1^-(\theta) \equiv G_0^-(\theta) \tag{10}$$

holds. Here

$$G_j^-(\theta) = \sum_{k=5}^{\infty} a_{-kj}\zeta^{-k},$$

if

$$F_j(\theta) = \sum_{k=-\infty}^{\infty} a_{kj}\zeta^k, \quad \zeta = e^{i\theta} \in \Gamma.$$

Theorem 3 *We suppose, that four roots of the characteristic Eq. (2) are equal to imaginary unit* i *and* $\lambda_k \neq i$ *and* $\Im\lambda_k > 0\,(k = 5, 6)$*,* $\lambda_5 \neq \lambda_6$*. In this case the general solution of the homogeneous Dirichlet problem (1), (5) may be represented in the form*

$$U_0(x, y) = (1 - z\bar{z})^3 \Phi(z), \tag{11}$$

where the function Φ *is an arbitrary analytic function. Supposing* $|\mu_5| \leq |\mu_6|$*, we get, that in-homogeneous Dirichlet problem (1), (5) has a solution if and only if the boundary functions* F_j $(j = 0, 1, 2)$ *belong to the class* $A^{(2,\alpha)}(|\mu_5|)$ *and equality:*

$$E_0^-(\theta) - (\mu_5 + \mu_6)E_1^-(\theta) \equiv -\mu_5\mu_6 E_2^-(\theta) \tag{12}$$

holds. Here

$$E_j^-(\theta) = \sum_{k=4}^{\infty} a_{-kj}\zeta^{-k},$$

if

$$F_j(\theta) = \sum_{k=-\infty}^{\infty} a_{kj}\zeta^k, \quad \zeta = e^{i\theta} \in \Gamma.$$

Theorem 4 *Let's suppose, that three roots of the characteristic Eq. (2) are equal to imaginary unit* i *and* $\lambda_j \neq i$*,* $\lambda_j \neq \lambda_k$*,* $\Im\lambda_j > 0\,(j = 4, 5, 6,\ j \neq k)$*. If we have*

$|\mu_4| \le |\mu_5| \le |\mu_6|$ and the boundary functions $F_j \in A^{(2,\alpha)}(|\mu_4|)$, then the Dirichlet problem (1), (5) is uniquely solvable.

We investigate also one case, when the imaginary unit is a double root of the characteristic Eq. (2).

Theorem 5 *Let's suppose that two roots of the characteristic Eq. (2) are equal to imaginary unit i and $\lambda_2 = \lambda_3 = \lambda_4 \neq \lambda_1$, $\lambda_k \neq i$, $\Im\lambda_k > 0\,(k = 1, 2)$. If the boundary functions $F_j \in A^{(2,\alpha)}(|\mu_2|)$ and $\sigma = \mu_1\mu_2^{-1}$, then the problem (1), (5) is uniquely solvable if and only if the conditions*

$$S_{l-3}(\sigma) = \sum_{m=0}^{l-3}(m+1)(m+2)\sigma^m \neq 0, \quad l = 4, 5, \ldots \tag{13}$$

hold. If these conditions fail for some l_0 then the homogeneous problem (1), (5) has a non-zero solution, which is a polynomial of order $l_0 + 2$ and the corresponding in-homogeneous problem has a solution if the boundary functions satisfy one linearly independent condition. The conditions (13) may fail for one value of l only, therefore, the defect numbers of the problem (1), (5) may be equal to zero or one.

Remark 1 The formulated theorems show, that the multiplicity of the root $\pm i$ plays an important role in investigation of the problem (1), (5) in the unit disc (may be because of exclusive position of zero point in the unit disc). It will be interesting to investigate the "dynamics"—variation of the defect numbers of the Dirichlet problem (1), (5) when the roots of the characteristic Eq. (2) vary in the neighborhood of the imaginary unit.

2 Proofs of the Formulated Results

In this section we will prove all formulated results. First, we formulate the auxiliary lemmas, what will be necessary for the proof of the main theorems.

2.1 Some Preliminary Results

The main tool we use for the proof of formulated propositions is a representation of the general solution of the Eq. (1). First, we consider the case when Eq. (1) is the polyanalytic equation, that is all roots of the Eq. (2) are equal i. Let's consider an equation

$$\left(\frac{\partial}{\partial \bar{z}}\right)^m u(x, y) = 0, \quad (x, y) \in D, \quad z = x + iy. \tag{14}$$

Lemma 1 *The general solution of the Eq. (14) for $m \geq 2$ may be represented in the form:*

$$u(x, y) = \sum_{k=0}^{m-1}(1 - z\bar{z})^k \Phi_k(z) + \sum_{k=0}^{m-2}(1 - z\bar{z})^k \bar{z} P_{m-2-k}(\bar{z}), \tag{15}$$

where Φ_k are arbitrary functions, analytic in D, P_j are polynomials of order j.

Remark 2 The form of representation (15) (where second term is a polynomial of z and $\bar{z}$) for $m = 2$ was introduced in [4] and for $m > 2$ was successfully used by A. V. Bitsadze and N. E. Tovmasyan [6], so the slight modification of the second summand in the representation (15) is new.

Proof The proof will be done by induction. For $m = 2$ the formula (15) is true (A. V. Bitsadze representation). We suppose, that the formula (15) holds for m and consider the case $m + 1$. In this case the Eq. (14) may be written in the form:

$$\left(\frac{\partial}{\partial \bar{z}}\right)\left(\frac{\partial^m u}{\partial \bar{z}^m}\right) = 0.$$

Therefore, the equation is equivalent to following:

$$\frac{\partial^m u}{\partial \bar{z}^m} = \Psi(z), \tag{16}$$

where Ψ is an analytic function. The solution of the Eq. (16) may be written in the form $u = u_0 + u_*$, where u_0 is a general solution of the homogeneous equation and u_* is a special solution of the in-homogeneous Eq. (16). By the induction supposition we have:

$$u_0(x, y) = \sum_{k=0}^{m-1}(1 - z\bar{z})^k \Phi_k(z) + \sum_{k=0}^{m-2}(1 - z\bar{z})^k \bar{z} P_{m-2-k}(\bar{z}). \tag{17}$$

The special solution of the inhomogeneous problem we will seek in the form:

$$u_*(x, y) = (1 - z\bar{z})^m \Phi_m(z) + \sum_{k=0}^{m-1}(1 - z\bar{z})^k a_{m-1-k} \bar{z}^{m-k}. \tag{18}$$

Here a_{m-1-k} are some complex constants. Substituting this solution to the Eq. (16), we get

$$\Psi(z) = (-1)^m m! z^m \Phi_m(z) + \sum_{k=0}^{m-1} m!(-1)^k z^k a_{m-1-k}. \tag{19}$$

From this equality we uniquely determine the function Φ_m and the constants a_{m-1-k} by the function Ψ. Finally, adding the functions (17) and (18) we get the sought representation for $m+1$. Lemma is proved. □

If the roots of Eq. (2) are not equal to $\pm i$, then we will use the representation, obtained in [14]:

Lemma 2 *The general solution of the equation*

$$\prod_{k=1}^{M}\left(\frac{\partial}{\partial \bar{z}} - \mu_k \frac{\partial}{\partial z}\right)^{m_k} \prod_{p=1}^{P}\left(\frac{\partial}{\partial z} - \nu_p \frac{\partial}{\partial \bar{z}}\right)^{s_p} u(x, y) = 0, \tag{20}$$

where $0 < |\mu_k| < 1$, $0 < |\nu_p| < 1$, $\mu_p \neq \mu_l$, $\nu_p \neq \nu_l$ $(p \neq l)$ *may be represented in the form:*

$$u(x, y) = \sum_{k=1}^{M}\sum_{j=0}^{m_k-1}\left(\frac{\partial}{\partial \theta}\right)^{j} \Phi_{kj}(z + \mu_k \bar{z}) + \sum_{p=1}^{P}\sum_{j=0}^{s_p-1}\left(\frac{\partial}{\partial \theta}\right)^{j} \Psi_{pj}(\bar{z} + \nu_p z), \tag{21}$$

where Φ_{kj} *and* Ψ_{pj} *are arbitrary functions, analytic in* $D(\mu_k) = \{z + \mu_k \bar{z} : z \in D\}$ *and* $D_1(\nu_p) = \{\bar{z} + \nu_p z : z \in D\}$ *correspondingly,* $\frac{\partial}{\partial \theta}$ *is a differentiation by the argument of complex number* ($z = re^{i\theta}$).

If the characteristic equation has multiple roots $\pm i$ and the roots, which differ from $\pm i$, then we must get the sum of the representations (15) and (21).

We must use also the representations of analytic in $D(\mu)$ and $D_1(\nu)$ functions $\Phi(z + \mu\bar{z})$ and $\Psi(\bar{z} + \nu z)$ in a neighborhood of the boundary by analytic functions in D. It was proved in [17], that if μ and ν are complex numbers, such that $0 < |\mu| < 1$ and $0 < |\nu| < 1$, then, for $|z| = 1$, we have

$$\Phi(z + \mu\bar{z}) = \omega(z) + \omega(\mu\bar{z}), \quad \Psi(\bar{z} + \nu z) = \chi(\bar{z}) + \chi(\nu z), \tag{22}$$

where ω and χ are analytic functions in the unit disc. Given the functions ω and χ, we can recover Φ and Ψ by the formulas:

$$\Phi(z + \mu\bar{z}) = \omega\left(\frac{z + \mu\bar{z} + \sqrt{(z + \mu\bar{z})^2 - 4\mu}}{2}\right) + \omega\left(\frac{z + \mu\bar{z} - \sqrt{(z + \mu\bar{z})^2 - 4\mu}}{2}\right),$$

$$\Psi(\bar{z} + \nu z) = \chi\left(\frac{\bar{z} + \nu z + \sqrt{(\bar{z} + \nu z)^2 - 4\nu}}{2}\right) + \chi\left(\frac{\bar{z} + \nu z - \sqrt{(\bar{z} + \nu z)^2 - 4\nu}}{2}\right),$$

where $|z| < 1$. Here we take the branch of $\sqrt{\zeta^2 - 4\mu}$ ($\sqrt{\zeta^2 - 4\nu}$), which is continuous outside the segment $[-2\sqrt{\mu}, 2\sqrt{\mu}]$ ($[-2\sqrt{\nu}, 2\sqrt{\nu}]$) and satisfies the condition $\zeta^{-1}\sqrt{\zeta^2 - 4\mu} \longrightarrow 1$ ($\zeta^{-1}\sqrt{\zeta^2 - 4\nu} \longrightarrow 1$), when $\zeta \longrightarrow \infty$.

We will use also the representation of the polynomials in the unit disc from [27].

Lemma 3 ([27], Theorem 5.1) *Let Q_n be a polynomial, which on the $\Gamma = \{\zeta : |\zeta| = 1\}$ satisfy the condition $Q_n\big|_\Gamma = g(\zeta)$. Then this polynomial may be represented in the form:*

$$Q_n(z, \bar{z}) = (1 - z\bar{z})Q_{n-2}(z, \bar{z}) + H(z, \bar{z}), \quad |z| < 1, \tag{23}$$

where Q_{n-2} is a polynomial of order $n - 2$ and H is a harmonic polynomial, what satisfy the condition $H\big|_\Gamma = g(\zeta)$.

2.2 The Case, When the Multiplicity of the i Is Grater Than Two

Proof (*Proof of the Theorem* 1) Let's suppose that all roots of the Eq. (2) are equal to i, that is the Eq. (1) is reduced to the form

$$\left(\frac{\partial}{\partial \bar{z}}\right)^6 U = 0.$$

Using Lemma 1, we can represent the solution of this equation in the form:

$$U(x, y) = \sum_{k=0}^{5}(1 - z\bar{z})^k \Phi_k(z) + \sum_{k=0}^{4}(1 - z\bar{z})^k \bar{z} P_{4-k}(\bar{z}), \tag{24}$$

Substituting this function in the boundary conditions (3), we get:

$$\Phi_0(e^{i\theta}) + e^{-i\theta} P_4(e^{-i\theta}) = f_0(\theta),$$

$$\zeta\Phi_0'(\zeta) + \bar{\zeta}(\bar{\zeta}P_4(\bar{\zeta}))' - 2\Phi_1(\zeta) - 2\bar{\zeta}P_3(\bar{\zeta}) = f_1(\theta), \ \ \zeta = e^{i\theta}, \tag{25}$$

$$\zeta^2\Phi_0''(\zeta) + \bar{\zeta}^2(\bar{\zeta}P_4(\bar{\zeta}))'' - 2\Phi_1(\zeta) - 2\bar{\zeta}P_3(\bar{\zeta}) - 2\zeta\Phi_1'(\zeta) - 2\bar{\zeta}(\bar{\zeta}P_3(\bar{\zeta}))' +$$

$$+ 8\Phi_2(\zeta) + 8\bar{\zeta}P_2(\bar{\zeta}) = f_2(\theta), \quad \zeta = e^{i\theta}.$$

First, we consider homogeneous problem (1), (3). If in (25) we have $f_j \equiv 0$, then from the first equation of (25) it follows $\Phi_0 \equiv P_4 \equiv 0$. Then, using these equalities from the second equation (25) we get $\Phi_1 \equiv P_3 \equiv 0$ and, after that $\Phi_2 \equiv P_2 \equiv 0$. Thus, the general solution of the homogeneous problem (1), (3) is represented in the form:

$$U(x, y) = \sum_{k=3}^{5}(1 - z\bar{z})^k \Phi_k(z) + \sum_{k=3}^{4}(1 - z\bar{z})^k \bar{z} P_{4-k}(\bar{z}). \tag{26}$$

Here Φ_k are arbitrary analytic functions and P_0, P_1 are arbitrary polynomials of order 0 and 1 correspondingly. Hence the general solution is a function (7) and the first part of the Theorem 1 is proved.

For the in-homogeneous problem (1), (3), using Eqs. (25), and taking into account, that Φ_j are analytic functions, we get that for the solvability of the problem it is necessary, that the functions $e^{5i\theta} f_j(\theta)$ will be analytic in D. It means, that if we represent the boundary functions by Fourier series:

$$f_j(\theta) = \sum_{k=-\infty}^{\infty} f_{kj} e^{ki\theta}, \quad j = 0, 1, 2 \tag{27}$$

then for the solvability of the in-homogeneous problem (1), (3) it is necessary $f_{kj} = 0$, for $k \leq -6$ and $j = 0, 1, 2$. Further, if we compare the Fourier coefficients of the right and left parts of the Eqs. (25) we will get the following conditions, necessary and sufficient for the solvability of the in-homogeneous problem (1), (3):

$$20 f_{-50} = 4 f_{-51} = f_{-52}, \;\; 5 f_{-41} - 8 f_{-40} = f_{-42}, \;\; f_{kj} = 0, \;\; k \leq -6. \tag{28}$$

Theorem 1 is proved.

Now, we suppose, that one of the roots of the characteristic equation differs from i, that is the Eq. (1) is reduced to the form:

$$\left(\frac{\partial}{\partial \bar{z}}\right)^5 \left(\frac{\partial}{\partial \bar{z}} - \mu_6 \frac{\partial}{\partial z}\right) U(x, y) = 0. \tag{29}$$

Here $0 < |\mu_6| < 1$.

Proof (*Proof of the Theorem* 2) The general solution of the Eq. (29) has the following form (see Lemmas 1 and 2):

$$U(x, y) = \sum_{k=0}^{4} (1 - z\bar{z})^k \Phi_k(z) + \sum_{k=0}^{3} (1 - z\bar{z})^k \bar{z} P_{3-k}(\bar{z}) + \Psi(z + \mu_6 \bar{z}),$$

where Φ_k are analytic in D functions and Ψ is analytic in $D(\mu_6) = \{z + \mu_6 \bar{z} : z \in D\}$. We will use the boundary conditions (5), therefore, it is more convenient to represent this function in the form:

$$U(x, y) = U_0(x, y) + \sum_{k=0}^{2} (1 - z\bar{z})^k \Phi_k(z) + \sum_{k=0}^{2} z^k \bar{z}^{k+1} Q_{3-k}(\bar{z}) + \Psi(z + \mu_6 \bar{z}), \tag{30}$$

where Φ_k are analytic functions and Q_j are polynomials of order j. The function U_0 is represented in the form

$$U_0(x, y) = (1 - z\bar{z})^3 (\Phi_3(z) + (1 - z\bar{z})\Phi_4(z) + C\bar{z}) \equiv (1 - z\bar{z})^3 v_2(x, y), \tag{31}$$

where v_2 is arbitrary bi-analytic function. Let's substitute the function (30) in the boundary conditions (5). We get:

$$2\zeta^2\Phi_2(\zeta)+\sum_{k=0}^{2}\zeta^k(\bar{\zeta}^{k+1}Q_{3-k}(\bar{\zeta}))''+\mu_6^2\Psi''(\zeta+\mu_6\bar{\zeta})=F_0(\theta), \tag{32}$$

$$-(\zeta\Phi_1(\zeta))'+2\Phi_2(\zeta)+\sum_{k=1}^{2}k\zeta^{k-1}(\bar{\zeta}^{k+1}Q_{3-k}(\bar{\zeta}))'+\mu_6\Psi''(\zeta+\mu_6\bar{\zeta})=F_1(\theta), \tag{33}$$

$$\Phi_0''(\zeta)-\bar{\zeta}\Phi_1'(\zeta)+2\bar{\zeta}^2\Phi_2(\zeta)+2\bar{\zeta}^3Q_1(\bar{\zeta})+\Psi''(\zeta+\mu_6\bar{\zeta})=F_2(\theta). \tag{34}$$

Here $\zeta=e^{i\theta}\in\Gamma$. Now, we use the representation (22) for the function Ψ'':

$$\Psi''(\zeta+\mu_6\bar{\zeta})=\psi(\zeta)+\psi(\mu_6\bar{\zeta}),\quad \zeta\in\Gamma, \tag{35}$$

where ψ is analytic in D, and the Fourier series for the functions F_j:

$$F_j(\theta)=0.5a_{0j}+\sum_{k=1}^{\infty}a_{kj}\zeta^k+0.5a_{0j}+\sum_{k=1}^{\infty}a_{-kj}\zeta^{-k}\equiv F_j^+(\theta)+F_j^-(\theta). \tag{36}$$

Substituting these expansions in the boundary Eqs. (32)–(34) and equating functions which are analytic outside $\bar{D}$, we get:

$$(\bar{\zeta}Q_3(\bar{\zeta}))''+\zeta(\bar{\zeta}^2(Q_2(\bar{\zeta})-Q_2(0)))''+\mu_6^2\psi(\mu_6\bar{\zeta})=F_0^-(\theta),$$
$$(\bar{\zeta}^2Q_2(\bar{\zeta}))'+2\zeta(\bar{\zeta}^3Q_1(\bar{\zeta}))'+\mu_6\psi(\mu_6\bar{\zeta})=F_1^-(\theta),$$
$$2\bar{\zeta}^3Q_1(\bar{\zeta})-\Phi_1'(0)\bar{\zeta}+2\bar{\zeta}^2(\Phi_2(0)+\Phi_2'(0)\zeta)+\psi(\mu_6\bar{\zeta})=F_2^-(\theta). \tag{37}$$

First, we consider homogeneous problem (1), (3). In this case, solving the system (37) we see, that the function ψ is a polynomial of order at most two and $Q_1\equiv 0$. After that, solving the system (32)–(34) we get, that the function $U-U_0$ is a polynomial of order at most five. Using homogeneous boundary conditions (3) and Lemma 3, we have that this polynomial is divisible by $(1-z\bar{z})^3$, that is polynomial of order not less than six. Therefore, $U-U_0\equiv 0$, and the general solution of the homogeneous problem is the function (31). First part of the Theorem 2 is proved.

Let's consider in-homogeneous problem (1), (3). We denote

$$G_j^-(\zeta)=\sum_{k=5}^{\infty}a_{-kj}\zeta^{-k}\equiv F_j^-(\theta)-0.5a_{0j}-\sum_{k=1}^{4}a_{-kj}\zeta^{-k}.$$

Then, from the system (37) we have

$$\mu_6^2 G_2^-(\theta) \equiv \mu_6 G_1^-(\theta) \equiv G_0^-(\theta).$$

These functions are equal to $\mu_6^2\psi(\mu_6\bar{\zeta})$, therefore, the boundary functions F_j must be analytic in the annulus $\{|\mu_6| < |z| < 1\}$. Theorem 2 is proved.

Proof (Proof of the Theorem 3) The proof of this theorem is similar to the proof of the Theorem 2, so we will provide only a sketch of the proof, omitting some details. The general solution of the equation is represented in the form:

$$U(x, y) = U_0(x, y) + \sum_{k=0}^{2}(1 - z\bar{z})^k \Phi_k(z)+$$

$$+\sum_{k=0}^{2} z^k \bar{z}^{k+1} Q_{2-k}(\bar{z}) + \Psi_5(z + \mu_5\bar{z}) + \Psi_6(z + \mu_6\bar{z}), \tag{38}$$

where Ψ_k are analytic functions and Q_j are polynomials of order j and the function U_0 is a function (11). Substituting this function in the boundary condition (5) we get:

$$2\zeta^2\Phi_2(\zeta) + \sum_{k=0}^{2} \zeta^k(\bar{\zeta}^{k+1} Q_{2-k}(\bar{\zeta}))'' + \sum_{k=5}^{6} \mu_k^2 \Psi_k''(\zeta + \mu_k\bar{\zeta}) = F_0(\theta), \tag{39}$$

$$2\Phi_2(\zeta) - (\zeta\Phi_1(\zeta))' + \sum_{k=1}^{2} k\zeta^{k-1}(\bar{\zeta}^{k+1} Q_{2-k}(\bar{\zeta}))' + \sum_{k=5}^{6} \mu_k \Psi_k''(\zeta + \mu_k\bar{\zeta}) = F_1(\theta), \tag{40}$$

$$\Phi_0''(\zeta) - \bar{\zeta}\Phi_1'(\zeta) + 2\bar{\zeta}^2\Phi_2(\zeta) + 2\bar{\zeta}^3 Q_0(\bar{\zeta}) + \sum_{k=5}^{6} \Psi_k''(\zeta + \mu_k\bar{\zeta}) = F_2(\theta). \tag{41}$$

Here $\zeta = e^{i\theta} \in \Gamma$. We use the representation (22) for the function Ψ_k'':

$$\Psi_k''(\zeta + \mu_k\bar{\zeta}) = \psi_k(\zeta) + \psi_k(\mu_k\bar{\zeta}), \quad \zeta \in \Gamma, \ \ k = 5, 6, \tag{42}$$

where ψ_k are analytic in D, and the Fourier expansion (36) for the functions F_j. Substituting these expansions in the boundary Eqs. (39)–(41) and equating functions which are analytic outside $\bar{D}$, we get:

$$(\bar{\zeta} Q_2(\bar{\zeta}))'' + \sum_{k=5}^{6} \mu_k^2 \psi_k(\mu_k\bar{\zeta}) = F_0^-(\theta),$$

$$(\bar{\zeta}^2 Q_1(\bar{\zeta}))' + 2\zeta(\bar{\zeta}^3 Q_0(\bar{\zeta}))' + \sum_{k=5}^{6} \mu_k \psi_k(\mu_k \bar{\zeta}) = F_1^-(\theta),$$

$$2\bar{\zeta}^3 Q_0(\bar{\zeta}) - \Phi_1'(0)\bar{\zeta} + 2\bar{\zeta}^2(\Phi_2(0) + \Phi_2'(0)\zeta) + \sum_{k=5}^{6} \psi_k(\mu_k \bar{\zeta}) = F_2^-(\theta). \quad (43)$$

From these equations as in the previous theorem, we derive, that the function U_0 is a general solution of the homogeneous problem (1), (3). Let's consider inhomogeneous problem. Using notations of the theorem we represent the boundary functions in the form

$$F_j(\theta) = E_j^-(\theta) + \sum_{k=-3}^{\infty} a_{kj}\zeta^k.$$

Then, equating the coefficients for the powers of ζ less than -3, we get:

$$\sum_{k=5}^{6} \mu_k^{2-j} \widehat{\psi_k}(\mu_k \bar{\zeta}) = E_j^-(\theta), \quad j = 0, 1, 2.$$

Here $\widehat{\psi_k}(z) = \psi_k(z) - \psi_k(0) - \psi_k'(0)z - \frac{1}{2}\psi_k''(0)z^2 - \frac{1}{6}\psi_k^{(3)}(0)z^3$. This system shows, that the functions E_j, and therefore F_j, belong to the class $A^{(2,\alpha)}(|\mu_5|)$ and the equality (12) holds. Theorem is proved.

The proof of the Theorem 4 is similar to the previous proofs, so will be omitted.

2.3 The Case, When the Multiplicity of the i Is Equal Two

In this subsection we consider the case, when multiplicity of i—the root of characteristic Eq. (2), is equal to two.

Proof (Proof of the Theorem 5) We suppose, that the roots of characteristic Eq. (2) satisfy the condition:

$$\lambda_1 \neq \lambda_k, \ k \neq 1, \ \lambda_2 = \lambda_3 = \lambda_4 \neq i, \ \lambda_5 = \lambda_6 = i, \ \Im\lambda_j > 0.$$

In this case the Eq. (4) is reduced to the form:

$$\left(\frac{\partial}{\partial \bar{z}}\right)^2 \left(\frac{\partial}{\partial \bar{z}} - \mu_1 \frac{\partial}{\partial z}\right) \left(\frac{\partial}{\partial \bar{z}} - \mu_2 \frac{\partial}{\partial z}\right)^3 U = 0. \quad (44)$$

The general solution of this equation by the Lemmas 1 and 2 has following form:

$$U(x, y) = \Phi_0(z) + (1 - z\bar{z})\Phi_1(z) + C\bar{z} +$$

$$+\,\Psi_0(z+\mu_1\bar{z})+\sum_{k=1}^{3}\left(\frac{\partial}{\partial\theta}\right)^{k-1}\Psi_k(z+\mu_2\bar{z}), \tag{45}$$

where Φ_k are analytic in D functions, Ψ_0 and Ψ_k $(k=1,2,3)$ are analytic functions in the domains $D(\mu_j)=\{z+\mu_j\bar{z}:z\in D\}$ for $j=1,2$ correspondingly. Using operator identity $\frac{\partial^{n+m}}{\partial\bar{z}^n\partial z^m}\frac{\partial^k}{\partial\theta^k}=\left(\frac{\partial}{\partial\theta}+i(m-n)I\right)^k\frac{\partial^{n+m}}{\partial\bar{z}^n\partial z^m}$ [14], after substituting the solution (45) to the boundary conditions (5) we get the following equations:

$$\mu_1^2\Psi_0''(\zeta+\mu_1\bar{\zeta})+\sum_{k=1}^{3}\mu_2^2\left(\frac{\partial}{\partial\theta}-2iI\right)^{k-1}\Psi_k''(\zeta+\mu_2\bar{\zeta})=F_0(\theta),\quad \zeta=e^{i\theta},$$

$$-\,(\zeta\Phi_1(\zeta))'+\mu_1\Psi_0''(\zeta+\mu_1\bar{\zeta})+\sum_{k=1}^{3}\mu_2\left(\frac{\partial}{\partial\theta}\right)^{k-1}\Psi_k''(\zeta+\mu_2\bar{\zeta})=F_1(\theta), \tag{46}$$

$$\Phi_0''(\zeta)-\bar{\zeta}\Phi_1'(\zeta)+\Psi_0''(\zeta+\mu_1\bar{\zeta})+\sum_{k=1}^{3}\left(\frac{\partial}{\partial\theta}+2iI\right)^{k-1}\Psi_k''(\zeta+\mu_2\bar{\zeta})=F_2(\theta),$$

Applying formulas (22) to the functions Ψ_k'', we get:

$$\Psi_0''(\zeta+\mu_1\bar{\zeta})=\psi_0(\zeta)+\psi_0(\mu_1\bar{\zeta})=\sum_{l=0}^{\infty}A_{0l}\zeta^l+\sum_{l=0}^{\infty}A_{0l}\mu_1^l\bar{\zeta}^l,\quad \zeta=e^{i\theta},$$

$$\Psi_k''(\zeta+\mu_2\bar{\zeta})=\psi_k(\zeta)+\psi_k(\mu_2\bar{\zeta})=\sum_{l=0}^{\infty}A_{kl}\zeta^l+\sum_{l=0}^{\infty}A_{kl}\mu_2^l\bar{\zeta}^l,\quad k=1,2,3. \tag{47}$$

Here we expand the functions ψ_k in Taylor series, which converges in D. These expansions hold on Γ, hence, we can differentiate them by θ. Let's substitute these expansions, and representation of the functions F_j in a Fourier series (36) in the first Eq. (46).

$$\mu_1^2\sum_{l=0}^{\infty}\left(A_{0l}\zeta^l+A_{0l}\mu_1^{l+2}\zeta^{-l}\right)+\sum_{l=0}^{\infty}\sum_{k=1}^{3}\left(A_{kl}(il-2i)^{k-1}\mu_2^2\zeta^l+\right.$$

$$\left.+A_{kl}(-il-2i)^{k-1}\mu_2^{l+2}\zeta^{-l}\right)=\sum_{l=-\infty}^{\infty}a_{l0}\zeta^l. \tag{48}$$

Equating parts of the second and third Eqs. (46) analytic in the complementary of $\bar{D}$, we get:

$$\sum_{l=0}^{\infty}\left(A_{0l}\mu_1^{l+1}+\sum_{k=1}^{3}A_{kl}(-il)^{k-1}\mu_2^{l+1}\right)\zeta^{-l}=\sum_{l=0}^{\infty}a_{-l1}\zeta^{-l},$$

$$-\Phi_1'(0)\zeta^{-1}+\sum_{l=0}^{\infty}\left(A_{0l}\mu_1^{l}+\sum_{k=1}^{3}A_{kl}(-il+2i)^{k-1}\mu_2^{l}\right)\zeta^{-l}=\sum_{l=0}^{\infty}a_{-l2}\zeta^{-l}. \quad (49)$$

Fourier series in Eqs. (48), (49) are equal for arbitrary $\zeta = e^{i\theta}$, therefore, coefficients for the same powers ζ^l are equal. For $l \geq 2$, we have system of four equations for determination four unknowns A_{kl}:

$$\mu_1^{l+2}A_{0l}+\sum_{k=1}^{3}A_{kl}(-il-2i)^{k-1}\mu_2^{l+2}=a_{-l0},\quad \mu_1^{l+1}A_{0l}+\sum_{k=1}^{3}A_{kl}(-il)^{k-1}\mu_2^{l+1}=a_{-l1},$$

$$\mu_1^{l}A_{0l}+\sum_{k=1}^{3}A_{kl}(2i-il)^{k-1}\mu_2^{l}=a_{-l2},\quad \mu_1^{2}A_{0l}+\sum_{k=1}^{3}A_{kl}(il-2i)^{k-1}\mu_2^{2}=a_{l0}. \quad (50)$$

If $l = 0, 1$ we get the similar rectangular systems, which is always solvable. We don't write these systems, because the coefficients A_{kl} for $l \leq 2$ do not affect the defect numbers of the problem (1), (3).

Let's consider the determinant of the main matrix of the system (50)

$$\Delta_l = \det\begin{pmatrix} \mu_1^{l+2} & \mu_2^{l+2} & -i(l+2)\mu_2^{l+2} & -(l+2)^2\mu_2^{l+2} \\ \mu_1^{l+1} & \mu_2^{l+1} & -il\mu_2^{l+1} & -l^2\mu_2^{l+1} \\ \mu_1^{l} & \mu_2^{l} & -i(l-2)\mu_2^{l} & -(l-2)^2\mu_2^{l} \\ \mu_1^{2} & \mu_2^{2} & i(l-2)\mu_2^{2} & -(l-2)^2\mu_2^{2} \end{pmatrix}.$$

Taking out of each row corresponding powers of μ_2, such that in the second column remain only units, out of third column $-i$ and of fourth column -1, we get, denoting $\sigma = \mu_1\mu_2^{-1}$:

$$\Delta_l = \mu_2^{3l+5}i\begin{vmatrix} \sigma^{l+2} & 1 & l+2 & (l+2)^2 \\ \sigma^{l+1} & 1 & l & l^2 \\ \sigma^{l} & 1 & l-2 & (l-2)^2 \\ \sigma^{2} & 1 & -(l-2) & (l-2)^2 \end{vmatrix} \equiv \mu_2^{3l+5}i\Omega_l. \quad (51)$$

Calculating the determinant Ω_l, we get:

$$\Omega_l = 8\sigma^2(\sigma-1)^3 S_{l-3}. \quad (52)$$

This equality shows that the determinant Δ_l for $l \geq 4$ is non-zero if and only if the conditions (13) hold.

First, let's suppose that conditions (13) hold. Let's consider homogeneous problem (1), (5). The solution of this problem is constructed using the coefficients A_{kl} for $l \geq 2$ which are determined from the homogeneous systems (50). Therefore, we get, that $A_{kl} = 0$ for $l > 3$. The coefficients A_{k3} are also equal to zero, because Δ_3 is a generalized Vandermonde determinant with different terms, and therefore, not equal to zero. Thus, the non-zero solution of the homogeneous problem (1), (5) may be polynomial of order at most four. But an arbitrary non-zero polynomial, which satisfies the homogeneous conditions (3) on Γ is divisible by $(1 - z\bar{z})^3$ (Lemma 3), which means that the homogeneous problem (1), (5) has only trivial solution.

Now, we return to in-homogeneous problem (1), (5). We consider the in-homogeneous system (50). First, taking into account, that $\Delta_l \neq 0$ for $l \geq 3$, we see that in this case coefficients A_{kl} are determined uniquely. In the system (50) for $l = 2$ left parts of the third and fourth equations are the same and the right parts are also the same, because relations (6) imply equality $a_{20} = a_{-22}$, hence the coefficients A_{k2} may also be found (not uniquely). Taking into account the equality $a_{10} = a_{-11}$ one may show, that coefficients A_{k1} and A_{k0} may be determined also not uniquely. After determination of the functions Ψ_k we will find the functions Φ_k from the Eqs. (46). Summing up, we can say that the in-homogeneous problem (1), (5) has a solution.

We must show, that this solution belongs to the class $C^{(2,\alpha)}(D \cup \Gamma)$ if the boundary functions are from the class $A^{(2,\alpha)}(|\mu_2|)$. For this goal, we solve the system (50). First, we suppose, that $|\sigma| < 1$. Then, denoting

$$T = 0.25\left(a_{-l0} - 2a_{-l1}\mu_2 + a_{-l2}\mu_2^2\right)\mu_2^{-l-2},$$

we get following asymptotic estimations for the solution:

$$A_{0l} \sim -\frac{l^2 T}{2\sigma^2}, \quad A_{1l} \sim -\frac{l^2 T}{8}, \quad A_{2l} \sim -i\frac{lT}{4}, \quad A_{3l} \sim -\frac{T}{8}, \tag{53}$$

Now, we use the estimations, which show a connection between rate of growth of the coefficients Fourier of the function and rate of smoothness of corresponding series (see [28], p. 210). If the boundary functions F_j belong to the class $A^{(2,\alpha)}(|\mu_2|)$, then we have

$$a_{-lm} \sim l^{-2}\gamma_l \mu_2^{l+2}, \quad m = 0, 1, 2$$

where γ_l are coefficients Fourier of Hölder continuous function. These relations and (53) imply , that all components of the solution are from the class $C^{(2,\alpha)}(\overline{D})$. Thus, the in-homogeneous problem (1), (5) has a solution in the class $C^{(2,\alpha)}(\overline{D})$ and this solution is unique. Thus, for $|\sigma| < 1$ the Theorem 5 is proved. In the case $|\sigma| \geq 1$ we get the same result, simply the estimations (53) must be replaced by

$$A_{0l} \sim \frac{T_1}{\sigma^{l-4}} + \frac{E_1}{\sigma^{l-4}(l-1)}, \quad A_{1l} \sim -\frac{(\sigma-1)^2 E_1 (l-1)}{8},$$

$$A_{2l} \sim \frac{a_{-l2}\mu_2^{-l} - T_1 - a_{l0}\mu_2^{-2}}{2l}, \quad A_{3l} \sim \frac{(\sigma-1)^2 E_1}{8(l-1)},$$

where

$$T_1 = \frac{8\left(a_{-l0} - 2a_{-l1}\mu_2 + a_{-l2}\mu_2^2\right)}{(\sigma-1)^2\mu_2^{l+2}}, \quad E_1 = \frac{2(a_{-l2}\mu_2 - a_{-l1})\mu_2^{-l-1} - 2(\sigma-1)T_1}{(\sigma-1)^2}.$$

Now, we suppose that conditions (13) fail. Let's suppose, that for some $p \geq 4$ $S_{p-3} = 0$ and, therefore, $\Delta_p = 0$. Then, the homogeneous system (50) for $l = p$ has one linearly independent solution A_{kp}, because the rank of the main matrix of this system is equal to three. Getting another coefficients A_{kl} for $l \neq p$ and the functions Φ_m equal to zero, and substituting these coefficients in formulas (45), we determine the linearly independent solution of the homogeneous problem (1), (5), which is a polynomial of order $p+2$. For example, if $S_1(\sigma) = 0$, or $1 + 3\sigma = 0$, then the function $(1 - z\bar{z})^3$ is a non-trivial solution of the homogeneous problem (1), (5). For in-homogeneous problem let's consider the in-homogeneous system (50). If the condition (13) fail for some l_0, then the rank of the main matrix of the corresponding system is equal three, so for the solvability of this system it is necessary that some linear combination of the right parts of the equations must be equal to zero, that is for the solvability of this system one linearly independent condition for the boundary functions F_k is necessary. If the boundary functions belong to the class $A^{(2,\alpha)}(|\mu_2|)$ then this condition is also sufficient.

For the completeness of the proof of Theorem 5, it remains to prove, that for the fixed σ the condition (13) may fail for one value of l only. For this goal let's consider the polynomial S_k from (13). By the Eneström–Kakeya theorem (see [29]) the roots of the polynomial $Q_n(z) = \sum_{j=0}^{n} q_j z^j$ with positive coefficients q_j belong to the annulus

$$\min_j \left(\frac{q_j}{q_{j+1}}\right) \leq |z| \leq \max_j \left(\frac{q_j}{q_{j+1}}\right). \tag{54}$$

If we apply this theorem to the polynomial (13), we find that the roots of polynomial S_{l-3} belong to the annulus.

$$\frac{1}{3} \leq |z| \leq \frac{l-3}{l-1}. \tag{55}$$

Let's suppose, that for some δ we have $S_{l-3}(\delta) = S_{l+m-3}(\delta) = 0$. Then the number δ is a root of polynomial S_{l-3} and the polynomial

$$V(z) \equiv S_{l+m-3}(z) - S_{l-3}(z) = z^{l-2}\sum_{j=0}^{m-1}(j+l-1)(j+l)z^j \equiv z^{l-2}R(z).$$

As $\delta \neq 0$, then δ is a root of the polynomial R. From other side, using the inequality (54), we get that the roots of the polynomial R are in the domain:

$$\frac{l-1}{l} \leq |z| \leq \frac{l+m-3}{l+m-2}. \tag{56}$$

The domains (55) and (56) do not intersect, hence, the polynomials S_{l-3} and R and, therefore, S_{l-3} and S_{l+m-3} can't have the same roots. This contradiction proves the proposition. Theorem 5 is proved.

Remark 3 We must mention, that from (55) we have, that for $|\sigma| < \frac{1}{3}$ and for $|\sigma| > 1$ the condition (13) holds, that is for $3|\mu_1| < |\mu_2|$ or $|\mu_1| > |\mu_2|$ the problem (1), (3) is uniquely solvable.

References

1. Lions, J.-L., Magenes, E.: Problémes Aux Limites Non Homogénes at Applications, vol. 1. Dunod, Paris (1968)
2. Gohberg, I., Krein, M.: Fundamental aspects of defect numbers, root numbers and indexes of linear operators. Uspekhi Math. Nauk. [Russian Math. Surveys] **12**(2)(74), 43–118 (1957)
3. Bicadze, A.V.: On a Uniqueness of the solution of the Dirichlet problem for partial differential equations of elliptic type. Uspekhi Math. Nauk. [Russian Math. Surveys] **3**(6) 211–212 (1948)
4. Bicadze, A.V.: Boundary Value Problems for the Second Order Differential Equations. Nauka, Moscow (1966). (Russian)
5. Tovmasyan, N.E.: The general boundary value problem for second order elliptic systems with constants cefficients. Differencialnye Uravneniya **2**(1) 3–23 (1966). (Russian)
6. Tovmasyan, N.E.: The general boundary value problem for second order elliptic systems with constants cefficients II. Differencialnye Uravneniya **2**(2), 163–171 (1966). (Russian)
7. Soldatov, A.P.: A function theory method in boundary value problems in the plane. I. The smooth case. Math. of the USSR, Izvestiya **39**(2), 1033–1061 (1992)
8. Soldatov, A.P.: A function theory method in elliptic problems in the plane. II. The piece-wise smooth case. Russ. Acad. Sci. Izv. Math. **40**(3), 529–563 (1993)
9. Volpert, A.I.: On an index and normal solvability of the boundary value problems for elliptic systems of differential equations on the plane. Trudy Moskovskogo Matematicheskogo Obschestva [Proceedings of Moscow Mathematical Society] **10**, 41–87 (1961). (Russian)
10. Lopatinskii, Ya.B.: On a method of reduction of the boundary value problem for the elliptic type system of the partial differential equations to the regular equations. Ukr. Math. J. **5**, 123–151 (1953). (Russian)
11. Tovmasyan, N.E.: Dirichlet problem for properly elliptic equation in the multiply connected domains. Izvestiya NAN Armenii, Mathematika (J. Contemp. Math. Anal.) **3**(6), 5–40 (2002)
12. Babayan, A.H.: On unique solvability of the Dirichlet problem for one class of properly elliptic equations. Topics in Analysis and its Applications. NATO Sciences Series, series 2, vol. 147, pp. 287–295. Kluwer Academic Publishers, Dordrecht (2004)
13. Burskii, V.P., Burjachenko, E.A.: Some aspects of the non-trivial solvability of homogeneous Dirichlet problem for linear equation of arbitrary even order in the unit disk. Math. Notes **77**(3), 461–470 (2005)
14. Babayan, A.H.: Dirichlet Problem for Properly Elliptic Equation in the Unit Disk. Izvestiya NAN Armenii, Matematika (J. Contemp. Math. Anal.) **38**(6), 39–48 (2003)

15. Soldatov, A.P.: The Neumann problem for elliptic systems on a plane. Sovremennaya matematika. Fundamentalnye napravleniya. **48**, 120–133 (2013). (Russian) [J. Math. Sci.] (2014), **202**(6), 897–910
16. Begehr, H., Kumar, A.: Boundary value problems for the in-homogeneous poly-analytic equation. I, Analysis. **25**(1), 55–71 (2005)
17. Tovmasyan, N.E.: New statement and investigation of the I, the II and the III boundary problems for two second order elliptic strongly bounded systems of differential equations. Izvestiya AN ArmSSR. Matematika **3**(6), 498–521 (1968). (Russian)
18. Tovmasyan, N.E.: Non-Regular Differential Equations and Calculations of Electromagnetic Fields. World Scientific, Singapore (1998)
19. Babayan, A.H.: O zadache Dirihle dlya nepravilno ellipticheskogo uravneniya chetvertogo poryadka. [On a Dirichlet problem for fourth order improperly elliptic equation]. Neklassicheskije uravnenija matematicheskoj fiziki [Nonclassical equations of mathematical physics]. Izd. Mathem. Inst., Novosibirsk, pp. 56–69 (2007). (Russian)
20. Babayan, A.H.: On a Dirichlet problem for fourth order partial differential equation in the case of double roots of characteristic equation. Math. Montisnigri **32**, 66–80 (2015). (Russian)
21. Babayan, A.H., Mohammadi, M.H.: On a Dirichlet Problem for One Properly Elliptic Equation in the Unit Disk. Rep. NAS Armen. **117**(3), 192–199 (2017). (Russian)
22. Bourgin, D.G., Duffin, R.: The Dirichlet problem for the vibrating string equation. Bull. AMS **45**, 851–859 (1939)
23. John, F.: The Dirichlet problem for hyperbolic equation. Am. J. Math **63**(1), 141–155 (1941)
24. Aleksandryan, R.A.: Spectral properties of the operators generated by the S.L. Sobolev type systems of differential equations, Trudy Moskovskogo Matematicheskogo Obschestva [Proceedings of Moscow Mathematical Society] **9**, 455–505 (1960). (Russian)
25. Zhura, N.A., Soldatov, A.P.: A boundary value problem for a first order hyperbolic system in a two-dimensional domain. Izvesiya: Mathematics **81**(3), 542–567 (2017)
26. Babayan, A.H., Babayan, V.A.: Defect numbers of the Dirichlet problem for higher order partial differential equations in the unit disc. Casp. J. Comput. Math. Eng. (CJCME) (1), 4–19 (2016). https://doi.org/10.22039/cjcme.2016.01
27. Axler, S., Bourdon, B., Ramey, W.: Harmonic Function Theory. Springer, New York (2001)
28. Bari, N.K.: Trigonometric Series. Nauka, Moscow (1961). (Russian)
29. Anderson, N., Saff, E.B., Varga, R.S.: On the Eneström-Kakeya theorem and its sharpness. Linear Algebr. Appl. **28**, 5–16 (1979)

On the 1-dim Defocusing NLS Equation with Non-vanishing Initial Data at Infinity

Nikolaos Gialelis and Ioannis G. Stratis

Abstract We show global well-posedness of certain type of strong-in-time and weak-in-space solutions for the Cauchy problem of the 1-dimensional nonlinear Schrödinger equation, in various cases of open sets, bounded and unbounded. These solutions do not vanish at the boundary or at infinity.

Keywords NLS equation · Cauchy problem · Strong-in-time solutions · Non-vanishing solutions · Global well-posedness · Zhidkov space

1 Introduction

The 1-dimensional nonlinear Schrödinger equation (NLS) emerges as a first order model in a variety of fields—from high intensity laser beam propagation, to Bose-Einstein condensation, to water waves theory, etc. The NLS is completely integrable, hence solvable, in one dimension on the infinite line, or with periodic boundary conditions.

In this work we consider the one-dimensional defocusing NLS equation

$$i v_t + v_{xx} - |v|^{\alpha} v = 0, \ \forall\, (t, x) \in J^* \times U, \tag{1.1}$$

where $v : J \times \overline{U} \to \mathbb{C}$, with J an interval $\subseteq \mathbb{R}$ such that $0 \in J$, U an open set $\subseteq \mathbb{R}$ and $\alpha > 0$.

We are interested in solutions with a prescribed initial condition on $\{t = 0\} \times \overline{U}$. Moreover, v is either not necessarily equal to zero on $J \times \partial U$, or not necessarily such

N. Gialelis · I. G. Stratis (✉)
Department of Mathematics, National and Kapodistrian University of Athens, Panepistimioupolis, 15784 Zographou (Athens), Greece
e-mail: istratis@math.uoa.gr

N. Gialelis
e-mail: ngialelis@math.uoa.gr

A. Karapetyants et al. (eds.), *Modern Methods in Operator Theory and Harmonic Analysis*, Springer Proceedings in Mathematics & Statistics 291,
https://doi.org/10.1007/978-3-030-26748-3_19

that $\lim_{|x|\to\infty} v=0$ on $J\times\overline{U}$. In the case that U is unbounded we assume that v has a constant amplitude at infinity.

Since we are interested in all possible cases of open sets, U could be bounded (e.g., a ball) or unbounded—with or without empty boundary (e.g., $\mathbb{R}_+$ or $\mathbb{R}$, respectively). Let us recall that when $U=\mathbb{R}$, the existence of many such solutions is well-known, e.g. the black soliton $v(t,x)=e^{-it}\tanh\left(2^{-\frac{1}{2}}x\right)$, for $\alpha=2$. In the present work we look for solutions of the form $v(t,x)=e^{irt}\,(u(t,x)+\zeta(x))$, for $r\in\mathbb{R}$ and u, ζ complex-valued functions over $J\times\overline{U}$ and $\overline{U}$, respectively, such that u vanishes at the boundary and at infinity, but ζ, in contrast, survives. The arising problem then becomes

$$\begin{cases} iu_t+(u+\zeta)_{xx}-(|u+\zeta|^{\alpha}+r)\,(u+\zeta)=0, \ \forall\,(t,x)\in J^*\times U \\ u=u_0, \text{ on } \{t=0\}\times\overline{U} \\ u=0, \text{ on } J\times\partial U \text{ and } u \xrightarrow{|x|\to\infty} 0, \text{ on } J\times\overline{U}, \end{cases} \tag{1.2}$$

for given r, ζ and also $u_0:\overline{U}\to\mathbb{C}$ which vanishes at the boundary and at infinity.

The problem (1.2) for $U=\mathbb{R}$ and

$$\alpha=2\tau \text{ for } \tau\in\mathbb{N}:=\{1,2,\dots\} \text{ and } r=-\rho^{\tau} \text{ for } \rho>0, \tag{1.3}$$

has been studied in [8]. There it is stated that if $\zeta\in C_b^2(\mathbb{R})$, $D\zeta\in H^2(\mathbb{R})$, and additionally $\left(|\zeta|^2-\rho\right)\in L^2(\mathbb{R})$, then (1.2) is globally well-posed in $H^1(\mathbb{R})$ and the energy of the solution is conserved.

Recently, in [10], the above result is extended not only by weakening the assumptions on ζ but also by considering more general cases of $U\subseteq\mathbb{R}$, other than the Euclidean space itself. Namely, it is shown that the problem (1.2) is globally and uniquely solvable in $H_0^1(U)$ for any open $U\subseteq\mathbb{R}$, if $\zeta\in X^1(U)$, the Zhidkov space over U (see the notations below) and additionally if α, r are as in (1.3); for unbounded U it is further assumed that $\left(|\zeta|^2-\rho\right)\in L^2(U)$.

In this work we introduce sufficient conditions on ζ that establish the continuous dependence on the initial data, as well as the conservation of energy. We show that in bounded sets there is no need for extra assumptions on ζ. On the other hand, in unbounded sets the assumptions on ζ are stronger than the ones for the bounded case, yet still, they remain weaker than the ones in [8] for the case $U=\mathbb{R}$ and they ascertain the rigorous proof of the well-posedness of the problem.

The present paper is organized as follows: the problem is formulated for two different sets of assumptions in Sect. 2, where the necessary notation is also introduced. The well-posedness of the problem for bounded U—with "minimal" assumptions on ζ—is treated in Sect. 3. We note that both the strong H_0^1-regularity and the continuous dependence of the solution on the initial data require the conservation of energy; the latter is established without any additional assumption in the case of bounded U. However, as shown in Sect. 4, for unbounded U the energy is proved to be conserved under stronger assumptions on ζ. The underlying reason for this, is that in the case of unbounded U the backward-in-time existence is not guaranteed by the "minimal" assumptions of Sect. 3; to surpass this obstacle we employ an approximation by regular solutions for which the energy is actually conserved.

2 Preliminaries

We start with some notation used throughout the paper:

1. J denotes any bounded interval such that $0 \in J$, $J_\pm := J \cap \mathbb{R}_\pm$ and U for any open $\subseteq \mathbb{R}$.
2. If $p, q, r \in [1, \infty]$ and $k, m \in \mathbb{N}_0 := \{0, 1, 2, \ldots\}$, we write

$$|\cdot|_{m,r,U} := \|\cdot\|_{W^{m,r}(U)}, \ |\cdot|_{-m,U} := \|\cdot\|_{H^{-m}(U)}$$
$$|\cdot|_{k,p,J;m,r,U} := \|\cdot\|_{W^{k,p}(J;W^{m,r}(U))}, \ |\cdot|_{K,p,J;-m,U} := \|\cdot\|_{W^{k,p}(J;H^{-m}(U))}.$$

   We omit $p = \infty$, $J = \mathbb{R}$ and $U = \mathbb{R}$ from the notation.
3. If $m \in \mathbb{N}$, then $X^m(U)$ stands for the Zhidkov space over U, defined as

$$X^m(U) := \left\{ u \in L^\infty(U) \mid D^k u \in L^2(U), \ \forall k \in \{1, \ldots, m\} \right\}$$

   and equipped with its natural norm $\|\cdot\|_{X^m(U)} := |\cdot|_{0,\infty,U} + \sum_{k=1}^m \left| D^k \cdot \right|_{0,2,U}$. The first version of such spaces over $\mathbb{R}$ is introduced in [13] and a generalization for higher dimensions (along with certain modifications) is done in [7–9, 14]. In this work, however, we consider X^m over any open set.
4. Let $\mathcal{F}(U_1; \mathbb{C})$ be a function space over $U_1 \subset U_2 \subseteq \mathbb{R}$ and $f \in \mathcal{F}(U_1)$. We denote by $\mathcal{E}_{U_2} f$ its extension by zero in $U_2 \backslash U_1$ and $\mathcal{E}_{U_2}\mathcal{F}(U_1) := \left\{ \mathcal{E}_{U_2} f \mid f \in \mathcal{F}(U_1) \right\}$. We omit $U_2 = \mathbb{R}$ from these notations. Moreover, if $g \in \mathcal{F}(U_2)$, we denote by $\mathcal{R}_{U_1} g$ and $\mathcal{R}_{U_1}\mathcal{F}(U_2)$ the restriction of g in U_1 and the set of these restricted functions, respectively.
5. We write C and c for any non-negative constant factor and exponent, respectively. These constants may be explicitly calculated in terms of known quantities and may change from line to line and also within a certain line in a given computation. We also employ the letter K for any increasing function $K : [0, \infty)^n \to [0, \infty)$. When J and U appear as subscripts in an element, they denote that this depends on them, while their absence designates independence.
6. If $u : J \times U \to \mathbb{C}$, with $u(t, \cdot) \in \mathcal{F}(U)$ for each $t \in J$, then, following the notation of, e.g., [6, 11], we associate with u the mapping $\mathbf{u} : J \to \mathcal{F}(U; \mathbb{C})$, defined by $[\mathbf{u}(t)](x) := u(t, x)$, for every $x \in U$ and $t \in J$.

Next, recall Hölder's inequality: let $U \subseteq \mathbb{R}^n$, $m \in \mathbb{N} \backslash \{1\}$, $\{p_k\}_{k=1}^m \subset [1, \infty]$, such that $\frac{1}{p_1} + \ldots + \frac{1}{p_m} = 1$ and $u_k \in L^{p_k}(U)$ for $k = 1, \ldots, m$. Then

$$\int_U |u_1 \cdots u_m| \, dx \leq \prod_{k=1}^m |u_k|_{0,p_k,U}. \tag{2.1}$$

From (2.1) for $p_1 = \frac{\alpha+2}{\alpha+1}$ and $p_2 = \alpha+2$, $\alpha \geq 0$, it clearly holds that

$$\int_U |u|^{\alpha+1}\, |v|\, dx \leq |u|^c_{0,\alpha+2,U} |v|_{0,\alpha+2,U}\,, \tag{2.2}$$

for $u,\, v \in L^{\alpha+2}(U)$.

In the sequel, we assume that $\alpha > 0$, $\zeta \in L^{\alpha+2}(U)$ and $r \in \mathbb{R}$.
From (2.2) and the scaling invariant embedding $H_0^1(U) \hookrightarrow L^{\alpha+2}(U)$, we define $g : H_0^1(U) \to Y_\alpha := L^{\frac{\alpha+2}{\alpha+1}}(U) + L^2(U) \hookrightarrow H^{-1}(U)$ to be the nonlinear and bounded operator such that

$$\langle g(u;\, \alpha, \zeta, r)\,,\, v\rangle := \int_U (|u+\zeta|^\alpha + r)\,(u+\zeta)\,\bar{v} dx,\ \text{ for } v \in H_0^1(U)\,.$$

For the above operator we have the following estimate.

Proposition 1 *Let $u,\, v \in H_0^1(U)$. Then*

$$\begin{aligned} \|g(u) - g(v)\|_{Y_\alpha} \leq K\big(|u|_{1,2,U}\,, |v|_{1,2,U}\,, |\zeta|_{0,\alpha+2,U}\big) \times \\ \times \big(|u-v|_{0,\alpha+2,U} + |u-v|_{0,2,U}\big)\,. \end{aligned} \tag{2.3}$$

Proof For $u,\, v \in L^{\alpha+2}(U)$, we have

$$\big||u|^\alpha u - |v|^\alpha v\big|_{0,\frac{\alpha+2}{\alpha+1},U} \leq C\,\big(|u|^c_{0,\alpha+2,U} + |v|^c_{0,\alpha+2,U}\big)\, |u-v|_{0,\alpha+2,U}\,. \tag{2.4}$$

This inequality follows by direct application of

$$\big||z_1|^\alpha z_1 - |z_2|^\alpha z_2\big| \leq C\, |z_1 - z_2|\, (|z_1|^\alpha + |z_2|^\alpha)\,,\ z_1, z_2 \in \mathbb{C}, \tag{2.5}$$

(2.1) for $p_1 = \alpha+1$ and $p_2 = \frac{\alpha+1}{\alpha}$, as well as $|z_1+z_2|^p \leq C\,(|z_1|^p + |z_2|^p)$, $p>0$. From (2.4) and the scaling invariant embedding $H_0^1(U) \hookrightarrow L^{\alpha+2}(U)$ we get

$$\begin{aligned} \|g(u) - g(v)\|_{Y_\alpha} \leq C\,\big(|u|^c_{1,2,U} + |v|^c_{1,2,U} + |\zeta|^c_{0,\alpha+2,U}\big)\, |u-v|_{0,\alpha+2,U} + \\ + C|u-v|_{0,2,U} \end{aligned}$$

and the result follows.

Now, we further assume that $\zeta \in H^1(U)$ and we define $\mathcal{N}[\cdot,\cdot] : \big(H_0^1(U)\big)^2 \to \mathbb{C}$ to be the form which is associated with the operator $D^2(\cdot + \zeta) - g$, such that $\mathcal{N}[u,\, v] := \big\langle D^2(u+\zeta)\,,\, v\big\rangle - \langle g(u)\,,\, v\rangle$, for every $u,\, v \in H_0^1(U)$.

We then restate the problem (1.2): we seek a solution $\mathbf{u}_J \in L^\infty\big(J;\, H_0^1(U)\big) \cap W^{1,\infty}\big(J;\, H^{-1}(U)\big)$ of

$$\begin{cases} i\,\big\langle \mathbf{u}'_J,\, v\big\rangle + \mathcal{N}[\mathbf{u}_J,\, v] = 0,\ \forall v \in H_0^1(U)\,,\ \text{a.e. in } J \\ \mathbf{u}_J(0) = u_0. \end{cases} \tag{2.6}$$

We also provide an estimate for the form $\mathcal{N}$.

Proposition 2 *Let $u, v \in H^1_0(U)$. Then*

$$|\mathcal{N}[u, v]| \leq K\left(|u|_{1,2,U}, |v|_{1,2,U}, |\zeta|_{1,2,U}, |\zeta|_{0,\alpha+2,U}\right). \tag{2.7}$$

Proof From (2.1) ($p_1 = p_2 = 2$), (2.2) and the scaling invariant embedding $H^1_0(U) \hookrightarrow L^{\alpha+2}(U)$, we get $|\mathcal{N}[u, v]| \leq C|D(u+\zeta)|_{0,2,U}|v|_{1,2,U} + C|u+\zeta|^c_{0,\alpha+2,U}|v|^c_{1,2,U}$, hence the result follows.

We further define the energy functional $E : H^1_0(U) \to \mathbb{R}_+ \cup \{\infty\}$ by

$$E(\cdot; \alpha, \zeta, r) := \frac{1}{2}|D(\cdot+\zeta)|^2_{0,2,U} + G(\cdot; \alpha, \zeta, r),$$

where $G : H^1_0(U) \to \mathbb{R}_+ \cup \{\infty\}$, with

$$G(\cdot; \alpha, \zeta, r) := \int_U V(|\cdot+\zeta|; \alpha, r)\, dx,$$

where $V : [0, \infty) \to [0, \infty)$ is defined as

$$V(x; \alpha, r) := \frac{1}{\alpha+2}x^{\alpha+2} + \frac{1}{2}rx^2 + \frac{\alpha}{2(\alpha+2)}|r|^{\frac{\alpha+2}{\alpha}}. \tag{2.8}$$

It easily follows that for every constant $C_\alpha > \alpha+2$ there exists a constant $A_\alpha > 0$, such that

$$x^{\alpha+2} \leq C_\alpha V(x), \ \forall x \geq A_\alpha. \tag{2.9}$$

For the functional G we have the following estimates.

Proposition 3 *Let $u, v \in H^1_0(U)$. If $(G(u) - G(v)) \in \mathbb{R}$, then*

$$\begin{aligned} |G(u) - G(v)| \leq K\left(|u|_{1,2,U}, |v|_{1,2,U}, |\zeta|_{1,2,U}, |\zeta|_{0,\alpha+2,U}\right) \times \\ \times \left(|u-v|_{0,\alpha+2,U} + |u-v|_{0,2,U}\right) \end{aligned} \tag{2.10}$$

and

$$G(u) \leq K\left(|u|_{1,2,U}, |\zeta|_{1,2,U}, |\zeta|_{0,\alpha+2,U}, |U|\right). \tag{2.11}$$

Proof From

$$\begin{aligned} G(u) - G(v) = \int_0^1 \frac{d}{ds}G(su + (1-s)\, v)\, ds = \\ = \int_0^1 \mathrm{Re}\langle g(su + (1-s)\, v), u - v\rangle\, ds, \end{aligned} \tag{2.12}$$

(2.2) and the scaling invariant embedding $H_0^1(U) \hookrightarrow L^{\alpha+2}(U)$ we get

$$|G(u)-G(v)| \leq C\left(|u|_{1,2,U}^c+|v|_{1,2,U}^c+|\zeta|_{0,\alpha+2,U}^c+|\zeta|_{1,2,U}^c\right) \times \\ \times\left(|u-v|_{0,\alpha+2,U}+|u-v|_{0,2,U}\right).$$

As for (2.11), we notice that

$$G(0)=\int_U V(|\zeta|)\,dx=\frac{1}{\alpha+2}|\zeta|_{0,\alpha+2,U}^{\alpha+2}+\frac{1}{2}r|\zeta|_{0,2,U}^2+\frac{\alpha}{2(\alpha+2)}|r|^{\frac{\alpha+2}{\alpha}}|U| \leq \\ \leq K\left(|\zeta|_{1,2,U},|\zeta|_{0,\alpha+2,U},|U|\right).$$

Then the result follows from (2.10) and the triangle inequality.

Let us now assume that $\zeta \in L^{\alpha+2}(U) \cap L^{\infty}(U)$. Two fine properties concerning the operator g follow.

Proposition 4 *Let $u, v \in H_0^1(U)$. Then $(g(u)-g(v)) \in L^2(U)$ with*

$$|g(u)-g(v)|_{0,2,U} \leq K\left(|u|_{1,2,U},|v|_{1,2,U},|\zeta|_{0,\infty,U}\right)|u-v|_{0,2,U}. \tag{2.13}$$

Proof By simple application of (2.5), we get

$$\int_U |g(u)-g(v)|^2 dx \leq \int_U\left(|u|^{2\alpha}+|v|^{2\alpha}\right)|u-v|^2 dx+ \\ +\left(|\zeta|_{0,\infty,U}^c+C\right)|u-v|_{0,2,U}^2.$$

We then employ the scaling invariant embedding $H_0^1(U) \hookrightarrow L^{\infty}(U)$.

Proposition 5 *Let $u, v \in H_0^1(U)$. If either $|U|<\infty$, or $|U|=\infty$, α and r be as in (1.3), as well as $\left(|\zeta|^2-\rho\right) \in L^2(U)$, then g maps to $L^2(U)$ and*

$$|g(u)|_{0,2,U} \leq \begin{cases} K_U\left(|u|_{1,2,U},|\zeta|_{0,\infty,U}\right), & \text{if } |U|<\infty \\ K\left(|u|_{1,2,U},|\zeta|_{0,\infty,U},\left||\zeta|^2-\rho\right|_{0,2,U}\right), & \text{otherwise.}\end{cases} \tag{2.14}$$

Proof We notice that $g(0)=(|\zeta|^{\alpha}+r)\zeta$, which belongs to $L^2(U)$. Indeed, for $|U|<\infty$ this is straightforward. For $|U|=\infty$, by expanding via

$$a^n-b^n=(a-b)\left(a^{n-1}+a^{n-2}b+\ldots+ab^{n-2}+b^{n-1}\right), \tag{2.15}$$

we get $|g(0)|_{0,2,U} \leq K\left(|\zeta|_{0,\infty,U},\left||\zeta|^2-\rho\right|_{0,2,U}\right)$. The result then follows from (2.13) and the triangle inequality.

Let us now notice that ζ being in $L^{\alpha+2}(U)$ plays no essential role at any of the above results. Hence, for

$$\alpha, r \text{ as in (1.3) and } \zeta \in L^{\infty}(U) \text{ with } \big(|\zeta|^2-\rho\big) \in L^2(U),$$

we define $g_s : H_0^1(U) \to L^2(U)$ by

$$\langle g_s(u;\tau,\zeta,\rho), v\rangle := \int_U \big(|u+\zeta|^{2\tau}-\rho^{\tau}\big)(u+\zeta)\,\bar{v}\,dx, \text{ for } v \in H_0^1(U),$$

which satisfies the above estimates.

Now, we further assume that $\zeta \in X^1(U)$ and we define $\mathcal{N}_s[\cdot,\cdot] : \big(H_0^1(U)\big)^2 \to \mathbb{C}$ to be the form which is associated with the operator $D^2(\cdot+\zeta)-g_s$, such that $\mathcal{N}_s[u,v] := \big\langle D^2(u+\zeta), v\big\rangle - \langle g_s(u), v\rangle$, for every $u, v \in H_0^1(U)$. We note that apart from belonging to $\mathcal{L}\big(H^1(U); H^{-1}(U)\big)$, $D^2 \in \mathcal{L}\big(X^1(U); H^{-1}(U)\big)$ also, with its usual definition. Now, the problem (1.2) becomes: find a solution $\mathbf{u}_J \in L^{\infty}\big(J; H_0^1(U)\big) \cap W^{1,\infty}\big(J; H^{-1}(U)\big)$ of

$$\begin{cases} i\,\big\langle \mathbf{u}_J', v\big\rangle + \mathcal{N}_s[\mathbf{u}_J, v] = 0, \ \forall v \in H_0^1(U), \text{ a.e. in } J \\ \mathbf{u}_J(0) = u_0. \end{cases} \tag{2.16}$$

From (2.14) and (2.1) (for $p_1 = p_2 = 2$), we derive the following estimate

$$|\mathcal{N}_s[u,v]| \le K\Big(|u|_{1,2,U}, |v|_{1,2,U}, \|\zeta\|_{X^1(U)}, \big||\zeta|^2-\rho\big|_{0,2,U}\Big) \tag{2.17}$$

for every $u, v \in H_0^1(U)$.

We also define the respective energy functional $E_s : H_0^1(U) \to \mathbb{R}_+ \cup \{\infty\}$ by

$$E_s(\cdot;\tau,\zeta,\rho) := \frac{1}{2}|D(\cdot+\zeta)|_{0,2,U}^2 + G_s(\cdot;\tau,\zeta,\rho),$$

where $G_s : H_0^1(U) \to \mathbb{R}_+ \cup \{\infty\}$, with

$$G_s(\cdot;\tau,\zeta,\rho) := \int_U V(|\cdot+\zeta|;\tau,\rho)\,dx,$$

for which we have

$$\begin{aligned} |G_s(u)-G_s(v)| \le K\Big(|u|_{1,2,U}, |v|_{1,2,U}, |\zeta|_{0,\infty,U}, \big||\zeta|^2-\rho\big|_{0,2,U}\Big) \times \\ \times |u-v|_{0,2,U}, \end{aligned} \tag{2.18}$$

from (2.12) and (2.14). Moreover, $G_s(0) < K\left(|\zeta|_{0,\infty,U}, \left||\zeta|^2-\rho\right|_{0,2,U}\right)$, which is obtained easily from

$$a^{n+1} - a\,(n+1)\,b^n + nb^{n+1} = (a-b)^2\left(a^{n-1} + 2a^{n-2}b + \ldots + (n-1)ab^{n-2} + nb^{n-1}\right).$$

Hence, from (2.18) and the triangle inequality we get

$$G_s(u) \leq K\left(|u|_{1,2,U}, |\zeta|_{0,\infty,U}, \left||\zeta|^2-\rho\right|_{0,2,U}\right), \tag{2.19}$$

for every $u \in H_0^1(U)$ and so $E_s, G_s : H_0^1(U) \to \mathbb{R}_+$.

We also need the following results.

Proposition 6 *Let $\mathcal{H}$ be a Hilbert space and $\mathcal{F}$ be a Banach space with the Radon-Nikodym property with respect to the Lebesgue measure in $(J, \mathscr{B}(J))$.*

1. *Let $\{\mathbf{u}_k\}_{k=1}^{\infty} \subset L^\infty(J;\mathcal{H})$ and $\mathbf{u} : J \to \mathcal{H}$ with $\mathbf{u}_k(t) \rightharpoonup \mathbf{u}(t)$ in $\mathcal{H}$, for a.e. $t \in J$. If $\|\mathbf{u}_k\|_{L^\infty(J;\mathcal{H})} \leq C$ uniformly for all $k \in \mathbb{N}$, then $\mathbf{u} \in L^\infty(J;\mathcal{H})$ with $\|\mathbf{u}\|_{L^\infty(J;\mathcal{H})} \leq C$, where C is the same in both inequalities.*
2. *Let $\{\mathbf{u}_k\}_{k=1}^{\infty} \cup \{\mathbf{u}\} \subset L^\infty(J;\mathcal{F}^*)$ with $\mathbf{u}_k \overset{*}{\rightharpoonup} \mathbf{u}$ in $L^\infty(J;\mathcal{F}^*)$.*[1] *If $\|\mathbf{u}_k\|_{L^\infty(J;\mathcal{F}^*)} \leq C$ uniformly for all $k \in \mathbb{N}$, then $\|\mathbf{u}\|_{L^\infty(J;\mathcal{F}^*)} \leq C$, where C is the same in both inequalities.*
3. *Let $p \in [1,\infty)$ and $\{\mathbf{u}_k\}_{k=1}^{\infty} \cup \{\mathbf{u}\} \subset L^p(J;\mathcal{H})$ with $\mathbf{u}_k \rightharpoonup \mathbf{u}$ in $L^p(J;\mathcal{H})$. If $\|\mathbf{u}_k\|_{L^\infty(J;\mathcal{H})} \leq C$ uniformly for all $k \in \mathbb{N}$, then $\|\mathbf{u}\|_{L^\infty(J;\mathcal{H})} \leq C$, where C is the same in both inequalities.*

Proof 1. We derive that $\|\mathbf{u}(t)\|_{\mathcal{H}} \leq C$, for a.e. $t \in J$, from the (sequentially) weak lower semi-continuity of the norm. Then, the result follows directly.

2. Let $v \in \mathcal{F}$ be such that $\|v\|_{\mathcal{F}} \leq 1$ and set $\mathbf{v} : J \to \mathcal{F}$ the constant function with $\mathbf{v}(t) := v$, for all $t \in J$. We have

$$\int_s^{s+h} \langle \mathbf{u}_k, \mathbf{v}\rangle \, dt \leq Ch, \text{ for every } s \in J^\circ \text{ and every sufficiently small } h > 0.$$

Letting $k \to \infty$, dividing both parts by h and then letting $h \to 0$, we get $\langle \mathbf{u}(s), v\rangle \leq C$, for every $s \in J^\circ$. Since v arbitrary, the proof is complete.

3. We deal as in 2.

Lemma 1 *Let $\mathbf{u} : \overline{J} \to H^m(U)$ be such that $\mathbf{u} \in C\left(\overline{J}; L^2(U)\right)$. Then $\mathbf{u}$ is weakly continuous (as a function to $H^m(U)$).*

Proof Let $t_0 \in \overline{J}$ and $\{t_n\}_{n=1}^{\infty} \subset \overline{J}$ be such that $t_n \to t_0$. Let also $v \in C_c^\infty(U)$. Then

---

[1] That is, $\mathbf{u}_k \overset{*}{\rightharpoonup} \mathbf{u}$ in $\sigma\left(L^\infty(J;\mathcal{F}^*), L^1(J;\mathcal{F})\right)$. Note that $L^\infty(J;\mathcal{F}^*) \cong \left(L^1(J;\mathcal{F})\right)^*$ (see, e.g., [5] Theorem 1, Sect. IV.1).

$$\left|(\mathbf{u}(t_n)-\mathbf{u}(t_0)\,,\,v)_{H^m(U)}\right| = \left|\sum_{k=0}^{m}\int_U D^k(\mathbf{u}(t_n)-\mathbf{u}(t_0))\,D^k\bar{v}dx\right| =$$

$$= \left|\sum_{k=0}^{m}(-1)^k\int_U (\mathbf{u}(t_n)-\mathbf{u}(t_0))\,D^{2k}\bar{v}dx\right| \leq$$

$$\overset{2.1}{\leq} |\mathbf{u}(t_n)-\mathbf{u}(t_0)|_{0,2,U}\sum_{k=0}^{m}\left|D^{2k}v\right|_{0,2,U} \to 0,$$

hence, the result follows from the density argument and the fact that $t_0 \in \overline{J}$ is arbitrary.

Lemma 2 *Let $\mathcal{H}$ be a Hilbert space and $\mathbf{u} : \overline{J} \to \mathcal{H}$ be weakly continuous. If $\|\mathbf{u}\|_{\mathcal{H}} \in C(\overline{J})$, then $\mathbf{u} \in C(\overline{J};\mathcal{H})$.*

Proof Let $t_0 \in \overline{J}$ and $\{t_n\}_{n=1}^{\infty} \subset \overline{J}$ be such that $t_n \to t_0$. Then

$$\|\mathbf{u}(t_n)-\mathbf{u}(t_0)\|_{\mathcal{H}}^2 = \|\mathbf{u}(t_n)\|_{\mathcal{H}}^2 - (\mathbf{u}(t_0)\,,\,\mathbf{u}(t_n))_{\mathcal{H}} -$$
$$-\,(\mathbf{u}(t_n)-\mathbf{u}(t_0)\,,\,\mathbf{u}(t_0))_{\mathcal{H}} \to 0,$$

hence, the result follows since $t_0 \in \overline{J}$ is arbitrary.

Lemma 3 *Let $U_1 \subset U_2 \subseteq \mathbb{R}$, $m \in \mathbb{N}_0$ and $\{u_k\}_{k=1}^{\infty} \cup \{u\} \subset H^m(U_2)$ such that $u_k \rightharpoonup u$ in $H^m(U_2)$. Then $\mathcal{R}_{U_1}u_k \rightharpoonup \mathcal{R}_{U_1}u$ in $H^m(U_1)$.*

Proof Let $v \in C_c^{\infty}(U_1)$. Then

$$\left(\mathcal{R}_{U_1}u_k - \mathcal{R}_{U_1}u,\,v\right)_{H^m(U_1)} = \sum_{k=0}^{m}\int_{U_1} D^k\left(\mathcal{R}_{U_1}u_k - \mathcal{R}_{U_1}u\right)D^k\bar{v}dx =$$

$$= \sum_{k=0}^{m}\int_{U_2} D^k(u_k-u)\,D^k\mathcal{E}_{U_2}\bar{v}dx = \left(u_k-u,\,\mathcal{E}_{U_2}v\right)_{H^m(U_2)} \to 0,$$

hence, the result follows from the density argument.

3 Solutions in Bounded Sets

In this section, we assume that $U \subset \mathbb{R}^n$ is bounded.

3.1 A General Result for $r \in \mathbb{R}$

Theorem 1 *Let $u_0 \in H_0^1(U)$. Then for every J, there exists a solution $\mathbf{u}_J \in L^{\infty}\left(J;\,H_0^1(U)\right) \cap W^{1,\infty}\left(J;\,H^{-1}(U)\right)$ of (2.6), such that*

$$|\mathbf{u}_J|_{0,J;1,2,U}+|\mathbf{u}'_J|_{0,J;-1,U}\leq\mathcal{K}, \tag{3.1}$$

where

$$\mathcal{K}:=\begin{cases}K\left(|u_0|_{1,2,U},|\zeta|_{1,2,U},|\zeta|_{0,\alpha+2,U}\right), & \text{if } r\geq 0\\ K_U\left(|u_0|_{1,2,U},|\zeta|_{1,2,U},|\zeta|_{0,\alpha+2,U}\right), & \text{if } r<0\end{cases}$$

and also

$$E(\mathbf{u}_J)\leq E(u_0)\,,\ \text{everywhere in } J. \tag{3.2}$$

Moreover, if u_0 and ζ are real-valued, then $\mathbf{u}_J(t)=\overline{\mathbf{u}_J(-t)}$, for all $t\in J$ with $|t|\leq\operatorname{dist}(0,\partial J)$.

Proof STEP 1: We make use of the standard Faedo-Galerkin method. It holds true that $H_0^1(U;\mathbb{C})\hookrightarrow\hookrightarrow L^2(U;\mathbb{C})$, hence there exists a countable subset of $H_0^1(U;\mathbb{R})\cap C^\infty\left(\overline{U};\mathbb{R}\right)$, which is an orthogonal basis of $L^2(U;\mathbb{C})$, e.g., the complete set of eigenfunctions for the operator $-D^2$ in $H_0^1(U;\mathbb{C})$.[2] Let $\{w_k\}_{k=1}^\infty\subset H_0^1(U;\mathbb{R})\cap C^\infty\left(\overline{U};\mathbb{R}\right)$ be that basis, appropriately normalized so that $\{w_k\}_{k=1}^\infty$ be an orthonormal basis of $L^2(U;\mathbb{C})$. Fixing any $m\in\mathbb{N}$, we define $\mathbf{d}_m\in C^\infty(J_m;\mathbb{C}^m)$, with $\mathbf{d}_m(t):=\left[d_m^1(t),\ldots,d_m^m(t)\right]^{\mathrm{T}}$, to be the unique maximal solution of the initial-value problem

$$\begin{cases}\mathbf{d}_m{}'(t)=F_m(\mathbf{d}_m(t))\,,\ \forall t\in J_m^*\\ \mathbf{d}_m(0)=[(u_0,w_1),\ldots,(u_0,w_m)]^{\mathrm{T}},\end{cases}$$

where $F_m\in C^\infty\left(\mathbb{R}^{2m};\mathbb{C}^m\right)$ with

$$F_m^k(\mathbf{z}):=i\mathcal{N}[\sum_{l=1}^m z_l w_l,\,w_k],\ \text{for all } \mathbf{z}\in\mathbb{C}^m,\ \text{with } \mathbf{z}:=[z_1,\ldots,z_m]^{\mathrm{T}},$$

for all $k\in\{1,\ldots,m\}$. Now, we define $\mathbf{u}_m\in C^\infty\left(J_m;H_0^1(U;\mathbb{C})\cap C^\infty\left(\overline{U};\mathbb{C}\right)\right)$, with

$$\mathbf{u}_m(t):=\sum_{k=1}^m d_m^k(t)\,w_k.$$

It is then trivial to verify that

$$i\left(\mathbf{u}'_m,w_k\right)+\mathcal{N}[\mathbf{u}_m,w_k]=0,\ \text{everywhere in } J_m, \tag{3.3}$$

[2] This specific subset is an orthogonal basis of both $H_0^1(U;\mathbb{C})$ and $L^2(U;\mathbb{C})$.

for all $k\in\{1,\ldots,m\}$. Note that $u_{0m}:=u_m(0,\cdot)=\mathbf{u}_m(0)\to u_0$ in $L^2(U)$ and $|u_{0m}|_{0,2,U}\le|u_0|_{0,2,U}$. Furthermore, $|u_{0m}|_{1,2,U}\le|u_0|_{1,2,U}$. Indeed, since $\sum_{k=1}^m a_k w_k\in$ span $\{w_k\}_{k=1}^m$ for some $\{a_k\}_{k=1}^m\subset\mathbb{C}$ we have that $\big(D^2u_{0m},u_{0m}\big)=\big(D^2u_{0m},u_0\big)$, hence we get

$$|Du_{0m}|_{0,2,U}^2=-\big(D^2u_{0m},u_{0m}\big)=(Du_{0m},Du_0)\le$$
$$\le\frac{1}{2}|Du_{0m}|_{0,2,U}^2+\frac{1}{2}|Du_0|_{0,2,U}^2.$$

Therefore $|Du_{0m}|_{0,2,U}\le|Du_0|_{0,2,U}$.

STEP 2: We multiply the variational equation in (3.3) by $-\overline{d_m^k}'(t)$, sum for $k=1,\ldots,m$, and take real parts of both sides, and thus obtain

$$\frac{d}{dt}E(\mathbf{u}_m)=0,\ \text{that is } E(\mathbf{u}_m)\le E(u_0),\ \text{everywhere in } J_m, \tag{3.4}$$

hence, if $r\ge 0$, we have that $|\mathbf{u}_m|_{1,2,U}\le\mathcal{K}$ and thus $J_m\equiv\mathbb{R}$. Since $m\in\mathbb{N}$ is arbitrary, we get $|\mathbf{u}_m|_{1,2,U}\le\mathcal{K}$, for all $m\in\mathbb{N}$. Hence, we conclude that $\{\mathbf{u}_m\}_{m=1}^\infty$ is uniformly bounded in $L^\infty\big(\mathbb{R};H_0^1(U)\big)$, with

$$|\mathbf{u}_m|_{0;1,2,U}\le\mathcal{K},\ \forall m\in\mathbb{N}. \tag{3.5}$$

If $r<0$, from (2.11) we have that $|D\mathbf{u}_m|_{0,2,U}\le E(\mathbf{u}_m)\le\mathcal{K}$ and thus $J_m\equiv\mathbb{R}$. Therefore, from the Poincaré inequality, we also get $|\mathbf{u}_m|_{0,2,U}\le\mathcal{K}$ and thus (3.5) follows.

STEP 3: We fix an arbitrary $v\in H_0^1(U)$ with $|v|_{1,2,U}\le 1$ and write $v=\mathcal{P}v\oplus(\mathcal{I}-\mathcal{P})v$, where $\mathcal{P}$ is the projection in span $\{w_k\}_{k=1}^m$. Since $\mathbf{u}_m'\in$ span $\{w_k\}_{k=1}^m$ and $\mathcal{N}[h,g]$ is (conjugate) linear for g, from the variational equation in (3.3) we get that $\langle\mathbf{u}_m',v\rangle=\big(\mathbf{u}_m',v\big)=\big(\mathbf{u}_m',\mathcal{P}v\big)=i\mathcal{N}[\mathbf{u}_m,\mathcal{P}v]$. Applying (2.7) we derive $\big|\langle\mathbf{u}_m',v\rangle\big|\le\mathcal{K}$. Hence $\big\{\mathbf{u}_m'\big\}_{m=1}^\infty$ is uniformly bounded in $L^\infty\big(\mathbb{R};H^{-1}(U)\big)$, with

$$\big|\mathbf{u}_m'\big|_{0;-1,U}\le\mathcal{K},\ \forall m\in\mathbb{N}. \tag{3.6}$$

STEP 4α: We fix an arbitrary J. From (3.5), (3.6), Theorem 1.3.14 (i) in [4] and **Proposition** 6 1., there exist a subsequence $\big\{\mathbf{u}_{m_l}\big\}_{l=1}^\infty\subseteq\{\mathbf{u}_m\}_{m=1}^\infty$ and a function $\mathbf{u}_J\in L^\infty\big(J;H_0^1(U)\big)\cap W^{1,\infty}\big(J;H^{-1}(U)\big)$, such that

$$\mathbf{u}_{m_l}(t)\rightharpoonup\mathbf{u}_J(t)\ \text{in } H_0^1(U),\ \text{for every } t\in\overline{J}\ \text{and also } |\mathbf{u}_J|_{0,J;1,2,U}\le\mathcal{K}. \tag{3.7}$$

STEP 4β: $H^{-1}(U)$ is separable since $H_0^1(U)$ is separable, hence by the Dunford-Pettis theorem (see, e.g., [5], Theorem 1, Sect. III.3) we have $L^\infty\big(J;H^{-1}(U)\big)\cong\big(L^1\big(J;H_0^1(U)\big)\big)^*$. In virtue of the above, from (3.6), the Banach-Alaoglu-Bourbaki theorem (see, e.g., [3], Theorem 3.16) and **Proposition** 6 2., there exist a subsequence of $\big\{\mathbf{u}_{m_l}\big\}_{l=1}^\infty$, which we still denote as such and a function $\mathbf{h}\in L^\infty\big(J;H^{-1}(U)\big)$, such that

$$\mathbf{u}'_{m_l} \overset{*}{\rightharpoonup} \mathbf{h} \text{ in } L^\infty\big(J; H^{-1}(U)\big) \text{ and also } |\mathbf{h}|_{0,J;-1,U} \leq \mathcal{K}. \tag{3.8}$$

From the convergence in (3.7), Lemma 1.1, Chap. 3 in [11], along with the Leibniz rule, we can derive that

$$\int_J \big(\mathbf{u}'_{m_l}, \psi v\big)\, dt \to \int_J \big\langle \mathbf{u}'_J, \psi v\big\rangle\, dt, \ \forall\, \psi \in C^1_c(J^\circ)\,,\, v \in H^1_0(U)\,,$$

hence $\mathbf{h} \equiv \mathbf{u}'_J$.

STEP 5α: Since U is bounded, $H^1_0(U) \hookrightarrow\hookrightarrow L^2(U) \hookrightarrow H^{-1}(U)$. Hence, from (3.5), (3.6) and the Aubin–Lions–Simon lemma (see [2], Theorem II.5.16), there exist a subsequence of $\{\mathbf{u}_{m_l}\}_{l=1}^\infty$, which we still denote as such and a function $\mathbf{y} \in C\big(\overline{J}; L^2(U)\big)$, such that

$$\mathbf{u}_{m_l} \to \mathbf{y} \text{ in } C\big(\overline{J}; L^2(U)\big)\,. \tag{3.9}$$

From the convergence in (3.7), we deduce that $\mathbf{y} \equiv \mathbf{u}_J$.

STEP 5β: From (3.5), (3.9) and the Gagliardo–Nirenberg inequality (see, e.g., Theorem 1.3.7 in [4]) $|u|_{0,\alpha+2,U} \leq C |Du|^c_{0,2,U} |u|^c_{0,2,U}$, we have

$$\mathbf{u}_{m_l} \to \mathbf{u}_J \text{ in } C\big(\overline{J}; L^{\alpha+2}(U)\big)\,. \tag{3.10}$$

STEP 5γ: From (2.3), (3.5), the bound in (3.7), (3.9) and (3.10) we get

$$g\big(\mathbf{u}_{m_l}\big) \to g(\mathbf{u}_J) \text{ in } C\big(\overline{J}; Y_\alpha\big)\,. \tag{3.11}$$

STEP 6α: Let now $\psi \in C^\infty_c(J^\circ)$ and fix $N \in \mathbb{N}$. We choose m_l such that $N \leq m_l$ and $v \in \operatorname{span}\{w_k\}_{k=1}^N$, hence, by the linearity of the inner product, we get from (3.3) that

$$\int_J i\, \big(\mathbf{u}'_{m_l}, \psi v\big) + \mathcal{N}\big[\mathbf{u}_{m_l}, \psi v\big]\, dt = 0.$$

We then pass to the weak, $*$-weak and strong limits (since $\psi v \in L^1\big(J; H^1_0(U)\big)$) to get

$$\int_J i\, \langle \mathbf{u}'_J, \psi v\rangle + \mathcal{N}[\mathbf{u}_J, \psi v]\, dt = 0.$$

Since ψ is arbitrary, $\mathbf{u}_J$ satisfies the variational equation in (2.6) for every $v \in \operatorname{span}\{w_k\}_{k=1}^N$. By the linear and continuous dependence on v, we get the desired result, after letting $N \to \infty$.

STEP 6β: For the initial condition, we fix an arbitrary $t_0 \in J^*$. Let $v \in H^1_0(\mathrm{U})$ be arbitrary and $\phi \in C^1\big(\overline{J}\big)$ such that $\phi(0) \neq 0$ and $\phi(t_0) = 0$. We then have from [11], Lemma 1.1, Chap. 3, along with the Leibniz rule, that

$$\int_0^{t_0} \left(\mathbf{u}'_{m_l}, \phi v\right) dt = -\int_0^{t_0} \left(\mathbf{u}_{m_l}, \phi' v\right) dt - \left(u_{0m_l}, \phi(0) v\right),$$
$$\int_0^{t_0} \left\langle \mathbf{u}'_J, \phi v\right\rangle dt = -\int_0^{t_0} \left(\mathbf{u}_J, \phi' v\right) dt - \left(\mathbf{u}_J(0), \phi(0) v\right).$$

Passing to the $*$-weak limits in the first equality, using that $u_{0m} \to u_0$ in $L^2(U)$ and the fact that $v \in H_0^1(U)$ is arbitrary, we derive that $\mathbf{u}_J(0) = u_0$.

STEP 7α: Now, for (3.2), we first derive from (2.10), (3.9) and (3.10) that $G\left(\mathbf{u}_{m_l}\right) \to G(\mathbf{u}_J)$, everywhere in J. On the other hand, from the convergence in (3.7), (3.9), the fact that if $\mathbf{u}_{m_l} \rightharpoonup \mathbf{u}_J$ in $H^1(U)$ and $\mathbf{u}_{m_l} \rightharpoonup \mathbf{u}_J$ in $L^2(U)$ then $D\mathbf{u}_{m_l} \rightharpoonup \mathbf{u}_J$ in $L^2(U)$, as well as the weak lower semi-continuity of the L^2-norm we get $\left|D\mathbf{u}_{m_l}\right|_{0,2,U} \leq |D\mathbf{u}_J|_{0,2,U}$, everywhere in J. Combining these two results, we get (3.2) from (3.4).

STEP 7βFinally, if ζ is real-valued, then $\overline{F_m(\mathbf{z})} = F_m(\bar{\mathbf{z}})$, for all $\mathbf{z} \in \mathbb{C}^m$ and if u_0 is real-valued, then $\mathbf{d}_m(0) \in \mathbb{R}^m$. Hence, under these two assumptions, it easily follows that $\mathbf{u}_m(t) = \overline{\mathbf{u}_m(-t)}$, for all $t \in \mathbb{R}$. Now, the symmetry $\mathbf{u}_J(t) = \overline{\mathbf{u}_J(-t)}$, for all $t \in J$ with $|t| \leq \operatorname{dist}(0, \partial J)$, follows directly from the respective symmetry $\mathbf{u}_m(t) = \overline{\mathbf{u}_m(-t)}$ for all $t \in \mathbb{R}$, $m \in \mathbb{N}$ and the convergence in (3.7).

3.2 *Uniqueness and Globality*

It is obvious that the uniqueness of the extracted local solutions implies the "globality" of those solutions.

Proposition 7 *The solution* $\mathbf{u}_J$ *of* ***Theorem*** *1 is unique everywhere in* J.

Proof Let $u_{0,1} = u_{0,2}$ and $\mathbf{u}_{J,1}$, $\mathbf{u}_{J,2}$ be the corresponding solutions. Setting $\mathbf{w} := \mathbf{u}_{J,1} - \mathbf{u}_{J,2}$, we have

$$i\mathbf{w}' + \Delta\mathbf{w} - \left(g\left(\mathbf{u}_{J,1}\right) - g\left(\mathbf{u}_{J,2}\right)\right) \overset{H^{-1}(U)}{=} 0, \text{ a.e. in } J. \tag{3.12}$$

We apply the functional of (3.12) on $\mathbf{w}(t)$, for arbitrary $t \in J$ and take the imaginary parts of both parts to get us

$$|\mathbf{w}|_{0,2,U}^2 \leq C\left|\int_0^t \left|\left\langle g\left(\mathbf{u}_{J,1}\right) - g\left(\mathbf{u}_{J,2}\right), \mathbf{w}\right\rangle\right| ds\right|, \forall t \in J.$$

Since $H^1(U) \hookrightarrow L^\infty(U)$, from (2.13) we deduce that

$$|\mathbf{w}|_{0,2,U}^2 \leq C\left|\int_0^t |\mathbf{w}|_{0,2,U}^2 ds\right| \leq C|t|\,|\mathbf{w}|_{0,[-t,t];0,2,U}^2,$$

hence

$$|\mathbf{w}|^2_{0,[-t,t];0,2,U} \le C|t||\mathbf{w}|^2_{0,[-t,t];0,2,U}.$$

For $|t|$ sufficiently small we have $\mathbf{w} \equiv 0$. Now, we show that $\mathbf{w} \equiv 0$ in J_+ and in an analogous fashion we can have that $\mathbf{w} \equiv 0$ in J_-. Since $\mathbf{w} \in C(\overline{J}, L^2(U))$, we set

$$s := \sup\left\{t_0 \in J \;\middle|\; \mathbf{w} = 0 \text{ for all } t \in [0, t_0]\right\}.$$

If $s \neq \sup J$, then there exists $\delta > 0$ such that $[s, s+\delta] \subset J$. Then the continuity of $\mathbf{w}$ implies $\mathbf{w}(s) = 0$ and, by dealing as above, we deduce that $\mathbf{w} = 0$ for "a little further" than s, which is a contradiction to the definition of s.

3.3 Conservation of Energy and Well-Posedness

Here, we utilize the existence backwards in time as well as the uniqueness of the solution, in order to complete the puzzle of the well-posedness of the problem. First, we show the following result.

Proposition 8 *The energy of the unique solution* $\mathbf{u}_J$ *of* ***Theorem*** *1 is conserved, that is*

$$E(\mathbf{u}_J) = E(u_0)\,, \text{ everywhere in } J. \tag{3.13}$$

Proof We show that the energy is conserved in J_+ and in an analogous fashion we can get conservation of the energy in J_-. Let $t_0 \in J_+$. We set $\widehat{J} := [-t_0, 0]$ and we define as $\mathbf{v}_{\widehat{J}} \in L^\infty(\widehat{J};\, H_0^1(U)) \cap W^{1,\infty}(\widehat{J};\, H^{-1}(U))$ a solution of

$$\begin{cases} i\left\langle \mathbf{v}'_{\widehat{J}},\, v\right\rangle + \mathcal{N}\left[\mathbf{v}_{\widehat{J}},\, v\right] = 0, \;\; \forall v \in H_0^1(U)\,, \text{ a.e. in } \widehat{J} \\ \mathbf{v}_{\widehat{J}}(0) = \mathbf{u}_J(t_0)\,, \end{cases}$$

which **Theorem** 1 provides us. From the uniqueness of the solution we have that $\mathbf{v}_{\widehat{J}}(\widehat{t}) = \mathbf{u}_J(t)$, for all $\widehat{t} \in [-t_0, 0]$ and all $t \in [0, t_0]$. Moreover, from (3.2) we have that

$$E(\mathbf{u}_J(t_0)) \le E(u_0) \text{ and } E\left(\mathbf{v}_{\widehat{J}}(-t_0)\right) \le E(\mathbf{u}_J(t_0))$$

and applying $\mathbf{v}_{\widehat{J}}(t - t_0) = \mathbf{u}_J(t)$, for all $t \in [0, t_0]$, i.e. an equivalent formulation of the above equality, we obtain

$$E(\mathbf{u}_J(t_0)) \le E(u_0) \text{ and } E(u_0) \le E(\mathbf{u}_J(t_0))\,.$$

Since $t_0 \in J_+$ is arbitrary, we deduce (3.13) with J_+ instead of J.

Corollary 1 *The unique solution* $\mathbf{u}_J$ *of* **Theorem** *1 is a strong* H_0^1*-solution in* J*, i.e.* $\mathbf{u}_J \in C(\overline{J};\, H_0^1(U)) \cap C^1(\overline{J};\, H^{-1}(U))$*, and is also continuously dependent on the initial datum.*

Proof For the regularity, since $\mathbf{u}_J \in C(\overline{J};\, L^2(U))$, we deduce that $\mathbf{u}_J$ is weakly continuous from **Lemma** 1 and also that $|\mathbf{u}_J|_{0,2,U} \in C(\overline{J})$ by the triangle inequality. Moreover, from (2.10) we also deduce that $G(\mathbf{u}_J) \in C(\overline{J})$. Therefore, from (3.13) we get that $|\mathbf{u}_J|_{1,2,U} \in C(\overline{J})$ and thus, from **Lemma** 2, we obtain that $\mathbf{u}_J \in C(\overline{J};\, H_0^1(U))$ and also, by the variational equation, that $\mathbf{u}'_J \in C(\overline{J};\, H^{-1}(U))$.

As far as the continuous dependence is concerned, we fix an arbitrary $u_0 \in H_0^1(U)$. Let $\{u_{0,m}\}_{m=1}^{\infty} \subset H_0^1(U)$ be such that $u_{0,m} \to u_0$ in $H_0^1(U)$. We write as $\mathbf{u}_J$ and $\mathbf{u}_{J,m}$, for $m \in \mathbb{N}$, the unique corresponding solutions of the problem (2.6). We deduce that $\{\mathbf{u}_J\} \cup \{\mathbf{u}_{J,m}\}_{m=1}^{\infty} \subset C(\overline{J};\, H_0^1(U))$ from above. We fix an arbitrary $m_0 \in \mathbb{N}$ and then there exists a constant C_{m_0} such that

$$|u_{0,m}|_{1,2,U} \leq |u_0|_{1,2,U} + C_{m_0}, \quad \text{for all } m \in \mathbb{N} \text{ such that } m \geq m_0.$$

From (3.1), the above estimate, as well as the increasing property of K we have

$$|\mathbf{u}_{J,m}|_{0,J;1,2,U} + |\mathbf{u}'_{J,m}|_{0,J;-1,U} \leq \mathcal{K},$$

for all m as above. Hence, by dealing as in the proof of **Theorem** 1 from *STEP* 4α to *STEP* 6β, there exist a subsequence $\{\mathbf{u}_{J,m_l}\}_{l=1}^{\infty} \subseteq \{\mathbf{u}_{J,m}\}_{m=m_0}^{\infty}$ and a function $\mathbf{y} \in L^{\infty}(J;\, H_0^1(U)) \cap W^{1,\infty}(J;\, \mathrm{H}^{-1}(\mathrm{U}))$, such that $\mathbf{y}$ solves the problem (2.6) and also $\mathbf{u}_{J,m_l} \to \mathbf{y}$ in $C(\overline{J};\, L^2(U) \cap L^{\alpha+2}(U))$. In view of **Proposition** 7, we deduce that $\mathbf{y} \equiv \mathbf{u}_J$. Moreover, from (2.10), (3.1), the latter convergence, and (3.13), we obtain that $|\mathbf{u}_{J,m_l}|_{1,2,U} \to |\mathbf{u}_J|_{1,2,U}$ uniformly in $\overline{J}$. Hence, from Proposition 1.3.14 (iii) in [4] we get that $\mathbf{u}_{J,m_l} \to \mathbf{u}_J$ in $C(\overline{J};\, H_0^1(U))$. Since $\{u_{0,m}\}_{m=1}^{\infty}$ is arbitrary we deduce that for every $\{u_{0,m}\}_{m=1}^{\infty} \subset H_0^1(U)$ such that $u_{0,m} \to u_0$ in $H_0^1(U)$, there exists a subsequence $\{u_{0,m_l}\}_{l=1}^{\infty} \subset \{u_{0,m}\}_{m=1}^{\infty}$ such that $\mathbf{u}_{J,m_l} \to \mathbf{u}_J$ in $C(\overline{J};\, H_0^1(U))$. Hence, $\mathbf{u}_{J,m} \to \mathbf{u}_J$ in $C(\overline{J};\, H_0^1(U))$ and since $u_0 \in H_0^1(U)$ is arbitrary we conclude that the map $u_0 \mapsto \mathbf{u}_J$ is continuous.

3.4 Regularity

Here, we provide a regularity result, which is useful for the next section. We do not intend to exhaust the whole subject, thus, we only show weak H^2-regularity for a particular type of the extracted solutions.

Theorem 2 *Let* $\mathbf{u}_J$ *be the unique, energy conserving, continuously dependent on the initial datum, strong* H_0^1*-solution of (2.6) in* $U = \bigcup_{j=1}^{\infty} U_j$*, for pairwise disjoint* *open (and bounded) intervals U_j with $|U_j| \geq \delta > 0$, for every $j \in \mathbb{N}$. If $u_0 = H_0^1(U) \cap H^2(U)$, $\zeta \in H^4(U)$ and α is as in (1.3), then $\mathbf{u}_J \in L^\infty\big(J_+;\, H_0^1(U) \cap H^2(U)\big) \cap W^{1,\infty}\big(J_+;\, L^2(U)\big)$, with*

$$|\mathbf{u}_J|_{0,J_+;2,2,U} + |\mathbf{u}_J{}'|_{0,J_+;0,2,U} \leq \widehat{\mathcal{K}}, \tag{3.14}$$

where $\widehat{\mathcal{K}} := K_{J_+,U}\big(|u_0|_{2,2,U},\, |\zeta|_{4,2,U}\big)$.

Proof STEP 1: Let $\{\mathbf{u}_m\}_{m=1}^\infty$ be as in the proof of **Theorem** 1. We have that $|u_{0m}|_{2,2,U} \leq |u_0|_{2,2,U}$. Indeed, since $D^4 u_{0m} \in \operatorname{span}\{w_l\}_{l=1}^m$ we have $\big(D^4 u_{0m},\, u_{0m}\big) = \big(D^4 u_{0m},\, u_0\big)$, hence we get that

$$\begin{aligned}\big|D^2 u_{0m}\big|_{0,2,U}^2 &= \big(D^4 u_{0m},\, u_{0m}\big) = \big(D^2 u_{0m},\, D^2 u_0\big) \leq \\ &\leq \frac{1}{2}\big|D^2 u_{0m}\big|_{0,2,U}^2 + \frac{1}{2}\big|D^2 u_0\big|_{0,2,U}^2,\end{aligned}$$

therefore $\big|D^2 u_{0m}\big|_{0,2,U} \leq \big|D^2 u_0\big|_{0,2,U}$.

STEP 2: We multiply the variational equation in (3.3) by $\lambda_l^2 \overline{d_m^l}(t)$, where λ_l the lth eigenvalue of $-D^2$ on $H_0^1(U)$, sum for $l = 1, \ldots, m$ and take imaginary parts of both sides to find

$$\begin{aligned}&\frac{1}{2}\frac{d}{dt}\big|D^2\mathbf{u}_m\big|_{0,2,U}^2 - \operatorname{Im}\big(D^4\zeta,\, D^2\mathbf{u}_m\big) - \\ &- \operatorname{Im}\big(D^2\big[\big(|\mathbf{u}_m+\zeta|^{2\tau}+r\big)\,(\mathbf{u}_m+\zeta)\big],\, D^2\mathbf{u}_m\big) = 0.\end{aligned} \tag{3.15}$$

Note that

$$\begin{aligned}&\big(D^2\big(|\mathbf{u}_m+\zeta|^{2\tau}\,(\mathbf{u}_m+\zeta)\big),\, D^2\mathbf{u}_m\big) = \\ &= \left(\sum_{q_1+\cdots+q_{2\tau+1}=2}\prod_{l=1}^{\tau+1} D^{q_l}(\mathbf{u}_m+\zeta)\prod_{l=\tau+2}^{2\tau+1} D^{q_l}\overline{(\mathbf{u}_m+\zeta)},\, D^2\mathbf{u}_m\right),\end{aligned}$$

hence we deduce, by application of (2.1) for $p_1 = p_2 = 2$ that

$$\begin{aligned}&\big|\big(D^2\big(|\mathbf{u}_m+\zeta|^{2\tau}\,(\mathbf{u}_m+\zeta)\big),\, D^2\mathbf{u}_m\big)\big| \leq \\ &\leq \left|\sum_{q_1+\cdots+q_{2\tau+1}=2}\prod_{l=1}^{\tau+1} D^{q_l}(\mathbf{u}_m+\zeta)\prod_{l=\tau+2}^{2\tau+1} D^{q_l}\overline{(\mathbf{u}_m+\zeta)}\right|_{0,2,U} \times \\ &\times \big|D^2\mathbf{u}_m\big|_{0,2,U}\end{aligned}$$

and again for $p_l = \frac{2}{q_l}$, where $l = 1, \ldots, 2\tau+1$, to get

$$\left|\left(D^2\left(|\mathbf{u}_m+\zeta|^{2\tau}(\mathbf{u}_m+\zeta)\right), D^2\mathbf{u}_m\right)\right| \le$$
$$\le \sum_{q_1+\cdots+q_{2\tau+1}=2} \prod_{l=1}^{2\tau+1} \left|D^{q_l}(\mathbf{u}_m+\zeta)\right|_{0,\frac{4}{q_l},U} \left|D^2(\mathbf{u}_m+\zeta)\right|_{0,2,U}.$$

Moreover, from the Gagliardo-Nirenberg inequality, in view of Theorem 8.6 in [3],[3] we have

$$\left|D^j u\right|_{0,\frac{4}{j},I} \le K\left(\frac{1}{|I|}\right)\left|D^2 u\right|_{0,2,I}^{\frac{j}{2}} |u|_{0,\infty,I}^{\frac{2-j}{2}}, \text{ for } j=0,\,1,\,2,$$

I being an interval (bounded or not) and $u \in H^2(I)$. Hence

$$\left|D^j u\right|_{0,\frac{4}{j},U} \le C(\delta)\left|D^2 u\right|_{0,2,U}^{\frac{j}{2}} |u|_{0,\infty,U}^{\frac{2-j}{2}}, \text{ for } j=0,\,1,\,2.$$

From the above inequality, the embedding $H^1(U) \hookrightarrow L^\infty(U)$ and (3.1), we then have

$$\begin{aligned} &\left|\left(D^2\left(|\mathbf{u}_m+\zeta|^{2\tau}(\mathbf{u}_m+\zeta)\right), D^2\mathbf{u}_m\right)\right| \le \\ &\le C\left|(\mathbf{u}_m+\zeta)\right|_{0,\infty,U}^{2\tau}\left|D^2(\mathbf{u}_m+\zeta)\right|_{0,2,U}^2 \le \\ K_U&\left(|u_0|_{1,2,U}, |\zeta|_{2,2,U}\right)\left(1+\left|D^2\mathbf{u}_m\right|_{0,2,U}^2\right). \end{aligned} \tag{3.16}$$

Combining (3.15) and (3.16), we derive

$$\left|D^2\mathbf{u}_m\right|_{0,2,U}^2 \le K_{J_+,U}\left(|u_0|_{2,2,U}, |\zeta|_{4,2,U}\right),$$

everywhere in J_+, from which, along with the estimates of **Theorem** 1, we obtain that $\{\mathbf{u}_m\}_{m=1}^\infty$ is bounded in $C\left(\overline{J_+};\, H^2(U)\right)$, with

$$|\mathbf{u}_m|_{0,J_+;2,2,U} \le \widehat{\mathcal{K}}, \ \forall m \in \mathbb{N}. \tag{3.17}$$

STEP 3: We set $m = m_l$. Obviously, $C\left(\overline{J_+};\, H^2(U)\right) \hookrightarrow L^2\left(J_+;\, H^2(U)\right)$. Therefore, applying Theorem 3, Sect. D.4 in [6] and **Proposition** 6 3., we get from (3.17) that there exist a subsequence of $\left\{\mathbf{u}_{m_l}\right\}_{l=1}^\infty$, which we still denote as such and a function $\mathbf{v} \in L^\infty\left(J_+;\, H^2(U)\right)$, such that

---

[3]We can modify the reflection technique used for the proof of this result, in order to cover the case of the extension of H^2-functions. In particular, we can apply the reflection technique used for Theorem 5.19 in [1].

$$\mathbf{u}_{m_l} \rightharpoonup \mathbf{v} \text{ in } L^2\big(J_+;\, H^2(U)\big) \text{ and also } |\mathbf{v}|_{0,J_+;2,2,U} \leq \widehat{\mathcal{K}}. \tag{3.18}$$

From (3.7), we can easily derive that $\mathbf{v} \equiv \mathbf{u}_J$.

STEP 4: From the embedding $H^1(U) \hookrightarrow L^\infty(U)$ and (2.14), we get that $g(\mathbf{u}_J(\cdot))$ maps to $L^2(U)$. Hence, from (3.18) and the variational equation in $H_0^1(U)$, we deduce that $\mathbf{u}_J$ solves the equation in $L^2(U)$ and

$$\left|\mathbf{u}'_J\right|_{0,J_+;0,2,U} \leq \widehat{\mathcal{K}}. \tag{3.19}$$

3.5 A Special Case of Solutions

We notice that problem (2.16) allows us to consider ζ which do not vanish at infinity, if U is unbounded, hence the $\cdot_s$-formulation is crucial for those sets. Before we proceed with the study of the unbounded case, we provide the next result.

Theorem 3 *Let $\mathbf{u}_J$ be the unique, energy conserving, continuously dependent on the initial datum, strong H_0^1-solution of (2.16). Then*

$$|\mathbf{u}_J|_{0,J_+;1,2,U} + \left|\mathbf{u}'_J\right|_{0,J_+;-1,U} \leq \mathscr{K}, \tag{3.20}$$

where $\mathscr{K} := K_{J_+}\left(|u_0|_{1,2,U},\, \|\zeta\|_{X^1(U)},\, \left||\zeta|^2 - \rho\right|_{0,2,U}\right)$.

Moreover, if $U = \bigcup_{j=1}^{\infty} U_j$, for pairwise disjoint open (and bounded) intervals U_j with $\left|U_j\right| \geq \delta > 0$, for every $j \in \mathbb{N}$, $u_0 \in H_0^1(U) \cap H^2(U)$ and $\zeta \in X^4(U)$, then

$$|\mathbf{u}_J|_{0,J_+;2,2,U} + \left|\mathbf{u}'_J\right|_{0,J_+;0,2,U} \leq \widehat{\mathscr{K}}, \tag{3.21}$$

where $\widehat{\mathscr{K}} := K_{J_+}\left(|u_0|_{2,2,U},\, \|\zeta\|_{X^4(U)},\, \left||\zeta|^2 - \rho\right|_{0,2,U}\right)$.

Proof Let $\{\mathbf{u}_m\}_{m=1}^{\infty}$ be as in the proof of **Theorem** 1. From (3.4) and (2.19) we get

$$|D\mathbf{u}_m|_{0,2,U} \leq K\left(|u_0|_{1,2,U},\, \|\zeta\|_{X^1(U)},\, \left||\zeta|^2 - \rho\right|_{0,2,U}\right), \quad \forall m \in \mathbb{N}. \tag{3.22}$$

Then, we multiply the variational equation in (3.3) by $\overline{d_m^k}(t)$, sum for $k = 1, \ldots, m$ and take imaginary parts of both sides, and thus obtain

$$\begin{aligned} &\frac{1}{2}\frac{d}{dt}|\mathbf{u}_m|_{0,2,U}^2 - \operatorname{Im}(D\zeta,\, D\mathbf{u}_m) - \\ &- \operatorname{Im}\left(\left(|\mathbf{u}_m + \zeta|^{2\tau} - \rho^\tau\right)(\mathbf{u}_m + \zeta)\,,\, \mathbf{u}_m\right) = 0. \end{aligned} \tag{3.23}$$

Applying (3.22) and expanding in view of (2.15), we deduce

$$\frac{d}{dt}|\mathbf{u}_m|_{0,2,U}^2 - K\left(|u_0|_{1,2,U},\, \|\zeta\|_{X^1(U)},\, \left||\zeta|^2-\rho\right|_{0,2,U}\right)\times \\ \times\left(1+|\mathbf{u}_m|_{0,2\tau+1,U}^{2\tau+1}\right)\leq 0. \tag{3.24}$$

In order to estimate the non constant term inside the parenthesis, we imitate the technique which has already been developed for the proof of Lemma 3.3 in [8]. We set $B=\left(A_{2\tau}+|\zeta|_{0,\infty,U}+1\right)^2$, where $A_{2\tau}$ is as in (2.9), $Q:=\left\{x\in U \mid |\mathbf{u}_m+\zeta|\leq\sqrt{B}\right\}$ and $R:=Q^c\cap U$. Then

$$\begin{aligned}|\mathbf{u}_m|_{0,2\tau+1,U}^{2\tau+1} &= \int_Q |\mathbf{u}_m|^2|\mathbf{u}_m|^{2\tau-1}dx+\int_R |\mathbf{u}_m|^{2\tau+1}dx\leq \\ &\leq \int_{\left\{x\in U \mid |\mathbf{u}_m|\leq\sqrt{B}+|\zeta|_{0,\infty,U}\right\}}|\mathbf{u}_m|^2|\mathbf{u}_m|^{2\tau-1}dx+ \\ &\quad +C\int_R|\zeta|^{2\tau+1}+|\mathbf{u}_m+\zeta|^{2\tau+1}dx\leq \\ &\overset{(2.9)}{\leq}\left(\sqrt{B}+|\zeta|_{0,\infty,U}\right)^c|\mathbf{u}_m|_{0,2,U}^2+C|\zeta|_{0,\infty,U}^c\int_R dx+CG_s(\mathbf{u}_m)\leq \\ &\leq\left(\sqrt{B}+|\zeta|_{0,\infty,U}\right)^c|\mathbf{u}_m|_{0,2,U}^2+ \\ &\quad +\frac{C|\zeta|_{0,\infty,U}^c}{\left(\sqrt{B}-|\zeta|_{0,\infty,U}\right)^c}|\mathbf{u}_m|_{0,2,U}^2+CG_s(\mathbf{u}_m)\leq \\ &\overset{(2.19)}{\leq} K\left(|u_0|_{1,2,U},\, \|\zeta\|_{X^1(U)},\, \left||\zeta|^2-\rho\right|_{0,2,U}\right)\left(1+|\mathbf{u}_m|_{0,2,U}^2\right).\end{aligned} \tag{3.25}$$

From (3.24) and (3.25), we derive that

$$|\mathbf{u}_m|_{0,2,U}\leq\mathscr{K} \text{ in } J_+,\ \forall m\in\mathbb{N}. \tag{3.26}$$

From (3.22) and (3.26) we conclude that $\{\mathbf{u}_m\}_{m=1}^{\infty}$ is uniformly bounded in $C\left(\overline{J_+};\, H_0^1(U)\right)$, with

$$|\mathbf{u}_m|_{0,J_+;1,2,U}\leq\mathscr{K},\ \forall m\in\mathbb{N}. \tag{3.27}$$

In addition, we make use of (2.17) to get that $\left\{\mathbf{u}_m{}'\right\}_{m=1}^{\infty}$ is uniformly bounded in $C\left(\overline{J_+};\, H^{-1}(U)\right)$, with

$$\left|\mathbf{u}_m'\right|_{0,J_+;-1,U}\leq\mathscr{K},\ \forall m\in\mathbb{N}. \tag{3.28}$$

We then deal as in **Theorem** 1 in order to obtain (3.20).

As far as the estimate (3.21) is concerned, we deal exactly as in **Theorem** 2, employing (3.20) instead of (3.1), as well as the scaling invariant embedding $H_0^1(U)\hookrightarrow L^\infty(U)$ instead of the scaling dependent $H^1(U)\hookrightarrow L^\infty(U)$.

4 Solutions in Unbounded Sets

Here, we assume $U \subseteq \mathbb{R}$ to be unbounded.

Theorem 4 *Let $u_0 \in H_0^1(U)$. Then there exists a unique and global solution $\mathbf{u} \in L^\infty_{loc}\big(0, \infty;\, H_0^1(U)\big) \cap W^{1,\infty}_{loc}\big(0, \infty;\, H^{-1}(U)\big)$ of (2.16) in the positive time ray, such that*

$$\left|\mathbf{u}_{J_+}\right|_{0,J_+;1,2,U} + \left|\mathbf{u}'_{J_+}\right|_{0,J_+;-1,U} \leq \mathscr{K}, \ \forall J_+, \tag{4.1}$$

where $\mathbf{u}_{J_+} := \mathcal{R}_{J_+}\mathbf{u}$ and also

$$E(\mathbf{u}) \leq E(u_0), \ \text{everywhere in } \mathbb{R}_+. \tag{4.2}$$

Moreover, if $U = \bigcup_{j=1}^\infty U_j$, for pairwise disjoint open (bounded or not) intervals U_j with $\left|U_j\right| \geq \delta > 0$, for every $j \in \mathbb{N}$., $u_0 \in H_0^1(U) \cap H^2(U)$ and $\zeta \in X^4(U)$, then $\mathbf{u} \in L^\infty_{loc}\big(0, \infty;\, H_0^1(U) \cap H^2(U)\big) \cap W^{1,\infty}_{loc}\big(0, \infty;\, L^2(U)\big)$, with

$$\left|\mathbf{u}_{J_+}\right|_{0,J_+;2,2,U} + \left|\mathbf{u}'_{J_+}\right|_{0,J_+;0,2,U} \leq \widehat{\mathscr{K}}, \ \forall J_+. \tag{4.3}$$

and also

$$E(\mathbf{u}) = E(u_0), \ \text{everywhere in } \mathbb{R}_+. \tag{4.4}$$

Proof We only show local existence in $H_0^1(U)$. H^2-regularity follows analogously. Equation (4.4) is a direct result of the fact that the regular solution $\mathbf{u}$ satisfies the equation in $L^2(U)$. We get uniqueness and globality exactly as in **Proposition** 7.

STEP 1: Since U is open, we fix an arbitrary $B_\varrho(x_0) \subset U$. Let $u_{0,k} := \mathcal{R}_U \eta_k u_0$, for all $k \in \mathbb{N}$, where $\{\eta_k\}_{k=1}^\infty$ is defined as follows: let $f \in C^\infty(\mathbb{R})$ with

$$f(t) := \begin{cases} e^{-\frac{1}{t}}, & t > 0 \\ 0, & t \leq 0, \end{cases}$$

and $\{a_k\}_{k=1}^\infty \subset \mathbb{R}_+$ increasing, such that $a_k > \varrho$ for all $k \in \mathbb{N}$ and $a_k \nearrow \infty$. We define $\{\eta_k\}_{k=1}^\infty \subset C_c^\infty(\mathbb{R})$ by

$$\eta_k(x;\, x_0, a_{k-1}, a_k) := \frac{f\,(a_k - |x - x_0|)}{f\,(|x - x_0| - a_{k-1}) + f\,(a_k - |x - x_0|)}, \ \forall\, x \in \mathbb{R}, k \in \mathbb{N} \setminus \{1\}$$

and

$$\eta_1\big(x;\, B_\varrho(x_0)\,, a_1\big) := \frac{f(a_1 - |x - x_0|)}{f(|x - x_0| - \varrho) + f(a_1 - |x - x_0|)}, \ \forall x \in \mathbb{R}.$$

It is trivial to show that

$$\eta_k(x)=\begin{cases}1, & x\in\overline{B_{a_{k-1}}(x_0)}\\ 0, & x\in B_{a_k}(x_0)^{\mathsf{C}},\end{cases} \quad \forall k\in\mathbb{N}\setminus\{1\} \text{ and } \eta_1(x)=\begin{cases}1, & x\in\overline{B_{\varrho}(x_0)}\\ 0, & x\in B_{a_1}(x_0)^{\mathsf{C}}.\end{cases}$$

If, in addition, $a_{k+1}-a_k=a_1-\varrho=C$ uniformly for all $k\in\mathbb{N}$ (i.e. C is independent of k), then $\left|D^{\beta}\eta_k\right|_{0,\infty}\le C_m$, for some $\{C_m\}_{m=0}^{\infty}\subset\mathbb{R}_+$, uniformly for all $k\in\mathbb{N}$ and every multi-index β such that $|\beta|=m$. In particular, $C_0=1$.

Therefore, for all $k\in\mathbb{N}$, we have that

$$\left|u_{0,k}\right|_{1,2,U}\le C|u_0|_{1,2,U}\,. \tag{4.5}$$

We also notice that $u_{0,k}=0$, in $B_{a_k}(x_0)^{\mathsf{C}}\cap U$, hence, by setting $B_k:=B_{a_k}(x_0)\cap U$, for every $k\in\mathbb{N}$, we obtain that $\left\{\mathcal{R}_{B_k}u_{0,k}\right\}_{k=1}^{\infty}\subset H_0^1(B_k)$. Moreover,

$$u_{0,k}\to u_0 \text{ in } L^2(U)\,. \tag{4.6}$$

Indeed,

$$\left|u_{0,k}-u_0\right|_{0,2,U}=|(\eta_k-1)u_0|_{0,2,U}\le|u_0|_{0,2,B_{a_{k-1}}(x_0)^{\mathsf{C}}\cap U}\to 0.$$

STEP 2α: Let J_+ be arbitrary. Fixing any $k\in\mathbb{N}$, we consider (2.16) in $U=B_k$, where we take $\mathcal{R}_{B_k}u_{0,k}$ as our initial datum and we set $\mathbf{u}^k\in L^{\infty}\left(J_+;\,H_0^1(B_k)\right)\cap W^{1,\infty}\left(J_+;\,H^{-1}(B_k)\right)$ to be the solution that **Theorem** 3 provides. From its proof, it follows that there exist $\left\{\mathbf{u}_m^k\right\}_{m=1}^{\infty}\subset C^{\infty}\left(\overline{J_+};\,H_0^1(U)\cap C^{\infty}\left(\overline{U}\right)\right)$, such that

$$\begin{aligned}\left|\mathbf{u}_m^k\right|_{0,J_+;1,2,B_k}+\left|\mathbf{u}_m^{k\,\prime}\right|_{0,J_+;-1,B_k}&\le\\ \le K_{J_+}\left(\left|u_{0,k}\right|_{1,2,B_k},\,\|\zeta\|_{X^1(B_k)},\,\left||\zeta|^2-\rho\right|_{0,2,B_k}\right),&\quad\forall m\in\mathbb{N}\end{aligned} \tag{4.7}$$

and

$$\begin{aligned}\mathbf{u}_m^k(t)\rightharpoonup\mathbf{u}^k(t) \text{ in } H_0^1(B_k)\,,&\text{ for every } t\in\overline{J_+},\\ \mathbf{u}_m^{k\,\prime}\overset{*}{\rightharpoonup}\mathbf{u}^{k\prime} \text{ in } L^{\infty}\left(J_+;\,H^{-1}(B_k)\right).&\end{aligned} \tag{4.8}$$

From (4.5), (4.7) and the increasing property of K we deduce that

$$\left|\mathbf{u}_m^k\right|_{0,J_+;1,2,B_k}+\left|\mathbf{u}_m^{k\,\prime}\right|_{0,J_+;-1,B_k}\le\mathscr{K},\ \forall m\in\mathbb{N}. \tag{4.9}$$

STEP 2β: Since $\mathcal{R}_{\partial B_k\setminus\partial U}\mathbf{u}_m^k=0$, the extensions by zero $\mathbf{v}_m^k:=\mathcal{E}_U\mathbf{u}_m^k$, for all $m\in\mathbb{N}$,[4] are continuous in $\partial B_k\setminus\partial U$ and thus $\left\{\mathbf{v}_m^k\right\}_{m=1}^{\infty}\subset C^{\infty}\left(\overline{J_+};\,H_0^1(U)\right)$. Evidently,

[4]For the H^2-regularity, we define $\mathbf{v}_m^k:=\eta_k\mathcal{E}_U\mathbf{u}_m^k$, for all $m\in\mathbb{N}$.

$$\left|\mathbf{v}_m^k\right|_{0,J_+;1,2,U}=\left|\mathbf{u}_m^k\right|_{0,J_+;1,2,B_k}$$
$$\left|\mathbf{v}_m^{k\,\prime}\right|_{0,J_+;-1,U}=\left|\mathbf{u}_m^{k\,\prime}\right|_{0,J_+;-1,B_k},$$

hence, from (4.9), we get that

$$\left|\mathbf{v}_m^k\right|_{0,J_+;1,2,U}+\left|\mathbf{v}_m^{k\,\prime}\right|_{0,J_+;-1,U}\le\mathcal{K},\ \forall m\in\mathbb{N}.$$

STEP 2γ: Dealing as in *STEP* 4 of the proof of **Theorem** 3, there exist a subsequence $\left\{\mathbf{v}_{m_l}^k\right\}_{l=1}^\infty\subseteq\left\{\mathbf{v}_m^k\right\}_{m=1}^\infty$ and a function $\mathbf{v}^k\in L^\infty\left(J_+;\,H_0^1(U)\right)\cap W^{1,\infty}\left(J_+;\,H^{-1}(U)\right)$, such that

$$\begin{aligned}\mathbf{v}_{m_l}^k(t)\rightharpoonup\mathbf{v}^k(t)\ \text{ in } H_0^1(U)\,,\ \text{for every } t\in\overline{J_+},\\ \mathbf{v}_{m_l}^{k\,\prime}\overset{*}{\rightharpoonup}\mathbf{v}^{k\prime}\ \text{in } L^\infty\left(J_+;\,H^{-1}(U)\right),\\ \left|\mathbf{v}^k\right|_{0,J_+;1,2,U}+\left|\mathbf{v}^{k\prime}\right|_{0,J_+;-1,U}\le\widetilde{\mathcal{K}}.\end{aligned}\tag{4.10}$$

Since $k\in\mathbb{N}$ is arbitrary, $\left\{\mathbf{v}^k\right\}_{k=1}^\infty\subset L^\infty\left(J_+;\,H_0^1(U)\right)\cap W^{1,\infty}\left(J_+;\,H^{-1}(U)\right)$ and the above estimate is satisfied for each $k\in\mathbb{N}$.

STEP 3α: Dealing again as before, there exist a subsequence $\left\{\mathbf{v}^{k_l}\right\}_{l=1}^\infty\subseteq\left\{\mathbf{v}^k\right\}_{k=1}^\infty$ and a function $\mathbf{u}_{J_+}\in L^\infty\left(J_+;\,H_0^1(U)\right)\cap W^{1,\infty}\left(J_+;\,H^{-1}(U)\right)$, such that

$$\begin{aligned}\mathbf{v}^{k_l}(t)\rightharpoonup\mathbf{u}_{J_+}(t)\ \text{ in } H_0^1(U)\,,\ \text{for every } t\in\overline{J_+},\\ \mathbf{v}^{k_l\prime}\overset{*}{\rightharpoonup}\mathbf{u}_{J_+}'\ \text{ in } L^\infty\left(J_+;\,H^{-1}(U)\right),\\ \left|\mathbf{u}_{J_+}\right|_{J_+;1,2,U}+\left|\mathbf{u}_{J_+}'\right|_{J_+;-1,U}\le\widetilde{\mathcal{K}}.\end{aligned}\tag{4.11}$$

STEP 3β: From (2.13), the estimate in (4.10) and Lemma 3.3.6 in [4] we deduce that $\left\{g_s\left(\mathbf{v}^{k_l}\right)\right\}_{l=1}^\infty$ is bounded in $C^{0,\frac{1}{2}}\left(\overline{J_+};\,L^2(U)\right)$. Hence, from Proposition 1.1.2 in [4], there exist a subsequence of $\left\{\mathbf{v}^{k_l}\right\}_{l=1}^\infty$, which we still denote as such, and a function $\mathbf{f}\in C\left(\overline{J_+};\,L^2(U)\right)$, such that

$$g_s\left(\mathbf{v}^{k_l}(t)\right)\rightharpoonup\mathbf{f}(t)\ \text{ in } L^2(U)\,,\ \forall t\in\overline{J_+}.\tag{4.12}$$

STEP 4α: Let Ω be any bounded open interval $\subset U$. For $k\in\mathbb{N}$ big enough so that $\Omega\subseteq B_k$, we have

$$\begin{aligned}\left(\mathbf{v}^k,\,\mathcal{E}_U v\right)=\left(\mathbf{u}^k,\,\mathcal{E}_{B_k}v\right),\ \ \left(g_s\left(\mathbf{v}^k\right),\,\mathcal{E}_U v\right)=\left(g_s\left(\mathbf{u}^k\right),\,\mathcal{E}_{B_k}v\right)\\ \text{and } \left\langle\mathbf{v}^{k\prime},\,\mathcal{E}_U v\right\rangle=\left\langle\mathbf{u}^{k\prime},\,\mathcal{E}_{B_k}v\right\rangle,\end{aligned}\tag{4.13}$$

for every $v\in C_c^\infty(\Omega)$. Indeed, for the first equality, we get from (4.10)

$$\int_U \mathbf{v}_{m_l}^k \mathcal{E}_U \bar{v} dx \to \int_U \mathbf{v}^k \mathcal{E}_U \bar{v} dx$$

and from (4.8)

$$\int_U \mathbf{v}_{m_l}^k \mathcal{E}_U \bar{v} dx = \int_{B_k} \mathcal{R}_{B_k} \mathbf{v}_{m_l}^k \mathcal{E}_{B_k} \bar{v} dx \to \int_{B_k} \mathbf{u}^k \mathcal{E}_{B_k} \bar{v} dx.$$

The second equality follows similarly. The third equality follows from the first one and Lemma 1.1, Chap. 3, in [11]. Now, since $\mathbf{u}^k$ is a solution in B_k,

$$i\left\langle \mathbf{u}^{k\prime}, \mathcal{E}_{B_k} v\right\rangle + \mathcal{N}_s\left[\mathbf{u}^k, \mathcal{E}_{B_k} v\right] = 0, \ \forall v \in C_c^\infty(\Omega), \text{ a.e. in } J_+,$$

hence, from (4.13),

$$i\left\langle \mathbf{v}^{k\prime}, \mathcal{E}_U v\right\rangle + \mathcal{N}_s\left[\mathbf{v}^k, \mathcal{E}_U v\right] = 0, \ \forall v \in C_c^\infty(\Omega), \text{ a.e. in } J_+. \tag{4.14}$$

STEP 4β: From the first convergence in (4.11), the weak lower semi-continuity of the H^1-norm and the compact embedding $H^1(\Omega) \hookrightarrow\hookrightarrow L^2(\Omega)$, we obtain that there exist a subsequence of $\{\mathbf{v}^{k_l}\}_{l=1}^\infty$, which we still denote as such, for which we have

$$\mathbf{v}^{k_l}(t) \to \mathbf{u}_{J_+}(t) \text{ in } L^2(\Omega), \ \forall t \in \overline{J_+}. \tag{4.15}$$

We set $k = k_l$ in (4.14) and we pass to the limit $l \to \infty$. From (4.11), (4.12) and (4.15), we deduce that

$$\int_{J_+} \left(i\left\langle \mathbf{u}_{J_+}', \mathcal{E}_U v\right\rangle + \left\langle \Delta \mathbf{u}_{J_+}, \mathcal{E}_U v\right\rangle + \langle \mathbf{f}, \mathcal{E}_U v\rangle\right) \bar{\psi} dt = 0,$$

for every $v \in C_c^\infty(\Omega)$ and $\psi \in C_c^\infty(J_+)$, hence

$$i\left\langle \mathbf{u}_{J_+}', \mathcal{E}_U v\right\rangle + \left\langle \Delta \mathbf{u}_{J_+}, \mathcal{E}_U v\right\rangle + \langle \mathbf{f}, \mathcal{E}_U v\rangle = 0, \ \forall v \in C_c^\infty(\Omega), \text{ a.e. in } J_+. \tag{4.16}$$

STEP 4γ: From (4.12) and **Lemma** 3 we have

$$g_s\left(\mathcal{R}_\Omega \mathbf{v}^{k_l}(t)\right) = \mathcal{R}_\Omega g_s\left(\mathbf{v}^{k_l}(t)\right) \rightharpoonup \mathcal{R}_\Omega \mathbf{f}(t) \text{ in } L^2(U), \ \forall t \in \overline{J_+}. \tag{4.17}$$

On the other hand, from (4.15) and **Lemma** 3,

$$\mathcal{R}_\Omega \mathbf{v}^{k_l}(t) \to \mathcal{R}_\Omega \mathbf{u}_{J_+}(t) \text{ in } L^2(\Omega), \ \forall t \in \overline{J_+}.$$

From (2.13) we get

$$g_s\left(\mathcal{R}_\Omega \mathbf{v}^{k_l}(t)\right) \to g_s\left(\mathcal{R}_\Omega \mathbf{u}_{J_+}(t)\right) = \mathcal{R}_\Omega g_s\left(\mathbf{u}_{J_+}(t)\right) \text{ in } L^2(U), \ \forall t \in \overline{J_+}. \quad (4.18)$$

From (4.17) and (4.18) we derive $\mathcal{R}_\Omega g_s(\mathbf{u}) \equiv \mathcal{R}_\Omega \mathbf{f}$ and so (4.16) gets the form

$$i\left\langle \mathbf{u}'_{J_+}, \mathcal{E}_U v\right\rangle + \mathcal{N}_s\left[\mathbf{u}_{J_+}, \mathcal{E}_U v\right] = 0, \ \forall v \in C_c^\infty(\Omega), \text{ a.e. in } J_+.$$

Since Ω is arbitrary, $\mathbf{u}_{J_+}$ satisfies the variational equation in (2.16).

STEP 5: As far as the initial condition is concerned, let t_0, v, ϕ be as in *STEP* 6β of the proof of **Theorem** 3. Then

$$\begin{aligned} \int_0^{t_0} \left(\mathbf{v}_m^{k\,\prime}, \phi v\right) dt &= -\int_0^{t_0} \left(\mathbf{v}_m^k, \phi' v\right) dt - \left(\mathbf{v}_m^k(0), \phi(0)\, v\right), \\ \int_0^{t_0} \left\langle \mathbf{u}'_{J_+}, \phi v\right\rangle dt &= -\int_0^{t_0} \left(\mathbf{u}_{J_+}, \phi' v\right) dt - \left(\mathbf{u}_{J_+}(0), \phi(0)\, v\right). \end{aligned} \quad (4.19)$$

Moreover, $\left(\mathbf{v}_m^k(0), \phi(0)\, v\right) = \left(\mathbf{u}_m^k(0), \phi(0)\, \mathcal{R}_{B_k} v\right)$, hence, by setting $m = m_l$ and letting $l \to 0$, we get

$$\int_0^{t_0} \left\langle \mathbf{v}^{k\prime}, \phi v\right\rangle dt = -\int_0^{t_0} \left(\mathbf{v}^k, \phi' v\right) dt - \left(\mathcal{R}_{B_k} u_{0k}, \phi(0)\, \mathcal{R}_{B_k} v\right).$$

Since $\left(\mathcal{R}_{B_k} u_{0k}, \phi(0)\, \mathcal{R}_{B_k} v\right) = \left(u_{0k}, \phi(0)\, v\right)$, we set $k = k_l$ and we pass to the limit as $l \to \infty$, applying (4.6), to get

$$\int_0^{t_0} \left\langle \mathbf{u}'_{J_+}, \phi v\right\rangle dt = -\int_0^{t_0} \left(\mathbf{u}_{J_+}, \phi' v\right) dt - \left(u_0, \phi(0)\, v\right). \quad (4.20)$$

From the second equation in (4.19) and (4.20), we conclude to $\mathbf{u}_{J_+}(0) = u_0$.

In fact, for $U = \mathbb{R}$, we need a weaker assumption on ζ, in order for the H^2-regularity result of **Theorem** 4 to hold. Indeed, in view of **Theorem** 4, (2.13) and (2.14), the following is a direct application of Theorem 5.3.1 and Remark 5.3.2 in [4].

Theorem 5 *If* $v_0 \in H^2(\mathbb{R})$ *and* $\zeta \in X^2(\mathbb{R})$, *then there exists a unique and global solution* $\mathbf{u} \in L^\infty_{loc}\left(0, \infty; H_0^1(U) \cap H^2(U)\right) \cap W^{1,\infty}_{loc}\left(0, \infty; L^2(U)\right)$ *of (2.16), with* $E(\mathbf{u}) = E(u_0)$ *everywhere in* $\mathbb{R}_+$.

Proposition 9 *Let* $U = \bigcup_{j=1}^\infty U_j$, *for pairwise disjoint open (bounded or not) intervals* U_j *with* $\left|U_j\right| \geq \delta > 0$, *for every* $j \in \mathbb{N}$, $u_0 \in H_0^1(U)$,

$$\zeta \in \begin{cases} X^2(U), & \text{if } U = \mathbb{R} \\ X^4(U), & \text{otherwise} \end{cases}$$

and $\mathbf{u}$ *the corresponding weak* H_0^1*-solution of (2.16). Then the energy of* $\mathbf{u}$ *is conserved,* $\mathbf{u}$ *is a strong* H_0^1*-solution, continuously dependent on the initial datum.*

Proof It suffices to show that the energy is conserved. Let J_+ be arbitrary and $\{u_{0,m}\}_{m=1}^{\infty} \subset C_c^{\infty}(U)$ be such that $u_{0,m} \to u_0$ in $H_0^1(U)$. We write as $\mathbf{u}_{J_+}$ and $\mathbf{u}_{J_+,m}$, for $m \in \mathbb{N}$, the unique corresponding solutions of the problem (2.16). In view of **Theorems** 4 and **5**, we have that $\mathbf{u}_{J_+,m}$ satisfies the differential equation in $L^2(U)$, for all $m \in \mathbb{N}$, hence, we can easily derive that the energy of every $\mathbf{u}_{J_+,m}$ is conserved, i.e.

$$E\left(\mathbf{u}_{J_+,m}\right) = E\left(u_{0,m}\right), \ \forall m \in \mathbb{N}, \ \text{everywhere in } J_+. \tag{4.21}$$

Moreover, we have that $\mathbf{u}_{J_+,m} \to \mathbf{u}_{J_+}$ in $C\left(\overline{J_+};\, L^2(U)\right)$. Indeed, dealing as in the proof of **Proposition** 7, we have that

$$|\mathbf{w}_m|_{0,2,U}^2 \leq C \int_0^t |\mathbf{w}_m|_{0,2,U}^2 ds, \ \forall t \in J_+,$$

where $\mathbf{w}_m := \mathbf{u}_{J_+,m} - \mathbf{u}_{J_+}$, for all $m \in \mathbb{N}$, therefore, the convergence follows from the Grönwall inequality and the fact that $u_{0,m} \to u_0$ in $L^2(U)$. Now, we deal as in the proof of **Theorem** 1 from *STEP* 4α to *STEP* 6β, minding to exclude *STEP* 5 and apply the above extracted convergence as well as (2.13), instead. Hence, there exist a subsequence $\{\mathbf{u}_{J_+,m_l}\}_{l=1}^{\infty} \subseteq \{\mathbf{u}_{J_+,m}\}_{m=m_0}^{\infty}$ and a function $\mathbf{y} \in L^{\infty}\left(J_+;\, H_0^1(U)\right) \cap W^{1,\infty}\left(J_+;\, \mathrm{H}^{-1}(\mathrm{U})\right)$, such that $\mathbf{y}$ solves the problem (2.6) and also $\mathbf{u}_{J_+,m_l} \to \mathbf{y}$ in $C\left(\overline{J_+};\, L^2(U)\right)$. From the uniqueness of the solution, we deduce that $\mathbf{y} \equiv \mathbf{u}_{J_+}$. Moreover, from (2.10), (4.1), the latter convergence and (4.21), we obtain that $\left|\mathbf{u}_{J_+,m_l}\right|_{1,2,U} \to \left|\mathbf{u}_{J_+}\right|_{1,2,U}$ uniformly in $\overline{J_+}$. Applying the aforementioned convergences, we then easily get from (4.21) that $E\left(\mathbf{u}_{J_+}\right) = E(u_0)$, everywhere in J_+.

Acknowledgements – **N. G.** acknowledges that this research has been co-financed—via a programme of State Scholarships Foundation (IKY)—by the European Union (European Social Fund—ESF) and Greek national funds through the action entitled "Strengthening Human Resources Research Potential via Doctorate Research" (contract number: 2016–*ΕΣΠΑ*–050-0502-5534) in the framework of the Operational Program "Human Resources Development Program, Education and Lifelong Learning" of the National Strategic Reference Framework (NSRF) 2014–2020.
– **I. G. S.** acknowledges that this work was made possible by NPRP grant #[8-764-160] from Qatar National Research Fund (a member of Qatar Foundation).
– The findings achieved herein are solely the responsibility of the authors.

References

1. Adams, R.A., Fournier, J.J.F.: Sobolev Spaces. Pure and Applied Mathematics, vol. 140, 2nd edn. Academic Press, Oxford (2003)
2. Boyer, F., Fabrie, P.: Mathematical Tools for the Study of the Incompressible Navier-Stokes Equations and Related Models, Applied Mathematical Sciences, vol. 183. Springer, New York (2013)

3. Brezis, H.: Functional Analysis, Sobolev Spaces and Partial Differential Equations. Universitext. Springer, New York (2011)
4. Cazenave, T.: Semilinear Schrödinger Equations. Lecture Notes, vol. 10. American Mathematical Society, Providence, Rhode Island (2003)
5. Diestel, J., Uhl, J. jr.: Vector Measures, Mathematical Surveys and Monographs, vol. 15. American Mathematical Society, Providence, Rhode Island (1977)
6. Evans, L.C.: Partial Differential Equations. Graduate Studies in Mathematics, vol. 19, 2nd edn. American Mathematical Society, Providence, Rhode Island (2010)
7. Gallo, C.: Schrödinger group on Zhidkov spaces. Adv. Differ. Equ. **9**(5–6), 509–538 (2004)
8. Gallo, C.: The Cauchy problem for defocusing nonlinear Schrödinger equations with non-vanishing initial data at infinity. Commun. Part. Differ. Equ. **33**(5), 729–771 (2008)
9. Gérard, P.: The Cauchy problem for the Gross-Pitaevskii equation. Annales de l'Institut Henri Poincaré C, Analyse Non Linéaire **23**(5), 765–779 (2006)
10. Gialelis, N., Stratis, I.G.: Non-vanishing at spatial extremity solutions of the defocusing nonlinear Schrödinger equation. Math. Methods Appl. Sci. 1–18. https://doi.org/10.1002/mma.5074
11. Temam, R.: Navier-Stokes Equations. Studies in Mathematics and its Applications, vol. 2, revised edn. North Holland, Amsterdam-New York (1979)
12. Temam, R.: Infinite-dimensional Dynamical Systems in Mechanics and Physics. Applied Mathematical Sciences, vol. 68, 2nd edn. Springer, New York (2012)
13. Zhidkov P.E.: The Cauchy problem for a nonlinear Schrödinger equation, JINR Communications Dubna, R5-87-373, (1987) (18 pages) (in Russian)
14. Zhidkov, P.E.: Korteweg-de Vries and Nonlinear Schrödinger Equations: Qualitative Theory. Lecture Notes in Mathematics, vol. 1756. Springer, Berlin-Heidelberg (2001)

On Time-Global Solutions of SDE Having Nowhere Vanishing Initial Densities

Yuri E. Gliklikh

Abstract We deal with stochastic differential equations (SDE) on the non-compact finite-dimensional manifolds and consider their solutions such that the initial values have densities nowhere equal to zero. First, we obtain a sufficient condition under which all such solutions exist on the entire half-axis $[0, \infty)$. Further, we introduce the notion of the system, generated by the above equation, continuous at infinity. A sufficient condition the latter property to be satisfied is found. Then we find the necessary and sufficient condition for existence of the above solutions on $[0, \infty)$ for the case where the corresponding system is continuous at infinity.

Keywords Stochastic differential equations on manifolds · Non-zero density of initial value · Continuity at infinity · Existence of solutions on $[0, \infty)$

1 Introduction

The paper is devoted to the problem under which conditions all the solutions of a stochastic differential equation (SDE) are time-global provided that the initial value of the solution has a nowhere vanishing density, i.e., exist on the entire half-axis $[0, \infty)$. This question is not an abstract one. There are some problems where the existence of solution is proved only under the assumption that the initial value has nowhere equal to zero density (see, e.g., [1–3]). We search for the conditions more general than the well-known Ito condition or some of their analogues. The obtained sufficient conditions are close to the necessary and sufficient ones. Moreover, under some additional assumptions (of "continuity at the infinity" type) we do arrive at the necessary and sufficient condition. We consider the general case of stochastic differential equations on the non-compact finite-dimensional manifolds.

Besides this introduction, the paper consists of three sections. In Sect. 2, the problem is set up. In particular, we introduce the necessary notions and notation as well

Y. E. Gliklikh (✉)
Voronezh State University, Universitetskaya pl. 1, Voronezh 394018, Russia
e-mail: yeg@math.vsu.ru

A. Karapetyants et al. (eds.), *Modern Methods in Operator Theory and Harmonic Analysis*, Springer Proceedings in Mathematics & Statistics 291,
https://doi.org/10.1007/978-3-030-26748-3_20

as make general assumptions. In Sect. 3, we prove a sufficient condition for the existence of a solution of the afore-mentioned type, introduce the notion of continuity at infinity and prove a sufficient condition under which it holds. In Sect. 4, we prove the necessary and sufficient condition under the assumption that the property of continuity at infinity is satisfied.

Everywhere in the paper, we use the Einstein summation convention, i.e., if in a certain monomial there is the same upper and lower index this means the summation in this index from 1 to n where n is the dimension of the manifold under consideration.

By E we denote the mathematical expectation and by P the probability.

The research is supported in part by RFBR Grant 18-01-00048.

2 Setting Up the Problem

Let M be a finite-dimensional non-compact manifold, $\dim M = n$. By $(q^1, \ldots, q^n)$ we denote the local coordinates in charts on M and by $\frac{\partial}{\partial q^i}$ the corresponding basis vectors in tangent spaces. They are also considered as first order differential operators.

Consider a stochastic differential equation on M in the Itô or in the Stratonovich form with smooth coefficients. Since the coefficients are smooth, we can pass from the Stratonovich equation to that of Itô and vice versa. Recall that the equation in the Stratonovich form is given as a section of natural generalization of tangent bundle while the equation in the Ito form is a section of the special bundle called the Ito bundle (see, e.g., [2, 4]).

From both the Stratonovich and Ito forms of equation, one can derive its generator that we denote by $\mathcal{L}$. Recall that in the local coordinates $\mathcal{L}$ has the form $\mathcal{L} = a^i \frac{\partial}{\partial q^i} + \frac{1}{2}\alpha^{ij}\frac{\partial^2}{\partial q^i \partial q^j}$ where the coordinates a^i and the symmetric matrix (α^{ij}) are constructed from the coefficients of the equation. By the above assumption, all a^i and the α^{ij} are smooth. Our additional assumption is that the matrix (α^{ij}) is non-degenerate, i.e., the operator $\mathcal{L}$ is strictly elliptic.

Consider the one-point compactification $M\bigcup\{\infty\}$ of M, where the system of open neighborhoods of $\{\infty\}$ consists of the complements to all compact sets of M. Following standard terminology, we call the general solution with deterministic initial conditions (i.e., the initial values are points of M) of the equation under consideration the stochastic flow (flow for short) and denote it by $\xi(s) : M \to M\bigcup\{\infty\}$. For any point $m \in M$ and time t, the orbit $\xi_{t,m}(s)$ of this flow is the unique solution of the above-mentioned equation with initial conditions $\xi_{t,m}(t) = m$.

In order to distinguish between the above flow and the general solution in our case of initial values with densities nowhere equal to zero, the latter general solution is called *the general system of solutions*. This means that here we consider all solutions with such initial values. For the solution with initial value ξ_0 (having nowhere vanishing density) and initial time instant t, we keep the name "orbit" and notation $\xi_{t,\xi_0}(s)$.

The point $\{\infty\}$ is the "cemetery" to where the orbits (defined on a random time interval) in both cases get after explosion.

The generator $\mathcal{L}$ will be used as the generator both for the flow and for the general system of solutions.

To deal with a solution having initial values with densities nowhere equal to zero, we have to modify the notion of local solution. Since the matrices (α^{ij}) are non-degenerate, we can recover the coefficients of equation from the generator. This is the reason why we explain this notion in terms of generators. For simplicity, we consider the case where M is a linear space. The general case of a manifold we leave to the reader as a simple exercise.

Consider in R^n an expanding sequence of compacts V_i with smooth boundaries, such that $V_i \subset V_{i+1} \bigcup\limits_{i=1}^{\infty} V_i = M$. We construct a system of smooth bell-shaped functions φ_i, equal to one in V_i, zero outside V_{i+1}, and having uniformly bounded first partial derivatives in all $V_{i+1} \backslash V_i$. Let α^* be a constant symmetric non-degenerate matrix. Consider the sequence of generators, where the drift has the form $\varphi_i a$ and the matrix of coefficients at the second order derivatives takes the form $\varphi_i(\alpha) + (1 - \varphi_i)(\alpha^*)$. We note that the equations with such generators satisfy the conditions introduced above. Since the coefficients of these equations are smooth and bounded, they possess unique solutions. We call these solutions the local solutions of equations under consideration.

Our main problem is to find conditions (just sufficient, and necessary and sufficient) that make possible to extend all local solutions to the global ones. If it is the case, we call the general systems of solutions complete.

3 Sufficient Condition for Completeness of General Systems of Solutions and for Continuity of the Systems at Infinity

Let us recall the following definition.

Definition 1 Let X be a topological space. A function $\varphi : X \to R$ is called proper, if the preimage of every relatively compact set in R is relatively compact in X.

Let $\mathcal{L}$ be the generator corresponding to the equation in question.

Theorem 1 *Let there exist a smooth positive proper function φ on M such that $\mathcal{L}(t, m)\varphi < C$ for a certain $C > 0$ for all $t \in R$ and $m \in M$. Then for every initial value ξ_0 having density nowhere equal to zero, the solution is well-posed on $[0, \infty)$.*

Proof For positive integers $1, 2, \ldots, k, \ldots$, consider the family of sets $W_k = \varphi^{-1}([0, k))$ in M. Since φ is proper, these sets are relatively compact and $\bigcup\limits_k W_k = M$.

Besides, by the construction $W_i \subset W_{i+1}$ $i = 1, 2, \ldots, k, \ldots$.

Specifying an arbitrary $t \in [0, +\infty)$ and a random initial value ξ_0 with the density nowhere equal to zero, we consider the solution $\xi_{t,\xi_0}(s)$. We describe $\varphi(\xi_{t,\xi_0}(s))$ by the Ito formula. Since the Ito integral is a martingale, its expectation equals zero. Hence $E\varphi(\xi_{t,\xi_0}(s)) = E\varphi(\xi_0) + E\int_t^s(\mathcal{L}\varphi)(\nu, \xi_{t,\xi_0}(\nu))d\nu < E\varphi(\xi_0) + Cs$, $\mathcal{L}(t,m)\varphi < C$.

We introduce the set $\Omega^k = \{\omega \in \Omega \mid \xi_{0_\omega} \in W_k\}$ and notation $P(\Omega_k) = \lambda_k$. Since for every $\varepsilon > 0$ there exists a compact $K_\varepsilon \subset M$ such that $P(\omega \mid \xi_{0_\omega} \in K_\varepsilon) > 1 - \varepsilon$, it is clear that $\lim_{k\to\infty} \lambda_k \to 1$.

Consider the set $\Omega^k_s = \{\omega \in \Omega^k \mid \xi_{0_\omega} \in W_k, s < \tau_k\}$, where τ_k is the time of the first hit of the sample trajectory to the boundary of W_k. Evidently, $k(\lambda_k - \mathsf{P}(\Omega^k_s) < \int_{\Omega^k}\varphi(\xi_{t,\xi_{0_\omega}}(s))d\mathsf{P}$. Indeed, for $\omega \in \Omega^k \backslash \Omega^k_s$, we obtain $\xi_{t,\xi_0}(s \wedge \tau_k, \omega) = \xi_{t,\xi_0}(\tau_k, \omega)$, i.e., $\varphi(\xi_{t,\xi_0}(s \wedge \tau_k, \omega)) = k$. $\int_{\Omega^k}\varphi(\xi_{t,\xi_{0_\omega}}(s))d\mathsf{P} < E\varphi(\xi_{t,\xi_0}(s)) < E\varphi(\xi_0) + Cs$. Thus,

$$\lambda_k - P(\Omega^k_s) < \frac{E\varphi(\xi_0) + Cs}{k}. \tag{1}$$

Hence $\lim_{k\to\infty}(\lambda_k - \mathsf{P}(\Omega^k_s)) = 0$. Since $\lim_{k\to\infty}\lambda_k = 1$ (see above), we have $\lim_{k\to\infty}\mathsf{P}(\Omega^k_s) = 1$. However, by construction $\lim_{k\to\infty}\Omega^k_s = \bigcup_{i=1}^{\infty}\Omega^i_s = \Omega$, i.e., for all specified $s \geq t$ the value $\xi_{t,\xi_0}(s)$ does exist in M with probability 1.

Theorem 1 is a generalization of [5, Theorem IX. 6A].

Consider a random element $\eta : \Omega \to M$ and a compact $K \subset M$. We introduce the measurable set $\Omega_K \subset \Omega$ by the relation $\Omega_K = \{\omega \in \Omega \mid \eta(\omega) \in K\}$. It is clear that $\mathsf{P}(\Omega_K) = \lambda > 0$ for some λ.

For the proper function φ on M, we construct the family of relatively compact sets $V_k = \varphi^{-1}([0,k))$, where k is a positive integer. Obviously, $V_k \subset V_{k+1}$ for every k and $\bigcup_{k=1}^{\infty} V_k = M$. We observe that K belongs to all V_k for k large enough. Consider the orbit $\xi_{0,\eta}(s)$ of $\xi(s)$ with random initial value η.

Choose an arbitrary $T > 0$. The following lemma is a slight modification of Theorem 1.

Lemma 1 *Let there exist a smooth proper positive function $\varphi : M \to R$ such that for all t and m, the inequality $\mathcal{L}\varphi < C$ holds for a certain real constant $C > 0$, independent of t and m, where $\mathcal{L}$ is the generator of $\xi(s)$. Then the probability $\mathsf{P}\{\omega \in \Omega_K \mid \xi_{0,\eta_\omega}(T) \notin V_k\}$ tends to zero as $k \to \infty$. If there is a sequence of general systems of the solution, for which this inequality holds with the same φ, C, T, K and λ, the corresponding probabilities tend to zero uniformly for all such processes.*

Proof Since φ is smooth, its values on the compact K are uniformly bounded by a certain constant $C_K > 0$: $\sup_{x\in K}\varphi(m) < C_K$.

It follows from the construction of compacts V_k that for k greater than some N all V_k contain K. Below we deal only with such k.

Consider the real-valued process $\varphi(\xi_{0,\eta}(s))$ and describe it by the Ito formula. In this description, the Ito integral is a martingale, therefore its expectation equals zero. Thus, it follows from the Ito formula and condition $\mathcal{L}\varphi < C$ that

$$E(\varphi(\xi_{0,\eta}(T)) - \varphi(\eta)) = E\int_0^T \mathcal{L}\varphi(\xi_{0,\eta}(\nu))d\nu < CT. \tag{2}$$

Taking into account the fact that $\sup_{x\in K}\varphi(m) < C_K$ (see above), we derive from (2) that

$$\int_{\Omega_K} \varphi(\xi_{0,\eta}(T))d\mathsf{P} < \int_{\Omega_K} (\varphi(\eta))d\mathsf{P} + E\int_0^T \mathcal{L}\varphi(\xi_{0,\eta})(\nu))d\nu < \lambda C_K + CT. \tag{3}$$

We introduce the sets $\Omega_T^k \subset \Omega_K$ by the relations

$$\Omega_T^k = \{\omega \in \Omega_K \mid T < \tau_k\},$$

where τ_k is the time of the first hit of the sample trajectory $\xi_{0,\eta_\omega}(s)$ to the boundary of compact V_k. Then

$$k(\lambda - \mathsf{P}(\Omega_T^k)) < \int_{\Omega_K} (\varphi(\eta))d\mathsf{P} + \int_{\Omega_K}\int_0^T A\varphi(\nu, \xi_{0,\eta}^{(n)}(\nu))d\nu d\mathsf{P},$$

since, if $\omega \in \Omega_K$ but $\omega \notin \Omega_T^k$, we obtain $\xi_{0,\eta}^{(k)}(T \wedge \tau_k)_\omega = \xi_{0,\eta}^{(k)}(\tau_k)_\omega = k$. Thus, it follows from (3) that for every k

$$\lambda - \mathsf{P}(\Omega_T^k) < \frac{\lambda C_K + CT}{k}. \tag{4}$$

Hence $\lim_{k\to\infty}(\lambda - \mathsf{P}(\Omega_T^k(k))) = 0$. The uniform convergence follow from the fact that estimate (4) is the same for all the flows under consideration.

Let η_k be a sequence of random elements in M.

Definition 2 We say that $\eta_k \to \infty$ as $k \to \infty$ if there exists a certain proper positive function ψ on M such that $E\psi(\eta_k) \to \infty$ as $k \to \infty$.

Definition 3 The general system of solutions $\eta(s)$ is continuous at infinity on the interval $[0, T] \subset R$ if for all $0 \le t \le T$ and for every compact $K \subset M$ the inequality

$$\lim_{\eta_{0_i}\to+\infty} \mathsf{P}(\eta_{t,\eta_i}(T)) \in K) = 0, \tag{5}$$

holds for every orbit $\eta_{t,\eta_i}(s)$. The flow is continuous at infinity if this property holds for every $T > 0$.

Definition 3 is a modification of the notion introduced by Schwartz [6, 7]).

Theorem 2 *Let on M there exist a smooth positive proper function u such that $\tilde{\mathcal{L}}u < C$ for a certain constant $C > 0$, where $\tilde{\mathcal{L}}$ is the generator of the backward general system of solutions $\tilde{\eta}(t)$. Then the forward general system of solutions $\eta(t)$ is continuous at infinity on* $[0, T]$.

Proof Suppose the contrary, i.e., that the forward general system of solutions is not continuous at infinity on $[0, T] \subset R$. This means that there exists a sequence of random variables η_i having the property $\eta_i \to \infty$, such that for a certain compact K, the relation

$$\lim_{\eta_i \to \infty} \mathsf{P}(\eta_{t,\eta_i}(T) \in K) \neq 0 \tag{6}$$

holds. Thus, there exists $\lambda > 0$ such that

$$\mathsf{P}(\eta_{t,\eta_i}(T) \in K) > \lambda. \tag{7}$$

for i large enough. Recall that, by the hypothesis, there exists a smooth positive proper function u such that $\tilde{\mathcal{L}}u < C$ on M for some positive constant C. By Theorem 3, this implies that for the general system of solutions $\tilde{\eta}(s)$, its orbits are a.s. well-posed at any time instant $t \in [0, T]$. Consider the orbits $\tilde{\eta}_{(T,\eta_{t,\eta_i}(T))}(s)$ of $\tilde{\eta}(s)$ with initial conditions $\tilde{\eta}_{(T,\eta_{t,\eta_i}(T))}(T) = \eta_{t,\eta_i}(T)$. By Lemma 1,

$$\lim_{k \to \infty} \mathsf{P}\{\omega \in \Omega \mid \tilde{\eta}_{T,\eta_i(T)}(T)_\omega \in K, \tilde{\eta}_{T,\eta_i(T)}(t)_\omega \notin V_k\} = 0,$$

for $k \to \infty$ uniformly in i, i.e., for every $\varepsilon > 0$, there exists $N(\varepsilon)$ such that for $k > N(\varepsilon)$, we obtain

$$\mathsf{P}\{\omega \in \Omega \mid \tilde{\eta}_{T,\eta_i(T)}(T)_\omega \in K, \tilde{\eta}_{T,\eta_i(T)}(t)_\omega \notin V_k\} < \varepsilon.$$

But from (7), it follows that for every i large enough there exists a set $\{\omega \in \Omega \mid \tilde{\eta}_{T,\eta_i(T)}(T)_\omega \in K, \tilde{\eta}_{T,\eta_i(T)}(t)_\omega \notin V_k\}$, with probability greater than λ, for which $\tilde{\eta}_{T,\eta_{i(T)}}(t) = \eta_{t,\eta_i}(t)$ lies outside V_k. This contradiction finishes the proof.

4 Necessary and Sufficient Condition for Completeness of the General Systems of Solutions

Theorem 3 ([2, 8]) *For every general system of solutions $\xi(s)$ on M, complete on the interval $[0, T]$, there exists a proper positive smooth function θ on M such that for all $t \in [0, T]$, $x \in M$ and $s \in [t, T]$, the inequality $E\theta(\xi_{t,x}(s)) < \infty$ holds.*

Note that in the construction of the proper function mentioned in Theorem 3, the orbits of flow with deterministic initial values (points) are used.

Specify an arbitrary $T > 0$ and consider the direct product $M_+ = [0, T] \times M$. Denote by $\pi_+ : M_+ \to M$ the natural projection: $\pi_+(t, m) = m$.

Theorem 4 ([2, 8]) *Let the generator $\mathcal{L}$ of a complete general system of solutions be smooth and strictly elliptic. Then the function $u(t, m) = E\theta(\xi_{t,x}(T))$ on M_+ is smooth and satisfies the equation*

$$(\frac{\partial}{\partial t} + \mathcal{L})u = 0. \tag{8}$$

Theorem 5 *If the complete general system of solutions $\xi(s)$ is continuous at infinity, then the function $u(t, m) = E\theta(\xi_{t,x}(T))$ on M_+ is proper.*

Proof Let the general system of solutions $\xi(s)$ be continuous at infinity in the mean of Definition 3. Then it is obvious that the restriction of this general system of solutions to the orbits with deterministic initial conditions is continuous at infinity as well. Now, in order to prove the properness of function $u(t, m)$, it suffices to show that from $u(t, x_i) \to \infty$ it follows that $\theta(x_i) \to \infty$, i.e., that for every $C > 0$ there exists $\Xi > 0$ such that from $\theta(x_i) > \Xi$ it follows that $u(t, x_i) > C$ for every $t \in [0, T]$. Since θ is proper, $K = \theta^{-1}([0, 2C])$ is compact. It follows from formula (5) that for every $t \in [0, T]$ there exists Ξ such that $\mathrm{P}(\xi_{t,x_i}(T) \notin K) > \frac{1}{2}$ for $\theta(x_i) > \Xi$. Then $u(t, m) = E\theta(\xi_{t,x}(T)) > 2C \cdot \frac{1}{2} = C$. Since t belongs to the compact set$[0, T]$ and $u(t, m)$ is continuous in t, this finishes the proof.

Consider on M_+ the general system of solutions $\eta(s) = (s, \xi(s))$. Evidently, for $(t, m) \in M_+$, the trajectories of $\eta_{(t,m)}(s)$ satisfy the relation

$$\pi_+(\eta_{(t,m)}(s)) = \xi_{t,x}(s).$$

It is clear that the generator $\mathcal{L}_+$ of $\eta(s)$ is of the form

$$\mathcal{L}_{+(t,m)} = \frac{\partial}{\partial t} + \mathcal{L}(t, m). \tag{9}$$

Theorem 6 *The general system of solutions $\xi(s)$ on M, with smooth strictly elliptic generator and continuous at infinity, is complete on $[0, T]$ if and only if there exists a smooth positive proper function $u_+ : M_+ \to R$ such that $\mathcal{L}_+u_+ < C$ for a certain constant $C > 0$ at all points $(t, m) \in M_+$.*

Proof Let there exist a smooth positive proper function $u_+(t, m)$ on M_+ such that $\mathcal{L}_+u_+ < C$ at all points of M_+. Then from Theorem 1 it follows that $\eta(s)$ is complete. Hence $\xi(s) = \pi\eta(s)$ is also complete.

Let $\xi(s)$ be complete. Consider the above mentioned function $\theta(m)$ on M_+, and the function $u_+(t, m) = E\theta(\xi_{t,x}(T))$ on M_+. Since $\xi(s)$ is continuous at infinity,

$u_+(t, m)$ is a proper function by Theorem 5. By Theorem 4, it is also smooth and satisfies the equality $(\frac{\partial}{\partial t} + \mathcal{L})u_+ = \mathcal{L}_+ u_+ = 0$.

Corollary 1 *The general system of solutions $\xi(s)$ on M as in Theorem 6 is complete if and only if for every $T > 0$ there exists a positive proper function $u_+ : M_+ \to R$ that is C^1-smooth in $t \in [0, T]$, C^2-smooth in $m \in M$ and such that $\mathcal{L}_+ u(t, m) < C$ for a certain constant $C > 0$ at all points $(t, m) \in M_+$.*

References

1. Gliklikh, Y.E., Zastawniak, T.J.: Solution of the Newton-Nelson equation with random initial data. Appendix in: Yuri E. Gliklikh. Ordinary and Stochastic Differential Geometery as a Tool for Mathematical Physics, pp. 166–174. Kluwer (1996)
2. Gliklikh, YuE: Global and Stochastic Analysis with Applications to Mathematical Physics. Springer, London (2011)
3. Azarina, S.V., Gliklikh, YuE: On the solvability of nonautonomous stochastic differential equations with current velocities. Math. Notes **100**(1), 3–10 (2016). https://doi.org/10.1134/S0001434616070014
4. Belopolskaya, Y.I., Dalecky, Y.L.: Stochastic Processes and Differential Geometry. Kluwer Academic Publishers, Dordrecht (1989)
5. Elworthy, K.D.: Stochastic differential equations on manifolds. Lecture Notes of London Mathematical Society, vol. 70. Cambridge University Press, Cambridge (1982)
6. Schwartz, L.: Processus de Markov et desingration regulieres. Annales de l'Institut Fourier de l'Université de Grenoble **27**, 211–277 (1977)
7. Schwartz L.: Le semigroupe d'une diffusion en liaison avec les trajectories. Séminair de Probabilités XXIII/Lecture Notes in Mathematics, vol. 1372, 326–342 (1989)
8. Gliklikh, YuE: Necessary and sufficient conditions for global in time existence of solutions of ordinary, stochastic, and parabolic differential equations. Abstr. Appl. Anal. Article ID **39786**, 1–17 (2006). https://doi.org/10.1155/AAA/2006/39786

On Transmutation Operators and Neumann Series of Bessel Functions Representations for Solutions of Linear Higher Order Differential Equations

Flor A. Gómez and Vladislav V. Kravchenko

Abstract A new representation for solutions of linear higher order ordinary differential equations with a spectral parameter is obtained in terms of Neumann series of Bessel functions. The result is based on a Fourier-Legendre series representation for the Borel transform of the solution with respect to the spectral parameter. Estimates for the coefficients and for the convergence of the representation are derived. Numerical illustrations of the applicability of the obtained formulae are presented.

Keywords Linear differential equations · Borel transform · Neumann series

1 Introduction

Transmutation operators for linear differential equations of order $n > 2$ have been subject of dozens of publications reflecting the efforts of mathematicians to obtain a satisfactory generalization of this concept, well understood and developed in the case $n = 2$. We refer to a historical review of these efforts in [4, 12]. Here we use an idea of such a transmutation operator for $n > 2$ developed in [9] and based on the Borel transform of entire functions. The formula for the inverse transform as was pointed out in [9] in fact represents a natural transmutation operator transmuting solutions of an elementary equation into solutions of the equation with variable coefficients. Moreover, as was noticed in [9, p. 59], this operator reduces to the usual transmutation operator in the case $n = 2$.

V. V. Kravchenko on sabbatical leave from Cinvestav, Mexico.

F. A. Gómez
Departamento de Matemáticas, Cinvestav, Unidad Querétaro, Juriquilla, Mexico
e-mail: florgomezz85@gmail.com

V. V. Kravchenko (✉)
Regional Mathematical Center of Southern Federal University, Rostov-on-Don, Russia
e-mail: vkravchenko@math.cinvestav.edu.mx

A. Karapetyants et al. (eds.), *Modern Methods in Operator Theory and Harmonic Analysis*, Springer Proceedings in Mathematics & Statistics 291,
https://doi.org/10.1007/978-3-030-26748-3_21

Based on the transmutation operator from [9] in the present work we obtain a representation of solutions of linear differential equations of order $n > 2$ in terms of Neumann series of Bessel functions. This is an extension of the recent results of [6] onto the case $n > 2$.

The main result of the present paper consists in a representation of a solution of a linear differential equation of the form

$$y^{(n)} + p_2(x)y^{(n-2)} + ... + p_n(x)y = \omega^n y, \quad x \in (0, b)$$

as a sum of four Neumann series of Bessel functions. The formulas for the coefficients of the series are derived and an estimate for the approximation of the solution by the partial sums of the series is obtained. The result is obtained by representing in the form of a Fourier-Legendre series of the Borel transform γ of the solution y along the boundary of a square centered at the origin and containing all singularities of γ.

Additionally we show that the representation obtained is applicable to numerical calculation.

The paper is structured as follows. In Sect. 2 we introduce two representations of the solution y, the spectral parameter power series (SPPS) representation and the integral representation as an inverse Borel transform. The relation between them allows us to obtain an auxiliary series representation of the solution. In Sect. 3 we develop a Fourier-Legendre series representation for the Borel transform γ and obtain formulas and estimates for the coefficients involved. Section 4 presents the Neumann series of Bessel functions representation for the solution y. To illustrate the application of the representation in Sect. 5 we first develop an example of a fourth order equation of a particular form for which an especially simple procedure from [5] allows one to compute the SPPS coefficients required. We give some numerical illustrations showing the applicability of the representation to computation of solutions of initial value and spectral problems. In Sect. 6 we apply the new representation to an equation of fifth order using the procedure for computing the SPPS coefficients from [9, p. 59] and giving a numerical illustration of the applicability of the new representation to equations of higher orders.

2 A Transmutation Operator and a Representation for the Solution

Let $y(\omega, x)$ denote the solution of the following Cauchy problem

$$L[y] = y^{(n)} + p_2(x)y^{(n-2)} + ... + p_n(x)y = \omega^n y, \quad x \in (0, b), \tag{1}$$

$$y(\omega, 0) = 1, \quad y'(\omega, 0) = \omega, \dots, y^{(n-1)}(\omega, 0) = \omega^{n-1} \tag{2}$$

where $p_2,\ldots, p_n$ are assumed to be complex valued continuous functions on $[0, b]$, $0 < b < \infty$ and $\omega \in \mathbb{C}$ is a spectral parameter.

The following two representations for the solution $y(\omega, x)$ will be used throughout the paper.

Theorem 1 [9, Sect. 3] *The solution $y(\omega, x)$ of the problem (1), (2) admits the following spectral parameter power series (SPPS) representation*

$$y(\omega, x) = \sum_{m=0}^{\infty} \frac{\alpha_m(x)\omega^m}{m!} \tag{3}$$

where the coefficients α_m can be computed using the simple recursive integration procedure from [9, Sect. 3] or from [7]. The series converges uniformly with respect to $x \in [0, b]$ and uniformly on any compact subset of the complex plane with respect to ω.

The solution $y(\omega, x)$ admits the following representation

$$y(\omega, x) = \frac{1}{2\pi i} \int_C \gamma(x, t)e^{\omega t} dt \tag{4}$$

where for any $x \in [0, b]$ fixed the function $\gamma(x, t)$ as a function of the variable t is analytic outside a regular n-sided polygon Π_x with center at the origin and one of whose vertices being the point $t = x$. Moreover, $\gamma(x, t)$ is continuous up to the boundary of the polygon Π_x.

Remark 1 The function $\gamma(x, t)$ is nothing but the Borel transform of the function $y(\omega, x)$ which is entire with respect to ω (see, e.g., [2, 9]).

Remark 2 Instead of (4) the following representation can be used as well

$$y(\omega, x) = e^{\omega x} + \frac{1}{2\pi i} \int_C \widetilde{\gamma}(x, t)e^{\omega t} dt$$

where the function $\widetilde{\gamma}$ enjoys the same properties as those of the function γ, formulated in the previous theorem.

Both integral representations can be regarded as transmutations of the solution of the elementary Cauchy problem

$$v^{(n)} = \omega^n v, \quad x \in (0, b),$$

$$v(\omega, 0) = 1, \quad v'(\omega, 0) = \omega, \ldots, v^{(n-1)}(\omega, 0) = \omega^{n-1}$$

into the solution of (1), (2).

Proposition 1 *Outside Π_x the function $\gamma(x,t)$ admits the following series representation*

$$\gamma(x,t)=\sum_{m=0}^{\infty}\frac{\alpha_m(x)}{t^{m+1}} \tag{5}$$

where α_m are the coefficients from (3).

Proof Let C be a circle centered at the origin with a radius greater than x. Then outside C the function $\gamma(x,t)$ admits the representation $\gamma(x,t)=\sum_{m=0}^{\infty}\frac{b_m(x)}{t^{m+1}}$. Substitution of this series into (4) gives us the equalities

$$2\pi i y(\omega,x)=\sum_{m=0}^{\infty}b_m(x)\int_C\frac{e^{\omega t}}{t^{m+1}}dt=2\pi i\sum_{m=0}^{\infty}b_m(x)\frac{\omega^m}{m!}.$$

From here and from (3) we obtain (5).

In what follows by C_δ we denote the counterclockwise oriented boundary of the square centered at the origin and with one of the vertex being $t_1=(x+\delta)+i(x+\delta)$ with $\delta>0$.

Theorem 2 *The solution $y(\omega,x)$ of the problem (1), (2) admits the following representation*

$$\begin{aligned}y(\omega,x)=\sum_{k=1}^{\infty}\frac{\alpha_{k-1}(x)}{2\pi i}\Bigg(&e^{-i\omega(x+\delta)}\int_{-(x+\delta)}^{x+\delta}\frac{e^{\omega\tau}}{(\tau-i(x+\delta))^k}d\tau+\\&+ie^{\omega(x+\delta)}\int_{-(x+\delta)}^{(x+\delta)}\frac{(-i)^ke^{i\omega\tau}}{(\tau-i(x+\delta))^k}d\tau-e^{i\omega(x+\delta)}\int_{-(x+\delta)}^{x+\delta}\frac{(-1)^ke^{-\omega\tau}}{(\tau-i(x+\delta))^k}d\tau\\&-ie^{-\omega(x+\delta)}\int_{-(x+\delta)}^{(x+\delta)}\frac{i^ke^{-i\omega\tau}}{(\tau-i(x+\delta))^k}d\tau\Bigg)\end{aligned} \tag{6}$$

where for any ω and δ fixed each of the four series converges absolutely and uniformly for $x\in[0,b]$.

Proof From (4) by a natural parametrization of contour C_δ we obtain the equality

$$\begin{aligned}y(\omega,x)=&\,e^{-i\omega(x+\delta)}\int_{-x-\delta}^{x+\delta}\gamma_1(x,\tau)e^{\omega\tau}d\tau+ie^{\omega(x+\delta)}\int_{-x-\delta}^{x+\delta}\gamma_2(x,\tau)e^{i\omega\tau}d\tau\\&-e^{i\omega(x+\delta)}\int_{-x-\delta}^{x+\delta}\gamma_3(x,\tau)e^{-\omega\tau}d\tau-ie^{-\omega(x+\delta)}\int_{-x-\delta}^{x+\delta}\gamma_4(x,\tau)e^{-i\omega\tau}d\tau\end{aligned} \tag{7}$$

where $\gamma_1(x,\tau):=\frac{1}{2\pi i}\gamma(x,\tau-i(x+\delta))$, $\gamma_2(x,\tau):=\frac{1}{2\pi i}\gamma(x,(x+\delta)+i\tau)$, $\gamma_3(x,\tau):=\frac{1}{2\pi i}\gamma(x,-\tau+i(x+\delta))$ and $\gamma_4(x,\tau):=\frac{1}{2\pi i}\gamma(x,-(x+\delta)-i\tau)$ for $\tau\in[-x-\delta,x+\delta]$. Due to Proposition 1 for the functions γ_j we have the following series representations

$$\gamma_1(x,\tau) = \frac{1}{2\pi i}\sum_{k=0}^{\infty}\frac{\alpha_k(x)}{(\tau - i(x+\delta))^{k+1}}, \quad \gamma_2(x,\tau) = \frac{1}{2\pi i}\sum_{k=0}^{\infty}\frac{(-i)^{k+1}\alpha_k(x)}{(\tau - i(x+\delta))^{k+1}},$$
$$\gamma_3(x,\tau) = \frac{1}{2\pi i}\sum_{k=0}^{\infty}\frac{(-1)^{k+1}\alpha_k(x)}{(\tau - i(x+\delta))^{k+1}}, \quad \gamma_4(x,\tau) = \frac{1}{2\pi i}\sum_{k=0}^{\infty}\frac{i^{k+1}\alpha_k(x)}{(\tau - i(x+\delta))^{k+1}}. \tag{8}$$

Let us prove that for any $x \in [0, b]$ fixed the series converge absolutely and uniformly with respect to τ on $[-x-\delta, x+\delta]$.

Consider the series for $\gamma_1(x,\cdot)$. We have

$$\left|\frac{a_k(x)}{(\tau - i(x+\delta))^{k+1}}\right| = \frac{|a_k(x)|}{(\tau^2 + (x+\delta)^2)^{\frac{k+1}{2}}} \le \frac{|a_k(x)|}{(x+\delta)^{k+1}}$$

Therefore,

$$\sum_{k=0}^{\infty}\left|\frac{a_k(x)}{(\tau - i(x+\delta))^{k+1}}\right| \le \sum_{k=0}^{\infty}\frac{|a_k(x)|}{(x+\delta)^{k+1}}. \tag{9}$$

Since for any $x \in [0, b]$, $y(\omega, x)$ is an entire function with respect to ω of order one and exponential type x [9, Sect. 3], we have that $x = \limsup_{k\to\infty} |y^{(k)}(\omega, x)|^{1/k}$ where ω is any fixed complex number, see [2, Chap. 2, Formula 2.2.12]. From the series representation (3) it is obtained that $|\alpha_k(x)| = |y^{(k)}(0, x)|$ and so $\limsup_{k\to\infty} |\alpha_k(x)|^{1/k} = x$. Therefore, by Cauchy's criterion the series on the right hand side of (9) converges for any $x \in [0, b]$. By Weierstrass M-test, we conclude that for any $x \in [0, b]$ the series in (8) are uniformly absolutely convergent on $[-x-\delta, x+\delta]$. Substitution of the function series for γ_j into equality (7) leads then to the representation (6).

In order to prove that the function series (6) converge absolutely for all x we first obtain the following estimates

$$\left|\int_{-x-\delta}^{x+\delta}\frac{e^{\omega\tau - i\omega(x+\delta)}}{(\tau - i(x+\delta))^k}d\tau\right| \le 2e^{\mathrm{Im}(\omega)(x+\delta)}\frac{\sinh(\mathrm{Re}(\omega)(x+\delta))}{\mathrm{Re}(\omega)(x+\delta)^k},$$
$$\left|\int_{-(x+\delta)}^{(x+\delta)}\frac{e^{i\omega\tau + \omega(x+\delta)}}{(\tau - i(x+\delta))^k}d\tau\right| \le 2e^{\mathrm{Re}(\omega)(x+\delta)}\frac{\sinh(\mathrm{Im}(\omega)(x+\delta))}{\mathrm{Im}(\omega)(x+\delta)^k},$$
$$\left|\int_{-x-\delta}^{x+\delta}\frac{e^{-\omega\tau + i\omega(x+\delta)}}{(\tau - i(x+\delta))^k}d\tau\right| \le 2e^{-\mathrm{Im}(\omega)(x+\delta)}\frac{\sinh(\mathrm{Re}(\omega)(x+\delta))}{\mathrm{Re}(\omega)(x+\delta)^k},$$
$$\left|\int_{-(x+\delta)}^{(x+\delta)}\frac{e^{-i\omega\tau - \omega(x+\delta)}}{(\tau - i(x+\delta))^k}d\tau\right| \le 2e^{-\mathrm{Re}(\omega)(x+\delta)}\frac{\sinh(\mathrm{Im}(\omega)(x+\delta))}{\mathrm{Im}(\omega)(x+\delta)^k}.$$

Considering the first of them we observe that the function

$$\frac{2e^{\mathrm{Im}(\omega)(x+\delta)}\sinh(\mathrm{Re}(\omega)(x+\delta))}{\mathrm{Re}(\omega)}$$

attains a maximum on the interval $[0, b]$. Denote

$$c_1(\omega, \delta) := \max_{x\in[0,b]} \frac{2e^{\mathrm{Im}(\omega)(x+\delta)} \sinh(\mathrm{Re}(\omega)(x+\delta))}{\mathrm{Re}(\omega)}.$$

Hence we have the following estimate for the first series

$$\sum_{k=1}^{\infty} \left| \alpha_{k-1}(x) \int_{-x-\delta}^{x+\delta} \frac{e^{\omega\tau - i\omega(x+\delta)}}{(\tau - i(x+\delta))^k} d\tau \right| \leq c_1(\omega, \delta) \sum_{k=1}^{\infty} \frac{|\alpha_{k-1}(x)|}{(x+\delta)^k}.$$

The series on the right converges for all $x \in [0, b]$ because $\limsup |\alpha_{k-1}(x)|^{1/k} = x$.

The convergence of the other three series is proved in a similar way.

3 A Fourier-Legendre Series Representation for the Kernels γ_j

Denote $c_{1,k} = 1$, $c_{2,k} = (-i)^k$, $c_{3,k} = (-1)^k$ and $c_{4,k} = i^k$. Then the kernels γ_j, $j = 1, 2, 3, 4$ can be written as follows

$$\gamma_j(x, \tau) = \frac{1}{2\pi i} \sum_{k=1}^{\infty} \frac{c_{j,k}\alpha_{k-1}(x)}{(\tau - i(x+\delta))^k}. \tag{10}$$

Proposition 2 *The functions $\gamma_j(x, \tau)$ admit the following representations*

$$\gamma_j(x, \tau) = \sum_{n=0}^{\infty} \frac{\beta_n^{(j)}(x; \delta)}{x+\delta} P_n\left(\frac{\tau}{x+\delta}\right) \tag{11}$$

where P_n stands for the Legendre polynomial of order n, and the coefficients β_n are defined by the equality

$$\beta_n^{(j)}(x; \delta) = \frac{(2n+1)}{2\pi i\, (n+1)_{n+1}} \times$$
$$\sum_{k=1}^{\infty} \frac{(-1)^k c_{k,j}\alpha_{k-1}(x)\, (k)_n\, (1-i)^{n+k}}{(x+\delta)^{k-1} 2^k} {}_2F_1\left(n+k, n+1; 2\,(n+1); \frac{2}{1+i}\right)$$

where $(k)_n$ is the Pochhammer symbol and ${}_2F_1$ is the Gauss hypergeometric function.

The series in (11) converges uniformly and for any $x \in [0, b]$ the following estimate is valid

$$|\beta_n^{(j)}(x; \delta)| \leq \sqrt{2n+1}(x+\delta)^{1/2} \left\| \gamma_j(x, \tau) \right\|_{L_2[-x-\delta, x+\delta]}. \tag{12}$$

Proof Since $\gamma(x,t)$ as a function of the variable t is analytic outside Π_x the functions $\gamma_j(x,\cdot)$ admit the uniformly convergent Fourier-Legendre series representations of the form (11). Multiplying (11) by $P_m\left(\frac{\tau}{x+\delta}\right)$ and integrating we obtain

$$\frac{2\beta_n^{(j)}(x;\delta)}{2n+1}=\int_{-x-\delta}^{x+\delta}\gamma_j(x,\tau)P_n\left(\frac{\tau}{x+\delta}\right)d\tau.$$

Now substitution of (10) leads to the equalities

$$\begin{aligned}\beta_n^{(j)}(x;\delta)&=\frac{2n+1}{4\pi i}\sum_{k=1}^{\infty}c_{j,k}\alpha_{k-1}(x)\int_{-x-\delta}^{x+\delta}\frac{P_n\left(\frac{\tau}{x+\delta}\right)}{(\tau-i(x+\delta))^k}d\tau\\&=\frac{2n+1}{4\pi i}\sum_{k=1}^{\infty}\frac{c_{j,k}\alpha_{k-1}(x)}{(x+\delta)^{k-1}}\int_{-1}^{1}\frac{P_n(s)}{(s-i)^k}ds\\&=\frac{(2n+1)\,2^n}{2\pi i\,(n+1)_{n+1}}\\&\times\sum_{k=1}^{\infty}\frac{(-1)^k\,c_{j,k}\alpha_{k-1}(x)}{(x+\delta)^{k-1}}\frac{(k)_n}{(1+i)^{n+k}}\,{}_2F_1\left(n+k,n+1;2\,(n+1)\,;\frac{2}{1+i}\right)\end{aligned}$$

where for calculating the integrals

$$p_{n,k}:=\int_{-1}^{1}\frac{P_n(s)}{(s-i)^k}ds$$

formula 2.17.1 (12) from [11] was used.

The estimate (12) follows from the Cauchy-Schwarz inequality,

$$\begin{aligned}|\beta_n^{(j)}(x;\delta)|=&\frac{2n+1}{2}\left|\int_{-x-\delta}^{x+\delta}\gamma_j(x,\tau)P_n\left(\frac{\tau}{x+\delta}\right)d\tau\right|\\\le&\frac{2n+1}{2}\left\|\gamma_j(x,\tau)\right\|_{L_2[-x-\delta,x+\delta]}\left\|P_n\left(\frac{\tau}{x+\delta}\right)\right\|_{L_2[-x-\delta,x+\delta]}\\=&\sqrt{2n+1}(x+\delta)^{1/2}\left\|\gamma_j(x,\tau)\right\|_{L_2[-x-\delta,x+\delta]}.\end{aligned}$$

4 The Neumann Series of Bessel Functions Representation of the Solution

Theorem 3 *The solution $y(\omega,x)$ of (1), (2) admits the representation*

$$y(\omega,x)=2\sum_{n=0}^{\infty}\left(\beta_n^{(1)}(x;\delta)e^{-i\omega(x+\delta)}+(-1)^{n+1}\,\beta_n^{(3)}(x;\delta)e^{i\omega(x+\delta)}\right)i_n(\omega(x+\delta))$$

$$+2\sum_{n=0}^{\infty}\left(i^{n+1}\beta_n^{(2)}(x;\delta)e^{\omega(x+\delta)}+(-i)^{n+1}\beta_n^{(4)}(x;\delta)e^{-\omega(x+\delta)}\right)j_n(\omega(x+\delta)) \quad (13)$$

with the coefficients $\beta_n^{(k)}$ *from Proposition 2. Here* j_n *stands for the spherical Bessel function of the first kind of order n, and* i_n *is the modified spherical Bessel function of the first kind of order n (see the definition, e.g., in [10, Chap. 10]).*

For the approximate solution

$$y_M(\omega,x)=2\sum_{n=0}^{M}\left(\beta_n^{(1)}(x;\delta)e^{-i\omega(x+\delta)}+(-1)^{n+1}\beta_n^{(3)}(x;\delta)e^{i\omega(x+\delta)}\right)i_n(\omega(x+\delta))$$

$$+2\sum_{n=0}^{M}\left(i^{n+1}\beta_n^{(2)}(x;\delta)e^{\omega(x+\delta)}+(-i)^{n+1}\beta_n^{(4)}(x;\delta)e^{-\omega(x+\delta)}\right)j_n(\omega(x+\delta)) \quad (14)$$

the following estimate holds

$$\begin{aligned}|y(\omega,x)-y_M(\omega,x)| <& \frac{2\sqrt{\pi}\left(M+\frac{1}{2}+2e\right)}{\Gamma(M+\frac{3}{2})}(x+\delta)\left\|\gamma_j(x,\tau)\right\|_{L_2[-x-\delta,x+\delta]}\\ &\sqrt{|\omega|}\left(\cosh((x+\delta)\operatorname{Im}(\omega))e^{(x+\delta)|\operatorname{Re}(\omega)|}+\right.\\ &\left.+\cosh((x+\delta)\operatorname{Re}(\omega))e^{(x+\delta)|\operatorname{Im}(\omega)|}\right).\end{aligned}$$

Proof Substitution of (11) into (7) with the aid of [10, formula 18.17.19] leads to (13).

Consider

$$\begin{aligned}&2\sum_{n=0}^{\infty}\left(\left|\beta_{n+M+1}^{(1)}(x;\delta)e^{-i\omega(x+\delta)}\right|+\left|\beta_{n+M+1}^{(3)}(x;\delta)e^{i\omega(x+\delta)}\right|\right)\left|i_{n+M+1}(\omega(x+\delta))\right|\\ &+2\sum_{n=0}^{\infty}\left(\left|\beta_{n+M+1}^{(2)}(x;\delta)e^{\omega(x+\delta)}\right|+\left|\beta_{n+M+1}^{(4)}(x;\delta)e^{-\omega(x+\delta)}\right|\right)\left|j_{n+M+1}(\omega(x+\delta))\right|\\ &\geq|y(\omega,x)-y_M(\omega,x)|\end{aligned}$$

Since

$$|J_\nu(z)|\leq\frac{|z|e^{|Im(z)|}}{2\Gamma(\nu+1)},\quad \left(\nu\geq-\frac{1}{2}\right)$$

(see formula 9.1.62 from [1]) we get

$$\left|j_{M+1+n}(\omega(x+\delta))\right|\leq\frac{\sqrt{\pi(x+\delta)}\sqrt{|\omega|}e^{(x+\delta)|\operatorname{Im}(\omega)|}}{2\sqrt{2}\Gamma(M+\frac{3}{2}+n)},$$

$$\left|i_{M+1+n}(\omega(x+\delta))\right|\leq\frac{\sqrt{\pi(x+\delta)}\sqrt{|\omega|}e^{(x+\delta)|\operatorname{Re}(\omega)|}}{2\sqrt{2}\Gamma(M+\frac{3}{2}+n)}.$$

Using (12) we obtain

$$\begin{aligned}|y(\omega,x)-y_M(\omega,x)| \leq & 2\sqrt{\pi}(x+\delta)\left\|\gamma_j(x,\tau)\right\|_{L_2[-x-\delta,x+\delta]}\sqrt{|\omega|}\\ & \times\left(\cosh((x+\delta)\operatorname{Im}(\omega))e^{(x+\delta)|\operatorname{Re}(\omega)|}+\right.\\ & \left.+\cosh((x+\delta)\operatorname{Re}(\omega))e^{(x+\delta)|\operatorname{Im}(\omega)|}\right)\sum_{n=0}^{\infty}\frac{\sqrt{M+n+3/2}}{\Gamma(M+n+\frac{3}{2})}.\end{aligned}$$

Notice that the series on the right hand side admits the following chain of relations

$$\begin{aligned}\sum_{n=0}^{\infty}\frac{\sqrt{M+n+\frac{3}{2}}}{\Gamma(M+n+\frac{3}{2})} &\leq \sum_{n=0}^{\infty}\frac{M+n+\frac{3}{2}}{\Gamma(M+n+\frac{3}{2})}\\ &=\sum_{n=0}^{\infty}\frac{M+n+\frac{1}{2}}{\Gamma(M+n+\frac{3}{2})}+\sum_{n=0}^{\infty}\frac{1}{\Gamma(M+n+\frac{3}{2})}\\ &=\sum_{n=0}^{\infty}\frac{1}{\Gamma(M+n+\frac{1}{2})}+\sum_{n=0}^{\infty}\frac{1}{\Gamma(M+n+\frac{3}{2})}\\ &=\frac{1}{\Gamma(M+\frac{1}{2})}+2\sum_{n=0}^{\infty}\frac{1}{\Gamma(M+n+\frac{3}{2})}.\end{aligned}$$

Note that

$$\sum_{n=0}^{\infty}\frac{1}{\Gamma(M+n+\frac{3}{2})}=\frac{1}{\Gamma(M+\frac{3}{2})}\sum_{n=0}^{\infty}\frac{1}{\left(M+\frac{3}{2}\right)_n},$$

and since $\frac{1}{(1)_n}>\frac{1}{\left(M+\frac{3}{2}\right)_n}$ for any $M\geq 0$, we have that

$$\sum_{n=0}^{\infty}\frac{1}{\left(M+\frac{3}{2}\right)_n}<\sum_{n=0}^{\infty}\frac{1}{n!}=e.$$

Thus,

$$\sum_{n=0}^{\infty}\frac{\sqrt{M+n+\frac{3}{2}}}{\Gamma(M+n+\frac{3}{2})}<\frac{1}{\Gamma(M+\frac{1}{2})}+\frac{2e}{\Gamma(M+\frac{3}{2})}=\frac{M+\frac{1}{2}+2e}{\Gamma(M+\frac{3}{2})}.$$

5 An Example of Application to Fourth Order Ordinary Differential Equations

In this section we give some numerical illustrations of the representation (13) in the case when the equation of the problem (1), (2) has the following form

$$y^{(4)} + \left(p(x)y'\right)' = \omega^4 y, \quad x \in (0, b), \tag{15}$$

$$y(\omega, 0) = 1, \quad y'(\omega, 0) = \omega, \quad y''(\omega, 0) = \omega^2, \quad y'''(\omega, 0) = \omega^3 \tag{16}$$

where $p \in C^1[0, b]$. We chose this particular equation due to the fact that for the construction of corresponding formal powers and hence of the SPPS representation of the solution an especially simple procedure was developed in [5].

5.1 The SPPS Representation

Let f be a particular solution (in general a complex-valued one) of the equation $v'' + pv = 0$ such that f and $1/f \in C[0, b]$. Then following [5] let us define a system of recursive integrals,

$$X_1^{(0)} \equiv X_2^{(0)} \equiv X_3^{(0)} \equiv X_4^{(0)} \equiv 1$$

$$X_1^{(m)} = \begin{cases} m\int_0^x X_1^{(m-1)}(s)f(s)ds, & m = 4k+4, \\ m\int_0^x X_1^{(m-1)}(s)\frac{1}{f^2(s)}ds, & m = 4k+3, \\ m\int_0^x X_1^{(m-1)}(s)f(s)ds, & m = 4k+2, \\ m\int_0^x X_1^{(m-1)}(s)ds, & m = 4k+1, \end{cases}$$

$$X_2^{(m)} = \begin{cases} m\int_0^x X_1^{(m-1)}(s)\frac{1}{f^2(s)}ds, & m = 4k+4 \\ m\int_0^x X_1^{(m-1)}(s)f(s)ds, & m = 4k+3 \\ m\int_0^x X_1^{(m-1)}(s)ds, & m = 4k+2 \\ m\int_0^x X_1^{(m-1)}(s)f(s)ds, & m = 4k+1 \end{cases}$$

$$X_3^{(m)} = \begin{cases} m\int_0^x X_1^{(m-1)}(s)f(s)ds, & m = 4k+4, \\ m\int_0^x X_1^{(m-1)}(s)ds, & m = 4k+3, \\ m\int_0^x X_1^{(m-1)}(s)f(s)ds, & m = 4k+2, \\ m\int_0^x X_1^{(m-1)}(s)\frac{1}{f^2(s)}ds, & m = 4k+1, \end{cases}$$

$$X_4^{(m)} = \begin{cases} m \int_0^x X_1^{(m-1)}(s)ds, & m = 4k+4 \\ m \int_0^x X_1^{(m-1)}(s) f(s)ds, & m = 4k+3 \\ m \int_0^x X_1^{(m-1)}(s) \frac{1}{f^2(s)}ds, & m = 4k+2 \\ m \int_0^x X_1^{(m-1)}(s) f(s)ds, & m = 4k+1 \end{cases}$$

Proposition 3 *The solution of the problem (15), (16) admits the following representation*

$$y(\omega, x) = \sum_{m=0}^{\infty} \frac{\alpha_m(x)}{m!}\omega^m \tag{17}$$

where $\alpha_0 \equiv 1$,

$$\alpha_m(x) = \begin{cases} \frac{1}{f(0)} X_2^{(m)}(x) - f'(0)\frac{X_3^{(m+1)}(x)}{(m+1)} - \frac{f''(0)}{f(0)}\frac{X_4^{(m+2)}(x)}{(m+1)(m+2)}, & m = 4k+1, \\ f(0)X_3^{(m)}(x), & m = 4k+2, \\ X_4^{(m)}(x), & m = 4k+3, \\ X_1^{m}(x), & m = 4k+4, \end{cases} \tag{18}$$

$k = 0, 1, \ldots$ *and the series (17) converges uniformly with respect to* x *on* $[0, b]$ *and with respect to* ω *on any compact subset of the complex plane.*

Proof The proof consists in application of Theorem 3 from [5] to the problem (15), (16). The solution of (15), (16) is obtained as a linear combination of the four linearly independent solutions from [5],

$$u_1(x) = \sum_{m=0}^{\infty} \frac{X_1^{(4m)}(x)}{(4m)!}\omega^{4m} \quad u_2(x) = \sum_{m=0}^{\infty} \frac{X_2^{(4m+1)}(x)}{(4m+1)!}\omega^{4m}$$

$$u_3(x) = \sum_{m=0}^{\infty} \frac{X_3^{(4m+2)}(x)}{(4m+2)!}\omega^{4m} \quad u_4(x) = \sum_{m=0}^{\infty} \frac{X_4^{(4m+3)}(x)}{(4m+3)!}\omega^{4m}.$$

The initial conditions (16) are fulfilled by a linear combination of these solutions with the corresponding constants chosen as

$$c_1 = 1, \quad c_2 = \frac{\omega}{f(0)}, \quad c_3 = \omega^2 f(0) - \omega f'(0), \quad c_4 = \omega^3 - \omega\frac{f''(0)}{f(0)}.$$

5.2 Numerical Illustrations

By $\beta_{n,K}^j$ we denote the approximation of the coefficient β_n^j for $j = 1, 2, 3, 4$, defined by

$$\beta_{n,K}^{j}(x;\delta) := \frac{2n+1}{4\pi i}\sum_{k=1}^{K}\frac{c_{j,k}\alpha_{k-1}(x)}{(x+\delta)^{k-1}}\int_{-1}^{1}\frac{P_n(s)}{(s-i)^k}ds. \tag{19}$$

The coefficients α_k defined by (18) are computed using the numerical integration approach explained in [8, Sect. 7], although other approaches (for example, that based on splines [3] though considerably slower is also applicable). We emphasize that the computation of a couple of hundreds of the coefficients α_k does not represent any difficulty and can be performed with a remarkable accuracy. The integrals in (19) were computed numerically as well. All the numerical results in this section are obtained with $K = 150$.

Solution of Initial Value Problems

Example 1 Consider the equation

$$y^{(4)} + 2y'' = \omega^4 y$$

subject to conditions (16). Figure 1 shows the absolute error of the approximate solution corresponding to the value $\delta = 0$ and computed for several distinct values of ω. It shows that meanwhile for the end point of the interval the accuracy is reasonably good it deteriorates rapidly for the values of x closer to the origin, and the representation (13) is not applicable in the vicinity of $x = 0$.

The situation is completely different for $\delta > 0$ as shown on Fig. 2 where the absolute error of the approximate solution is depicted in the case $\delta = 1$. The solution is sufficiently accurate on the whole interval, the accuracy deteriorates to the right end of the interval, and according to the estimate from Theorem 3 the accuracy is better for smaller values of $|\omega|$.

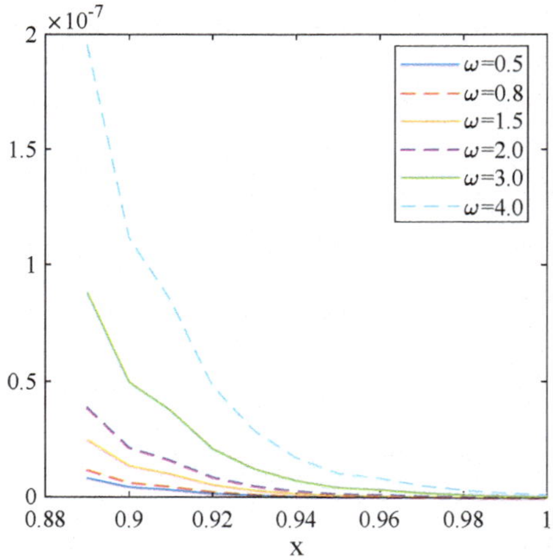

Fig. 1 The absolute error of the approximate solution (14) computed with $\delta = 0$ for several values of ω

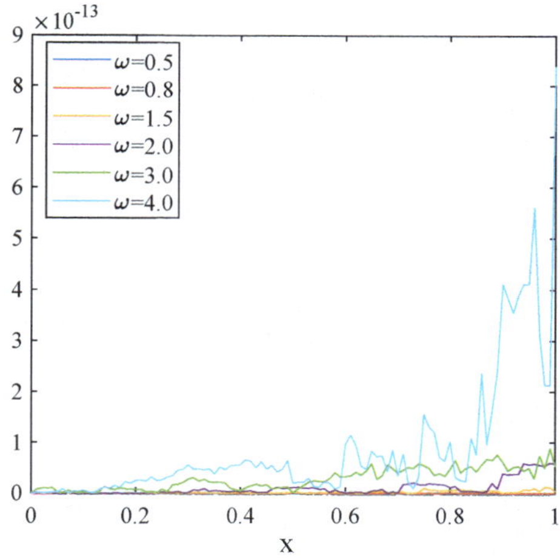

Fig. 2 The absolute error of the approximate solution (14) computed with $\delta = 1$ for several values of ω

Solution of Eigenvalue Problems

Example 2 Consider the following eigenvalue problem

$$u^{(4)} = w^4 u(x), \quad u(\omega, 0) = 0 = u''(\omega, 0) \quad u(\omega, 1) = 0 = u'(\omega, 1).$$

The exact characteristic equation of the problem has the form

$$(e^{\omega} - e^{-\omega}) \cos(\omega) = (e^{\omega} + e^{-\omega}) \sin(\omega).$$

In terms of the solution $y(\omega, x)$ satisfying conditions (16) the dispersion equation can be written as follows

$$\begin{aligned} &y(-\omega, 1)y'(\omega, 1) - y(\omega, 1)y'(-\omega, 1) + y(\omega, 1)y'(-i\omega, 1) \\ &-y(-i\omega, 1)y'(\omega, 1) + y(-i\omega, 1)y'(-\omega, 1) - y(-\omega, 1)y'(-i\omega, 1) \\ &+y(i\omega, 1)y'(-i\omega, 1) - y(-i\omega, 1)y'(i\omega, 1) = 0. \end{aligned}$$

The first six eigenvalues computed by solving this equation with the aid of the representation (13) are presented in the table below together with their corresponding absolute error. The eigenvalues increase rapidly, and similarly to the SPPS [5] the representation (13) allows one accurate computation of several lower index eigenvalues.

$N = 40, \quad k = 150 \quad \text{and} \quad \delta = 0.25$		
n	NSM	$E(Abs.error)$
0	237.721067584372	5.3×10^{-8}
1	2496.48744387544	6.0×10^{-6}
2	10867.5824450557	2.2×10^{-4}
3	31780.0997391812	3.2×10^{-3}
4	74000.9108835559	6.1×10^{-2}
5	148630.254375198	4.2

6 Application to Order $n > 4$

Example 3 Consider the following equation of order five

$$y^{(5)}(x) + y(x) = \omega^5 y(x), \quad x \in (0, 1)$$

with the initial conditions (2). Here we compute the coefficients $A_m(x) = \frac{\alpha_m(x)}{m!}$ using the method from [9, p. 31]. Thus, the coefficients are defined by the formula

$$A_{m+5}(x) = \sum_{k=1}^{5} y_k(x)(-1)^{k+1} \int_0^x A_m(t) \frac{\Delta_k(t)}{\Delta(t)} dt$$

where $y_1, \ldots, y_5$ is a fundamental system of the equation $y^{(5)}(x) + y(x) = 0$, $\Delta(x)$ is their Wronskian and $\Delta_k(x)$ is the cofactor of the $(5, k)$ entry of $\Delta(x)$.

The absolute error of the approximate solution computed with the aid of (13) with $\delta = 1$ and $N = 20$ is reported on Fig. 3 for several different values of ω.

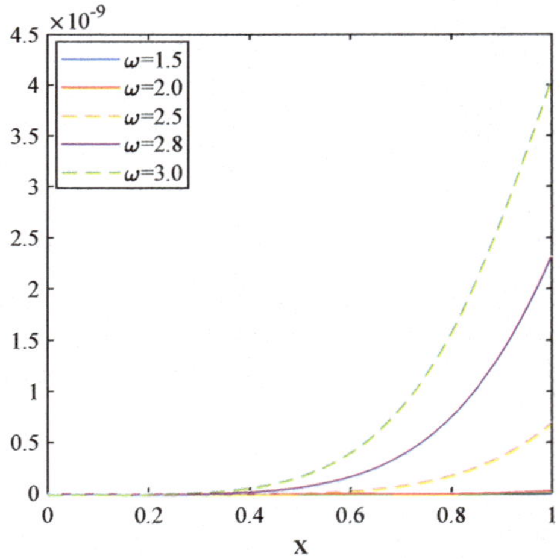

Fig. 3 The absolute error of the approximate solution computed with $\delta = 1$ and $N = 20$ for several different values of ω

7 Conclusions

A representation for solutions of linear differential equations with spectral parameter of order greater than two is obtained in terms of four Neumann series of Bessel functions. The formulas for the coefficients of the series are derived. An estimate for the approximation of the solution by the partial sums of the series is obtained. The numerical illustrations are presented which show the applicability of the representation for numerical solution of initial value and spectral problems.

Acknowledgements Research was supported by CONACYT, Mexico via the project 284470.

References

1. Abramovitz, M., Stegun, I.A.: Handbook of Mathematical Functions. Dover, New York (1972)
2. Boas, R.P.: Entire Functions. Academic, New York (1954)
3. Castillo, R., Kravchenko, V.V., Oviedo, H., Rabinovich, V.S.: Dispersion equation and eigenvalues for quantum wells using spectral parameter power series. J. Math. Phys. **52**(4), 043522(10 pp.) (2011)
4. Katrakhov, V.V., Sitnik, S.M.: The transmutation method and boundary value problems for singular differential equations. Contemp. Math. Fundam. Dir. **64**(2), 211–428 (2018). (in Russian)
5. Khmelnytskaya, K.V., Kravchenko, V.V., Baldenebro-Obeso, J.A.: Spectral parameter power series for fourth-order Sturm-Liouville problems. Appl. Math. Comput. **219**, 3610–3624 (2012)
6. Kravchenko, V.V., Navarro, L.J., Torba, S.M.: Representation of solutions to the one-dimensional Schrödinger equation in terms of Neumann series of Bessel functions. Appl. Math. Comput. **314**(1), 173–192 (2017)
7. Kravchenko, V.V., Porter, R.M., Torba, S.M.: Spectral parameter power series for arbitrary order linear differential equations. Math. Methods Appl. Sci. https://doi.org/10.1002/mma.4769
8. Kravchenko, V.V., Torba, S.M.: Analytic approximation of transmutation operators and applications to highly accurate solution of spectral problems. J. Comput. Appl. Math. **275**, 1–26 (2015)
9. Leontiev, A.F.: Generalizations of Exponential Series. Nauka, Moscow (1981). (in Russian)
10. Olver, F.W.J., Lozier, W., Boisvert, R.F., Clark, C.W.: NIST Handbook of Mathematical Functions. Cambridge University Press, New York (2010)
11. Prudnikov, A.P., Brychkov, YuA, Marichev, O.I.: Integrals and Series, vol. 2. Special Functions, p. 750. Gordon & Breach Science Publishers, New York (1986)
12. Sitnik, S.M.: Transmutations and applications: a survey, arXiv:1012.3741v1. (Originally published in the book) In: Korobeinik, Y.F., Kusraev, A.G. (eds.) Advances in Modern Analysis and Mathematical Modeling, pp. 226–293. Vladikavkaz Scientific Center of the Russian Academy of Sciences and Republic of North Ossetia–Alania, Vladikavkaz (2008)

On a Boundary Value Problem with Infinite Index

H. M. Hayrapetyan

Abstract The generalized Riemann boundary value problem for analytic functions is investigated in the weighted space. It is supposed that the weight function has infinitely many zeroes on the unit circumference. It is proved that the homogeneous problem has an infinite number of linearly independent solutions and under some additional conditions on the order of zeroes of the weight function these solutions determined in explicit form.

Keywords Boundary value problem · Index · Weighted space

1 Introduction. Formulation of the Problem

Let G^+ be a simply connected domain in the complex plane bounded by a rectifiable curve L, and G^- be the complement of the set $G^+ \cup L$. Riemann boundary value problem is formulated as (cf. [1–3]): find analytic in $G^\pm$ functions $\varphi^\pm$, $(\varphi^-(\infty) = 0)$ such that the equality

$$\varphi^+(t) - a(t)\varphi^-(t) = f(t), t \in L,$$

where a and f are given on L functions is satisfied. Investigation of this problem in the classes C^α, $L^p(1 < p < \infty)$ is based on the fact that Cauchy type integral is a bounded operator in the spaces C^α and L^p $(1 < p < \infty)$.

For the investigation of the Riemann boundary value problem in the class L^1 another formulation of this problem is proposed, which in the case of the unit circle has the following form [4]: find analytic in $D^+ \cup D^-$, $D^+ = \{z; |z| < 1\}$, $D^- = \{z; |z| > 1\}$ functions $\varphi^\pm$, $(\varphi^-(\infty) = 0)$ such that the equality

$$\lim_{r \to 1-0} \|\varphi^+(rt) - a(t)\varphi^-(r^{-1}t) - f(t)\|_{L^1} = 0$$

H. M. Hayrapetyan (✉)
Yerevan State University, Alek Manukyan st. 1, Yerevan 0025, Armenia
e-mail: hhayrapet@gmail.com

A. Karapetyants et al. (eds.), *Modern Methods in Operator Theory and Harmonic Analysis*, Springer Proceedings in Mathematics & Statistics 291,
https://doi.org/10.1007/978-3-030-26748-3_22

is satisfied. In similar settings the Riemann boundary value problem is investigated in spaces C, L^{∞}, W. (cf. [5–8]). In [9–11] similar problems are investigated in the weighted spaces, where the weight function admits zero value in a finite set of boundary points. In the present paper we investigate the Riemann boundary value problem in the unit circle D^+ formulated as follows: find analytic in $D^+ \cup D^-$ function φ, $(\varphi(\infty) = 0)$ such that

$$\lim_{r \to 1-0} \|\varphi^+(rt) - a(t)\varphi^-(r^{-1}t) - f(t)\|_{L^1(\rho)} = 0, \tag{1}$$

$\rho(t) = \prod_{k=1}^{\infty} |t_k - t|^{\delta_k}$, $1 > \delta_k > 0$, $\sum_{k=1}^{\infty} \delta_k < \infty$, $t_k \in T$, $T = \{z; |z| = 1\}$. We suppose that $t_k = e^{i\theta_k}$, $\theta_k \downarrow 0$.

In the present paper is proved that the homogeneous problem (1) has an infinite number of linearly independent solutions and if additional condition on $\{t_k\}_1^{\infty}$, $\{\delta_k\}$ is imposed, then the solution of the homogeneous problem (particularly, if $a(t) \equiv 1$) may be represented as

$$\varphi(z) = \sum_{k=1}^{\infty} \frac{A_k}{t_k - z},$$

where $\{A_k\}_1^{\infty} \in l^1$.

2 Some Auxiliary Results

Here and in the sequel we consider problem (1) when $a(t) \equiv 1$.

Lemma 1 *Function $\rho(t) = \prod_{k=1}^{\infty} |t_k - t|^{\delta_k}$ is continuous at 1 and $\rho(1) = 0$, if and only if*

$$\sum_{k=1}^{\infty} \delta_k \ln|1 - t_k| = -\infty \tag{2}$$

Proof The sufficiency of condition (2) is evident, as $\rho(t_k) = 0$, $k = 1, 2, \ldots$. Consider the function

$$R(t) = \sum_{k=1}^{\infty} \delta_k \ln|t_k - t|.$$

We prove now that $\lim_{t \to 0} R(t) = -\infty$. Let $A > 0$ be an arbitrary positive number. Chose $N > 0$ such that

$$\sum_{k=1}^{N} \delta_k \ln|1 - t_k| < -A.$$

A number $\delta > 0$ may be chosen such that at $|1 - t| < \delta$ the inequality

$$\sum_{k=1}^{N} \delta_k \ln |t - t_k| < -\frac{A}{2}$$

is satisfied. Then we have

$$\sum_{k=1}^{\infty} \delta_k \ln |t - t_k| < -\frac{A}{2},$$

at $|1 - t| < \delta$. Lemma 1 is proved.

Lemma 2 *Let the sequence* $\{t_k\}_1^\infty$ *satisfy the following conditions:*

$$\sum_{k=1}^{\infty} \delta_k \ln |1 - t_k| > -\infty; \quad |t_k - t_j| > c|t_k - 1|, \quad j \neq k. \tag{3}$$

for some fixed number $c > 0$. *Then*

$$\inf \rho_i = \rho_0 > 0, \quad i = 1, 2, ..., \tag{4}$$

where

$$\rho_i = \prod_{k \neq i}^{\infty} |t_k - t_i|^{\delta_k}$$

Proof Let $t_k = e^{i\theta_k}$, $k = 1, 2, ...$. Then $|t_k - t_i| > |\theta_k - \theta_i| > c\theta_k$, $|t_k - t_j| > c|t_k - 1|$, where $c > 0$ does not depend on k and j. Hence $|t_k - t_j|^{\delta_k} > c^{\delta_k}|t_k - 1|^{\delta_k}$ and

$$\prod_{k \neq i}^{\infty} |t_k - t_i|^{\delta_k} > \prod_{k=1}^{\infty} c^{\delta_k} \prod_{k=1}^{\infty} |t_k - 1|^{\delta_k}$$

Note that condition (3) implies that

$$\rho(1) = \delta > 0. \tag{5}$$

Lemma 2 is proved.

Let $t_k = e^{i\theta_k}$, $\theta'_0 = 2\pi$, $\theta'_k = 2^{-1}(\theta_k + \theta_{k+1})$, $t'_k = e^{i\theta'_k}$, $k = 0, 1, ...,$ $T_k = [t'_{k-1}, t'_k)$, $k = 1, 2, ...$ It is clear that $T_k \cap T_{k+1} = \emptyset$, $\cup_{k=1}^{\infty} T_k = T$.
We put

$$\delta_k(t) = \prod_{k \neq j}^{\infty} |t_j - t|^{\delta_j}$$

Lemma 3 *Let the sequence of points* $\{t_k\}_1^\infty$ *satisfy condition (3). Then there exists* $\delta > 0$ *such that for any* $k = 1, 2, ...$

$$\inf_{t\in T_k} \delta_k(t) > \delta > 0.$$

Proof Let $t \in (t'_{k-1}, t_k)$, then $|t_j - t| \geq |t_j - t_k|$, at $j \geq k+1$. If $j < k$ we have $|t_j - t| > |t_j - t_{k-1}|$ and

$$\prod_{j<k} |t_j - t|^{\delta_j} > \prod_{j<k-1} |t_j - t_{k-1}|^{\delta_j} > \delta.$$

Hence $\delta_k(t) \geq \delta^2, t \in (t'_{k-1}, t_k)$. Let now $t \in (t_k, t'_k)$, then at $j < k$ we have $|t_j - t| > |t_j - t_k|$ and

$$\prod_{j=1}^{k-1} |t_j - t|^{\delta_j} \geq \prod_{j=1}^{k-1} |t_j - t_k|^{\delta_j} > \delta.$$

At $j \geq k+1$ we have $|t_j - t| > c|t_j - t_{k-1}|$ where $c > 1$ does not depend on j, k. We get

$$\prod_{j>k} |t_j - t|^{\delta_j} > \prod_{j>k} c^{\delta_j} \prod_{j>k} |t_j - t_{k-1}|^{\delta_j} > c\delta.$$

Hence $\delta_k > c\delta^2$. Lemma 3 is proved.

Let's denote

$$f_k(t) = \begin{cases} f(t), & t \in T_k \\ 0, & t \in T \backslash T_k \end{cases}.$$

Define the function $\varphi_{1k}(z)$ as follows

$$\varphi_{1k}(z) = \frac{1}{2\pi i (t_k - z)} \int_{T_k} \frac{f_k(t)(t_k - t)dt}{t - z}, k = 1, 2, ... \tag{6}$$

Lemma 4 *The estimate*

$$\|\varphi_{1k}^+(rt) - \varphi_{1k}^-(r^{-1}t)\|_{L^1(\rho)} \leq C\|f_k(t)\|_{L^1(\rho)}, \tag{7}$$

where the constant C is independent of k and r, is true. The limit relation

$$\lim_{r\to 1-0} \|\varphi_{1k}^+(rt) - \varphi_{1k}^-(r^{-1}t) - f_k(t)\|_{L^1(\rho)} = 0.$$

also holds.

Proof Note that

$$\varphi_{1k}^{+}(rt) - \varphi_{1k}^{-}(r^{-1}t) = I_1(r,t,k) + I_2(r,t,k).$$

where

$$I_1(r,t,k) = \frac{1}{2\pi i}\frac{\bar{t}_k(1-r^2)}{|t_k - rt|^2}\int_{T_k}\frac{f_k(\tau)(t_k-\tau)d\tau}{\tau - rt},$$

$$I_2(r,t,k) = \frac{1}{2\pi i}\frac{1}{t_k - r^{-1}t}\int_{T_k}\frac{f_k(\tau)(t_k-\tau)(1-r^2)|d\tau|}{|\tau - rt|^2}.$$

Further

$$I_1(r,t,k)|t_k - t|^{\delta_k} \le \frac{(1-r^2)^{1-\delta_k}}{|t_k - rt|^{2-\delta_k}}\Big(\int_{T_k}|f_k(\tau)||t_k-\tau|^{\delta_k}|d\tau|$$

$$+\int_{T_k}|f_k(\tau)||t_k-\tau|^{\delta_k}\left|\frac{t_k - rt}{\tau - rt}\right|^{1-\delta_k}|d\tau|\Big) = I_1'(r,t,k) + I_2'(r,t,k)$$

$$\int_T I_1'(r,t,k)|dt| \le \int_{T_k}|f_k(\tau)||t_k-\tau|^{\delta_k}|d\tau|\int_T\frac{(1-r^2)^{1-\delta_k}}{|t_k - rt|^{2-\delta_k}}|dt|$$

$$\le C_1\int_{T_k}|f_k(\tau)||t_k-\tau|^{\delta_k}|d\tau|$$

$$\int_T I_1''(r,t,k)|dt| \le \int_{T_k}|f_k(\tau)||t_k-\tau|^{\delta_k}|d\tau|\int_T\frac{(1-r^2)^{1-\delta_k}}{|t_k - rt||\tau - rt|^{1-\delta_k}}|dt|$$

$$\le C_2\int_{T_k}|f_k(\tau)||t_k-\tau|^{\delta_k}|d\tau|$$

It may be noted that the constants C_1, C_2 do not depend on r and k. Therefore

$$\|I_1(r,t,k)\|_{L^1(\rho)} \le M_1\|f_k\|_{L^1(\rho)},$$

where M_1 is a constant independent of r and k.

Further we have

$$I_2(r,t,k)|t_k - t|^{\delta_k} \le \frac{|t_k - t|^{\delta_k}|1-r|^{1-\delta_k}}{|t_k - r^{-1}t|}\int_{T_k}\frac{|f_k(\tau)||t_k-\tau|^{\delta_k}(1-r)^2}{|\tau - rt|^2}|d\tau|$$

and as

$$\int_T\frac{|t_k - t|^{\delta_k}|1-r|^{1-\delta_k}}{|t_k - r^{-1}t|}|dt| \le c_1$$

we get

$$\|I_2(r,t,k)\|_{L^1(\rho)} \leq M_2\|f_k\|_{L^1(\rho)},$$

where M_2 is a constant independent of r and k.

Therefore

$$\|\varphi_{1k}(rt) - \varphi_{1k}^{-}(r^{-1}t)\|_{L^1(\rho)} \leq M\|f_k\|_{L^1(\rho)}$$

where $M = \max\{M_1, M_2\}$. From this estimate immediately follows the proof of the lemma.

3 Main Theorems

Theorem 1 *Let the sequence $\{t_k\}_1^\infty$ satisfy condition (3). Then the general solution of the homogeneous problem*

$$\lim_{r\to 1-0} \|\varphi^+(rt) - \varphi^-(r^{-1}t)\|_{L^1(\rho)} = 0, \tag{8}$$

may be represented as

$$\Phi_0(z) = \sum_{k=1}^{\infty} \frac{A_k}{t_k - z}, \tag{9}$$

where $\{A_k\} \in l^1$.

Proof Sufficiency. Let $\Phi_{0k} = A_k(t_k - z)^{-1}$, $|z| \neq 1$, $k = 1, 2, \ldots$. As

$$\Phi_{0k}(rt) - \Phi_{0k}(r^{-1}t) = A_k \frac{(1-r^2)\bar{t}_k}{|t_k - rt|^2},$$

then

$$\int_T \frac{1-r^2}{|t_k - rt|^2}|t_k - t|^{\delta_k}|dt| \to 0$$

Further we have

$$\Phi_0(rt) - \Phi_0(r^{-1}t) = \sum_{k=1}^{\infty} \frac{A_k(1-r^2)}{|t_k - rt|^2}$$

As this series converges uniformly at $r \in (0, 1)$, then

$$\int_T |\Phi_0(rt) - \Phi_0(r^{-1}t)|\rho(t)|dt| \leq C \sum_{k=1}^{\infty} A_k \int_T \frac{(1-r^2)|t - t_k|^{\delta_k}}{|t_k - rt|^2}|dt| \leq$$

$$\leq C\sum_{k=1}^{\infty}|A_k|(1-r)^{1-\delta_k}.$$

Recalling that

$$\sum_{k=1}^{\infty}|A_k|(1-r)^{1-\delta_k}\to 0$$

at $r\to 1-0$ we get the proof of the sufficiency of this theorem.

We prove now the necessity of this condition, i.e. that any solution of the homogeneous problem (8) may be represented as in (9). Let $\Phi_0(z)$ be a solution of this problem. It is clear that the points $\{t_k\}_1^{\infty}$ are first order poles of this function and if $r_k(z)=A_k(t_k-z)^{-1}$ is the principal part of Laurent series about this point, then it is sufficient to establish that

$$\sum_{k=1}^{N}|A_k|<S \tag{10}$$

where the number S does not depend on N. Put $t_k=e^{i\theta_k}$, $\theta_k'=2^{-1}(\theta_k+\theta_{k+1})$, $t_k'=e^{i\theta_k'}$, $T_{Nk}=[t_k',t_{k+1}')$, $T_N=\sum_{k=1}^{N}T_{Nk}$.

Let

$$M=\sup\int_{T_N}|\Phi_0(rt)-\Phi_0(r^{-1}t)|\rho(t)|dt|,\ 0<r<1.$$

We have

$$\int_{T_N}|R_N(rt)-R_N(r^{-1}t)|\rho(t)|dt|\leq 2M,\ r>r_0 \tag{11}$$

where

$$R_N(z)=\sum_{k=1}^{N}\frac{A_k}{t_k-z},$$

$r\in(r_0,1)$, $r_0>0$. Estimate (11) follows from the fact that the function $\Phi_0(z)-R_N(z)$ is analytic on the set $\overline{T}_N$. For any $r>r_0$ we have

$$\begin{aligned}I_k(r)&=\int_{T_{Nk}}|R_N(rt)-R_N(r^{-1}t)||t_k-t|^{\delta_k}|dt|\\&>|A_k|\int_{T_{Nk}}\frac{(1-r)|t-t_k|^{\delta_k}}{|t_k-rt|^2}|dt|\\&-\sum_{j\neq k}^{N}\int_{T_{Nk}}|A_j|\frac{(1-r)|t-t_k|^{\delta_k}}{|t_j-rt|^2}|dt|\end{aligned}$$

As $|t_k - rt|^2 = (1-r)^2 + r(\theta - \theta_k)^2 > 2\sqrt{2}(1-r)(\theta - \theta_k)$, then

$$\int_{T_{Nk}} \frac{(1-r)|t-t_k|^{\delta^k}}{|t_k - rt|^2}|dt| \geq \frac{c_0(\theta_k' - \theta_k)^{\delta_k}}{1+\delta_k} > c_0' {\theta_k}^{\delta_k},$$

where $c_0' > 0$ does not depend on k, r, N. Taking into account that according to the condition $|\theta_j - \theta_k| > c_1\theta_k$, where c_1 does not depend on k, j we get

$$\int_{T_{Nk}} \frac{(1-r)|t-t_k|^{\delta_k}}{|t_j - rt|^2}|dt| \leq \int_{T_{Nk}} \frac{(1-r)|t-t_k|^{\delta_k}}{2\sqrt{2}|\theta_j - \theta_k|}|dt|$$

$$\leq \frac{(1-r)^2(1-r)^{\delta_k}}{(1-r)|\theta_j - \theta_k|} = \frac{(1-r)^{1+\delta_k}}{2|\theta_j - \theta_k|}.$$

Hence from $|\theta_k' - \theta_k| > c\theta_k$ we get

$$I_k(r) \geq |A_k|c_0'{\theta_k}^{\delta_k} - \sum_{j\neq k}^{N} \int_{T_{Nk}} |A_j| \frac{(1-r)|t-t_k|^{\delta_k}}{|t_j - rt|}|dt|.$$

As

$$\sum_{j\neq k}^{N} |A_j| \frac{(1-r)|t-t_k|^{\delta_k}}{|t_j - rt|}|dt| \to 0$$

uniformly at $t \in T_{Nk}$, from (4) we get

$$\lim_{r\to 1} I_k(r) \geq c_0''|A_k|$$

where c_0'' is a positive constant independent of r and k. As by (10)

$$\sum_{k=1}^{N} I_k(r) < \frac{2M}{c_0''}$$

we get

$$\sum_{k=1}^{N} |A_k| < \frac{2M}{c_0''}$$

Theorem is proved.

Remark 1 It is necessary to note that the Problem (8) in $L^p(\rho)(1 < p < \infty)$ has the finite number solutions.

Theorem 2 *Let the sequence* $\{t_k\}_1^\infty$ *satisfy the condition (2). We denote*

$$\Phi_0(z) = \frac{A_0}{1-z} + \sum_{k=1}^{\infty} \frac{A_k}{t_k - z},$$

where $\{A_k\} \in l^1$, A_0 *is an arbitrary complex number. Then the function* Φ_0 *is a solution of the homogeneous problem (8).*

Proof It is sufficient to check that the function $\Phi_{00}(z) = A_0(1-z)^{-1}$ is a solution of the Problem (8). Indeed

$$\Phi_{00}(rt) - \Phi_{00}(r^{-1}t) = A_0 \frac{1-r^2}{|1-rt|^2},$$

and according to Lemma 1 the function $\rho(t)$ is continuous on T and $\rho(1) = 0$, therefore,

$$\|\Phi_{00}(rt) - \Phi_{00}(r^{-1}t)\|_{L^1(\rho)} \to 0, \quad r \to 1-0.$$

Now, we consider the problem (1) in the case where $a(t)$ is not constant and $a(t) \in C^\alpha(T)$. Using standard factorization [1] we get $a(t) = S^+(t)(S_-(t))^{-1}$, where

$$S^+(z) = exp\Big\{\frac{1}{2\pi i}\int_{-\infty}^{+\infty} \frac{\ln(t^{-\kappa}a(t))}{t-z}dt\Big\}, \qquad z \in D^+$$

$$S^-(z) = z^{-\kappa} exp\Big\{\frac{1}{2\pi i}\int_{-\infty}^{+\infty} \frac{\ln(t^{-\kappa}a(t))}{t-z}dt\Big\}, \qquad z \in D^-$$

$\kappa = inda(t), t \in T.$

By applying Lemmas 1, 3 and 4, it is not difficult to establish the following theorems.

Theorem 3 *Let the sequence* $\{t_k\}_1^\infty$ *satisfy the condition (3). If* $\kappa \geq 0$*, then the general solution of the homogeneous problem (1) may be represented as*

$$\varphi_0(z) = S(z)\Big(P_{\kappa-1}(z) + \sum_{k=1}^{\infty} \frac{A_k}{t_k - z}\Big),$$

where $P_{\kappa-1}(z)$ *is an arbitrary polynomial of degree* $\kappa - 1$*, and the sequence* $\{A_k\}_{k=1}^\infty \in l^1$.
(b) If $\kappa < 0$*, then the general solution of the homogeneous problem (1) may be represented as*

$$\varphi_0(z) = S(z)\Big(\sum_{k=1}^{\infty} \frac{A_k}{t_k - z}\Big),$$

where $\{A_k\}_{k=1}^{\infty} \in l^1$, $A_{-\kappa+1}, A_{-\kappa+2}, \ldots$ *are arbitrary, and* $A_1, A_2, \ldots, A_\kappa$ *are uniquely determined from the system of linear equations*

$$A_1 t_1^j + A_2 t_2^j + \ldots + A_\kappa t_\kappa^j = -\sum_{k=\kappa+1}^{\infty} A_k t_k^j,$$

where $j = 0, 1, \ldots, \kappa - 1$.

Taking into account Lemma 3 we can prove following result.

Theorem 4 *Let the sequence* $\{t_k\}_1^\infty$ *satisfy the condition (3). Then the following assertions hold: (a) If* $\kappa \geq 0$, *then the general solution of the in-homogeneous problem (1) may be represented as* $\varphi(z) = \varphi_0(z) + \varphi_1(z)$, *where* φ_0 *is the general solution of the homogeneous equation and*

$$\varphi_1(z) = \sum_{k=1}^{\infty} \varphi_{1k}(z) \tag{12}$$

where

$$\varphi_{1k}(z) = \frac{S(z)}{t_k - z} \frac{1}{2\pi i} \int_{T_k} \frac{f(t)(t_k - t)}{S^+(t)(t - z)} dt.$$

b) If $\kappa < 0$, *then the general solution of the homogeneous problem (1) may be represented as* $\varphi(z) = \varphi_0(z) + \varphi_1(z)$, *where*

$$\varphi_0(z) = \sum_{k=1}^{\infty} \frac{A_k}{t_k - z}$$

and $\{A_k\}_{k=1}^{\infty} \in l^1$, $A_{-\kappa+1}, A_{-\kappa+2}, \ldots$ *are arbitrary and the numbers* $A_1, A_2, \ldots, A_\kappa$ *are uniquely defined by the system of linear equations*

$$\sum_{k=1}^{\infty} A_k = 0$$

$$\sum_{k=1}^{\infty} A_k t_k = -\sum_{k=1}^{\infty} B_{k1}$$

$$\sum_{k=1}^{\infty} A_k t_k^2 = -\sum_{k=1}^{\infty} (B_{k2} + t_k B_{k1})$$

$$\ldots$$

$$\sum_{k=1}^{\infty} A_k t_k^{\kappa-1} = -\sum_{k=1}^{\infty} (B_{k\kappa-2} + t_k B_{k\kappa-3} + \ldots + t_k^{\kappa-3} B_{k1})$$

where

$$B_{kj} = -\int_{T_k} \frac{f(t)(t_k - t)}{S^+(t)} t^{j-1} dt, \; j = 1, 2, \ldots .$$

Acknowledgements Research was conducted with support of SCS MES RA in the scope of joint scientific project NYSU-SFU-16/1 financed based on the results of international competition "SCS MES RA-YSU-SFU RF-2016".

References

1. Gakhov, F.D.: Boundary Value Problems. Nauka, Moscow (1963)
2. Muskhelishvili, N.I.: Singular Integral Equations. Nauka, Moscow (1968)
3. Khvedelidze, B.V.: Methods of Cauchy type integrals in discontinuous boundary value problems for holomophic functions of one complex variable. J. Sov. Math. **7**(3), 309–414 (1977)
4. Hayrapetyan, H.M.: Discontinuous Riemann-Privalov problem with a shift. (Izvestiya NAN Armenii) Mathematics **25**, 1 (1990)
5. Hayrapetyan, H.M., Babayan, V.A.: On the Riemann problem in the space of functions, continuous with a weight. Sci. Rep. Belgorod Univ. **17**(112) (24), 5–15 (2011)
6. Hayrapetyan, H.M., Poghosyan, L.V.: Riemann problem in the sense of weak convergence. In: International Conference of Young Scientists, Nalchik (2011)
7. Hayrapetyan, H.M., Poghosyan, L.V.: Riemann-Hilbert boundary value problem for Bitzadze equation with boundary conditions from subspace of measures. Doklady NAN Armenii **112**, 127–132 (2013)
8. Kazaryan, K.S., Soria, F., Spitkovsky, I.M.: Riemann boundary value problem in the spaces with a weight, admitting singularities. Doklady RAN **353**(N6), 717–719 (1997)
9. Hayrapetyan, H.M.: On the solvability of Dirichlet problem for boundary functions belonging to a space with a weight. Math. Notes **76**(5), 599–605 (2004)
10. Hayrapetyan, H.M., Meliksetyan, P.E.: Hilbert boundary value problem in the half plane in the weighted spaces. Izvestija NAN Armenii Mathematics **38**(6), 17–32 (2003)
11. Hayrapetyan, H.M.: Dirichlet Problem in the Half Plane for RO-Varying Weight Functions. NATO Sciences Series. Series 2, Topics in Analysis and Applications, vol. 147, pp. 311–317, Kluwer Academic Publishers, Dordrecht (2002)

A Numerical Realization of the Wiener–Hopf Method for the Kolmogorov Backward Equation

Oleg Kudryavtsev and Vasily Rodochenko

Abstract We propose a new numerical method for a certain type of boundary value problems for 3-dimensional partial differential equations, which are related to first passage time distributions of Itô diffusions. We consider the Kolmogorov backward equation, which arises in various applications including mathematical finance. The technique presented is based on a probabilistic interpretation of the problem, which involves a Markov chain approximation, and a Wiener–Hopf factorization. First, we use Carr's time randomization and approximate the second component of the related diffusion process with a Markov chain. As the result, we reduce the original problem to a sequence of 1-dimensional differential equations with suitable boundary conditions, associated with Gaussian processes, whose constant coefficients are defined by the Markov chain constructed. We also suggest an improvement for the approximation procedure, which lowers the number of nodes used. Then we express an analytic solution to each problem in terms of a probabilistic form of Wiener–Hopf factorization. We implement explicit and approximate factorization formulae numerically using the Fast Fourier Transform algorithm and provide the results of numerical experiments to illustrate the performance of the method developed.

Keywords Partial differential equations · First passage problems · Wiener–Hopf factorization · Kolmogorov equations · Numerical methods

---

O. Kudryavtsev (✉)
Russian Customs Academy Rostov branch, Budennovskiy 20, Rostov-on-Don 344002, Russia
e-mail: koe@donrta.ru

V. Rodochenko
Institute for Mathematics, Mechanics, and Computer Science in the name of I.I. Vorovich (Southern Federal University), Milchakova str. 8a, Rostov-on-Don 344090, Russia
e-mail: vrodochenko@gmail.com

A. Karapetyants et al. (eds.), *Modern Methods in Operator Theory and Harmonic Analysis*, Springer Proceedings in Mathematics & Statistics 291,
https://doi.org/10.1007/978-3-030-26748-3_23

1 Introduction

It has been well known since Dynkin, Feinman, and Katz published their fundamental works, that the solution to a diffusion equation in a given domain can be interpreted as an expected value of a function of Brownian motion at the first exit point from the domain. Itô diffusions are widely used for modeling different physical processes.

The idea that a geometric Brownian motion can be efficiently used for the purpose of modeling asset prices behavior, along with other vital assumptions, gave rise to a famous Black–Scholes model, first introduced in [4]. The Black–Scholes model has been the most popular model for derivative pricing and hedging for decades, although it has shown severe problems with explaining the "volatility smile" phenomena.

As the result, new financial market models with more than one source of randomness were invented. There is a family of models known as the "stochastic volatility models", where the market volatility is governed by a diffusion process. The Heston model [24] with a variance driven by Cox-Ingersoll-Ross (CIR) [20] process is perhaps the most popular stochastic volatility model, which allow one to obtain solutions in closed form to initial value problems related to pricing European options. Fast numerical methods for valuating European options in Heston model and calibrating this model can be found in [14, 45], respectively.

Pricing exotic options in Heston model is more complicated due to the fact that the problem involves solving complex partial differential equations known as the Kolmogorov backward equations, subject to certain initial and boundary conditions specified by an option type. As a rule, there are no known analytical methods for solving such equations, which brings a necessity to use methods of numerical analysis.

The equation we analyze in this paper has numerous applications in physics. In finance, its solution, with suitable boundary conditions, can be interpreted as a price of a derivative contract.

There are three main approaches to solve boundary value problems for the Kolmogorov backward equation numerically.

The first one is based on a Monte Carlo simulation technique (see e.g. [1]). Typically Monte Carlo methods provide relatively good speed and accuracy away from a barrier. The most significant drawback of the methods is low computational performance, especially for the case of exotic options, which comes from a huge number of simulations needed to achieve a good accuracy.

The second approach uses finite-difference schemes [17, 47] to discretize the 3-dimensional Kolmogorov equation. An overview of these methods is provided in [15]. Notice that finite difference schemes will produce sufficiently good results on a very fine grids which makes them computationally expensive. Some authors apply splitting operator method [28, 30] to improve computational efficiency of finite difference methods. Notice that tree-based methods can be considered as explicit finite difference schemes. In [38], a combined tree for both diffusion processes (stock prices and volatilities) was constructed. The method is relatively fast and robust, but the number of tree nodes grows quadratically in the number of time steps.

The third approach could be described as hybrid methods. A common idea is to build an approximation for the variance process (see, for example, [11, 18] or [2]) and proceed from a model with stochastic variance to a regime switching model, some advantages of which are covered in [26]. In fact, by reducing 1 dimension of the equation, we obtain a system of partial differential equations. A tree approximation from [2] has a good robustness and converges to an original CIR process relatively fast.

As a theoretical background behind the procedure, methods of the third group use the method of horizontal lines [16] which includes a time discretization while a space variable remains continuous. It can be shown (see e.g. [6, 32]) that such approach is equivalent to a time randomization technique suggested by Carr in [16]. After that, a sequence of certain stationary boundary problems for differential equations on a line arises. To solve them, one may apply either finite-difference methods like in [15, 31] or Wiener–Hopf factorization method (see, for example, [34]).

The Wiener–Hopf method is a general tool for solving integro-differential equations with convolution-type kernels on a half-line. In application to finance the Wiener–Hopf method was widely used to solve 2-dimensional initial boundary value problems for pricing path-dependent options under Lévy models. The value of such options depends on the trajectory of the underlying asset and a frequency of its price monitoring. In the case of continuously monitored options, one can reduce the initial Kolmogorov backward equation to the Wiener–Hopf one using the method of lines, as we mention above, or apply the Laplace transform in time variable (see e.g. [7–9, 13, 32, 33, 36, 40]). A discrete monitoring case can be treated analogously by means of z-transform (see e.g. [22, 23, 40, 41]). There are a few papers which deal with higher dimensions and use the Wiener–Hopf approach to solve free boundary problems (see [10, 11]) and initial-boundary value problems (see [12]) for the Kolmogorov backward equation with variable coefficients.

In the current paper, we confine ourselves to a certain type of boundary value problems for the Kolmogorov backward equation related to a first passage time distribution of a stochastic process known as the Heston model. We develop a hybrid scheme to solve the problem by means of Wiener–Hopf factorization method. By using a time randomization and approximating the CIR process by an appropriate Markov chain, we reduce the original problem to a sequence of 1-dimensional boundary problems associated with Gaussian processes, whose constant coefficients are defined by the constructed Markov chain. An analytic solution to each problem can be expressed in terms of a Wiener–Hopf factorization in probabilistic form. We implement explicit and approximate factorization formulae using a Fast Fourier Transform algorithm. In order to improve the speed of our method we suggest a modification of the Markov chain approximation from [2], which allows to get accurate results faster by using less nodes of the tree for CIR process than we used with the original procedure. Preliminary results obtained with similar methodology, concerning equations arising for barrier options in the Heston model, were presented at the International Scientific Conference "Modern Methods, Problems and Applications of Operator Theory and Harmonic Analysis" in 2017 and 2018.

The rest of the paper is organized as follows. In Sect. 2 we give a general problem setup for the Kolmogorov backward equation and it's probabilistic interpretation as a first passage problem of a 2-dimensional Itô diffusion. Section 3 specifies the problem for the case known in mathematical finance as the Heston model. Section 4 describes the steps of our method to reduce the original problem to a sequence of 1-dimensional differential equations with suitable boundary conditions: a substitution that simplifies the operator, Carr's randomization and a tree approximation for the CIR process. In Sect. 5 we recall necessary facts on the Wiener–Hopf factorization method and suggest two approaches for solving the problems arisen by means of explicit and approximate factorization formulae. Numerical experiments are placed to Sect. 6. Section 7 concludes.

2 Problem Setup

Let us consider a 3-dimensional partial differential equation for a function $u = u(x,t)$ with $x = (x_1, x_2) \in \mathbf{R}^2$, $t \in \mathbf{R}$:

$$\Big(\frac{\partial}{\partial_t} + L\Big)u = 0, \tag{1}$$

where

$$L = \sum_i \mu_i(x)\frac{\partial}{\partial x_i} + \frac{1}{2}\sum_{i,j}(\sigma\sigma^T)_{ij}(x)\frac{\partial^2}{\partial x_i \partial x_j},$$

$i = 1, 2$, $j = 1, 2$, the functions $\mu(x) = (\mu_1(x), \mu_2(x))$ and

$$\sigma(x) = \begin{pmatrix} \sigma_{11}(x) & \sigma_{12}(x) \\ \sigma_{21}(x) & \sigma_{22}(x) \end{pmatrix}$$

satisfy the conditions in Theorem 5.2.1 (existence and uniqueness theorem for stochastic differential equations) [39]:

Let $T > 0$ and $\mu : \mathbf{R}^2 \to \mathbf{R}^2$, $\sigma : \mathbf{R}^2 \to \mathbf{R}^{2\times 2}$ be measurable functions satisfying

$$|\mu(x)| + |\sigma(x)| \le C(1 + |x|); \qquad x \in \mathbf{R}^2$$

for some constant C (where $|\sigma|^2 = \sum |\sigma_{i,j}|^2$) and such that

$$|\mu(x) - \mu(y)| + |\sigma(x) - \sigma(y)| \le D|(x - y)|; \qquad x, y \in \mathbf{R}^2$$

for some constant D.

Recall that μ and σ—(or sometimes $\frac{1}{2}(\sigma\sigma^T)$) are often called the drift coefficient and the diffusion coefficient, respectively.

Being essentially a diffusion equation (and the Kolmogorov backward equation), (1) describes an evolution of a 2-dimensional process in time and arises in numerous applications.

Let $H > 0$ be an absorbing barrier. We consider equation (1) subject to the following initial (terminal, to be specific) and boundary conditions.

$$\begin{cases} (\frac{\partial}{\partial t} + L)u = 0, & x_1 > H, x_2 \in \mathbf{R}, t < T, \\ \\ u(x_1, x_2, T) = 0, & x_1 > H, x_2 \in \mathbf{R}, \\ u(x_1, x_2, t) = 1, & x_1 \leq H, x_2 \in \mathbf{R}, t \leq T. \end{cases} \tag{2}$$

A review of existence and uniqueness conditions for different problems for the Kolmogorov equation can be found in [21, 39] and, for the less general case we focus on, in [17, 25]. Although general analytic solutions to the problem are not known, one may develop an appropriate numerical technique by taking the behavior of $\mu_i(x)$ and $\sigma(x)$ into account.

We suggest an approach which uses a probabilistic interpretation of this equation. Let $B_1(t)$, $B_2(t)$ be independent Brownian motions [39]. It can be shown (see e.g. [39]) that the function u can be represented as a conditional expectation of an indicator-function for the event of the first entrance of two-dimensional stochastic process X_t into the region $x_1 \leq H$, where $X_t = (X_1(t), X_2(t))$ is defined by a system of two stochastic differential equations in Itô's form:

$$\begin{aligned} dX_1(t) &= \mu_1 dt + \sigma_{11} dB_1(t) + \sigma_{12} dB_2(t), \\ dX_2(t) &= \mu_2 dt + \sigma_{21} dB_1(t) + \sigma_{22} dB_2(t). \end{aligned} \tag{3}$$

In mathematical physics, the solution of (3) may be interpreted as the mathematical description of small particle motion in a moving fluid. For that reason, such stochastic processes are called (Itô) diffusions.

Notice that in probabilistic terms, problem (2) is closely related to the first-passage time T_H of $(X_1(t), X_2(t))$ to the lower half-plane $x_1 \leq H$:

$$T_H := \inf\{t > 0 | X_1(t) \leq H\},$$

where $H < x_1^0 (= X_1(0))$. Hence, we have

$$u(x_1, x_2, 0) = \mathbf{P}(T_H \leq T) = \mathbf{P}(\underline{X}_1(T) \leq H), \tag{4}$$

where $\underline{X}_1(T) = \inf_{0 \leq s \leq T} X_1(s)$.

In next sections we demonstrate our approach based on a numerical Wiener–Hopf factorization technique for a special case of (3) which arises in mathematical finance.

3 Heston Model Case

Let us now focus on a less general form of Eq. (1), which arises from a model for asset prices in financial applications. First, we define positive constants $\kappa_V, \theta_V, \sigma_V$; and $\rho \in (-1, 1)$. Then set $\hat{\rho} = \sqrt{1-\rho^2}$ and assume $\sigma_{22} = 0$. Denote $X_1(t)$ as S_t and $X_2(t)$ as V_t. Putting $\mu_1 = 0$, $\mu_2 = \kappa_V(\theta_V - V_t)$ and

$$\sigma = \begin{pmatrix} \rho\sqrt{V_t}S_t & \hat{\rho}\sqrt{V_t}S_t \\ \sigma_V\sqrt{V_t} & 0 \end{pmatrix},$$

from system (3) we obtain:

$$\begin{aligned} dS_t &= \sqrt{V_t}S_t(\rho dB_1(t) + \hat{\rho} dB_2(t)), \\ dV_t &= \kappa_V(\theta_V - V_t)dt + \sigma_V\sqrt{V_t}dB_1(t). \end{aligned} \tag{5}$$

The generator L in Eq. (1) in terms of functions $u = u(S, v, t)$ (we replace x_1 by S and x_2 with v) can be rewritten as follows:

$$L = \frac{1}{2}S^2 v\frac{\partial^2}{\partial S^2} + \rho\sigma_V v S\frac{\partial^2}{\partial S\partial v} + \frac{1}{2}\sigma_V^2 v\frac{\partial^2}{\partial v^2} + \kappa_V(\theta_V - v)\frac{\partial}{\partial v}. \tag{6}$$

System (5) consists of two diffusional random processes, which, from a financial viewpoint, can be interpreted as a dynamics of an underlying asset price S_t and its variance V_t (governed by a widely used CIR process [20]), where parameters are positive and have the following sense:

- parameter κ_V defines the speed of "regression" of variance process to a "long-run" variance average value θ_V,
- parameter θ_V is a long-run average value,
- parameter σ_V is referred to as a "volatility" of variance (in a sense that the variance of variance equals to σ_V^2).

Notice that $S_t > 0$ and $V_t \geq 0$. If $V(0) > 0$ and the Feller condition $2\kappa_V\theta_V \geq \sigma_V^2$ is satisfied, it could be shown that the variance process V_t is strictly positive [20]. Processes $(\rho dB_1(t) + \hat{\rho} dB_2(t))$ and $B_1(t)$ are Brownian motions with correlation coefficient ρ. It is also assumed that S_t is a martingale process under a suitable risk-neutral measure. With all these assumptions, (5) describes the Heston model [24].

Replacing V_t for all $t > 0$ with a positive constant V_0 makes S_t a Gaussian process, and (5) turns into a geometric Brownian motion model.

Now we may rewrite problem (2) in new notations as follows:

$$\begin{cases} (\frac{\partial}{\partial t} + \frac{1}{2}S^2 v \frac{\partial^2}{\partial S^2} + \rho\sigma_V v S \frac{\partial^2}{\partial S \partial v} + \frac{1}{2}\sigma_V^2 v \frac{\partial^2}{\partial v^2} + \kappa_V(\theta_V - v)\frac{\partial}{\partial v})u = 0, \\ \qquad\qquad S > H, v \geq 0, t < T, \\ u(S, v, T) = 0, \quad S > H, v \geq 0, \\ u(S, v, t) = 1, \quad 0 < S \leq H, v \geq 0, t \leq T. \end{cases} \tag{7}$$

As in [2, 15, 17], we do not impose the Feller condition, hence $v \geq 0$.

A 0-time solution to problem (7) can be interpreted as the following conditional expectation:

$$u(S, v, 0) = E[\mathbf{1}_{[0,T]}(T_H)|S_0 = S, V_0 = v], \tag{8}$$

where T_H is the first entrance time of S_t into $(0, H]$ (crossing the barrier H from below). In the case with the barrier H from above, we flip the inequality sign in the relationships between S and H in (7).

$$\begin{cases} (\frac{\partial}{\partial t} + \frac{1}{2}S^2 v \frac{\partial^2}{\partial S^2} + \rho\sigma_V v S \frac{\partial^2}{\partial S \partial v} + \frac{1}{2}\sigma_V^2 v \frac{\partial^2}{\partial v^2} + \kappa_V(\theta_V - v)\frac{\partial}{\partial v})u = 0, \\ \qquad\qquad 0 < S < H, v \geq 0, t < T, \\ u(S, v, T) = 0, \quad 0 < S < H, v \geq 0, \\ u(S, v, t) = 1, \quad S \geq H, v \geq 0, t \leq T. \end{cases} \tag{9}$$

The correspondent solution can be found as the expectation (8) again, but with T_H being the first entrance time of S_t into $[H, +\infty)$.

From the financial viewpoint, the solution $u(S, v, 0)$ can be interpreted as the price of a first touch digital option which is a contract which pays 1$ to its holder as stock price S_t for the first time crosses the barrier H (a positive constant) any time between $t = 0$ and $t = T$ (an expiration date). If this event does not take place before the expiration date, at $t = T$ the option becomes worthless. First touch digital options are important for risk management and risk evaluation. They belong to a class of path-dependent options—i.e. contracts with a value depending on a whole trajectory of an underlying asset price.

A numerical solution to 3-dimensional problem (7) with a standard finite difference technique is rather time consuming. In particular, there are computational difficulties caused by errors from an approximation of the mixed partial derivative in the operator L defined in (6) which is due to the non-zero correlation between S_t and V_t in (5). In the next section we make a substitution to avoid the correlation effect and reduce our problem to a family of 1-dimensional problems on a half-line.

4 Reduction to a Family of One-Dimensional Problems

Section 4 describes the techniques and methods we use to simplify the original problem, make the approximation for the variance process V_t and obtain a sequence of 1-dimensional problems. In Sect. 4.1 we introduce a substitution which modifies the behavior of the barrier, but allows to rewrite L with no mixed partial derivatives. Carr's randomization technique, which is used to reduce the original problem to a sequence of problems on small time intervals, is described in Sect. 4.2. The Markov chain approximation technique originally presented in [2] is applied to the case we consider in Sect. 4.3. Section 4.3 continues by deriving a resulting sequence of 1-d problems with constant variance and introducing a modification which allows to reduce the number of computations the method needs.

4.1 Substitution

An ability to reflect the effect of correlation between the price and its variance is an important feature of the model. There are a number of ways to take it into account. The method we use is similar to the one in [15], and aimed at constructing a suitable substitution affecting correlation between the model processes. Let us introduce a process

$$Y_t = \ln \frac{S_t}{H} - \frac{\rho}{\sigma_V} V_t.$$

In these terms,

$$S_t = H \exp(Y_t + \frac{\rho}{\sigma_V} V_t).$$

First of all, it allows to change the S scale to a logarithmic scale and make a normalization with respect to the barrier H, which is convenient for computations. By applying Itô's lemma, the system (5) can be rewritten in terms of Y_t and V_t as follows (see [15]):

$$\begin{aligned} dY_t &= \mu_Y(V_t)dt + \hat{\rho}\sqrt{V_t} dB_2(t), \\ dV_t &= \mu_V(V_t)dt + \sigma_V \sqrt{V_t} dB_1(t), \end{aligned} \tag{10}$$

where the drifts $\mu_Y(v)$ and $\mu_V(v)$ in the system (10) are as follows:

$$\begin{aligned} \mu_Y(V_t) &= -\tfrac{1}{2} V_t - \tfrac{\rho}{\sigma_V} \kappa_V(\theta_V - V_t), \\ \mu_V(V_t) &= \kappa_V(\theta_V - V_t). \end{aligned}$$

Notice that Y_t and V_t are independent, hence the associated generator L has the following, more convenient, form without mixed derivatives:

$$L = \frac{1}{2}\hat{\rho}^2 v \frac{\partial^2}{\partial y^2} + \frac{1}{2}\sigma_V^2 v \frac{\partial}{\partial v^2} + \mu_Y(v)\frac{\partial}{\partial y} + \mu_V(v)\frac{\partial}{\partial v}.$$

The substitution however changes the behavior of the barrier. Set $f(y, v, t) = u\Big(H \exp(y + \frac{\rho}{\sigma_V} v), v, t\Big)$, then the problem (7) can be written as follows.

$$\begin{cases} \left(\frac{\partial}{\partial_t} + \frac{1}{2}\hat{\rho}^2 v \frac{\partial^2}{\partial y^2} + \frac{1}{2}\sigma_V^2 v \frac{\partial}{\partial v^2} + \mu_Y(v)\frac{\partial}{\partial y} + \mu_V(v)\frac{\partial}{\partial v}\right) f = 0, \\ \qquad\qquad y + \frac{\rho}{\sigma_V} v > 0, v \geq 0, t < T, \\ f(y, v, T) = 0, \quad y + \frac{\rho}{\sigma_V} v > 0, v \geq 0, \\ f(y, v, t) = 1, \quad y + \frac{\rho}{\sigma_V} v \leq 0, v \geq 0, t \leq T. \end{cases} \tag{11}$$

Then T_H can be interpreted as the first time the process $Y_t + \frac{\rho}{\sigma} V_t$ enters a region $(-\infty, 0]$:

$$T_H = \inf_{t \geq 0}\{t : Y_t + \frac{\rho}{\sigma_V} V_t \leq 0\}.$$

Hence, we can rewrite the expectation (8) in terms of f:

$$f(y, v, 0) = E^{y,v}[\mathbf{1}_{[0,T]}(T_H)]. \tag{12}$$

Recall that the shorthand notation $E^{y,v}[\cdot]$ means that we calculate an expectation conditioned on $Y_0 = y$, $V_0 = v$.

4.2 Carr's Randomization

The procedure of Carr's randomization, introduced in [16] for the case of American call options, is similar to the method of horizontal lines (MOL) [43]. Later it was generalized for a wider class of stochastic control problems in paper [5]. In [7, 32] the technique was successfully used for pricing first touch digital options in Lévy models.

The core idea is to substitute the deterministic time T in expectation (12) with a suitable random variable T' whose distribution can be concentrated arbitrarily close around T. This allows one to reduce the initial problem to a sequence of stationary problems.

Let's define $N \in \mathbf{N}$, $\Delta\tau = \frac{T}{N}$ and for $i = 0, \ldots, N$ set $t_i = i\Delta\tau$. Consider a sequence $\{\tau_i\}_{i=1}^N$ of independent identically exponentially distributed variables (and also independent of Y_t, V_t) with an average value equal to $\Delta\tau$. Next step, we set $T' = \sum_{i=1}^N \tau_i$, which is equal in law to a Gamma random variable $\Gamma(n, q)$ with parameters $n = N$ and $q = N/T$. Since $T' \sim \Gamma(N, N/T)$, then $E[T'] = T$ and it's variance $Var[T'] = \frac{T^2}{N} \to 0$ as $N \to \infty$.

Let us focus on $f(y, v, 0)$. Let's denote

$$Z_t = Y_t + \frac{\rho}{\sigma_V} V_t.$$

At this point we replace T in (12) with a gamma distributed random variable $T' \sim \Gamma(N, N/T)$, independent of Y_t and V_t.

Taking into account (4) we can see that $f(y, v, 0)$ can be approximated by:

$$\begin{aligned} f_0(y, v) &= E^{y,v}[\mathbf{1}_{[0,T']}(T_H)] \\ &= E^{y,v}[\mathbf{1}_{(-\infty,0]}(\underline{Z}_{T'})], \end{aligned} \tag{13}$$

where

$$\underline{Z}_t = \inf_{0 \le s \le t} \{Z_s\}$$

is the infimum process of Z_t. It was shown in [5] that we can iteratively calculate an approximation for $f_0(y, v) = f(y, v, 0)$. Let us denote $f_n(y, v)$ a Carr's approximation of $f(y, v, t_n)$ as

$$f_n(y, v) = E[\mathbf{1}_{(-\infty,0]}(\underline{Z}_{T'}) | Y_{T'_n} = y, V_{T'_n} = v], \tag{14}$$

where $T'_n = \sum_{i=1}^n \tau_i,$. Then we have

$$f_N(y, v) = \begin{cases} 0, \; y > -\frac{\rho}{\sigma_V} v, \\ 1, \; y \le -\frac{\rho}{\sigma_V} v. \end{cases}$$

Further, for $n = N-1, N-2, \ldots, 0$ we can express $f_n(y, v)$ via $f_{n+1}(y, v)$:

$$\begin{aligned} f_n(y, v) &= E\Big[\mathbf{1}_{(-\infty,0]}(\underline{Z}_{T'_{n+1}}) \mathbf{1}_{(-\infty,0]}(\underline{Z}_{T'}) | Y_{T'_n} = y, V_{T'_n} = v\Big] + \\ &+ E\Big[\mathbf{1}_{(0,+\infty)}(\underline{Z}_{T'_{n+1}}) \mathbf{1}_{(-\infty,0]}(\underline{Z}_{T'}) | Y_{T'_n} = y, V_{T'_n} = v\Big] = \\ &= E[\mathbf{1}_{(-\infty,0]}(\underline{Z}_{T'_{n+1}}) | Y_{T'_n} = y, V_{T'_n} = v] + \\ &+ E[\mathbf{1}_{(0,+\infty)}(\underline{Z}_{T'_{n+1}}) f_{n+1}(Y_{T'_{n+1}}, V_{T'_{n+1}}) | Y_{T'_n} = y, V_{T'_n} = v]. \end{aligned} \tag{15}$$

Generally speaking, for $l \in \mathbf{N}, n + l \le N$, we can calculate $f_n(y, v)$ using $f_{n+l}(y, v)$:

$$\begin{aligned} f_n(y, v) &= E[\mathbf{1}_{(-\infty,0]}(\underline{Z}_{T'_{n+l}}) | Y_{T'_n} = y, V_{T'_n} = v] + \\ &+ E[\mathbf{1}_{(0,+\infty)}(\underline{Z}_{T'_{n+l}}) f_{n+l}(Y_{T'_{n+l}}, V_{T'_{n+l}}) | Y_{T'_n} = y, V_{T'_n} = v]. \end{aligned} \tag{16}$$

According to the Markov property of Itô diffusion, one can rewrite (15) as

$$f_n(y, v) = E^{y,v}[\mathbf{1}_{(-\infty,0]}(\underline{Z}_{\tau_{n+1}})] + E^{y,v}[\mathbf{1}_{(0,+\infty)}(\underline{Z}_{\tau_{n+1}}) f_{n+1}(Y_{\tau_{n+1}}, V_{\tau_{n+1}})], \tag{17}$$

and (16) as

$$f_n(y,v) = E^{y,v}[\mathbf{1}_{(-\infty,0]}(\underline{Z}_{T_l'})] + E^{y,v}[\mathbf{1}_{(0,+\infty)}(\underline{Z}_{T_l'}) f_{n+l}(Y_{T_l'}, V_{T_l'})]. \quad (18)$$

Note that the random time T_l' is a gamma distributed random variable, which is equal in law to $\Gamma(l, N/T)$.

Let us justify the convergence in time of the scheme. Proofs of the Carr's randomization convergence in the case of similar problems for Lévy processes are presented in [6, 32]. Notice that in the current paper we apply a different technique.

Denote $F(y,v,T) = f(y,v,0)$, with an emphasis on the fact that the solution $f(y,v,0)$ depends on a deterministic time $T > 0$. We have

$$\begin{aligned} F(y,v,T) &= E[\mathbf{1}_{[0,T]}(T_H)|Y_0 = y, V_0 = v] \\ &= E[\mathbf{1}_{(-\infty,0]}(\underline{Z}_T)|Y_0 = y, V_0 = v] \end{aligned} \quad (19)$$

Applying Laplace transform to $F(y,v,T)$ with respect to time T, we obtain

$$\begin{aligned} \hat{F}(y,v,q) &= \int_0^{+\infty} e^{-qt} E^{y,v}\big[\mathbf{1}_{(-\infty,0]}(\underline{Z}_t)\big]dt \\ &= q^{-1} E^{y,v}[\mathbf{1}_{(-\infty,0]}(\underline{Z}_{T_q})], \end{aligned} \quad (20)$$

where $T_q \sim Exp(q)$. Differentiating both parts of the Eq. (20) $n-1$ times and multiplying it on $\frac{(-1)^{n-1}q^n}{(n-1)!}$, we obtain

$$\begin{aligned} \frac{(-1)^{n-1}q^n}{(n-1)!}\partial_q^{n-1}\hat{F}(y,v,q) &= \frac{q^n}{(n-1)!}\int_0^{+\infty} t^{n-1}e^{-qt}E^{y,v}\big[\mathbf{1}_{(-\infty,0]}(\underline{Z}_{T_q})\big]dt \\ &= E^{y,v}\big[\mathbf{1}_{(-\infty,0]}(\underline{Z}_{\Gamma(n,q)})\big]. \end{aligned} \quad (21)$$

Notice that at $q = N/T$ (21) coincides with (13). Hence, the convergence of the method depends on the convergence of the left hand side of (21) to $F(y,v,T)$.

If $f(T)$ is a function of a non-negative variable T and it's Laplace transform $\tilde{f}(q) = \int_0^\infty e^{-qT} f(T)\,dT$ is known, then the approximate Post–Widder formula (see e.g. [3]) for $f(T)$ can be written as follows.

$$\begin{aligned} f(T) &= \lim_{N\to\infty} f_N(T); \\ f_N(T) &:= \frac{(-1)^N}{N!}\left(\frac{N}{T}\right)^{N+1} \tilde{f}^{(N)}\left(\frac{N}{T}\right), \end{aligned}$$

where $\tilde{f}^{(N)}(q)$—Nth derivative of the function $\tilde{f}$ at q.

As a result, we conjecture that the Carr's approximation (13) to the solution to the problem (12) always converges to the actual value for the Heston model. Moreover, according to the properties of the Post–Widder formula (see [3]) we obtain the order of the convergence: $O(N^{-1})$, where N is the number of time steps and we can apply Richardson extrapolation to accelerate the convergence, as it has been suggested in [32, 36].

4.3 Markov Chain Approximation and Recurrent Calculations

We can now introduce an approximation for the CIR process V_t with a process on a lattice. A detailed reasoning why process V_t on a sufficiently small time interval $(t, t + \Delta s]$ ($\Delta s \in \mathbf{R}$ small) can be replaced by a suitable deterministic continuous function (namely, its constant value in $t + \Delta s$) for this purpose can be found in [15].

For performance reasons it is convenient to make an approximating tree "computationally simple", with a number of nodes growing at most linearly on the number of time intervals. The notion was introduced in [37], although independently implemented by Hull and White, as it is mentioned in [19]. A recombining binomial tree (with "shared" nodes, where consequent moves up and down leads to the same node as down and up ones) belongs to this class. We build such tree and define transitional probabilities following the procedure from [2]. Among the advantages of the scheme we can mention a fast convergence towards the parent process and an ability to remain adequate in a wide range of parameters: for example, as it has been shown in numerical experiments provided in [2, 15], the parameters of CIR process are not restricted by Feller condition $2\kappa\theta \geq \sigma_V^2$.

As it was noted in Sect. 4.2, Post–Widder formula has the order of convergence $O(N^{-1})$, where N is the number of time steps. On the other hand, the convergence speed of the Markov chain proposed in [2] (to an original CIR process) is faster than, for example, in [37, 46] or [38], which makes constructing a tree with a large N redundant.

To make use of this, we may define constants $l \in \mathbf{N}$ and $N_\upsilon \in \mathbf{N}$, set $N = N_\upsilon l$, and group $\{\tau_n\}_{n=1}^N$ into N_υ subsequences of length l. The sum, for $\sum_{i=jl}^{(j+1)l} \tau_i$, $j = 0, 1, \ldots, N_\upsilon$, is a gamma-distributed random variable with a law $\Gamma(l, \Delta s)$, where $\Delta s = l\Delta\tau$. For $\{\tau_n\}_{n=1}^l, \{\tau_n\}_{n=l+1}^{2l}, \ldots$ we can assume the variance process to be constant and only allow transitions at T'_{jl}, for $j = 0, 1, \ldots, N_\upsilon$.

Consider a diffusion process: $dV_t = \kappa_V(\theta_V - V_t)dt + \sigma_V\sqrt{V_t}dW_t$. Its direct discretization is not possible (see [37] for details), because the tree does not recombine. Let's first define a transformation proposed in [37]:

$$R_V(\upsilon) = \frac{2\sqrt{\upsilon}}{\sigma_V}, \tag{22}$$

and set $V_0 = \upsilon$. A binomial tree with nodes $x_V(n, k)$, for $n' = 0, 1, \ldots N_\upsilon$, $k = 0, 1 \ldots n'$, may be therefore defined by formula:

$$x_V(n', k) = x_V(0, 0) + (2k - n')\sqrt{\Delta s}. \tag{23}$$

Now we can set a tree with nodes denoted as $V_{n',k}$ using a substitution:

$$V_{n',k} = \begin{cases} \frac{\sigma_V^2 x_V^2(n',k)}{4}, x_V(n', k) > 0, \\ 0, x_V \leq 0. \end{cases} \tag{24}$$

From (23)–(24) we can derive a formula $V_{n',k}$:

$$V_{n',k} = \left(\sqrt{V_0} + \tfrac{\sigma_V}{2}(2k - n')\sqrt{\Delta s}\right)^2 \mathbf{1}_{(0,+\infty)}\left(\sqrt{V_0} + \tfrac{\sigma_V}{2}(2k - n')\sqrt{\Delta s}\right), \tag{25}$$

and easily see that the value V_0 of equals to $V_{0,0}$ in the approximating tree

An important idea from [27, 46] is that transitions from $V_{n',k}$ are not only allowed to $V_{n'+1,k}$ and $V_{n'+1,k+1}$. Starting from $V_{n',k}$, it can either move:

- "upwards"—to a node $V_{n'+1,k_u(n',k)}$,
- "downwards"—to $V_{n'+1,k_d(n',k)}$,

where values of $k_u(n', k)$ and $k_d(n', k)$ are calculated as in [2] to fit the drift $\mu_V(V_{n',k})$. First, we define two sets:

$$\begin{aligned} K_u(n', k) &:= \{k^* : k + 1 \leq k^* \leq n' + 1, \quad V_{n',k} + \mu_V(V_{n',k})\Delta s \leq V_{n'+1,k^*}\}, \\ K_d(n', k) &:= \{k^* : 0 \leq k^* \leq k, \quad V_{n',k} + \mu_V(V_{n',k})\Delta s \geq V_{n'+1,k^*}\}. \end{aligned}$$

Then, from $K_u(n', k)$ and $K_d(n', k)$, we may calculate, $k_u(n', k)$ and $k_d(n', k)$ as follows:

$$k_u(n', k) = \begin{cases} \min\{K_u(n', k)\}, & K_u(n', k) \neq \emptyset, \\ n' + 1, & K_u(n', k) = \emptyset; \end{cases}$$

$$k_d(n', k) = \begin{cases} \max\{K_d(n', k)\}, & K_u(n', k) \neq \emptyset, \\ 0, & K_d(n', k) = \emptyset. \end{cases}$$

Transitions from $V_{n',k}$ happen with probabilities $p_u(n', k)$ and $p_d(n', k)$ to $V_{n'+1,k_u}$ and $V_{n'+1,k_d}$, respectively:

$$\begin{aligned} p_u(n', k) &= \frac{(\mu_V(V_{n',k})\Delta s + V_{n',k}) - V_{n'+1,k_d(n',k)}}{V_{n'+1,k_u(n',k)} - V_{n'+1,k_d(n',k)}}, \\ p_d(n', k) &= 1 - p_u(n', k). \end{aligned}$$

Finally, let us define restricting rules to prevent negative probabilities from appearing for any nodes of the lattice:

$$p_u(n',k) := \begin{cases} 1, & p_u(n',k) > 1; \\ p_u(n',k), & p_u(n',k) \in [0,1]; \\ 0, & p_{k_u}(n',k) < 0; \end{cases}$$

$$p_d(n',k) := 1 - p_{k_u}(n',k).$$

This gives rise to a Markov chain that weakly converges, as $\Delta s \to 0$, to the diffusion process V_t. In terms of the Markov chain, we will approximate $f_n(y, v)$ in (14) at points $v = V_{n',k}$, where $n = n'l$.

We will use shorthands $k_u := k_u(n',k)$, $k_d := k_d(n',k)$ for indices, and $p_u := p_{k_u(n',k)}$, $p_d := p_{k_d(n',k)}$, for probabilities, to simplify the notation.

Let's define a process $Y_t^{n',k}$ as Y_t with coefficients generated by $V_{n',k}$, and define

$$Z_t^{n',k} = Y_t^{n',k} + \frac{\rho}{\sigma} V_{n',k}.$$

An indicator function $\mathbf{1}_{(-\infty,0]}(\underline{Z}_t)$ can be decomposed as:

$$\mathbf{1}_{(-\infty,0]}(\underline{Z}_t) = \mathbf{1}_{(-\infty,0]}(\inf_{0<s\leq t}\{Z_s\}) \cdot \mathbf{1}_{(-\infty,0]}(Z_0). \tag{26}$$

The same modification is applicable for $\mathbf{1}_{(0,+\infty)}(\underline{Z}_t)$. Let us for the rest of the section redefine $\underline{Z}_t^{n,k} = \inf_{0<s\leq t}\{Z_s^{n,k}\}$ and denote

$$h^+(y)\big(= h^+_{n',k}(y)\big) = \mathbf{1}_{(0,+\infty)}(y + \frac{\rho}{\sigma_V} V_{n',k}),$$

and

$$h^-(y)\big(= h^-_{n',k}(y)\big) = \mathbf{1}_{(-\infty,0]}(y + \frac{\rho}{\sigma_V} V_{n',k}).$$

Using these notations and the fact that we now allow the variation process to change its value at moments T_j', $j = n'l$, from $V_{n',k}$ to either $V_{n'+1,k_u}$ or $V_{n'+1,k_d}$, the expectations (18) can be calculated in terms of the Markov chain for $n' = N_v, .., 0$, and $k = 0, ..., n'$ as:

$$\begin{aligned} &f_{n'l}(y, V_{n',k}) = \\ &= E^{y,V_{n',k}}\big[h^-(y) \cdot \mathbf{1}_{(-\infty,0]}(\underline{Z}_{T_l'})\big] + E^{y,V_{n',k}}\big[h^+ \cdot \mathbf{1}_{(0,+\infty)}(\underline{Z}_{T_l'}) f_{n'l+l}(Y_{T_l'}, V_{T_l'})\big] \\ &= p_u f^u_{n',k}(y) + p_d f^d_{n',k}(y), \end{aligned} \tag{27}$$

where

$$f^u_{n',k}(y) = h^-(y) \cdot E^y\big[\mathbf{1}_{(-\infty,0]}(Y^{n'+1,k_u}_{T'_l} + \tfrac{\rho}{\sigma_V} V_{n'+1,k_u})\big] +$$

$$+h^+(y) \cdot E^y\big[\mathbf{1}_{(0,+\infty)}(Y^{n'+1,k_u}_{T'_l} + \tfrac{\rho}{\sigma_V} V_{n'+1,k_u}) f_{n'l+l}(Y^{n'+1,k_u}_{T'_l}, V_{n'+1,k_u})\big],$$

$f^d_{n',k}(y)$ is defined by analogy. We may then calculate $f^u_{n',k}(y)$ and $f^d_{n',k}(y)$ recursively as expectations over exponentially distributed variables. For $m = l-1, l-2, ..., 0$ we define:

$$\begin{aligned} w^u_{m,k}(y) &= h^-(y) \cdot E^y[\mathbf{1}_{(-\infty,0]}(Y^{n'+1,k_u}_{\tau_{m+1}} + \tfrac{\rho}{\sigma_V} V_{n'+1,k_u})] \\ &+h^+(y) \cdot E^y[\mathbf{1}_{(0,+\infty)}(Y^{n'+1,k_u}_{\tau_{m+1}} + \tfrac{\rho}{\sigma_V} V_{n'+1,k_u}) w^u_{m+1,k}(Y^{n'+1,k_u}_{\tau_{m+1}})], \end{aligned} \tag{28}$$

where $w^u_{l,k}(y) = f_{n'l+l}(y, V_{n'+1,k_u})$. Then we approximate $f^u_{n',k}(y)$ with $w^u_{0,k}(y)$. Calculations for $f^d_{n',k}(y)$ are analogous.

This modification, for $l > 1$, replicates a tree from [15] and allows to N be large enough to obtain good convergence for the Laplace transform inversion while keeping the number of nodes respectively low. Numerical experiments (see Tables 2 and 8) support this argument.

5 Wiener–Hopf Factorization Approach

In this section we deal with a Wiener–Hopf factorization method, which is the main tool to compute the expectations (28) in our approach. More precisely, in Sect. 5.1 we recall general factorization formulae in several forms and apply them in the operator form for computing the expectations (28), while in Sect. 5.2 we give practical explicit and approximate expressions for the symbols of the Wiener–Hopf operators. Section 5.3 provides a numerical technique based on the Fast Fourier Transform which allows to implement approximate Wiener–Hopf factorization formulae efficiently for our problem and for more general cases admitting jumps.

5.1 General Formulae

It is well-known that a Gaussian process X_t (and a more general Lèvy process [44]) has a characteristic exponent $\psi(\xi)$, which is a continuous function $\psi : \mathbf{R} \to \mathbf{C}$ satisfying conditions $\psi(0) = 0$ and:

$$E[e^{i\xi X_t}] = e^{-t\psi(\xi)}, \qquad \forall \xi \in \mathbf{R}, t \geq 0.$$

A characteristic exponent of $Y_t^{n',k}$ is defined by:

$$\psi(\xi) = \frac{\sigma_{n',k}^2}{2}\xi^2 - i\gamma_{n',k}\xi, \tag{29}$$

where $\sigma_{n',k} = \hat{\rho}\sqrt{V_{n',k}}$, $\gamma_{n',k} = \mu_Y(V_{n',k})$. We use the characteristic exponents (29) to solve our problems (28) by means of the Wiener–Hopf factorization.

There are several forms of Wiener–Hopf factorization (see details in [9, 13, 44]). The Wiener–Hopf factorization identity used in probability reads:

$$E[e^{i\xi X_\tau}] = E[e^{i\xi \overline{X}_\tau}]E[e^{i\xi \underline{X}_\tau}] \quad \forall \xi \in \mathbf{R}, \tag{30}$$

where $\tau \sim \mathrm{Exp}\, q$, and $\overline{X}_\tau = \sup_{0\le s\le \tau} X_s$, and $\underline{X}_\tau = \inf_{0\le s\le \tau} X_s$ are supremum and infimum processes of a Lévy process X_t. There is a useful formula for $\underline{X}_t$ and $\overline{X}_t$:

$$\overline{X}_t \overset{d}{\sim} X_t - \underline{X}_t.$$

With notations:

$$\begin{aligned} \phi_q^+(\xi) &= qE\left[\int_0^\infty e^{-qt}e^{i\xi\overline{X}_\tau}dt\right] = E\left[e^{i\xi\overline{X}_\tau}\right], \\ \phi_q^-(\xi) &= qE\left[\int_0^\infty e^{-qt}e^{i\xi\underline{X}_\tau}dt\right] = E\left[e^{i\xi\underline{X}_\tau}\right], \end{aligned} \tag{31}$$

we can represent (30) as

$$\frac{q}{q+\psi(\xi)} = \phi_q^+(\xi)\phi_q^-(\xi). \tag{32}$$

The functions $\frac{q}{q+\psi(\xi)}$, $\phi_q^\pm(\xi)$ are the symbols of expected present value operators $\mathcal{E}_q$, $\mathcal{E}_q^\pm$ (EPVs), introduced in [9]:

$$\mathcal{E}_q g(x) = E\left[\int_0^\infty qe^{-qt}g(X_t)dt|X_0 = x\right] = E[g(x+X_\tau)], \tag{33}$$

$$\mathcal{E}_q^+ g(x) = E\left[\int_0^\infty qe^{-qt}g(\overline{X_t})dt|X_0 = x\right] = E[g(x+\overline{X}_\tau)], \tag{34}$$

$$\mathcal{E}_q^- g(x) = E\left[\int_0^\infty qe^{-qt}g(\underline{X_t})dt|X_0 = x\right] = E[g(x+\underline{X}_\tau)]. \tag{35}$$

From a theory of pseudodifferential operators (PDO) it is known that the product of symbols corresponds to the composition of operators. It can be shown (see [9]) that in appropriate functional spaces:

$$\mathcal{E}_q = \mathcal{E}_q^+\mathcal{E}_q^- = \mathcal{E}_q^-\mathcal{E}_q^+.$$

Expectations from (28), $w^u_{n,k}(y)$ and $w^d_{n,k}(y)$, can be calculated in terms of $\mathcal{E}^\pm$ (see [7] for details). By definition of a PDO, operators $\mathcal{E}^\pm$ can be written in the following form:

$$\mathcal{E}_q^\pm = \frac{1}{2\pi}\int_{-\infty}^{+\infty} e^{ix\xi}\phi^\pm(\xi)\hat{f}(\xi)d\xi, \tag{36}$$

where $\hat{f}(\xi) = \int_{-\infty}^{+\infty} e^{-ix\xi} f(x)d\xi$. In a numerical implementation the integrals can be represented as discrete sums and calculated via the FFT algorithm.

Let $\mathcal{E}_q^\pm$ be the EPVs for the supremum and infimum processes of $Y_t^{n'+1,k_u}$. By applying them for $w^u_{m,k}(y)$ in (28) with $q = \Delta\tau^{-1}$, we obtain:

$$w^u_{m,k}(y) = h^-(y)\cdot E^y[\mathbf{1}_{(-\infty,0]}(\underline{Y}^{n'+1,k_u}_{\tau_{m+1}} + \tfrac{\rho}{\sigma_V}V_{n'+1,k_u})]$$

$$+h^+(y)\cdot E^y[\mathbf{1}_{(0,+\infty)}(\underline{Y}^{n'+1,k_u}_{\tau_{m+1}} + \tfrac{\rho}{\sigma_V}V_{n'+1,k_u})w^u_{m+1,k}(y + Y^{n'+1,k_u}_{\tau_{m+1}})] =$$

$$= \mathbf{1}_{(-\infty,0]}(y + \tfrac{\rho}{\sigma_V}V_{n',k})\cdot\mathcal{E}_q^-\mathbf{1}_{(-\infty,0]}(y + \tfrac{\rho}{\sigma_V}V_{n'+1,k_u}) +$$

$$+\mathbf{1}_{(0,+\infty)}(y + \tfrac{\rho}{\sigma_V}V_{n',k})\cdot\mathcal{E}_q^-\mathbf{1}_{(-\frac{\rho}{\sigma_V}V_{n'+1,k_u},+\infty)}(\cdot)\mathcal{E}_q^+ w^u_{m+1,k}(y),$$

and calculate $w^d_{m,k}(y)$ by analogy. After having calculated the expectations with respect to the Markov chain, we use the fact that $v = V_{0,0}$ in (12) and obtain the approximate value of $f(y, v, 0)$ as $f_0(y, V_{0,0})$, and therefore solve the problem.

5.2 *Approximate and Explicit Factorization*

The characteristic exponent (29) of the process $Y_t^{n',k}$ allows an analytical factorization of the expression $q/(q+\psi(\xi))$ (see e.g. [13]):

$$\phi_q^+(\xi) = \frac{\beta^+_{n',k}}{\beta^+_{n',k} - i\xi}, \quad \phi_q^-(\xi) = \frac{-\beta^-_{n',k}}{-\beta^-_{n',k} + i\xi},$$

where

$$\beta^-_{n',k} = -\frac{\gamma_{n',k} + \sqrt{\gamma^2_{n',k} + 2\sigma^2_{n',k}q}}{\sigma^2_{n',k}},$$

$$\beta^+_{n',k} = -\frac{\gamma_{n',k} - \sqrt{\gamma^2_{n',k} + 2\sigma^2_{n',k}q}}{\sigma^2_{n',k}}.$$

If the process S_t admits Lévy jumps, one may use general approximate formulae for the factors introduced in [32] for a wide class of Lévy models. The function $\phi_q^+(\xi)$ allows analytical continuation to a half-plane $\Im\xi > \omega_-$ and can be represented as follows:

$$\phi_q^+(\xi) = \exp\left[i\xi F^+(0) - \xi^2 \hat{F}^+(\xi)\right],$$
$$F^+(x) = \mathbf{1}_{(-\infty,0]}(x)(2\pi)^{-1} \int_{-\infty+i\omega_-}^{+\infty+i\omega_-} e^{ix\eta} \frac{\ln(q+\psi(\eta))}{\eta^2} d\eta,$$
$$\hat{F}^+(\xi) = \int_{-\infty}^{+\infty} e^{-ix\xi} F^+(x)dx.$$

Analogously, $\phi_q^-(\xi)$ allows an analytical continuation to a halfplane $\Im\xi < \omega_+$ and can be represented as follows:

$$\phi_q^-(\xi) = \exp\left[-i\xi F^-(0) - \xi^2 \hat{F}^-(\xi)\right],$$
$$F^-(x) = \mathbf{1}_{[0,+\infty)}(x)(2\pi)^{-1} \int_{-\infty+i\omega_+}^{+\infty+i\omega_+} e^{ix\eta} \frac{\ln(q+\psi(\eta))}{\eta^2} d\eta;$$
$$\hat{F}^-(\xi) = \int_{-\infty}^{+\infty} e^{-ix\xi} F^-(x)dx.$$

The constants $\omega_- < 0 < \omega_+$ depend on Lévy process parameters (see details in [32]). To implement the $\mathcal{E}_q^\pm$ operators numerically (see details in [32, 33]) it is vital to present the expressions for $\phi_q^\pm(\xi)$ in a computationally efficient form.

5.3 Fast Fourier Transform Implementation

The algorithm for calculating the integral via the FFT for the case when the closed formulae for the factors exist is described in [33]. Let us construct the scheme of calculating an approximate value of $\phi_q^\pm(\xi)$. For the sake of brevity, we restrict ourself to the case of $\phi_q^-(\xi)$. The symbol $\phi_q^-(\xi)$ has a form:

$$\phi_q^-(\xi) = exp[-i\xi F^-(0) - \xi^2 \hat{F}^-(\xi)]$$

We may change the integration line for $F^-(x)$ and use Jordan's lemma to obtain the following representation:

$$F^-(x) = 1_{[0,+\infty)}(x) e^{-x\omega_+} \frac{1}{2\pi} \int_{-\infty}^{+\infty} e^{ix\upsilon} \frac{\ln(1+\frac{\psi(\upsilon+i\omega_+)}{q})}{(\upsilon+i\omega_+)^2} d\upsilon.$$

Let's define constants M (an integer and a power of 2) and Δ_x, which plays the role of a discretization instant for the x variable. Let's set $M+1$ equally spaced points:

$$\{\upsilon_l\}_{l=0}^{M}: \quad \upsilon_l = -\frac{\pi}{\Delta_x} + \frac{2\pi l}{M\Delta_x}$$

and denote $\Delta_\upsilon = \upsilon_{l+1} - \upsilon_l = \frac{2\pi}{M\Delta_x}$. We can use a simple trapezoid rule to calculate the integral numerically for $\{x_k\}_{k=0}^{M-1}$:

$$F^-(x_k) = 1_{[0,+\infty)}(x_k)e^{-x_k\omega_+}\frac{1}{2\pi}\Delta_\upsilon \sum_{\upsilon_l=-\frac{\pi}{\Delta_x}}^{\frac{\pi}{\Delta_x}} e^{ix_k\upsilon_l}\frac{\ln(1+\frac{\psi(\upsilon_l+i\omega_+)}{q})}{(\upsilon_l+i\omega_+)^2}.$$

Let us choose $\{x_k\}_{k=0}^{M-1}: \quad x_k = k\Delta_x$. Having performed the multiplications, we simplify the expression in the power of the exponent, and obtain:

$$F^-(k\cdot\Delta_x) = \frac{(-1)^k\cdot e^{-k\Delta_x\omega_+}}{M\Delta_x}\sum_{l=0}^{M-1} e^{\frac{2\pi ikl}{M}}\frac{\ln(1+\frac{\psi(-\frac{\pi}{\Delta_x}+\frac{2\pi l}{M\Delta_x}+i\omega_+)}{q})}{(-\frac{\pi}{\Delta_x}+\frac{2\pi l}{M\Delta_x}+i\omega_+)^2}$$

Let's denote $F_k := F^-(k\cdot\Delta_x), k = 0, 1..., M-1$. Now we can calculate $F(\xi)$. Let's use an array F_k as an input data. We can get the values of $F(\xi)$ in M points, namely $\{\xi_l\}_{l=0}^{M-1}: \quad \xi_l = -\frac{\pi}{\Delta_x} + \frac{2\pi l}{M\Delta_x}$ Let us use a trapezoid rule again:

$$\hat{F}^-(\xi_l) = \Delta_x\sum_{k=0}^{M}\frac{e^{-ik\Delta_x\cdot\xi_l}F_k^- + e^{-i(k+1)\Delta_x\cdot\xi_l}F_{k+1}^-}{2}.$$

Although values of F_0^- and F_M^- are not equal, we can safely assume F_M^- to be zero. Analogous to the previous case, we can write down a discrete sum as

$$\hat{F}^-(\xi_l) = -\frac{\Delta_x F_0^-}{2} + \\ +\frac{1}{M}\sum_{k=0}^{M-1} e^{-\frac{2\pi ikl}{M}}\left(e^{-k\Delta_x\omega_+}\sum_{l=0}^{M-1} e^{\frac{2\pi ikl}{M}}\frac{\ln(1+\frac{\psi(-\frac{\pi}{\Delta_x}+\frac{2\pi l}{M\Delta_x}+i\omega_+)}{q})}{(-\frac{\pi}{\Delta_x}+\frac{2\pi l}{M\Delta_x}+i\omega_+)^2}\right) \tag{37}$$

This sum can easily be calculated via the FFT. Most libraries offer fast calculation of the following sum for the IFFT: $IDFT_k(a_k) = \frac{1}{M}\sum_{l=0}^{M-1}\exp(\frac{2\pi ikl}{M})a_k$ and FFT: $DFT_k(c_k) = \sum_{l=0}^{M-1}\exp(-\frac{2\pi ikl}{M})c_k$ Having found the values of expression

$$\frac{\ln(1+\frac{\psi(-\frac{\pi}{\Delta_x}+\frac{2\pi l}{M\Delta_x}+i\omega_+)}{q})}{(-\frac{\pi}{\Delta_x}+\frac{2\pi l}{M\Delta_x}+i\omega_+)^2}), \text{ for } \quad l = 0, 1, \ldots, M-1$$

and saved them into array a_k, we can calculate $c_k = e^{\frac{2\pi ikl}{M}} IDFT_k(a_k)$. The only step remaining is to calculate the $DFT_k(c_k)$ and correct the values by substituting $\frac{\Delta_x F_0^-}{2}$. It should be noted that good results can be achieved using vectorized computations.

The values of characteristic exponents and their parameters should be calculated for each node. When the value of a symbol is known, a problem for each node can be solved similar to the case of the analytic factorization. We will check the performance of the approximate formulae in comparison with the explicit ones in Sect. 6.

6 Numerical Experiments

We provide the results of numerical experiments to demonstrate the convergence and accuracy of the methods proposed. We denote the approximate method, based on the Markov Chain approximation and the approximate Wiener–Hopf factors, as FWH-A. For analytical factorization with a classical lattice (where $l = 1$) we use a notation FWH-CF, and FWH-CFG for our modification with $l > 1$. It was shown that Monte Carlo methods (see e.g. [15, 17] are in good agreement with finite-difference methods for the problem considered. Since Monte Carlo methods are time consuming, we used the results calculated on a fine grid with the hybrid finite-difference (FD) method developed in [15] as a benchmark. The parameters of the FD-method which implemented in platform Premia [42] are N time steps and M spatial variable points. The parameter N also controls the number of nodes of the Markov chain.

We have solved the problem (9) in the Heston model, for the case of a barrier from above, with different sets of parameters. To perform the experiments we used a PC with a following configuration: Intel(R) Core(TM) i5-3470 CPU, 3.2 GHz, RAM 8 Gb, under OS MS Windows 10 ×64. We used scientific Python stack (numpy and scipy packages) to implement the program code.

The parameters for FWH-A and FWH-CF are as follows: N_v is the number of time frames where transitions in the Markov chain are allowed, and l is the number of time steps between them. To control the localization error of the FFT we use the coefficient L, (see details in [32]), and define an equally spaced grid for y on $[-L\ln(2), L\ln(2)]$. The parameter M defines a number of points in the interval.

Tables 1 and 2 report the results obtained by of all the methods mentioned above for the following parameters of the Heston model: $T = 0.5$, $H = 100$, $V_0 = 0.1$, $\kappa = 2.0$, $\theta = 0.1$, $\sigma = 0.2$, $\rho = 0$. The values obtained by FWH-A and the ones of FWH-CF are in good agreement, but the former method needs a denser grid to achieve a good accuracy. Note that it took less time for the FWH-CFG method to achieve better values of relative error than for the FWH-CF one.

In the rest of the section we use a more convenient parametrization. We choose the constant h as a spatial variable discretization instant and define an equally spaced grid for y on $[-L\ln(2), L\ln(2)]$. The parameter M is therefore derived from h.

Table 1 Solution values, $T = 0.5$, $V_0 = 0.1$, $\kappa = 2.0$, $\theta = 0.1$, $\sigma = 0.2$, $\rho = 0$

S	FD	FWH-CF		FWH-CFG		FWH-A
	$M = 1200$	$M = 2^{12}$	$M = 2^{12}$	$M = 2^{12}$	$M = 2^{12}$	$M = 2^{14}$
	$N = 400$	$N_v = 300$	$N_v = 50$	$N_v = 30$	$N_v = 30$	$N_v = 30$
		$L = 2$	$L = 2$	$L = 2$	$L = 2$	$L = 2$
				$l = 3$	$l = 5$	$l = 5$
60	0.017529	0.01749	0.01713	0.01725	0.01733	0.01737
65	0.043552	0.04333	0.04208	0.04266	0.04295	0.04290
70	0.092036	0.09174	0.08906	0.09034	0.09099	0.09079
75	0.170332	0.17003	0.16578	0.16786	0.16889	0.16847
80	0.282601	0.28242	0.27701	0.27968	0.28101	0.28029
85	0.428349	0.42842	0.42283	0.42561	0.42698	0.42593
90	0.602431	0.60289	0.59830	0.60059	0.60172	0.60032
95	0.796318	0.79723	0.79467	0.79595	0.79658	0.79487
Time (s)	11.6	70	2.71	2.27	3.68	22.7

Table 2 Relative error values (%), $T = 0.5$, $V_0 = 0.1$, $\kappa = 2.0$, $\theta = 0.1$, $\sigma = 0.2$, $\rho = 0$

S	FWH-CF		FWH-CFG		FWH-A
	$M = 2^{12}$	$M = 2^{12}$	$M = 2^{12}$	$M = 2^{12}$	$M = 2^{14}$
	$N_v = 300$	$N_v = 50$	$N_v = 30$	$N_v = 30$	$N_v = 30$
			$l = 3$	$l = 5$	$l = 5$
60	0.22	2.28	1.59	1.14	0.91
65	0.51	3.38	2.05	1.38	1.50
70	0.32	3.23	1.84	1.14	1.35
75	0.18	2.67	1.45	0.85	1.09
80	0.06	1.98	1.03	0.56	0.82
85	0.02	1.29	0.64	0.32	0.56
90	0.08	0.69	0.31	0.12	0.35
95	0.11	0.21	0.05	0.03	0.18
Time (s)	70.0	2.71	2.27	3.68	22.7

Tables 3 and 4 demonstrate the behavior of the FWH-CF method for the set of parameters from [17], namely: $T = 0.5$, $H = 100$, $V_0 = 0.1$, $\kappa = 2.0$, $\theta = 0.1$, $\sigma = 0.1$, $\rho = -0.5$.

For long time periods (and wider variance trees) it takes more iterations for the method to converge. Tables 5 and 6, demonstrate the behavior of the methods for a parameter set with a higher variance and expiration time: $T = 1.0$, $H = 100$, $V_0 = 1.0$, $\kappa = 2.0$, $\theta = 1.0$, $\sigma = 0.1$, $\rho = 0.5$.

On the contrary, shorter time periods allows to obtain good results faster and fully benefit from the use of FWH-CFG methods. Tables 5 and 6 present the results for a

Table 3 Solution values, $T = 0.5$, $V_0 = 0.1$, $\kappa = 2.0$, $\theta = 0.1$, $\sigma = 0.1$, $\rho = -0.5$

S	FD	FWH-CF ($h = 0.001$, $L = 3$)			
	$N = 400$				
	$M = 1200$	$N_v = 500$	$N_v = 300$	$N_v = 100$	$N_v = 50$
75	0.17033	0.16896	0.17306	0.17696	0.16200
80	0.28260	0.28521	0.29036	0.29520	0.29699
85	0.42834	0.43193	0.43995	0.44527	0.44652
90	0.60319	0.60863	0.61369	0.61868	0.61744
95	0.79631	0.79598	0.79985	0.80383	0.79325
Time (s)	34.3	338.4	132.8	16.7	4.27
S	FD	FWH-CF ($h = 0.002$, $L = 3$)			
	$N = 400$				
	$M = 1200$	$N_v = 500$	$N_v = 300$	$N_v = 100$	$N_v = 50$
75	0.17033	0.16890	0.16955	0.17354	0.17751
80	0.28260	0.28415	0.28632	0.29136	0.29623
85	0.42834	0.43286	0.43632	0.44187	0.44707
90	0.60319	0.60976	0.61207	0.61728	0.62184
95	0.79631	0.80159	0.80326	0.80699	0.80941
Time (s)	34.3	180.0	70.0	8.94	2.43

Table 4 Relative error (%), $T = 0.5$, $V_0 = 0.1$, $\kappa = 2.0$, $\theta = 0.1$, $\sigma = 0.1$, $\rho = -0.5$

S	FWH-CF ($h = 0.001$, $L = 3$)			
	$N_v = 500$	$N_v = 300$	$N_v = 100$	$N_v = 50$
75	0.62	0.81	1.60	3.89
80	0.39	0.92	2.75	4.46
85	0.95	0.84	2.71	3.95
90	0.90	0.90	1.74	2.57
95	0.08	0.04	0.44	0.94
Time (s)	338.4	132.8	16.7	4.27
S	FWH-CF ($h = 0.002$, $L = 3$)			
	$N_v = 500$	$N_v = 300$	$N_v = 100$	$N_v = 50$
75	0.84	0.46	1.88	4.22
80	0.55	1.32	3.10	4.82
85	1.06	1.86	3.16	4.37
90	1.09	1.47	2.34	3.09
95	0.66	0.87	1.34	1.65
Time (s)	180.0	70.0	8.94	2.43

Table 5 Solution values, $T = 1.0$, $V_0 = 1.0$, $\kappa = 2.0$, $\theta = 1.0$, $\sigma = 0.1$, $\rho = 0.5$

S	FD	FWH-CF ($h = 0.01$, $L = 4$)				
	$N = 1000$					
	$M = 2000$	$N_v = 3000$	$N_v = 2000$	$N_v = 1000$	$N_v = 500$	$N_v = 300$
70	0.58461	0.58024	0.57884	0.57600	0.57229	0.56832
75	0.65210	0.64837	0.64702	0.64447	0.64105	0.63731
80	0.72043	0.71747	0.71622	0.71405	0.71102	0.70760
85	0.78941	0.78726	0.78624	0.78455	0.78204	0.77901
90	0.85885	0.85747	0.85711	0.85591	0.85415	0.85136
95	0.92861	0.92798	0.92912	0.92808	0.92771	0.92437
Time (s)	151.0	2841.0	1256.0	234.3	89.2	32.8

Table 6 Relative errors (%), $T = 1.0$, $V_0 = 1.0$, $\kappa = 2.0$, $\theta = 1.0$, $\sigma = 0.1$, $\rho = 0.5$

S	FWH-CF ($h = 0.01$, $L = 4$)				
	$N_v = 3000$	$N_v = 2000$	$N_v = 1000$	$N_v = 500$	$N_v = 300$
70	0.75	0.99	1.47	2.11	2.79
75	0.57	0.78	1.17	1.69	2.27
80	0.41	0.59	0.89	1.31	1.78
85	0.27	0.40	0.62	0.93	1.32
90	0.16	0.20	0.34	0.55	0.87
95	0.07	0.05	0.06	0.10	0.46
Time (s)	2841.0	1256.0	234.3	89.2	32.8

parameter set with a short expiration time: $T = 0.1$, $H = 100$, $V_0 = 1.0$, $\kappa = 2.0$, $\theta = 1.0$, $\sigma = 0.1$, $\rho = 0.5$.

The method works reliably for a range of commonly used values of parameters and demonstrates good accuracy. The correlation coefficient ρ doesn't seem to have any significant effect on accuracy or speed (see Tables 7 and 8), although due to the structure of the substitution described in Sect. 4.1 sufficiently large value of $\frac{\rho}{\sigma_V}$ with large variance values may require to increase L.

Choosing small V_0 and θ_0 (< 0.001) also can make method work slower and less accurate. This effect is caused by large values which arise in the Wiener–Hopf factors. Notice that for small values of variance (0.01 and less) and large q (small step in time), the values of $\beta_q^{\pm}$ also become large, which affects the shape of $\phi_q^{\pm}(\xi)$ functions and reduces the convergence speed of the FFT-based procedure. In this case we recommend to use methods of contour deformation (see e.g. [8, 29] for details). The problem however only becomes pronounced for trees with sufficiently large amount of nodes $V_{n,k}$ with a small variance.

Table 7 Solution values, $T = 0.1$, $V_0 = 1.0$, $\kappa = 2.0$, $\theta = 1.0$, $\sigma = 0.1$, $\rho = 0.5$

S	FD	FWH-CF ($L = 2$)				FWH-CFG ($L = 2$)	
	$N = 1000$	$h = 0.005$	$h = 0.005$	$h = 0.005$	$h = 0.005$	$h = 0.005$	$h = 0.005$
	$M = 2000$	$N_v = 500$	$N_v = 300$	$N_v = 200$	$N_v = 100$	$N_v = 100$	$N_v = 100$
						$l = 2$	$l = 10$
70	0.21590	0.21479	0.21423	0.21366	0.21251	0.2134	0.21559
75	0.31227	0.31099	0.31031	0.30959	0.30827	0.30941	0.31227
80	0.42676	0.42544	0.42467	0.42383	0.42251	0.42385	0.42732
85	0.55643	0.55524	0.55444	0.55356	0.55244	0.55391	0.5579
90	0.69760	0.69679	0.69606	0.69525	0.69459	0.69613	0.70056
95	0.84633	0.84631	0.84563	0.84541	0.84563	0.8473	0.85221
Time (s)	151.0	85.5	30.9	13.5	3.57	5.92	25.4
S	FD	FWH-CF ($L = 2$)				FWH-CFG ($L = 2$)	
	$N = 1000$	$h = 0.01$	$h = 0.01$	$h = 0.01$	$h = 0.01$	$h = 0.01$	$h = 0.01$
	$M = 2000$	$N_v = 500$	$N_v = 300$	$N_v = 200$	$N_v = 100$	$N_v = 100$	$N_v = 100$
						$l = 2$	$l = 10$
70	0.21590	0.21474	0.21425	0.21374	0.21248	0.21342	0.21581
75	0.31227	0.31091	0.31034	0.30971	0.30818	0.30939	0.31251
80	0.42676	0.42531	0.4247	0.42401	0.42234	0.42377	0.42756
85	0.55643	0.55508	0.55449	0.5538	0.55213	0.5537	0.55808
90	0.69760	0.69659	0.69607	0.69547	0.69395	0.69558	0.70041
95	0.84633	0.84604	0.84562	0.84515	0.84387	0.84551	0.85064
Time (s)	151.0	57.4	20.3	10.5	2.74	4.31	15.0

It should be noted that for the method proposed, a smaller variance requires a finer grid (smaller h), while a higher one requires L and N_v to be sufficiently large. For a large expiration time T and/or σ_V, the performance of method suffers, because both conditions must be held at the same time. For these cases, finite difference methods might be a better and faster option. However, a highly volatile process which leads to the Heston model with large values of V_0, σ_V and θ, may often be described more accurately via a suitable jump-diffusion model. The results obtained with the approximate and closed formulae are in good agreement, the method proposed may be used for models with jumps with approximate factorization formulae. In this case the Kolmogorov backward equation has an integral component, which results in increasing computational difficulty for finite difference schemes, but not for our method.

Table 8 Relative errors (%), $T = 0.1$, $V_0 = 1.0$, $\kappa = 2.0$, $\theta = 1.0$, $\sigma = 0.1$, $\rho = 0.5$

S	FWH-CF ($L = 2$)				FWH-CFG ($L = 2$)	
	$h = 0.005$	$h = 0.005$	$h = 0.005$	$h = 0.005$	$h = 0.005$	$h = 0.005$
	$N_v = 500$	$N_v = 300$	$N_v = 200$	$N_v = 100$	$N_v = 100$	$N_v = 100$
					$l = 2$	$l = 10$
70	0.51	0.77	1.04	1.57	1.16	0.14
75	0.41	0.63	0.86	1.28	0.92	0.00
80	0.31	0.49	0.69	1.00	0.68	0.13
85	0.21	0.36	0.52	0.72	0.45	0.26
90	0.12	0.22	0.34	0.43	0.21	0.42
95	0.00	0.08	0.11	0.08	0.11	0.69
Time (s)	85.5	30.9	13.5	3.57	5.92	25.4
S	FWH-CF ($L = 2$)				FWH-CFG ($L = 2$)	
	$h = 0.01$	$h = 0.01$	$h = 0.01$	$h = 0.01$	$h = 0.01$	$h = 0.01$
	$N_v = 500$	$N_v = 300$	$N_v = 200$	$N_v = 100$	$N_v = 100$	$N_v = 100$
					$l = 2$	$l = 10$
70	0.54	0.76	1.00	1.58	1.15	0.04
75	0.44	0.62	0.82	1.31	0.92	0.08
80	0.34	0.48	0.64	1.04	0.70	0.19
85	0.24	0.35	0.47	0.77	0.49	0.30
90	0.14	0.22	0.31	0.52	0.29	0.40
95	0.03	0.08	0.14	0.29	0.10	0.51
Time (s)	57.4	20.3	10.5	2.74	4.31	15.0

7 Conclusion

A new numerical method for solving the Kolmogorov backward equation with variable coefficients, which arises in first-passage problems for diffusions, has been developed. We used a probabilistic interpretation of the equation and analyzed the problem in terms of expectations of stochastic processes. We reduced the initial 3-d problem to a sequence of 1-dimensional problems. First, we used a time randomization and approximated one of two diffusion's components by a discrete-time Markov chain; then we solved each of the arising 1-d problems by applying the Wiener–Hopf method. It should be noted that the methods constructed in the paper have common ingredients with the existing methods from [10, 11, 15].

The approximation of diffusions using Markov chains is not a novel approach. According to [18, 35], for the diffusion process to be approximated, a continuous-time Markov chain is constructed for a given state-space, with probabilities that preserve the instantaneous drift and volatility structure. For any given state, only the neighboring states need to be reached, resembling a trinomial tree in continuous time. This idea was used in [10, 11] to solve more difficult free boundary 3-d and

4-d problems, respectively. Under the correspondent regime switching structure the authors in [10, 11] apply Carr's randomization with exponentially distributed time periods, and then a system of 1-dimensional problems with respect to space variable should be solved.

In contrast to the papers [10, 11], we applied Carr's randomization first, and then used the robust binomial tree approximation from [15] based on matching the local drift between the discrete-time Markov chain and the continuous diffusion model. Moreover, we approximated the second diffusion component by the binomial tree over gamma distributed time periods. In comparison to previous researches (see [15]), we improved the computational efficiency of hybrid methods by reducing the number of tree nodes with respect to the number of time steps.

According to the tree algorithm, we needed to solve 1-d problems in its nodes only, unlike the regime switching model, where all the states of the Markov chain should be handled on each time step. However, we found the solutions to the 1-d problems using the Wiener–Hopf method similar to [10, 11], instead of a finite-difference scheme as in [15].

If in (3), the process $X_1(t)$ had constant coefficients, $X_1(t)$ and $X_2(t)$ were independent or the correlation coefficient was small, and the coefficients of $X_2(t)$ depended on $X_2(t)$ only, then the problem (2) for the Kolmogorov backward equation could be most efficiently solved with the eigenfunction expansion method developed in [12].

Using the algorithm of approximate Wiener–Hopf factorization (see Sect. 5.2), the method proposed in the paper allows a generalization for a wide class of models admitting jumps whose variance is governed by the CIR process. In this case, the correspondent Kolmogorov backward equation has a partial integro-differential form, but the computational complexity of the method does not increase significantly.

Acknowledgements The reported study was funded by the Russian Foundation for Basic Research according to the research project No 18-01-00910.

References

1. Alfonsi, A.: High order discretization schemes for the CIR process: application to affine term structure and Heston models. Math. Comput. **79**, 209–237 (2010)
2. Apolloni, E., Caramellino, L., Zanette, A.: A robust tree method for pricing American options with CIR stochastic interest rate. IMA J. Manag. Math. **26**(4), 377–401 (2015)
3. Abate, J., Whitt, W.: A unified framework for numerically inverting Laplace transforms. INFORMS J. Comput. **18**(4), 408–421 (2006)
4. Black, F., Scholes, M.: The pricing of options and corporate liabilities. J. Polit. Econ. **81**(3), 637–654 (1973)
5. Bouchard, B., El Karoui, N., Touzi, N.: Maturity randomization for stochastic control problems. Ann. Appl. Probab. **15**(4), 2575–2605 (2005)
6. Boyarchenko, M.: Carr's randomization for finite-lived Barrier options: proof of convergence. Working Paper, Available at SSRN: http://papers.ssrn.com/abstract=1275666 (2008)
7. Boyarchenko, M., Levendorskiĭ, S.: Prices and sensitivities of barrier and first-touch digital options in Levy-driven models. Int. J. Theor. Appl. Financ. **12**(08), 1125–1170 (2009)

8. Boyarchenko, M., Levendorskiĭ, S.: Ghost calibration and pricing Barrier options and credit default swaps in spectrally one-sided Lévy models: the parabolic laplace inversion method. Quant. Financ. **15**(3), 421–441 (2015)
9. Boyarchenko, S.I., Levendorskiĭ, S.Z.: American options: the EPV pricing model. Ann. Financ. **1**(3), 267–292 (2005)
10. Boyarchenko, S.I., Levendorskiĭ, S.Z.: American options in Lévy models with stochastic interest rates. J. Comput. Financ. **12**(4) (2009)
11. Boyarchenko, S., Levendorskiĭ, S.: American options in the Heston model with stochastic interest rate and its generalizations. Appl. Math. Financ. **20**(1), 26–49 (2013)
12. Boyarchenko, S., Levendorskiĭ, S.: Efficient pricing barrier options and CDS in Lévy models with stochastic interest rate. Math. Financ. (2016). https://doi.org/10.1111/mafi.12121
13. Boyarchenko, S.I., Levendorskiĭ, S.Z.: Non-Gaussian Merton-Black-Scholes Theory. Advanced series on statistical science and applied probability, vol. 8. World Scientific Publishing Co, Singapore (2002)
14. Boyarchenko, S., Levendorskiĭ, S.: SINH-acceleration: efficient evaluation of probability distributions, option pricing, and Monte-Carlo simulations (2018). https://doi.org/10.2139/ssrn.3129881
15. Briani, D.M., Caramellino, L., Zanette, A.: A hybrid approach for the implementation of the Heston model. IMA J. Manag. Math. **28**(4), 467–500 (2017)
16. Carr, P.: Randomization and the American put. Rev. Financ. Stud. **11**, 597–626 (1998)
17. Chiarella, C., Kang, B., Meyer, G.H.: The evaluation of barrier option prices under stochastic volatility. Comput. Math. Appl. **64**, 2034–2048 (2010)
18. Chourdakis, K.: Levy processes driven by stochastic volatility. Asia Pac. Financ. Mark. **12**, 333–352 (2005)
19. Costabile, M., Leccadito, A., Massabó, I.: Computationally simple lattice methods for option and bond pricing. Decis. Econ. Financ. **32**, 161–181 (2009). https://doi.org/10.1007/s10203-009-0092-9
20. Cox, J.C., Ingersoll, J.E., Ross, S.A.: A theory of the term structure of interest rates. Econometrica **53**, 385–408 (1985)
21. Eström, E., Tysk, J.: The Black-Scholes equation in stochastic volatility models. J. Math. Anal. Appl. **368**, 498–507 (2010)
22. Fusai, G., Germano, G., Marazzina, D.: Spitzer identity, Wiener-Hopf factorization and pricing of discretely monitored exotic options. Eur. J. Oper. Res. **251**(1), 124–134 (2016)
23. Green, R., Fusai, G., Abrahams, I.D.: The Wiener-Hopf technique and discretely monitored path-dependent option pricing. Math. Financ. Int. J. Math., Stat. Financ. Econ. **20**(2), 259–288 (2010)
24. Heston, L.A.: Closed-form solution for options with stochastic volatility with applications to bond and currency options. Rev. Financ. Stud. **6**, 327–343 (1993)
25. Heston, S.L., Loewenstein, M., Willard, G.A.: Options and bubbles. Rev. Financ. Stud. **20**(2), 359–390 (2006). https://doi.org/10.1093/rfs/hhl005
26. Hieber, P.: Pricing exotic options in a regime switching economy: a Fourier transform method. Rev. Deriv. Res. **21**, 231–252 (2018)
27. Hilliard, J.E., Schwartz, A.L., Tucker, A.L.: Bivariate binomial pricing with generalized interest rate processes. J. Financ. Res. **XIX4**, 585–602 (1996)
28. Ikonen, S., Toivanen, J.: Componentwise splitting methods for pricing American options under stochastic volatilityInt. J. Theor. Appl. Financ. **10**, 331–361 (2007)
29. de Innocentis, M., Levendorskiĭ, S.: Calibration Heston model for credit risk. Risk. 90–95 (2017)
30. Itkin, A.: Pricing Derivatives Under Levy Models. Birkhauser, Basel (2017)
31. Kudryavtsev, O.: Finite difference methods for option pricing under Levy processes: Wiener-Hopf factorization approach. Sci. World J. (Article ID 963625), 12 (2013)
32. Kudryavtsev, O.: Advantages of the Laplace transform approach in pricing first touch digital options in Lévy-driven models. Boletin de la Sociedad Matematica Mexicana **22**(2), 711–731 (2016)

33. Kudryavtsev, O., Levendorskiĭ, S.: Fast and accurate pricing of barrier options under Lévy processes. Financ. Stoch. **13**(4), 531–562 (2009)
34. Kudryavtsev, O., Rodochenko, V.: A Wiener-Hopf factorization approach for pricing barrier options in the Heston model. Appl. Math. Sci. **11**(2), 93–100 (2017)
35. Kushner, H.J.: Numerical methods for stochastic control problems in continuous time. SIAM J. Control. Optim. **28**(5), 999–1048 (1990)
36. Levendorskiĭ, S.: Convergence of Carr's randomization approximation near barrier. SIAM FM **2**(1), 79–111 (2011)
37. Nelson, D.B., Ramaswamy, K.: Simple binomial processes as diffusion approximations in financial models. Rev. Financ. Stud. **3**(3), 393–430 (1990)
38. Nieuwenhuis, H., Vellekoop, M.: A tree-based method to price American options in the Heston model. J. Comput. Financ. **13**, 1–21 (2009)
39. Oksendal, B.: Stochastic Differential Equations. Springer, New York (2012)
40. Phelan, C.E., Marazzina, D., Fusai, G., Germano, G.: Fluctuation identities with continuous monitoring and their application to price barrier options. Eur. J. Oper. Res. **271**(1), 210–223 (2018)
41. Phelan, C.E., Marazzina, D., Fusai, G., Germano, G.: Hilbert transform, spectral filters and option pricing. Ann. Oper. Res. 1–26 (2018)
42. Premia: a platform for pricing financial derivatives. https://www.rocq.inria.fr/mathfi/Premia
43. Rothe, E.: Zweidimensionale parabolische Randwertaufgaben als Grenzfall eindimensinaler Randwerttaufgaben. Math. Ann. **102** (1930)
44. Sato, K.: Lévy processes and infinitely divisible distributions. Cambridge University Press, Cambridge (1999)
45. Yiran, C., Rollin, d.B., Guido, S.G.: Full and fast calibration of the Heston stochastic volatility model. Eur. J. Oper. Res. **263** (2015). https://doi.org/10.1016/j.ejor.2017.05.018
46. Wei, J.Z.: Valuing American equity options with a stochastic interest rate: a note. J. Financ. Eng. **2**, 195–206 (1996)
47. Zvan, R., Forsyth, P., and Vetzal, K.: A penalty method for American options with stochastic volatility. J. Comput. Appl. Math. **92**, 199–218 (1998)

On Waves Processes in Transversally-Inhomogeneous Waveguides

Alexander Vatulyan and Victor Yurov

Abstract Different types of problems on wave propagation in transversally inhomogeneous cylindrical waveguides are considered. General properties of the dispersion set of inhomogeneous cylindrical waveguides are presented. Asymptotic formulas for the dispersion set branches in the vicinity of radial resonance points are constructed. A condition of solvability of inhomogeneous problems is built. Different models of materials are used for numerical testing of the asymptotic formulas. Dispersion relations for elastic, viscoelastic, electro-elastic and prestressed elastic waveguides are analyzed. The formulas allowing to estimate the influence of residual stresses on the radial resonances points are presented. The effect of various types of homogeneous boundary conditions on the structure of dispersion set is analysed.

Keywords Inhomogeneity · Waveguides · Attenuation · Residual stresses · Forced oscillations

1 Introduction

Mathematical aspects of wave propagation in inhomogeneous waveguides require studying operator sheaf with two spectral parameters. In [1], polynomial operator sheaves are fundamentally studied. It is possible to construct dispersion relations in analytical form for homogeneous waveguides with canonical cross-section shape. One may perform only asymptotic or numerical analysis for operators with variable properties. In [2], the authors present a brief review of dispersion characteristics of

A. Vatulyan · V. Yurov (✉)
Southern Federal University, Rostov-on-Don 344090, Russia
e-mail: vitja.jurov@yandex.ru

Southern Mathematical Institute of the Vladikavkaz Scientific Centre of the RAS, Vladikavkaz 362027, Russia

A. Vatulyan
e-mail: vatulyan@math.rsu.ru

A. Karapetyants et al. (eds.), *Modern Methods in Operator Theory and Harmonic Analysis*, Springer Proceedings in Mathematics & Statistics 291,
https://doi.org/10.1007/978-3-030-26748-3_24

normal modes in an elastic slab and a cylinder. Key topics of the 125-year history of the problem and its modern reflection in the global information space are elucidated.

The relevance of studies for inhomogeneous waveguides with ring or circular cross-section is due to wide practical applications; here let us mention elastic pipelines with multilayer coatings, piezo-waveguides, cylindrical structures immersed in the soil, waveguides in the presence of polymer-composite coatings. In [3], the propagation of shear waves in a coupled plate consisting of a piezoelectric layer and an elastic layer with initial stress is investigated. In [4], the authors analyze an isotropic planarly layered waveguide consisting of an inhomogeneous core that is enclosed between two homogeneous layers forming the cladding, where the spectral parameter power series method is applied. Often, to take into account the inhomogeneity of the functionally graded cylinder, they use a layered model for which the inhomogeneity is replaced by a set of perfectly connected homogeneous layers with different elastic moduli. For example, in [5], the problem on a thick radially inhomogeneous cylinder is solved. In [6], the oscillations of a cylinder inhomogeneous in the circumferential and radial directions are studied. In [7], a numerical method for calculating wave propagation in an infinite laminated cylinder is presented.

Acoustic monitoring of pipelines immersed in the ground or a liquid generates problems on submerged waveguides. In [8], features of axisymmetric wave motions in a fluid-filled elastic cylinder surrounded by a fluid medium are discussed. A technique for determining real and complex roots of the dispersion equation is proposed. For buried/submerged elastic waveguides, the surrounding medium is modelled by a perfectly matched layer (PML) [9].

In [10], authors provide information on various aspects of the dispersion set structure, free and forced oscillations in a cross-inhomogeneous anisotropic layer in 2D case. The comparison of different principles of selecting a single solution is carried out.

The work [11] is devoted to the study of waves in irregular waveguides based on the method of homogeneous solutions to investigate the processes of propagation and reflection of waves in longitudinally inhomogeneous waveguides with an arbitrary cross section. Solutions union for cylindrical waveguides of different diameters including those with broken line is performed. Based on the perturbation method and the Saint-Venant solutions, a structure of the lower modes of the dispersion set is investigated; special attention is paid to the waveguides with a circular cross section.

When studying wave processes in thin-walled cylinders, using the shells theory is common practice. In [12], low-frequency analysis of propagating modes in an orthotropic cylindrical shell is presented. A short dispersion equation, providing a good approximation to the exact dispersion equation for the lower three branches is given.

To conduct natural experiments, simplified models are required. In the book [13], ultrasonic waves in waveguides are studied with reference to problems of non-destructive testing and structural health monitoring of elements.

2 General Problem Statement

Consider dynamic processes in an infinite cylinder with an annular cross section and denote the inner radius $r = a$, the outer radius $r = b$ and a dimensionless coordinate $x = r/b$, where $\xi \le x \le 1$, $\xi = a/b$.

All the problems for waveguides can be divided into two classes: problems of free vibrations and problems of forced oscillations under the action of a local axisymmetric load. Next, we will consider several types of problems differing in type of inhomogeneity, the characteristics of the medium and the boundary conditions.

3 Free Vibrations

The problems of free oscillations in radially inhomogeneous cylindrical waveguides are reduced to finding nontrivial solutions of the first order matrix differential operators of the form

$$\mathbf{X}' = \mathbf{A}\mathbf{X} \tag{1}$$

with homogeneous boundary conditions

$$\mathbf{M}_1\mathbf{X}|_{x=\xi} = 0 \text{ and } \mathbf{M}_2\mathbf{X}|_{x=1} = 0, \tag{2}$$

where $\mathbf{A}$ is the matrix of dimension $N \times N$ with variable coefficients. The constant matrixes $\mathbf{M}_1$, $\mathbf{M}_2$ have dimension $N/2 \times N$. $\mathbf{X}$ is a vector of the amplitudes of the basic variables for boundary value problems of dimension N (displacements, stresses, electric potential, induction, etc.).

In this section two types of problems are considered, which are conventionally divide it into problems without attenuation (elastic or electoelastic body) (type 1) and problems with attenuation (type 2), which is modeled in the framework of the concept of complex modules [14, 15] (for models of viscoelasticity or model of electroelasticity with attenuation).

For the problems of the first type, the matrix of coefficients of the operator can be represented as a quadratic sheaf

$$\mathbf{A} = \mathbf{A}_0 - \kappa^2\mathbf{A}_{01} + \gamma\mathbf{A}_1 + \gamma^2\mathbf{A}_2 \tag{3}$$

where κ is the dimensionless frequency parameter, γ is the dimensionless wave number. The coefficients of the square matrices $\mathbf{A}_0$, $\mathbf{A}_{01}$, $\mathbf{A}_1$, $\mathbf{A}_2$ have dimensions $N \times N$, depend on the radial coordinate x and do not depend on the spectral parameters. When $N = 6$, the matrixes $\mathbf{A}_0$, $\mathbf{A}_1$, $\mathbf{A}_2$ have the following structure:

$$\mathbf{A}_j = \begin{pmatrix} a_{11}^j & 0 & a_{13}^j & 0 & a_{15}^j & a_{16}^j \\ 0 & a_{22}^j & 0 & a_{24}^j & 0 & 0 \\ a_{31}^j & 0 & a_{33}^j & 0 & a_{35}^j & a_{36}^j \\ 0 & a_{42}^j & 0 & a_{44}^j & 0 & 0 \\ a_{51}^j & 0 & a_{53}^j & 0 & a_{55}^j & a_{56}^j \\ a_{61}^j & 0 & a_{63}^j & 0 & a_{65}^j & a_{66}^j \end{pmatrix}, \; j = 0, 2, \; \mathbf{A}_1 = \begin{pmatrix} 0 & a_{12}^1 & 0 & a_{14}^1 & 0 & 0 \\ a_{21}^1 & 0 & a_{23}^1 & 0 & a_{25}^1 & a_{26}^1 \\ 0 & a_{32}^1 & 0 & a_{34}^1 & 0 & 0 \\ a_{41}^1 & 0 & a_{43}^1 & 0 & a_{45}^1 & a_{46}^1 \\ 0 & a_{52}^1 & 0 & a_{54}^1 & 0 & 0 \\ 0 & a_{62}^1 & 0 & a_{64}^1 & 0 & 0 \end{pmatrix}$$

For $N = 4$, their structure can be described by the following rule: $a_{ij}^m = 0$, if the number $i + j$ is odd; $a_{ij}^m \neq 0$, if the number $i + j$ is even, where $m = 0, 2$; $a_{ij}^1 = 0$ if the number $i + j$ is even; $a_{ij}^1 \neq 0$, if the number $i + j$ is odd.

For the second type of problems, the matrix of coefficients of the operator can be represented as a sheaf

$$\mathbf{A} = \mathbf{A}_0(\kappa) - \kappa^2 \mathbf{A}_{01} + \gamma \mathbf{A}_1(\kappa) + \gamma^2 \mathbf{A}_2(\kappa) \tag{4}$$

In this case, the coefficients of the complex-valued matrices $\mathbf{A}_0(\kappa), \mathbf{A}_1(\kappa), \mathbf{A}_2(\kappa)$ depend on the radial coordinate and are rational functions of the frequency parameter κ. The introduced matrices have real limit values at $\kappa = 0$ and $\kappa = \infty$; these limit cases can be analysed on the basis of consideration of the problems of the first type.

Problems of both types can be formulated for four-dimensional ($N = 4$, problems 1A, 2A) $\mathbf{X} = (X_1, X_2, X_3, X_4)^T$ and six-dimensional ($N = 6$, problems 1B, 2B) $\mathbf{X} = (X_1, X_2, X_3, X_4, X_5, X_6)^T$ vector of unknowns.

Let us study the dispersion set of the problem (1)–(2) as a set of pairs of spectral parameters (κ, γ) in which the problem has at least one nontrivial solution. Note that the problem (1) with homogeneous boundary conditions (2) has a trivial solution for any combination of spectral parameters. In a further study, we assume that the spectral parameter $\kappa \geq 0$.

4 Dispersion Equation Structure

1. At $\kappa = 0$, $\gamma = 0$, the vector differential Eq. (1) takes the degenerate form $\mathbf{X}' = \mathbf{A}_0\mathbf{X}$, and the corresponding boundary conditions (2) remain the same because they do not depend on spectral parameters. In the study of the degenerate boundary value problem, the two following situations are possible:
   (a) The degenerate boundary value problem has only a trivial solution, then the point $\kappa = 0$, $\gamma = 0$ does not belong to the dispersion set.
   (b) The degenerate boundary value problem has nontrivial solution $\mathbf{X}_0$ (one or more), then the point under consideration belongs to the dispersion set. In accordance with the principles of analytical continuation, a branch of the dispersion set (one or more) [10] comes out of it; its characteristics can be found by perturbation theory methods. Suppose that, for small γ and κ, the

spectral parameters have a linear relationship $\kappa = t\gamma$. The solution decomposition of the problem by the parameter γ is performed. For this purpose let us assume $\mathbf{X} = \mathbf{X}_0 + \gamma\mathbf{X}_1 + \gamma^2\mathbf{X}_2 + \ldots$ and, equating the coefficients at the same powers γ, the sequence of boundary value problems are obtained.

$$\mathbf{X}_0' = \mathbf{A}_0\mathbf{X}_0$$

$$\mathbf{X'}_1 = \mathbf{A}_0\mathbf{X}_1 + \mathbf{F}_1, \quad \mathbf{F}_1 = \mathbf{A}_1\mathbf{X}_0$$

$$\mathbf{X'}_2 = \mathbf{A}_0\mathbf{X}_2 + \mathbf{F}_2, \quad \mathbf{F}_2 = \mathbf{A}_1\mathbf{X}_1 + \mathbf{A}_2\mathbf{X}_0 - t^2\mathbf{A}_{01}\mathbf{X}_0$$

$$\mathbf{M}_1\mathbf{X}_j\big|_{x=\xi} = 0 \text{ and } \mathbf{M}_2\mathbf{X}_j\big|_{x=1} = 0,\ j = 0, 1, 2$$

from which the parameter t can be defined.

2. For $\kappa = 0$, $\gamma \neq 0$ (static case)

$$\mathbf{X}' = \left(\mathbf{A}_0 + \gamma\mathbf{A}_1 + \gamma^2\mathbf{A}_2\right)\mathbf{X} \tag{5}$$

where the matrixes $\mathbf{A}_j$ are real even in the presence of attenuation (for all the introduced types of problems) there is a countable set of complex roots, which are quadruples: $\gamma,\ -\gamma,\ \bar{\gamma},\ -\bar{\gamma}$ on the complex plane Re (γ), Im (γ).
Proof Consider the spectral boundary value problem of finding a nontrivial vector solution $\mathbf{X}$ satisfying the Eq. (5) and boundary conditions (2). Suppose that it has a nontrivial solution $\mathbf{X}$ for $\gamma = \gamma_0$. We apply a complex conjugation to the Eq. (5) and obtain $\bar{\mathbf{X}}' = \left(\mathbf{A}_0 + \bar{\gamma}_0\mathbf{A}_1 + \bar{\gamma}_0^2\mathbf{A}_2\right)\bar{\mathbf{X}}$, from which it follows that the problem also has a nontrivial solution $\bar{\mathbf{X}}$ when $\gamma = \bar{\gamma}_0$. Let us prove the second part of the statement for the case $N = 4$. Let us introduce two subsets: R_1^4—contains vector functions of the form $\mathbf{Z}_1 = (X_1, 0, X_3, 0)$ and R_2^4—contains vector functions of the form $\mathbf{Z}_2 = (0, X_2, 0, X_4)$. The matrixes structure $\mathbf{A}_0$, $\mathbf{A}_1$, $\mathbf{A}_2$ give the following properties.
If $\mathbf{X} \in R_1^4$, then $\mathbf{A}_0\mathbf{X} \in R_1^4$, $\mathbf{A}_1\mathbf{X} \in R_2^4$, $\mathbf{A}_2\mathbf{X} \in R_1^4$
If $\mathbf{X} \in R_2^4$, then $\mathbf{A}_0\mathbf{X} \in R_2^4$, $\mathbf{A}_1\mathbf{X} \in R_1^4$, $\mathbf{A}_2\mathbf{X} \in R_2^4$
Using these properties, we will look for a solution in the form $\mathbf{X} = \mathbf{X}_1 + \mathbf{X}_2$, where $\mathbf{X}_1 \in R_1^4$, $\mathbf{X}_2 \in R_2^4$. Then the equation $\mathbf{X}' = \left(\mathbf{A}_0 + \gamma\mathbf{A}_1 + \gamma^2\mathbf{A}_2\right)\mathbf{X}$ is converted to a system of the following form

$$\mathbf{X'}_1 = \mathbf{A}_0\mathbf{X}_1 + \gamma\mathbf{A}_1\mathbf{X}_2 + \gamma^2\mathbf{A}_2\mathbf{X}_1$$
$$\mathbf{X'}_2 = \mathbf{A}_0\mathbf{X}_2 + \gamma\mathbf{A}_1\mathbf{X}_1 + \gamma^2\mathbf{A}_2\mathbf{X}_2$$

Now it is easy to verify that if $\mathbf{X} = \mathbf{X}_1 + \mathbf{X}_2$ is a solution for $\gamma = \gamma_0$ then $\mathbf{X} = \mathbf{X}_1 - \mathbf{X}_2$ is a solution for $\gamma = -\gamma_0$.
The case of existence of a solution for $\gamma = -\bar{\gamma}_0$ is considered similarly.

3. When $\gamma = 0$ (standing waves) equation (1) takes the following form $\mathbf{X}' = \left(\mathbf{A}_0 - \kappa^2 \mathbf{A}_{01}\right) \mathbf{X}$. Note that the problems of the first type (problem 1A and 1B) in this case have a set of real eigenvalues [10]. The sheaf often splits into sheaves of smaller dimension (usually 2). Let $K = K_1 \cup K_2$ have a set of eigenvalues $\kappa_j \geq 0$ at which the problem has a nontrivial solution. Eigenvalues of κ_j characterize the radial resonances of the waveguide; in virtue of the analytic nature of the dispersion set [10] from these points the dispersion set branches come out. Branches are divided into families according to belonging κ_j to the set K_1 or K_2.
   Note that the sets K_1, K_2 can have the same elements that generate multiple roots (this case is not considered in the future, since multiple situations are easily destroyed by moving the parameters of the problem).
   In the presence of attenuation (task 2A and 2B) dispersion set is significantly rebuilt. Problem $\mathbf{X}' = \left(\mathbf{A}_0(\kappa) - \kappa^2 \mathbf{A}_{01}\right) \mathbf{X}$ has only the trivial solution, with the exception, perhaps, of the case $\kappa = 0$. The branches of the dispersion set have a complex structure.

5 Properties of Equation Solution

For further purposes, let us introduce the scalar product of complex-valued vector functions $(\mathbf{s}(x), \mathbf{v}(x)) = \int\limits_{\xi_0}^{1} \left(\sum\limits_{i=1}^{N} s_i(x) \bar{v}_i(x) \right) x dx$, where N is the dimension of vectors, $\bar{v}_i$ are components of a complex-conjugate vector of a complex-conjugate vector to $\mathbf{v}(x)$.

Property 1 *Let* $\mathbf{X}_0$ *be a nontrivial solution of the boundary value problem for the vector differential Eq. (1) with boundary conditions (2). Then the corresponding boundary-value problem for the vector equation*

$$\mathbf{X}' = \mathbf{A}\mathbf{X} + \mathbf{F} \tag{6}$$

with boundary conditions (2) is solvable if the solvability condition

$$(\mathbf{F}, \mathbf{Y}) = 0 \tag{7}$$

is satisfied, where $\mathbf{Y}$ *is the solution of the conjugate problem*

$$-(x\mathbf{Y})' = x\mathbf{A}^T \mathbf{Y} \tag{8}$$

$$\mathbf{M}_1^0 \mathbf{Y}\big|_{x=\xi} = 0 \textit{ and } \mathbf{M}_2^0 \mathbf{Y}\big|_{x=1} = 0 \tag{9}$$

where $\mathbf{A}^T$ is the transposed matrix. The matrices $\mathbf{M}_1^0$ and $\mathbf{M}_2^0$ should be chosen so that $(\mathbf{X}_0 \cdot \mathbf{Y})\, x|_\xi^1 = 0$, where $(\mathbf{X}_0 \cdot \mathbf{Y}) = \sum_{i=1}^{N} X_{0i}\bar{Y}_i$ is ordinary scalar product of complex-valued vectors.

Proof Let us multiply the vector Eq. (6) by the trial vector function $\mathbf{Y}$ on the right. Integration by parts is performed on the left-hand side, and then the property of the scalar product is used on the right-hand side of equation

$$(\mathbf{X} \cdot \mathbf{Y})\, x|_\xi^1 - \int_\xi^1 \left(\mathbf{X} \cdot (x\mathbf{Y})'\right) dx = \int_\xi^1 \left(\mathbf{X} \cdot \mathbf{A}^T\mathbf{Y}\right) dx + (\mathbf{F}, \mathbf{Y})$$

It is required that the vector function $\mathbf{Y}$ is a solution of the conjugate Eq. (8) with boundary conditions (9) satisfying the relation $(\mathbf{X} \cdot \mathbf{Y})\, x|_\xi^1 = 0$. Thus we obtain the required condition (7).

6 Asymptotic Analysis

Let us investigate the dispersion curves structure for the problems 1A and 1B in the neighborhood of the line $\gamma = 0$. Let $\kappa_0 \in K$ and κ_0 is not a multiple eigenvalue, i.e. κ_0 does not belong simultaneously to K_1 and K_2. Let us seek a decomposition in the form

$$\kappa^2 = {\kappa_0}^2 + a_1\gamma + a_2\gamma^2 + \ldots \tag{10}$$

and the solution of the problem (1)–(2) in the form of a regular γ expansion

$$\mathbf{X} = \mathbf{X}_0 + \gamma\mathbf{X}_1 + \gamma^2\mathbf{X}_2 + \ldots \tag{11}$$

Let us formulate the problems for the same powers of γ

$$\mathbf{X}_0' = \left(\mathbf{A}_0 - {\kappa_0}^2\mathbf{A}_{01}\right)\mathbf{X}_0 \tag{12}$$

$$\mathbf{X}_1' = \left(\mathbf{A}_0 - {\kappa_0}^2\mathbf{A}_{01}\right)\mathbf{X}_1 + \mathbf{F}_1, \text{ where } \mathbf{F}_1 = (\mathbf{A}_1 - a_1\mathbf{A}_{01})\,\mathbf{X}_0 \tag{13}$$

$$\mathbf{X}_2' = \left(\mathbf{A}_0 - {\kappa_0}^2\mathbf{A}_{01}\right)\mathbf{X}_2 + \mathbf{F}_2, \text{ where } \mathbf{F}_2 = (\mathbf{A}_1 - a_1\mathbf{A}_{01})\,\mathbf{X}_1 + (\mathbf{A}_2 - a_2\mathbf{A}_{01})\,\mathbf{X}_0 \tag{14}$$

$$\mathbf{M}_1\mathbf{X}_j\big|_{x=\xi} = 0,\ \mathbf{M}_2\mathbf{X}_j\big|_{x=1} = 0, \quad j = 0, 1, 2 \tag{15}$$

The boundary conditions (2) do not depend on the spectral parameters, and therefore they will have the same form (15) for each degree of the γ expansion of the problem.

Equations (13) and (14) are equations with the right side with respect to (12). In the situation where (12) has a nontrivial solution, the solvability condition is imposed on the right-hand sides of Eqs. (13) and (14) in accordance with the property 1.

Let us apply the solvability condition to problem (13)–(15) and (14)–(15) with respect to problem (12)–(15). We obtain that $(\mathbf{F}_i, \mathbf{Y}) = 0, \quad i = 1, 2$, where $\mathbf{Y}$ is a solution of the problem

$$-(x\mathbf{Y})' = x\left(\mathbf{A}_0 - {\kappa_0}^2\mathbf{A}_{01}\right)^T\mathbf{Y},\ \mathbf{M}_1^0\mathbf{Y}\big|_{x=\xi} = 0,\ \mathbf{M}_2^0\mathbf{Y}\big|_{x=1} = 0$$

Let us rewrite the vectors of the right-hand sides $\mathbf{F}_1, \mathbf{F}_2$ in the following form $\mathbf{F}_1 = \mathbf{F}_{10} - a_1\mathbf{F}_{11},\ \mathbf{F}_2 = \mathbf{F}_{20} - a_2\mathbf{F}_{21}$, where $\mathbf{F}_{10} = \mathbf{A}_1\mathbf{X}_0,\ \mathbf{F}_{11} = \mathbf{F}_{21} = \mathbf{A}_{01}\mathbf{X}_0,\ \mathbf{F}_{20} = (\mathbf{A}_1 - a_1\mathbf{A}_{01})\,\mathbf{X}_1 + \mathbf{A}_2\mathbf{X}_0$

We obtain

$$a_1 = (\mathbf{F}_{10}, \mathbf{Y})\,(\mathbf{F}_{11}, \mathbf{Y})^{-1},\ a_2 = (\mathbf{F}_{20}, \mathbf{Y})\,(\mathbf{F}_{21}, \mathbf{Y})^{-1} \tag{16}$$

Definition 1 Let $(\mathbf{F}_{11}, \mathbf{Y}) = (\mathbf{F}_{21}, \mathbf{Y}) \neq 0$ and $a_1 = 0$. We say that the branch of the dispersion set leaving the point $(\kappa_0, 0)$ has a normal dispersion if $a_2 > 0$ and has an abnormal dispersion if $a_2 < 0$.

7 Problem Statements

Let us consider the examples of specific problems for different types of waveguides.

Problems 1AS, 1A and 1Aβ

Problem 1AS on the longitudinal-radial vibrations of a hollow cylinder with residual stresses (RS) can be reduced to the problem (1)–(2), where the sheaf matrices have the following nonzero components

$\mathbf{A}_0$: $a_{11}^0 = \dfrac{-g_1}{x\,(G+S_1)},\quad a_{13}^0 = \dfrac{1}{G+S_1}\quad a_{24}^0 = \dfrac{1}{g_2+S_1},\quad a_{31}^0 = \dfrac{G+S_2}{x^2} - \dfrac{g_1^2}{x^2\,(G+S_1)},\ a_{33}^0 = -\dfrac{2g_2+S_1}{x\,(G+S_1)},\ a_{44}^0 = -\dfrac{1}{x}.$

$\mathbf{A}_{01}$: $a_{31}^{01} = a_{42}^{01} = p\,(x).$

$\mathbf{A}_1$: $a_{12}^1 = \dfrac{g_1}{G+S_1},\quad a_{21}^1 = -\dfrac{g_2}{g_2+S_1},\quad a_{32}^1 = -\dfrac{g_1\,(2g_2+S_1)}{x\,(G+S_1)},\quad a_{34}^1 = \dfrac{g_2}{g_2+S_1},$
$a_{41}^1 = -\dfrac{g_1\,(2g_2+S_1)}{x\,(G+S_1)},\ a_{43}^1 = -\dfrac{g_1}{G+S_1}.$

$\mathbf{A}_2$: $a_{31}^2 = \dfrac{g_2S_3 + S_1\,(g_2+S_3)}{g_2+S_1},\ a_{42}^2 = G + S_3 - \dfrac{g_1^2}{G+S_1}.$

Boundary conditions (2) corresponding to stress-free boundaries are characterized by matrices

$$\mathbf{M}_1 = \mathbf{M}_2 = \begin{pmatrix} 0\,0\,1\,0 \\ 0\,0\,0\,1 \end{pmatrix}$$

Let us list the dimensionless parameters and variables used below.

The values of the shear modulus and the density averaged over the wall thickness $\mu_0 = (b-a)^{-1} \int_a^b \mu(r)\,dr$, $\rho_0 = (b-a)^{-1} \int_a^b \rho(r)\,dr$. The vector of unknowns $\mathbf{X}$ contains the displacement components $u_r = bX_1, u_z = ibX_2$ and components of the additional Piola stress tensor $t_r = \mu_0 X_3, t_{rz} = i\mu_0 X_4$ [16]. g_1, g_2 are nondimentional Lamé parameters, S_1, S_2, S_3 are nondimentional components of the RS tensor, where $\sigma_r^0 = \mu_0 S_1,\ \sigma_\phi^0 = \mu_0 S_2,\ \sigma_z^0 = \mu_0 S_3,\ p$ is nondimentional density. $g_1 + 2g_2 = G$, $\kappa^2 = \rho_0\omega^2 b^2/\mu_0$, γ is dimensionless wave number.

Problem 1A corresponds to vibration of hollow cylinder without RS (Let $S_1 = S_2 = S_3 = 0$ be, the boundary conditions do not change).

Problem 1Aβ is actually Problem 1A with non-canonical boundary conditions corresponding to the matrices

$$\mathbf{M}_1 = \begin{pmatrix} 0 & 0 & 1 & 0 \\ 0 & 0 & 0 & 1 \end{pmatrix} \text{ and } \mathbf{M}_2 = \begin{pmatrix} \beta_1 & 0 & 1 & 0 \\ 0 & \beta_2 & 0 & 1 \end{pmatrix}$$

containing two real positive parameters $\beta_1 > 0,\ \beta_2 > 0$. Boundary conditions of this type relate the displacements and stresses of the waveguide outer boundary and model the contact with the external elastic medium, and for $\beta_1 = \beta_2 = 0$, the boundary conditions correspond to the free external boundary, the case $\beta_1, \beta_2 = \infty$ corresponds to anchorage of the waveguide outer boundary, and the case $\beta_1 = \infty,\ \beta_2 = 0$ corresponds to a waveguide that is in a rigid holder without friction.

Problem 1B

Problem 1B on the longitudinal-radial vibrations of an electroelastic hollow cylinder can be reduced to the problem (1)–(2), where the sheaf matrices have the following nonzero components

$\mathbf{A}_0$: $a_{11}^0 = -\dfrac{\Delta_4}{x\Delta_1}$, $a_{13}^0 = \dfrac{\epsilon_{33}}{\Delta_1}$, $a_{16}^0 = \dfrac{\tau_{33}}{\Delta_1}$, $a_{24}^0 = \dfrac{1}{c_{44}}$, $a_{31}^0 = \dfrac{c_{11}}{x^2} + \dfrac{\Delta_2}{x^2\Delta_1}$, $a_{33}^0 = \dfrac{\Delta_4}{x\Delta_1} - \dfrac{1}{x}$, $a_{36}^0 = \dfrac{\Delta_3}{x\Delta_1}$, $a_{44}^0 = -\dfrac{1}{x}$, $a_{51}^0 = \dfrac{(c_{33}\tau_{13} - \tau_{33}c_{13})}{x\Delta_1}$, $a_{53}^0 = a_{16}^0$, $a_{56}^0 = -\dfrac{c_{33}}{\Delta_1}$, $a_{66}^0 = -\dfrac{1}{x}$.

$\mathbf{A}_{01}$: $a_{31}^{01} = a_{42}^{01} = p(x)$.

$\mathbf{A}_1$: $a_{12}^1 = \dfrac{\Delta_4}{\Delta_1}$, $a_{21}^1 = -1$, $a_{25}^1 = -\dfrac{\tau_{15}}{c_{44}}$, $a_{32}^1 = -\dfrac{1}{x}\left(c_{12} + \dfrac{\Delta_2}{\Delta_1}\right)$, $a_{34}^1 = 1$, $a_{41}^1 = a_{32}^1$, $a_{43}^1 = -\dfrac{\Delta_4}{\Delta_1}$, $a_{46}^1 = -\dfrac{\Delta_3}{\Delta_1}$, $a_{52}^1 = \dfrac{\Delta_3}{\Delta_1}$, $a_{64}^1 = \dfrac{\tau_{15}}{c_{44}}$.

$\mathbf{A}_2$: $a_{42}^2 = c_{11} + \dfrac{\Delta_2}{\Delta_1}$, $a_{65}^2 = -\epsilon_{11} - \dfrac{{\tau_{15}}^2}{c_{44}}$.

and boundary conditions (2) corresponding to stress-free and electrically shorted boundaries are characterized by matrices

$$\mathbf{M}_1 = \mathbf{M}_2 = \begin{pmatrix} 0\,0\,1\,0\,0\,0 \\ 0\,0\,0\,1\,0\,0 \\ 0\,0\,0\,0\,1\,0 \end{pmatrix}$$

Let us introduce characteristic values of the amplitudes: the electric potential ϕ_0, the vector of electrical induction D_0, the elastic moduli μ_0, the piezoelectric constants e_0, and the permittivity at constant deformations ϵ_0. The dimensioning was carried out, we introduce the following parameters and functions $\kappa^2 = \rho_0\omega^2 b^2/\mu_0$, $u_r = bX_1$, $u_z = ibX_2$, $\sigma_r = \mu_0 X_3$, $\sigma_{rz} = i\mu_0 X_4$, $C_{ij} = \mu_0 c_{ij}$, $e_{ij} = e_0\tau_{ij}$, $\phi = \phi_0 X_5$, $\epsilon_{ii} = \epsilon_0\epsilon_{ii}$, $D_r = D_0 X_6$; The following relation is assumed $\phi_0 e_0 = b\mu_0$. Also the following relations are introduced $\Delta_1 = \tau_{33}^2 + \epsilon_{33}c_{33}$, $\Delta_2 = \tau_{31}c_{33}\tau_{13} - \tau_{31}\tau_{33}c_{13} - c_{13}\tau_{33}\tau_{13} - c_{13}\epsilon_{33}c_{13}$, $\Delta_3 = c_{13}\tau_{33} - \tau_{31}c_{33}$, $\Delta_4 = \tau_{31}\tau_{33} + c_{13}\epsilon_{33}$.

Problems 2A and 2B

To take into account the attenuation in problems 1A and 1B, we replace g_1, g_2 and c_{ij} with their complex analogs $g_1(\kappa)$, $g_2(\kappa)$, $c_{ij}(\kappa)$. We will be guided by the concept of complex modules and apply the model of a standard viscoelastic inhomogeneous body, where the frequency dependence has the form $(N(x) - in\kappa M(x))(1 - in\kappa)^{-1}$, where $N(x)$ is the analog of the long-term module, $M(x)$ is the analog of the instantaneous module, and n is the dimensionless relaxation time.

Property 2 *For the case $N = 4$ Let us introduce the block matrix* $\mathbf{J} = \begin{pmatrix} 0 & -\mathbf{E} \\ \mathbf{E} & 0 \end{pmatrix}$, *where* $\mathbf{E}$ *is unit matrix of dimension* 2×2. *For the case* $N = 6$ - $\mathbf{J} = \begin{pmatrix} 0 & -\mathbf{E} & 0 \\ \mathbf{E} & 0 & 0 \\ 0 & 0 & \mathbf{H} \end{pmatrix}$, *where* $\mathbf{H} = \begin{pmatrix} 0 & -1 \\ 1 & 0 \end{pmatrix}$

Due to the fact that for the above problems (1A, 1AS, 1Aβ and 1B) the property $\mathbf{J}\mathbf{A}^T\mathbf{J} - \frac{1}{x}\mathbf{E} = \mathbf{A}$ *is performed, then the conjugate vector for a problem of the form (6) has a specific form* $\mathbf{Y} = \mathbf{J}\mathbf{X}_0$, *where* $\mathbf{X}_0$ *is the solution of the problem (1)–(2).*

Proof The conjugate equation $-(x\mathbf{Y})' = x\mathbf{A}^T\mathbf{Y}$ is transformed under the substitution $\mathbf{Y} = \mathbf{J}\mathbf{X}_0$ to the form $-(x\mathbf{J}\mathbf{X}_0)' = x\mathbf{A}^T\mathbf{J}\mathbf{X}_0$. Let us multiply the conjugate equation by $\mathbf{J}$ from the left and take into account that $\mathbf{J}\mathbf{J}\mathbf{X} = -\mathbf{X}$. We get $x\mathbf{X}'_0 = x\mathbf{J}\mathbf{A}^T\mathbf{J}\mathbf{X}_0 - \mathbf{X}_0$, then regroup $\mathbf{X}'_0 = \left(\mathbf{J}\mathbf{A}^T\mathbf{J} - \frac{1}{x}\mathbf{E}\right)\mathbf{X}_0$ ($\mathbf{E}$ is the identity matrix of dimension $N \times N$). The resulting problem will have a nontrivial solution if $\mathbf{J}\mathbf{A}^T\mathbf{J} - \frac{1}{x}\mathbf{E} = \mathbf{A}$, since the problem $\mathbf{X}' = \mathbf{A}\mathbf{X}$ has the nontrivial solution $\mathbf{X}_0$ by the condition of property 1.

Property 3 *For the problems introduced above, the expansion coefficient a_1, determined by formula (16) in Sect. 4, is reduced to zero.*

Proof Let us carry out the proof for the case $N = 4$. By the condition of Section 4, $\kappa_0 \in K$ and is not a multiple eigenvalue. Due to the structure of the matrix

$\left(\mathbf{A}_0 - \kappa_0^2 \mathbf{A}_{01}\right)$, the solution $\mathbf{X}_0$ of the problem (12), (15) belongs to one of the two subsets R_1^4 or R_2^4, introduced above. Taking into account property 2, we obtain that $a_1 = (\mathbf{A}_1\mathbf{X}_0, \mathbf{J}\mathbf{X}_0)(\mathbf{A}_{01}\mathbf{X}_0, \mathbf{J}\mathbf{X}_0)^{-1}$. It is easy to verify that $\mathbf{A}_1\mathbf{X}_0 \in R_2^4$, $\mathbf{J}\mathbf{X}_0 \in R_1^4$, if $\mathbf{X}_0 \in R_1^4$ and $\mathbf{A}_1\mathbf{X}_0 \in R_1^4$, $\mathbf{J}\mathbf{X}_0 \in R_2^4$, if $\mathbf{X}_0 \in R_2^4$. In both cases, the scalar product $(\mathbf{A}_1\mathbf{X}_0, \mathbf{J}\mathbf{X}_0)$ is identically equal to zero.

Corollary 1 *The second expansion coefficient, determined by (16), is calculated the from formula* $a_2 = (\mathbf{A}_1\mathbf{X}_1 + \mathbf{A}_2\mathbf{X}_0, \mathbf{J}\mathbf{X}_0)(\mathbf{A}_{01}\mathbf{X}_0, \mathbf{J}\mathbf{X}_0)^{-1}$. *The result is a corollary of the properties 2 and 3, with reference to (16).*

8 Properties of Problems 1AS, 1A, 1Aβ and 1B

Problem 1AS

For an elastic waveguide with RS, problem (12), (15) is divided into two subproblems differing in the kinematics of nontrivial (homogeneous) solutions.

$$\begin{cases} X_{01}' = -\dfrac{g_1}{x\,(G+S_1)}X_{01} + \dfrac{1}{(G+S_1)}X_{03} \\ X_{03}' = \left(-\kappa^2 p + \dfrac{G+S_2}{x^2} - \dfrac{g_1^2}{x^2\,(G+S_1)}\right)X_{01} - \dfrac{2g_2+S_1}{x\,(G+S_1)}X_{03} \\ X_{03}(\xi) = X_{03}(1) = 0 \end{cases} \tag{17}$$

subproblem 1S

$$\begin{cases} X_{02}' = \dfrac{1}{g_2+S_1}X_{04} \\ X_{04}' = -\kappa^2 p X_{02} - \dfrac{X_{04}}{x} \\ X_{04}(\xi) = X_{04}(1) = 0 \end{cases} \tag{18}$$

subproblem 2S

Let $K_{1,s}$ and $K_{2,s}$ be sets of eigenvalues of subproblems 1S and 2S.

Property 4 $\kappa = 0$ *belongs to the set* $K_{2,s}$, *and the non-trivial solution of the subproblem 2S has the form* $X_{02} = 1,\ X_{04} = 0$.

Property 5 *If* $\kappa_1 \in K_{1,s}$ *and* X_{01}, X_{03} *are non-trivial solutions of the problem (17), then the following relation holds*

$$\kappa_1^2 = \left(\int_\xi^1 \left(\frac{G+S_2}{x} - \frac{g_1^2}{x\,(G+S_1)}\right)X_{01}^2 dx + \int_\xi^1 \frac{xX_{03}^2}{G+S_1}dx\right)\left(\int_\xi^1 xpX_{01}^2 dx\right)^{-1} \tag{19}$$

Proof Let us multiply the second equation of the subproblem 1S by $X_{01}(x)\,x$ and integrate over the interval $[\xi, 1]$. We integrate by parts and substitute X'_{01} from the first Eq. (17). After the isolation of κ_1^2, we get (19).

Property 6 *If $\kappa_2 \in K_{2,S}$ and X_{02}, X_{04} are non-trivial solutions of the problem (18), then the following relation holds*

$$\kappa_1^2 = \left(\int\limits_{\xi}^{1} \frac{xX_{04}^2}{g_2 + S_1} dx\right)\left(\int\limits_{\xi}^{1} xpX_{02}^2 dx\right)^{-1} \tag{20}$$

Property 7 *For an elastic waveguide with RS, the coefficient a_2 in the expansion (16) has the form $a_2 = \int_{\xi}^{1} P_j(x)\,xdx \left(\int\limits_{\xi}^{1} X_{0j}^2\, p\, x\, dx\right)^{-1}$, where $j = 1$, if $\kappa_0 \in K_{1,S}$ (subproblem 1S) and $j = 2$, if $\kappa_0 \in K_{2,S}$ (subproblem 2S). The introduced functions $P_j(x)$ are defined by the following formulas*

$$P_1(x) = \left(S_3 + \frac{S_1 g_2}{g_2 + S_1}\right) X_{01}^2 - \frac{g_1}{G_1 + S_1} X_{12} X_{03} - \frac{g_1(2g_2 + S_1)}{x(G_1 + S_1)} X_{12} X_{01} + \\ + \frac{g_2}{g_2 + S_1} X_{14} X_{01} \tag{21}$$

$$P_2(x) = \left(G_1 + S_3 - \frac{g_1^2}{G_1 + S_1}\right) X_{02}^2 + \frac{g_2}{g_2 + S_1} X_{11} X_{04} - \\ - \frac{g_1(2g_2 + S_1)}{x(G_1 + S_1)} X_{11} X_{02} - \frac{g_1}{G_1 + S_1} X_{13} X_{02} \tag{22}$$

Low-frequency asymptotics for the problem 1AS. For $\kappa = \gamma = 0$, the subproblem 2S has a non-trivial solution $X_{02} = 1$, $X_{04} = 0$ (property 4). It is easy to obtain the slope of the dispersion curve $t = \sqrt{a_2}$ leaving the origin, where

$$a_2 = \tilde{p}^{-1} \int_{\xi}^{1} \left[\left(G_1 + S_3 - \frac{g_1^2}{G_1 + S_1}\right) - \frac{g_1(2g_2 + S_1)}{x(G_1 + S_1)} X_{11} - \frac{g_1}{G_1 + S_1} X_{13}\right] xdx \tag{23}$$

where $\tilde{p} = \int\limits_{\xi}^{1} p(x)\,x\,dx$. In the case of constants g_1, g_2, p and RS, corresponding to the action of internal pressure, one can construct a solution X_{11}, X_{13} and formula (23) takes the form

$$a_2 = \frac{1}{p}\left[\frac{g_2(3g_1 + 2g_2)}{(g_1 + g_2)}\right] + \frac{1}{p}\left[\frac{g_1^2}{2(g_1 + g_2)^2} + \frac{g_1}{(g_1 + g_2)}\right]\varepsilon + O(\varepsilon^2) \tag{24}$$

Problem 1A

For an elastic waveguide problem (12), (15) is divided into two subproblems.

$$\begin{cases} X'_{01} = -\dfrac{g_1}{xG}X_{01} + \dfrac{1}{G}X_{03} \\ X'_{03} = \left(-\kappa^2 p + \dfrac{4g_2\,(g_1+g_2)}{x^2 G}\right) X_{01} - \dfrac{2g_2}{xG}X_{03} \\ X_{03}\,(\xi) = X_{03}\,(1) = 0 \end{cases} \tag{25}$$

subproblem 1

$$\begin{cases} X'_{02} = \dfrac{1}{g_2}X_{04} \\ X'_{04} = -\kappa^2 p X_{02} - \dfrac{X_{04}}{x} \\ X_{04}\,(\xi) = X_{04}\,(1) = 0 \end{cases} \tag{26}$$

subproblem 2

Let K_1 and K_2 be sets of eigenvalues of subproblems 1 and 2.

Property 8 *If $\kappa_1 \in K_1$ and X_{01}, X_{03} are non-trivial solutions of the problem (25), then the following relation holds*

$$\kappa_1^2 = \left(\int\limits_{\xi}^{1} \left(\frac{G^2 - g_1^2}{xG}\right) X_{01}^2 dx + \int\limits_{\xi}^{1} \frac{xX_{03}^2}{G} dx\right)\left(\int\limits_{\xi}^{1} xpX_{01}^2 dx\right)^{-1} \tag{27}$$

Property 9 *If $\kappa_2 \in K_2$ and X_{02}, X_{04} are non-trivial solutions of the problem (26), then the following relation holds*

$$\kappa_2^2 = \left(\int\limits_{\xi}^{1} \frac{xX_{04}^2}{g_2} dx\right)\left(\int\limits_{\xi}^{1} xpX_{02}^2 dx\right)^{-1} \tag{28}$$

Property 10 *In this case the coefficient a_2 of expansion (16) is also calculated separately for two subproblems*

$$a_2 = \tilde{p}_1^{-1} \int_{\xi}^{1} \left[X_{14}X_{01} - \frac{g_1}{G}X_{12}X_{03} - \frac{2g_1 g_2}{xG}X_{12}X_{01}\right] x dx \tag{29}$$

(subproblem 1)

$$a_2 = \tilde{p}_2^{-1} \int_\xi^1 \left[\left(G - \frac{g_1^2}{G} \right) X_{02}^2 + X_{11} X_{04} - \frac{2 g_1 g_2}{x G} X_{11} X_{02} - \frac{g_1}{G} X_{13} X_{02} \right] x dx \tag{30}$$

(subproblem 2)

where $\tilde{p}_j = \int\limits_\xi^1 X_{0j}^2 \, p \, x \, dx$.

Low-frequency asymptotics for the problem 1A. The solution of the subproblem 2S (18) is also a solution of the subproblem 2 (26). Similarly, we get that

$$a_2 = \int_\xi^1 \left[\left(G - \frac{g_1^2}{G} \right) - \frac{2 g_2 g_1}{x G} X_{11} - \frac{g_1}{G} X_{13} \right] x dx \left(\int\limits_\xi^1 p \, x \, dx \right)^{-1} \tag{31}$$

In the case of constants g_1, g_2, p we get

$$a_2 = \frac{g_2 (3 g_1 + 2 g_2)}{p (g_1 + g_2)}, \text{ where } X_{11}(x) = \frac{g_1 x}{2 (g_1 + g_2)}, \; X_{13}(x) = 0 \tag{32}$$

Problem 1Aβ

For an elastic waveguide with modified boundary conditions problem (12), (15) is divided into two subproblems.

$$\begin{cases} X_{01}' = -\dfrac{g_1}{x G} X_{01} + \dfrac{1}{G} X_{03} \\ X_{03}' = \left(-\kappa^2 p + \dfrac{4 g_2 (g_1 + g_2)}{x^2 G} \right) X_{01} - \dfrac{2 g_2}{x G} X_{03} \\ X_{03}(\xi) = 0, \quad X_{03}(1) = -\beta_1 X_{01}(1) \end{cases} \tag{33}$$

subproblem 1β

$$\begin{cases} X_{02}' = \dfrac{1}{g_2} X_{04} \\ X_{04}' = -\kappa^2 p X_{02} - \dfrac{X_{04}}{x} \\ X_{04}(\xi) = 0, \quad X_{04}(1) = -\beta_2 X_{02}(1) \end{cases} \tag{34}$$

subproblem 2β

Let $K_{1,\beta}$ and $K_{2,\beta}$ be sets of eigenvalues of subproblems 1β and 2β.

Property 11 *If $\kappa_1 \in K_{1,\beta}$ and X_{01}, X_{03} are non-trivial solutions of the problem (33), then the following relation holds*

$$\kappa_1^2 = \left(\beta_1 X_{01}^2 (1) + \int\limits_{\xi}^{1} \left(\frac{G^2 - g_1^2}{xG} \right) X_{01}^2 dx + \int\limits_{\xi}^{1} \frac{x X_{03}^2}{G} dx \right) \left(\int\limits_{\xi}^{1} x p X_{01}^2 dx \right)^{-1} \tag{35}$$

Property 12 *If $\kappa_2 \in K_{2,\beta}$ and X_{02}, X_{04} are non-trivial solutions of the problem (34), then the following relation holds*

$$\kappa_2^2 = \left(\beta_2 X_{02}^2 (1) + \int\limits_{\xi}^{1} \frac{x X_{04}^2}{g_2} dx \right) \left(\int\limits_{\xi}^{1} x p X_{02}^2 dx \right)^{-1} \tag{36}$$

Property 13 *The formulas for the expansion coefficient a_2 are identical in structure to the formulas given in property 10, but are calculated through the solutions of subproblems 1β and 2β.*

The low-frequency asymptotics can be constructed only in case $\beta_2 = 0$, since $\beta_2 \neq 0$, $0 \notin K_{2,\beta}$.

$$a_2 = \int_{\xi}^{1} \left[\left(G - \frac{g_1^2}{G} \right) - \frac{2 g_2 g_1}{xG} X_{11} - \frac{g_1}{G} X_{13} \right] x dx \left(\int\limits_{\xi}^{1} p\, x\, dx \right)^{-1} \tag{37}$$

In the case of constants g_1, g_2, p, the formula (37) is transformed to the form

$$a_2 = \frac{g_2 \left(2 \left(3 g_1 g_2 + 2 g_2^2 \right) \left(\xi_0^2 - 1 \right) - \left(\left(3 g_1 + 2 g_2 \right) \xi_0^2 + g_1 + 2 g_2 \right) \beta_1 \right)}{p \left(2 \left(g_1 g_2 + g_2^2 \right) \left(\xi_0^2 - 1 \right) - \left(\left(g_1 + g_2 \right) \xi_0^2 + g_2 \right) \beta_1 \right)} \tag{38}$$

$$X_{11}(x) = \frac{1}{2x} \frac{g_1 \left(\beta_1 \xi_0^2 + x^2 \left(2 \xi_0^2 g_2 - 2 g_2 - \beta_1 \xi_0^2 \right) \right)}{2 \left(g_1 g_2 + g_2^2 \right) \left(\xi_0^2 - 1 \right) - \beta_1 \left(\left(g_1 + g_2 \right) \xi_0^2 + g_2 \right)},$$

$$X_{13}(x) = \frac{1}{x^2} \frac{g_1 g_2 \beta_1 \left(x^2 - \xi_0^2 \right)}{2 \left(g_1 g_2 + g_2^2 \right) \left(\xi_0^2 - 1 \right) - \beta_1 \left(\left(g_1 + g_2 \right) \xi_0^2 + g_2 \right)}$$

The slope t of the dispersion set branch emerging from the origin is analyzed. For a homogeneous waveguide, it is revealed that $t = \sqrt{a_2 (\beta_1)}$ monotonically depends on β_1 and changes in a narrow range. More detailed results are given in [17].

Problem 1B

Consideration 1B problem is began with the case of $\gamma = 0$. The problem (12), (15) is divided into two subproblems, besides, the first subproblem does not contain the electric field component.

$$\begin{cases} X_{01}' = \dfrac{1}{\Delta_1}\left[\tau_{33}X_{06} + \epsilon_{33}X_{03} - \Delta_4\dfrac{X_{01}}{x}\right] \\ X_{03}' = \left(\dfrac{c_{11}}{x^2} + \dfrac{\Delta_2}{x^2\Delta_1} - p\kappa^2\right)X_{01} + \left(\dfrac{\Delta_4}{x\Delta_1} - \dfrac{1}{x}\right)X_{03} + \dfrac{\Delta_3}{x\Delta_1}X_{06} \\ X_{05}' = \dfrac{1}{\Delta_1}\left[-\Delta_3\dfrac{X_{01}}{x} + \tau_{33}X_{03} - c_{33}X_{06}\right] \\ X_{06}' = -\dfrac{X_{06}}{x} \\ X_{03}(\xi) = X_{03}(1) = X_{05}(\xi) = X_{05}(1) = 0 \end{cases} \tag{39}$$

subproblem 1B

$$\begin{cases} X_{02}' = \dfrac{1}{c_{44}}X_{04} \\ X_{04}' = -p\kappa^2 X_{02} - \dfrac{1}{x}X_{04} \\ X_{04}(\xi) = X_{04}(1) = 0 \end{cases} \tag{40}$$

subproblem 2B

Property 14 $\kappa = 0$ *belongs to the set of eigenvalues of (40).*

Property 15 *For an electroelastic waveguide, the coefficient* a_2 *in the expansion (16) has the form* $a_2 = \int_\xi^1 P_j(x)\,x dx \left(\int\limits_\xi^1 (X_{0j})^2 p\,x\,dx\right)^{-1}$, *where the notation is introduced according to property 7.*

$$P_1(x) = Q_1 X_{05}^2 - \frac{\Delta_4}{\Delta_1}X_{12}X_{03} + Q_2X_{12}X_{01} + X_{14}X_{01} - \frac{\Delta_3}{\Delta_1}X_{12}X_{06} + \frac{\tau_{15}}{c_{44}}X_{14}X_{05} \tag{41}$$

$$P_2(x) = Q_3X_{02}^2 + X_{11}X_{04} + \frac{\tau_{15}}{c_{44}}X_{15}X_{04} + Q_2X_{11}X_{02} - \frac{\Delta_4}{\Delta_1}X_{13}X_{02} - \frac{\Delta_3}{\Delta_1}X_{16}X_{02} \tag{42}$$

where $Q_1 = \left(-\epsilon_{11} - \frac{(\tau_{15})^2}{c_{44}}\right)$, $Q_2 = -\frac{1}{x}\left(c_{12} + \frac{\Delta_2}{\Delta_1}\right)$, $Q_3 = \left(c_{11} + \frac{\Delta_2}{\Delta_1}\right)$.

Low-frequency asymptotics for the problem 1B. Similarly, we get that

$$a_2 = \int_\xi^1 \left[Q_3 + Q_2X_{11} - \frac{\Delta_4}{\Delta_1}X_{13} - \frac{\Delta_3}{\Delta_1}X_{16}\right]xdx\left(\int\limits_\xi^1 p(x)\,x\,dx\right)^{-1} \tag{43}$$

Comparative Results

Let us compare sets K_S, K, K_β

Property 16 *$\kappa_{i,\beta} > \kappa_i$, if $\kappa_{i,\beta} \in K_\beta$, $\kappa_i \in K$ and $\beta_1 > 0$, $\beta_2 > 0$. The proof is given in the article [17].*

Property 17 *Elements of the sets $K_{1,S}$ and $K_{2,S}$ can be obtained through the elements K_1 and K_2 and the corresponding eigenfunctions.*

The problems (17), (18) contain three components of the tensor RS S_1, S_2, S_3, which are small in comparison with g_1, g_2 and their dimensionless amplitude ε normally lies in the interval $10^{-5} \sim 10^{-3}$, then you can expand the problem in a Taylor series for ε and get the problem for ε^0 and ε^1. The application of the solvability condition to the problem for ε^1 yields a formula that reflects the change in the frequency parameter κ as a function of the components of the tensor RS S_1, S_2, S_3.

Let $\kappa^{(j)} = \kappa_0^{(j)} + \kappa_1^{(j)} + o\left(\kappa_1^{(j)}\right)$, where $\kappa^{(j)} \in K_S$, $\kappa_0^{(j)} \in K$, $\kappa_1^{(j)}$—a small correction, then $\kappa_1^{(j)} = \int\limits_\xi^1 F(x)\, x dx \left(2\kappa_0^{(j)} \int\limits_\xi^1 \left(X_{10}^2(x) + X_{20}^2(x)\right) p(x)\, x dx\right)^{-1}$, where

$$F(x) = \left(\frac{S_2}{x^2} + \left(\frac{g_1}{xG}\right)^2 S_1\right) X_{10}^2(x) + X_{30}^2(x)\frac{S_1}{G^2} + X_{40}^2(x)\frac{S_1}{g_2^2} - X_{10}(x)\, X_{30}(x)\frac{2g_1 S_1}{xG^2}$$

where $(X_{01}, X_{02}, X_{03}, X_{04})$ is solution of the problem 1A calculated at $\kappa_0^{(j)}$ and $\gamma = 0$.

In [16], calculations were made for a homogeneous cylindrical waveguide with RS, which correspond to the action of internal pressure. Calculations have shown that $\kappa_1^{(j)}$ can take positive and negative values if $\kappa_0^{(j)} \in K_1$. This is due to the fact that under the integral there is a quadratic form that is not sign-definite. If $\kappa_0^{(j)} \in K_2$ and S_1 does not change the sign to $[\xi, 1]$, then the signs $\kappa_0^{(j)}$ and S_1 are the same.

9 The Numerical Solution Scheme

Using the false position method, the numerical analysis of the dispersion relations is produced. The solution of the problem (1)–(2) for $N = 6$ is sought as a linear combination of three (two at $N = 4$) vector functions $\mathbf{X}^{(1)}$, $\mathbf{X}^{(2)}$, $\mathbf{X}^{(3)}$.

$$X_1^{(1)}(\xi) = 1, \quad X_2^{(1)}(\xi) = X_3^{(1)}(\xi) = X_4^{(1)}(\xi) = X_5^{(1)}(\xi) = X_6^{(1)}(\xi) = 0 \tag{44}$$

$$X_1^{(2)}(\xi) = 0, \quad X_2^{(2)}(\xi) = 1, \quad X_3^{(2)}(\xi) = X_4^{(2)}(\xi) = X_5^{(2)}(\xi) = X_6^{(2)}(\xi) = 0 \tag{45}$$

$$X_1^{(3)}(\xi) = X_2^{(3)}(\xi) = X_3^{(3)}(\xi) = X_4^{(3)}(\xi) = X_5^{(3)}(\xi) = 0, \quad X_6^{(3)}(\xi) = 1 \tag{46}$$

where $\mathbf{X}^{(1)}$ is a solution to the Cauchy problem (1)–(44), $\mathbf{X}^{(2)}$ is a solution (1)–(45) and $\mathbf{X}^{(3)}$ is a solution (1)–(46). Vector function $\mathbf{X} = p_1\mathbf{X}^{(1)} + p_2\mathbf{X}^{(2)} + p_3\mathbf{X}^{(3)}$ automatically satisfies the condition $\mathbf{M}_1\mathbf{X}|_{x=\xi} = 0$ and is a solution of (1).

Requiring the condition $\mathbf{M}_2\mathbf{X}|_{x=1} = 0$, a homogeneous algebraic system is obtained and equating its determinant to zero, the dispersion set points are obtained.

10 Forced Oscillations. Residue Calculation

Let us consider axisymmetric waves in a cylindrical waveguide with an annular cross section $a \leq r \leq b$, inhomogeneous in thickness. Let the internal and external boundaries of the waveguide be free from stresses; on the external boundary of the waveguide $r = b$ there is a periodic in time with the frequency ω concentrated normal load $\mu_0\delta(z)\exp(-i\omega t)$. We will consider the steady-state mode of oscillations in the axisymmetric case, assuming that the components of the physical fields do not depend on ϕ.

We apply the Fourier integral transform along the axial coordinate to the boundary value problem, having previously dimensioned it by the formula $z = yb$.

$$\tilde{X}_n(x,\alpha) = \int_{-\infty}^{\infty} X_n(x,y)\exp(i\alpha y)\,dy, \quad n = 1..4 \tag{47}$$

Then with respect to the transformants we have an operator bundle with two spectral parameters (κ, α) of the following form

$$\tilde{\mathbf{X}}' = \left(\mathbf{A}_0 - \kappa^2\mathbf{A}_{01} - \alpha\mathbf{A}_1 + \alpha^2\mathbf{A}_2\right)\tilde{\mathbf{X}}, \text{ where } \tilde{\mathbf{X}} = \left(\tilde{X}_1, \tilde{X}_2, \tilde{X}_3, \tilde{X}_4\right) \tag{48}$$

The vector Eq. (48) coincides with Eq. (1) if we assume $\alpha = -\gamma$, $\tilde{\mathbf{X}} = \mathbf{X}$.

The boundary conditions for the types of waveguides introduced above assume the form

$$\tilde{X}_k(\xi_0) = 0, \quad \tilde{X}_k(1) = -\delta_{3k} \tag{49}$$

where $k = 3, 4$ in the case of $N = 4$ and $k = 3, 4, 5$ in the case $N = 6$.

The problem (48)–(49) has a solution for any (κ, α), except for the points of the dispersion set and can be solved by the method of alignment.

It remains only to find the inverse Fourier transform by the formula

$$X_j(x,y) = \frac{1}{2\pi}\int_{\Gamma} \tilde{X}_j(x,\alpha)\exp(-i\alpha y)\,d\alpha, \quad j = 1, 2, 3, 4 \tag{50}$$

where the integration is carried out along the contour Γ, which coincides with the real axis everywhere except for the real poles of the transform and envelopes them in accordance with the principle of limit absorption [10]; in the regular case, the positive poles are enveloped below, and the negative ones from above; in the irregular case, the smallest pole in magnitude is enveloped in the opposite way.

We give a scheme of calculating primitives of the problem for $N = 4$, which can be generalized to the case $N = 6$.

We calculate the Fourier inverse integral transform for all the components of the wave field assuming that $y > 0$. The integrands $\tilde{X}_j(x, \alpha)$ are meromorphic, have singularities of the real and complex poles type, which are determined by the dispersion equation, and can be found numerically with the aid of the scheme described above. We also note that with the growth of α, the solution of the boundary value problem (48)–(49) begins to possess a boundary layer structure, and the error of the solution found by the Runge–Kutta method grows.

Let us find (50), applying the theory of residues. Consider the following contour integral

$$\frac{1}{2\pi}\int\limits_{L} \tilde{X}_j(x, \alpha)\exp(-i\alpha y)\,d\alpha = \\ = \frac{1}{2\pi}\int\limits_{L_0} \tilde{X}_j(x, \alpha)\exp(-i\alpha y)\,d\alpha + \frac{1}{2\pi}\int\limits_{L_R} \tilde{X}_j(x, \alpha)\exp(-i\alpha y)\,d\alpha \tag{51}$$

Here L is a closed contour in the complex plane of the parameter α consisting of two parts $L = L_0 \cup L_R$, where L_0 is the arc of the semicircle $\operatorname{Im}\alpha \leq 0$ with radius R and center at $\alpha = 0$, L_R is a part of the contour Γ, which is bounded by the segment $[-R, R]$.

Applying the theory of residues for the integral along the contour L, we obtain on the boundary of the waveguide $x = 1$.

$$\frac{1}{2\pi}\int\limits_{L} \tilde{X}_j(x, \alpha)\exp(-i\alpha y)\,d\alpha = -i\sum_{n=1}^{m}\operatorname{Res}\Big(\tilde{X}_j(1, \alpha_n)\exp(-i\alpha_n y)\Big),\ j = 1, 2, 3, 4 \tag{52}$$

where α_n are poles of the first order inside the contour L, and m their number.

In accordance with the Jordan lemma, as $R \to \infty$, the integral along the contour L_0 tends to zero, the integral in L_R tends to the integral (50) on the contour Γ, $m \to \infty$. Thus, in accordance with (52), in the regular case it is necessary to sum the residues over all negative real poles and over all complex poles with negative imaginary part.

$$\tilde{X}_1^{(1)}(\xi) = 1,\ \tilde{X}_2^{(1)}(\xi) = \tilde{X}_3^{(1)}(\xi) = \tilde{X}_4^{(1)}(\xi) = 0 \tag{53}$$

$$\tilde{X}_1^{(2)}(\xi) = 0, \tilde{X}_2^{(2)}(\xi) = 1,\ \tilde{X}_3^{(2)}(\xi) = \tilde{X}_4^{(2)}(\xi) = 0 \tag{54}$$

To find the residue at the pole α_n we use the following scheme. While searching for the solution of the problem by the method of alignment in the form $\tilde{\mathbf{X}} = c_1\tilde{\mathbf{X}}^{(1)} + c_2\tilde{\mathbf{X}}^{(2)}$ where $\tilde{\mathbf{X}}^{(1)}, \tilde{\mathbf{X}}^{(2)}$ are solutions (48)–(53) and (48)–(54). Satisfying the boundary conditions for $x = 1$, we obtain

$$\mathbf{B}(1,\alpha)\,\mathbf{C}(\alpha)=\mathbf{F} \tag{55}$$

where $\mathbf{B}(1,\alpha)=\begin{pmatrix}\tilde{X}_3^{(1)}(1,\alpha)\ \tilde{X}_3^{(2)}(1,\alpha)\\ \tilde{X}_4^{(1)}(1,\alpha)\ \tilde{X}_4^{(2)}(1,\alpha)\end{pmatrix}$, $\mathbf{C}(\alpha)=\begin{pmatrix}c_1(\alpha)\\ c_2(\alpha)\end{pmatrix}$, $\mathbf{F}=\begin{pmatrix}-1\\ 0\end{pmatrix}$

Let us decompose the vector function $\mathbf{C}(\alpha)$ and the matrix-valued function $\mathbf{B}(1,\alpha)$ into the Laurent series in the neighborhood of the pole α_n

$$\mathbf{C}(\alpha)=\mathbf{C}_{-1}\frac{1}{\alpha-\alpha_n}+\mathbf{C}_0+\ldots$$
$$\mathbf{B}(1,\alpha)=\mathbf{B}_0(1,\alpha_n)+\mathbf{B}_1(1,\alpha_n)(\alpha-\alpha_n)+\ldots,\ \text{where}\ \mathbf{B}_1(1,\alpha_n)=\left.\frac{\partial\mathbf{B}}{\partial\alpha}\right|_{\alpha=\alpha_n} \tag{56}$$

In accordance with the introduced expansions, the solution of the problem (48)–(49) can also be represented as a Laurent series. The coefficient of the Laurent series at $(\alpha-\alpha_n)^{-1}$: $\left[c_{-1}^1\tilde{X}_j^{(1)}(1,\alpha_n)+c_{-1}^2\tilde{X}_j^{(2)}(1,\alpha_n)\right]$ defines the residue at the first-order pole and allows you to switch from (52) to the following formula

$$X_j(1,y)=-i\sum_{n=1}^{\infty}\left[c_{-1}^1(\alpha_n)\,\tilde{X}_j^{(1)}(1,\alpha_n)+c_{-1}^2(\alpha_n)\,\tilde{X}_j^{(2)}(1,\alpha_n)\right]\exp(-i\alpha_n y),$$
$$j=1,2,3,4 \tag{57}$$

To find the coefficients used in (57), we use the representation (55), (56), where we equate terms with identical degrees of expansion, similarly to [18].

$$\begin{cases}\mathbf{B}_0(1,\alpha_n)\,\mathbf{C}_{-1}=0\\ \mathbf{B}_0(1,\alpha_n)\,\mathbf{C}_0+\mathbf{B}_1(1,\alpha_n)\,\mathbf{C}_{-1}=\mathbf{F}\end{cases} \tag{58}$$

To find the unknown matrix components $\mathbf{B}_1(1,\alpha_n)$, we form two auxiliary vector functions $\tilde{\mathbf{Z}}_1=\left(\tilde{\mathbf{X}}^{(1)},\tilde{\mathbf{W}}^{(1)}\right)^T$ and $\tilde{\mathbf{Z}}_2=\left(\tilde{\mathbf{X}}^{(2)},\tilde{\mathbf{W}}^{(2)}\right)^T$.

$$\begin{cases}\tilde{\mathbf{X}}'=\left(\mathbf{A}_0-\kappa^2\mathbf{A}_{01}+\alpha\mathbf{A}_1+\alpha^2\mathbf{A}_2\right)\tilde{\mathbf{X}}\\ \tilde{\mathbf{W}}'=\left(\mathbf{A}_0-\kappa^2\mathbf{A}_{01}+\alpha\mathbf{A}_1+\alpha^2\mathbf{A}_2\right)\tilde{\mathbf{W}}+(\mathbf{A}_1+2\alpha\mathbf{A}_2)\,\tilde{\mathbf{X}},\ \text{where}\ \tilde{\mathbf{W}}=\dfrac{\partial\tilde{\mathbf{X}}}{\partial\alpha}\end{cases} \tag{59}$$

The vector functions $\tilde{\mathbf{Z}}_1$ and $\tilde{\mathbf{Z}}_2$ are solutions of the Eq. (59) under the following boundary conditions

$$\tilde{X}_j^{(1)}(\xi)=\delta_{1j},\quad \tilde{W}_j^{(1)}(\xi)=0,\quad j=1,2,3,4 \tag{60}$$

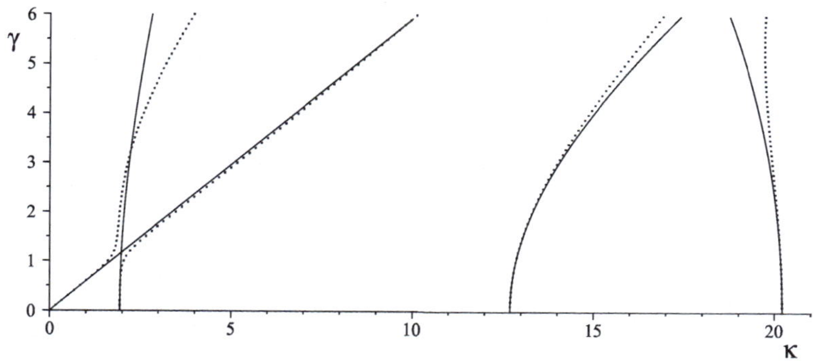

Fig. 1 Numerical dispersion curves and their quadratic asymptotics

$$\tilde{X}_j^{(2)}(\xi) = \delta_{2j}, \quad \tilde{W}_j^{(2)}(\xi) = 0, \quad j = 1, 2, 3, 4 \tag{61}$$

Solving the Cauchy problem (59), (60), and (59), (61), we obtain

$$\mathbf{B}_1(1, \alpha_n) = \begin{pmatrix} \tilde{W}_3^{(1)}(1, \alpha_n) & \tilde{W}_3^{(2)}(1, \alpha_n) \\ \tilde{W}_4^{(1)}(1, \alpha_n) & \tilde{W}_4^{(2)}(1, \alpha_n) \end{pmatrix} \tag{62}$$

Find $c_{-1}^1(\alpha_n)$ and $c_{-1}^2(\alpha_n)$ up to a constant multiplier from the first equation of the system (58). This equation is homogeneous and defines an eigenvector of the form: $\mathbf{C}_{-1} = q\mathbf{M}_0$, where $\mathbf{M}_0$ is the unit eigenvector of the matrix $\mathbf{B}_0(1, \alpha_n)$. We substitute this solution in the second equation and find the constant q. To do this, we multiply the second equation from (58) by the vector $\bar{\mathbf{M}}_0$, which is the eigenvector of the conjugate matrix $\bar{\mathbf{B}}_0(1, \alpha_n)$.

$$\left(\mathbf{B}_0\mathbf{C}_0, \bar{\mathbf{M}}_0\right) + q\left(\mathbf{B}_1\mathbf{M}_0, \bar{\mathbf{M}}_0\right) = \left(\mathbf{F}, \bar{\mathbf{M}}_0\right) \tag{63}$$

The first term on the left-hand side of (63) vanishes by the property of an eigenvector and for an unknown coefficient we have

$$q = \left(\mathbf{F}, \bar{\mathbf{M}}_0\right) / \left(\mathbf{B}_1\mathbf{M}_0, \bar{\mathbf{M}}_0\right) \tag{64}$$

11 Numerical Test

Let us consider the problem 1A. Let $\xi = 0.75$, $g_1(x) = 1.5\,th(50x - 49) + 2$, $g_2(x) = th(50x - 49) + 2$ be.

In Fig. 1, the first four real dispersion curves are shown. Dispersion curves calculated numerically are presented as dot line. The quadratic approximations of the dispersion curves calculated by the formulas (31) and (32) at the points $\kappa = 0$,

$\kappa = 1.93005, \kappa = 12.70087$ and $\kappa = 20.21128$ are presented by a solid line. Approximations allow us to determine the cases of normal dispersion for the first three curves and of abnormal dispersion for the fourth curve.

12 Conclusion

The four types of boundary value problems on free and forced oscillations of an inhomogeneous cylindrical waveguide with an arbitrary inhomogeneity are considered. The general properties of the dispersion set in the long-wave range are studied. The formulas allowing to construct a quadratic approximation of the dispersion set branches in a neighbourhood of the radial resonances points are given. The formulas for specific waveguides are obtained, and the solution of the problem on forced oscillations is presented.

This work is supported by the Russian Science Foundation, project # 18-11-00069.

References

1. Markus, A. S.: Vvedenie v spektral'nuyu teoriyu polinomial'nykh operatornykh puchkov [Introduction to the Spectral Theory of Polynomial Operator Sheaf], 260 p. Shtiintsa, Kishinev (1986)
2. Meleshko, V.V., Bondarenko, A.A., Dovgiy, S.A., et al.: Elastic waveguides: history and the state of the art I. J. Math. Sci. **162**, 99–120 (2009). https://doi.org/10.1007/s10958-009-9623-8
3. Son, M.S., Kang, Y.J.: The effect of initial stress on the propagation behavior of SH waves in piezoelectric coupled plates. Ultrasonics **51**(4), 489–495 (2011). https://doi.org/10.1016/j.ultras.2010.11.016
4. Figueroa, V.B., Kravchenko, V.V., Rabinovich, V.S.: Spectral parameter power series analysis of isotropic planarly layered waveguides. Appl. Anal. **93**(4), 729–755 (2014). https://doi.org/10.1080/00036811.2013.794940
5. Hasheminejad, S.M., Ahamdi-Savadkoohi, A.: Vibro-acoustic behavior of a hollow FGM cylinder excited by on-surface mechanical drives. Compos. Struct. **92**, 86–96 (2010). https://doi.org/10.1016/j.compstruct.2009.06.014
6. El-Raheb, M.: Transient waves in an inhomogeneous hollow infinite cylinder. Int. J. Solids Struct. **42**, 5356–5376 (2005). https://doi.org/10.1016/j.ijsolstr.2005.02.030
7. Elmaimouni, L., Lefebvre, J.E., Zhang, V., Gryba, T.: Guided waves in radially graded cylinders: a polynomial approach. NDT&E Int. **38**, 344–353 (2005). https://doi.org/10.1016/j.ndteint.2004.10.004
8. Grinchenko, V.T., Komissarova, G.L.: Properties of normal waves of a composite elastic-fluid waveguide placed to a fluid. Acoust. Bull. **9**(4), 17–34 (2006). (Russian)
9. Kalkowski, M.K., Muggleton, J.M., Rustighi, E.: Axisymmetric semi-analytical finite elements for modelling waves in buried/submerged fluid-filled waveguides. Comput. Struct. **196**, 327–340 (2018). https://doi.org/10.1016/j.compstruc.2017.10.004
10. Vorovich, I.I., Babeshko, V.A.: Dinamicheskie smeshannye zadachi teorii uprugosti dlya neklassicheskikh oblastej [Dynamic Mixed Elastic Problems for Nonclassical Regions], 320 p. Nauka, Moskva (1979) (Russian)

11. Getman, I.P., Ustinov, Y.A.: Matematicheskaya teoriya neregulyarnykh tverdykh volnovodov [Mathematical Theory of Irregular Solid Waveguides] 144 p. Izdat. RGU, Rostov-on-Don (1993)
12. Kaplunov, J., Nobili, A.: A robust approach for analysing dispersion of elastic waves in an orthotropic cylindrical shell. J. Sound Vib. **401**, 23–35 (2017). https://doi.org/10.1016/j.jsv.2017.04.028
13. Rose, J.L.: Ultrasonic guided waves in solid media, 512 p. Cambridge University Press, Cambridge (2014). https://doi.org/10.1017/CBO9781107273610
14. Vatulyan, A.O., Yurov, V.O.: Dispersion Properties of an Inhomogeneous Piezoelectric Waveguide with Attenuation. Acoust. Phys. **63**(4), 369–377 (2017). https://doi.org/10.1134/S1063771017040133
15. Vatulyan, A.O., Yurov, V.O.: On the dispersion relations for an inhomogeneous waveguide with attenuation. Mech. Solids. **51**(5), 576–582 (2016). https://doi.org/10.3103/S0025654417010137 (https://doi.org/10.3103/S0025654416050101)
16. Vatulyan, A.O., Yurov, V.O.: Wave processes in a hollow cylinder in an inhomogeneous prestress field. J. Appl. Mech. Tech. Phy. **57**(4), 731–739 (2016). https://doi.org/10.1134/S0021894416040180
17. Vatulyan A. O., Yurov V. O.: On the properties of the dispersion set for an inhomogeneous cylindrical waveguide. Vladikavkazskij matematicheskij zhurnal [Vladikavkaz Math. J.] **20**(1), 50–60 (2018). https://doi.org/10.23671/VNC.2018.1.11397 (Russian)
18. Gusakov, D.V., Vatul'yan, A.O.: Dispersion properties of inhomogeneous poroelastic layer. ZAMM Zeitschrift fur Angewandte Mathematik und Mechanik (4), 532-541 (2018)

Inverse Spectral Problems for Differential Systems

Vjacheslav Anatoljevich Yurko

Abstract Inverse problems of spectral analysis for non-selfadjoint systems of ordinary differential equations are studied. We establish properties of the spectral characteristics, give statements of the inverse problems, prove uniqueness theorems, obtain algorithms for the solutions of the inverse problems and provide necessary and sufficient conditions for their solvability.

Keywords Differential systems · Spectral characteristics · Inverse problems · Method of spectral mappings

AMS Classification 34A55 · 34B40 · 34L05 · 47E05

1 Differential Systems on the Half-Line

Consider the Following System of Differential Equations on the Half-Line

$$\ell Y(x) := Q_0 Y'(x) + Q(x)Y(x) = \rho Y(x), \quad x > 0. \tag{1.1}$$

Here $Y = [y_k]^t_{k=\overline{1,n}}$ is a column-vector (t is the sign for the transposition), ρ is the spectral parameter, $Q_0 = \mathrm{diag}[q_k]_{k=\overline{1,n}}$, $Q(x) = [q_{kj}(x)]_{k,j\overline{1,n}}$, where $q_k \neq 0,\ k = \overline{1,n}$ are different complex numbers, and $q_{kk}(x) \equiv 0$. The matrix $Q(x)$ is called the potential. By W_N we denote the set of functions $f(x),\ x > 0$, such that the functions $f^{(\nu)}(x),\ \nu = \overline{0, N-1}$ are absolutely continuous and $f^{(\nu)}(x) \in L(0,\infty), \nu = \overline{0, N}$. We shall say $\ell \in V_N$, if $q_{kj}(x) \in W_N, k, j = \overline{1,n}$. We will consider the operator ℓ in the classes $V_N,\ N \geq 1$.

In the present paper we study inverse spectral problems for system (1.1) on the half-line (Sect. 1) and on a finite interval (Sect. 2) for arbitrary complex numbers

V. A. Yurko (✉)
Department of Mathematics, Saratov State University,
Astrakhanskaya 83, Saratov 410026, Russia
e-mail: yurkova@info.sgu.ru

A. Karapetyants et al. (eds.), *Modern Methods in Operator Theory and Harmonic Analysis*, Springer Proceedings in Mathematics & Statistics 291,
https://doi.org/10.1007/978-3-030-26748-3_25

q_k and for arbitrary behavior of the spectrum. Inverse spectral problems consist in recovering operators from their spectral characteristics. Such problems often appear in mathematics and different branches of natural sciences and engineering.

Inverse problems for systems of the form (1.1) in different statements were studied by many authors. Most of the works are devoted to the so-called degenerate case when all characteristic numbers q_k are real (or, which is the same, are situated on a line containing the origin). For example, Dirac and AKNS systems are degenerate. The main tool for these investigations was the transformation operator method, and the obtained results are similar to the results for the Sturm-Liouville operator (see [1–3]). However for system (1.1) of general form with arbitrary complex q_k and with integrable $Q(x)$, the inverse problem is essentially more difficult for studying. In particular, in the general case the transformation operator method does not work.

In order to study inverse problems for system (1.1) in the general non-degenerate case we will use here another method, namely: the method of spectral mappings [4]. As the main spectral characteristics we introduce and investigate the so-called Weyl matrix which is an analog of the classical Weyl function for the Sturm-Liouville operator and an analog of the Weyl matrix introduced in [5] for higher-order differential operators. Developing the ideas of the method of spectral mappings for system (1.1) and using analytic and structural properties of the Weyl matrix we prove the uniqueness theorem for the solution of the inverse problem from the Weyl matrix, obtain a constructive procedure for the solution and give necessary and sufficient conditions for the solvability of this nonlinear inverse problem. For more details see [6–9].

Some words about the structure of the paper. In Sect. 1 we study the system on the half-line. Section 2 is devoted to the case of a finite interval. For simplicity, in these two sections we confine ourselves to the case when all characteristic roots are simple. In Sects. 3 and 4 we consider the general case of multiplied roots and show how to deal with it.

We note that the solution of the inverse problem for system (1.1) on the line is an unsolved problem. Some particular results for the case of the line see in [10, 11].

Denote $\beta_k = 1/q_k$. Let for definiteness $\beta_k \neq \beta_j$ for $k \neq j$. It is known that the ρ- plane can be partitioned into sectors $S_j = \{\rho : \arg\rho \in (\theta_j, \theta_{j+1})\}$, $j = \overline{0, 2r-1}$, $0 \le \theta_0 < \theta_1 < \ldots < \theta_{2r-1} < 2\pi$, in each of which there exists a permutation $i_k = i_k(S_j)$ of the numbers $1, \ldots, n$, such that for the numbers $R_k = R_k(S_j)$ of the form $R_k = \beta_{i_k}$ one has

$$\operatorname{Re}(\rho R_1) < \ldots < \operatorname{Re}(\rho R_n), \quad \rho \in S_j. \tag{1.2}$$

Let a matrix $h = [h_{\xi\nu}]_{\xi,\nu=\overline{1,n}}$, $\det h \neq 0$ be given, where $h_{\xi\nu}$ are complex numbers. We introduce the linear forms $U(Y) = [U_\xi(Y)]^t_{\xi=\overline{1,n}}$ by the formula $U(Y) = hY(0)$, i.e. $U_\xi(Y) = [h_{\xi 1}, \ldots, h_{\xi n}]Y(0)$. Denote $\Omega^0_{mk}(j_1, \ldots, j_m) = \det[h_{\xi, j_\nu}]_{\xi=\overline{1,m-1},k;\ \nu=\overline{1,m}}$. Assume that

$$\Omega^0_{mm}(i_1, \ldots, i_m) \neq 0, \quad m = \overline{1, n-1}, \quad , j = \overline{0, 2r-1}, \tag{1.3}$$

where $i_k = i_k(S_j)$ is the above-mentioned perturbation for the sector S_j. Condition (1.3) is called the information condition for the pair $L = (\ell, U)$. Systems, which do not satisfy the information condition, possess qualitatively different properties for investigating inverse problems, and are not considered in this paper. Without loss of generality we assume that the following normalizing conditions are fulfilled: $\det h = 1$, and for a fixed sector (for definiteness, for the sector S_0) one has $\Omega^0_{mm}(i_1, \dots, i_m) = 1$, $m = \overline{1, n-1}$.

Let the vector-functions $\Phi_m(x, \rho) = [\Phi_{km}(x, \rho)]^t_{k=\overline{1,n}}$, $m = \overline{1, n}$, be solutions of system (1.1) under the conditions $U_\xi(\Phi_m) = \delta_{\xi m}$, $\xi = \overline{1, m}$, and $\Phi_m(x, \rho) = O(\exp(\rho R_m x))$, $x \to \infty$, $\rho \in S_j$ in each sector S_j with the property (1.2). Here and in the sequel, $\delta_{\xi m}$ is the Kronecker delta. These conditions determine uniquely the solutions $\Phi_m(x, \rho)$. Let $M_{m\xi}(\rho) = U_\xi(\Phi_m)$, $\xi > m$, $M(\rho) = [M_{m\xi}(\rho)]_{m,\xi=\overline{1,n}}$, $M_{m\xi}(\rho) = \delta_{\xi m}$ for $\xi \le m$, $\Phi(x, \rho) = [\Phi_1(x, \rho), \dots, \Phi_n(x, \rho)] = [\Phi_{km}(x, \rho)]_{k,m=\overline{1,n}}$. The functions $\Phi_m(x, \rho)$ and $M_{m\xi}(\rho)$ are called the Weyl solutions and the Weyl functions respectively. The matrix $M(\rho)$ is called the Weyl matrix or the spectrum of $L = (\ell, U)$.

Fix Q_0, i.e. the numbers $\beta_k = 1/q_k$, $k = \overline{1, n}$, are known and fixed. The inverse problem is formulated as follows: *given* $M(\rho)$, *construct the pair* $L = (\ell, U)$.

1.2. Denote $\Phi^*(x, \rho) = (Q_0 \Phi(x, \rho))^{-1}$, $\mathcal{N}(\rho) = M^t(\rho)$, $\mathcal{N}^*(\rho) = (\mathcal{N}(\rho))^{-1}$. For $\xi = \overline{0, n-2}$ we introduce the functions $B^\xi_{mk}(\rho)$, $m = \overline{1, n-\xi-1}$, $k = \overline{m+\xi+1, n}$ by the recurrent formulae

$$B^0_{mk}(\rho) = M_{mk}(\rho), \quad B^\xi_{mk}(\rho) = B^{\xi-1}_{mk}(\rho) - B^{\xi-1}_{m,m+\xi}(\rho) B^0_{m+\xi,k}(\rho).$$

Denote $\Gamma_j = \{\rho : \arg \rho = \theta_j\}$, $j = \overline{0, 2r-1}$, $\Gamma_{2r} := \Gamma_0$. We cut the ρ- plane along the rays Γ_j and denote by $\Gamma^\pm_j = \{\rho : \arg \rho = \theta_j \pm 0\}$ the sides of the cuts. We put $\overline{S}_j = S_j \cup \Gamma^+_j \cup \Gamma^-_{j+1}$ and denote by $\Sigma = \mathbf{C} \setminus \left(\bigcup_{j=0}^{2r-1} \Gamma_j \right) = \bigcup_{j=0}^{2r-1} S_j$ the ρ- plane without the cuts along the rays Γ_j, and denote by $\overline{\Sigma} = \bigcup_{j=0}^{2r-1} \bar{S}_j$ the closure of Σ (we differ the sides of the cuts).

Fix $j = \overline{0, 2r-1}$. For $\rho \in \Gamma_j$, strict inequalities from (1.2) in some places become equalities. Let $m_i = m_i(j)$, $p_i = p_i(j)$, $i = \overline{1, s}$ be such that for $\rho \in \Gamma_j$: $\operatorname{Re}(\rho R_{m_i - 1}) < \operatorname{Re}(\rho R_{m_i}) = \ldots = \operatorname{Re}(\rho R_{m_i + p_i}) < \operatorname{Re}(\rho R_{m_i + p_i + 1})$, $i = \overline{1, s}$, where $R_k = R_k(S_j)$. Let $N_j := \{m : m = \overline{m_1, m_1 + p_1 - 1}, \dots, \overline{m_s, m_s + p_s - 1}\}$, $J_m := \{j : m \in N_j\}$, $\gamma_m = \bigcup_{j \in J_m} \Gamma_j$, and let $\Sigma_m = \mathbf{C} \setminus \gamma_m$ be the ρ- plane without the cuts along the rays from γ_m, and $\overline{\Sigma}_m$ be the closure of Σ_m (we differ the sides of the cuts). The domain $\Sigma_m = \bigcup_\nu S_{m\nu}$ consists of the sectors $S_{m\nu}$, each of which is a union of several sectors S_j with the same set $\{R_\xi\}_{\xi=\overline{1,m}}$. Denote by $\mathcal{M}$ the set of functions $M(\rho) = [M_{mk}(\rho)]_{m,k=\overline{1,n}}$ with the properties:

(1) $M_{mk}(\rho) \equiv \delta_{mk}$ for $m \ge k$;

(2) $M_{mk}(\rho)$, $k > m$ are analytic in Σ_m with the exception of an at most countable

bounded set Λ'_m of poles, and are continuous in $\overline{\Sigma}_m$ with the exeption of a bounded set Λ_m (in general the sets Λ'_m and Λ_m are different for each matrix $M(\rho)$ from $\mathcal{M}$);
(3) $B^{m-\nu}_{\nu k}(\rho)$ are analytic on $\Gamma_j \setminus \Lambda'_m$ for $j \notin J_m$, $1 \le \nu \le m \le n-1$, $m+1 \le k \le n$;
(4) for $|\rho| \to \infty$, $\rho \in \overline{S}_j$,

$$M_{mk}(\rho) = \mu^0_{mk}(S_j) + O(\rho^{-1}), \quad \mu^0_{mk}(S_j) = \frac{\Omega^0_{mk}(i_1, \ldots, i_m)}{\Omega^0_{mm}(i_1, \ldots, i_m)}, \quad i_k = i_k(S_j).$$

Theorem 1.1 *If $M(\rho)$ is the Weyl matrix for a pair $L = (\ell, U)$, then $M(\rho) \in \mathcal{M}$.*

Theorem 1.2 *The specification of the Weyl matrix $M(\rho)$ uniquely determines the potential $Q(x)$ and the matrix h.*

1.3. Together with $L = (\ell, U)$ we consider the pair $\tilde{L} = (\tilde{\ell}, \tilde{U})$ of the same form but with different matrices $\tilde{Q}, \tilde{h}$. Everywhere below if a symbol α denotes an object related to L, then $\tilde{\alpha}$ will denote the analogous object related to $\tilde{L}$, and $\hat{\alpha} := \alpha - \tilde{\alpha}$. Let the pair $\tilde{L} = (\tilde{\ell}, \tilde{U})$ be chosen such that

$$\hat{M}(\rho) = O(\rho^{-1}), \quad |\rho| \to \infty, \tag{1.4}$$

i.e. $\mu^0_{mk}(S_j) = \tilde{\mu}^0_{mk}(S_j)$ for all m, k, j. We note that condition (1.4) is equivalent to the condition $h = \tilde{h}$. For example, one can take $\tilde{Q}(x) \equiv 0$.

In the ρ- plane we consider the contour $\omega = \omega^0 \cup \omega^+ \cup \omega^-$ (with a counterclockwise circuit), where ω^0 is a bounded closed contour encircling the set $\Lambda \cup \tilde{\Lambda} \cup \{0\}$ (i.e. $\Lambda \cup \tilde{\Lambda} \cup \{0\} \subset \operatorname{int}\omega^0$), and $\omega^{\pm} = \bigcup_{j=0}^{2r-1} \omega^{\pm}_j$, $\omega^{\pm}_j = \Gamma^{\pm}_j \setminus \operatorname{int}\omega^0$.

We define the matrices $A(\rho) = [a_{mk}(\rho)]_{m,k=\overline{1,n}}$ and $\tilde{A}(\rho) = [\tilde{a}_{mk}(\rho)]_{m,k=\overline{1,n}}$ by the formulae $A(\rho) = \mathcal{N}^*(\rho)\hat{\mathcal{N}}(\rho)$, $\tilde{A}(\rho) = \tilde{\mathcal{N}}^*(\rho)\hat{\mathcal{N}}(\rho)$, where $\mathcal{N}(\rho) = (M(\rho))^t$, $\mathcal{N}^*(\rho) = (M^*(\rho))^t$. Fix $j = \overline{0, 2r-1}$. Let $N_j = \{m_1, m_1 + p_1 - 1, \ldots, m_s, m_s + p_s - 1\}, m_i - 1, m_i + p_i \notin N_j$, $i = \overline{1, s}$. Consider the matrices $A^{(j)}(\rho) = [a^{(j)}_{k\xi}(\rho)]_{k,\xi=\overline{1,n}}$, $\tilde{A}^{(j)}(\rho) = [\tilde{a}^{(j)}_{k\xi}(\rho)]_{k,\xi=\overline{1,n}}$, $\rho \in \Gamma^{\pm}_j$, where $a^{(j)}_{k\xi}(\rho) = a_{k\xi}(\rho)$, $\tilde{a}^{(j)}_{k\xi}(\rho) = \tilde{a}_{k\xi}(\rho)$ for $m_i \le \xi < k \le m_i + p_i$, $i = \overline{1, s}$, and $a^{(j)}_{k\xi}(\rho) = \tilde{a}^{(j)}_{k\xi}(\rho) = 0$ otherwise. For $\rho \in \omega$ we introduce the matrices $A_0(\rho)$ and $\tilde{A}_0(\rho)$ by the formulae

$$A_0(\rho) = \begin{cases} A^{(j)}(\rho), & \rho \in \omega^{\pm}_j, \\ A(\rho), & \rho \in \omega^0, \end{cases} \quad \tilde{A}_0(\rho) = \begin{cases} \tilde{A}^{(j)}(\rho), & \rho \in \omega^{\pm}_j, \\ \tilde{A}(\rho), & \rho \in \omega^0. \end{cases}$$

In the contour ω it is convenient to stick together the sides of the cuts. Therefore, in the ρ- plane we consider the contour $\omega^* := \omega^0 \cup \omega^1$, where $\omega^1 = \bigcup_{j=0}^{2r-1} \omega^1_j$, $\omega^1_j := \{\rho : \rho \in \Gamma_j \setminus \omega^0\}$ (with the orientation towards the growth of the modulus of ρ). For $\rho \in \omega^*$ we define the matrices $\varphi(x, \rho)$, $\tilde{\varphi}(x, \rho)$, $\tilde{G}(x, \rho)$, $\tilde{S}(\rho)$, $\tilde{r}(x, \mu, \rho)$ and $D(x, \rho)$ by the formulae

$$\varphi(x,\rho) = \begin{cases} [\Phi^+(x,\rho), \Phi^-(x,\rho)], \\ \Phi(x,\rho), \end{cases} \quad \tilde{\varphi}(x,\rho) = \begin{cases} [\tilde{\Phi}^+(x,\rho), \tilde{\Phi}^-(x,\rho)], & \rho \in \omega^1, \\ \tilde{\Phi}(x,\rho), & \rho \in \omega^0, \end{cases}$$

$$\tilde{G}(x,\rho) = \begin{bmatrix} -(A_0(\rho)\tilde{\Phi}^*(x,\rho))^+ Q_0 \\ (A_0(\rho)\tilde{\Phi}^*(x,\rho))^- Q_0 \end{bmatrix}, \ \rho \in \omega^1,$$

$$\tilde{G}(x,\rho) = A_0(\rho)\tilde{\Phi}^*(x,\rho)Q_0, \ \rho \in \omega^0,$$

$$\tilde{S}(\rho) = \begin{bmatrix} E - \frac{1}{2} A_0^+(\rho) & \frac{1}{2}\Big(A_0(\rho)\tilde{\mathcal{N}}^*(\rho)\Big)^+ \tilde{\mathcal{N}}^-(\rho) \\ \frac{1}{2}\Big(A_0(\rho)\tilde{\mathcal{N}}^*(\rho)\Big)^- \tilde{\mathcal{N}}^+(\rho) & E - \frac{1}{2} A_0^-(\rho) \end{bmatrix}, \ \rho \in \omega^1,$$

$$\tilde{S}(\rho) = E - \frac{1}{2} A_0(\rho), \ \rho \in \omega^0, \quad \tilde{r}(x,\mu,\rho) = \frac{\tilde{G}(x,\mu)\tilde{\varphi}(x,\rho)}{\mu - \rho}, \ \rho, \mu \in \omega^*,$$

$$D(x,\rho) = \operatorname{diag}[D_k(x,\rho)]_{k=\overline{1,n}}, \ D_k(x,\rho) = \exp(-\rho R_k x), \text{ for } \rho \in \omega^0,$$

$$D(x,\rho) = \operatorname{diag}[D_k(x,\rho)]_{k=\overline{1,2n}}, \ D_k(x,\rho) = \exp(-\rho R_k x) \ (k \le n),$$

$$D_k(x,\rho) = \exp(-\rho R_{k-n} x) \ (k > n), \ R_k = R_k(S_j) \text{ for } \rho \in \omega_j^1,$$

where $f^\pm := f_{|\omega^\pm}$. We consider the Banach space $\mathcal{B}_p := \{f(\rho) : f(\rho)\rho^{-1} \in L_p(\omega^*)\}$, $p > 1$ with the norm $\|f\|_{\mathcal{B}_p} := \|f(\rho)\rho^{-1}\|_{L_p(\omega^*)}$.

Theorem 1.3 *Let $M(\rho)$ be the Weyl matrix for the pair $L = (\ell, U)$. Let the pair $\tilde{L} = (\tilde{\ell}, \tilde{U})$ be chosen such that (1.4) holds. Then the following relation is valid*

$$\tilde{\varphi}(x,\rho) = \varphi(x,\rho)\tilde{S}(\rho) + \frac{1}{2\pi i}\int_{\omega^*} \varphi(x,\mu)\tilde{r}(x,\mu,\rho)\,d\mu, \quad \rho \in \omega^*. \tag{1.5}$$

For each fixed $x \ge 0$, Eq. (1.5) has the unique solution $\varphi(x,\rho)$ in the class $\varphi(x,\rho)D(x,\rho) \in \mathcal{B}_p$ for each $p > 1$; moreover $\sup_{x\ge0} \|\varphi(x,\rho)D(x,\rho)\|_{\mathcal{B}_p} < \infty$.

Equation (1.5) is called the main equation of the inverse problem. Let us now formulate necessary and sufficient conditions for the solvability of the inverse problem.

Theorem 1.4 *For a matrix $M(\rho) \in \mathcal{M}$ to be the Weyl matrix for a pair $L = (\ell, U)$, $\ell \in V_N$, it is necessary and sufficient that the following conditions are fulfilled:*
(1) (asymptotics) there exists a pair $\tilde{L} = (\tilde{\ell}, \tilde{U})$ such that (1.4) holds;
(2) (condition P) for each fixed $x \ge 0$, the main Eq. (1.5) has a unique solution $\varphi(x,\rho)$ in the class $\varphi(x,\rho)D(x,\rho) \in \mathcal{B}_p$, $p > 1$, and $\sup_{x\ge0} \|\varphi(x,\rho)D(x,\rho)\|_{\mathcal{B}_p} < \infty$;
(3) $\varepsilon(x) \in W_N$, where

$$\varepsilon(x) = \frac{1}{2\pi i}\int_{\omega} \Big(\Phi(x,\mu)A_0(\mu)\tilde{\Phi}^*(x,\mu)Q_0 - Q_0\Phi(x,\mu)A_0(\mu)\tilde{\Phi}^*(x,\mu)\Big)\,d\mu.$$

Under these conditions the pair $L = (\ell, U)$ is constructed by the formulae

$$Q(x) = \tilde{Q}(x) + \varepsilon(x), \quad h = \tilde{h}.$$

2 Differential Systems on a Finite Interval

We Consider the Following System of Differential Equations on a Finite Interval

$$\ell Y(x) := Q_0 Y'(x) + Q(x)Y(x) = \rho Y(x), \quad 0 \le x \le T. \tag{2.1}$$

Here $Y = [y_k]^t_{k=\overline{1,n}}$ is a column-vector, $Q_0 = \operatorname{diag}[q_k]_{k=\overline{1,n}}$, where $q_k \neq 0$ are different complex numbers, and $Q(x) = [q_{kj}(x)]_{k,j=\overline{1,n}}$, where $q_{kj}(x) \in L(0,T)$ are complex-valued functions, and $q_{kk}(x) \equiv 0$. The matrix $Q(x)$ is called the potential. Denote $B_0 = Q_0^{-1}$, hence $B_0 = \operatorname{diag}[\beta_k]_{k=\overline{1,n}}$, where $\beta_k = 1/q_k$. In this section we study the inverse spectral problem for system (2.1) with arbitrary complex q_k and with an arbitrary integrable potential $Q(x)$.

Let matrices $h = [h_{\xi\nu}]_{\xi,\nu=\overline{1,n}}$ and $H = [H_{\xi\nu}]_{\xi,\nu=\overline{1,n}}$ be given, where $h_{\xi\nu}$, $H_{\xi\nu}$ are complex numbers, and $\det h \neq 0$, $\det H \neq 0$. We introduce the linear forms $U(Y) = [U_\xi(Y)]^t_{\xi=\overline{1,n}}$, $V(Y) = [V_\xi(Y)]^t_{\xi=\overline{1,n}}$ by the formulae $U(Y) = hY(0)$, $V(Y) = HY(T)$, i.e. $U_\xi(Y) = [h_{\xi 1}, \dots, h_{\xi n}]Y(0)$, $V_\xi(Y) = [H_{\xi 1}, \dots, H_{\xi n}]Y(0)$. Let the vector-functions $\Phi_m(x,\rho) = [\Phi_{\nu m}(x,\rho)]^t_{\nu=\overline{1,n}}$, $m = \overline{1,n}$ be solutions of system (2.1) under the conditions $U_\xi(\Phi_m) = \delta_{\xi m}$, $\xi = \overline{1,m}$, $V_\eta(\Phi_m) = 0$, $\eta = \overline{1,n-m}$. Let $M_{m\xi}(\rho) = U_\xi(\Phi_m)$, $M(\rho) = [M_{m\xi}(\rho)]_{m,\xi=\overline{1,n}}$, $\Phi(x,\rho) = [\Phi_1(x,\rho), \dots, \Phi_n(x,\rho)] = [\Phi_{\nu m}(x,\rho)]_{\nu,m=\overline{1,n}}$. The functions $\Phi_m(x,\rho)$ and $M_{\xi m}(\rho)$ are called the *Weyl solutions* and the *Weyl functions* respectively. The matrix $M(\rho)$ is called the *Weyl matrix* for system (2.1). Since $M_{m\xi}(\rho) = \delta_{m\xi}$ for $\xi \le m$, the Weyl matrix is triangular, and $\det M(\rho) \equiv 1$.

Fix Q_0, h, H, i.e. the numbers $\beta_k = 1/q_k$, $h_{\xi\nu}$, $H_{\xi\nu}$, $k, \xi, \nu = \overline{1,n}$ are known and fixed. The inverse problem is formulated as follows: *given the Weyl matrix $M(\rho)$*, it construct the potential $Q(x)$, $0 < x < T$.

It is easy to see that in the general case, the specification of the Weyl matrix does not determine uniquely the matrices of the linear forms h and H. It is possible to point out particular cases (when the matrices h and H are normalized and have a special structure) for which the specification of the Weyl matrix uniquely determines not only the potential but also the matrices of the linear forms. For simplicity, we confine ourselves here to the most principal and difficult problem of recovering the potential $Q(x)$.

2.2. Let us study the properties of the Weyl matrix. By definition we have

$$U(\Phi) = h\Phi(0) = \mathcal{N}(\rho), \quad \text{where} \quad \mathcal{N}(\rho) = M^t(\rho).$$

Let $C(x,\rho) = [C_{km}(x,\rho)]_{k,m=\overline{1,n}}$ be a matrix-solution of system (2.1) under the initial condition $U(C) = hC(0,\rho) = E$ (here and below E denotes the identity matrix of the corresponding dimension or the identity operator in the corresponding Banach space). In other words, the column-vectors $C_m(x,\rho) = [C_{km}(x,\rho)]^t_{k=\overline{1,n}}$, $m = \overline{1,n}$, are solutions of (2.1) under the initial conditions $U_\xi(C_m) = \delta_{\xi m}$, $\xi, m = \overline{1,n}$. The functions $C_{km}(x,\rho)$ are entire in ρ for each fixed x. Clearly,

$$\Phi(x,\rho) = C(x,\rho)\mathcal{N}(\rho) \quad \text{or} \quad \Phi_m(x,\rho) = C_m(x,\rho) + \sum_{k=m+1}^{n} M_{mk}(\rho)C_k(x,\rho), \tag{2.2}$$

$$\det C(x,\rho) = \det \Phi(x,\rho) = \exp(\rho(\beta_1 + \cdots + \beta_n)x). \tag{2.3}$$

It follows from (2.2) to (2.3) that the vectors $\{C_m(x,\rho)\}_{m=\overline{1,n}}$ and $\{\Phi_m(x,\rho)\}_{m=\overline{1,n}}$ form fundamental systems of solutions for system (2.1). Denote

$$\Delta_{mk}(\rho) := (-1)^{m+k} \det[V_\xi(C_\nu)]_{\xi=\overline{1,n-m};\nu=\overline{m,n}\setminus k}, \quad 1 \le m \le k \le n, \ \Delta_{nn}(\rho) := 1.$$

It is easy to verify that

$$\Phi_m(x,\rho) = (\Delta_{mm}(\rho))^{-1} \det[C_\nu(x,\rho), V_1(C_\nu), \ldots, V_{n-m}(C_\nu)]_{\nu=\overline{m,n}}, \tag{2.4}$$

$$M_{mk}(\rho) = (\Delta_{mm}(\rho))^{-1}\Delta_{mk}(\rho), \quad 1 \le m < k \le n. \tag{2.5}$$

It is known that the ρ- plane can be partitioned into sectors $S_j = \{\rho : \arg\rho \in (\theta_j, \theta_{j+1})\}$, $j = \overline{0, 2r-1}$, $0 \le \theta_0 < \theta_1 < \ldots < \theta_{2r-1} \le 2\pi$ in which there exist permutations $i_k = i_k(S_j)$ of the numbers $1, \ldots, n$, such that for the numbers $R_k = R_k(S_j)$ of the form $R_k = \beta_{i_k}$ one has

$$\mathrm{Re}(\rho R_1) < \ldots < \mathrm{Re}(\rho R_n), \quad \rho \in S_j. \tag{2.6}$$

We put $\theta_{j+2kr} := \theta_j$, $S_{j+2kr} := S_j$, $k \in \mathbf{Z}$, and denote $\Gamma_j = \{\rho : \arg\rho = \theta_j\}$. Clearly, $\Gamma_{j+2kr} := \Gamma_j$, $k \in \mathbf{Z}$. We note that the number $2r$ of the sectors S_j depends on the location of the numbers $\{\beta_k\}_{k=\overline{1,n}}$ on the complex-plane, and $1 \le r \le n(n-1)/2$. For example, if all β_k lie on a line containing the origin, then $r = 1$. Denote

$$\Omega^0_{mk}(j_1, \ldots, j_m) := \det[h_{\xi, j_\nu}]_{\xi=\overline{1,m-1},k;\ \nu=\overline{1,m}}, \quad 1 \le m \le k \le n,$$

$$\Omega^0_m(j_1, \ldots, j_m) := \Omega^0_{mm}(j_1, \ldots, j_m), \quad \Omega^0_0 := 1,$$

$$\Omega^1_m(j_{m+1}, \ldots, j_n) := \det[H_{\xi, j_\nu}]_{\xi=\overline{1,n-m};\ \nu=\overline{m+1,n}}.$$

Let

$$\Omega_m^0(i_1,\ldots,i_m)\neq 0,\quad \Omega_m^1(i_{m+1},\ldots,i_n)\neq 0,\quad m=\overline{1,n-1},\quad j=\overline{0,2r-1}, \tag{2.7}$$

where $i_k=i_k(S_j)$ is the above-mentioned perturbation for the sector S_j. Condition (2.7) is called the *regularity condition*. Systems, which do not satisfy the regularity condition, possess qualitatively different properties for investigating inverse problems, and are not considered in this paper.

Fix $j=\overline{0,2r-1}$. For $\rho\in\Gamma_j$, strict inequalities from (2.6) in some places become equalities. Let $m_i=m_i(j)$, $p_i=p_i(j)$, $i=\overline{1,s}$, be such that for $\rho\in\Gamma_j$,

$$\mathrm{Re}(\rho R_{m_i-1})<\mathrm{Re}(\rho R_{m_i})=\ldots=\mathrm{Re}(\rho R_{m_i+p_i})<\mathrm{Re}(\rho R_{m_i+p_i+1}),\quad i=\overline{1,s},$$

where $R_k=R_k(S_j)$. Denote $N_j:=\{m:\ m=\overline{m_1,m_1+p_1-1,\ldots,m_s,m_s+p_s-1}\}$, $J_m:=\{j:\ m\in N_j\}$, $\gamma_m=\bigcup_{j\in J_m}\Gamma_j$, $\Sigma_m=\mathbf{C}\setminus\gamma_m$ is the ρ- plane without the cuts along the rays from γ_m. Clearly, the domain $\Sigma_m=\bigcup_\nu S_{m\nu}$ consists of sectors $S_{m\nu}$, each of which is a union of several sectors S_j with the same collection $\{R_\xi\}_{\xi=\overline{1,m}}$. Let $\Gamma_{j,\sigma}:=\{\rho:\ \mathrm{dist}\,(\rho,\Gamma_j)\leq\sigma\}$, $\sigma>0$, be a strip along the ray Γ_j, and let $\gamma_{m,\sigma}:=\bigcup_{j\in J_m}\Gamma_{j,\sigma}=\{\rho:\ \mathrm{dist}\,(\rho,\gamma_m)\leq\sigma\}$. Denote $\Gamma_\sigma^1:=\bigcup_{j=0}^{2r-1}\Gamma_{j,\sigma}$, $\Gamma^1:=\bigcup_{j=0}^{2r-1}\Gamma_j$.

Example 2.1 Let $n=4$, $\beta_1=1$, $\beta_2=2$, $\beta_3=3$, $\beta_4=-i$. Let $\rho=\sigma+i\tau$. Then $r=4$, $\Gamma_0=\{\rho:\ \tau=\sigma,\ \sigma>0\}$, $\Gamma_1=\{\rho:\ \tau=2\sigma,\ \sigma>0\}$, $\Gamma_2=\{\rho:\ \tau=3\sigma,\ \sigma>0\}$, $\Gamma_3=\{\rho:\ \sigma=0,\ \tau>0\}$, $\Gamma_4=\{\rho:\ \tau=\sigma,\ \sigma<0\}$, $\Gamma_5=\{\rho:\ \tau=2\sigma,\ \sigma<0\}$, $\Gamma_6=\{\rho:\ \tau=3\sigma,\ \sigma<0\}$, $\Gamma_7=\{\rho:\ \sigma=0,\ \tau<0\}$, $N_0=N_6=\{1\}$, $N_1=N_5=\{2\}$, $N_2=N_4=\{3\}$, $N_3=\{1,2\}$, $N_7=\{2,3\}$, $J_1=\{0,3,6\}$, $J_2=\{1,3,5\}$, $J_3=\{2,4,7\}$, $\gamma_1=\Gamma_0\cup\Gamma_3\cup\Gamma_6$, $\gamma_2=\Gamma_1\cup\Gamma_3\cup\Gamma_5$, $\gamma_1=\Gamma_2\cup\Gamma_4\cup\Gamma_7$.

Denote by $\Lambda_m=\{\rho_{lm}\}_{l\geq 1}$, $m=\overline{1,n-1}$, the set of zeros (with multiplicities) of the entire function $\Delta_{mm}(\rho)$, and put $\Lambda:=\bigcup_{m=1}^{n-1}\Lambda_m$. The numbers $\{\rho_{lm}\}_{l\geq1}$ coincide with the eigenvalues of the boundary value problem L_m for system (1.1) under the conditions $U_\xi(Y)=V_\eta(y)=0$, $\xi=\overline{1,m}$, $\eta=\overline{1,n-m}$. Denote $G_{\delta,m}=\{\rho:\ |\rho-\rho_{lm}|\geq\delta,\ l\geq1\}$, $G_\delta=\bigcap_{m=1}^{n-1}G_{\delta,m}$. Let $\Lambda_m^0=\{\rho_{lm}^0\}_{l\geq1}$, $m=\overline{1,n-1}$, be the eigenvalues of the "simplest" boundary value problems L_m^0 for system (2.1) with $Q(x)\equiv 0$.

Theorem 2.1 *(1) There exists $\sigma>0$ such that $\Lambda_m\subset\gamma_{m,\sigma}$. Moreover, $\Lambda_m=\bigcup_{j\in J_m}\Lambda_{mj}$, where $\Lambda_{mj}\in\Gamma_{j,\sigma}$ is the subsequence of Λ_m located in the strip $\Gamma_{j,\sigma}$, $j\in J_m$.*

(2) The number n_{ma} of zeros of $\Delta_{mm}(\rho)$ in the domain $\gamma_{m,\sigma}^a:=\{\rho:\ \rho\in\gamma_{m,\sigma},\ |\rho|\in[a,a+1]\}$ is bounded with respect to a.

(3) There exist positive numbers $r_N \to \infty$ such that for sufficiently small $\delta > 0$, the circles $|\rho| = r_N$ lie in G_δ for all N.

(4) For $m = \overline{1, n-1}$, one has $\rho_{lm} = \rho^0_{lm} + O(l^{-1}),\ l \to \infty$.

Theorem 2.2 *The Weyl functions $M_{mk}(\rho)$, $k > m$ are meromorphic in ρ with the set $\Lambda_m = \{\rho_{lm}\}_{l\geq 1}$ of poles. For $|\rho| \to \infty$, $\arg \rho = \theta \in (\theta_j, \theta_{j+1})$,*

$$M_{mk}(\rho) = M^0_{mk} + O(\rho^{-1}),$$

where $M^0_{mk} = (\Omega^0_m(i_1, \dots, i_m))^{-1}\Omega^0_{mk}(i_1, \dots, i_m)$. Moreover, $|M_{mk}(\rho)| \leq C_\delta$, $\rho \in G_{\delta,m}$,

$$|\Phi_{\mu m}(x, \rho)| \leq C_\delta |\exp(\rho R_m x)|, \quad \rho \in \overline{S_j} \cap G_{\delta,m},\ x \in [0, T],\ \mu, m = \overline{1, n}, \tag{2.8}$$

$$\Phi_{\mu m}(x, \rho) = (\alpha_{\mu m j} + O(\rho^{-1})) \exp(\rho R_m x), \quad |\rho| \to \infty,\ \arg \rho = \theta \in (\theta_j, \theta_{j+1}), \tag{2.9}$$

where the numbers $\alpha_{\mu m j}$ do not depend on x and ρ; they depend only on h, H.

We consider the differential system

$$\ell^* Z(x) := -Z'(x)Q_0 + Z(x)Q(x) = \rho Z(x), \tag{2.10}$$

where $Z = [z_k]_{k=\overline{1,n}}$ is a row-vector. Put $U^*(Z) = Z(0)h^*$, where $h^* = [h^*_{k\xi}]_{k,\xi=\overline{1,n}}$ $:= Q_0 h^{-1}$. Then $U^*(Z) = [U^*_n(Z), \dots, U^*_1(Z)]$, where $U^*_{n-\xi+1}(Z) = Z(0)$ $[h^*_{k\xi}]^t_{k=\overline{1,n}}$. Clearly,

$$Z(x)\ell Y(x) - \ell^* Z(x) Y(x) = \frac{d}{dx}\Big(Z(x)Q_0 Y(x)\Big), \tag{2.11}$$

$$Z(0)Q_0Y(0) = U^*(Z)U(Y) = \sum_{\xi=1}^n U^*_{n-\xi+1}(Z)U_\xi(Y).$$

Moreover, it follows from (2.11) that if $\ell Y(x, \rho) = \rho Y(x, \rho)$, $\ell^* Z(x, \mu) = \mu Z(x, \mu)$, then

$$(\rho - \mu) Z(x, \mu) Y(x, \rho) = \frac{d}{dx}\Big(Z(x, \mu) Q_0 Y(x, \rho)\Big).$$

Denote $R^*_m := -R_{n-m+1}$. Let vector-functions $\Phi^*_m(x, \rho) = [\Phi^*_{km}(x, \rho)]_{k=\overline{1,n}}$, $m = \overline{1, n}$, be solutions of (2.10) satisfying the conditions $U^*_\xi(\Phi^*_m) = \delta_{\xi m}$, $\xi = \overline{1, m}$, $V^*_\eta(\Phi^*_m) = 0$, $\eta = \overline{1, n-m}$. We put $\Phi^*(x, \rho) = [\Phi^*_{n-m+1}(x, \rho)]^t_{m=\overline{1,n}} = [\Phi^*_{n-m+1,k}(x, \rho)]_{m,k=\overline{1,n}}$,

$$M^*_{mk}(\rho) = U_k(\Phi^*_m), \quad M^*(\rho) = [M^*_{n-\xi+1, n-k+1}(\rho)]_{k,\xi=\overline{1,n}}, \quad \mathcal{N}^*(\rho) = (M^*(\rho))^t.$$

We put $C^*(x,\rho) = [C^*_{n-m+1}(x,\rho)]^t_{m=\overline{1,n}}$ where the vectors $C^*_m(x,\rho) = [C^*_{mk}(x,\rho)]_{k=\overline{1,n}}$ are solutions of (2.10) under the initial conditions $U^*_\xi(C^*_m) = \delta_{\xi m},\ \xi,\ m = \overline{1,n}$ (i.e. $U^*(C^*) = C^*(0,\rho)h^* = E$). The functions $C^*_{mk}(x,\rho)$ are entire in ρ for each fixed x, and

$$\Phi^*(x,\rho) = \mathcal{N}^*(\rho)C^*(x,\rho), \quad M^*_{mk}(\rho) = (\Delta^*_{mm}(\rho))^{-1}\Delta^*_{mk}(\rho),$$

$$\det C^*(x,\rho) = \det \Phi^*(x,\rho) = (\det Q_0)^{-1}\exp(-\rho(\beta_1 + \cdots + \beta_n)x),$$

where

$$\Delta^*_{mk}(\rho) := (-1)^{m+k}\det[V^*_\xi(C_\nu)^*]_{\xi=\overline{1,n-m};\,\nu=\overline{m,n}\setminus k}, \quad 1 \le m \le k \le n.$$

The following relations hold

$$\Phi^*(x,\rho) = (Q_0\Phi(x,\rho))^{-1}, \tag{2.12}$$

$$M^*(\rho) = M^{-1}(\rho), \quad \mathcal{N}^*(\rho) = \mathcal{N}^{-1}(\rho), \tag{2.13}$$

$$C(x,\rho) = \Phi(x,\rho)\mathcal{N}^*(\rho), \quad C^*(x,\rho) = \mathcal{N}(\rho)\Phi^*(x,\rho). \tag{2.14}$$

2.3. In this subsection we provide a constructive solution of the inverse problem of recovering the potential matrix $Q(x)$ from the given Weyl matrix $M(\rho)$. For this purpose we reduce our nonlinear inverse problem to the solution of the so-called *main equation*, which is a linear equation in a corresponding Banach space of sequences. We give a derivation of the main equation, and prove its unique solvability. Using the solution of the main equation we provide an algorithm for the solution of the inverse problem. First we formulate the following uniqueness theorem for the solution of the inverse problem.

Theorem 2.3 *The specification of the Weyl matrix $M(\rho)$ uniquely determines the potential matrix $Q(x)$.*

Now we go on to the construction of the solution of the inverse problem. For simplicity, in the sequel we confine ourselves to the case when the functions $\Delta_{mm}(\rho)$, $m = \overline{1,n-1}$, have only simple zeros (the general case requires minor technical modifications). Then the matrices $M(\rho)$ and $M^*(\rho)$ have only simple poles.

For $\rho_0 \in \Lambda$ we define the matrix $F(\rho_0) = [F_{jk}(\rho_0)]_{j,k=\overline{1,n}}$ via

$$F(\rho_0) = -(N_{\langle 0\rangle}(\rho_0))^{-1}N_{\langle -1\rangle}(\rho_0).$$

Here and below $f_{\langle k\rangle}(\rho_0) = (f(\rho))^{\langle k\rangle}_{|\rho=\rho_0}$ denotes the kth Laurent's coefficient for the function $f(\rho)$ at the point $\rho = \rho_0$. In particular, $f_{\langle -1\rangle}(\rho_0) = \operatorname*{Res}_{\rho=\rho_0} f(\rho)$.

Since $M_{mk}(\rho) = \delta_{mk},\ m \ge k$, it follows that $F_{jk}(\rho_0) = 0$ for $j \le k$. One has

$$\left.\begin{array}{c} F(\rho_0)F(\rho_0)=0, \\ \Phi_{\langle -1\rangle}(x,\rho_0)=-\Phi_{\langle 0\rangle}(x,\rho_0)F(\rho_0),\ \ \Phi^*_{\langle -1\rangle}(x,\rho_0)=F(\rho_0)\Phi^*_{\langle 0\rangle}(x,\rho_0). \end{array}\right\} \quad (2.15)$$

It follows from (2.4) and $\ell\Phi_m(x,\rho)=\rho\Phi_m(x,\rho)$ that

$$V_{n-m+1}(\Phi_m(x,\rho))=(-1)^{n-m}(\Delta_{mm}(\rho))^{-1}\Delta_{m-1,m-1}(\rho),$$

$$\left.\begin{array}{c} \ell\Phi_{m,\langle -1\rangle}(x,\rho_0)=\rho_0\Phi_{m,\langle -1\rangle}(x,\rho_0), \\ \ell\Phi_{m,\langle 0\rangle}(x,\rho_0)=\rho_0\Phi_{m,\langle 0\rangle}(x,\rho_0)+\Phi_{m,\langle -1\rangle}(x,\rho_0). \end{array}\right\}$$

We provide two important properties of the Weyl matrix. Define $\Lambda_0=\Lambda_n=\emptyset$.

Theorem 2.4 *If* $\rho_0\notin\Lambda_m$, *then* $F_{m+1,j}(\rho_0)=\ldots=F_{nj}(\rho_0)=0$, $j=\overline{1,m}$. *If, moreover,* $\nu<m-1$, $\rho_0\notin\Lambda_\nu$, $\rho_0\in\Lambda_{\nu+1}\bigcap\ldots\bigcap\Lambda_{m-1}$, $\rho_0\notin\Lambda_m$, $1\le\nu+1<m\le n$, *then* $F_{m,\nu+1}(\rho_0)\ne 0$.

Denote $A_s(\rho_0)=[F_{\nu j}(\rho_0)]_{j=\overline{1,n-s},\ \nu=\overline{n-s,n}}$, $s=\overline{1,n-1}$.

Theorem 2.5 *The following relation holds*

$$\mathit{rank}\, A_s(\rho_0)\le 1,\quad s=\overline{1,n-1}.$$

Let the Weyl matrix $M(\rho)$ for system (2.1) be given. We take an arbitrary off-diagonal potential $\tilde{Q}(x)$. For example, one can take $\tilde{Q}(x)\equiv 0$. We agree that everywhere below if a symbol α denotes an object related to Q, then $\tilde{\alpha}$ will denote the analogous object related to $\tilde{Q}$, and $\hat{\alpha}:=\alpha-\tilde{\alpha}$. Denote

$$\xi_l:=\sum_{k=2}^{n}\Big(|\rho_{l,k-1}-\tilde{\rho}_{l,k-1}|+\sum_{s=1}^{k-1}|\gamma_{skl}-\tilde{\gamma}_{skl}|\Big),\quad l\ge 1,$$

where $\gamma_{skl}:=F_{ks}(\rho_{l,k-1})$, $\tilde{\gamma}_{skl}:=\tilde{F}_{ks}(\tilde{\rho}_{l,k-1})$, $s=\overline{1,k-1}$, $k=\overline{2,n}$, $l\ge 1$. It follows from (2.7) and the properties of the Weyl matrix that

$$\xi_l=O(l^{-1}),\quad l\to\infty. \quad (2.16)$$

Choose $\sigma>0$ such that $\Lambda^1:=\Lambda\cup\tilde{\Lambda}\subset\Gamma^1_\sigma$. Let ω be the contour (with a counterclockwise circuit) which is the boundary of Γ^1_σ, i.e. $\omega:=\{\rho:\ \mathrm{dist}(\rho,\Gamma^1)=\sigma\}$. Denote $J:=\{\rho:\ \mathrm{dist}(\rho,\Gamma^1)>\sigma\}$. Using (2.8), (2.9), (2.12) and Cauchy's integral formula, we get

$$\tilde{\Phi}(x,\rho)=\Phi(x,\rho)+\frac{1}{2\pi i}\int_\omega\Phi(x,\mu)\tilde{\Phi}^*(x,\mu)Q_0\tilde{\Phi}(x,\rho)\frac{d\mu}{\mu-\rho},\quad \rho\in J, \quad (2.17)$$

$$\frac{\Phi^*(x,\theta)Q_0\Phi(x,\rho)}{\rho-\theta}-\frac{\tilde{\Phi}^*(x,\theta)Q_0\tilde{\Phi}(x,\rho)}{\rho-\theta}$$

$$= \frac{1}{2\pi i}\int_\omega \Phi^*(x,\theta)Q_0\Phi(x,\mu)\tilde{\Phi}^*(x,\mu)Q_0\tilde{\Phi}(x,\rho)\,\frac{d\mu}{(\mu-\theta)(\rho-\mu)},\quad \rho,\theta\in J. \tag{2.18}$$

For $\rho_0 \in \Lambda^1$ we introduce the matrix $\tilde{D}(x,\rho_0,\rho) = [\tilde{D}_{sm}(x,\rho_0,\rho)]_{s,m=\overline{1,n}}$ by the formula

$$\tilde{D}(x,\rho_0,\rho) := \Big(\frac{\tilde{\Phi}^*(x,\mu)Q_0\tilde{\Phi}(x,\rho)}{\rho-\mu}\Big)^{\langle 0\rangle}_{|\mu=\rho_0}$$

$$= \Big(\frac{\tilde{\Phi}^*_{\langle -1\rangle}(x,\rho_0)}{(\rho-\rho_0)^2} + \frac{\tilde{\Phi}^*_{\langle 0\rangle}(x,\rho_0)}{\rho-\rho_0}\Big)Q_0\tilde{\Phi}(x,\rho).$$

Taking (2.15) into account we calculate

$$\Phi_{\langle -1\rangle}(x,\rho_0)\tilde{\Phi}^*_{\langle -1\rangle}(x,\rho_0) = -\Phi_{\langle 0\rangle}(x,\rho_0)F(\rho_0)\tilde{F}(\rho_0)\tilde{\Phi}^*_{\langle 0\rangle}(x,\rho_0)$$

$$-\Phi_{\langle 0\rangle}(x,\rho_0)\hat{F}(\rho_0)\tilde{F}(\rho_0)\tilde{\Phi}^*_{\langle 0\rangle}(x,\rho_0) = -\Phi_{\langle 0\rangle}(x,\rho_0)\hat{F}(\rho_0)\tilde{\Phi}^*_{\langle -1\rangle}(x,\rho_0),$$

$$\Phi_{\langle -1\rangle}(x,\rho_0)\tilde{\Phi}^*_{\langle 0\rangle}(x,\rho_0) + \Phi_{\langle 0\rangle}(x,\rho_0)\tilde{\Phi}^*_{\langle -1\rangle}(x,\rho_0) = -\Phi_{\langle 0\rangle}(x,\rho_0)\hat{F}(\rho_0)\tilde{\Phi}^*_{\langle 0\rangle}(x,\rho_0).$$

Since

$$\operatorname*{Res}_{\mu=\rho_0}\frac{\Phi(x,\mu)\tilde{\Phi}^*(x,\mu)}{\rho-\mu} = \frac{\Phi_{\langle -1\rangle}(x,\rho_0)\tilde{\Phi}^*_{\langle -1\rangle}(x,\rho_0)}{(\rho-\rho_0)^2}$$

$$+\frac{\Phi_{\langle -1\rangle}(x,\rho_0)\tilde{\Phi}^*_{\langle 0\rangle}(x,\rho_0) + \Phi_{\langle 0\rangle}(x,\rho_0)\tilde{\Phi}^*_{\langle -1\rangle}(x,\rho_0)}{\rho-\rho_0}$$

$$= -\frac{\Phi_{\langle 0\rangle}(x,\rho_0)\hat{F}(\rho_0)\tilde{\Phi}^*_{\langle -1\rangle}(x,\rho_0)}{(\rho-\rho_0)^2} - \frac{\Phi_{\langle 0\rangle}(x,\rho_0)\hat{F}(\rho_0)\tilde{\Phi}^*_{\langle 0\rangle}(x,\rho_0)}{\rho-\rho_0},$$

it follows from (2.17) and the residue theorem that

$$\tilde{\Phi}(x,\rho) = \Phi(x,\rho) + \sum_{\rho_0\in\Lambda^1}\Phi_{\langle 0\rangle}(x,\rho_0)\hat{F}(\rho_0)\tilde{D}(x,\rho_0,\rho),$$

or in the coordinates

$$\tilde{\Phi}_{im}(x,\rho) = \Phi_{im}(x,\rho) + \sum_{\rho_0\in\Lambda^1}\Big(\sum_{k=2}^{n}\Phi_{ik,\langle 0\rangle}(x,\rho_0)\sum_{s=1}^{k-1}\hat{F}_{ks}(\rho_0)\tilde{D}_{sm}(x,\rho_0,\rho)\Big). \tag{2.19}$$

The series in (2.19) converges "with brackets"': $\sum_{\rho_0\in\Lambda^1} = \lim_{N\to\infty}\sum_{\rho_0\in\Lambda^1_N}$, where $\Lambda^1_N := \Lambda^1\cap\{\rho:\ |\rho|<r_N\}$. From (2.19), using the structural properties of the Weyl matrix, we infer

$$\tilde{\Phi}_{im}(x,\rho) = \Phi_{im}(x,\rho) + \sum_{l\ge 1}\Big(\sum_{k=2}^{n}\Phi_{ik,\langle 0\rangle}(x,\rho_{l,k-1})\sum_{s=1}^{k-1}\gamma_{skl}\tilde{D}_{sm}(x,\rho_{l,k-1},\rho)$$

$$-\sum_{k=2}^{n}\Phi_{ik,\langle 0\rangle}(x,\tilde{\rho}_{l,k-1})\sum_{s=1}^{k-1}\tilde{\gamma}_{skl}\tilde{D}_{sm}(x,\tilde{\rho}_{l,k-1},\rho)\Big),\ i,m=\overline{1,n}. \qquad (2.20)$$

The series in (2.20) converges "with brackets": $\sum_{l\ge 1} = \lim_{N\to\infty}\sum_{l\in I_N}$, where $I_N := \{l:\ |\rho^0_{lm}| < r_N,\ \forall m = \overline{1,n-1}\}$.

Denote $\rho_{lk0} = \rho_{lk},\ \rho_{lk1} = \tilde{\rho}_{lk},\ \gamma^{sk}_{l0} = \gamma_{skl},\ \gamma^{sk}_{l1} = \tilde{\gamma}_{skl}$. We introduce the matrices

$$\varphi_{l\varepsilon}(x) = [\varphi^{ik}_{l\varepsilon}(x)]_{i=\overline{1,n},\,k=\overline{2,n}},\quad \tilde{\varphi}_{l\varepsilon}(x) = [\tilde{\varphi}^{ik}_{l\varepsilon}(x)]_{i=\overline{1,n},\,k=\overline{2,n}},$$

$$P_{l\varepsilon}(x,\rho) = [P^{km}_{l\varepsilon}(x,\rho)]_{k=\overline{2,n},\,m=\overline{1,n}},\quad \tilde{P}_{l\varepsilon}(x,\rho) = [\tilde{P}^{km}_{l\varepsilon}(x,\rho)]_{k=\overline{2,n},\,m=\overline{1,n}},$$

$$G_{(l\varepsilon)(l_0\varepsilon_0)}(x) = [G^{km}_{(l\varepsilon)(l_0\varepsilon_0)}(x)]_{k,m=\overline{2,n}},\ \tilde{G}_{(l\varepsilon)(l_0\varepsilon_0)}(x) = [\tilde{G}^{km}_{(l\varepsilon)(l_0\varepsilon_0)}(x)]_{k,m=\overline{2,n}},$$

$$l, l_0 > 1;\quad \varepsilon,\varepsilon_0 = 0,1,$$

by the formulae

$$\varphi^{ik}_{l\varepsilon}(x) = \Phi_{ik,\langle 0\rangle}(x,\rho_{l,k-1,\varepsilon}),\quad \tilde{\varphi}^{ik}_{l\varepsilon}(x) = \tilde{\Phi}_{ik,\langle 0\rangle}(x,\rho_{l,k-1,\varepsilon}),$$

$$P^{km}_{l\varepsilon}(x,\rho) = \sum_{s=1}^{k-1}\gamma^{sk}_{l\varepsilon}D_{sm}(x,\rho_{l,k-1,\varepsilon},\rho),\ \tilde{P}^{km}_{l\varepsilon}(x,\rho) = \sum_{s=1}^{k-1}\gamma^{sk}_{l\varepsilon}\tilde{D}_{sm}(x,\rho_{l,k-1,\varepsilon},\rho),$$

$$G^{km}_{(l\varepsilon)(l_0\varepsilon_0)}(x) = P^{km}_{l\varepsilon,\langle 0\rangle}(x,\rho_{l_0,m-1,\varepsilon_0}),\ \tilde{G}^{km}_{(l\varepsilon)(l_0\varepsilon_0)}(x) = \tilde{P}^{km}_{l\varepsilon,\langle 0\rangle}(x,\rho_{l_0,m-1,\varepsilon_0}).$$

Denote $u_{lm(x)} := |\exp(\tilde{\rho}_{l,m-1}R_m x)|,\ l\in\tilde{\Lambda}_{m-1,j},\ j\in J_{m-1},\ R_m = R_m(S_j)$. We rewrite (2.20) in the following form

$$\tilde{\Phi}_{im}(x,\rho) = \Phi_{im}(x,\rho) + \sum_{l\ge 1}\sum_{k=2}^{n}\Big(\varphi^{ik}_{l0}(x)\tilde{P}^{km}_{l0}(x,\rho) - \varphi^{ik}_{l1}(x)\tilde{P}^{km}_{l1}(x,\rho)\Big),$$

or in the matrix form

$$\tilde{\Phi}(x,\rho) = \Phi(x,\rho) + \sum_{l\ge 1}\Big(\varphi_{l0}(x)\tilde{P}_{l0}(x,\rho) - \varphi_{l1}(x)\tilde{P}_{l1}(x,\rho)\Big). \qquad (2.21)$$

It follows from (2.21) that

$$\tilde{\varphi}_{l_0\varepsilon_0}(x) = \varphi_{l_0\varepsilon_0}(x) + \sum_{l\ge 1}\Big(\varphi_{l0}(x)\tilde{G}_{(l0)(l_0\varepsilon_0)}(x) - \varphi_{l1}(x)\tilde{G}_{(l1)(l_0\varepsilon_0)}(x)\Big), \qquad (2.22)$$

$$l_0 > 1; \quad \varepsilon_0 = 0, 1.$$

Remark 2.1 For each fixed $x \in [0, T]$, relation (2.22) can be considered as a system of linear equations with respect to $\varphi_{l\varepsilon}(x),\ l \geq 1,\ \varepsilon = 0, 1$. But the series in (2.22) converges only "with brackets". Therefore, it is not convenient to use (2.22) as a main equation of the inverse problem. Below we will transfer (2.22) to a linear equation in a corresponding Banach space of sequences (see (2.26)).

For this purpose we introduce the matrices

$$\psi_{l\varepsilon}(x) = [\psi_{l\varepsilon}^{ik}(x)]_{i=\overline{1,n},\, k=\overline{2,n}},\ H_{(l\varepsilon)(l_0\varepsilon_0)}(x) = [H_{(l\varepsilon)(l_0\varepsilon_0)}^{km}(x)]_{k,m=\overline{2,n}},\ l, l_0 > 1;\ \varepsilon, \varepsilon_0 = 0, 1,$$

by the formulae

$$\psi_{l0}^{ik}(x) = (\xi_l u_{lk}(x))^{-1}(\varphi_{l0}^{ik}(x) - \varphi_{l1}^{ik}(x)),\ \psi_{l1}^{ik}(x) = (u_{lk}(x))^{-1}\varphi_{l1}^{ik}(x), \tag{2.23}$$

$$H_{(l0)(l_0 0)}^{km}(x) = \xi_l u_{lk}(x)(\xi_{l_0} u_{l_0 m}(x))^{-1}\Big(G_{(l0)(l_0 0)}^{km}(x) - G_{(l0)(l_0 1)}^{km}(x)\Big),$$

$$H_{(l0)(l_0 1)}^{km}(x) = \xi_l u_{lk}(x)(u_{l_0 m}(x))^{-1}\, G_{(l0)(l_0 1)}^{km}(x),$$

$$H_{(l1)(l_0 0)}^{km}(x) = u_{lk}(x)(\xi_{l_0} u_{l_0 m}(x))^{-1}\Big(G_{(l0)(l_0 0)}^{km}(x) - G_{(l0)(l_0 1)}^{km}(x) - G_{(l1)(l_0 0)}^{km}(x) + G_{(l1)(l_0 1)}^{km}(x)\Big),$$

$$H_{(l1)(l_0 1)}^{km}(x) = u_{lk}(x)(u_{l_0 m}(x))^{-1}\Big(G_{(l0)(l_0 1)}^{km}(x) - G_{(l1)(l_0 1)}^{km}(x)\Big).$$

Similarly we define the matrices $\tilde{\psi}_{l\varepsilon}(x)$ and $\tilde{H}_{(l\varepsilon)(l_0\varepsilon_0)}(x)$. One has,

$$|\psi_{l\varepsilon}^{ik}(x)| \leq C,\ |\tilde{\psi}_{l\varepsilon}^{ik}(x)| \leq C, \tag{2.24}$$

$$|H_{(l\varepsilon)(l_0\varepsilon_0)}^{km}(x)|,\ |\tilde{H}_{(l\varepsilon)(l_0\varepsilon_0)}^{km}(x)| \leq \frac{C\xi_l}{|l - l_0| + 1}. \tag{2.25}$$

Let V be a set of indices $u = (l, \varepsilon),\ l \geq 1,\ \varepsilon = 0, 1$. For each fixed $x \in [0, T]$, we define the vector

$$\psi(x) = [\psi_u(x)]_{u \in V} = [\psi_{l0}(x), \psi_{l1}(x)]_{l \geq 1} = [\psi_{10}, \psi_{11}, \psi_{20}, \psi_{21}, \ldots],$$

and the block matrix

$$H(x) = [H_{uv}(x)]_{u,v \in V} = \begin{bmatrix} H_{(l0)(l_0 0)}(x) & H_{(l0)(l_0 1)}(x) \\ H_{(l1)(l_0 0)}(x) & H_{(l1)(l_0 1)}(x) \end{bmatrix}_{l, l_0 \geq 1},$$

$$u = (l, \varepsilon), \quad v = (l_0, \varepsilon_0), \quad \varepsilon, \varepsilon_0 = 0, 1.$$

Analogously we define the matrices $\tilde{\psi}(x)$ and $\tilde{H}(x)$. In view of our notations, relation (2.22) transforms to the form

$$\tilde{\psi}_{l_0\varepsilon_0}(x) = \psi_{l_0\varepsilon_0}(x) + \sum_{l,\varepsilon} \psi_{l\varepsilon}(x)\tilde{H}_{(l\varepsilon)(l_0\varepsilon_0)}(x), \quad l, l_0 \geq 1, \quad \varepsilon, \varepsilon_0 = 0, 1,$$

or, which is the same,

$$\tilde{\psi}(x) = \psi(x)(E + \tilde{H}(x)). \tag{2.26}$$

According to (2.16), (2.24) and (2.25), the series in (2.26) converges absolutely and uniformly in $x \in [0, T]$.

Starting from (2.17) we arrived at (2.26). By similar arguments, starting from (2.18) one can get the relation

$$(E + \tilde{H}(x))(E - H(x)) = E.$$

Interchanging places for Q and $\tilde{Q}$, we obtain

$$(E - H(x))(E + \tilde{H}(x)) = E.$$

Let us consider the Banach space m of bounded sequences $\alpha = [\alpha_u]_{u\in V}$ with the norm $\|\alpha\| = \sup_{u\in V} |\alpha_u|$. It follows from (2.25) that for each fixed $x \in [0, T]$, the operators $E + \tilde{H}(x)$ and $E - H(x)$, acting from m to m, are linear bounded operators, and

$$\|H(x)\|, \ \|\tilde{H}(x)\| \leq C \sup_{l\geq 1} \sum_{l_0 \geq 1} \frac{\xi_l}{|l - l_0| + 1} < \infty.$$

Thus, we have proved the following theorem

Theorem 2.6 *For each fixed $x \in [0, T]$, the vector $\psi(x) \in m$ satisfies the Eq. (2.26) in the Banach space m. Moreover, the operator $E + \tilde{H}(x)$ has a bounded inverse operator, i.e. Eq. (2.26) is uniquely solvable.*

Equation (2.26) is called the *main equation* of the inverse problem. Solving (2.26) we find the vector $\psi(x)$, and consequently, the functions $\varphi_{l\varepsilon}(x)$. Then, by (2.21) we calculate $\Phi(x, \rho)$. Since $\Phi(x, \rho)$ is the solution of (2.21), we can construct the potential matrix $Q(x)$. Thus, we obtain the following algorithm for the solution of inverse problem

Algorithm 2.1 *Given the Weyl matrix $M(\rho)$.*
(1) *Choose $\tilde{Q}(x)$, and construct $\tilde{\psi}(x)$ and $\tilde{H}(x)$.*
(2) *Find $\psi(x)$ by solving Eq. (2.26).*
(3) *Calculate $\Phi(x, \rho)$ via (2.21), where $\varphi_{l\varepsilon}(x)$ is constructed from (2.23).*
(4) *Construct $Q(x)$ by the formula $Q(x) = -Q_0\Phi'(x, \rho)(\Phi(x, \rho))^{-1} + \rho E$.*

Remark 2.2 Using the method of spectral mappings, one can show that the properties of the Weyl matrix, established in Theorems 2.1, 2.2 and 2.4, 2.5, 2.6, are not only necessary but also sufficient for the solvability of the inverse problem.

Remark 2.3 *Inverse problem from a system of spectra.* The zeros $\Lambda_{mk} := \{\rho_{lmk}\}$ of the entire functions $\Delta_{mk}(\rho)$ coincide with the eigenvalues of the boundary value problems L_{mk} for system (2.1) with the boundary conditions

$$U_1(Y) = \ldots = U_{m-1}(Y) = U_k(Y) = V_1(Y) = \ldots = V_{n-m}(Y) = 0.$$

The inverse problem of recovering the potential from the system of spectra is formulated as follows: given the spectra Λ_m and Λ_{mk} of the boundary value problems L_m and L_{mk} $(m = \overline{1, n-1},\ k > m)$, construct the potential $Q(x)$. Since the functions $\Delta_{mk}(\rho)$ are uniquely determined by their zeros, it follows from (2.7) that this inverse problem can be reduced to the inverse problem from the Weyl matrix.

3 Differential Systems on the Half-Line with Multiplied Roots of the Characteristic Polynomial

Consider the Following System of Differential Equations on the Half-Line

$$Y'(x) + B(x)Y(x) = \rho B_0 Y(x), \quad x > 0. \tag{3.1}$$

Here $Y = [y_k]^t_{k=\overline{1,n}}$ is a column-vector, B_0 is a constant diagonal matrix with arbitrary complex numbers $\beta_i^0 \neq 0, i = \overline{1, p}$ of multiplicities n_i $(p > 1,\ n_1 + \cdots + n_p = n)$ on the main diagonal, and $B(x) = [B_{k\nu}(x)]_{k,\nu=\overline{1,n}}$, where $B_{k\nu}(x) \in L(0, \infty)$ are complex-valued functions. The matrix $B(x)$ is called the *potential*.

The presence of multiplied roots produces additional interesting qualitative points in the investigation of the inverse problem. In the case of multiplied roots even the construction of the Weyl matrix is not a trivial task. In particular, it is turned out that the structure of the constructed Weyl matrix depends essentially on the multiplicities n_i and also on the distribution of the numbers β_i^0 on the complex plane. This implies serious qualitative modifications in the investigation of the inverse problem. It is also interesting that for system (3.1) on a *finite interval* the definition and the structure of the Weyl matrix are different from those on the half-line (unlike the case of simple roots).

We renumber β_i^0, $i = \overline{1, p}$, taking their multiplicities into account, i.e. we rewrite the matrix B_0 in the form $B_0 = \mathrm{diag}[\beta_k]_{k=\overline{1,n}}$, where $\beta_k := \beta_i^0$ for $k = \overline{n_{i-1} + 1, n_i}$, $i = \overline{1, p}$, $n_0 := 1$. It is convenient to write the potential $B(x)$ in the following block form: $B(x) = [B_{ki}^0(x)]_{k,i=\overline{1,p}}$, where $B_{ki}^0(x)$ are $n_k \times n_i$ matrices. Without loss of generality we assume that $B_{kk}^0(x) \equiv 0$, $k = \overline{1, p}$.

It is known that the ρ- plane can be partitioned into sectors $S_j = \{\rho : \arg\rho \in (\theta_j, \theta_{j+1})\}$, $j = \overline{0, 2r-1}$, $0 \le \theta_0 < \theta_1 < \ldots < \theta_{2r-1} < 2\pi$, $\theta_{2r} := \theta_0 + 2\pi$, in which there exist permutations $i_k^0 = i_k^0(j)$, $k = \overline{1, p}$ of the numbers $1, \ldots, p$, such that

$$\operatorname{Re}(\rho\beta_{1j}^0) < \ldots < \operatorname{Re}(\rho\beta_{pj}^0) \quad \text{for} \quad \rho \in S_j, \quad \beta_{kj}^0 := \beta_{i_k^0(j)}^0. \tag{3.2}$$

We note that the number $2r$ of sectors S_j depends on the distribution of the numbers $\{\beta_i^0\}_{i=\overline{1,p}}$ on the complex plane, and $1 \le r \le p(p-1)/2$. For example, if all of the numbers β_i^0 lie on a line containing the origin, then $r = 1$. This case is called *degenerate*. Let $m_{kj} := n_{i_k^0(j)}$, $k = \overline{1, p}$ be the permutation of the numbers $n_1, \ldots, n_p$, corresponding to the sector S_j. Put $m_{0j} := 1$. In each sector S_j we also renumber the quantities $\{\beta_k\}_{k=\overline{1,n}}$ respectively, namely, we consider the numbers $\beta_{kj} := \beta_{ij}^0$ for $k = \overline{m_{i-1,j} + 1, m_{ij}}$, $i = \overline{1, p}$. In other words, $\beta_{kj} = \beta_{i_k(j)}$, where $i_k(j)$, $k = \overline{1, n}$ is the corresponding permutation of the numbers $1, \ldots, n$. Denote $s_{kj} = \sum_{i=1}^{q} m_{ij}$ for $m_{0j} + \cdots + m_{q-1,j} \le k \le m_{1j} + \cdots + m_{qj}$, $q = \overline{1, p}$, Clearly, $s_{kj} \ge k$, $s_{k+1,j} \ge s_{kj}$, $s_{nj} = n$. For example, in the case of simple roots of the characteristic polynomial (when $n_i = 1$ for all i) one has $s_{kj} = k$ for all j. Other examples see below.

Let a matrix $h = [h_{\xi\nu}]_{\xi,\nu=\overline{1,n}}$, where $h_{\xi\nu}$ are complex numbers, be given. We introduce linear forms $U(Y) = [U_\xi(Y)]^t_{\xi=\overline{1,n}}$ by the formula $U(Y) = hY(0)$, i.e. $U_\xi(Y) = [h_{\xi 1}, \ldots, h_{\xi n}]Y(0)$. Denote $\Omega_{kj}^0 := \det[h_{\xi, i_\mu(j)}]_{\xi,\mu=\overline{1,s_{kj}}}$, $j = \overline{0, 2r-1}$, $k = \overline{1, n}$. We assume that

$$\Omega_{kj}^0 \ne 0, \quad j = \overline{0, 2r-1}, \quad k = \overline{1, n}. \tag{3.3}$$

Condition (3.3) is called the *information condition*. Systems, which do not satisfy the information condition, possess qualitatively different properties for investigating inverse problems, and are not considered in this paper. Everywhere below we shall assume that the information condition is fulfilled.

Let vector-functions $\Phi_k(x, \rho) = [\Phi_{\nu k}(x, \rho)]^t_{\nu=\overline{1,n}}$, $k = \overline{1, n}$ be solutions of (3.1) satisfying the conditions $U_\xi(\Phi_k) = \delta_{\xi k}$, $\xi = \overline{1, s_{kj}}$, and $\Phi_k(x, \rho) = O(\exp(\rho\beta_{kj}x))$, $x \to \infty$, $\rho \in S_j$ in each sector S_j with property (3.2). It will be shown below that these conditions uniquely determine the solutions $\Phi_k(x, \rho)$. Denote $M_{\xi k}(\rho) := U_\xi(\Phi_k)$, $\xi > s_{kj}$, $\rho \in S_j$, $M(\rho) := [M_{\xi k}(\rho)]_{\xi,k=\overline{1,n}}$, $M_{\xi k}(\rho) := \delta_{\xi k}$ for $\xi \le s_{kj}$, $\rho \in S_j$, $\Phi(x, \rho) := [\Phi_1(x, \rho), \ldots, \Phi_n(x, \rho)] = [\Phi_{\nu k}(x, \rho)]_{\nu,k=\overline{1,n}}$. The functions $\Phi_k(x, \rho)$ and $M_{\xi k}(\rho)$ are called the *Weyl solutions* and the *Weyl functions* respectively. The matrix $M(\rho)$ is called the *Weyl matrix*. We note that the Weyl matrix is triangular, and $\det M(\rho) \equiv 1$.

Fix B_0, h, i.e. the numbers β_i^0 and $h_{\xi\nu}$, are known and fixed. The inverse problem is formulated as follows: *given the Weyl matrix* $M(\rho)$, *construct the potential* $B(x)$, $x > 0$.

Clearly, the specification of the Weyl matrix does not determine uniquely the matrix h. It is possible to point out particular cases (when the matrix h has a special structure) for which the specification of the Weyl matrix uniquely determines not only the potential but also the matrix of linear forms. For simplicity, we confine ourselves to the most interesting and difficult problem of recovering the potential $B(x)$.

We note that the Weyl functions $M_{\xi k}(\rho)$ are piecewise analytical, and its structure depends essentially on the multiplicities n_i and also on the distribution of the numbers β_i^0 on the complex plane (see below Theorem 3.1 and Examples 3.1, 3.2, 3.3).

Theorem 3.1 *The Weyl functions $M_{\xi k}(\rho)$ are analytic in each sector S_j, $j = \overline{0, 2r-1}$ with the exception of an at most countable bounded set Λ'_{kj} of poles and are continuous in each closed sector $\overline{S_j}$ with the exception of a bounded set Λ_{kj}. For $|\rho| \to \infty$, $\rho \in \overline{S_j}$, $\xi > s_{kj}$,*

$$M_{\xi k}(\rho) = \mu_{\xi kj} + O(\rho^{-1}),$$

where the numbers $\mu_{\xi kj}$ depend only on h.

3.2. In this subsection we provide the uniqueness theorem for the solution of the inverse problem of recovering the potential from the given Weyl matrix. For this purpose we agree that together with $B(x)$ we consider a potential $\tilde{B}(x)$ of the same form. Everywhere below if a symbol α denotes an object related to $B(x)$, then $\tilde{\alpha}$ will denote the analogous object related to $\tilde{B}(x)$.

Theorem 3.2 *If $M(\rho) = \tilde{M}(\rho)$, then $B(x) = \tilde{B}(x)$ a.e. for $x > 0$. Thus, the specification of the Weyl matrix $M(\rho)$ uniquely determines the potential $B(x)$.*

Remark 3.1 Using the method of spectral mappings [4] and the properties of the Weyl matrix, obtained above, one can get a constructive procedure for the solution of the inverse problem along with necessary and sufficient conditions for its solvability.

Let us give several examples in order to illustrate the introduced notions and to show the dependence of the structure of the Weyl matrix on the multiplicities n_i and on the distribution of the numbers β_i^0 on the complex plane.

Example 3.1 Let $n = 4$, $p = 2$, $n_1 = n_2 = 2$, i.e.

$$B_0 = \begin{bmatrix} \beta_1^0 & 0 & 0 & 0 \\ 0 & \beta_1^0 & 0 & 0 \\ 0 & 0 & \beta_2^0 & 0 \\ 0 & 0 & 0 & \beta_2^0 \end{bmatrix}, \qquad B(x) = \begin{bmatrix} 0 & 0 & B_{13}(x) & B_{14}(x) \\ 0 & 0 & B_{23}(x) & B_{24}(x) \\ B_{31}(x) & B_{32}(x) & 0 & 0 \\ B_{41}(x) & B_{42}(x) & 0 & 0 \end{bmatrix}.$$

Then $S_0 = \{\rho : \operatorname{Re}(\rho\beta_1^0) < \operatorname{Re}(\rho\beta_2^0)\}$, $S_1 = \{\rho : \operatorname{Re}(\rho\beta_2^0) < \operatorname{Re}(\rho\beta_1^0)\}$, $\beta_1 = \beta_2 = \beta_1^0$, $\beta_3 = \beta_4 = \beta_2^0$. For $\rho \in S_j$, $j = 0, 1$ one has

$$s_{1j} = s_{2j} = 2, \quad s_{3j} = s_{4j} = 4,$$

$$U_1(\Phi_1) = 1, \; U_2(\Phi_1) = 0, \; \Phi_1(x, \rho) = O(\exp(\rho\beta_{j+1}^0 x)), \; x \to \infty,$$

$$U_1(\Phi_2) = 0,\ U_2(\Phi_2) = 1,\ \Phi_2(x, \rho) = O(\exp(\rho\beta_{j+1}^0 x)),\ x \to \infty,$$

$$U_1(\Phi_3) = U_2(\Phi_3) = 0,\ U_3(\Phi_3) = 1,\ U_4(\Phi_3) = 0,$$

$$U_1(\Phi_4) = U_2(\Phi_4) = U_3(\Phi_4) = 0,\ U_4(\Phi_4) = 1,$$

$$M_{31}(\rho) = U_3(\Phi_1),\ M_{41}(\rho) = U_4(\Phi_1),$$

$$M_{32}(\rho) = U_3(\Phi_2),\ M_{42}(\rho) = U_4(\Phi_2),$$

$$M(\rho) = \begin{bmatrix} 1 & 0 & 0 & 0 \\ 0 & 1 & 0 & 0 \\ M_{31}(\rho) & M_{32}(\rho) & 1 & 0 \\ M_{41}(\rho) & M_{42}(\rho) & 0 & 1 \end{bmatrix}.$$

The information condition (3.3) takes the form

$$\begin{vmatrix} h_{11} & h_{12} \\ h_{21} & h_{22} \end{vmatrix} \neq 0, \quad \begin{vmatrix} h_{13} & h_{14} \\ h_{23} & h_{24} \end{vmatrix} \neq 0, \quad \det h \neq 0.$$

Thus, in this case the structure of the Weyl matrix does not depend on the sector.

Example 3.2 Let $n = 4,\ p = 2,\ n_1 = 1,\ n_2 = 3$, i.e.

$$B_0 = \begin{bmatrix} \beta_1^0 & 0 & 0 & 0 \\ 0 & \beta_2^0 & 0 & 0 \\ 0 & 0 & \beta_2^0 & 0 \\ 0 & 0 & 0 & \beta_2^0 \end{bmatrix}, \qquad B(x) = \begin{bmatrix} 0 & B_{12}(x) & B_{13}(x) & B_{14}(x) \\ B_{21}(x) & 0 & 0 & 0 \\ B_{31}(x) & 0 & 0 & 0 \\ B_{41}(x) & 0 & 0 & 0 \end{bmatrix}.$$

Then $S_0 = \{\rho :\ \mathrm{Re}(\rho\beta_1^0) < \mathrm{Re}(\rho\beta_2^0)\}$, $S_1 = \{\rho :\ \mathrm{Re}(\rho\beta_2^0) < \mathrm{Re}(\rho\beta_1^0)\}$, $\beta_1 = \beta_1^0$, $\beta_2 = \beta_3 = \beta_4 = \beta_2^0$. In this case the structure of the Weyl matrix depends on the sector; hence we consider the sectors separately. For $\rho \in S_0$ one has

$$s_{10} = 1, \quad s_{20} = s_{30} = s_{40} = 4,$$

$$U_1(\Phi_1) = 1,\ \Phi_1(x, \rho) = O(\exp(\rho\beta_1^0 x)),\ x \to \infty,$$

$$U_1(\Phi_2) = 0,\ U_2(\Phi_2) = 1,\ U_3(\Phi_2) = U_4(\Phi_2) = 0,$$

$$U_1(\Phi_3) = U_2(\Phi_3) = 0,\ U_3(\Phi_3) = 1,\ U_4(\Phi_3) = 0,$$

$$U_1(\Phi_4) = U_2(\Phi_4) = U_3(\Phi_4) = 0,\ U_4(\Phi_4) = 1,$$

$$M_{21}(\rho) = U_2(\Phi_1),\ M_{31}(\rho) = U_3(\Phi_1),\ M_{41}(\rho) = U_4(\Phi_1),$$

$$M(\rho) = \begin{bmatrix} 1 & 0\,0\,0 \\ M_{31}(\rho) & 1\,0\,0 \\ M_{31}(\rho) & 0\,1\,0 \\ M_{41}(\rho) & 0\,0\,1 \end{bmatrix}.$$

For $\rho \in S_1$ one has

$$s_{11} = s_{21} = s_{31} = 3, \; s_{41} = 4,$$

$$U_1(\Phi_1) = 1, \; U_2(\Phi_1) = U_3(\Phi_1) = 0, \; \Phi_1(x, \rho) = O(\exp(\rho\beta_2^0 x)), \; x \to \infty,$$

$$U_1(\Phi_2) = 0, \; U_2(\Phi_2) = 1, \; U_3(\Phi_2) = 0, \; \Phi_2(x, \rho) = O(\exp(\rho\beta_2^0 x)), \; x \to \infty,$$

$$U_1(\Phi_3) = U_2(\Phi_3) = 0, \; U_3(\Phi_3) = 1, \; \Phi_3(x, \rho) = O(\exp(\rho\beta_2^0 x)), \; x \to \infty,$$

$$U_1(\Phi_4) = U_2(\Phi_4) = U_3(\Phi_4) = 0, \; U_4(\Phi_4) = 1,$$

$$M_{41}(\rho) = U_4(\Phi_1), \; M_{42}(\rho) = U_4(\Phi_2), \; M_{43}(\rho) = U_4(\Phi_3),$$

$$M(\rho) = \begin{bmatrix} 1 & 0 & 0 & 0 \\ 0 & 1 & 0 & 0 \\ 0 & 0 & 1 & 0 \\ M_{41}(\rho) & M_{42}(\rho) & M_{43}(\rho) & 1 \end{bmatrix}.$$

The information condition (3.3) takes the form

$$h_{11} \neq 0, \quad \begin{vmatrix} h_{12} & h_{13} & h_{14} \\ h_{22} & h_{23} & h_{24} \\ h_{32} & h_{33} & h_{34} \end{vmatrix} \neq 0, \quad \det h \neq 0.$$

Example 3.3 Let $n = 6, \; p = 3, \; n_1 = 1, \; n_2 = 2, \; n_3 = 3$.

Case 1. Let $\beta_k^0 = \exp(2\pi i(k-1)/3), \; k = 1, 2, 3$. Here $r = 3$, i.e. we have 6 sectors: $S_j = \{\rho : \; \arg\rho \in (j\pi/3, (j+1)\pi/3)\}, \; j = \overline{0, 5}$. In each sector the Weyl matrix has the different structure since

$$s_{10} = s_{20} = 2, \; s_{30} = s_{40} = s_{50} = 5, \; s_{60} = 6,$$

$$s_{11} = s_{21} = 2, \; s_{31} = 3, \; s_{41} = s_{51} = s_{61} = 6,$$

$$s_{12} = 1, \; s_{22} = s_{32} = 3, \; s_{42} = s_{52} = s_{62} = 6,$$

$$s_{13} = 1, \; s_{23} = s_{33} = s_{43} = 4, \; s_{53} = s_{63} = 6,$$

$$s_{14} = s_{24} = s_{34} = 3, \; s_{44} = 4, \; s_{54} = s_{64} = 6,$$

$$s_{15} = s_{25} = s_{35} = 3, \; s_{45} = s_{55} = 5, \; s_{65} = 6.$$

Case 2. Let β_k^0 are real and $\beta_1^0 < \beta_2^0 < \beta_3^0$. Here $r = 1$, i.e. we have 2 sectors: $S_0 = \{\rho : \ \mathrm{Re}\,\rho > 0\}$ and $S_1 = \{\rho : \ \mathrm{Re}\,\rho < 0\}$. Clearly,

$$s_{10} = 1, \ s_{20} = s_{30} = 3, \ s_{40} = s_{50} = s_{60} = 6,$$

$$s_{11} = s_{21} = s_{31} = 3, \ s_{41} = s_{51} = 5, \ s_{61} = 6,$$

Thus, the structure of the Weyl matrix depends not only on the multiplicities n_i, but also depends on the distribution of the numbers β_i^0 on the complex plane.

4 Differential Systems on a Finite Interval with Multiplied Roots of the Characteristic Polynomial

We Consider the Following System of Differential Equations on a Finite Interval

$$Y'(x) + B(x)Y(x) = \rho B_0 Y(x), \quad 0 < x < T. \tag{4.1}$$

Here $Y = [y_k]^t_{k=\overline{1,n}}$ is the column-vector, B_0 is a constant diagonal matrix with arbitrary complex numbers $\beta_i^0 \neq 0$, $i = \overline{1,p}$ of multiplicities n_i $(p > 1, \ n_1 + \cdots + n_p = n)$ on the main diagonal, and $B(x) = [B_{k\nu}(x)]_{k,\nu=\overline{1,n}}$, where $B_{k\nu}(x) \in L(0,T)$ are complex-valued functions.

We renumber β_i^0, $i = \overline{1,p}$, taking their multiplicities into account, i.e. we rewrite the matrix B_0 in the form $B_0 = \mathrm{diag}[\beta_k]_{k=\overline{1,n}}$, where $\beta_k := \beta_i^0$ for $k = \overline{n_{i-1}+1, n_i}$, $i = \overline{1,p}$, $n_0 := 1$. It is convenient to write the potential $B(x)$ in the following block form: $B(x) = [B_{ki}^0(x)]_{k,i=\overline{1,p}}$, where $B_{ki}^0(x)$ are $n_k \times n_i$ matrices. Without loss of generality we assume that $B_{kk}^0(x) \equiv 0$, $k = \overline{1,p}$.

It is known that the ρ- plane can be partitioned into sectors $S_j = \{\rho : \ \arg\rho \in (\theta_j, \theta_{j+1})\}$, $j = \overline{0, 2r-1}$, $0 \le \theta_0 < \theta_1 < \ldots < \theta_{2r-1} < 2\pi$, $\theta_{2r} := \theta_0 + 2\pi$, in each of which there exists a permutation $i_k^0 = i_k^0(j)$, $k = \overline{1,p}$ of the numbers $1, \ldots, p$, such that $\mathrm{Re}(\rho\beta_{1j}^0) < \ldots < \mathrm{Re}(\rho\beta_{pj}^0)$ for $\rho \in S_j$, $\beta_{kj}^0 := \beta_{i_k^0(j)}^0$. We note that the number $2r$ of sectors S_j depends on the distribution of the numbers $\{\beta_i^0\}_{i=\overline{1,p}}$ on the complex plane, and $1 \le r \le p(p-1)/2$. For example, if all of the numbers β_i^0 lie on a line containing the origin, then $r = 1$. This case is called degenerate. Let $m_{kj} := n_{i_k^0(j)}$, $k = \overline{1,p}$ be the permutation of the numbers $n_1, \ldots, n_p$, corresponding to the sector S_j. Put $m_{0j} := 1$. In each sector S_j we also renumber the quantities $\{\beta_k\}_{k=\overline{1,n}}$ respectively, namely, we consider the numbers $\beta_{kj} := \beta_{ij}^0$ for $k = \overline{m_{i-1,j}+1, m_{ij}}$, $i = \overline{1,p}$. In other words, $\beta_{kj} = \beta_{i_k(j)}$, where $i_k(j)$, $k = \overline{1,n}$ is the corresponding permutation of the numbers $1, \ldots, n$. Denote $s_{kj} = \sum_{i=1}^{q} m_{ij}$ for $m_{0j} + \cdots + m_{q-1,j} \le k \le m_{1j} + \cdots + m_{qj}$, $q = \overline{1,p}$, and put $s_k := \min\limits_j s_{kj}$, $k = \overline{1,n}$. Clearly, $s_k \ge k$, $s_{k+1} \ge s_k$, $s_n = n$. For example, in the case of simple roots of the characteristic polynomial (when $n_i = 1$ for all i) one has $s_k = k$ for all j. We note that the numbers s_k depend not only on the multiplicities $n_1, \dots, n_p$, but also on the distribution of the numbers $\beta_i^0, i = \overline{1, p}$ on the complex plane (see Examples 4.1, 4.2).

4.2. Let matrices $h = [h_{\xi\nu}]_{\xi,\nu=\overline{1,n}}$, $H = [H_{\xi\nu}]_{\xi,\nu=\overline{1,n}}$ be given, where $h_{\xi\nu}$, $H_{\xi\nu}$ are complex numbers, $\det h \neq 0$, $\det H \neq 0$. We introduce the linear forms $U(Y) = [U_\xi(Y)]^t_{\xi=\overline{1,n}}$, $V(Y) = [V_\xi(Y)]^t_{\xi=\overline{1,n}}$ by the formulae $U(Y) = hY(0)$, $V(Y) = HY(T)$, i.e. $U_\xi(Y) = [h_{\xi 1}, \dots, h_{\xi n}]Y(0)$, $V_\xi(Y) = [H_{\xi 1}, \dots, H_{\xi n}]Y(0)$. Let the vector-functions $\Phi_k(x, \rho) = [\Phi_{\nu k}(x, \rho)]^t_{\nu=\overline{1,n}}$, $k = \overline{1, n}$ be solutions of system (4.1) under the conditions $U_\xi(\Phi_k) = \delta_{\xi k}$, $\xi = \overline{1, s_k}$, $V_\eta(\Phi_k) = 0$, $\eta = \overline{1, n - s_k}$. Let $M_{\xi k}(\rho) = U_\xi(\Phi_k)$, $\xi > s_k$, $M(\rho) = [M_{\xi k}(\rho)]_{\xi,k=\overline{1,n}}$, $M_{\xi k}(\rho) = \delta_{\xi k}$ $(\xi \leq s_k)$, $\Phi(x, \rho) = [\Phi_1(x, \rho), \dots, \Phi_n(x, \rho)] = [\Phi_{\nu k}(x, \rho)]_{\nu,k=\overline{1,n}}$. Thus, $M(\rho) = h\Phi(0, \rho)$. The functions $\Phi_k(x, \rho)$ and $M_{\xi k}(\rho)$ are called the *Weyl solutions* and the *Weyl functions* respectively. The matrix $M(\rho)$ is called the *Weyl matrix* for system (4.1). We note that the Weyl matrix is triangular, and $\det M(\rho) \equiv 1$.

Fix B_0, h, H. The inverse problem is formulated as follows: *given the Weyl matrix $M(\rho)$, construct the potential $B(x)$, $0 < x < T$.*

Denote $\Omega_k^0(\rho) := \det[h_{1\nu}, \dots, h_{s_k,\nu}, H_{1\nu}\exp(\rho\beta_\nu T), \dots, H_{n-s_k,\nu}\exp(\rho\beta_\nu T)]_{\nu=\overline{1,n}}$. For $|\rho| \to \infty$, $\rho \in S_j$, $\arg\rho = \theta \in (\theta_j, \theta_{j+1})$ there exist the finite limits $\Omega_{kj}^0 := \lim \Omega_k^0(\rho)\exp(-\rho\omega_{kj}T)$, where $\omega_{kj} := \sum_{\mu=s_k+1}^n \beta_{\mu j}$, and the numbers Ω_{kj}^0 depend only on $h_{\xi\nu}$, $H_{\eta\nu}$, $\xi = \overline{1, s_k}$, $\eta = \overline{1, n - s_k}$. Suppose that

$$\Omega_{kj}^0 \neq 0, \quad k = \overline{1, n}, \ j = \overline{0, 2r - 1}. \tag{4.2}$$

Condition (4.2) is called the information condition. Systems, which do not satisfy the information condition, possess qualitatively different properties for investigating inverse problems, and are not considered in this paper. Everywhere below we shall assume that the information condition is fulfilled.

Let $C(x, \rho) = [C_{\nu k}(x, \rho)]_{\nu,k=\overline{1,n}}$ be the matrix solution of system (4.1) under the initial condition $U(C) = hC(0, \rho) = E$ (here and below by E we denote the identity matrix of the corresponding dimension). In other words, the column-vectors $C_k(x, \rho) = [C_{\nu k}(x, \rho)]^t_{\nu=\overline{1,n}}$ are solutions of (1) under the initial conditions $U_\xi(C_k) = \delta_{\xi k}$, $\xi, k = \overline{1, n}$. The functions $C_{\nu k}(x, \rho)$ are entire in ρ for each fixed x, and

$$\Phi(x, \rho) = C(x, \rho)M(\rho), \quad \det C(x, \rho) = \det\Phi(x, \rho) = (\det h)^{-1}\exp(\rho(\beta_1 + \dots + \beta_n)x). \tag{4.3}$$

Let $\{\rho_{l\xi k}\}$ be the zeros of the entire functions

$$\Delta_{\xi k}(\rho) = \det[U_1(C_\mu), \dots, U_{k-1}(C_\mu), U_\xi(C_\mu), U_{k+1}(C_\mu), \dots, U_{s_k}(C_\mu), V_1(C_\mu), \dots, V_{n-s_k}(C_\mu)]_{\mu=\overline{1,n}}.$$

Denote $\Delta_k(\rho) := \Delta_{kk}(\rho)$, $\rho_{lk} := \rho_{lkk}$, $G_\delta := \{\rho : |\rho - \rho_{lk}| \geq \delta\}$, $\delta > 0$. Obviously,

$$M_{\xi k}(\rho) = \Delta_{\xi k}(\rho)(\Delta_k(\rho))^{-1}, \quad \xi > s_k\,. \tag{4.4}$$

Now we provide the uniqueness theorem for the solution of the inverse problem of recovering the potential from the given Weyl matrix. For this purpose we agree that together with $B(x)$ we consider a potential $\tilde{B}(x)$ of the same form (we remind that the matrices B_0, h and H are known a priori and fixed). Everywhere below if a symbol α denotes an object related to $B(x)$, then $\tilde{\alpha}$ will denote the analogous object related to $\tilde{B}(x)$.

Theorem 4.1 *If $M(\rho) = \tilde{M}(\rho)$, then $B(x) = \tilde{B}(x)$ a.e. for $(0, T)$. Thus, the specification of the Weyl matrix $M(\rho)$ uniquely determines the potential $B(x)$.*

4.3. Now we consider the inverse problem from a system of spectra. The zeros $\{\rho_{l\xi k}\}$ of the entire functions $\Delta_{\xi k}(\rho)$ coincide with the eigenvalues of the boundary value problems $Q_{\xi k}$ for system (4.1) with the boundary conditions $U_1(Y) = \ldots = U_{k-1}(Y) = U_\xi(Y) = U_{k+1}(Y) = \ldots = U_{s_k}(Y) = V_1(Y) = \ldots = V_{n-s_k}(Y) = 0$. Put $Q_k := Q_{kk}$. We note that some of these boundary value problems can be identical when $s_{k+1} = s_k$, and can be degenerate (Cauchy problems) when $s_k = n$.

The inverse problem from a system of spectra is formulated as follows: given the spectra $\{\rho_{lk}\}$ and $\{\rho_{l\xi k}\}, \xi > s_k$ of the boundary value problems Q_k and $Q_{\xi k}$, construct the potential $B(x)$, $0 < x < T$. The number of these spectra depend on the distribution of the quantities β_i^0, $i = \overline{1, p}$ on the complex plane and on the multiplicities $n_1, \ldots, n_p$ (see Examples 4.1, 4.2).

Theorem 4.2 *If $\rho_{lk} = \tilde{\rho}_{lk}$; $\rho_{l\xi k} = \tilde{\rho}_{l\xi k}$, $\xi > s_k$ for all l, k, then $B(x) = \tilde{B}(x)$ a.e. on $(0, T)$. Thus, the specification of the spectra of the boundary value problems Q_k and $Q_{\xi k}$, $\xi > s_k$ uniquely determines the potential $B(x)$.*

Theorem 4.2 is a corollary of Theorem 4.1. Indeed, the entire functions $\Delta_k(\rho)$ is uniquely determined by their zeros $\{\rho_{lk}\}$. Analogously one can prove that the entire functions $\Delta_{\xi k}(\rho)$ is uniquely determined by their zeros $\{\rho_{l\xi k}\}$. Hence, using (4.4) and the conditions of Theorem 4.2, we obtain that $M(\rho) = \tilde{M}(\rho)$, and consequently, $B(x) = \tilde{B}(x)$ a.e. on $(0, T)$ by Theorem 4.1.

Remark 4.1 The structure of the Weyl matrix $M(\rho)$ depends on distribution of the numbers β_i^0, $i = \overline{1, p}$ on the complex plane and on their multiplicities n_i. In particular, a part of the elements of the Weyl matrix, situated below the main diagonal, can be equal to zero. Together with the Weyl matrix $M(\rho)$ in some problems it is natural to consider also the so-called *complete Weyl matrix* $M^0(\rho) = [M_{\xi k}^0(\rho)]_{\xi,k=\overline{1,n}}$ of the form $M^0(\rho) := h\Phi^0(0, \rho)$, where $\Phi^0(x, \rho) = [\Phi_k^0(x, \rho)]_{k=\overline{1,n}}^t$, and the column-vectors $\Phi_k^0(x, \rho)$ are solutions of system (4.1) under the conditions $U_\xi(\Phi_k^0) = \delta_{\xi k}$, $\xi = \overline{1, k}$, $V_\eta(\Phi_k^0) = 0$, $\eta = \overline{1, n-k}$. The matrix $M^0(\rho)$ is an analog of the Weyl matrix introduced for system (1.1) on the half-line. One can study the inverse problem of recovering the potential $B(x)$ from the complete Weyl matrix $M^0(\rho)$. This inverse problem can be reduced to the inverse problem from the Weyl matrix $M(\rho)$, since $\Phi_k^0(x, \rho) = \sum_{i=k}^{s_k} \Phi_i(x, \rho) M_{ik}^0(\rho)$, $M_{\xi k}^0(\rho) = \sum_{i=k}^{s_k} M_{\xi i}(\rho) M_{ik}^0(\rho)$, and consequently, the matrix $M(\rho)$ is uniquely determined from the matrix $M^0(\rho)$.

Let us give examples in order to illustrate the dependence of the structure of the Weyl matrix $M(\rho)$ and the number of the prescribed spectra on the distribution of the numbers β_i^0, $i = \overline{1, p}$ on the complex plane and on their multiplicities $n_1, \ldots, n_p$.

Example 4.1 Let $n = 4,\ p = 2,\ n_1 = 1,\ n_2 = 3$. Then $s_1 = 1,\ s_2 = s_3 = 3,\ s_4 = 4$,

$$\Delta_1(\rho) = \det[U_1(C_\mu), V_1(C_\mu), V_2(C_\mu), V_3(C_\mu)]_{\mu=\overline{1,4}},$$

$$\Delta_2(\rho) = \Delta_3(\rho) = \det[U_1(C_\mu), U_2(C_\mu), U_3(C_\mu), V_1(C_\mu)]_{\mu=\overline{1,4}},$$

$\Delta_4(\rho) \equiv 1,\ M_{32}(\rho) \equiv 0$. The potential $B(x)$ is uniquely determined by the specification of 7 spectra of the boundary value problems generated by the boundary conditions $U_k(Y) = V_1(Y) = V_2(Y) = V_3(Y) = 0, k = \overline{1, 4}$ and $U_1(Y) = U_\xi(Y) = U_\eta(Y) = V_1(Y) = 0, 2 \le \xi < \eta \le 4$.

Let now $n = 4,\ p = 2,\ n_1 = n_2 = 2$. Then $s_1 = s_2 = 2,\ s_3 = s_4 = 4$,

$$\Delta_1(\rho) = \Delta_2(\rho) = \det[U_1(C_\mu), U_2(C_\mu), V_1(C_\mu), V_2(C_\mu)]_{\mu=\overline{1,4}}, \quad \Delta_3(\rho) = \Delta_4(\rho) \equiv 1,$$

$M_{21}(\rho) = M_{43}(\rho) \equiv 0$. The potential $B(x)$ is uniquely determined by the specification of 5 spectra of the boundary value problems generated by the boundary conditions $U_\xi(Y) = U_\eta(Y) = V_1(Y) = V_2(Y) = 0, \xi = 1, 2, 1 \le \xi < \eta \le 4$.

Example 4.2 Let $n = 7,\ p = 3, n_1 = 1,\ n_2 = 2,\ n_3 = 4$.
(i) Let β_k^0 be real, $\beta_1^0 < \beta_2^0 < \beta_3^0$. Then $s_1 = 1,\ s_2 = s_3 = 3,\ s_4 = 4,\ s_5 = s_6 = 6,\ s_7 = 7,\ M_{32}(\rho) = M_{65}(\rho) \equiv 0$, and the potential $B(x)$ is uniquely determined by the specifiation of 23 spectra.
(ii) Let $\beta_k^0 = \exp(2\pi(k-1)i/3), k = 1, 2, 3$. Then $s_k = k$, $M(\rho) \equiv M^0(\rho)$, and the potential $B(x)$ is uniquely determined by the specifiation of 27 spectra.

Acknowledgements This work was supported in part by Grant 1.1660.2017/4.6 of the Russian Ministry of Education and Science and by Grant 19-01-00102 of Russian Foundation for Basic Research.

References

1. Marchenko, V.A.: Sturm-Liouville Operators and Their Applications. Naukova Dumka, Kiev (1977). English translation: Birkhauser (1986)
2. Levitan B.M.: Inverse Sturm-Liouville Problems. Nauka, Moscow (1984). English translation: VNU Sci. Press, Utrecht (1987)
3. Freiling, G., Yurko, V.A.: Inverse Sturm-Liouville Problems and their Applications. NOVA Science Publishers, New York (2001)
4. Yurko, V.A.: Method of Spectral Mappings in the Inverse Problem Theory. Inverse and Ill-posed Problems Series. VSP, Utrecht (2002)
5. Yurko V.A.: Recovery of nonselfadjoint differential operators on the half-line from the Weyl matrix. Mat. Sbornik **182**(3), 431–456 (1991) (Russian). English translation: Math. USSR Sbornik **72**(2), 413–438 (1992)

6. Yurko V.A.: An inverse spectral problem for singular non-selfadjoint differential systems. Matem. Sbornik **195**(12), 123–156 (2004) (Russian). English translation: Sbornik Math. **195**(12), 1823–1854 (2004)
7. Yurko, V.A.: Inverse spectral problems for differential systems on a finite interval. Results Math. **48**(3–4), 371–386 (2005)
8. Yurko, V.A.: An inverse problem for differential systems with multiplied roots of the characteristic polynomial. J. Inverse Ill-Posed Probl. **13**(5), 503–512 (2005)
9. Yurko V.A.: An inverse problem for differential systems on a finite interval in the case of multiple roots of the characteristic polynomial. Differ. Uravnenija **41**(6), 781–786 (2005) (Russian). English translation: Differ. Equ. **41**(6), 818–823 (2005)
10. Beals, R., Coifman, R.R.: Scattering and inverse scattering for first order systems. Comm. Pure Appl. Math. **37**, 39–90 (1984)
11. Zhou, X.: Direct and inverse scattering transforms with arbitrary spectral singularities. Comm. Pure Appl. Math. **42**, 895–938 (1989)

Printed by Printforce, the Netherlands